Advances in Intelligent Systems and Computing

Volume 1070

The series "Advances in Intelligent Systems and Computing" contains publications on theory, applications, and design methods of Intelligent Systems and Intelligent Computing. Virtually all disciplines such as engineering, natural sciences, computer and information science, ICT, economics, business, e-commerce, environment, healthcare, life science are covered. The list of topics spans all the areas of modern intelligent systems and computing such as computational intelligence, soft computing including neural networks, fuzzy systems, evolutionary computing and the fusion of these paradigms, social intelligence, ambient intelligence, computational neuroscience, artificial life, virtual worlds and society, cognitive science and systems, Perception and Vision, DNA and immune based systems, self-organizing and adaptive systems, e-Learning and teaching, human-centered and human-centric computing, recommender systems, intelligent control, robotics and mechatronics including human-machine teaming, knowledge-based paradigms, learning paradigms, machine ethics, intelligent data analysis, knowledge management, intelligent agents, intelligent decision making and support, intelligent network security, trust management, interactive entertainment, Web intelligence and multimedia.

The publications within "Advances in Intelligent Systems and Computing" are primarily proceedings of important conferences, symposia and congresses. They cover significant recent developments in the field, both of a foundational and applicable character. An important characteristic feature of the series is the short publication time and world-wide distribution. This permits a rapid and broad dissemination of research results.

** Indexing: The books of this series are submitted to ISI Proceedings, EI-Compendex, DBLP, SCOPUS, Google Scholar and Springerlink **

More information about this series at http://www.springer.com/series/11156

Kohei Arai · Rahul Bhatia · Supriya Kapoor
Editors

Proceedings of the Future Technologies Conference (FTC) 2019

Volume 2

 Springer

Editors
Kohei Arai
Faculty of Science and Engineering
Saga University
Saga, Japan

Rahul Bhatia
The Science and Information
(SAI) Organization
Bradford, West Yorkshire, UK

Supriya Kapoor
The Science and Information
(SAI) Organization
Bradford, West Yorkshire, UK

ISSN 2194-5357   ISSN 2194-5365   (electronic)
Advances in Intelligent Systems and Computing
ISBN 978-3-030-32522-0   ISBN 978-3-030-32523-7   (eBook)
https://doi.org/10.1007/978-3-030-32523-7

This Springer imprint is published by the registered company Springer Nature Switzerland AG
The registered company address is: Gewerbestrasse 11, 6330 Cham, Switzerland

Editor's Preface

After the success of three FTCs, the Future Technologies Conference (FTC) 2019 was held on October 24–25, 2019, in San Francisco, USA, a city of immense beauty—be it the charming neighborhoods such as Castro/Upper Market or landmarks such as the Golden Gate Bridge—and even denser fog. FTC 2019 focuses on technological breakthroughs in the areas of computing, electronics, AI, robotics, security, and communications.

The ever-changing scope and rapid development of computer technologies create new problems and questions, resulting in the real need for sharing brilliant ideas and stimulating good awareness of this important research field. The aim of this conference is to provide a worldwide forum, where the international participants can share their research knowledge and ideas on the recent and latest research and map out the directions for future researchers and collaborations.

For this conference proceedings, researchers, academics, and technologists from leading universities, research firms, government agencies, and companies from 50+ countries submitted their latest research at the forefront of technology and computing. After the double-blind review process, we finally selected 143 full papers including 6 poster papers to publish.

We would like to express our gratitude and appreciation to all of the reviewers who helped us maintain the high quality of manuscripts included in this conference proceedings. We would also like to extend our thanks to the members of the organizing team for their hard work. We are tremendously grateful for the contributions and support received from authors, participants, keynote speakers, program committee members, session chairs, steering committee members, and others in their various roles. Their valuable support, suggestions, dedicated commitment, and hard work have made FTC 2019 a success.

We hope that all the participants of FTC 2019 had a wonderful and fruitful time at the conference and that our overseas guests enjoyed their sojourn in San Francisco!

Kind Regards,
Kohei Arai

Contents

Mutually Authenticated Group Key Management Protocol for Healthcare IoT Networks 1
Firdous Kausar, Waqas Aman, and Dawood Al-Abri

Enhancing the Linguistic Landscape with the Proper Deployment of the Internet of Things Technologies: A Case Study of Smart Malls 13
Fahad Algarni, Azmat Ullah, and Khalid Aloufi

Design and Fabrication of Smart Bandeau Bra for Breathing Pattern Measurement 40
Rafiu King Raji, Xuhong Miao, Ailan Wan, Li Niu, Yutian Li, and Andrews Boakye

Electrical Internet of Things - EIoT: A Platform for the Data Management in Electrical Systems 49
Santiago Gil, Germán D. Zapata-Madrigal, and Rodolfo García-Sierra

Infrastructure-Aided Networking for Autonomous Vehicular Systems 66
Izhak Rubin and Yulia Sunyoto

3D Objects Indexing Using Chebyshev Polynomial 87
Y. Oulahrir, F. Elmounchid, S. Hellam, A. Sadiq, and S. Mbarki

Machine Learning Classification and Segmentation of Forest Fires in Wide Area Motion Imagery 100
Melonie Richey, David Blum, Meredith Gregory, Kendall Ocasio, Zachary Mostowsky, and Jordan Chandler

A Fast Multi-phases Demon Image Registration for Atlas Building 112
Youshan Zhang

Hand Sign Language Feature Extraction Using Image Processing 122
Abdelkader Chabchoub, Ali Hamouda, Saleh Al-Ahmadi, Wahid Barkouti,
and Adnen Cherif

Butterfly, Larvae and Pupae Defects Detection Using
Convolutional Neural Network and Apriori Algorithm 132
Jerwin M. Montellano

Understanding Dynamic Logic Using 2D Images 162
John Promersberger, Sudhanshu Kumar Semwal, and Leonid Perlovsky

Screens Affordances in Image Consumption. 176
Ema Rolo, Helena Nobre, and Vania Baldi

Photofeeler-D3: A Neural Network with Voter Modeling
for Dating Photo Impression Prediction 188
Agastya Kalra and Ben Peterson

Human-Augmented Robotic Intelligence (HARI)
for Human-Robot Interaction. 204
Vishwas Mruthyunjaya and Charles Jankowski

Kinematic Analysis of the Bio-inspired Design of a Wearable
Compliant Gait Rehabilitation Robotic System for Smart
and Connected Health 224
S. M. Mizanoor Rahman

Grasp Rehabilitation of Stroke Patients Through Object
Manipulation with an Intelligent Power Assist Robotic System 244
S. M. Mizanoor Rahman

Reasonably Optimal Utilisation Through Evolution (ROUTE)
in Airspace Design 260
Thomas Lawson and Yanyan Yang

Autonomous Robot Navigation with Fuzzy Logic
in Uncontrolled Environments 275
Ryan Reck and Sherine Antoun

The Future of Socially Assistive Robotics:
Considering an Exploratory Governance Framework (EGF) 284
Hock Chuan Lim

A Novel Power Management System for Autonomous Robots 293
Mamadou Doumbia, Xu Cheng, and Haixian Chen

Some Cyberpsychology Techniques to Distinguish Humans
and Bots for Authentication 306
Ishaani Priyadarshini, Haining Wang, and Chase Cotton

**Multi Criteria Method for Determining the Failure Resistance
of Information System Components** 324
Askar Boranbayev, Seilkhan Boranbayev, Assel Nurusheva,
Yerzhan Seitkulov, and Askar Nurbekov

**Time-Invariant Cryptographic Key Generation
from Cardiac Signals** 338
Sarah Alharbi, Md Saiful Islam, and Saad Alahmadi

**Privacy Implication and Technical Requirements Toward
GDPR Compliance** 353
Ching-Chun (Jim) Huang and Zih-shiuan (Spin) Yuan

STRIDE-Based Threat Modeling for MySQL Databases 368
James Sanfilippo, Tamirat Abegaz, Bryson Payne, and Abi Salimi

**Efficient RSA Crypto Processor Using Montgomery
Multiplier in FPGA** 379
Lavanya Gnanasekaran, Anas Salah Eddin, Halima El Naga,
and Mohamed El-Hadedy

New Residue Signed-Digit Addition Algorithm 390
Shugang Wei

**Study and Evaluation of Unsupervised Algorithms Used
in Network Anomaly Detection** 397
Juliette Dromard and Philippe Owezarski

**Recurrent Binary Patterns and CNNs for Offline
Signature Verification** 417
Mustafa Berkay Yılmaz and Kağan Öztürk

**An Improvement of Compelling Graphical Confirmation Plan
and Cryptography for Upgrading the Information Security
and Preventing Shoulder Surfing Assault** 435
Norman Dias and S. R. Reeja

**Application of Siamese Neural Networks for Fast Vulnerability
Detection in MIPS Executable Code** 454
Roman Demidov and Alexander Pechenkin

**Where Are We Looking for Security Concerns? Understanding
Android Security Static Analysis** 467
Suzanna Schmeelk

Decentralized Autonomous Video Copyright Protection 484
Qifeng Chen, Shiyu Zhang, and Wilson Wei

A ZigBee Based Architecture for Public Safety Communication
in Hurricane Scenario 493
Imtiaz Parvez, Yemeserach Mekonnen, and Arif I. Sarwat

An Efficient Singularity Detector Network for Fingerprint Images 511
Geetika Arora, C. Jinshong Hwang, Kamlesh Tiwari, and Phalguni Gupta

An Implementation of Smart Contracts by Integrating BIM
and Blockchain. 519
Alireza Shojaei, Ian Flood, Hashem Izadi Moud, Mohsen Hatami,
and Xun Zhang

Are Esports the Next Big Varsity Draw? An Exploratory
Study of 522 University e-Athletes 528
Thierry Karsenti

Using Maximum Weighted Cliques in the Detection
of Sub-communities' Behaviors in OSNs 542
Izzat Alsmadi and Chuck Easttom

Gamification in Enterprise Systems: A Literature Review 552
Changiz Hosseini and Moutaz Haddara

An Experiment with Link Prediction in Social Network:
Two New Link Prediction Methods 563
Ahmad Rawashdeh

Dynamic User Modeling in the Context of Customer Churn
Prediction in Loyalty Program Marketing 582
Ishani Chakraborty

Buddha Bot: The Exploration of Embodied Spiritual Machine
in Chatbot 589
Pat Pataranutaporn, Bank Ngamarunchot, Korakot Chaovavanich,
Sornchai Chatwiriyachai, Potiwat Ngamkajornwiwat,
Nutchanon Ninyawee, and Werasak Surareungchai

My Personal Brand and My Web Presence: Mining Digital
Footprints and Analyzing Personas in the World of IOT
and Digital Citizenry 596
Fawzi BenMessaoud, Taryn Elizabeth Husted, Dwight William Hall,
Holly Nichole Handlon, and Niranjan Valmik Kshirsagar

Formalizing Graph Database and Graph Warehouse
for On-Line Analytical Processing in Social Networks 605
Frank S. C. Tseng and Annie Y. H. Chou

Blending Six Sigma and Software Development 619
Hassan Pournaghshband

**Toys That Mobilize: Past, Present and Future of Phygital
Playful Technology** 625
Katriina Heljakka and Pirita Ihamäki

Unified and Stable Project: "Ushering in the Future" 641
Mohamed E. Fayad, Gaurav Kuppa, and David Hamu

Global Logistics in the Era of Industry 4.0 652
Ayodeji Dennis Adeitan, Clinton Aigbavboa,
and Emmanuel Emem-Obong Agbenyeku

**Enabling Empirical Research: A Corpus of Large-Scale
Python Systems** 661
Safwan Omari and Gina Martinez

**M-Health Android Application Using Firebase, Google APIs
and Supervised Machine Learning** 670
Charu Nigam, Mahima Narang, Nisha Chaurasia, and Aparajita Nanda

**ICTs Adoption in SMEs Located in Rural Regions: Case Study
of Northern of Portugal and Castela and Leão (Spain)** 682
João Paulo Pereira, Valeriia Ostritsova, and João Pereira

**Kaizen 4.0 Towards an Integrated Framework
for the Lean-Industry 4.0 Transformation** 692
Peter Burggräf, Carolin Lorber, Andreas Pyka, Johannes Wagner,
and Tim Weißer

**Energy Efficient Power Management Modes for Smartphone
Battery Life Conservation** 710
Evelyn Sowells-Boone, Rushit Dave, Brinta Chowdhury,
and DeWayne Brown

**Providing Some Minimum Guarantee for Real-Time Secondary
Users in Cognitive Radio Sensor Networks** 717
Changa Andrew, Tonny Bulega, and Michael Okopa

**Sparse Signal Reconstruction by Batch Orthogonal
Matching Pursuit** 731
Lichun Li and Feng Wei

**Performance of Cooperative System Based on LDPC Codes
in Wireless Optical Communication** 745
Ibrahima Gueye, Ibra Dioum, K. Wane Keita, Idy Diop, Papis Ndiaye,
Moussa Diallo, and Sidi Mohamed Farssi

**Studying the Impacts of the Renewable Energy Integration
in Telecommunication Systems: A Case Study in Lome** 758
Koffi A. Dotche, Adekunlé A. Salami, Koffi M. Kodjo, François Sekyere,
and Koffi-Sa Bedja

Effect of the Silicon Substrate in the Response of MIS Transistor
Sensor for Nano-Watts Light Signal........................... 781
J. Hernández-Betanzos, A. A. Gonzalez-Fernandez, J. Pedraza,
and M. Aceves-Mijares

Energy Efficient Balanced Tree-Based Routing Protocol
for Wireless Sensor Network (EEBTR) 795
Rafla Ghoul, Jing He, Ammar Hawbani, and Sana Djaidja

Effect of Traffic Generator and Density on the Performance
of Protocols in Mobile Wireless Networks 823
Amine Kada, Hassan Echoukairi, Khalid Bouragba,
and Mohammed Ouzzif

Development of P2P Educational Service in Russia 833
Sergey Avdoshin and Elena Pesotskaya

Assessing Collaborative Learning with E-Tools in Engineering
and Computer Science Programs 848
Steven Billis and Oscar Cubenas

School Leadership Preparation and Technology Implementation:
Ensure Successful Change Processes Through Transformative
Mind Shifts... 855
Jeff Faust and Ted Price

Developing Active Personal Learning Environments
on Smart Mobile Devices 871
Brian Whalley, Derek France, Julian Park, Alice Mauchline,
and Katharine Welsh

Effective and Innovative Interactives for icseBooks 890
Y. Daniel Liang

Learning Analytics as a Sociotechnical System................... 904
Marcel Simonette, Mario Magalhães, and Edison Spina

Cluster and Sentiment Analyses of YouTube Textual Feedback
of Programming Language Learners to Enhance Learning
in Programming 913
Rex P. Bringula, John Noel Victorino, Marlene M. De Leon,
and Ma. Regina Estuar

Method for Estimation of Multiple Reflection, Scattering
and Absorption in Mountainous Areas of Remote Sensing
Satellite Data 925
Kohei Arai

**A Method to Input Secret Information Using an Eye
Tracking Device** 936
Hazuki Owada, Daiki Kamitai, Chinayo Shonen Inoue,
and Manabu Okamoto

**Web-Based Learning for Enhancing CSL Learners'
Language Proficiency in Singapore** 944
Liu May

**Deep Siamese Networks with Bayesian Non-parametrics
for Video Object Tracking** 950
Anthony D. Rhodes and Manan Goel

**A Distributed Ledger Based Cyber-Physical Architecture
to Enforce Social Contracts: Paper Cup Recycling** 959
Tarun Goel, Yingqi Gu, Francesco Pilla, and Robert Shorten

**3D Design and Manufacturing Analysis of Liquid
Propellant Rocket Engine (LPRE) Nozzle** 968
Samuel O. Alamu, Marc J Louise Caballes, Yulai Yang, Orlyse Mballa,
and Guangming Chen

Author Index 981

Mutually Authenticated Group Key Management Protocol for Healthcare IoT Networks

Firdous Kausar[1]([⊠]), Waqas Aman[2], and Dawood Al-Abri[1]

[1] Department of Electrical and Computer Engineering, College of Engineering,
Sultan Qaboos University, Muscat, Sultanate of Oman
{firdous, alabrid}@squ.edu.om
[2] Department of Information System, College of Economics and Political
Science, Sultan Qaboos University, Muscat, Sultanate of Oman
waqas@squ.edu.om

Abstract. Healthcare is one of the most promising application of IoT. Because of the critical nature of health-related data, it is important to transfer it securely on the network and allow only legitimate IoT devices to participate in the network. This paper proposed a mutually authenticated group key management protocol (MAGKMP) for healthcare applications of IoT. Each IoT device is properly authenticated before allowing it to join the network. It also provides the facility to IoT devices to authenticate the trusted servers and smart e-health gateways before sharing the session and group keys with them. For secure multicast communication, group keys are distributed securely after mutual authentication between different participating devices in the network. The analysis of proposed MAGKMP shows that it is secure against different types of attacks and provides mutual authentication and forward and backward secrecy in the group communication. It also offers load balancing among gateways by exploiting the context profile of gateways.

Keywords: Mutual authentication · Group key management · Secure multicast · Healthcare · Internet of Things

1 Introduction

Internet of Things (IoT) is the interconnection of varied kind of devices communicating and interacting by exploiting different networking technologies. IoT has become widely used in many aspects of our daily life. Healthcare is one of the most widely used application of IoT which allows continuous and remote monitoring of patients in hospitals or living alone at home and providing timely medical assistance. Healthcare IoT applications are much concerned about the security and privacy of health-related data of patients that collected through different type of health sensors and communicated on wireless network.

The conventional security solutions are not applicable for healthcare IoT applications because of the limited computation capability of medical sensor nodes. In order to provide end-to-end secure communication, it needs to use the cryptographic keys. It is

© Springer Nature Switzerland AG 2020
K. Arai et al. (Eds.): FTC 2019, AISC 1070, pp. 1–12, 2020.
https://doi.org/10.1007/978-3-030-32523-7_1

also important to mutually authenticate each party before sharing keys with each other. Secure group key distribution is an important service for secure broadcast or multicast communication in healthcare IoT network. Most of the current protocol provide pairwise key sharing for unicast communication but not emphasis on group key negotiation.

This paper provides mutually authenticated group key management protocol (MAGKMP) to provide secure group communication among nodes in health care IoT network. Our proposed architecture consists of three types of participating nodes in the network: (1) patient sensor nodes (PSN), (2) smart eHealth gateways (GW), and (3) context aware authentication server (CAAS). All nodes are sharing unicast and multicast keys after mutually authenticating each other without ever exposing the pre-shared secret keys on network. CAAS also provide load balancing among GWs.

Rest of the paper is organized as follows. Section 2 discusses the state-of-the-art work in the same area. Section 3 presents the network architecture. Section 4 provides the details of proposed MAGKMP. Section 5 performs the security analysis of proposed MAGKMP. Section 6 concludes the paper.

2 Related Work

IoT is mainly composed of resource-constrained devices. Conventional security measures may not be suitable for it. To address such issues, Datagram Transport Layer Security (DTLS) [1] offers a number of services including key management, authentication, and encryption. It can be utilized in four different settings; no security, symmetric, asymmetric, and certificate-based. The major drawbacks of DTLS is that it does not support IP multicasting, which could be useful in group keying scenarios. Moreover, since its handshake has many exchanges, it creates high communication overhead, particularly when deployed in dynamic IoT environment.

Multimedia Internet Keying (MIKEY) [2] is another standard protocol that establishes security secrets among the communicating parties. However, it has high-energy consumption concerns in symmetric key settings and scalability issues in asymmetric configurations. To make it suitable for the low-end devices in the IoT, a distributed MIKEY mode for eHealth application is presented in [3]. It uses the cooperation of third parties to relieve the heavy computations from the constrained devices. Symmetric settings are adopted in constrained region while asymmetric mode is utilized in the resourceful region. Similar approach can also be found in [4] where the authors have suggested a signcryption-based key distribution solution for MIKEY.

Farash et al. [5] have proposed a password-based key management scheme. It is a 2-party authentication scheme that authenticates a client to a server. The server performs the key management procedure, which establishes and exchanges the session key. The underlying cryptosystem allows the client to use either a symmetric key, an open key, or the server public key settings. Anonymity of the client is also protected. However, the scheme is complex and requires considerable computational cost to be realized.

A multiparty password authentication key exchange (M-PAKE) is proposed in [6]. Each server holds the password of each user, which uses it for user authentication and further use it to generate and distribute session keys. The scheme uses Elliptic Curve Cryptography (ECC) for keys generation. A similar attempt is also implemented in [7], where the authors have utilized a 3-party PAKE (P3-PAKE) and Diffie-Hellman to establish symmetric session keys. In the context of IoT, such schemes could be very complex as nodes may enters and leaves a particular network (server) dynamically. Moreover, they are computationally costly and have communication overheads thus, may not be feasible for low-end devices in the IoT.

In [8], the authors have presented a 3-party authentication scheme for health record exchanges. Whenever two medical institutes intend to exchange medical records, they are authenticated by a trusted server. After authentication, the server performs a key agreement process that establishes a shared session key that is used to protect the exchange. The scheme uses more than 15 exponentiations and does not protect the identity of the medical institutes. Therefore, poses a threat to privacy and tracking attacks.

Sanaah et al. [9] have described a key management scheme for smart homes. Two modes are used to protect information exchanges. The KEYEncrypt mode protects the exchange of the shared secret, whereas the DATAEncrypt is used to protect the data using the shared secret. The scheme does not address the authentication perspective, which is usually the preliminary step in key management processes.

In a more recent attempt, Zahid et al. [10] have proposed a distributed multiparty key establishment scheme. They utilized Chebyshev polynomials and Chaotic maps in their cryptosystem. Their scheme mainly consists of two phases. In Phase I, it verifies the group heads (IoT gateways) to a trusted server and creates a session key between them. In Phase II, member nodes are registered as well as inter and intra group sessions keys are established.

A lightweight key management protocol has been presented in [11] for IoT eHealth applications. The protocol uses the collaboration of the end nodes and trusted third parties. A symmetric key is generated to protect the communication channel. The key agreement phase has intensive public-key operations, including encryption, signing, and verification. Although, the computational cost is minimized by cooperating with a third party, the scheme has considerable communication overhead as it requires multiple entities to communicate the protocol messages from the nodes to the intermediate parties and then to the main server.

In [12], a group key management model is proposed for the IoT. The model uses a Context Aware Security Server (CASS) that manages and distributes multicast session and group keys. Considering that IoT environment changes dynamically, the CASS collects contextual information from the key distribution servers (KDS) and sensors, and assesses it to select an optimal KDS for the sensors. Context includes information such as location, status, supported groups and their registered members. The authors presented abstract level performance analysis, however it is unclear how the proposed model secure the model may defend against the fundamental threats associated with key management.

3 Network Architecture

Our proposed network architecture consists of three different types of entities i.e. patient sensor nodes (PSN), smart e-Health gateways (GW), and context aware authentication server (CAAS).

3.1 Patient Sensor Nodes

Patients are equipped with different type of health sensors e.g. body temperature, glucometer, position, accelerometer, pulse oximetry, ECG, blood pressure etc. All these sensors send their telematics to the central wearable microcontroller node. Patient sensor node (PSN) consists of different type of health sensors and central wearable microcontroller node on the patient body as shown in Fig. 1.

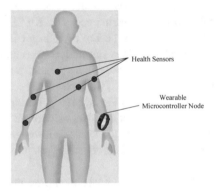

Fig. 1. Patient sensor node

3.2 Smart e-Health Gateways

Smart e-Health Gateway are responsible for gathering patients' health related data from PSNs in its zone, perform necessary processing on it and sends those records to the local switch and/or Internet.

3.3 Context Aware Authentication Server

The CAAS server performs as the trusted entity that is responsible for authenticating the PSNs before allowing them to join network/zone. It also acts as key distribution center that securely provide the session keys for communication between PSNs and GWs.

We consider CAAS as context aware in the sense that it also collects information about the environment from PSNs and GWs such as their locations, number of PSNs in each zone i.e. load on GWs, which PSNs are currently participating in which zone, accessible GWs in the proximity of each PSN and store that information in its database.

3.4 Network Model

We consider that a hospital area is divided into different zones as shown in Fig. 2. Each zone consists of a number of PSNs. Further, each zone is having its own gateway (GW) which is responsible for providing the services to all the PSNs in its zone. Each PSN periodically sends its health-related information securely to the GW of its zone. When the patient is moving across the hospital from one area to another, PSN joins one zone while leaving the other one. There is one central context aware authentication server (CAAS) which is responsible for authenticating each newly joined PSN, key distribution, and allocating it to the most appropriate zone with respect to PSN location and load on different GWs.

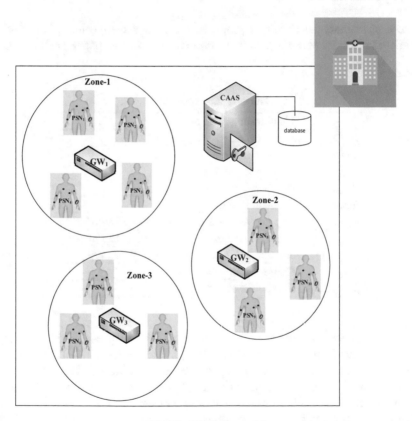

Fig. 2. Network model

3.5 System Requirements

- Secure: An eavesdropper should not be capable of getting any information about the patient telematics data transfer on the network between PSN and GW.
- Scalable: The overall system should be able to support large numbers of PSNs and GWs in the network.

- Reliable: The system should be distributed in nature and by failure of one GW should not result in the unavailability of supported services in the network.
- Mutual Authentication: Each PSN should be properly authenticated by CAAS and GW, and each CAAS and GW is properly authenticated by PSN before sharing secret key with each other.
- Context Awareness: The system should be aware of the context of environment. Contains profile for context information of GWs and PSNs with respect to their location information, load on GWs, and PSNs participated in each zone.

4 Proposed MAGKMP

This section presents the details of proposed mutually authenticated group key management protocol (MAGKMP). List of symbols use in the paper are described in Table 1 below.

Table 1. List of symbols

ID_{PSN_i}	Identification of PSN_i
ID_{GW_x}	Identification of GW_x
$nonce$	Random number
SK_{PSN_i}	Pre-shared secret key of PSN_i
SK_{GW_x}	Pre-shared secret key of GW_x
$location_{PSN_i}$	Location of PSN_i
K_{PSN_i}	Key derived from pre shared secret key of PSN_i
K_{GW_x}	Key derived from pre shared secret key of GW_x
$K_{PSN_i-GW_x}$	Session Key shared between PSN_i and GW_x
GK_{z_x}	Group key of zone x
$sessionID_{Z_x}$	Session ID of zone x

Each Patient Sensor node (PSN_i) is configured with its pre-shared secret key i.e. SK_{PSN_i} and unique identification number (ID_{PSN_i}). Each Gateway (GW_x) is configured with its pre-shared secret key i.e. SK_{GW_x} and unique identification number (ID_{GW_x}).

Context aware authentication server (CAAS) saves the information of all PSNs and GWs in its database. Each record of database consists of the unique identification number and cryptographic hash of the corresponding pre-shared secrete key of GW_x or PSN_i.

A newly arrived PSN_i when joins a network, it needs to discover its nearby gateway. It broadcasts the *Discover_GW* message on the network. The exchange of messages between PSN$_i$, GW$_x$ and CAAS is shown in Fig. 3. *Discover_GW* message consists of the ID of PSN_i and its location information.

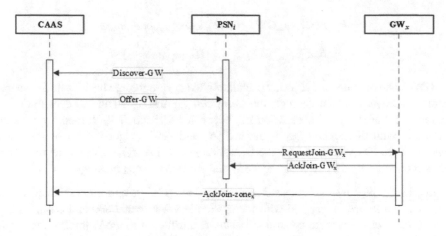

Fig. 3. MAGKMP message flow

$$Discover_{GW} = ID_{PSN_i}, location_{PSN_i}, nonce_1$$

CAAS when receives this *Discover_GW* message, it first searches the ID of *PSN_i* in its database. If it exists in the database then it replies with Offer-GW message after searching the most appropriate gateway in the proximity of *PSN_i* that would provide it with best service while controlling the load balancing on GW.

- *Offer_GW* message is encrypted with the cryptographic hash of the PSN pre shared secret key.
- *Offer_GW* message consists of ID of GW_x, session key between PSN_i and GW_x, and $Certificate_{GW_x}$.
- $Certificate_{GW_x}$ consists of ID of PSN_i and session key between PSN_i and GW_x encrypted with the key derived by calculating the hash on the GW_x pre-shared secret key i.e. SK_{GW_x}.

$$Offer\_GW = E_{K_{PSN}}(ID_{GW_x}, K_{PSN_i-GW_x}, nonce_1, Certificate_{GW_x})$$
$$Certificate_{GW_x} = E_{K_{GW}}(ID_{PSN_i}, K_{PSN_i-GW_x})$$
$$K_{PSN_i} = hash(SK_{PSN_i})$$
$$K_{GW_x} = hash(SK_{GW_x})$$

PSN_i first computes the K_{PSN_i} by calculating the cryptographic hash of its pre-shared secret key. It then decrypts the *Offer_GW* message with K_{PSN_i} and retrieve the id of GW_x, session key between PSN_i and GW_x and $Certificate_{GW_x}$.

It then sends the *RequestJoin_GW_x* message to GW_x that consists of the $Certificate_{GW_x}$ and $badge_{PSN_i}$.

- $badge_{PSN_i}$ consists of ID of PSN_i and random number ($nonce_2$) encrypted with the session key between PSN_i and GW_x.

$$RequestJoin\_GW_x = Certificate_{GW_x}, badge_{PSN_i}$$

$$badge_{PSN_i} = E_{K_{PSN_i-GW_x}}(ID_{PSN_i}, nonce_2)$$

GW_x first computes the K_{GW_x} by calculating the cryptographic hash of its pre-shared secret key SK_{GW_x}. It then decrypts the $Certificate_{GW_x}$ message and retrieves the information of ID of PSN_i and the session key between PSN_i and GW_x. It then decrypts the $badge_{PSN_i}$ with the session key between PSN_i and GW_x to get the ID of PSN_i and $nonce_2$. If the ID of PSN_i in both the $Certificate_{GW_x}$ and $badge_{PSN_i}$ are same then GW_x sends the $AckJoin\_GW_x$ message to PSN_i and $AckJoin\_zone_x$ message to CAAS.

- $AckJoin\_GW_x$ message consists of a group key (GK_{z_x}), current session id of the $zone_x$ and $nonce_2$ encrypted with the session key between PSN_i and GW_x.
- $AckJoin\_zone_x$ message consists of status confirmation of PSN_i joining the $zone_x$, total number of PSNs in the $zone_x$, and message authentication code (MAC) calculated on both these values by using the K_{GW_x}.

$$AckJoin\_GW_x = E_{K_{PSN_i-GW_x}}(GK_{z_x}, sessionID_{Z_x}, nonce_2)$$

$$AckJoin\_zone_x = Status, NumofPSN_{Z_x}, MAC_{K_{GW_x}}(Status, NumofPSN_{Z_x})$$

PSN_i decrypts the $ACKJoin\_GW_x$ message using the already shared session key between PSN_i and GW_x and accept the group key of $zone_x$ if $nonce_2$ of $ACKJoin\_GW_x$ matches with the $nonce_2$ of the $badge_{PSN_i}$ message sent by PSN_i. PSN_i then successfully join the $zone_x$ and participate in communication within $zone_x$ using the GK_{z_x}.

CAAS after receiving the $AckJoin\_zone_x$ perform message authentication by recalculating the MAC on $Status$ and $NumofPSN_{Z_x}$ and matching it with the received MAC, if both are same then it update its database with current number of PSN in $zone_x$.

4.1   Addition or Removal of PSN in the Zone

Whenever a new PSN joins or an old PSN leaves the $zone_x$, the group key of $zone_x$ has to be updated. For this purpose, GW_x generate a random number $(nonce_i)$ and compute the cryptographic hash of the current GK_{z_x} concatenating with the $nonce_i$ as follows:

$$Update\_GK_{z_x} = hash(GK_{z_x}, nonce_i)$$

GW then unicast the $Update\_GK_{z_x}$ message to each PSNs in the $zone_x$ by encrypting it with $K_{PSN_i-GW_x}$ as shown in Fig. 4.

GW also informs the CAAS in case if PSN leaves the zone by sending $AckLeave\_GW_{zone_x}$ message which consists of status confirmation of the PSN leaving the $zone_x$, total number of PSNs in the $zone_x$, and message authentication code (MAC) calculated on both these values by using the K_{GW}.

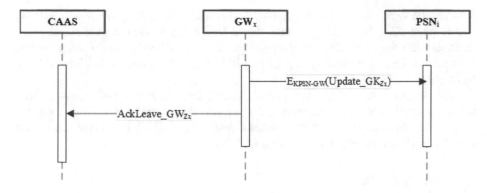

Fig. 4. Message flow for addition or removal of PSN

$$AckLeave\_GW_{zone_x} = Status, NumofPSN_{Z_x}, MAC_{K_{GW}}(Status, NumofPSN_{Z_x})$$

4.2 Periodic Rekeying

GW of each zone also periodically broadcasts the $Update\_GK_{z_x}$ message in their corresponding zones to update their group key. After receiving $Update\_GK_{z_x}$, all the nodes in the $zone_x$ updates the GK_{z_x} in the same manner as discussed above in Sect. 4(A).

5 Security Analysis

This section discusses the security analysis of proposed MAGKMP.

5.1 Mutual Authentication

Our proposed MAGKMP provides a method for mutual authentication where not only PSN is authenticated by CAAS and GW, but PSN also authenticate both the CAAS and GW before sharing a session and group key with them.

When PSN_i sends request to CAAS to discover GW, the CAAS responds with $Offer\_GW$ message, which is encrypted with the key derived from the pre-shared secret of PSN_i. As only CAAS knows the information of pre-shared secret of PSN_i, therefore PSN_i can authenticate CAAS if $Offer\_GW$ message is decrypted successfully and include the same value of $nonce_2$ as in $Discover\_GW$ message. CAAS authenticate PSN by making it sure that only legitimate PSN_i with valid pre-shared secret key can decrypt the $Offer\_GW$ message and can retrieve the session key between PSN_i and GW_x.

When PSN_i sends request to join GW_x, this message consists of $Certificate_{GW}$ issued by CAAS to PSN_i and ID_{PSN} and nonce encrypted with the session key K_{PSN-GW} provided by CAAS. As only legitimate PSNs can recover the session key between PSN_i and GW_x after decrypting successfully the *Offer_GW* message. GW_x authenticate PSN_i by making it sure that ID of PSN_i in decrypted $Certificate_{GW_x}$ matches with the ID of PSN_i in decrypted $badge_{PSN_i}$ in *RequestJoin_GW$_x$* message. This also make it sure that a $Certificate_{GW_x}$ is presented by the same PSN_i to which it has been issue by CAAS.

PSN authenticate GW if it successfully decrypts the *ACKJoin_GW$_x$* message with the session key between PSN_i and GW_x because only legitimate GW can get the session key between PSN_i and GW_x from $Certificate_{GW_x}$ which is encrypted with the key derived from the pre-shared secret of GW_x.

5.2   Replay Attack

Inclusion of $nonce_1$ in *er_GW*, *Offer_GW* messages, and $nonce_2$ in *RequestJoin_GW$_x$*, *ACKJoin_GW$_x$* prevents from replay attack. If an attacker capture *discover_GW* message and replay it after PSN_i leaves the network, it won't be accepted by CAAS because $nonce_1$ should not be repeated. If attacker capture the *Offer_GW* message and replay it later, PSN_i won't accept it because of repetition of $nonce_1$. If attacker replays either *RequestJoin_GW$_x$* or *ACKJoin_GW$_x$* messages, it won't be accepted by GW_x or PSN_i because of repetition of $nonce_2$.

5.3   Pre-shared Secret Key Security

Both PSN_i and GW_x successfully shared the pairwise session keys with each other and CAAS and group key of $zone_x$ without ever revealing the pre-shared secret keys on the network.

5.4   Eavesdropping Attack

Our MAGKMP is secure against eavesdropping attack. If attacker is capturing all the communication among PSNs, CAAS, and GWs, it will not be able to deduce any of the secret key. *Offer_GW* message is encrypted with the secrete key derived from the pre-shared secret of PSN_i which is only know to PSN_i and CAAS. Attacker can perform the brute force attack on it to get the session key between PSN_i and GW_x but It will take so long that PSN_i have already moved to new zone. Similarly, attacker cannot get any information from encrypted $Certificate_{GW_x}$. Further the group key of $zone_x$ is included in the encrypted *ACKJoin_GW$_x$* message and attacker does not know the encryption key i.e. $K_{PSN_i-GW_x}$ so cannot find GK_{z_x}.

5.5   Forward and Backward Secrecy

Every time when a new PSN_i joins or old PSN_i leaves the $zone_x$, group key of $zone_x$ is updated so that newly joined PSN_i should not be able to access the previous communication in the $zone_x$ or old PSN_i should not be able to see the future communication in the $zone_x$.

5.6 Central Administration of CAAS

We consider that our network model is distributed in nature with many gateways and each gateway is responsible of administrating its zone. In such a distributed environment by storing the pre-shared secret keys of PSNs on each GW poses a security risk and make it more difficult to manage. MAGKMP introduce the concept of centralized database stored on CAAS, which is responsible for authentication of PSNs and GWs before allowing them to join the network or zone. Although overall security of network is relying on securing this central database of CAAS, but it is much simpler to harden the security of single system instead of distributing the pre- shared secrets on many gateways.

5.7 Load Balancing

CAAS acts as load balancer also in the network. It is aware of the number of PSNs in each zone and assigns the PSN to a zone after considering the load on all nearby zones of this PSN. This way it manages the load on GWs in each zone.

6 Conclusion

Secure group communication is vital in healthcare IoT applications. A mutual authentication is crucial of each entity before participation in secure communication in the network. We propose a mutually authenticated group key management protocol for healthcare IoT networks. Patient sensor nodes successfully share the group key with their gateways while moving across the hospital. The security analysis of proposed MAGKMP proves that it provides mutual authentication, forward and backward secrecy, protection against replay and eavesdropping attacks. Further, it also provides central administration with the help of trusted CAAS in a highly distributed network and load balancing among GWs. As a future work, context awareness will be added in the current MAGKMP by deploying some deep learning techniques in healthcare IOT environment.

References

1. Rescorla, E., Nagendra M.: Datagram transport layer security version 1.2, No. RFC 6347 (2012)
2. Arkko, J., Carrara, E., Lindholm, F., Norrman, K., Naslund, M.: Mikey: multimedia internet keying, No. RFC 3830 (2004)
3. Abdmeziem, M.R., Tandjaoui, D., Romdhani, I.: A new distributed MIKEY mode to secure e-health applications. In: Proceedings of the International Conference on Internet of Things and Big Data, IoTBD, vol. 1, pp. 88–95 (2016)
4. Nguyen, K.T., Oualha, N., Laurent, M.: Novel lightweight signcryption-based key distribution mechanisms for MIKEY. In: IFIP International Conference on Information Security Theory and Practice, pp. 19–34. Springer, Cham (2016)

5. Farash, M.S., Attari, M.A., Kumari, S.: Cryptanalysis and improvement of a three-party password-based authenticated key exchange protocol with user anonymity using extended chaotic maps. Int. J. Commun. Syst. **30**(1), e2912 (2017)
6. Lu, C.F.: Multi-party password-authenticated key exchange scheme with privacy preservation for mobile environment. KSII Trans. Internet Inf. Syst. **9**(12) (2015)
7. Kwon, J.O., Jeong, I.R., Lee, D.H.: Practical password-authenticated three-party key exchange. KSII Trans. Internet Inf. Syst. **2**(6) (2008)
8. Lin, T.H., Lee, T.F.: Secure verifier-based three-party authentication schemes without server public keys for data exchange in telecare medicine information systems. J. Med. Syst. **38**(5), 30 (2014)
9. Al Salami, S., Baek, J., Salah, K., Damiani, E.: Lightweight encryption for smart home. In: 2016 11th International Conference on Availability, Reliability and Security (ARES), pp. 382–388. IEEE (2016)
10. Mahmood, Z., Ullah, A., Ning, H.: Distributed multi-party key management for efficient authentication in the Internet of Things. IEEE Access **6**, 29460–29473 (2018)
11. Abdmeziem, M.R., Tandjaoui, D.: An end-to-end secure key management protocol for e-health applications. Comput. Electr. Eng. **44**, 184–197 (2015)
12. Harb, H., William, A., El-Mohsen, O.A., Mansour, H.A: Multicast security model for Internet of Things based on context awareness. In: 13th International Computer Engineering Conference (ICENCO), Cairo, pp. 303–309 (2017)

Enhancing the Linguistic Landscape with the Proper Deployment of the Internet of Things Technologies: A Case Study of Smart Malls

Fahad Algarni[1]([⊠]), Azmat Ullah[2], and Khalid Aloufi[3]

[1] University of Bisha, Bisha, Saudi Arabia
Fahad.alqarni@ub.edu.sa
[2] La Trobe University, Melbourne, Australia
A.ullah@latrobe.edu.au
[3] Taibah University, Madinah, Saudi Arabia
Koufi@taibahu.edu.sa

Abstract. Major advances in the Internet of Things (IoT) technologies are offering unprecedented automated services to various sectors. IoT has shown great promise in transforming buildings and institutions into smart buildings and institutions, making traditional task management more efficient. In particular, there has been remarkable progress in successfully integrating IoT into various contexts, such as smart homes, businesses and cities. Success in such technological transformations is not only due to the proper deployment and implementation of IoT but also to the effective employment of related technologies, such as big data and cloud computing technologies, and making informed decisions on the inclusion of the required hardware and software for all the elements involved in the IoT environment. While a significant amount of IoT research in various contexts has been proliferated in recent years, attention given to its deployment for linguistic landscapes (LLs) is still very rare. The aim of this paper is to propose a smart Al-Noor Mall LL which takes into account the appropriate deployment of IoT tools and relevant technologies. The paper also thoroughly discusses the background of the field, covering IoT technologies, cloud computing, and big data studies, followed by a review of the recent publications involving the successful deployment of IoT technologies combined with big data and cloud computing in various contexts. The literature also covers the success factors that contribute to the robustness of the required real-time operating system (RTOS) in an IoT-based environment in order to enable IoT devices to manage their resources efficiently. Based on a detailed analysis of the literature, a three-level framework (i.e. the object, communications and application levels) together with the elements that are required for IoT-based smart LL solutions is proposed. A case study of Al-Noor Mall is used to validate the proposed framework and the results indicate that IBM Bluemix can help Al-Noor Mall management and customers to locate any particular shop in the huge shopping mall, which thereby makes the customers' shopping experience more satisfying. Finally, the challenges (i.e. energy efficiency, security, and intelligent data analysis) facing the successful deployment of IoT technologies are discussed with proposals for future research directions in the IoT research field.

© Springer Nature Switzerland AG 2020
K. Arai et al. (Eds.): FTC 2019, AISC 1070, pp. 13–39, 2020.
https://doi.org/10.1007/978-3-030-32523-7_2

Keywords: IoT · Linguistic landscapes · Smart environment · Energy efficiency · Big data · RTOS · Digital transformation

1 Introduction

Over the past decades, scholars have produced a significant body of research pertinent to language studies. The research on linguistic landscapes (LLs) covers the science of public signage and focuses on all of the linguistic objects that mark certain places such as schools, airports, businesses, shopping malls, parks, educational institutions, etc. It comes under the linguistic discipline as a subdivision of broad language research. An LL was initially defined by Landry and Bourhis [1], who stated "the language of public road signs, advertising billboards, street names, place names, commercial shop signs, and public signs on government buildings combines to form the linguistic landscape of a given territory, region, or urban agglomeration".

The LL as a field of study has been proven to be a significant area for examining the dynamics of the main characteristics of social life due to the fact that it both shapes and is shaped by social and cultural associations [2]. Signs which form the LL are characteristically categorized as either official, such as guidance and official warning signage created by governmental bodies such as councils, or non-official signage such as that found in commercial locales such as advertising for products and services, which is intended to attract customers' attention [3]. Both types of signage tend to disseminate various messages in symbolic and/or informational forms to the target audience.

The IoT is generating an increasing amount of discussion both inside and outside the context of a shopping centre. It is a notion that not only has the potential to influence how people live but also how they work in the workplace. It is defined as a system of interconnected computing devices, software, digital and mechanical machines and people all of which have a unique identifier and the system has the capability of transmitting data over a network without any internet. This transmission does not require any computer-to-human interaction or human-to-human interaction. Therefore, the IoT is very useful in the context of LLs in shopping centres. The aim of this paper is to propose suitable IoT tools and techniques to make Al-Noor shopping centre smart, as it is common for shoppers in large shopping centres to experience difficulties in locating particular shops.

2 Context of Our Work

This section summarises the context of this research, which encompasses Al-Noor Mall, the IoT and smart LLs.

2.1 Al-Noor Mall

Saudi Arabia is a country, which accommodates diverse and multiethnic population groups from various cultural and linguistic backgrounds [4]. Al Madinah is the second most important holy city in Saudi Arabia after Makkah, the most important holy city, both attracting over seven million pilgrims/religious visitors every year from all over the globe to participate in specific activities like prayers and visiting the Prophet's mosque [5]. Hosting such a huge number of visitors, both from within Saudi and from other countries, even for a short period of time, is a huge challenge for the relevant authorities to ensure facilities are equipped with the required support. According to the latest population statistics published by the Saudi General Authority of Statistics, the country had a total population of 32,552,336 people in 2017, with Al Madinah having a total of 1,265,561 Saudi residents and 516,172 non-Saudi residents [6]. The non-Saudi population comes from different cultural and linguistic backgrounds. The Al Madinah region is shown in Fig. 1.

Fig. 1. Al-Madinah region highlighted in bold black

Fig. 2. The Al-Noor Mall (highlighted by the red location marker) (source: Google Maps)

Al-Noor Mall is one of the major marketplaces located on King Abdullah Road, Al-Madinah, approximately 11 km north of the Al-Madinah city centre, Saudi Arabia (see Fig. 2). The mall comprises more than 100 businesses and services in a single location. Al-Noor Mall houses a variety of restaurants, cafes, clothing and fashion shops, gifts shops and fresh food stores and other professional businesses including insurance companies, public services, clinics and banks. The mall is managed by the Alhokair Group, which has been one of the leading property management groups in Saudi for decades. The group was established in 1975 to invest in the sectors of entertainment and hospitality under the leadership of Abdulmohsin Alhokair. Over the last five decades, the group's projects extended to incorporate 91 entertainment centres and 35 hotels spread across the Kingdom of Saudi Arabia and the United Arab Emirates [7].

The Al Madinah Regional Authority (MRM), the government body presiding over the region, has a wide range of responsibilities involving but not limited to: preserving the identity and privacy of the Madinah area, ensuring the safety and cleanliness of the community, achieving the required quality of public satisfaction, providing business support including business start-up advice, along with the delivery of customer care programs, preventing pollution, noise and disease [8], and other services that also involve engaging in the creation and viewing approval of public/private signage. MRM recently announced that Al-Noor Mall is considered to be a significant landmark of the Al Madinah region, under the Al-Madinah Region Development Authority's (MRDA) initiative of cultural and historic awareness, as, being built in 2008, it was the first modern and comprehensive mall established within the city [9, 10] (see Fig. 6). Al-Noor Mall is the largest shopping mall in Al Madinah, comprising three levels. With the linguistically diverse nature of Al-Noor Mall's local and seasonal shoppers, different languages have dominated the written shop signs within the mall. The two languages that are most commonly used are Arabic and English, and some examples are presented in Fig. 3. This can be an obstacle for people who speak other languages, especially in Al Madinah, the city that hosts millions of people from different linguistic and cultural backgrounds every year [11].

Fig. 3. A landmark sign placed in front of Al-Noor Mall

Figure 2 shows the location of Al-Noor Mall and Fig. 3 shows an example of signage at the front of the mall. In previous LL research, studying a specific locale proved to be crucial, particularly where the residents living in the surrounding areas are from diverse backgrounds, such as immigrants from diverse countries with different linguistic and cultural backgrounds [12, 13]. To ensure the inclusiveness and rigor of the analysis, a particular section of the mall was chosen for this study, which can assist in focusing efforts on a specific area of a manageable scale, taking into account the limitations of time and resources allocated to the study. Figure 4 shows some examples of the Al-Noor Mall LLs, where the majority of signage uses a combination of Arabic and English with a mixture of symbols and/or pictures, with different colours and designs put together into a spatially definable frame.

Fig. 4. Commercial and guidance signs using Arabic and English at the Al-Noor Mall

2.2   IoT and Smart LLs

The use of contemporary technology and multimedia has contributed positively in the improvement of traditional LLs. For instance, signage is becoming more innovative, especially with the addition of dynamism, flexibility and interactive features, rather than traditional signs which are static. Dynamic signs have now evolved where the employment of small screens has replaced traditional signs. Such screens have dynamic functionalities due to the fact that they are connected to computers, which enables content to be updated remotely, based on specific needs. The deployments of small screens in commercial areas have increased exponentially and have contributed to the current shape of LLs [14]. In addition, technology has resulted in more flexible LLs and enables LL signage owners to update the content of their signs in an effective manner to meet their required objectives. Technologically advanced LLs are interactive, enabling customers to discover information or find directions by touching the screen. Digital screens are supported by a special antenna that connects to the data storage server/s for remote content update purposes and/or to add more interactive features.

Hyun and Ravishankar [15] recently designed a smart system that enables visually impaired people to navigate through buildings. The system involves three elements, namely an assistive long cane with a Bluetooth low energy receiver built in, a sign with a wireless connection and a smart device. When the visually impaired person approaches the sign, signals are exchanged between the system components and the required information and/or instructions are provided to deliver the required assistance. Therefore, the successful incorporation of contemporary technologies has the potential to effectively shape the LLs of various contexts in ways that significantly assist consumers with visual impairments or those with other special needs. Despite the continued efforts to utilise technologies for the development of LLs, no research, as yet, has focused on the deployment of IoT technologies to enhance the current LLs.

In this section, a cloud-centric, three-level framework comprising the object level, the communication level and the application level, is proposed to incorporate the IoT into a LL, based on a detailed review of the literature in similar contexts, namely smart homes [16], smart buildings [17], smart cities [18], and smart environments [19]. It is evident that the proper deployment of IoT technologies in these scenarios has resulted in more effective, efficient and productive task management. The following three sub-sections further explain each level of the proposed multi-level smart LL framework.

The Object Level

At this level, objects/things are fitted with sensing and/or actuating capabilities. For the LLs, every shop sign must be outfitted with sensors that are wirelessly connected to the relevant shop and the mall servers. These objects and/or things can then sense, actuate, process data, and communicate over the wireless networks, similar to those installed for the IoT in smart homes [20]. It is also imperative to consider that sensors at the object level are required to process the data collected at the initial phase prior to forwarding this to the communication-level components, such as the related hubs, gateways, and data storage in the cloud [21]. When IoT is effectively deployed in LLs, customers can be better served if they have their smart device equipped with a smart application. For example, when a shopper walks around a shop, sensors can detect the shopper and communicate specific information (translated if required) about products and/or services, provide personalized messages and/or directions, and display offers as appropriate.

In some scenarios, actuators may also need to be attached to the LLs, especially if certain dynamic functionalities are required. Therefore, Al-Noor Mall shop owners can gain an advantage from the proper deployment of IoT technologies (i.e. smart objects equipped with the appropriate RTOS as discussed in earlier sections) and have better opportunities to provide and distribute updated information about their services, goods and facilities to the cloud. This also can assist Mall management to analyse the customers' requirements over a certain period of time to be able to provide practical advice to shop owners to ensure better decision making for future arrangements. Overall, the quality of objects, things and other related peripheral devices must be considered in the early design phase to support the successful employment of IoT technologies. Other challenges such as battery dependency and security are discussed in latter sections. The object-level and its components are illustrated in Fig. 5.

The Communication Level

At the communication level, three components are taken into account, the hub, the cloud and big data. The hub is a device that collects raw and/or processed data from things (i.e. the object level) and forwards them to the cloud or vice versa for storage and analysis purposes. Hubs are normally required to perform some local data processing after the data has been gathered from the smart objects, things and peripheral devices to moderate the amount of data being sent to the cloud and to eliminate redundancies [21, 22]. Similar to a smart home setting, the hub directs commands to smart things, objects and peripheral devices to act as a local planner or regulator [20]. For the purpose of this paper which aims at developing smart LLS, the hub is called the smart hub, as shown in Fig. 5. Smart hubs, in this case, are required to be efficient in order to be able to fully understand the communication protocols used by these smart things, objects and peripheral devices and to effectively regulate the data flow between them or to the cloud. As a consequence, communication difficulties and expected challenges will need to be addressed and solved to ensure a seamless workflow among the IoT elements [23, 24].

The second component in the communication level is the cloud. As explained in the introductory parts of this paper, the cloud provides the required infrastructure to run IoT data effectively. In the proposed smart LL setting, illustrated in Fig. 8, three cloud models can be utilised as appropriate. These models are Software as a Service (SaaS), Platform as a Service (PaaS), and Infrastructure as a Service (IaaS). Each has its own benefits as well as disadvantages. The SaaS model enables customers to access software or applications from different client devices through the Internet and they do not need to manage or control underpinning cloud infrastructure, such as servers and/or operating systems, among others. The PaaS model enables customers to arrange the relevant cloud infrastructure and gives them the ability to create the required applications using programming languages and tools maintained by the cloud provider. The IaaS model provides customers with processing, storage, networks, and other fundamental computing resources from the cloud providers, offering the highest flexibility of the three models [25]. To choose the optimal model for this case study, it is necessary to understand the differences between SaaS, PaaS and IaaS to support successful deployment of IoT technologies at the Al-Noor Mall.

The third component in the communication level is big data. As the amount of data generated by smart things, objects and peripheral devices is huge, it forms the big data component within the relevant cloud that is responsible for storing and processing such data [26, 27]. In the proposed smart LL setting, customers who have a specific smart application installed on their smart devices are able to send inquiries and receive information on for example, directions, services, products and events in their preferred language from neighbouring shops, restaurants, businesses etc. Such timely communication comes from updated information stored in the cloud. Sensors embedded in neighbouring businesses, shops etc. sense and communicate information to the cloud through the smart LL hubs. For instance, a customer who only speaks Urdu may be interested in purchasing a suitcase from a shop in the Al-Noor Mall. In this case, nearly all business signs in the Mall are written in Arabic and English. Thus, the customer opens the smart application on arrival and types in the product name (suitcase) in Urdu and the search term is then internally translated and the smart LL system provides a list of available products in the mall. The list is translated into Urdu for the customer and the smart application provides the required information and guidance as appropriate. Figure 5 shows the communication level for the smart IoT-based Al-Noor Mall LLs, including the three components, namely the smart hub, the cloud and the big data.

The Application Level
The application level involves smart application/s for the mall's management staff, business owners, customers and other relevant stakeholders. In order to properly deploy IoT technologies for the mall's LLs, the developed application/s must be effective, multilingual, responsive and smart. Normally, the smart application is required to provide real-time solutions for end-users' demands. The application should be designed in a way that delivers a personalised, intelligent and seamless shopping experience. It

should also provide customers with the required instructions on how to utilize the developed technologies in an operative and efficient manner. In order to harness the anticipated potential of such advanced technology and provide the advantages of time saving and informed decision making for the users, the smart application must be predictive and have the ability to provide actionable data for future shopping experiences. On the other hand, customers must have an adequate understanding of how to use the smart application. For example, there are seven mobile phone service provider businesses in Al-Noor Mall, so when a customer wants to purchase a certain package (i.e. a specified speed within a specific budget), they can obtain a list of prices from all seven providers within the mall area by utilising the intelligently developed application and make an optimal choice instead of physically walking into each mobile phone service provider to request information on their preferred package. Comparably, in a smart IoT-based home scenario, the applications developed take into account the personalized advice based on the customers' requirements in order to deliver more efficient home management tasks with the proper deployment of IoT elements [28]. The application level is illustrated in Fig. 5, taking into account the possible applications that can be involved in order to better serve all stakeholders in the mall.

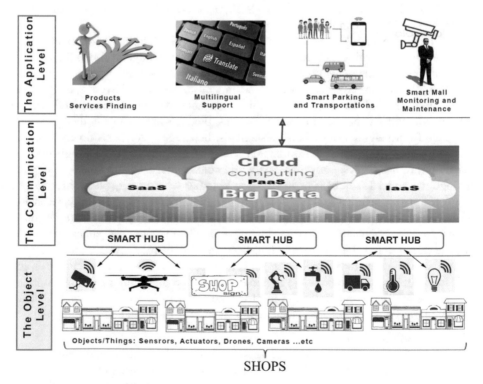

Fig. 5. IoT cloud-centric framework involving the required components for smart Al-Noor Mall LLs

3 What Is Required to Achieve the Smart Digital Transformation of Al-Noor Mall?

The following concepts, tools and techniques are required for the smart digital transformation of Al-Noor Mall.

3.1 IoT

The concept of the IoT involves a network of devices such as electronics, sensors, actuators and software, which enable different devices to exchange data and to connect. The IoT creates an opportunity for the direct integration of computer-based systems. As a consequence, several economic benefits are achieved due to a reduced level of human intervention [29]. The IoT has been adopted in shopping centres and public centres and is expected to be used in the development of infrastructure in the upcoming years [30]. The monitoring of public infrastructure also requires the IoT to ensure that safety is guaranteed and that emergency response coordination is achieved. The IoT can be used to transform various buildings into smart facilities.

3.2 IoT Operating Systems

Fundamentally, operating systems were defined as system software that regulates the roles for computerized devices (i.e. involving both hardware and software) and manages the encompassed resources in order to provide the desired services [31]. Different types of operating systems have been developed based on specific requirements. These involve, but are not limited to, single- and multi-tasking systems, distributed systems, and real-time operating systems (RTOS) (Tanenbaum 2009). For many IoT scenarios, RTOS is essential to handle sensitive tasks more effectively and efficiently.

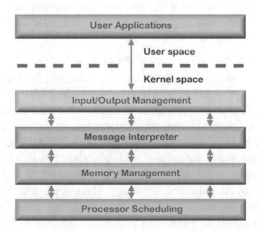

Fig. 6. Traditional OS architecture [32].

The differences between traditional operating systems, traditional operating systems comprise five layers: user applications, input/output management, message interpreter, memory management and processor scheduling, as shown in Fig. 6. RTOS, in which missions to be achieving goals and objective in a timely manner are critical. The RTOS architecture consists of task management, communication and synchronization, memory management, timer management, interrupt and event handling and device I/O management, as shown in Fig. 7.

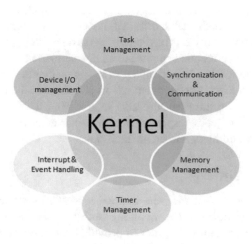

Fig. 7. RTOS architecture [33].

Proofread In recent years, there has been a significant shift toward the utilization of RTOS. This shift has occurred in both general purpose electronic devices and IoT smart devices as a result of the increased need for mission-critical tasks for daily life, such as heart monitoring and vital resources management, where operating systems must work efficiently with no or minimal errors.

Operating System Efficiency (Low-Power IoT Devices)
Operating systems are used to manage resources in the IoT. Operating systems fall into two categories: gateways and devices which use endpoint nodes [34]. End devices tend to be smaller in their capability compared to gateways which have a larger capability. Traditional devices which used to run without an operating system are now using a new form of operating system that is customized for the IoT [34]. This is mainly due to a higher demand for processing capabilities which is made possible by networks. The use of an operating system in the IoT promotes a revolution of technology and added constraints that need to be addressed. Currently, there is an existence of a close source of the IoT and an open source [34]. This is because the IoT has enhanced connectivity and the use of applications and hardware properties. Therefore, it is expected that the market will sustain multiple users of operating systems.

The devices used in the IoT usually require an operating system to ensure that they are able to carry out their functions, particularly because these devices tend to be complex. This complexity has increased particularly as gateways are increasingly adopting the use of more sensors which are required to perform functions to promote an increased connectivity and data processing capabilities [34]. In many cases, devices have more user interfaces which have face recognition, graphical displays and voice recognition [34]. Moreover, the devices that have an 8-bit operating system are currently adopting 32-bit architectures even as the complexity of these devices increases and costs decline significantly. Operating systems have been used to simplify tasks and to ensure that programmers are able to solve low-level challenges.

It is important that operating systems are efficient to ensure that resources are managed efficiently. This is essential for an effective memory management mechanism and so that an effective process is developed to ensure that operating systems run smoothly. The use of an efficient energy management system in operating systems is therefore considered to be significant [35]. To achieve this objective, it is essential for signal processing, communication design, data reception and data transmission to be executed efficiently to reduce energy consumption. Low-power IoT devices tend to send sensor data infrequently, especially if these devices are connected to a fixed power supply [36]. As such, the efficient management of energy in operating systems is necessary to ensure that strategies, mechanisms and effective data structures are created.

Low power IoT devices can be used to ensure that they use the lowest energy in endpoint functions, such as processing, capturing and sharing data online. Recent advancements in technology have allowed companies to design small-sized, low-power devices that can be connected to the Internet [37]. This mechanism enables these devices to cover a range of applications to enable them to be used in smart homes, healthcare monitoring and smart buildings. The reliability of IoT devices has also been enhanced by the development of low power devices.

Low power IoT devices are preferred since they are easy to use and incur low costs in their management. Furthermore, these devices are efficient and making it possible to connect to smartphones, personal computers and tablets [38]. Power management techniques can be applied to ensure low power IoT devices do not consume large amounts of energy. A wireless sensor node could be used to ensure that the battery power of these devices is saved. Most devices which are interconnected to the IoT use battery power [39]. New technologies have therefore been developed to ensure that the battery life of these devices can be extended without being detrimental to the performance of the device.

As shown in Fig. 8, six factors influence RTOS robustness: security, which refers to the security of the device through a secure boot; portability, which refers to the data transformation from one computer to another; connectivity, to support different connectivity protocols, for example Wi-Fi and Ethernets; scalability, which refers devices scalability; reliability, means, system should check reliabilities of network components; and energy-efficiency, refers to the management of energy uses.

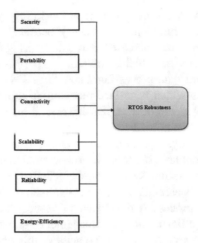

Fig. 8. Factors influencing RTOS robustness.

3.3   Cloud Computing

Cloud computing allows information technology to be accessed through a configured system of resources. It is through cloud computing that an organization can create a reliable system of sharing resources to ensure that efficiency and economies of scale are achieved in an organization [40]. Cloud computing allows companies to focus on their core businesses instead of using their resources on maintenance and computer infrastructure. The advantage of cloud computing is that it enables an organization to reduce its infrastructure costs by ensuring that are being able to run their applications rapidly [41]. Through the use of cloud computing, companies can use storage on a needs basis rather than incurring unnecessary costs. There are companies which offer cloud computing services and charge based on company usage.

Types of Cloud Computing
Cloud computing is defined based on its location or on the service that is offered by the cloud service providers. In relation to the site of cloud computing services, it can be categorized as a hybrid, public, private or a community cloud. In relation to the services that are offered, cloud computing is classified according to whether it offers software, infrastructure or whether it is a basic platform for data storage [42]. A public cloud describes computing infrastructure, which is located on the premises of a cloud. A private cloud ensures that the information which belongs to one organization is not shared with others. This form of cloud storage could either be located on the company premises or outside the premises.

Hybrid cloud storage is a type of cloud storage that uses both public and private clouds for information storage. Infrastructure as a service is another type of cloud storage that is made possible through storage disks and virtual servers. This is offered by companies such as Amazon [23]. Platform as a service cloud storage allows companies to use an operating system and a programming language in the storage of information. These services are offered by Microsoft Azure and the Google App engine. Software as a service is another type of cloud storage method that allows users to access information through the use of different software applications through different modes. These services are offered by Google Docs and Gmail.

3.4 Big Data

Big data is a descriptor of datasets which are complex and occupy large spaces which conventional data processing software is unable to handle. The challenges of managing big data arise from the difficulty of handling voluminous information [43]. Identifying the data source and ensuring that information privacy is guaranteed is considered to be a significant challenge in handling big data. Big data is analysed to identify business trends and to prevent crime [44]. Private organizations can use big data to monitor the consumption habits of their consumers. Big data is relevant to companies which handle a large number of transactions per day [45]. The IoT and big data work in tandem and the use of big data promotes a high degree of interconnectivity. The IoT gathers sensory data which is helpful for retail businesses to improve the shopping experience of customers.

3.5 Role of IoT in Transforming Buildings and Shopping Centers

The IoT has had a significant impact on building and construction. This is because it is used to transform multiple facets of buildings and how construction designs are implemented [46]. Using IoTs, a three-dimensional model of a building can be created through a building information model [47]. This is critical to ensure that builders can update their building plans based on their needs. IoT sensors can also be used after the completion of a building to monitor temperature, trends, energy usage or the movement of people in a building [48]. The information gathered from these models can then be analyzed to ensure that future building projects are improved.

The IoT is also relevant in the building industry as it enables building contractors to monitor the energy output of green buildings. Through the IoT, it is possible for companies to ensure that the building is more sustainable by automatically switching off devices when they are not needed [27]. As a consequence of the use of the IoT, the building process has become affordable [49]. Furthermore, heavy construction equipment is fitted with sensors to ensure that monitoring is possible as a way of making sure that maintenance issues such as excessive vibrations and temperature fluctuations are appropriately monitored [50]. Abnormal patterns can be monitored and maintenance efforts can be carried out before heavy and sensitive machines break down.

The IoT has also been known to improve the shopping experience as it improves customer engagement by improving communication capabilities and monitoring the products which are marketed to consumers [42]. The IoT has also been effective in allowing consumers to give their feedback to allow organizations to improve the products offered [51]. Through the use of location-based technologies such as Bluetooth or Wi-Fi or eBluetooth, retail stores can gather the behavioral trends of consumers which can be used to create a marketing strategy.

3.6 Peripheral Devices

Peripheral devices can be designed to detect environmental changes through the use of sensors which can be used in everyday objects such as touch-sensitive elevators [52]. Sensors have been used to monitor shopping trends in retail stores by installing smart lighting which is adaptive to the existing weather [42]. Sensors have also been used in smart lighting which adjusts itself according to the preferences of the customer.

Actuators are used to convert energy into motion. They are categorized according to the energy source. Hydraulic actuators use liquid to generate motion while thermal actuators use heat to generate motion. Pneumatic actuators use compressed air to generate motion whereas electric actuators use an external power to generate motion [53]. One of the most common examples of actuators is a brace which is wearable during exercise and is used to determine the heartbeat of an individual or their blood pressure.

Brands of Sensors used in Smart Contexts
There numerous brands of sensors that are currently deployed for smart contexts. Examples are as follows: Smart Video Doorbell manufactured by SkyBell that enables homeowners to see and talk to visitors who are at their front door, the Intelligent convection oven manufactured by June which allows users to monitor the cooking process and the temperature through the use of inbuilt cameras, the Smart Smoke Alarm manufactured by Roost has leak detection sensors and a smoke detection capability to alert users of rising smoke levels in the house, and the Await Glow manufactured by Await is an air monitoring device which enables users to monitor the level of toxins in the air and the chemicals present in a room [54].

Brands of Actuators used in Smart Contexts
There are many brands of actuators that have been used effectively in various smart contexts. An example is the Smart Light Switch manufactured by Deacon that employs Bluetooth connectivity so that homeowners are able to control the lighting of their rooms using their smartphones. Also, the Cloud-based IoT SaaS platform manufactured by Jasper is another example that is ideal for startup businesses and small and medium enterprises to assist them to manage and digitize their services. In addition, Samsara Company has manufactured the IoT data platform, which is used in fleet management where users are able to track the location of all vehicles, and monitor asset utilization and energy levels. Finally, the Smart Wi-Fi LED bulb by TP Link enables homeowners to control the lighting in their homes through the utilization of Wi-Fi capability. Using this actuator, users are able to track energy consumption, track the brightness of the bulbs and the colour of the bulbs in a room [54].

Advantages of Peripheral Devices
Peripheral devices are beneficial since they assist in file sharing across devices, enabling users to access information remotely if it is stored on connected devices [55]. Moreover, peripheral devices are used in resource sharing because they allow information and devices to be used by different users, helping organizations to save money and resources.

Peripheral devices can also be used in organizations to share an Internet connection to protect their systems. In effect, this allows the company to increase its storage capacity since they can store more information on this peripheral device [36]. Instead of relying on a computer to store information, it is possible for organizations to use peripheral devices to ensure that they develop an extra capacity at a lower cost.

4   Case Study with IBM Bluemix

In the business environment, there is a huge demand for cloud computing and the IoT. This is because of the need for decentralised data as well as the continuous monitoring of various devices integrated with businesses. Cloud computing enables access to database with no restrictions on time and location with elasticity ('pay as you go') in relation to the storage and the IoT incorporates sensor devices that are used in daily life such as smart phones, smart vehicles, etc. Consequently, the integration of cloud computing and the IoT has resulted in a huge growth in data analytics, machine learning and artificial intelligence.

IBM Bluemix has many features and services that help a user in creating projects and applications that can help businesses. It has the IoT Foundation that uses the IBM Watson IoT Platform. Early IoT applications performed three tasks: connecting objects, monitoring applications and managing and analysing the objectives.

This section discusses the case study of Al-Noor Mall in Saudi Arabia, which is an attractive shopping destination for diverse groups of visitors to Almadinah city, coming from various cultural and linguistic backgrounds. However, due to the expanding demand of shoppers, the mall is unable to deliver products and services equally to diverse customers. Further, the mall is facing the following problems:

- Firstly, it is clear that the specific demands of customers with various cultural and linguistic backgrounds require attention.
- Secondly, there are only two languages used in the mall's linguistic landscape.
- Thirdly, the use of contemporary technologies and smart applications has not been implemented to address the existing challenges.

The use of technology can provide possible solutions for the problems facing Al-Noor Mall. Bluemix can integrate the various services managed by application providers which will help Al-Noor Mall to improve the customer experience by using IBM Cloud (Bluemix) and Watson IoT technologies. Potential business solutions are as follows:

- It can be assumed that as business grows, there is a substantial requirement to integrate cloud technology in the business to store information. Furthermore, as the cloud provides 24×7 services and a dynamic database can be made accessible to all employees anywhere, this will enhance the services provided to customers and will help Al-Noor Mall management to efficiently address a wider range of customers.

- By implementing IoT technology and developing the relevant smart application, Al-Noor Mall customers will be able to find the location of their desired products/services in a timely manner. By using IoT's geospatial information and telemetry capability, customers can find the location of their desired shops. This will save time and overcome the language barriers.
- The IoT-enabled smart application also provides a notification system that can be used to notify Al-Noor Mall management and customers about any updates or offers. As a result, customers will be assured that they have received the latest offers and updates which are delivered in a fast and convenient manner. This will significantly improve the traditional shopping experience of each customer using Al-Noor Mall smart application.
- Shop owners and service providers must update the dynamic database constantly to ensure precise services are provided to their customers.

Appendix A below demonstrates how the proposed system works using IBM Bluemix.

5 Discussion

The employment of digital technologies for traditional signs has made them more effective and interactive. Such employment has been experienced with the use of digital signs which allow users to touch the screen to find out more about certain information, directions, and services [15]. The content of digital signs can be updated remotely when connected to a server or by attaching a special antenna that enables connectivity for updating purposes. The proper deployment of IoT-enabled technology can significantly improve the LL further for any shopping center, especially when they have been equipped with the right selections of hardware and software.

The IoT has the potential to improve customers' shopping experience. This is because the IoT allows companies to demonstrate that they are committed to overcome environmental and social issues in the society. If a business demonstrates a positive attitude to sustainability, the IoT will improve the sales of an organization by attracting customers who are attracted to the virtues of sustainability [56]. The IoT can also improve the shopping experience of customers by improving customer service which leads to improved sales. This is because this forum provides detailed data to demonstrate the flow of customers through their stores. Through the IoT, retail stores can monitor the movement of customers and their purchases on an online platform.

Shopping centres can be transformed into smart centres using consumer data to create an environment which makes the shopping centre comfortable. During winter, smart data can be used to ensure that the temperature inside the shopping centre is maintained at an appropriate level [42]. A certain kind of music could be played in the retail store if devices are installed to detect the customer's preferred genre of music.

Smart strategies and solutions can be used to transform shopping centres into smart centres. This can be done by ensuring that high energy consuming buildings and shopping centres use less energy to ensure that they are cost-effective [36]. The use of the IoT in shopping centres can be redesigned to ensure the connection of all shops in one shopping centre and their customer can locate any shop through app. This can be done by transforming the shopping centre into an entertainment destination and a meeting place.

Smart stores can be created by enhancing the customer experience by connecting with them through social media within the retail store and even when they are not in the retail store. User-generated content can also be used in a retail store to allow consumers to interact with social media accounts [30]. An interactive map could also be made available through a smartphone application to assist consumers to navigate the retail store. Augmented reality mirrors could also be installed in retail stores to give shoppers different viewing angles.

IoT helps shopping centres save energy, for example, analytical equipment maintenance is undertaken to save and manage energy, forecast equipment failure or sense other matters. For instance, each shop in the centre has a lot of composite equipment. If all these equipment's are connect to the central point and make it possible to monitor the health of all equipment's, could save the every and many other resources.

However, the problem with IoT is security. IoT is a general term that includes, different devices, people, network, cloud computing, intranet, etc. and its goal is to connect all stakeholders and establish communication among them. However, today, IoT devices are a hacker's dream as IoT devices are largely unsecure and provide simple access points for malicious action to be taken and for sensitive information, for example, a shopper's personal information, to be accessed.

6 Conclusion

The IoT has played an integral role in transforming many sectors into smart sectors, where task handling is more effective and efficient. The use of IoT approaches have contributed to creating new business models which, in turn, improve the shopping experience of consumers [23]. The IoT ensures that risks in new markets are sufficiently reduced and that business revenue is increased. In practice, the IoT has been used to support the implementation of vertical and horizontal solutions as a way of ensuring that the company can increase its profitability levels [57]. In addition, the IoT is considered to be critical in ensuring that an operational support system is provided in the marketplace and that high levels of efficiency are achieved especially when several infrastructural objects are connected to improve productivity.

Previous technologies deployed for traditional LLs have achieved some improvements. However, having a smart LLs that overcome linguistic barriers and efficiency have not been achieved as yet. The results of this paper have proven that there is much potential for the effective integration of IoT into smart LLs, especially from the perspective of smart application. Several tools, techniques and methods to enable the smart digital transformation of Al-Noor Mall have been used. An IoT-based framework for the smart transformation of Al-Noor Mall is presented, followed by the validation and simulation of the project utilizing IBM Bluemix. The results demonstrated the effectiveness of the proper deployment of IoT-enabled solutions towards the desired smart transformation of Al-Noor Mall. However, the paper is limited to the literature on IoT and other related technologies. Future research directions could shed light on additional deployments of IoT-enabled tasks for Al-Noor Mall to overcome the challenges related to energy consumption, predictive maintenance, and improved quality of services. Issues such as securing IoT devices should be considered along with addressing the challenges related to IoT devices operating system process management, communication protocols, and hardware architecture. Another future research direction could focus on tackling the issue of multilingualism, in contexts where people come from diverse linguistic backgrounds, with a proper incorporation of IoT technologies.

Appendix A

A.1 Step by Step Process: Implementation Through IBM Bluemix

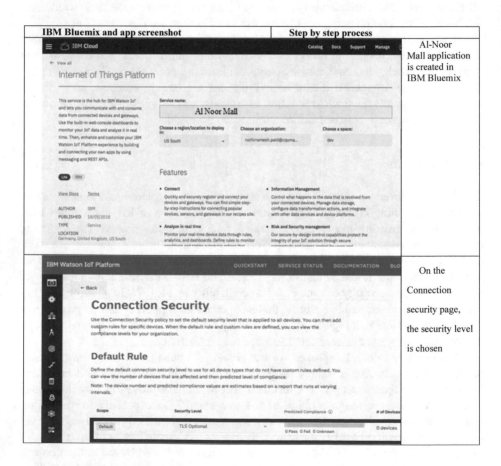

IBM Bluemix and app screenshot	Step by step process
	Al-Noor Mall application is created in IBM Bluemix
	On the Connection security page, the security level is chosen

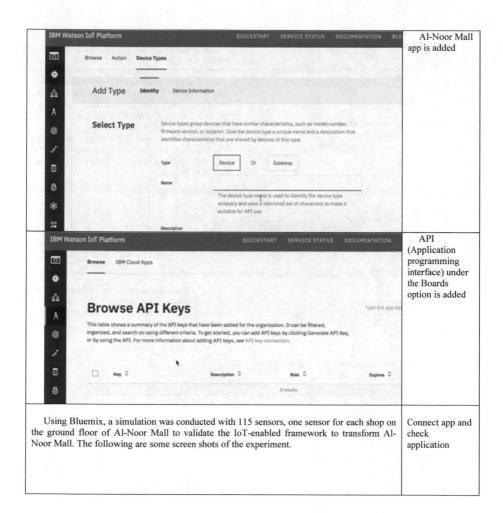

	Al-Noor Mall app is added
	API (Application programming interface) under the Boards option is added
Using Bluemix, a simulation was conducted with 115 sensors, one sensor for each shop on the ground floor of Al-Noor Mall to validate the IoT-enabled framework to transform Al-Noor Mall. The following are some screen shots of the experiment.	Connect app and check application

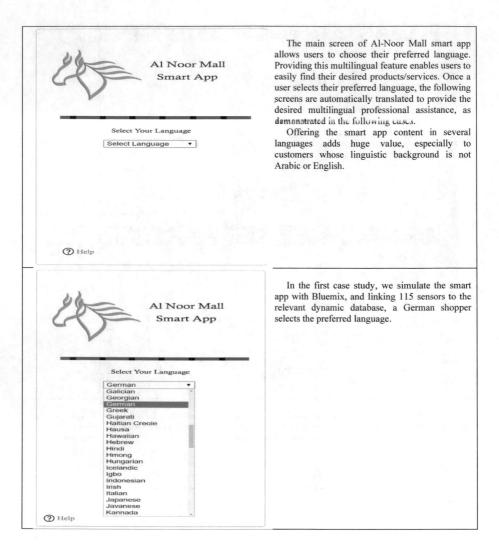

The main screen of Al-Noor Mall smart app allows users to choose their preferred language. Providing this multilingual feature enables users to easily find their desired products/services. Once a user selects their preferred language, the following screens are automatically translated to provide the desired multilingual professional assistance, as demonstrated in the following cases.

Offering the smart app content in several languages adds huge value, especially to customers whose linguistic background is not Arabic or English.

In the first case study, we simulate the smart app with Bluemix, and linking 115 sensors to the relevant dynamic database, a German shopper selects the preferred language.

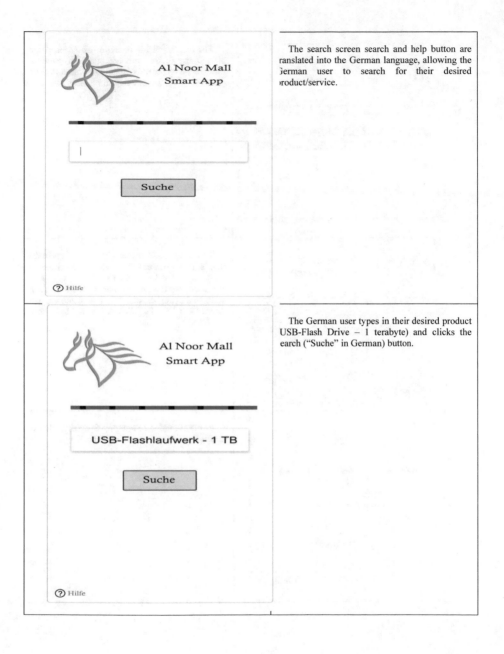

The search screen search and help button are translated into the German language, allowing the German user to search for their desired product/service.

The German user types in their desired product USB-Flash Drive – 1 terabyte) and clicks the search ("Suche" in German) button.

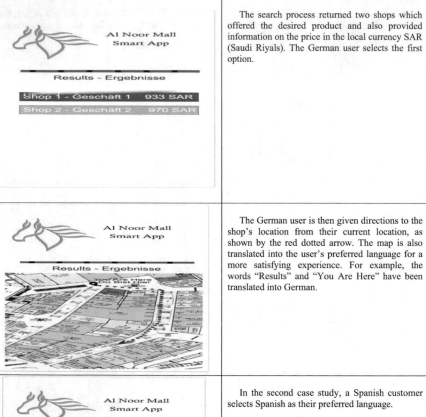

The search process returned two shops which offered the desired product and also provided information on the price in the local currency SAR (Saudi Riyals). The German user selects the first option.

The German user is then given directions to the shop's location from their current location, as shown by the red dotted arrow. The map is also translated into the user's preferred language for a more satisfying experience. For example, the words "Results" and "You Are Here" have been translated into German.

In the second case study, a Spanish customer selects Spanish as their preferred language.

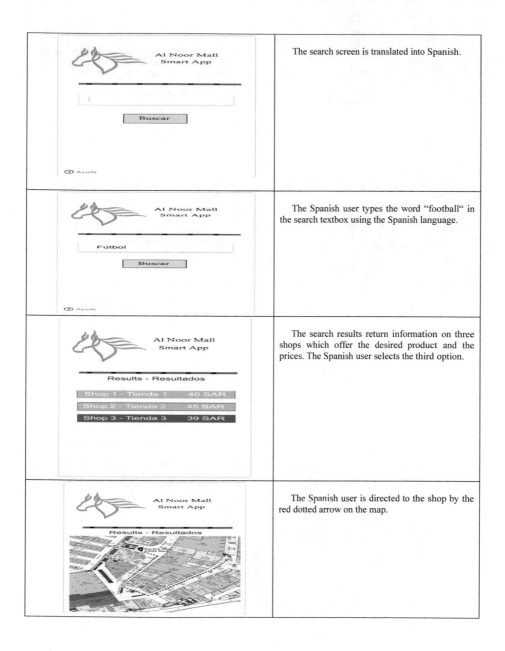

Al Noor Mall Smart App — Buscar — ⑦ Ayuda	The search screen is translated into Spanish.
Al Noor Mall Smart App — Fútbol — Buscar — ⑦ Ayuda	The Spanish user types the word "football" in the search textbox using the Spanish language.
Al Noor Mall Smart App — Results - Resultados — Shop 1 - Tienda 1 40 SAR — Shop 2 - Tienda 2 45 SAR — Shop 3 - Tienda 3 39 SAR	The search results return information on three shops which offer the desired product and the prices. The Spanish user selects the third option.
Al Noor Mall Smart App — Results - Resultados	The Spanish user is directed to the shop by the red dotted arrow on the map.

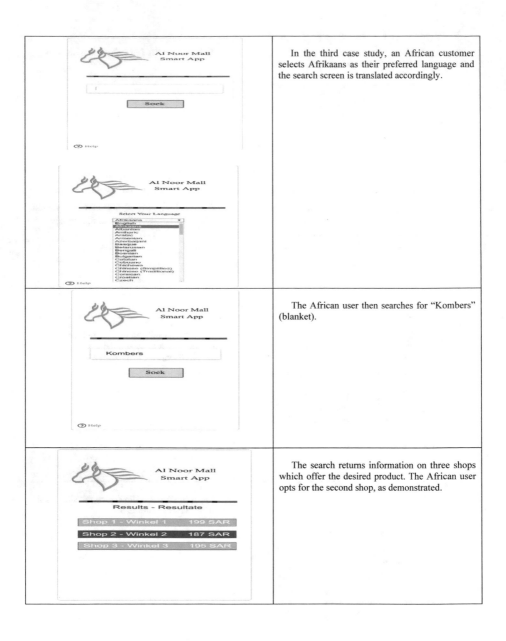

	In the third case study, an African customer selects Afrikaans as their preferred language and the search screen is translated accordingly.
	The African user then searches for "Kombers" (blanket).
	The search returns information on three shops which offer the desired product. The African user opts for the second shop, as demonstrated.

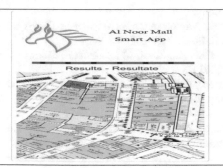

	The African user is directed to the location of the selected shop by the red dotted arrows.

References

1. Landry, R., Bourhis, R.Y.: Linguistic landscape and ethnolinguistic vitality an empirical study. J. Lang. Soc. Psychol. **16**, 23–49 (1997)
2. Li, S.: English in the linguistic landscape of Suzhou. Engl. Today **31**, 27–33 (2015)
3. Backhaus, P.: Linguistic Landscapes: A Comparative Study of Urban Multilingualism in Tokyo. Multilingual Matters, vol. 136 (2007)
4. Abdel-Latif, M.M., Abdel-Wahab, B.A.: Knowledge and awareness of adverse drug reactions and pharmacovigilance practices among healthcare professionals in Al-Madinah Al-Munawwarah, Kingdom of Saudi Arabia. Saudi Pharm. J. **23**, 154–161 (2015)
5. Brdesee, H.: Exploring factors impacting e-commerce adoption in tourism industry in Saudi Arabia. Doctorate, RMIT University, Melbourne, Australia (2013)
6. GASSA: The General Authority of Statistics in the Kingdom of Saudi Arabia. https://www.stats.gov.sa/sites/default/files/en-maddinah-pulation-by-gender-govnernorate-nationality_0.pdf. Accessed 13 Aug 2018
7. Group, A.-H.: Alhokair Group. https://www.alhokair.com/About. Accessed 13 Aug 2018
8. MRM: Al Madinah Regional Authority. http://www.amana-md.gov.sa/Pages/OpenData/DataLiberary.aspx. Accessed 14 Aug 2018
9. MRDA: Al Madinah Region Development Authority Initiatives. https://www.mda.gov.sa/MUD/10000. Accessed 14 Aug 2018
10. Soliman, S.K.: Marketing plan to launch a new brand in Saudi Arabia April/2014. Marketing **1009**, 11 (2014)
11. Abdellah, A., Ibrahim, M.: Towards developing a language course for Hajj guides in Al-Madinah Al-Munawwarah, a needs assessment. Int. Educ. Stud. **6**, 192–212 (2013)
12. Gaiser, L., Matras, Y.: The spatial construction of civic identities: a study of Manchester's linguistic landscapes. University of Manchester (2016)
13. Roeder, R., Walden, B.C.: The changing face of dixie: Spanish in the linguistic landscape of an emergent immigrant community in the New South. Ampersand **3**, 126–136 (2016)
14. Gorter, D.: Linguistic landscapes in a multilingual world. Annu. Rev. Appl. Linguist. **33**, 190–212 (2013)
15. Hyun, W.K., Ravishankar, S.: Smart signage: technology enhancing indoor location awareness for people with visual impairments. J. Technol. Persons Disabil., 204 (2016)

16. Risteska Stojkoska, B., Trivodaliev, K., Davcev, D.: Internet of Things framework for home care systems. Wirel. Commun. Mob. Comput. (2017)
17. Zafari, F., Papapanagiotou, I., Christidis, K.: Microlocation for Internet-of-Things-equipped smart buildings. IEEE Internet Things J. **3**, 96–112 (2016)
18. Zanella, A., Bui, N., Castellani, A., Vangelista, L., Zorzi, M.: Internet of Things for smart cities. IEEE Internet Things J. **1**, 22–32 (2014)
19. Ahmed, E., Yaqoob, I., Gani, A., Imran, M., Guizani, M.: Internet-of-Things-based smart environments: state of the art, taxonomy, and open research challenges. IEEE Wirel. Commun. **23**, 10–16 (2016)
20. Byun, J., Jeon, B., Noh, J., Kim, Y., Park, S.: An intelligent self-adjusting sensor for smart home services based on ZigBee communications. IEEE Trans. Consum. Electron. **58**(3), 794–802 (2012)
21. Viani, F., Robol, F., Polo, A., Rocca, P., Oliveri, G., Massa, A.: Wireless architectures for heterogeneous sensing in smart home applications: concepts and real implementation. Proc. IEEE **101**, 2381–2396 (2013)
22. Zhu, Q., Wang, R., Chen, Q., Liu, Y., Qin, W.: IoT gateway: bridgingwireless sensor networks into Internet of Things. In: 2010 IEEE/IFIP Proceedings of the 8th International Conference on Embedded and Ubiquitous Computing (EUC), pp. 347–352 (2010)
23. Gubbi, J., Buyya, R., Marusic, S., Palaniswami, M.: Internet of Things (IoT): a vision, architectural elements, and future directions. Future Gener. Comput. Syst. **29**, 1645–1660 (2013)
24. Heile, B.: Smart grids for green communications [Industry Perspectives]. IEEE Wirel. Commun. **17**(3), 4–6 (2010)
25. Hsu, P.-F., Ray, S., Li-Hsieh, Y.-Y.: Examining cloud computing adoption intention, pricing mechanism, and deployment model. Int. J. Inf. Manage. **34**, 474–488 (2014)
26. Zhou, J., Leppanen, T., Harjula, E., Ylianttila, M., Ojala, T., Yu, C., Jin, H., Yang, L.T.: CloudThings: a common architecture for integrating the Internet of Things with cloud computing. In: 2013 IEEE Proceedings of the 17th International Conference on Computer Supported Cooperative Work in Design (CSCWD), pp. 651–657 (2013)
27. Sun, Y., Song, H., Jara, A.J., Bie, R.: Internet of Things and big data analytics for smart and connected communities. IEEE Access **4**, 766–773 (2016)
28. Liu, C.H., Yang, B., Liu, T.: Efficient naming, addressing and profile services in Internet-of-Things sensory environments. Ad Hoc Netw. **18**, 85–101 (2014)
29. Li, W., Wang, B., Sheng, J., Dong, K., Li, Z., Hu, Y.: A resource service model in the industrial IoT system based on transparent computing. Sensors **18**, 981 (2018)
30. Chen, M., Mao, S., Liu, Y.: Big data: a survey. Mob. Netw. Appl. **19**, 171–209 (2014)
31. Tanenbaum, A.S.: Modern Operating System. Pearson Education Inc., London (2009)
32. Cristopher, J.: Operating System Architecture. Technology-UK (2009)
33. Gaitan, V.G., Gaitan, N.C., Ungurean, I.: CPU architecture based on a hardware scheduler and independent pipeline registers. IEEE Trans. Very Large Scale Integr. VLSI Syst. **23**, 1661–1674 (2015)
34. Arvind, P.: IoT Operating Systems, Devopedia. Version 9, 23 August. https://devopedia.org/iot-operating-systems. Accessed 03 Jan 2019
35. Silberschatz, A., Gagne, G., Galvin, P.B.: Operating System Concepts. Wiley, Hoboken (2018)
36. Sicari, S., Rizzardi, A., Grieco, L.A., Coen-Porisini, A.: Security, privacy and trust in Internet of Things: the road ahead. Comput. Netw. **76**, 146–164 (2015)
37. Lee, I., Lee, K.: The Internet of Things (IoT): applications, investments, and challenges for enterprises. Bus. Horiz. **58**, 431–440 (2015)

38. Riggins, F.J., Wamba, S.F.: Research directions on the adoption, usage, and impact of the Internet of Things through the use of big data analytics. In: Proceedings of the 48th Hawaii International Conference on System Sciences (HICSS), pp. 1531–1540 (2015)

39. Chen, F., Deng, P., Wan, J., Zhang, D., Vasilakos, A.V., Rong, X.: Data mining for the Internet of Things: literature review and challenges. Int. J. Distrib. Sens. Netw. **11**, 431047 (2015)

40. Sezer, O.B., Dogdu, E., Ozbayoglu, A.M.: Context-aware computing, learning, and big data in Internet of Things: a survey. IEEE Internet Things J. **5**, 1–27 (2018)

41. Pan, J., McElhannon, J.: Future edge cloud and edge computing for Internet of Things applications. IEEE Internet Things J. **5**, 439–449 (2018)

42. Al-Fuqaha, A., Guizani, M., Mohammadi, M., Aledhari, M., Ayyash, M.: Internet of Things: a survey on enabling technologies, protocols, and applications. IEEE Commun. Surv. Tutor. **17**, 2347–2376 (2015)

43. Rahmani, A.M., Gia, T.N., Negash, B., Anzanpour, A., Azimi, I., Jiang, M., Liljeberg, P.: Exploiting smart e-Health gateways at the edge of healthcare Internet-of-Things: a fog computing approach. Future Gener. Comput. Syst. **78**, 641–658 (2018)

44. Ferracuti, F., Freddi, A., Monteriù, A., Prist, M.: An integrated simulation module for cyber-physical automation systems. Sensors **16**, 645 (2016)

45. Aguirre, E., Lopez-Iturri, P., Azpilicueta, L., Astrain, J.J., Villadangos, J., Santesteban, D., Falcone, F.: Implementation and analysis of a wireless sensor network-based pet location monitoring system for domestic scenarios. Sensors **16**, 1384 (2016)

46. Kim, J., Hwangbo, H.: Sensor-based optimization model for air quality improvement in home IoT. Sensors **18**, 959 (2018)

47. Liu, G., Tan, Q., Kou, H., Zhang, L., Wang, J., Lv, W., Dong, H., Xiong, J.: A flexible temperature sensor based on reduced graphene oxide for robot skin used in Internet of Things. Sensors (Basel, Switzerland) **18**(5), 1400 (2018)

48. Jung, J., Lee, W., Kim, H.: Cooperative computing system for heavy-computation and low-latency processing in wireless sensor networks. Sensors **18**, 1686 (2018)

49. Wu, Y., Zhang, W., Shen, J., Mo, Z., Peng, Y.: Smart city with Chinese characteristics against the background of big data: idea, action and risk. J. Clean. Prod. **173**, 60–66 (2018)

50. Zhang, M., Cao, T., Zhao, X.: Applying sensor-based technology to improve construction safety management. Sensors **17**, 1841 (2017)

51. Botta, A., De Donato, W., Persico, V., Pescapé, A.: Integration of cloud computing and Internet of Things: a survey. Future Gener. Comput. Syst. **56**, 684–700 (2016)

52. Hossain, M.S., Muhammad, G.: Cloud-assisted industrial Internet of Things (IIoT)–enabled framework for health monitoring. Comput. Netw. **101**, 192–202 (2016)

53. Perera, C., Zaslavsky, A., Christen, P., Georgakopoulos, D.: Sensing as a service model for smart cities supported by Internet of Things. Trans. Emerg. Telecommun. Technol. **25**, 81–93 (2014)

54. Angela, R.: Internet of Things: 25 innovative IoT Companies and products you need to know. Entrepreneur: https://www.entrepreneur.com/article/298943. Accessed 13 Nov 2018

55. Mooring, D.J., Pallakoff, M.G.: File sharing between devices. Google Patents (2012)

56. Vermesan, O., Friess, P., Guillemin, P., Gusmeroli, S., Sundmaeker, H., Bassi, A., Jubert, I.S., Mazura, M., Harrison, M., Eisenhauer, M.: Internet of Things strategic research roadmap. In: Internet of Things-Global Technological and Societal Trends, vol. 1, pp. 9–52 (2011)

57. Díaz, M., Martín, C., Rubio, B.: State-of-the-art, challenges, and open issues in the integration of Internet of Things and cloud computing. J. Netw. Comput. Appl. **67**, 99–117 (2016)

Design and Fabrication of Smart Bandeau Bra for Breathing Pattern Measurement

Rafiu King Raji[1,4(✉)], Xuhong Miao[1], Ailan Wan[1,2,3], Li Niu[1],
Yutian Li[1], and Andrews Boakye[1]

[1] Engineering Research Center of Knitting Technology, Ministry of Education,
Jiangnan University, Wuxi, People's Republic of China
mrkingraji@outlook.com
[2] Key Laboratory of Advanced Textile Materials and Manufacturing
Technology, Ministry of Education, Zhejiang Sci-Tech University, Hangzhou,
People's Republic of China
[3] Key Laboratory of Textile Fiber and Product, Ministry of Education, Wuhan
Textile University, Wuhan, People's Republic of China
[4] Glorious Sun Guangdong Fashion College, Department of Fashion
Engineering, Huizhou University, 46 Yanda Avenue,
Huizhou, People's Republic of China

Abstract. Design and fabrication of a smart bandeau bra for breathing pattern mensuration is presented in this paper. The bra is knitted using Santoni's Top 2 seamless circular weft knitting machine. By connecting the bra to a customized portable signal acquisition; processing and transmission box, breathing signals emanating from the excursion of wearer's thorax is processed and transmitted to a computer interface. Five main respiration indices are recorded by this system and they include Respiratory rate, Respiratory cycle, Respiratory volume, Moment ventilation and Inspiratory duty cycles. Results show a successful application of the smart bra with result patterns being in congruence with studies conducted using other related respiratory measurement technologies and instruments.

Keywords: Bandeau bra · Seamless weft knitting · Respiratory pattern ·
Smart bra and signal processing

1 Introduction

Medically determining the general wellbeing of an individual is done by observing certain vital signs including body temperature, heart rate, oxygen level, respiratory rate (RR) or breathing rate (BR), blood pressure, and pain. BR is one of the most sensitive and vital clinical signs as it serve as a key predictor of health severity in emergency cases such as injury severity, respiratory infections and even death. Furthermore, athletes depend on their breathing as an indication of fitness, favoring slow but deep breaths during exercises.

A person's respiration rate therefore can be said to be a good reflection of his or her health status and physical condition. Several technologies for measuring breathing rate

© Springer Nature Switzerland AG 2020
K. Arai et al. (Eds.): FTC 2019, AISC 1070, pp. 40–48, 2020.
https://doi.org/10.1007/978-3-030-32523-7_3

exist, they include spirometry, thoracic impedance pneumography or respiratory inductance plethysmography, pulse oximetry or photoplethysmography, acoustic methods, capnography, accelerometry and derivations from other physiological measurements such as phonocardiography, ECG, heart rate, etc.

Test results however vary slightly depending on the equipment or technology used. Nevertheless in clinical practice, the most common method for respiration rate measurement is by physical assessment, either by counting chest wall movements or by auscultation of breath sounds with a stethoscope [1].

Breathing rate measurements are currently not restricted to the clinical setting as they are increasingly been embedded in personal health devices and smart apparels. Most of these sensors are primarily focused on quantitative measurements of breathing rate [2, 3]. The breathing rate gives an indication of the health of the lungs and breathing system as well as other physical and emotional indicators.

Smart apparels with respiration measuring capabilities are fabricated via a number of routes and are based on different sensing technologies including optical diffraction, piezotronic, photoelastic, and conductivity related properties such as piezoresistance, piezoelectricity, capacitance and inductance [4]. Some of the methods of fabrication includes coating piezo-resistive materials [5] on a fabric, directly knitting conductive fibers into fabrics or by stitching.

Wearable breathing sensors or smart apparels with breathing measurement functionality are usually designed such that, the sensors are positioned around the trunk of the user in the form of bands or shirts [6]; however, other areas of placement including attachment to backrest of chairs [7], to bed and the wrist have been experimented on.

Wrapping around the trunk however tends to be the dominant mode as it affords not only ambulatory measurements but also the stimuli provided by the trunk movement is also significant enough to be sensed by fabric sensors.

This paper presents a wearable respiration measurement system were the sensor is embedded in a bandeau bra. The system works by correlating the excursion of the user's thorax to respiration indicators per minute. The sensing system presented is characterized by good form factor comparable to commercial pneumographic or plethysmographic products and capable of ambulatory respiration measurement. Qualities that set this system apart from our previously published studies include the customized signal acquisition and processing software and hardware that affords untethered measurement of respiration data.

2 Experimental

2.1 Materials

The bra is circular weft knitted based on 1 × 1 mock rib structure. Santoni SM8-TOP2 circular weft knitting machine with the following parameters was used. Gauge number: E28, Diameter 15 in., RPM: 40–65 r/min. 1–8Feeds. (Santoni SPA, Italy). Materials used for the bra include silver plated yarns purchased Qingdao Hengtong X-Silver Specialty Textile Company, Nylon DTY and covered elastic yarn from Zhejiang Jinqi New Material Tech. Co., Ltd.

2.2   Methods

2.2.1   Smart Bra Design

The smart bra was fabricated at Engineering Research Center of Knitting Technology (KTC), Jiangnan University. The designed was using Photon and Digraph 3 software (Dinema S.P.A) described in our previous study [8]. The file was subsequently transferred to the knitting machine via a USB flash disk. Figure 1 shows image of the bra design using Digraph 3 software. Images of the prototype smart bra are shown in Fig. 2.

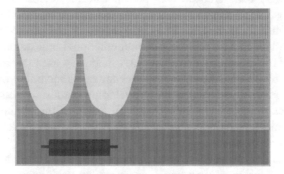

Fig. 1. Digraph 3, Quasar image of the bra design.

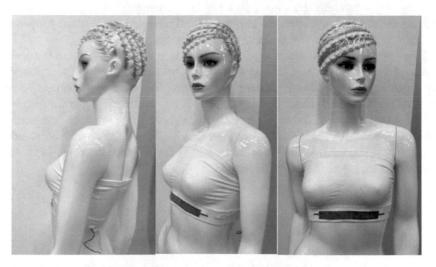

Fig. 2. Images of prototype smart bandeau bra

2.3   Systems and Methods for Respiratory Event Detection

2.3.1   System Design, Architecture and Functions

The system is enabled with a portable signal acquisition, processing and a bluetooth signal transmission capability referred to as a processing box. The processing box encompass a microprocessor and a memory, power unit, BLE wireless transmitter and receiver, rechargeable battery unit and a USB charging unit. A bluetooth dongle plugged into the computer receive signals transmitted to the computer. The main hardware components of the processing box are an ARM Cortex-M3 STM32F401RCT6 core board, designed power supply and signal acquisition circuit, HC-08 bluetooth 4.0 BLE Serial Port Module, two Li-Ion rechargeable batteries and a USB charging unit and connecting cables, were purchased online from Tmall.com. The electronic processing box is shown in Fig. 3.

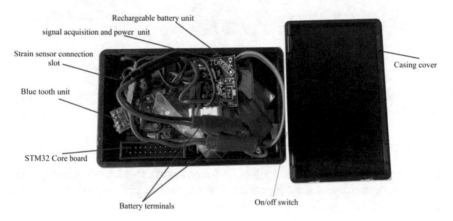

Fig. 3. Electronic processing box showing the various components (KTC Electro-box)

The digitized respiratory signals are filtered to limit non-respiratory components. This processing combines these signals, into a further signal proportional to actual, moment-by-moment lung volumes referred to as breath volume, respiratory frequency referred to as cpm and respiratory rate (rpm).

The test interface designed using Visual Studio 2012 is presented in Fig. 4; it entails a panel to enter name and age of the user, signal display range and the resistance of the sensor. A start button and a pause/resume button and a replay button to replay the latest test which is recorded automatically. All five respiration indicators are shown below the displayed pattern.

The indicators used include Respiratory cycle (RC) measured in cycles per minute (cpm), Respiratory rate (RR) measured in respiration per minute (rpm). Inspiratory Duty Cycle (DTCY) measured in percent (%), Breath Volume (BV) measured in ohms (Ω) likened to Tidal Volume (VT) and Moment Ventilation (MV) measured ohms/min (Ω/min) likened to Minute ventilation (VE) in standard spirometry, respectively. The

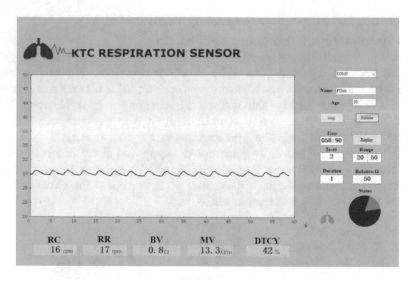

Fig. 4. Respiration sensor interface

interface software program apart from graphic display also automatically saves the original digitized signals and the derived parameters in an excel format strip-chart.

2.4   Testing Procedures

The user wears the bra just like any other ordinary bra after which two wires connected from both ends of the sensor are plugged into the signal acquisition system as shown in Fig. 5.

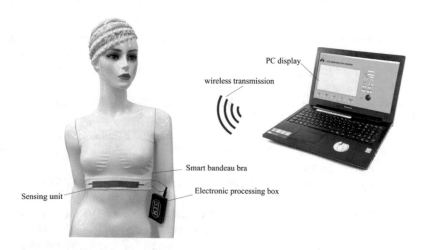

Fig. 5. Scheme of the smart bandeau bra sensing system

The computer or mobile device interface is then started to commence measurement of respiration. Signal emanating from the excursion of the user's thorax are wirelessly transmitted to the computer or related device for interpretation. The user is made to assume any of the stipulated postures or engage in the activities while information is being collected.

3 Results and Discussions

The respiration patterns for various postures and activities are shown in Figs. 6, 7, 8 and 9. Results indicate no significant difference between standing and sitting respiration cycles and rates of the user. However with regards to BV, standing and the sitting postures studied, standing posture tends to have produced higher values. Increased lung volumes of subjects or users in standing position compared to other positions can be attributed to an increase in the volume of the chest cavity [9].

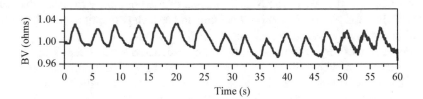

Fig. 6. Standing posture respiration pattern

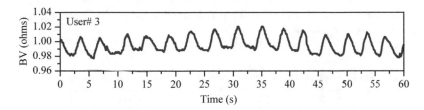

Fig. 7. Sitting posture respiration pattern

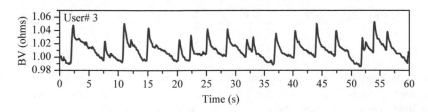

Fig. 8. Reading activity respiration pattern

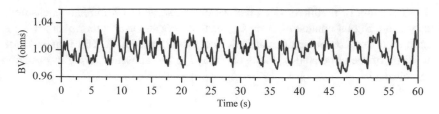

Fig. 9. Indoor walking activity respiration pattern

Reading activity patterns are also identical with patterns adduced by some voice recording sensors [10] The patterns are similar to those obtained using respitrace in a study and how speech production changes when people move [11]. This is in congruence with the assertion by Coyle *et al.* [12] that speaking episode, one of the most frequent types of physical activity and behavior, can alter a variety of physiological systems including breathing.

The results of the reading activity thus hints of the possibility of using this sensor for researches in the area of speech production and speech perception and adaptation of breathing during the perception of speech [10].

Comparison of respiration indices under physical activity was done. This is because during physical exercise, the muscles are strained beyond normal and calories are burnt. The muscles tend to require more oxygen than they normally use to burn these extra calories. The results showed higher values compared to those adduced when the user was stand and sitting confirming increased breathing levels after every physical activity.

The user's normal breathing rate at rest as can be seen in Table 1 was 14.1 ± 1.2 rpm, however indoor walking exercise increased the breathing rate up to 19.3 respirations per minute.

Table 1. Respiration results under different activities and postures

Activity/Posture	RC	RR	BV	MV	DTCY
Standing posture	14.9 ± 0.9	14.1 ± 1.2	1.2 ± 0.3	16.2 ± 3.0	35.4 ± 2.1
Sitting posture	15 ± 0.7	14.05 ± 1.1	0.65 ± 0.1	9.55 ± 1.9	37.55 ± 1.5
Reading activity	17.5 ± 0.8	18.95 ± 0.9	0.75 ± 0.2	14.5 ± 0.8	28.7 ± 1.2
Indoor walking	19 ± 2.8	18.9 ± 3.7	1.5 ± 0.2	26.9 ± 1.9	37.9 ± 2.6

Comparison between our device and others cited in literature and even some commercial products is presented in Table 2. Comparatively our device measured the most respiratory indicators. All systems that used digital multimeters also have the disadvantage of not permitting unrestricted or ambulatory measurements whilst all those designated as customized permit unrestricted measurements.

Table 2. A comparison of related respiration devices/systems and number/type of indices measured

Sensor type	Signal measurement system	Number and type of respiratory indices	
Proposed system (bandeau bra)	Customized (KTC Electro-box)	(5) Respiratory rate, Respiratory cycle, Breath volume, Moment ventilation and Inspiratory duty cycle	
Chest band	A digital multimeter (Agilent U3402A)	(1) respiratory rate	[13]
Infant apparel	Digital multimeter (APPA 505)	(1) respiratory rate	[14]
WEALTHY project (jump suit)	Customized (WEALTHY box/PC)	(3) electrocardiogram, respiration, activity	[5]
Respiration belt	Wheatstone bridge circuit	(1) Respiratory frequency	[15]
Respiration belt	bridge sensing circuit	(1) Respiratory rate	[3]
Respiration monitor belt (Vernier^{TM})	Customized (Vernier Gas Pressure Sensor)	(1) Respiratory rate	[16]

With the exception of [5] and other systems that are designed to measure multiple physiological events, systems designated for respiratory measurements usually measure only the quantitative respiratory rates or frequencies.

4 Conclusion

This paper presents a successful application of a wearable smart bandeau bra capable of measuring an array of respiration indices. The more expansive indices measured by our system sets it apart from other similar systems available in literature.

Projects of this nature serve to create a product that will assist in raising health consciousness by providing a cheaper means of untethered testing of respiration data, especially outside the hospital environment. It might also allow doctors and physicians to also remotely monitor their patients.

The results of our sensing system can also serve as primary data for other systems such as personal identification and human activity tracking systems. It is anticipated that further analysis, some of the basic data, will extend the practical, theoretical and non-clinical utility of respiration data.

References

1. Agnihotri, A.: Human body respiration measurement using digital temperature sensor with I2c interface. Int. J. Sci. Res. Publ. **3**(3), 1–8 (2013)
2. Guo, L., et al.: Knitted wearable stretch sensor for breathing monitoring application. In: Department of Signals and Systems. Ambience Conference 2011, Borås, Sweden, p. 33. Chalmers University of Technology (2012)

3. Huang, C.-T., et al.: A wearable yarn-based piezo-resistive sensor. Sens. Actuators A Phys. **141**(2), 396–403 (2008)
4. Raji, R.K., et al.: Knitted piezoresistive strain sensor performance, impact of conductive area and profile design. J. Ind. Text. 1–19 (2019)
5. Paradiso, R., Loriga, G., Taccini, N.: A wearable health care system based on knitted integrated sensors. IEEE Trans. Inf Technol. Biomed. **9**(3), 337–344 (2005)
6. Qureshi, W., et al.: Knitted wearable stretch sensor for breathing monitoring application. In: Ambience 2011, Borås, Sweden, pp. 1–5 (2011)
7. Griffiths, E., Saponas, T.S., Brush, A.J.B.: Health chair: implicitly sensing heart and respiratory rate. In: 2014 ACM International Joint Conference on Pervasive and Ubiquitous Computing, Association for Computing Machinery (ACM) Seattle WA USA, pp. 761–771 (2014)
8. Raji, R.K., et al.: Influence of rib structure and elastic yarn type variations on textile piezoresistive strain sensor characteristics. Fibers Text. Eastern Eur. **26**(5), 24–31 (2018)
9. Talaminos Barroso, A., et al.: Factors affecting lung function: a review of the literature. Archivos de Bronconeumología (English Edition) **54**(6), 327–332 (2018)
10. Rochet-Capellan, A., Fuchs, S.: Changes in breathing while listening to read speech: the effect of reader and speech mode. Front. Psychol. **2013**(4), 906 (2013)
11. Fuchs, S., Reichel, U., Rochet-Capellan, A.: Changes in speech and breathing rate while speaking and biking. In: International Congress of Phonetic Sciences ICPhS 2015, Glasgow Scotland, pp. 1–5 (2015)
12. Coyle, M., et al.: Systems and methods for respiratory event detection. In: Patents, E. (ed.) European Patents, pp. 1–49. Adidas AG, USA (2015)
13. Zhao, Y., et al.: A novel flexible sensor for respiratory monitoring based on in situ polymerization of polypyrrole and polyurethane coating. RSC Adv. **2017**(7), 49576–49585 (2017)
14. Jakubas, A., Łada-Tondyra, E.: A study on application of the ribbing stitch as sensor of respiratory rhythm in smart clothing designed for infants. J. Text. Inst. **109**(9), 1208–1216 (2018)
15. Atalay, O., Kennon, W.R., Demirok, E.: Weft-knitted strain sensor for monitoring respiratory rate and its electro-mechanical modeling. IEEE Sens. J. **15**(1), 111–122 (2015)
16. Vernier Software & Technology, L. Respiration Monitor Belt (2019). https://www.vernier.com/products/sensors/respiration-monitors/rmb/.last. Accessed 21 May 2019

Electrical Internet of Things - EIoT: A Platform for the Data Management in Electrical Systems

Santiago Gil[1(✉)], Germán D. Zapata-Madrigal[1],
and Rodolfo García-Sierra[2]

[1] Grupo T&T, Facultad de Minas, Universidad Nacional de Colombia,
Medellín, Colombia
{sagilar, gdzapata}@unal.edu.co
[2] Enel-Codensa S.A. E.S.P. (Enel Group), Bogotá, Colombia
rodolfo.garcia@enel.com

Abstract. The arrival of the IoT has benefited multiple industrial sectors; one of them is the electrical industry. It makes the IoT an attractive platform for the smart grid, because it improves the monitoring, analysis, availability, autonomy, and control of grid systems, from distribution to transmission. In this work, it is proposed the Electrical Internet of Things – EIoT, a platform for the data management in electrical systems. The EIoT platform relies on the LPWAN technologies to connect electrical things geographically distributed. This approach attempts to cover Fire Protection Systems in Substations, Control and Monitoring for the Demand Management, Electrical Asset Management, and Intelligent Power Management and Monitoring in Refrigerators applications, but other smart grid applications fit as well. The EIoT platform implemented LoRa, a LPWAN technology, and its LoRaWAN protocol to connect electrical elements remotely in a first stage. The prototyping phase of the EIoT platform was developed using Mbed-enabled LoRa nodes coupled to electrical elements with the corresponding instrumentation, a LoRaWAN network, and a custom application server; then, all the components of an IoT network were considered. Some results of the implementation are presented, which demonstrate the suitability of the EIoT platform to address several applications of the electrical industry. The results also expose promising opportunities to deal with data processing and analytics, integrate more IoT protocols, and consolidate a complete IoT ecosystem in the electrical industry and the smart grids in further researches.

Keywords: IoT · LPWAN · Electrical and power systems ·
Control and Monitoring · Fire protection systems · Smart grids

1 Introduction

The arrival of the Internet of Things (IoT) has benefited multiple industrial sectors; one of them is the electrical industry [1]. The IoT enables to connect devices such as sensors, controllers, actuators, etc. To monitor, control, and track the elements. It

© Springer Nature Switzerland AG 2020
K. Arai et al. (Eds.): FTC 2019, AISC 1070, pp. 49–65, 2020.
https://doi.org/10.1007/978-3-030-32523-7_4

makes the IoT suitable for from the smallest electrical systems such as microgrids, up to bigger ones such as transmission lines, generation plants, etc.

The IoT solutions can be used in several electrical applications like energy management systems for homes or for smart electricity metering. The IoT can be applied over sensors in the distributed network, smart substations, consumption profiles, energy efficiency tracking, electric vehicles tracking, demand management, among others [2]. The IoT is a key for the success of the smart grid and the Electrical Industry 4.0.

IoT represents a concept in which is possible to deploy multiple solutions, but what about the means to do it possible? Here, the communication protocols for IoT begin to play an important role in IoT solutions. Many technologies and protocols are addressing IoT applications, wired or wireless [3]; also many of them have been addressed specifically for applications in the electrical industry [4].

Recently, the Low Power Wide Area Network (LPWAN) technologies have emerged as a promising alternative to support IoT solutions at a wide range and low power consumption. These technologies can cover ranges up to some kilometers and they can meet requirements for industrial, health, agriculture, but also for energy and power applications [5, 6]. LPWAN technologies can operate over more than 10 years with a single battery due to their efficient performance [6], in comparison to ZigBee that can operate for more than 10 years too, it only covers distances at maximum 90–100 m, constraint that LPWAN technologies satisfy. Even with such suitable features, the LPWAN technologies have not been enough adopted in the electrical industry; many legacy communications technologies are still being used in current electrical, energy, and power systems projects and applications [7].

The main aim of this project is to exploit the benefits of the IoT for the electrical industry using the LPWAN technologies to create a consolidated platform for the data management in multiple applications of the electrical systems. The EIoT platform is designed to consolidate a cyber-physical ecosystem for electrical elements of the Electrical Industry 4.0 and the smart grids. The EIoT platform is composed by an IoT network using LPWAN nodes, a network server and a custom application server for the data management of the electrical systems. It has interfaces for the interaction of humans and things. The design and development consider scalable components to add further IoT resources in the platform which will enrich the electrical systems in the future.

The remainder of this paper is as follows:

- Section 2 presents a brief review on IoT platforms and how Colombia is in this subject.
- Section 3 details the design and development related to the EIoT platform.
- Section 4 presents an implementation of the EIoT platform.
- Section 5 presents some conclusions and future work.

2   Review on IoT Platforms and Applications for Electrical Systems

Ray made a deep survey on IoT-based architectures for various domains in [8]. He reviewed different aspects of IoT solutions such as wireless communication technologies, hardware platforms, public IoT cloud platforms, some application domains, and specific architectures for the domains proposed in the state of the art. The domain architectures presented cover from RFID, Service Oriented Architecture (SOA), Wireless Sensor Network (WSN), supply chain management and industry, health care, smart society, cloud service and management, social computing, to security.

Also Perera et al. did a review on the Internet of Things from the industrial market perspective [9]. They shared the importance of IoT solutions for new business approaches and how the energy market is one of the most benefited.

The review done by Sethi and Sarangi in [10] presents several relevant topics to consider when designing and implementing IoT applications, such as sensor technologies, fog computing, types of middleware, communication technologies, applications, and architectures. The LPWAN technologies are presented as suitable for constrained power and far away devices. Once more the electrical, energy, and grid applications are suggested to be addressed with IoT solutions.

Yun and Yuxin shared many application scenarios for the internet of things, but also emphasized on smart grid application approaches with IoT technologies in their review [2]. Some of the key characteristics in smart grid to satisfy with IoT are self-healing, participation of the users, distributed generation, demand response, fault detection, and asset management.

Wan et al. made a software approach for the Industrial Internet of Things in [11]. They proposed a software-defined architecture, an approach to Cyber-Physical Systems (CPSs) and suitable to be deployed in platforms composed by hardware and software elements.

A great perspective about how the IoT and the electrical systems (a big smart grid) can converge is provided by Collier in [12]. He stated how the current grid (a legacy grid) needs to be equipped with new technologies from electronics, communications, software, information, etc. This will enable the smart grid to support the future of the electrical systems: millions of distributed generators, distributed storage, Energy Management Systems (EMSs), appliances, lights, Electric Vehicles, etc. It makes the IoT an attractive platform for the smart grid, because it improves the monitoring, analysis, availability, autonomy, and control of grid systems, from distribution to transmission. That is why the smart grid is considered to be a largest and a pioneer example of IoT solution, and thus, the smart grids and the IoT need to be researched together for the development of the electrical systems of the Industry 4.0, as this work intends to.

Shahryari et al. [13] proposed a IoT-based Demand Side Management with planning and scheduling approach using LoRa technology and LoRaWAN standard. They considered an architecture with IoT nodes layer, fog computing layer, core network, and cloud data center. The fog computing represents the multiple microgrids domain (network in the edge) while cloud computing to storage the data generated by the multiple fogs. The LoRaWAN gateway acts as the interface for the two layers.

Ferrari et al. [14] implemented an application to integrate Electric Vehicles to main grid using LoRa technology. They relied on geographical and battery load data to characterize the power consumption that will be demanded by the vehicles because of the unpredictable nature of this variable. This approach could be extended for more applications.

Song et al. [15] proposed IoET – Internet of Energy Things, an infrastructure for Demand Side Management using LPWAN technologies. They stated a 3-layer architecture composed by a cloud platform layer, a LPWAN transmission layer, and an end device layer. The IoET is extensible for different applications such as smart appliances and smart home systems, microgrids and distributed energy systems, active distribution networks and aggregated demand response, electric vehicles and aggregated vehicle-to-grid operation, and energy distribution network and multi-energy systems. They stated a similar approach using a generic LPWAN architecture for the electrical things, but do not implemented application cases or customized platforms, which, in addition, are proposed in this work, with a perspective towards the intelligent data management in electrical systems.

Gharavi and Hu developed a Machine-to-Machine (M2M) communication network for distributed power grid monitoring [16]. They based on WLAN technologies to support real-time and high amount of data instead of LPWAN technologies which provide wider coverage and more battery lifetime but less data capacity transmission. The adoption of WLANs instead WANs, or in this case, LPWANs, limits the area coverage for the electrical things, which by nature are geographically distributed; it drops the chance to cover multiple applications of the electrical systems.

The cellular technologies have also taken an important role in the development of IoT solutions in electrical and power systems. An assessment of LTE technology for monitoring of energy distribution is done by Madueño et al. in [17]. The emerging LPWAN technology, NB-IoT, also promises to be suitable to address certain smart grid applications; a survey about this is carried out by Javadi and Seifvand in [18]. Unfortunately, this alternative is not available yet in Colombia, what is a restriction; so, other communication alternatives should be considered.

It is also very important to consider the standardized and consolidated reference architectural models for IoT applications, developed by Industrial Internet Consortium (IIC) - Industrial Internet Reference Architecture (IIRA) [19], Platform Industrie 4.0 – Reference Architectural Model Industrie 4.0 (RAMI) [20], IEEE Architectural Framework for IoT – The standard for an architectural framework for the Internet of Things (under the Project 2413), and other approaches [21]. These reference models can provide a basis for big and scalable IoT projects while involving challenges from different disciplines that are addressed.

Many approaches and efforts are being done for the development of the Internet of Things and for the Industry 4.0, which are related concepts. The electrical industry is one of the most benefited industries by these technological advances, thus, more research in these areas is required. No singular applications fit perfectly for all the IoT applications in the electrical industry, and that is why several approaches with different focus or objectives are done. The communication protocols for long range applications are necessary in the electrical industry and the standardization is very important too. The idea is to begin from the technological advances of the state of the art to propose a novel IoT solution for the electrical industry; the EIoT platform intends to cover the data acquisition and management for IoT applications in electrical systems: smart grids, demand management, asset tracking, etc. In the first stage by using the LPWAN technologies, but with a higher level of scalability to incorporate further resources to consolidate a powerful Electrical IoT platform.

3   EIoT: The Platform for Electrical Things

This section details the procedure to select the LPWAN technology, the design of the communication network, the IoT architecture for data and information integration, and the fundamental considerations about the instrumentation to carry out the Electrical IoT platform.

3.1   The LPWA Network

The EIoT platform was designed based on the LPWAN technology approach, in which there are multiple nodes (end devices) coupled to the electrical things, gateways to acquire the data of the nodes of the WAN and support the IP protocol, and then, the data is transmitted into web-based servers to be processed.

The review of requirements about LPWAN to select some alternative(s) to deploy the EIoT-LPWA network compared operation frequencies, range of coverage, battery lifetime, bidirectional communication, data transmission capability (data rate and payload), latency, Quality of Service (QoS), scalability, standardization, open source or proprietary technology-deployment, and local availability of technology criteria of the most known LPWAN technology alternatives: LoRa, SigFox, NB-IoT, Ingenu (RPMA), Telensa, Waviot (NB-Fi), EC-GSM, and eMTC. The review and the criteria based on [5, 6, 22–31] to select the best alternative(s), and some personal inquiries to know about the local availability in Colombia.

Without conflict of interests LoRa and the LoRaWAN protocol were selected as the means to deploy the LPWA network of the EIoT platform this time. The main criteria was the availability of the technology. It is very important to do the remark that the platform is in constant development and the idea is to converge as much IoT technologies and protocols as possible. The other LPWAN technologies are very promising as well, highlighting the 4.5G technologies (NB-IoT and LTE MTC Cat M1) for the future of IoT in electrical applications considering that the development of the communication infrastructure goes hand in hand with the development of energy infrastructure. The NB-IoT network, for example, is not available in Colombia yet, that is why considering the availability criteria is indispensable for the deployment of a LPWAN solution; the same occurs with operation frequencies regarding to regional and national regulations.

3.2 The Communication Architecture

The communication architecture was proposed considering a general LPWAN implementation, not based to a specific technology. This approach based on the LoRaWAN standard architecture stated by LoRa Alliance, that also shares similitudes with reference architectures stated by other LPWAN technologies, such as SigFox and Waviot. The architecture includes additional resources such as the integration with further web services that could be Google Cloud IoT Services or external analytics services. Figure 1 shows the communication architecture for the EIoT platform. It is a very common architecture, but this work focused on the development of a scalable application server to interact with more components and provide more communication interfaces for IoT networks.

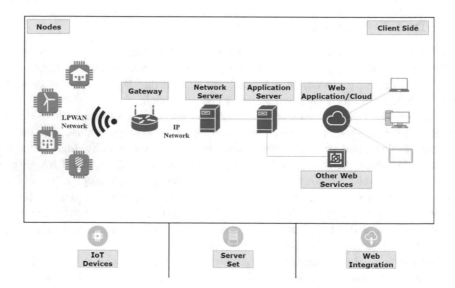

Fig. 1. Communication architecture for the EIoT platform.

3.3 The Logical Architecture for the World of Electrical Things

The platform structuration required a complete software design and development, subsequently embedded in the components of the communication architecture. It was necessary to state a logical architecture to manage the interactions and data flow.

In Fig. 2 is possible to notice the considerations about the logical architecture, communication interfaces for electrical things (not specific to a protocol) and the interface for humans to interact with the processes. Currently, the module for data processing and analytics is in development yet. This module pretends to cover the data analytics, data processing, and learning about consumption profiles; predict behaviors, recommendations of use, detect abnormal operation states or failures in the electrical things, etc. As stated in Sect. 3.2, the focus is done over the application server to provide additional data-based value-added tasks for the processes.

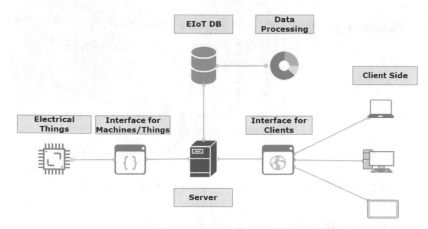

Fig. 2. Logical architecture for the interactions and data management in the EIoT platform.

3.4   The Instrumentation Diagram

Due to the electrical nature of the project, the instrumentation considerations are mandatory. The proposed diagram is a first approach to monitor some important variables in electrical elements. It considers the possibility of integrating different electrical systems such as three-phase or monophasic elements, DC loads, smart electrical meters, infrastructure of electrical systems, etc. with no limit. It depends on the scope and requirements of the project.

Figure 3 shows the proposed diagram to measure different variables in electrical things that belong to the EIoT platform. As stated in the "Other Sensors" legend, any appropriated sensor or meter could be integrated with the purpose of satisfying specific requirements of the application.

Legends

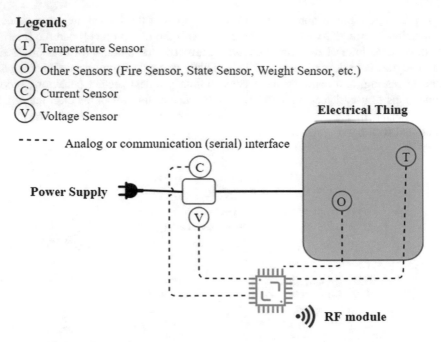

Ⓣ Temperature Sensor

Ⓞ Other Sensors (Fire Sensor, State Sensor, Weight Sensor, etc.)

Ⓒ Current Sensor

Ⓥ Voltage Sensor

Fig. 3. Proposed instrumentation diagram to measure EIoT elements.

3.5 The Purposes and Benefits of the Platform

The main idea of the platform is to provide continuous monitoring and control of electrical systems and their elements for energy service providers (generation, transmission, and distribution). It can also be applied for customers and final users to track and control the operation of their equipment. The platform represents a means to integrate the data of multiple devices. The first stage emphasizes on the deployment of the LPWA network, but after that, other IoT networks are expected to be integrated to consolidate a powerful IoT platform for the electrical industry.

Due to the capabilities of LPWAN technologies, in this case LoRaWAN, the platform can cover wide areas of EIoT nodes with few gateways, what makes the platform suitable to work with electrical systems which are distributed over wide areas in different locations, urban or rural; even with constrained devices.

In subsequent stages, the EIoT platform intends to cover several applications of the electrical industry, and thus, massify the data acquisition in multiple processes to carry out data analysis and convert these data into useful information. It will enable the performance of value-added tasks which enhance the electrical industry.

4 Implementation of the EIoT Platform

With the purpose of reaching an extensible IoT solution for the electrical industry, this project is looking for providing novel, not yet addressed, and very necessary applications. Thus, this project pretends to cover insights and objectives according to the trends related to smart grid management for energy service providers. In the prototyping phase of the project, some approaches to Fire Protection Systems in Substations, Control and Monitoring for the Demand Management, Electrical Asset Management, and Intelligent Power Management and Monitoring in Refrigerators were done.

4.1 The Application Cases

Regarding to the Intelligent Power Management and Monitoring in Refrigerators application, it is a special application that this project wants cover. It is due to industrial refrigeration represents a significant part of the power consumption of some companies. The idea consists of monitoring refrigerator variables, even the intervened by the users, like the opening or closing doors or consumables overload, to provide a service to the customers based on recommendations to optimize the power consumption in refrigerators and freezers. For this approach, a little home refrigerator was used. At this moment, current, power, temperature, and door state of the refrigerator were monitored, but the idea is to work on doing a completely instrumented approach, considering the weight and existence of products via image detection.

About Demand Management the most important variables to track are the related to the power consumption. Employing the EIoT platform for this purpose allows to track variable consumption profiles of loads and the subsequent implementation of data analysis, suitable for Demand Management applications. In this case, a low-cost current sensor was coupled to a hourly-based scheduled load to compare the profiles with a smart electricity meter.

The Fire Protection Systems in Substations is a quite relevant matter for energy providers; according to [32], the transformers are one of the most important assets to take care in fire protection systems in substations. By monitoring electrical variables as the stated along this paper, it is possible to determine normal and abnormal conditions in critical electrical elements, for example, the transformers, to keep aware of eventual fire.

The Electrical Asset Management is another relevant matter for energy service providers. A means to know the state of the multiple electrical assets that an energy service company owns becomes a useful way to give value-added information for the owners. It can help to schedule maintenance tasks, estimate useful life, inefficiencies, ensure the security of the physical assets, among other functions. For this application as many variables as selected can fit to manage information about the assets, depending on the established necessities of the companies. In this case, it was proposed the GPS and fire-detection monitoring in addition to the consumption variables. It can be used to provide information for the companies when assets operate as mobile elements or when assets must be under extreme care, avoiding material losses.

4.2   The Prototype Platform

For the implementation of the EIoT Prototype Platform some STMicroelectronics Mbed-enabled devices were used. The online Mbed compiler provides many resources to develop LoRa and LoRaWAN applications for several microcontrollers. The gateway was implemented using a LairdTech RG191, the network server was deployed using a local LoRa Server [33], and the custom application server using a Django Python based server. The strategy of using the LoRa Server and the Django server allowed a low-cost, embedded, and easy-to-deploy solution suitable to address IoT projects, considering that both of them can be assembled in Unix-based platforms, like a Raspberry PI. The data flow integration currently works under the HTTP protocol by means of the REST services, though the LoRa Server has implicit use of the MQTT protocol. The custom application server is used as a Graphical User Interface (GUI), but it also employs some Application Program Interfaces (APIs) to interact with other things and services using structured formats and it has additional features.

The development of the server and the interfaces based on the features of Python and the Django web framework. The database runs in the PostgreSQL engine. All the development of the servers was carried out over Unix-based OS, allowing the EIoT platform to be deployed on a Raspberry PI or similar embedded systems. At this moment, the GUI run under a prototype dashboard.

4.3   Results and Analysis

In this literal are presented some results about the collected data in the EIoT platform. The GUI and the collected data were used to present the graphs and tables and support the results. The platform provides a way to collect data from the different electrical things: the microcontrollers coupled to the electrical things run customized code to treat the physical variables, convert them to data, and send the data through the LoRaWAN protocol. The digital variables (Boolean) are represented as 0 or 1, like the door state or the fire detector variables; these variables allow the use of events to capture the changes of state and send messages when occur. The analog variables require a more complex process. For instance, the current variable is obtained by means of a low-cost current sensor that returns AC voltage. Then, a digital filter and a RMS calculation were needed to get the final RMS current value. This algorithm runs many times in the message transmission period to calculate mean values.

In Fig. 4 are shown the results of the current and temperature monitoring applied over the home refrigerator where it is possible to appreciate: (1) a stand by operation around 0 A, (2) followed by a steady state around 0.8 A, (3) and the variation of the refrigerator temperature between 7 and 7.5 °C. It represents the operation when the compressor is (1) turned off, (2) refrigerating, and (3) no door or temperature perturbances. This is a feature that the dashboard offers to allow the real-time monitoring of the interest variables. It also adds the visualization feature as a means to manage the data in electrical processes.

 mean_current

 Delete

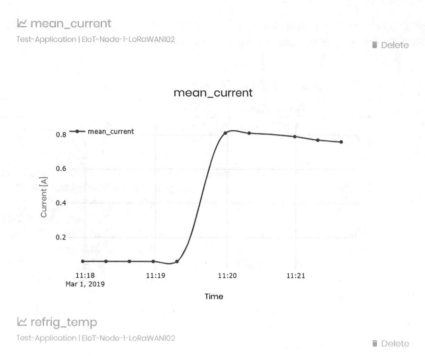

 refrig_temp

 Delete

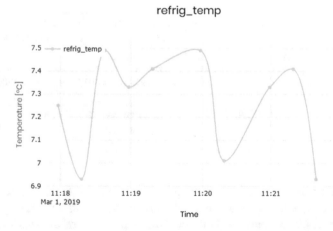

Fig. 4. Mean current - refrigerator temperature vs date time applied over the home refrigerator coupled to a EIoT node (1). Taken from dashboard.

In Table 1, some collected messages of the platform are shown. The messages contain the timestamp with milliseconds, what is extremely important for electrical systems because of their criticality. The identifier, this time the EUI of the LoRaWAN protocol, and the data. The data has a structure format of dictionaries or JSON objects which allow an easier way to deal with the data in subsequent processes. This type of

messages can be transmitted through the LoRaWAN protocol, though it is not recommended because of the length causes more battery consumption.

Table 1. Collected data from the EIoT platform.

Timestamp	Device EUI (ID)	Data
2019-04-24T13:22:33.455289-05:00	fd4ba37434271263	{"mean_current":0.06,"mean_power":7.13, "max_power":7.50,"refrig_temp":14.26, "freez_temp":72.85,"door_state":1}
2019-04-24T13:22:46.958067-05:00	fd4ba37434271263	{"mean_current":0.06,"mean_power":7.05, "max_power":7.29,"refrig_temp":14.59, "freez_temp":72.61,"door_state":0}
2019-04-24T13:23:33.680785-05:00	fd4ba37434271263	{"mean_current":0.06,"mean_power":7.01, "max_power":7.35,"refrig_temp":22.73, "freez_temp":72.77,"door_state":0}
2019-04-24T13:29:17.711484-05:00	c3674d88da03791f	{"mean_current":0.06,"mean_power":7.31, "max_power":159.04,"fire_detector":1}
2019-04-24T13:29:57.396177-05:00	c3674d88da03791f	{"mean_current":0.07,"mean_power":7.89, "max_power":8.66,"fire_detector":1}
2019-04-24T13:30:29.101972-05:00	c3674d88da03791f	{"mean_current":0.06,"mean_power":6.88, "max_power":6.88,"fire_detector":0}
2019-04-24T13:41:15.785953-05:00	e731127bf511a0cf	{"mean_current":0.30,"mean_power":35.95, "max_power":36.91,"fire_detector":0}
2019-04-24T13:43:36.131633-05:00	e731127bf511a0cf	{"mean_current":0.36,"mean_power":43.48, "max_power":44.81,"fire_detector":0}
2019-04-24T13:44:16.355487-05:00	e731127bf511a0cf	{"mean_current":0.45,"mean_power":53.32, "max_power":55.48,"fire_detector":0}

The messages contain more fields (not specified in table) like the device name or
the application name, which help to filter the data by keys. For instance, the application
of the substation contains some EIoT devices, the application of the solar-wind generation farm contains some others, etc.

In the presented data, the mean_current is the calculated mean RMS current (in
Amperes) along the transmission period; the mean_power multiplies the mean_current
to a voltage (for these tests, constant voltage), and the max_power considers the
maximum power value along the transmission period (both variables in Watts). The
refrig_temp and the freez_temp refer to the temperature in the fridge and freezer
temperature (2 measures, both in °C). The door_state refers to the open/closed state of
the door of the fridge; the value "1" represents a closed door, while the value "0"

represents an open door. Similar occurs in the fire_detector, it refers to the fire or smoke detection state; the value "0" represents no fire or smoke, while the value "1" represents that there is fire or smoke (abnormal operation).

With the data it is possible to detect correlations of variables. For example, in Table 1, the first 3 messages represent three almost consecutive messages of the refrigerator. The door was closed (door_state = 1) and the temperature of the refrigerator was low (14.26). Then, when the door was open (door_state = 0), the temperature began to increment (14.59 °C | 22.73 °C). The temperature of the freezer should be correlated as well, but according to the data, the temperature in the freezer (1) was higher than the refrigerator and (2) did not change when opening the door. It demonstrates that there is a problem in the temperature sensor of the freezer and should be addressed.

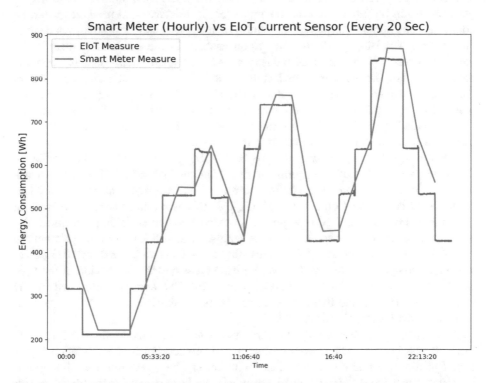

Fig. 5. Comparison of the EIoT sensor vs Smart Meter energy measurements for the demand management applications.

In Fig. 5, a test of the demand management application case was performed. In this case, it was compared the data of the energy measurement of an electricity smart meter (commercial use) and the estimated energy calculation by means of the current sensor in a node of the LoRaWAN platform. Both devices were connected to the same isolated hourly-scheduled load, and then, a controlled environment was guaranteed. The data of the EIoT node was collected every 20 s, while the smart meter collects data every 1 h.

From this test, it was possible to check the same behavior of the consumption profile in both measurements, with few (and not relevant for this purpose) errors in the measure. Even the more populated profile would serve better for very dynamic loads, though the amount of data could represent a problem when processing the data if many electrical things transmit messages so often. For the prototyping purposes, it worked perfectly.

As presented here, many tasks can be performed and some information can be inferred from the data sent by the Electrical Things in the EIoT platform. For instance, when transformers work at the top of their capacity, or when motors work in different speeds, or the temperature or gas are in an abnormal condition. The data can be used for multiple applications in the electrical systems, and the EIoT platform eases the way to deal with the data for those applications. The communication architecture and the design of the platform guarantee a scalable approach to integrate the data in several application cases of the electrical industry; later, when more communication protocols get involved in the EIoT platform, multiple electrical devices and machines could be connected one another despite of the nature of the elements – Industrial, residential, commercial. The EIoT platform works as an enabler for the data management in the electrical systems of the Electrical Industry 4.0 and the smart grids, contexts where the data is essential. The platform offers features to collect, store, visualize, manage, and, in the future, analyze the data from the electrical systems. The LoRaWAN protocol, which was used this time, fit the requirements to send data in wide areas with the required data length. Other LPWAN technologies which allow such data length in the messages (not too short) could fit for the purposes as well.

This proposal is aligned with the proposal of Song et al. [15], but in addition to define the generalities of a platform for the electrical Internet of Things, this work deploys the network and the platform for the data management with consistent applications cases. As Collier exposed in [12], this work intends to cover several requirements of the smart grid, proposing an extensible and scalable platform for the electrical systems to work in the internet of things. This work can also complement the approach of Gharavi and Hu [16] by providing an IoT network for distributed power grid monitoring and control while covering wide areas and the integration of multiple electrical applications in the same network-platform. The approach proposed in this work is aligned with the trends and requirements for the evolution of the electrical systems in the Electrical Industry 4.0.

This time, the electrical things to carry out the tests were monophasic AC and DC loads. It was because of the requirements of the project, but it is a limitation in the coverage for three-phase elements; this limitation can be overcome using the appropriated instrumentation. Even industrial elements can be connected to the platform with the correct integration through the industrial software applications or instrumentation interfaces.

This approach has a great potential for the development of the Internet of Things in the electrical systems. The consolidation of the smart grids requires the adoption of systems based on data and the exploitation of the operational value of the data. The EIoT platform provides a way to empower the electrical systems with the data and the further integration with information systems to improve the processes of the electrical industry, as a contribution for the consolidation of the smart grids.

5 Conclusions and Future Work

The project in a first stage was implemented using the LoRa technology with no LoRaWAN protocol. Then, LoRaWAN was adopted and it was appreciated the benefits of working under standardization in IoT projects. Other LPWAN technologies or communication protocols to consider in the future should be framed by standards and organizations that ensure the scalability of them.

The multidisciplinary knowledge requirements must be recognized as essential to deploy full IoT solutions, because they are composed by electronics, instrumentation, hardware in general, software, communications, and in this case, electrical engineering knowledge. The systems framed by the IoT show the evolution of the software engineering, the impact that it has generated in other disciplines, and its necessity for the process control and automation, benefiting the development of solutions with high value-added.

Although the EIoT platform was designed to work with electrical systems, any IoT solution can fit to applications in different industrial sectors by doing just minor modifications: major changes are due to notations and physical components. The EIoT platform can be eventually extended to cover multiple industrial sectors.

Another relevant topic is the adoption of protocols. In the era of the IoT is very important to comply the interoperability requirement. That is why protocol use is on the top of considerations in IoT or Industry 4.0 solutions. The application server of the EIoT platform was designed to work using the HTTP protocol by means of REST services, but it considers the integration of other important protocols for IoT communication such as MQTT, OPC UA, IEC 61850, and others later. It is also planned to integrate other LPWAN technologies and protocols into this solution and deploy similar IoT networks using the same platform. The NB-IoT technology, for example, should be covered later due to its suitability for the electrical industry [18, 34], when the networks be available in Colombia. The integration of multiple IoT protocols inside the platform is currently in working.

Up to now the EIoT platform performs data acquisition, storage, reading, and displaying, but it has great potential to perform hard tasks such as data analyzing, machine learning, autonomous control, among others, from the available data; it is on current deployment. It is important to state a future work to consolidate the Smart EIoT platform for the electrical systems, and other Smart IoT platforms for different industrial sectors. This approach can be subsequently integrated with other IoT platforms like Google Cloud IoT, Azure IoT, AWS IoT, and others. Such kind of integrations should be considered to take advantage of the distributed computing and to empower the features of the EIoT platform for more domain-specific tasks and the other platforms for the general purposes.

Acknowledgment. This project was developed and executed by Universidad Nacional de Colombia and funded by Enel-Codensa S.A. E.S.P. (Energy Service Provider) under the agreement "Plataforma EIoT - Electrical Internet of Things - para Sistemas de Energía". The authors would like to thank Universidad Nacional de Colombia, Enel-Codensa S.A. E.S.P. and the Administrative Department of Science, Technology and Innovation for their support in the development of this work.

References

1. Stankovic, J.A.: Research directions for the Internet of Things. IEEE Internet Things J. **1**, 3–9 (2014). https://doi.org/10.1109/JIOT.2014.2312291
2. Yun, M., Yuxin, B.: Research on the architecture and key technology of Internet of Things (IoT) applied on smart grid. In: International Conference on Advances in Energy Engineering, ICAEE 2010, pp. 69–72 (2010). https://doi.org/10.1109/ICAEE.2010.5557611
3. Wollschlaeger, M., Sauter, T., Jasperneite, J.: The future of industrial communication. automation networks in the era of the Internet of Things and Industry 4.0. IEEE Ind. Electron. Mag. **11**, 17–27 (2017). https://doi.org/10.1109/MIE.2017.2649104
4. Monteiro, K., Marot, M., Ibn-Khedher, H.: Review on microgrid communications solutions: a named data networking - fog approach. In: 16th Annual Mediterranean Ad Hoc Networking Workshop, Med-Hoc-Net (2017). https://doi.org/10.1109/MedHocNet.2017.8001656
5. Mekki, K., Bajic, E., Chaxel, F., Meyer, F.: A comparative study of LPWAN technologies for large-scale IoT deployment. ICT Express **5**, 1–7 (2018). https://doi.org/10.1016/j.icte.2017.12.005
6. Raza, U., Kulkarni, P., Sooriyabandara, M.: Low power wide area networks: an overview. IEEE Commun. Surv. Tutorials **19**, 855–873 (2017). https://doi.org/10.1109/COMST.2017.2652320
7. Andreadou, N., Guardiola, M.O., Fulli, G.: Telecommunication technologies for smart grid projects with focus on smart metering applications. Energies **9**, 375 (2016). https://doi.org/10.3390/en9050375
8. Ray, P.P.: A survey on Internet of Things architectures. J. King Saud Univ. Comput. Inf. Sci. **30**, 291–319 (2018). https://doi.org/10.1016/j.jksuci.2016.10.003
9. Perera, C., Liu, C.H.I.H., Jayawardena, S., Chen, M.: A survey on Internet of Things from industrial market perspective. IEEE Access **2**, 1660–1679 (2014). https://doi.org/10.1109/ACCESS.2015.2389854
10. Sethi, P., Sarangi, S.R.: Internet of Things: architectures, protocols, and applications. J. Electr. Comput. Eng. **2017**, 1–25 (2017). https://doi.org/10.1155/2017/9324035. Article ID 9324035, 25 pages
11. Wan, J., Tang, S., Shu, Z., Li, D., Wang, S., Imran, M., Vasilakos, A.V.: Software-defined industrial Internet of Things in the context of Industry 4.0. IEEE Sens. J. **16**, 7373–7380 (2016). https://doi.org/10.1109/JSEN.2016.2565621
12. Collier, S.E.: The emerging enernet. IEEE Ind. Appl. Mag. **23**, 12–16 (2016)
13. Shahryari, K., Anvari-Moghaddam, A.: Demand side management using the internet of energy based on fog and cloud computing. Kurdistan J. Appl. Res. **2**, 112–119 (2017)
14. Ferrari, P., Flammini, A., Rinaldi, S., Rizzi, M., Sisinni, E.: On the use of LPWAN for EVehicle to grid communication
15. Song, Y., Lin, J., Tang, M., Dong, S.: An internet of energy things based on wireless LPWAN. Engineering **3**, 460–466 (2017). https://doi.org/10.1016/J.ENG.2017.04.011
16. Gharavi, H., Hu, B.: Wireless infrastructure M2M network for distributed power grid monitoring. IEEE Netw. **31**, 122–128 (2017). https://doi.org/10.1109/MNET.2017.1700002
17. Madueno, G.C., Nielsen, J.J., Kim, D.M., Pratas, N.K., Stefanovic, C., Popovski, P.: Assessment of LTE wireless access for monitoring of energy distribution in the smart grid. IEEE J. Sel. Areas Commun. **34**, 675–688 (2016). https://doi.org/10.1109/JSAC.2016.2525639
18. Javadi, S., Seifvand, S.: NB-IoT applications in smart grid: survey and research challenges. Int. J. Commun. **11**, 74–77 (2017)

19. Lin, S.-W., Miller, B., Durand, J., Bleakley, G., Chigani, A., Martin, R., Murphy, B., Crawford, M.: The Industrial Internet of Things Volume G1: Reference Architecture. Ind. Internet Consort. 1.80, 1–7 (2017)
20. Plattform Industrie 4.0: Reference Architectural Model Industrie 4.0 (RAMI 4.0) - An Introduction. 0, 21 (2016)
21. Platform Industrie 4.0; Robot Revolution Initiative; 4.0, S.C.I.: The common strategy on international standardization in field of the Internet of Things/Industrie 4.0
22. WAVIOT: WAVIOT NB-FI LPWAN Technology: Product and Tech Description (Release 1.8), pp. 1–29 (2016)
23. On-Ramp Wireless Inc.: RPMA Technology for the Internet of Things, pp. 1–46 (2015)
24. LoRa Alliance, T.M.W.: What is it? (2015)
25. Sigfox: Sigfox technical overview 1, 26 (2017)
26. GSMA: 3GPP Low Power Wide Area Technologies (LPWA). GSMA White Paper, pp. 19–29 (2016)
27. UNB Wireless - Telensa. https://www.telensa.com/unb-wireless/
28. Boulogeorgos, A.-A.A., Diamantoulakis, P.D., Karagiannidis, G.K.: Low power wide area networks (LPWANs) for Internet of Things (IoT) applications: research challenges and future trends, pp. 1–15 (2016)
29. Centenaro, M., Vangelista, L., Zanella, A., Zorzi, M.: Long-range communications in unlicensed bands: the rising stars in the IoT and smart city scenarios. IEEE Wirel. Commun. 23, 60–67 (2016). https://doi.org/10.1109/MWC.2016.7721743
30. Qualcomm: Leading the LTE IoT evolution to connect the massive Internet of Things. Qualcomm Technol. Inc. 14 (2017)
31. Van Schalkwyk, T.: OEM Business Unit - Sierra Wireless - 3GPP LPWA Standards: LTE-M, NB-IoT & EC-GSM (2017)
32. Duarte, D.: A performance overview about fire risk management in the Brazilian hydroelectric generating plants and transmission network. J. Loss Prev. Process Ind. 17, 65–75 (2004). https://doi.org/10.1016/j.jlp.2003.09.007
33. Brocaar, O.: LoRa Server, open-source LoRaWAN network-server. https://www.loraserver.io/. Accessed 24 Apr 2019
34. Li, Y., Cheng, X., Cao, Y., Wang, D., Yang, L.: Smart choice for the smart grid: Narrowband Internet of Things (NB-IoT). IEEE Internet Things J. 4662, 1–11 (2017). https://doi.org/10.1109/JIOT.2017.2781251

Infrastructure-Aided Networking for Autonomous Vehicular Systems

Izhak Rubin and Yulia Sunyoto[✉]

Electrical and Computer Engineering Department,
University of California at Los Angeles (UCLA), Los Angeles, CA 90024, USA
rubin@ee.ucla.edu, yulia.sunyoto@ucla.edu

Abstract. We study infrastructure-aided data communication networking for autonomous transportation system. The infrastructure consists of Roadside Units (RSU) which are placed along the side of a highway segment and are connected by optical fiber. The infrastructure provides full communication coverage of the segment's vehicles. We present a data networking protocol enabling packet flows to be disseminated to all vehicles within a specified span. We study the performance behavior of the synthesized data network as a function of the number of employed RSU nodes. For each case, we configure the structure of the medium access control scheduling scheme, and configure the employed Modulation Coding Schemes (MCS) and corresponding data rates, the transmit power levels and the spatial-reuse factors. We aim to obtain high throughput rates under prescribed packet delay limits. In addition, we impose high packet successful reception rate requirements for assuring reliable dissemination of packet flows.

Keywords: Vehicle-to-Infrastructure · V2I · MAC · TDMA ·
IEEE 802.11p · Vehicular network · VANET · Autonomous vehicles ·
Autonomous transportation

1 Introduction

Autonomous driving can improve safety and enhance vehicular traffic efficiency. A fully-autonomous highway environment must make use of a reliable, robust and delay-aware communication system which enables data networking for the dissemination of both critical and non-critical messages to and from highway vehicles. The availability of a communications backbone infrastructure should be employed to enhance the performance of the highway vehicular communication system. In this paper, we assume the availability of a backbone network that consists of roadside unit (RSU) stations that are interconnected by a fiber optic backbone. We vary the density of the RSU nodes. Yet, for each RSU backbone configuration, the RSU stations are configured to provide full communications coverage of all vehicles traveling along the underlying segment of the highway. We study the data throughput performance behavior of this infrastructure-aided

© Springer Nature Switzerland AG 2020
K. Arai et al. (Eds.): FTC 2019, AISC 1070, pp. 66–86, 2020.
https://doi.org/10.1007/978-3-030-32523-7_5

autonomous vehicular system, subject to specified bounds on maximum packet delays and on minimum reception success rates. The fiber optic infrastructure consists of RSUs that are placed in linear formation along the highway. RSUs are equipped with wireless and fiber optic transceivers. The wireless radio is used by the RSU to transmit data packets downlink to covered vehicles (which form its cell members), and to receive uplink data transmissions from such vehicles. The fiber optic backbone network is used for the transmission of data packets among RSU stations. Vehicles are equipped with radios that are used for wireless communication with the infrastructure, enabling Vehicle-to-Infrastructure (V2I) and Infrastructure-to-Vehicle (I2V) transmissions. We present RSU-aided networking protocols that are employed for the dissemination of packets from source vehicles to all vehicles within a targeted dissemination span away from the corresponding sources. The infrastructure provides full communication coverage to all vehicles on the highway, and each vehicle can communicate with at least one RSU station (transmit to and receive from such an associated station) at any location along the highway. Source vehicles transmit packets uplink to a RSU, which then forwards the packets to the involved neighboring RSUs. The involved RSUs multicast these flow packets downlink to vehicles that are located within the targeted dissemination span. We characterize the performance behavior of the network system as a function of the density of the infrastructure backbone network, and thus the inter-RSU distance, while simultaneously properly setting MAC and PHY design parameters, impacting the employed scheduling schemes, spatial-reuse coloring indices, data rates, modulation coding schemes (MCS) and transmit power levels used by RSU nodes and by highway vehicles. In this paper, we focus on the development and study of an infrastructure aided Time-Division Multiple Access (TDMA) scheduling scheme that is employed for the sharing of the corresponding uplink and downlink wireless communication channels.

The major contributions of this work can be summarized as follows:

- We investigate the design of an infrastructure-aided communication networking system, assuming that the infrastructure provides full communication coverage of vehicles traveling along a segment of the highway under consideration. A two-layer hierarchical network architecture is synthesized and employed.
- We study and demonstrate the data throughput performance behavior of this infrastructure-aided autonomous vehicular system as we vary the inter-RSU distance (and thus the density of the RSU backbone layout). The design is subject to specified bounds on maximum packet delays and minimum reception success rates for disseminating packet flows from source vehicles to vehicles that travel within targeted dissemination spans.
- For each configured inter-RSU distance level, we determine the optimal joint setting of the network system parameters.
- We demonstrate the non linear behavior of the system throughput performance as the infrastructure backbone density is varied.
- We compare the data throughput performance attained by using the infrastructure aided communication system presented in this paper with the performance attained by a pure V2V system that employs no infrastructure-based

backbone network, assuming each one to use a TDMA-based scheduling protocol. There are several studies which discuss platoon-based V2V communications for autonomous vehicular systems, such as [1] and [2]. We use the V2V protocol and scheme presented in [2] for performance comparison to the infrastructure aided communication system presented in this paper.

The paper is organized as follows. In Sect. 2, we present an overview of related work on infrastructure-aided vehicular communication and TDMA-based MAC schemes. The network systems model, including network parameters and performance metrics, and the underlying network architecture and protocols used for data dissemination, are presented in Sect. 3. In Sect. 4, we present the infrastructure-aided TDMA based MAC scheme that is used to coordinate uplink and downlink transmissions. Its corresponding performance characteristics are presented in Sect. 5. In Sect. 6, we compare the performance behavior of several distinct infrastructure aided MAC schemes, including TDMA-based, IEEE 802.11p-based, and a hybrid of TDMA and IEEE 802.11p-based MAC schemes. In Sect. 7, we compare the data throughput performance of an infrastructure-aided V2I system with that exhibited by a platoon-based V2V system. Conclusions are drawn in Sect. 8.

2   Related Work

Several studies [3–5] develop cooperative routing strategies and cost-effective RSU placements [6] in environments where RSUs provide partial coverage to vehicles on the highway. The strategies discussed include multihop V2V operations. Performance metrics involved include data throughput [3–5] and message delay [3,5]. Such studies generally account for only uplink or only downlink data transmissions. The study in [6] aims to maximize the aggregate uplink throughput rate, including as parameters the vehicular data rates, the number of hops traversed from source vehicles to the corresponding destination RSUs, and inter-vehicular signal interference. We note that several of the above mentioned studies consider an infrastructure system that provides for only partial coverage of the highway segment. In turn, the study that we present in this paper focuses on the optimization of the performance of the synthesized data network system when the infrastructure system is capable to provide for full coverage of the highway segment, though cost considerations induce variations in the inter-RSU distance levels, as expected to be the practical case when considering an autonomous highway segment that resides in a busy urban environment.

The authors in [7] study the downlink data throughput performance of a mmWave infrastructure aided vehicular communication as a function of base station density (i.e. average inter base station distance). Beamforming is used for communication between a vehicle and a base station. The downlink data throughput performance is impacted by the stochastic occurrence of blockage scenarios. Some factors considered include resulting SINR between vehicle and base station, probability of coverage, and duration of maintaining connection during transmission. It considers transmission from each RSU to a single tracked vehicle. Unlike our study, [7] does not consider broadcasting packets from a source vehicle to all vehicles residing within a targeted geographical span, a scenario which is of critical importance for many classes of safety messages produced by vehicles traveling along the autonomous highway.

The study by [8] uses fractional frequency reuse (which is a hybrid of spatial reuse and frequency reuse) to mitigate interference among downlink transmissions by LTE base stations. However, the topological layout of the cellular base stations is non linear, impacting accordingly the interference signals that are produced.

Several studies consider TDMA (or FDMA) oriented MAC schemes that are used to support vehicular system applications, including RSU-based scheduling schemes [9–12], and ad hoc scheduling methods [13]. They often present methods for slot allocation for the purpose of accommodating contention free data transmissions by considering various performance objectives, including metrics that involve throughput maximization [9] and message delay bounds [9,10]. Several models also accommodate the transmission of high priority packets [9,10]. The scheme presented in [9] attaches higher weights for the support of vehicles that are associated with higher channel quality conditions, while also considering the position of vehicles for service fairness purposes. The paper uses demand assigned TDMA schemes, whereby the duration of a random access reservation period is adapted to ensure low collision rate levels for the transmission of reservation request messages. Both [9] and [10] allocate time slots based on EDCA (AC) factor, as is employed by IEEE 802.11e based systems, by attaching higher priority indicators to critical messages. Consequently, several of these studies also evaluate the performance efficiency of the TDMA MAC schemes during the contention free transmission period. However, [9,10] use TDMA scheduling within the neighborhood of a single RSU, and the slot scheduling scheme is tailored to yield performance efficiency within the RSU's region of coverage. Whereas, our study presents TDMA-based scheduling schemes which coordinate the scheduling of multiple RSUs jointly with the adaptive configuration of the employed data rate, MCS and spatial-reuse levels. Such an operation allows for interference mitigation, leading to system-wide throughput optimization.

In contrast, the authors in [11–13] propose a TDMA/FDMA based scheduling mechanisms that involve multiple RSUs. In [11], the authors propose a centralized RSU-centric TDMA time slots allocation scheme, in serving vehicles that reside within its coverage. The study assumes the use of two different frequency bands employed by adjacent RSUs, such that no signal interference is induced among transmissions executed in neighboring cells. The study [12] proposes installation of large number of base stations which are interconnected to a controller by optical fiber. Each base station covers a cell. The base stations are categorized into several groups, each forming a virtual cellular zone (VCZ), which consists of several base stations operating in the same frequency range. Only one base station can be active at a time to transmit downlink within a VCZ to prevent interference within a VCZ. Base stations transmissions within a VCZ are scheduled in a TDMA fashion. Adjacent VCZs also use disjoint frequency ranges. The study proposes a demand-assigned TDMA scheme within each VCZ. The authors in [13] propose a combination of using SDMA, OFDMA and TDMA. In using SDMA, the highway is divided into cells, whereby each cell is allocated different sets of subcarriers from adjacent cells. Within each cell, the subcarriers are shared in a TDMA fashion. It also specifies that there should be four minimum different frequencies used for adjacent cells to avoid time slots oriented overlap collisions. The setting in [11,13] however does not employ an optimized number of frequency bands for optimizing the spatial-reuse (coloring) configuration. Similarly, the scheme presented in [12] does not study methods for the optimization of the number of simultaneously active base stations employed within a VCZ. It also does not determine the best selection of the number of frequency bands that should be used at adjacent VCZs, as used to enhance throughput by mitigating signal interference. More TDMA protocols are discussed in [14]. In contrast with the setting of the joint uplink/downlink TDMA schemes employed by us in this paper, above noted papers which use TDMA MAC schemes tend to not provide for joint uplink/downlink scheduling, whereby the aim is to maximize the data throughput rate for the dissemination of packets flows over a targeted span.

The authors in [15,16] study the impact of RSU antenna designs and mounting on data throughput performance. In [17], a study of RSU-aided data networking performance is presented when considering several channel propagation models. Inter-RSU distances are adapted to ensure a higher degree of seamless coverage. These studies however do not consider MAC scheduling schemes that are employed for disseminating packet flows over specified spans, coupled with the effective setting of network parameters as performed in this paper.

To the best of our knowledge, none of the above studies propose and examine, as we present in this paper, the joint setting of data networking parameters for autonomous vehicular highway communication system that employs a backbone network that provides full coverage of highway segment vehicles, whereby the MAC scheduling structure and parameters are optimally configured, as performed in this paper, message flows are disseminated over specified spans, and the system is set to yield high throughput rates under specified delay and reception rates, in terms of the density level of the infrastructure's transmission system.

3   Network Systems Model

3.1   Network Parameters

RSUs are placed along the side of a linear multiple-lane highway of length L with inter-RSU distance of d_{RSU}, forming a RSU backbone network. The RSU nodes are interconnected by high speed fiber optic links. This infrastructure is used to aid communications networking along the vehicular highway system. Figure 1 illustrates the network system studied in this paper. Vehicles are assumed to be distributed along the highway in either a random fashion or organized into platoons, which are often used in each lane of the autonomous highway. In carrying out network performance evaluations in this paper, we assume, at no loss in generality, vehicles to be uniformly distributed along each lane.

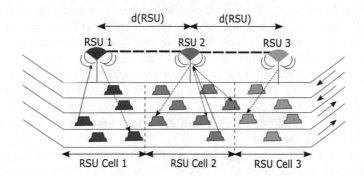

Fig. 1. Linear highway of length L with Fiber Optic Roadside Infrastructure

A RSU is equipped with both wireless and fiber optic transceivers. A vehicle's On-Board Unit (OBU) is equipped with a wireless radio. To illustrate the performance tradeoffs available to the designer under various network configuration options, our performance evaluations employ the following parameter values. We consider d_{RSU} to range from 100 m to 1000 m. The number of vehicles N admitted to a lane segment of length $L = 5$ km is set to 250. We note that simulations and analyses are performed on a longer highway length, such as 10 km, and data throughput performance is aggregated from the middle 5 km segment, to attenuate edge effects. Through our simulation and analysis based evaluations, we have found out that results presented for a single lane can be well used to reflect the performance behavior of the system in a multi lane environment. We focus here on performance results obtained when assuming vehicles to travel along a single lane. To illustrate the performance of data networking schemes, we use wireless communications radios that employ the 5.9 GHz frequency band, as often used for testing autonomous vehicular system technologies such as DSRC and C-V2X, based in the United States [18] or Europe [19].

Our networking schemes make use of subdivision of the spectrum into distinct communication channels, including a control channel (CCH) and multiple service channels (SCH). The CCH is used in a similar manner to that employed by the WAVE system [20] and by other architectures. It is used by station entities to periodically (e.g., every 100 ms) broadcast status and control messages [21]. This process allows RSUs and vehicles to exchange essential up-to-date status information, such as location and velocity states, which is used for the synthesis of our double-hierarchy based network architecture. The CCH can also be used for the transmission of reservation and assignment messages that serve to request and assign time slot resources [22]. The CCH can also be employed for the transmission of critical safety messages [14, 22].

A service channel (SCH) is used in our scheme for the transmission of data packet flows produced by a multitude of applications. To simplify, packet flows are categorized into two classes. Class 1 packets include critical (e.g., safety) messages and are granted higher priority. The throughput rate of such event-induced packet flows is relatively low. However, such packets should be disseminated across their spans to achieve a Packet Delivery Ratio (PDR) that is equal to at least 0.90. Furthermore, successfully delivered packets should complete their dissemination within a 90-percentile time delay requirement of $D_{P,max} = 50$ ms. Class 2 packet flows are of lower priority but often impose high throughput rate requirements. We assume in this study that packet flows can be generated by multiple source vehicles, and each packet must be disseminated over a targeted dissemination span of d_{span}, reaching vehicles travelling behind the source vehicle within such distance.

We aim to configure the communications network to achieve high performance, realizing a high data throughput rate while satisfying prescribed message delay (90-percentile packet delay capped at $D_{P,max}$) levels and successful delivery rate (PDR ≥ 0.90) objectives. We carry out our performance analysis and design evaluations under the assumption that vehicular nodes are highly loaded by packets that can either belong to Class 1 or Class 2. To illustrate, we assume $d_{span} = 300$ m. Our protocols and basic models can readily be extended to scenarios whereby the dissemination span is applied bi-directionally to cover vehicles that move behind and ahead of each source vehicle.

We aim to determine the best configuration of uplink and downlink MAC scheduling schemes and their associated parameter values, uplink and downlink MCS configurations and their ensuing data rates, the spatial-reuse (coloring) levels, and the transmit power levels used by radio modules at the vehicles and at the RSUs, under varying d_{RSU} range values. Noting that packet transmissions across the infrastructure are carried out along the fiber optic based backbone links, we consider the wireless communications channels that are used for respective uplink and downlink transmissions to be shared according to one of the following MAC protocols: Demand-assigned (DA) TDMA uplink - TDMA downlink (MAC 1), IEEE 802.11p uplink - TDMA downlink (MAC 2) and IEEE 802.11p uplink - IEEE 802.11p downlink (MAC 3). In this paper, we focus on the design and performance evaluation of a system that employs MAC 1. To carry out performance analyses when using TDMA-related MAC schemes, we combine the use of analytical models and simulation evaluations.

For our illustrative performance analyses, we assume data rates to be selected among 3 Mbps, 6 Mbps, 12 Mbps; noting that each corresponding MCS employs a rate 1/2 coding scheme. These data rates are recommended as mandatory by [23] for use in the 10 MHz channels of IEEE 802.11p. Accounting for the code rate, using these data rates equivalently corresponds to transmission rate values R of 6 Mbps, 12 Mbps and 24 Mbps respectively. We subsequently express the data throughput performance behavior as a function of the latter transmission rate levels. We denote the transmission rates employed across the uplink and downlink channels as R_{ul} and R_{dl}, respectively. The transmit power levels P_{tx}

(dBm) used by vehicles and RSUs are denoted as $P_{tx,v}$ and $P_{tx,r}$, respectively. For our illustrations, we assume transmit power values of 23 dBm and 33 dBm [18]. We set the communications channel propagation loss model and its associated parameter values to be the same as that used in [24]. We assume vehicles and RSUs to use omni-directional antennas. The signal power received at a node (either a vehicle or a RSU), when transmitted by another node that is located at a distance that is equal to d, is denoted as $P_{rx}(d)$.

$$P_{rx}(d) = \begin{cases} P_{rx}(d_0) - 10\gamma_1 log_{10}\frac{d}{d_0}, & \text{if } d_0 \le d \le d_c \\ P_{rx}(d_0) - 10\gamma_2 log_{10}\frac{d}{d_c} - 10\gamma_1 log_{10}\frac{d_c}{d_0}, & \text{if } d > d_c \end{cases} \tag{1}$$

For a given data rate and MCS, to assure an acceptably high PDR, a link-level packet transmission is considered to be successfully received only if a minimum receiver SINR threshold $SINR_t$ and a minimum receiver sensitivity (i.e. minimum received power) rv_t levels are satisfied. For example, when setting R to be 24 Mbps, we have $SINR_t = 20$ dB [25,26] and $rv_t = -77$ dBm [23]. The network parameters that we use in our system evaluations are summarized in Table 1.

Table 1. Network parameters

Parameter	Value	Parameter	Value
L	5 km	$P_{tx,v}, P_{tx,r}$	23 and 33 dBm[18]
d_{RSU}	100–1000 m	P_{noise}	–104 dBm
R_{ul}, R_{dl}	6, 12, 24 Mbps [23]	k_{ul}, k_{dl}	1, 2, 3 and 4
$SINR_t$	7, 11, 20 dB [25,26]	$D_{P,max}$	50 ms [2]
rv_t	–85, –82, –77 dBm [23]	PDR	≥ 0.90
P	3024 bit [27]	γ_1, γ_2	1.9, 3.8
d_{span}	300 m [2]	d_0, d_c	10, 80 m

3.2 Performance Metrics

We characterize the data throughput performance in terms of sole aggregate uplink throughput $TH_{ul,I}$, sole aggregate downlink throughput $TH_{dl,I}$ and joint aggregate data throughput of the system $TH_{ul,dl}$.

The aggregate throughput rate of data packets supported across an uplink channel when assuming that all system bandwidth is allocated for uplink operations (thus isolating its performance from that induced by downlink operations) is denoted as $TH_{ul,I}$. Similarly, assuming all system bandwidth to be used for downlink operations, the aggregate downlink data throughput is denoted as $TH_{dl,I}$. The aggregate joint system throughput $TH_{ul,dl}$ represents the total supported data rate of uplink packet flows originating from source vehicles. Included are only messages that are successfully disseminated to vehicles within d_{span} from their source vehicles. A flow of such packets is considered to be successfully disseminated if its packets are correctly received by at least 90% of vehicles within d_{span} (i.e., PDR ≥ 0.90), while incurring a packet delay level that is lower than $D_{\mathrm{P,max}} = 50$ msec for at least 90% of the packets. The aggregate data throughput metric is scaled by a factor that is proportional to the number of RSUs placed along the highway segment, N_{RSU}.

3.3 Network Architecture

Each RSU station manages a RSU cell, which is normally set to extend over a highway region whose boundary envelopes a range of $d_{RSU}/2$ from the RSU station covering vehicles travelling in both directions. For analytical simplification, we assume RSU cell regions to be disjoint, as illustrated in Fig. 1. In the double-hierarchy network architecture being formed, we assume that the number of RSUs forming the backbone network is sufficient to allow these RSUs to provide full coverage of all vehicles traveling along the underlying highway segment. A vehicle is assumed to associate with the RSU that manages the cell to which it belongs. Packet dissemination flows are thus carried out without resorting to the use of V2V communications.

A Highway System Manager (HSM) is employed. It exchanges status information with the RSUs, which includes vehicular status vectors, obtained from status message exchanges across the CCH. A status message contains information on vehicular location coordinates, speeds, destinations, channel quality indices, radio states, available processing rates and transmit power resource capabilities. The HSM also communicates with the RSUs to set system-wide data networking parameters, which are then announced by the RSUs to their mobiles. As it travels along the highway, a vehicle proceeds to associate with the RSU from which it receives the highest quality radio signals. The vehicle is then deemed to become a member of the selected RSU's cell.

3.4 Networking Protocol

During any period of time, packets are generated by certain vehicles that are traveling along the highway segment and happen to be stochastically engaged in the origination of packet flows. The packets produced by each source vehicle are disseminated to all vehicles located within d_{span} from the source vehicle, assuming here the dissemination to proceed in a direction that is opposite to the travel direction of the source vehicle. For this purpose, a source vehicle transmits its packets uplink to its associated RSU. The packets are then transported across the fiberoptic backbone to the other involved RSU nodes. The involved RSUs transmit these packets downlink across their cells, consequently reaching the vehicles which are located within the targeted dissemination d_{span}. We denote by n_{span} the number of RSUs, which are used to multicast these packets across their downlink channels. The average value of n_{span} is expressed as $E[n_{\mathrm{span}}] = 1 + \frac{d_{span}}{d_{RSU}}$.

It is noted that at times the selection of RSUs to forward packets may induce packet reception by vehicles that reside beyond the targeted span. Depending on the underlying application, such excess receptions may be acceptable (e.g., for critical safety messages). Otherwise, the receiving mobile may filter unwanted messages by using location or other identifiers.

4   Infrastructure-Aided MAC Schemes

Packet transmissions between RSUs are executed across a fiber-optic backbone network at a data rate that is much higher than that used for communications across the system's wireless channels. Hence, packets received across a wireless uplink at a RSU are effectively instantly transported to the other RSUs that need to receive such packets for dissemination to vehicles through their scheduling for downlink transmissions. The ensuing packet delays incurred across the fiber-optic links are therefore relatively negligible.

4.1   DA/TDMA Uplink - TDMA Downlink (MAC 1)

We assume that RSUs acquire time synchronization from a regional network manager. The RSUs transmit periodically beacons that provide time stamp information, which allows regional vehicles traveling along the underlying highway segment to be time synchronized, enabling the implementation of a TDMA mechanism across the system uplinks and downlinks. A Demand Assigned TDMA (DA/TDMA) scheme is used to allocate uplink access resources. Each time frame, denoted as T_f, contains periods used by mobiles to send reservation packets, as well as time periods used by mobiles and by RSUs to transmit uplink and dowlink packets. We denote the duration of the per-frame reservation period as Tp_{rsv}, the duration of the per-frame uplink data transmission period as Tp_{ul} and the duration of the per-frame downlink data transmission period as Tp_{dl} (see Fig. 2). Hence $T_f = Tp_{rsv} + Tp_{ul} + Tp_{dl}$.

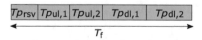

Fig. 2. Illustrative Time Slot formation in a MAC 1 Time Frame with $k_{ul} = 2$ and $k_{dl} = 2$. $Tp_{ul,1}$ and $Tp_{dl,1}$ denote uplink and downlink periods used by RSU cells of index 2n–1. $Tp_{ul,2}$ and $Tp_{dl,2}$ denote uplink and downlink periods used by RSU cells of index 2n ($n \geq 1$).

Only a single vehicle member of a RSU cell is assigned to execute uplink transmission while using a given time slot. To reduce the impact of interference signals induced by simultaneous uplink transmissions by mobiles that reside in neighboring RSU cells, an uplink coloring factor (i.e., spatial-reuse index) k_{ul} is employed. During a Tp_{ul} period, active vehicles that are members of a certain

RSU cell are set to transmit for a fraction of time that is equal to $\frac{1}{k_{ul}}Tp_{ul}$, so that during such a period only 1 out of k_{ul} consecutive RSU cells is scheduled to allow uplink transmissions. Each corresponding time slot assumes the duration $Ts_{ul} = \frac{P}{R_{ul}}$ for the transmission of a packet.

Through performance analyses, we determine the value for k_{ul} in aiming to mitigate inter-RSU cell interference signals, serving to assure a high uplink PDR level, PDR_{ul}, of at least 0.95, while attaining the highest feasible uplink throughput level. The coloring value to be properly configured depends on the employed R_{ul}. The value of k_{ul} is determined by using Monte Carlo simulation of uplink transmissions. For design purposes, we assume a conservative case whereby all RSU cells are assumed to be always active (i.e., vehicles are kept highly loaded with packets that they wish to disseminate) such that all RSU cells are kept busy during the entire Tp_{ul} and Tp_{dl} periods. When a higher R_{ul} is used, a higher reuse k_{ul} level is typically required due to the higher SINR threshold that is required at the intended receiver. For example, we have determined that, for the underlying scenario, under $R_{ul} = 6$ Mbps and $R_{ul} = 12$ Mbps, an optimal reuse level of $k_{ul} = 2$ should be employed. In turn, under $R_{ul} = 24$ Mbps, one should set an optimal reuse level of $k_{ul} = 3$. Similarly, we can determine optimal k_{dl} value, which depends on the employed R_{dl} by the RSUs. Overall, optimal coloring values k_{ul} and k_{dl} must be selected to ensure that at least 90% of the packets meet the corresponding required minimum SINR levels, and guarantee reception of packets across for both uplink and downlink channels at high success rates to reach their intended destination mobiles. For example, to simplify system design, we note that by guaranteeing $PDR_{ul} > 0.95$ and $PDR_{dl} > 0.95$, we assure a PDR level that is equal to a high value, which is often equal to at least 90.25%.

We coordinate the joint allocation of uplink and downlink bandwidth resources, noting that each RSU node must be able to process and transmit (downlink) the traffic that it receives (across its uplink) from vehicles in its RSU cell and from other RSU nodes (across the fiberoptic links), provided it is located within the traffic's backbone dissemination path. Included are packets received from $E[n_{span}]-1$ neighboring RSU nodes. Hence, the corresponding time periods Tp_{ul} and Tp_{dl} allocated in each frame are set to satisfy the following ratio:

$$\frac{Tp_{ul}}{Tp_{dl}} = \frac{R_{dl}PDR_{dl}k_{ul}}{E[n_{span}]R_{ul}PDR_{ul}k_{dl}} \tag{2}$$

Noting that our throughput analysis has been performed by assuming vehicles to be highly loaded in the production of packet flows, and hence inducing highly loaded RSU nodes, we deduce that the aggregate uplink throughput can be expressed as $TH_{ul,I} = N_{RSU}\frac{R_{ul}PDR_{ul}}{k_{ul}}$. The aggregate system throughput is expressed as follows (noting that all traffic flows are generated by mobiles):

$$TH_{ul,dl} = TH_{ul,I}\frac{Tp_{ul}}{T_f} \tag{3}$$

To assess the packet delay components, we note the following. The Frame Latency (FL) delay component represents the time elapsed between the arrival (or production) time of a packet at the source vehicle and the time instant at which the vehicle transmits its reservation packet. FL is incurred only by the first packet of a packet flow (or 'stream'); the underlying scheme can use the first data packet to include (piggyback) a reservation request. Subsequent flow packets will not experience such FL delays, as we assume that the flow is allocated slots at a rate that matches its requested rate. To state an upper bound on incurred packet delay level, we set a worst case frame latency to assume the value $FL = T_f$. The MAC 1 scheme is especially suitable for the scheduling of stream-oriented packet flows, which are characterized by periodic generation process at an application specific rate.

High priority Class 1 messages may use periods allocated within the CCH to assure a very high success rate in the transmission of short critical data packets and/or reservation packets to reserve TDMA time slots. Various other reservation schemes can be employed (for example the discussion in [14]). Clearly, a longer reservation period allocated within a SCH time-frame results in a lower residual capacity available to support the transmissions of data packets. In our performance analyses, we assume there that the targeted reservation latencies are met with the use of reservation schemes which adapt to the underlying traffic rate to ensure high probability of success.

To achieve a bounded delay level performance for admitted packets, we employ the following flow regulation and control scheme at each RSU node. All packets that reach a RSU node during a time frame period, either from its uplink channel or from its neighboring RSUs, are targeted by the RSU node for complete downlink transmission during the current time frame downlink period. Since the uplink and downlink time period setting uses an averaged n_{span} parameter, $E[n_{span}]$, to regulate the loading on each RSU under conditions whereby different flows may require dissemination along a different number of RSUs, we configure the system would to employ a flow control mechanism that serves to block the admission of flows which induce overloading of the downlink queue at RSU nodes. In this manner, one can assure an end-to-end packet delay of the order of $D_p \leq FL + T_f \leq 2T_f$ (supplemented by an ensuing reservation delay when applicable, normalized in relation to the average flow duration as only the first packet of the flow incurs the latter delay component).

A MAC 1 type scheme can be applied in a similar manner when resources are assigned in the joint time/frequency (TDMA/FDMA) two-dimensional domain, as often performed by 4G cellular systems.

4.2   DA/TDMA Uplink - TDMA Downlink (MAC 1) Under Adaptive Uplink Transmission Rate Selection

A mobile position-based adaptive uplink rate selection scheme can be implemented as follows. A vehicle adjusts its uplink data transmission rate R_{ul} in accordance with its current distance from the associated RSU. A corresponding higher R_{ul} value is set by a vehicle when it is located around the center of a cell, while a lower R_{ul} level is used by a vehicle that is located closer to the cell's boundary. An uplink coloring index k_{ul} is jointly configured in relation to the employed uplink rate R_{ul}, following the design method described above in Sect. 4.1. The aggregate uplink throughput $TH_{ul,I}$ determined for the adaptive uplink rate scheme is obtained by first averaging the attainable uplink data rate R_{ul} by assuming mobiles to be uniformly distributed over a single RSU cell area. We then scale the results by considering the totality of system RSU cells. For analytical simplicity in deriving an approximate expression for the system's throughput efficiency when using adaptive uplink rate, the Tp_{ul} and Tp_{dl} durations are apportioned by using the average values obtained for the system's R_{ul} and k_{ul} parameters. Note that due to the multicast nature of the message dissemination process, the downlink transmission rate R_{dl} values are configured to provide for reception by all vehicles residing within the RSU cell, regardless of their position. Hence, the values determined for the R_{dl} and k_{dl} parameters are selected in a fashion that is similar to that used when considering the non-adaptive rate scheme, as explained above in Sect. 4.1.

We note that such an adaptive uplink rate scheme can improve data throughput performance. An optimal selection of distance thresholds that are used to adapt the uplink transmission rate is yet to be investigated.

5   Performance Behavior

5.1   DA/TDMA Uplink - TDMA Downlink (MAC 1)

In Fig. 3, we show the variation of the aggregate system throughput rate, $TH_{ul,dl}$, as a function of the configured d_{RSU}. The $TH_{ul,dl}$ performance is obtained by using the mathematical model presented in Sect. 4.1. We specify the targeted overall PDR to be higher than 0.90 and the packet delay to be lower than 50 ms. For a given uplink rate R_{ul}, the downlink rate and coloring factor parameters, R_{dl} and k_{dl}, are selected such that the aggregate data throughput rate is maximized. For example, when vehicles and RSUs use a transmit power level $P_{tx,v} = P_{tx,r} = 23$ dBm at $d_{RSU} = 100$ m, and vehicles use $R_{ul} = 6$ Mbps, these values are paired with $R_{dl} = 24$ Mbps and $k_{dl} = 3$.

The highest throughput rate is noted to be attained when setting $R_{ul} = 24$ Mbps, $k_{ul} = 3$, $R_{dl} = 24$ Mbps, $k_{dl} = 3$ at $d_{RSU} = 100$ m. As d_{RSU} increases from 100 m to 800 m, n_{span} is reduced, which on its own serves to enhance throughput performance. However, we find that the global throughput $TH_{ul,dl}$ rate is actually reduced. This is attributed to the lower number of RSUs that are installed leading to an overall reduction in the aggregate spectral resources available and thus to the reduced level of available uplink and downlink transmission rate capacity. It is also noted that by increasing the uplink rate from 6 Mbps to 24 Mbps, the resulting throughput rate increases in a lower than proportional manner due to the ensuing increase in k_{ul}. In order to evaluate the performance of different coloring factors k_{ul} and k_{dl} for the uplink and downlink operation, we have carried out a Matlab-based Monte Carlo simulation. We note that as the d_{RSU} is further increased, say from 100 m to 400 m, R_{ul} and R_{dl} must be decreased to 12 Mbps. This is caused by the decrease in the received power level, which

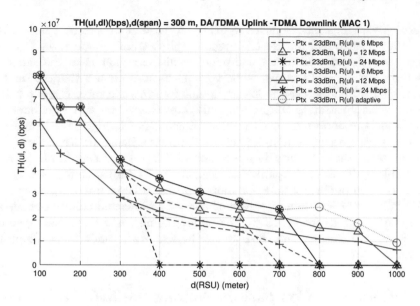

Fig. 3. $TH_{ul,dl}$ (bps) vs d_{RSU} (m) using MAC 1

may then be reduced below the required receiver sensitivity value, especially for vehicles located at the edge of the RSU cell. To resolve such an issue, we have studied a system that uses an increased transmission power level. We have set $P_{tx,r} = P_{tx,v} = 33$ dBm. We observed the increased transmit power to not lead to increased throughput rate performance when setting shorter inter-RSU range distances, such as $d_{RSU} = 100$ m to 300 m. However, an increase in the transmit power level yields a significant throughput upgrade under longer inter-RSU distances such as $d_{RSU} \geq 400$ m, as it enabled the system to meet the required minimum receiver sensitivity level of using high transmit rate $R_{ul} = 24$ Mbps and $R_{dl} = 24$ Mbps.

5.2 DA/TDMA Uplink - TDMA Downlink (MAC 1) Under Adaptive Uplink Rate Selection Scheme

As discussed in Sect. 4.2, an adaptive scheme can be used to command a vehicle to adjust its uplink transmission data rate and associated parameters in accordance with is location within a cell and its distance from the associated RSU.

Observing the results presented in Fig. 3, we conclude that when $P_{tx,r} = P_{tx,v} = 33$ dBm and vehicles adopt non-adaptive uplink rate selection, for $d_{RSU} < 800$ m, we should set $R_{ul} = 24$ Mbps. As the d_{RSU} range increases to $800 \leq d_{RSU} < 1000$ m, we set $R_{ul} = 12$ Mbps, and under $d_{RSU} \geq 1000$, we configure $R_{ul} = 6$ Mbps. Using these results, we have implemented a corresponding adaptive uplink rate setting scheme. A corresponding higher data rate is set by a vehicle when it is located around the center of a cell, while a lower uplink data rate is used by a vehicle that is located closer to the cell boundary.

Accordingly, to illustrate, vehicles which are located in a distance from the RSU that is lower than 365 m, use the highest uplink transmission rate, $R_{ul} = 24$ Mbps; those are located at a corresponding distance that is in the range 365 m–450 m, use $R_{ul} = 12$ Mbps, while vehicles located farther than 450 m away from their RSU nodes, use the lowest uplink transmission rate, $R_{ul} = 6$ Mbps.

Using this illustrative adaptive uplink rate scheme, we aim to enhance the data throughput performance of the scheduling scheme. In Fig. 3, we demonstrate that when using adaptive uplink transmission rate scheme, under the setting of $d_{RSU} = 800$ m, the aggregate data throughput $TH_{ul,dl}$ increases from 15.7 Mbps (attainable under a fixed uplink rate scheme) to 24 Mbps, thus resulting in a 52% increase in performance. For a more dense backbone, under $d_{RSU} = 100$ m to 700 m, all vehicles use uplink rate of $R_{ul} = 24$ Mbps and coloring of $k_{ul} = 3$, under the adaptive scheme, as well as under the non-adaptive scheme, so that there is no improvement attained under the use of the underlying adaptation mechanism.

6   Performance Comparison Among the MAC Schemes

In Fig. 4, we present the variation of the aggregate data throughput $TH_{ul,dl}$ attained by using the three MAC schemes, under delay and PDR constraints, as a function of the underlying inter-RSU distance d_{RSU}. Under each prescribed d_{RSU} value, for each scheme, we have set optimal values for the attained data rates and the respective coloring value to yield the best aggregate throughput performance for each scheme.

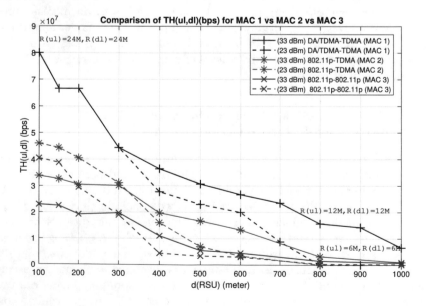

Fig. 4. Comparison between MAC 1, MAC 2, MAC 3: $TH_{ul,dl}$ (bps) vs d_{RSU} (m)

Under specified transmit power values, the MAC 1 scheme is noted to yield the highest throughput rate, under all d_{RSU} distance levels. It is followed by the MAC 2 scheme. The lowest throughput rates are attained when using the MAC 3 scheme. The highest throughput rate is attained by MAC 1 when configured at $R_{ul} = 24$ Mbps, $k_{ul} = 3$, $R_{dl} = 24$ Mbps, $k_{dl} = 3$ at $d_{RSU} = 100$ m, which is equal to about 80 Mbps). While generally MAC 2 achieves higher data throughput than MAC 3, the throughput differential is lower than that attained when comparing the throughput advantage realized by using MAC 1. For all the MAC schemes, we note the corresponding increase in system throughput that is realized as the number of installed RSU nodes is increased. Yet, the throughput variation depends in a non linear manner on the prescribed d_{RSU} level. As an example, we note that MAC 1's exhibited $TH_{ul,dl}$ values experience rapid degradation at shorter d_{RSU} ranges (i.e., shorter than 300 m) when compared with those incurred when longer d_{RSU} ranges are used (i.e., longer than 300 m). Especially, the use of a MAC 1 scheme is noted to be highly effective when implementing a high density backbone network whose density is incrementally increased (e.g., when the d_{RSU} is reduced from 300 m to 100 m).

7 Performance Comparison of Infrastructure-Aided V2I and V2V Systems Using TDMA

Two different communication networking architectural modes are considered in this section. On one hand, we consider the system studied above, which employs a backbone of RSUs that provide full coverage of vehicles traveling along the highway segment. The system uses V2I and I2V wireless transmissions as well as the underlying fiber-optic backbone network to disseminate packet flows. It is labeled here as the "V2I" system. In contrast, we consider a system that employs no infrastructure, using vehicle-to-vehicle communications for disseminating packet flows, which is labeled here as the "V2V" system. We assume that for the "V2V" system vehicles are admitted into each lane to form platoons. Platoons move along the autonomous highway at prescribed speeds and inter-vehicular distances. Each platoon is managed by a Platoon Leader (PL) that serves to control and synchronize its platoon members. Under the architecture developed in [2], an algorithm is used to elect certain PLs to act as Backbone Nodes (BNs).

The BNs form a vehicular backbone network (Bnet), whereby the inter-BN distance is denoted as d_{BN}. Vehicles associate with the nearest BN, or the one from which they receive the best signal, and are then identified as the clients of the latter BN. The client vehicles of a BN form the access network (Anet) of this BN. Packets produced by each vehicle are disseminated across this "V2V" network by first being transmitted across the corresponding Anet to its associated BN. Subsequently, packets are transmitted across the wireless Bnet to other BNs that reside in their dissemination span, and are simultaneously broadcasted for reception also by vehicles located in the corresponding Anets. The system forms a two layer hierarchical network that consists of a Bnet and several Anets.

However, the architectural topology is not fixed; as vehicles travel along, enter or exit the highway, Anet and Bnet formations are dynamically re-synthesized.

We compare the performance behavior of the two architectures as a function of the following parameters. For the "V2I" network, we vary the infrastructure's d_{RSU} distance levels as well as the uplink and downlink transmission rates, R_{ul} and R_{dl}, and the corresponding coloring values, k_{ul} and k_{dl}. For the "V2V" system, we vary the inter-BN distance d_{BN}, and the transmission rates across the Anet and Bnet sub-networks, R_{Anet} and R_{Bnet}, respectively.

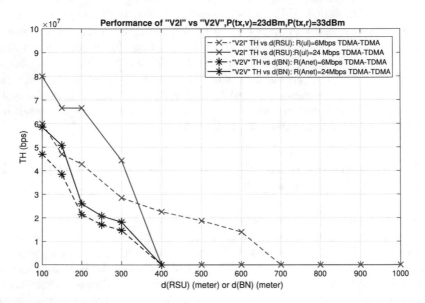

Fig. 5. Data Throughput TH (bps) vs d_{RSU} for "V2I" (m) or d_{BN} for "V2V", using TDMA MAC schemes

In Fig. 5, we present the aggregate data throughput performance attained when using TDMA-based MAC schemes under "V2I" (DA/TDMA Uplink-TDMA Downlink) and "V2V" (DA/TDMA Anet-TDMA Bnet) modes, as d_{RSU} and d_{BN} are varied respectively. We set a higher transmit power level at the RSU nodes, $P_{tx,r} = 33$ dBm, while the transmit power by a vehicular node is lower, $P_{tx,v} = 23$ dBm.

We observe that generally the "V2I" system exhibits uniformly better performance behavior than that attained by the "V2V" system, under various d_{RSU} and d_{BN} distance ranges. The "V2I" system is noted to achieve aggregate data throughput values that are about 30%–40% higher than those achieved by the "V2V" system, when setting for $d_{RSU} = d_{BN}$, assuming the inter-RSU distances to be configured to values in the 100 m–300 m range. For longer inter-RSU ranges, it is much more advantageous to employ a "V2I" system.

In aiming for a medium level data throughput of, say, 30 Mbps, when d_{RSU} is low, such as $d_{RSU} < 400$ m, such a data throughput level can be achieved by using the "V2I" system, by setting $R_{ul} = R_{dl} = 24$ Mbps. In turn, under a sparse RSU infrastructure with $d_{RSU} \geq 400$ m, the designer can institute a "V2V" operation by grouping vehicles into vehicular formations (or platoons) such that $d_{BN} < 200$ m. This can be readily realized when the highway segment is quite loaded by vehicular traffic. Yet, under such conditions, if the backbone network is expanded so that one can place RSUs at relatively short inter-RSU distances, a much enhanced operation is realized, as we capitalize on the high data rate capacity of the backbone network and on the availability of RSU nodes that are placed at fixed known locations.

A higher data throughput level, no lower than 75 Mbps, assuming the underlying scenario and parameter levels, can be achieved only by using the "V2I" system, with RSU nodes set to fully cover the highway segment at $d_{RSU} < 150$ m. Under the considered "V2V" system, when the d_{BN} range assumes widely different values in a range of 100 m to 1000 m, the system is noted to achieve a data throughput rate that is no higher than 60 Mbps.

8   Conclusions

In this paper, we develop and study data networking schemes for vehicular communications, including an autonomous transportation highway system, when aided by a roadside infrastructure which consists of road side units (RSUs) that are interconnected by a fiberoptic backbone. The study assumes that there is a sufficient number of RSU nodes to provide complete coverage of the highway segment under consideration. Induced by a multitude of applications, vehicles produce at random times packet flows which they aim to disseminate to all vehicles over a specified span. For this purpose, an active source vehicle transmits its packets uplink to its associated RSU. The RSU then forwards the received packets across the fiberoptic backbone network to other RSU nodes. The latter then transmit these packets downlink for reception by targeted vehicles.

We study the data throughput performance behavior of the infrastructure aided network system as a function of inter-RSU distance levels. We determine the optimal setting of the network system parameters, including the values set for the uplink and downlink data rates, configuration of the MAC scheduling schemes for the uplink and downlink wireless channels, spatial-reuse factor values and transmit power levels. In evaluating the ensuing data throughput performance behavior, we require the dissemination and scheduling scheme to meet prescribed limits for packet delay and packet success reception rates. We find that the system yields higher aggregate data throughput rates when shorter inter-RSU distance levels are employed, within limits. Yet, we demonstrate the corresponding dependence to exhibit a nonlinear functional behavior.

We also compare the performance of the underlying infrastructure-aided V2I system with that of a V2V-based vehicular backbone network. The enhanced performance extent achieved by the V2I system is exhibited and characterized. Such an advantage is realized when a sufficient number of RSU nodes are placed along the highway segment. Otherwise, V2I communications can be integrated with V2V operation, as the latter is employed over highway portions that do not provide for communications with RSU nodes. For highway segments that are not fully covered by RSU nodes, a hybrid system that employs both V2V and V2I networking components can be used. The cross layer design of such a system, as well as of a system that jointly supports autonomous and non-autonomous vehicular nodes, are topics of ongoing studies.

Acknowledgment. This work was supported in part by a research grant provided by the UCLA Institute of Transportation Studies under the support of California SB1.

References

1. Lin, Y., Rubin, I.: Throughput maximization under guaranteed dissemination coverage for VANET systems. In: 2015 Information Theory and Applications Workshop (ITA), pp. 313–318 (2015). https://doi.org/10.1109/ITA.2015.7309006
2. Rubin, I., Baiocchi, A., Sunyoto, Y., Turcanu, I.: Traffic management and networking for autonomous vehicular highway systems. Ad Hoc Netw. **83**, 125–148 (2019). https://doi.org/10.1016/j.adhoc.2018.08.018. http://www.sciencedirect.com/science/article/pii/S1570870518306164
3. Ni, Y., He, J., Cai, L., Bo, Y.: Data uploading in hybrid V2V/V2I vehicular networks: modeling and cooperative strategy. IEEE Trans. Veh. Technol. **67**(5), 4602–4614 (2018). https://doi.org/10.1109/TVT.2018.2796563
4. Chen, J., Mao, G., Li, C., Liang, W., Zhang, D.: Capacity of cooperative vehicular networks with infrastructure support: multiuser case. IEEE Trans. Veh. Technol. **67**(2), 1546–1560 (2018). https://doi.org/10.1109/TVT.2017.2753772
5. Ota, K., Dong, M., Chang, S., Zhu, H.: MMCD: cooperative downloading for highway VANETs. IEEE Trans. Emerg. Top. Comput. **3**(1), 34–43 (2015). https://doi.org/10.1109/TETC.2014.2371245
6. Wu, T., Liao, W., Chang, C.: A cost-effective strategy for road-side unit placement in vehicular networks. IEEE Trans. Commun. **60**(8), 2295–2303 (2012). https://doi.org/10.1109/TCOMM.2012.062512.100550
7. Giordani, M., Rebato, M., Zanella, A., Zorzi, M.: Coverage and connectivity analysis of millimeter wave vehicular networks. Ad Hoc Netw. **80**, 158–171 (2018). https://doi.org/10.1016/j.adhoc.2018.08.007. http://www.sciencedirect.com/science/article/pii/S1570870518305687
8. Chang, H., Rubin, I.: Optimal downlink and uplink fractional frequency reuse in cellular wireless networks. IEEE Trans. Veh. Technol. **65**(4), 2295–2308 (2016). https://doi.org/10.1109/TVT.2015.2425356
9. Zhang, R., Cheng, X., Yang, L., Shen, X., Jiao, B.: A novel centralized TDMA-based scheduling protocol for vehicular networks. IEEE Trans. Intell. Transp. Syst. **16**(1), 411–416 (2015). https://doi.org/10.1109/TITS.2014.2335746
10. Boulila, N., Hadded, M., Laouiti, A., Saidane, L.A.: QCH-MAC: a Qos-aware centralized hybrid MAC protocol for vehicular ad hoc networks. In: 2018 IEEE 32nd International Conference on Advanced Information Networking and Applications (AINA), pp. 55–62 (2018). https://doi.org/10.1109/AINA.2018.00021
11. Guo, W., Huang, L., Chen, L., Xu, H., Xie, J.: An adaptive collision-free MAC protocol based on TDMA for inter-vehicular communication. In: 2012 International Conference on Wireless Communications and Signal Processing (WCSP), pp. 1–6 (2012). https://doi.org/10.1109/WCSP.2012.6542833

12. Kim, H., Emmelmann, M., Rathke, B., Wolisz, A.: A radio over fiber network architecture for road vehicle communication systems. In: 2005 IEEE 61st Vehicular Technology Conference, vol. 5, pp. 2920–2924 (2005). https://doi.org/10.1109/VETECS.2005.1543881

13. Abdalla, G.M., Abu-Rgheff, M.A., Senouci, S.: Space-orthogonal frequency-time medium access control (SOFT MAC) for VANET. In: 2009 Global Information Infrastructure Symposium, pp. 1–8 (2009). https://doi.org/10.1109/GIIS.2009.5307071

14. Hadded, M., Muhlethaler, P., Laouiti, A., Zagrouba, R., Saidane, L.A.: TDMA-based MAC protocols for vehicular ad hoc networks: a survey, qualitative analysis, and open research issues. IEEE Commun. Surv. Tutorials 17(4), 2461–2492 (2015). https://doi.org/10.1109/COMST.2015.2440374

15. Shivaldova, V., Paier, A., Smely, D., Mecklenbräuker, C.F.: On roadside unit antenna measurements for vehicle-to-infrastructure communications. In: 2012 IEEE 23rd International Symposium on Personal, Indoor and Mobile Radio Communications - (PIMRC), pp. 1295–1299 (2012). https://doi.org/10.1109/PIMRC.2012.6362546

16. Shieh, W., Lee, W., Tung, S., Jeng, B., Liu, C.: Analysis of the optimum configuration of roadside units and onboard units in dedicated short-range communication systems. IEEE Trans. Intell. Transp. Syst. 7(4), 565–571 (2006). https://doi.org/10.1109/TITS.2006.884888

17. Kastell, K.: Analysis of planning constraints for wireless access in vehicular environments with respect to different mobility and propagation models. In: 2016 18th International Conference on Transparent Optical Networks (ICTON), pp. 1–4 (2016). https://doi.org/10.1109/ICTON.2016.7550366

18. Federal Communications Commission: Report and Order 03-324 (2004). https://docs.fcc.gov/public/attachments/FCC-03-324A1.pdf. Accessed 19 Feb 2019

19. 5G Automotive Association: Coexistence of C-V2X and 802.11p at 5.9 GHz. 5gaa.org/news/position-paper-coexistence-of-c-v2x-and-802-11p-at-5-9-ghz

20. IEEE Standard for Wireless Access in Vehicular Environments (WAVE) – Multi-Channel Operation. IEEE Std 1609.4-2016 (Revision of IEEE Std 1609.4-2010), pp. 1–94 (2016). https://doi.org/10.1109/IEEESTD.2016.7435228

21. van Eenennaam, M., Remke, A., Heijenk, G.: An analytical model for beaconing in VANETs. In: 2012 IEEE Vehicular Networking Conference (VNC), pp. 9–16 (2012). https://doi.org/10.1109/VNC.2012.6407451

22. Omar, H.A., Zhuang, W., Li, L.: VeMAC: a TDMA-based MAC protocol for reliable broadcast in VANETs. IEEE Trans. Mob. Comput. 12(9), 1724–1736 (2013). https://doi.org/10.1109/TMC.2012.142

23. IEEE Standard for Information technology–Telecommunications and information exchange between systems Local and metropolitan area networks–Specific requirements - Part 11: Wireless LAN Medium Access Control (MAC) and Physical Layer (PHY) Specifications. IEEE Std 802.11-2016 (Revision of IEEE Std 802.11-2012), pp. 1–3534 (2016). https://doi.org/10.1109/IEEESTD.2016.7786995

24. ETSI: Intelligent Transport Systems (ITS); STDMA recommended parameters and settings for cooperative ITS; Access Layer Part. Technical report 102 861 v1.1.1, ETSI (2012). https://www.etsi.org/deliver/etsi_tr/102800_102899/102861/01.01.01_60/tr_102861v010101p.pdf

25. Sepulcre, M., Gozalvez, J., Coll-Perales, B.: Why 6 Mbps is not (always) the optimum data rate for beaconing in vehicular networks. IEEE Trans. Mob. Comput. 16(12), 3568–3579 (2017). https://doi.org/10.1109/TMC.2017.2696533

26. Cassidy, W.G., Jaber, N., Ruppert, S.A., Toimoor, J., Tepe, K.E., Abdel-Raheem, E.: Interference modelling and SNR threshold study for use in vehicular safety messaging simulation. In: 26th Biennial Symposium on Communications (QBSC), pp. 52–55 (2012). https://doi.org/10.1109/QBSC.2012.6221350
27. Vehicle safety communications - applications (vsc-a) final report. U.S. Department of Transportation, pp. 1–102 (2011). https://www.nhtsa.gov/sites/nhtsa.dot.gov/files/811492a.pdf

3D Objects Indexing Using Chebyshev Polynomial

Y. Oulahrir[(⊠)], F. Elmounchid, S. Hellam, A. Sadiq, and S. Mbarki

Ibn Tofail University, Kénitra, Morocco
oulahrir@outlook.fr

Abstract. In this paper, we propose a new method for three-dimensional object indexing based on a new descriptor: Chebyshev Polynomial Descriptor (3D-CPD). For this end, we propose a numeric calculation of the coefficients of Chebyshev polynomial with maximum precision. The aim of this method is the search of similar 3D objects to a request object model and to minimize the processing time in the large database. Firstly we start by defining the new descriptor 3D-CPD using a new description of 3-D object. Then we define a new distance which will be tested and prove his efficiency in searching similar objects in the large databases in which we have objects with very various and important size.

Keywords: 3D object indexing · 3D shape descriptor ·
Chebyshev Polynomial · 3D object similarity

1 Introduction

The recognition of 3D objects is an active domain of research since decades. The search for similar forms in large databases has become a necessity in several areas: medicine (comparison of a sick organ to a healthy organ), paleontology (comparison of fossils), astronomy, security (crimes), material engineering and games.

Several basic mathematical concepts have been used and prove to be effective in image processing and 3D objects recognition of such as Markov models, neuron networks and Reeb graphs.

In this work, we propose a new descriptor 3D-CPD to describe 3D objects based on Calculation of the Chebyshev Polynomial Coefficients with numeric method using Simpson integration. For measurement of similarity between 3D objects, we define a new distance based on Chebyshev Polynomial coefficients. Finally, we have tested this distance on two family of 3D objects classes for a medium and important size.

This article is organized as follows: Sect. 2 beckons to the ancient researches that have proven their efficiency in the recognition of 3D objects. Then we present our new descriptor based on the Chebyshev polynomial coefficients and we explain the numerical method of calculation of these coefficients. The following section presents practical tests and results. Finally, the conclusion point out the limitations and some perspectives.

© Springer Nature Switzerland AG 2020
K. Arai et al. (Eds.): FTC 2019, AISC 1070, pp. 87–99, 2020.
https://doi.org/10.1007/978-3-030-32523-7_6

2 State of the Art

To evaluate the efficiency of the descriptor, the properties of invariance are required to eliminate differences due to translation, rotation and variation of scale.

In the literature of the 3D indexation, several types of shape descriptors were created and used: Local descriptors; Global shape descriptors; Graph based methods; Geometric methods based on 2D views of 3D models.

This approach [1] is based on creating a feature vector of the form of a 3D object by using the direction of its faces. The database of 3D object models was filled using a large number of scanned objects with 3D laser scanners. With this type of scanner, authors can simultaneously acquire the shape and color of a 3D object.

The main idea of [2] is to represent the signature of an object as a probability distribution of a form of global function measuring geometric properties of 3D objects.

Ding et al. [3] have recently developed a new method of recognizing 3D objects based on the skeleton. In this approach, the position of the human body is described by screw motions between 3D rigid bodies, this relation can be done between matrix of 3D rigid bodies.

Zaharia et al. in [4] have proposed 3D shape descriptor spectrum (SF3D). The 3D shape of the spectrum is a distribution of an aspect ratio which provides a representation of local geometric characteristics of the surface of a 3D object.

Osada et al. [5] used the Euclidian distances distribution between very large numbers of couples of points of the meshing. The Distribution of the elementary volumes bound to every facet of the meshing are also used.

Vranic and Saupe [6] propose to use the coefficients of the Fourier transform as 3D shape descriptor. The Fourier transform of 3D, allowed characterizing the information by frequency representations.

To benefit effectively from information on form, Turan et al. [7] studied the use of an advanced form descriptor that serves to classify emotional expressions. This descriptor is called the Angular Radial Transform (ART).

On the other hand, methods using 3D model views are motivated by psychophysical results that show that in the human visual system, a 3D object is represented by a set of 2D views rather than a 3D model [8].

In 3D recognition, Bicego et al. [9] have used the Hidden Markov Model (HMM) to deduct an element by having a part of that.

In this area of 3D indexing, Daoudi and his team have long worked on different approaches for indexing, analyzing and classifying 3D models: 2D/3D methods in [11], structural approach [10] and probabilistic approach in [12].

Each team has its own way of dealing with the problem of rotational invariance, for example, the Princeton team proposed to apply the spherical harmonic decomposition of the spherical functions characterizing the 3D object [13]. Few searchers used spherical harmonic coefficients [14–17], but in an earlier work, my team proposed a direct method for calculating spherical harmonic coefficients [18] and achieved the best result.

3   Chebyshev Polynomial Method

3.1   Invariant 3D Shape Descriptors

In order to ensure unique representation, it is designed to normalize the different objects in a coordinate. It's required to verify the properties of invariance to avoid differences due to translation, rotation, and the variation of scale. To eliminate these dependencies, we shall normalize the models by using the center of mass for translation, the root of the average square radius for scale and principal axis for rotation [8, 14, 26].

3.2   Representation of Chebyshev Polynomial

In the case of non-periodic problems, mathematicians advise using more appropriate basic functions: orthogonal polynomials, such as Chebychev polynomiales. The base of these polynomes is an appropriate alternative to the Fourier base. The extension of the Chebychev series can be considered a Fourier cosinated series, so that it possesses the valuable properties of the latter concerning, in particular, the convergence and possible use of the FFT. On the other hand, the extension of the Chebyshev series is exempt from the Gibbs phenomenon at the borders.

The differential equation:

$$(1 - x^2)\ddot{y} - x\dot{y} + n^2 y = 0 \tag{1}$$

Where n = 0, 1, 2, 3…
The solution of the equation (E) is given by:

$$T_n(x) = \cos(n \arccos x)$$
$$= x^n - \binom{n}{2} x^{n-2}(1 - x^2) + \binom{n}{4} x^{n-4}(1 - x^2)^2 - \cdots \tag{2}$$

Where T_n are functions of Chebyshev defined on the interval [−1, 1], with:

$$T_0(x) = 1 \tag{3}$$

$$T_1(x) = x \tag{4}$$

$$T_2(x) = 2x^3 - 1 \tag{5}$$

$$T_3(x) = 4x^3 - 3x \tag{6}$$

$$T_4(x) = 8x^4 - 8x^2 + 1 \tag{7}$$

$$T_5(x) = 16x^5 - 20x^3 + 5x \tag{8}$$

$$T_6(x) = 32x^6 - 48x^4 + 18x^2 - 1 \tag{9}$$

$$T_7(x) = 64x^7 - 112x^5 + 56x^3 - 7x \tag{10}$$

$$T_8(x) = 128x^8 - 256x^6 + 160x^4 - 32x^2 + 1 \tag{11}$$

Special Cases:

$$T_n(-x) = (-1)^n T_n(x) \tag{12}$$

$$T_n(-1) = (-1)^n \tag{13}$$

$$T_n(1) = 1 \tag{14}$$

$$T_{2n}(0) = (-1)^n \tag{15}$$

$$T_{2n+1}(0) = 0 \tag{16}$$

Relations of recurrence:

$$T_{n+1}(x) = 2xT_n(x) - T_{n-1}(x) \tag{17}$$

$$T_m(x)T_n(x) = T_{n+m}(x) + T_{n-m}(x)/n \geq m \tag{18}$$

Other relations:

$$T_n(x) = \frac{(-1)^n \sqrt{1 - x^2}}{(2n - 1)!!} \frac{d^n}{dx^n} \left(1 - x^2\right)^{n - \frac{1}{2}} \tag{19}$$

$$\frac{1 - t^2}{1 - 2xt + t^2} = T_0(x) + 2 \sum_{k=0}^{\infty} T_k(x)x^k \tag{20}$$

Primitive
For $n > 2$

$$\int T_n(x)dx = \frac{1}{2} \left[\frac{T_{n+1}(x)}{n+1} - \frac{T_{n-1}(x)}{n-1} \right] + cte \tag{21}$$

Orthogonality

$$\int_{-1}^{1} \frac{T_n(x)T_m(x)}{\sqrt{1 - x^2}} dx = \begin{cases} 0 : m \neq n \\ \pi : n = m = 0 \\ \frac{\pi}{2} : n = m \neq 0 \end{cases} \tag{22}$$

Graphical representation (Fig. 1) of T_n curves. Where n = 0, 1, 2, 3...

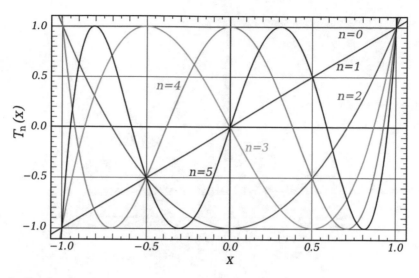

Fig. 1. The first few Chebyshev polynomials T_n of the first kind in the domain $-1 < x < 1$.

3.3   The Decomposition of a Function f in Chebyshev Polynomials (DTP)

In the appropriate Sobolev space, the set of Chebyshev polynomials form an orthonormal basis, so that a function in the same space can, on the interval $-1 \leq x \leq 1$ be expressed via the expansion [19–26].

$$f(x) = \frac{1}{2}A_0T_0(x) + A_1T_1(x) + A_2T_2(x) + \ldots \tag{23}$$

$$f(x) = \frac{1}{2}A_0T_0(x) + \sum_{k=1}^{\infty} A_kT_k(x) \tag{24}$$

Where

$$A_k = \frac{2}{\pi}\int_{-1}^{1} \frac{f(x)T_k(x)}{\sqrt{1-x^2}}\,dx \tag{25}$$

And

$$A_k = \frac{2}{\pi}\int_{0}^{\pi} f(\cos\theta)\cos(k\theta)\,d\theta, \quad k \geq 0 \tag{26}$$

In our case, f(x) is a characteristic function
We have

$$f(r) = \begin{cases} 1 \, si\, M \in \xi \\ 0 \, si\, M \notin \xi \end{cases} \tag{27}$$

Radial representation is shown in Fig. 2.

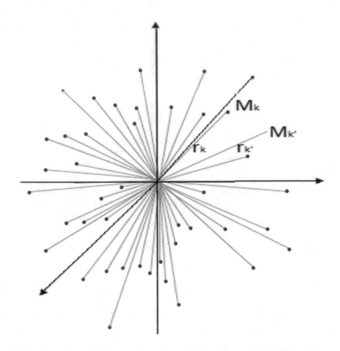

Fig. 2. The radial representation of the points of the 3D object.

$$A_k = \frac{2}{\pi} \int_0^{r_i} \frac{T_k(t)}{\sqrt{1 - t^2}} dt \tag{28}$$

4 Tests and Experimentation

We calculate the coefficients A_k with numeric method using Simpson integral. Thus, we define the matrix Mj describing the object Oj using A_k^j, the Chebyshev Polynomial Coefficients as:

$$M_j = \begin{bmatrix} A_1^{1,j} & \cdots & A_n^{1,j} \\ \vdots & \ddots & \vdots \\ A_1^{N,j} & \cdots & A_n^{N,j} \end{bmatrix} \tag{29}$$

For measurement of similarity between two objects Obj1 and Obj2, we define a new distance:

$$d = |T_1 - T_2| \tag{30}$$

Where:

$$T^j = \underset{1 \leq i \leq N}{Max} \frac{\sum_{k=1}^{n} \sqrt{|A_k^{i,j}|}}{(n+1)N} \tag{31}$$

4.1 Experimentation and Results

For testing our distance and new method we choice two objects families: Animals and Cars. In the following Table 1, we present objects of the first class, Animals.

Table 1. Class of Animals

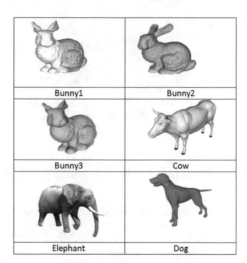

The following Table 2 shows the number of points for objects in class Animal.

Table 2. The number of points of the objects of the class Animal

The object	Number of points
Bunny 1	30516
Bunny 2	26002
Bunny 3	34834
Dog	4858
Cow	58006
Elephant	3667

In the following Table 3, we present objects of the second class, Cars.

Table 3. Class of Cars

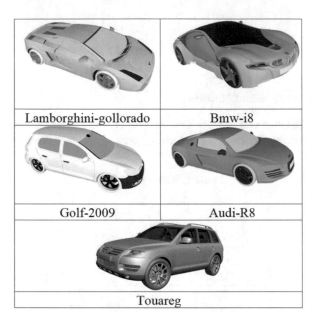

Lamborghini-gollorado	Bmw-i8
Golf-2009	Audi-R8
Touareg	

The following Table 4 presents the number of points per object in the class of cars.

Table 4. The number of points of the objects of the class Cars

Car	Name	Number of points
Car1	Lamborghini-gollorado	519036
Car2	Audi-R8	488189
Car3	Maclaren	563136
Car4	Golf-2009	327836
Car5	Touareg	279583

In the first test, we calculate the distance d between bunny1 and different animals of his class (Fig. 3). The following Table 5 shows the results.

Table 5. Distance between bunny1 and different objects of his class

K	bun1-bun3	bun1-bun2	bun1-eleph	bun1-dog	bun1-cow
1	2,29E−06	9,79E−06	2,54E−04	2,26E−04	7,98E−05
2	4,71E−06	1,37E−05	2,29E−04	1,67E−04	1,37E−04
3	3,69E−06	1,05E−05	1,73E−04	1,27E−04	1,03E−04
4	4,51E−06	1,05E−05	1,53E−04	1,13E−04	9,21E−05
5	5,66E−06	1,13E−05	1,46E−04	1,07E−04	8,81E−05
6	7,97E−06	1,38E−05	1,55E−04	1,14E−04	9,43E−05
7	9,87E−06	1,60E−05	1,63E−04	1,21E−04	9,99E−05
8	1,23E−05	1,90E−05	1,79E−04	1,33E−04	1,10E−04
9	1,59E−05	2,35E−05	2,06E−04	1,54E−04	1,28E−04
10	1,93E−05	2,79E−05	2,34E−04	1,75E−04	1,45E−04
11	2,47E−05	3,49E−05	2,82E−04	2,11E−04	1,76E−04
12	3,19E−05	4,45E−05	3,47E−04	2,60E−04	2,17E−04
13	3,70E−05	5,12E−05	3,92E−04	2,95E−04	2,46E−04
14	4,54E−05	6,23E−05	4,69E−04	3,53E−04	2,95E−04
15	6,15E−05	8,39E−05	6,20E−04	4,67E−04	3,90E−04
16	7,03E−05	9,55E−05	7,01E−04	5,28E−04	4,41E−04
17	8,95E−05	1,21E−04	8,82E−04	6,65E−04	5,56E−04
18	1,20E−04	1,61E−04	1,17E−03	8,79E−04	7,35E−04
19	1,39E−04	1,88E−04	1,35E−03	1,02E−03	8,53E−04
20	1,73E−04	2,33E−04	1,67E−03	1,26E−03	0,00106
21	2,24E−04	3,01E−04	2,15E−03	1,62E−03	0,00136
22	3,01E−04	4,04E−04	2,88E−03	2,17E−03	0,00182
23	3,66E−04	4,92E−04	3,50E−03	2,64E−03	0,00221
24	4,52E−04	6,07E−04	0,00432	0,00326	0,00273
25	5,73E−04	7,68E−04	0,00546	0,00412	0,00345
26	7,29E−04	9,77E−04	0,00694	0,00524	0,00438
27	9,56E−04	1,28E−03	0,0091	0,00687	0,00575
28	1,28E−03	1,72E−03	0,01222	0,00922	0,00772
29	0,00154	2,07E−03	0,01469	0,01109	0,00928
30	0,00199	0,00266	0,0189	0,01426	0,01193

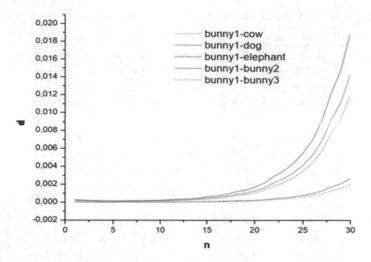

Fig. 3. Distance between bunny1 and the other objects in the same class

In the second test, we calculate the distance d between car1 and different cars of his class. The following Table 6 shows the results.

Table 6. Distance between car1 and different objects of his class

n	car1-car2	car1-car3	car1-car4	car1-car5
1	3,49E−07	1,96E−07	5,37E−07	9,06E−07
2	3,72E−06	3,19E−06	5,68E−06	1,41E−05
3	2,85E−06	2,45E−06	4,35E−06	1,08E−05
4	2,68E−06	2,31E−06	4,08E−06	1,02E−05
5	3,06E−06	2,64E−06	4,63E−06	1,15E−05
6	3,75E−06	3,24E−06	5,64E−06	1,41E−05
7	4,89E−06	4,22E−06	7,33E−06	1,83E−05
8	7,15E−06	6,19E−06	1,07E−05	2,67E−05
9	9,70E−06	8,40E−06	1,45E−05	3,63E−05
10	1,41E−05	1,22E−05	2,10E−05	5,25E−05
11	2,09E−05	1,81E−05	3,11E−05	7,80E−05
12	3,16E−05	2,74E−05	4,70E−05	1,18E−04
13	4,84E−05	4,20E−05	7,21E−05	1,81E−04
14	7,01E−05	6,08E−05	1,04E−04	2,62E−04
15	1,10E−04	9,56E−05	1,64E−04	4,12E−04
16	1,71E−04	1,49E−04	2,55E−04	6,39E−04
17	2,78E−04	2,41E−04	4,13E−04	0,00104
18	4,77E−04	4,13E−04	7,10E−04	0,00178
19	7,79E−04	6,75E−04	0,00116	0,00291

(*continued*)

Table 6. (*continued*)

n	car1-car2	car1-car3	car1-car4	car1-car5
20	0,00125	0,00109	0,00187	0,00467
21	0,0021	0,00182	0,00312	0,00782
22	0,00347	0,00301	0,00517	0,01295
23	0,00572	0,00496	0,00852	0,02136
24	0,00998	0,00865	0,01487	0,03724
25	0,01672	0,01449	0,02489	0,06236
26	0,02826	0,0245	0,04208	0,10542
27	0,04716	0,04088	0,07022	0,17591
28	0,08018	0,06951	0,1194	0,29912
29	0,1352	0,11721	0,20133	0,50436
30	0,23144	0,20063	0,34464	0,86334

The following figure (Fig. 4) shows the variation of distance d between car1 and other objects of his family.

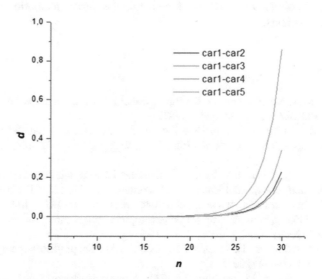

Fig. 4. Distance d between car1 and the others objects in class Cars.

This figure shows that car1, car2 and car3 are most similar to the other cars.

4.2 Discussion of Results

Both tests show clearly that the new method is very effective, whatever their density, it groups the objects which are alike. As in the other domains: the signal processing, the medicine, the geology, the security and others, the Chebyshev Polynomials show their efficiency and reliability in the indexation of the 3D objects.

In Figs. 3 and 4, we note that if n is less than 20, no object indexing effect but for n greater than 20, the curves of the objects are separated, similar objects gets down other curves and that indexing objects starts from n greater than 20. Similar objects are optimized by the distance chosen. For $n \geq 20$, if n increases, similar objects are slightly different from the other objects, it proves that two similar objects are not identical because their points distribution, points number and the position are not the same. It proves the efficiency of this method and that the objects are not identical hundred percent because of the same kind or group.

5   Conclusion and Future Work

In this work, we presented a new 3D shape descriptor. Numerical method for calculating Chebyshev Polynomial coefficients and the indexing process provide a 3D shape descriptor robust to translation, rotation and scale variations. The new descriptor which we had developed has allowed us to present the efficiency of indexing the 3D objects in large databases, this global efficiency acting even for objects presented by clouds at large number of points. Our future work concerns the direct calculation of Chebyshev Polynomial coefficients.

References

1. Cabral, B., Max, N., Springmeyer, R.: Bidirectional reflection functions from surface bump maps. In: SIGGRAPH, pp. 273–281 (1987)
2. D'Zmura, M.: Shading ambiguity: reflection and illumination. In: Landy, M.S., Movshon, J. A. (eds.) Computational Models of Visual Processing, pp. 187–207. MIT Press, Cambridge (1991)
3. Ding, W., Liu, K., Chen, H., Tang, F.: Human action recognition using similarity degree between postures and spectral learning. IET Comput. Vis. **12**, 110–117 (2018)
4. Zaharia, T., Preteux, F.: 3D shape-based retrieval within the MPEG- 7 framework. In: SPIE Conference 4304 on Nonlinear Image Processing and Pattern Analysis, San Jose, vol. XII, pp. 133–145 (2001)
5. Osada, R., Funkhouser, T., Chazelle, B., Dobkin, D.: Shape distributions. ACM Trans. Graph. **21**(4), 807–832 (2002)
6. Vranic, D.V., Saupe, D., Richter, J.: IEEE Fourth Workshop on Multimedia Signal Processing (Cat. No. 01TH8564), pp. 293–298 (2001)
7. Turan, M., Ekenel, H.K.: Shape-based facial expression classification using angular radial transform. In: 21st Signal Processing and Communications Applications Conference (SIU), pp. 1–4 (2013)
8. Ramamoorthi, R., Hanrahan, P.: An efficient representation for irradiance environment maps. In: SIG-GRAPH, pp. 497–500 (2001)
9. Bicego, M., Castellani, U., Murino, V.: A hidden Markov model approach for appearance-based 3D object recognition. Pattern Recogn. Lett. **26**(16), 2588–2599 (2005)

10. Daoudi, M., Ansary, T.F., Tierny, J., Vandeborre, J.P.: 3D-mesh models: view-based indexing and structural analysis. In: Thanos, C., Borri, F., Candela, L. (eds.) Digital Libraries: Research and Development, vol. 4877, pp. 298–307. Springer, Heidelberg (2007)
11. Bustos, B., et al.: Feature-based similarity search in 3D object databases. ACM Comput. Surv. **37**(4), 345–387 (2005)
12. Wang, M., Gao, Y., Lu, K., Rui, Y.: View-based discriminative probabilistic modeling for 3D object retrieval and recognition. IEEE Trans. Image Process. **22**(4), 1395–1407 (2013)
13. Vranic, D-V.: An improvement of rotation invariant 3D-shape based on functions on concentric spheres. In: 2003 International Conference on Image Processing, ICIP Proceedings, vol. 2, vol. 3, pp. III–757–60 (2003)
14. Mousa, M., Chaine, R., Akkouche, S.: Frequency-based representation of 3D models using spherical harmonics. In: Proceedings of the 14th International Conference in Central Europe on Computer Graphics, Visualization and Computer Vision, WSCG 2006, Plzen, Czech Republic, January 30–February 2 2006, vol. 14, pp. 193–200 (2006)
15. Liu, P., Wang, Y., Zhang, Z., Wang, Y.: Automatic and robust 3D face registration using multiresolution spherical depth map. In: Proceedings of 2010 IEEE 17th International Conference on Image Processing, Hong Kong, 26–29 September 2010
16. Dietmar, S., Dejan, V.-V.: 3D model retrieval with spherical harmonics and moments. In: Radig, B., Florczyk, S. (eds.) Pattern Recognition, DAGM, pp. 392–397. Springer, Heidelberg (2001)
17. Byerly, W.E.: Spherical Harmonics, Chap. 6, pp. 195–218. Dover, New York (1959)
18. Hellam, S., Oulahrir, Y., El Mounchid, F., Sadiq, A., Mbarki, S.: 3D objects indexing with a direct and analytical method for calculating the spherical harmonics coefficients. World Acad. Sci. Eng. Technol. Inter. J. Comput. Inf. Eng. **9**(2), 613–623 (2015)
19. Boyd, J.P.: Chebyshev and Fourier Spectral Methods, 2nd edn. Dover, Mineola (2001). ISBN 0-486-41183-4
20. Elliott, D.: The evaluation and estimation of the coefficients in the Chebyshev Series expansion of a function. Math. Comp **18**(86), 274–284 (1964). https://doi.org/10.1090/S0025-5718-1964-0166903-7. MR 0166903
21. Eremenko, A., Lempert, L.: An extremal problem for polynomials. Proc. Am. Math. Soc. **122**(1), 191–193 (1994)
22. Hernandez, M.A.: Chebyshev's approximation algorithms and applications. Comput. Math. Appl. **41**, 433–445 (2001)
23. Mason, J.C.: Some properties and applications of Chebyshev polynomial and rational approximation. Lecture Notes in Mathematics, vol. 1105, pp. 27–48 (1984)
24. Mason, J.C., Handscomb, D.C.: Chebyshev Polynomials. Taylor & Francis, Boca Raton (2002)
25. Mathar, R.J.: Chebyshev series expansion of inverse polynomials. J. Comput. Appl. Math. **196**, 596–607 (2006)
26. Koornwinder, T.H., Wong, R.S.C., Koekoek, R., Swarttouw, R.F.: Orthogonal Polynomials. In: Olver, F.W.J., Lozier, D.M., Boisvert, R.F., Clark, C.W. (eds.) NIST Handbook of Mathematical Functions. Cambridge University, New York (2010)

Machine Learning Classification and Segmentation of Forest Fires in Wide Area Motion Imagery

Melonie Richey$^{(\boxtimes)}$, David Blum, Meredith Gregory, Kendall Ocasio, Zachary Mostowsky, and Jordan Chandler

Next Tier Concepts, Inc., 8150 Leesburg Pike Suite 1400, Vienna, VA 22182, USA
melonie.richey@ntconcepts.com

Abstract. Numerous models and simulations exist for characterizing and predicting wildland fire behavior. The U.S. Forest Service (USFS) and other organizations have devoted decades of research to identifying the parameters that affect fire movement, rate of spread, and direction of spread across geographic terrain. While this research is invaluable to the firefighting community, due to computational constraints, these models do not run in real time or against imagery at time-of-collect, and therefore do little to assist the firefighter and first responders on the ground during a wildland fire event. We present the first part of a multi-step automated computational methodology to characterize fire behavior and rate of spread in real time across any geographic terrain. This first step is the classification and segmentation of the wildland fire in Wide Area Motion Imagery (WAMI) using Machine Learning (ML) methods. The continuation of this research, detailed herein, will involve training a more robust, purpose-built Recurrent Neural Network architecture incorporating many of the parameters the USFS has been studying for decades. The goal of this research is to deploy models using 'lite' ML frameworks on edge devices, mounted on collection platforms for real-time decision support for firefighting operations as imagery and other data are being collected during a wildland fire event.

Keywords: Machine learning · Forest fires · Wide Area Motion Imagery · Image classification · Image segmentation

1 Introduction

Over the past 20 years, wildland fires have become an increasingly destructive and costly problem in the United States. Nationwide data compiled by the National Interagency Fire Center (NIFC) indicates that in every year since 2000, an average of 73,200 wildfires burned an average of 6.9 million acres [1]. This figure is nearly double the average 3.3 million acres burned annually in the 1990s. Empirical models such as FARSITE [2] and Prometheus [3] exist for the quick estimation of fundamental wildland fire spread parameters. These include fire spread rate, flame length, and fireline intensity at a point of a specific fuel complex. Other three-dimensional physical models [4] join computational fluid dynamic models with the wildland fire component

© Springer Nature Switzerland AG 2020
K. Arai et al. (Eds.): FTC 2019, AISC 1070, pp. 100–111, 2020.
https://doi.org/10.1007/978-3-030-32523-7_7

and allow the fire to impact the microclimate in addition to the atmosphere existing as a parameter in the calculation of fire spread. Existing challenges in this field include the computational complexity of requisite models, the limited performance skill of weather-based and atmospheric models at a spatial resolution under 1 km, and the total absence of existing applications of deep learning techniques to the problem set.

To address these challenges, ML models create solutions that use multispectral, 0.5 m resolution, WAMI data in combination with meteorological, land cover, and topographical data sources. At this stage in our research, these solutions include fire segmentation and classification techniques that facilitate automated fire perimeter extraction and rate of spread calculations. The deeper purpose of the research is to work in concert with USFS-provisioned WAMI datasets to develop descriptive and predictive models of fire movement that run in constrained computational environments in order to support operational decision making. Ideally, edge devices such as Google's Edge TPU or Nvidia's AGX Xavier and RAPIDS package will facilitate these models running on the collection platform itself alongside the sensor, so as to deploy the models in real-time and communicate their results through a downlink to firefighters on the ground.

While in the process of developing a computationally-optimized ML model that characterizes and predicts full fire behavior in real time (Fig. 1, Output 1), the initial research to date includes classification of the existing burnt and burning area (Fig. 1, Outputs 4 & 3).

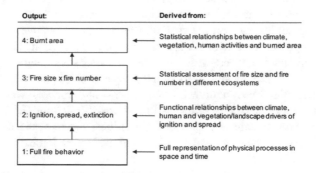

Fig. 1. Figure showing Outputs 1 to 4. (Source Hantson et al. 2016)

What follows is a brief overview of the remainder of this paper. It begins with a background on ML and how it is incorporated into prediction of natural weather events, followed by the approach that was taken to review the provided imagery, obtain supplemental data for prediction, create appropriate data preparation models, and, ultimately, perform segmentation, classification and prediction. The descriptive model will be implemented in order to convey information about the fire's location to the prediction model. Finally, the paper discusses the impact this work can have on current efforts to contain wildland fires particularly in the context of edge computing.

2   Background

Extant literature includes modeling and simulation approaches to complex environmental systems where spatio-temporal dynamics are highly relevant. These stochastic, empirical, and simulation models address the modeling of hurricanes [5, 6], typhoons [7], fires [2], earthquakes [8], and other natural disasters. For a comprehensive review of previously-developed fire models, see Hantson et al. 2016 [9].

Many existing wildland fire models, while scientifically valid, are constrained by the computational resources required to run them at scale. All are run retroactively and none, to our knowledge, are executed in real-time on a data collection platform. A gap exists between the scientific research community and the first-response operators such as firefighters and medics who deploy during an ongoing wildland fire event. While the models referenced above and in Hantson et al. yield greater insight into how and why fires behave the way they do, which enables effective policymaking, these models do little to assist the firefighter during an actual wildland fire emergency. For the requisite speed and computational scale, ML techniques and edge computing devices are leveraged. In support of this approach, our approach reviews (a) candidate descriptive and predictive ML model architectures and (b) geophysical and spatio-temporal factors affecting fire movement that the candidate model architecture must incorporate.

According to Hantson, "fire occurrence and spread depends not only on instantaneous climatic drivers and sources of ignition (such as humans, lightening, or both), but also on state variables, such as the state and amount of available fuel" [9]. Reichstein elaborates, "fire spread and thus the burned area depends not only on the local conditions of each pixel but also on the spatial arrangement and connectivity of fuel, its moisture, terrain properties, and of course wind speed and direction" [10]. To study the ground characteristics in concert with climactic characteristics, many types of models have been employed previously, most notably Dynamic Global Vegetation Models (DVGMs), Terrestrial Ecosystem Models (TEMs), and models with more global context such as those executed within Earth Systems Modelling (ESM) frameworks that account for atmospheric and climactic conditions. Agent-based models (ABMs) have also been used for the study and simulation of the interaction between burning fire and environment [11, 12].

The basic factors that influence fire movement have been widely researched over the decades by the USFS, including environmental, climatic, and physical factors [13]. Environmental factors include fuel and ignition sources (e.g. land use, land cover, fuel combustibility) in the target region, soil type, near-surface soil moisture, elevation, and slope [14–17]. Climatic variables are more difficult to calculate insofar as the fire, as it burns, affects the micro-climate just as the micro-climate affects fire spread through soot yield, smoke concentration, and sheer addition of heat to the local atmosphere. Physical factors of the burning fire include elements such as the Heat Release Rate (HRR), fire mass log (kg/second), oxygen consumption, and the Hot Gas Layer Temperature (HGL) [13].

3   Approach

3.1   Data

The data used in this research are 9 TB of WAMI collected by the USFS-owned, CRI-built LodeStar HDM sensor, which is housed in a 15" Gimbal. The spatial resolution of the data is 0.5 m and the spectral resolution is five bands (Blue, Green, and Red Electro-optical imagery (E/O), Mid-Wave Infrared (MWIR), and thermal) (see Fig. 2). The temporal resolution is that of traditional WAMI data, approximately 2–3 Hertz (Hz). In total, there are 3.1 flyover hours distributed across three U.S. wildland fires that burned in 2016 across Nevada, Arizona, and New Mexico. The number of flyover minutes per-fire is 53, 113, and 20, respectively.

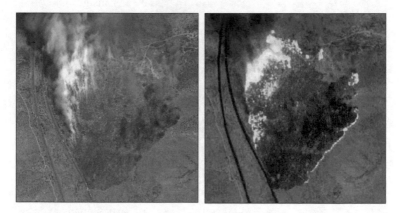

Fig. 2. Raw multi-spectral WAMI imagery chip (True Color Composite (RGB), left; normalized false color composite (Band 5, 4, & 3), right)

Prior to preprocessing, the WAMI is stored in GeoTiff format in Google Cloud Storage (GCS). The classification of the imagery takes place in Google Earth Engine (GEE), a cloud-based platform for performing large-scale geoprocessing and ML across multiple layers of both raster and vector data [18], requiring a pipeline be created between the two services.

3.2   Classification, Segmentation and Local Rate of Spread

In order to track the spread of wildfire using WAMI, it is necessary to extract unburnt, burnt, and burning areas from each individual orthorectified image comprising the WAMI. After the perimeters are extracted, frame over frame, they are used to define the local rate of spread. Fires burn very differently across different landscapes forcing the classification process to be manual. Manually extracting the perimeter for thousands of frames, however, is simply too time consuming. To circumvent this, DeepLab – a Deep Convolutional Neural Network (DCNN) – will be trained to segment the images [19].

A manually labeled subset of the WAMI will be used to train the model which will then segment the rest of the dataset.

Using Python's OpenCV module, the pixels of each raw image are clustered into 15 classes using a K-Means algorithm [20] on the middle infrared and thermal bands (bands 4 and 5). Each of the 15 classes are mapped to one of the three categories – unburnt, burnt, or burning – using a semi-automated rules-based approach (see Fig. 3). The rules themselves were constructed by comparing the clusters to the original optical image and determining that specific clusters were consistently associated with the burnt or burning regions. What remains is unburnt. The result of extracting unburnt, burnt, and burning regions from WAMI that allow for computing area-based and local rates of spread is shown in Fig. 4.

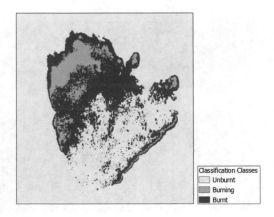

Fig. 3. Re-classified image (15 classes to 3)

While the K-Means classification method would be sufficient for a single fire, its categorical representations would not remain static across multiple fires and geographical regions. For this reason, it is imperative to train the DeepLab model to learn the features of various fire states equally and segment images into their unburnt, burnt, and burning regions. It is further described in Sect. 3.3 Automated Fire Perimeter Extraction.

The approach that was taken to compute the rate of wildfire spread is to use the area of burning regions (or the union of burnt and burning regions) and compute the rate of spread as the ratio of the area in one image to the corresponding area in the previous image. Depending on the particular decision to be supported, an area-based calculation may be relevant. For example, area-based calculations may adequately support decisions involving property loss or resource consumption which scale by area. Other decisions, however, are better supported by knowing the local rate of spread (e.g. the continual reallocation of firefighting resources). The local rate of spread is especially important because it, together with the location of a given perimeter segment, constitutes the "state" in any future State Space Model for predicting fire movement based on data like weather, temperature, topography, and fuel load.

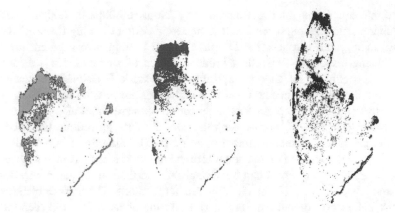

Fig. 4. Fire perimeter extraction frame over frame. Perimeters range from frame 1 to frame 1000.

Computing the local rate of spread requires that the perimeter be discretized into a set of compact rasters (hereafter perimeter segments). Next is to find the shortest distance between each perimeter segment and the perimeter segments from the previous image. The local rate of spread (v) measured at each perimeter segment in a given image (indexed by j:n) found at a given position (x) is the distance between it and the nearest perimeter segments from the previous image (indexed by j:n−1), multiplied by the WAMI sample rate (F). For a given perimeter segment we compute the local rate of spread along the longitudinal and latitudinal vectors (indexed by i=long, lat). See Eq. (1).

$$v_{i,j:n} = F * \min_{j:n-1}\left(x_{i,j:n} - x_{i,j:n-1}\right)$$ (1)

Note that with this approach, the direction of spread for a given perimeter segment is the opposite of its bearing to the nearest perimeter segment in the previous image.

3.3 Automated Fire Perimeter Extraction

If the only goal were simply to locate fire once, a reclassified unsupervised K-Means classification performs this task well. However, in order to automatically locate fire anywhere in real time across any geographic landscape, it is no longer viable as fires look different in different areas and at different times of the day. Instead, the use of a deep learning approach will overcome these challenges. To begin, training data needs to be created for the DeepLab model. A normalized false color composite showing the thermal, middle infrared, and red bands is generated from the WAMI image. The normalization process stretches the image, so the lowest pixel value becomes 0 and the highest pixel value becomes 255. An unsupervised K-Means classification produces 15 classes that a human then maps to three classes: burning, burnt and unburnt. This raster layer is now called the label. Next, each WAMI image and its corresponding label is chipped into equal sizes ranging from 300x by 300px to 600px by 600px randomly at 100px increments padding black pixels as necessary. After dividing the WAMI set into training, tuning, and test sets, the DeepLab model is trained to segment the WAMI images, using a Google Tensorflow implementation of DeepLab [21]. Additionally, testing has begun on an alternative Keras-based Mask R-CNN implementation to provide the flexibility to handle images at higher resolutions, up to 1024×1024, in addition to testing the DeepLab V3+ architecture for segmentation.

By training on WAMI images using the normalization procedure described above, the model learns to differentiate unburnt, burnt, and burning areas in a WAMI image consistently, regardless of fire location. Once the model has been trained and deployed, images captured by the sensor can be normalized and tiled in the same way described above and processed through the model in prediction mode. The reassembled result is a real time delineation of unburnt, burnt, and burning areas within the scene at 1 Hz. Additionally, this provides the input to the model answering the questions, how is the fire moving and where will it be going.

4   Future Work

4.1   Descriptive Model of Fire Perimeter Movement

The fire perimeter is also considered to be comprised of a set of segments that behave like objects, constrained kinematically in two spatial dimensions (without consideration of the vertical dimension as a kinematic constraint, though topography may still be a predictor of movement). For that reason, the local rate of movement (i.e., spread), in two spatial dimensions, together with the location of each perimeter segment, constitute the state that persists through time. For the movement of the fire perimeter, a discrete time State Space Model will be used with the following procedures iterated at each timestep:

1. For every fire perimeter segment, the rate of spread in both spatial dimensions from the previous timestep is updated based on exogenous factors.
2. The location of each fire segment is updated based on its rate of spread in the current timestep.
3. "White space" between fire perimeter segments that has been created by the outward movement of two fire perimeter segments is filled through the creation of new fire perimeter segments, whose location and rate of spread are inherited through linear interpolation from their two nearest neighbors.
4. Overlapping segments resulting from the inward movement of two fire perimeter segments are merged and their rate of spread averaged.

In consultation with experts at the USFS and through comparison of aggregated results to existing simulations like FARSITE and Prometheus, a determination must be made as to whether a model utilizing this algorithm adequately allows for realistic dynamics of fire. One open question is whether such a model will maintain a single contiguous fire rather than allow for isolated burning pockets to form. (The linear interpolation of location and rate of spread from the two nearest neighbors, as described in Step 3, may need to be replaced with an alternative method for inheriting state).

It is worth highlighting that Step 1, the updating step, allows for external inputs to the model. The changes in the local rate of spread at each perimeter segment are, in fact, the key parameter to be learned from data and eventually predicted. In other words, this updating step provides the linkage between descriptive and predictive models.

4.2   Prediction of Fire Perimeter Movement

In order to provide decision support for firefighting operations, a model capable of predicting total fire behavior is required. To achieve this purpose, a review and brief test of various candidate model architectures are being conducted alongside an in-depth study of fire movement parameters.

To predict the change in the local rate of spread in every timestep, the process began with gathering U.S. Geological Survey (USGS) topographic data, land cover data, weather data from the National Oceanic and Atmospheric Administration (NOAA), as well as temperature and weather data from mobile sensors fielded by the USFS, and the state and local firefighting organizations with whom it coordinates. Other data sources that have been compiled include USGS' 3DEP products for elevation, slope, and aspect; the National Land Cover Dataset (NCLD); the U.S. Department of Agriculture (USDA) Forest Service and Nature Conservancy fire fuel datasets; and the National Oceanic and Atmospheric Administration (NOAA) Palmer Drought Severity Index (PDSI), and other weather affects data. The ideal model for this prediction is one that can operate online, optimizes the predicted parameter values across time, and can be closely coupled with the descriptive model detailed in Sect. 4.1. This close coupling implies a model that is autoregressive in nature. It must also incorporate exogenous inputs in every timestep, which will vary both by the location of the perimeter segment whose rate of spread is being updated, and in some cases over time as well (as in the case of weather).

One leading candidate that appears to have these characteristics is the Nonlinear Auto-Regressive eXogenous recurrent neural network, or NARX. A generic NARX model is shown in Fig. 5 [22]. In addition to possessing the characteristics listed above, two other features of NARX make it attractive. First, a training procedure exists to decouple the feedback connections in the network [23]. The typical reason that researchers utilize this training procedure is to employ a classical static back propagation algorithm rather than the more computationally-intensive Back Propagation Through Time algorithm. More importantly, however, decoupling the feedback connections is attractive to us because the exogenous predictors vary by the location of each fire perimeter segment, which in turn depends on the true change in local rate of spread, and yet the algorithm allows for the creation and merging of segments, so reconstruction of an entire full "path" to train on would be complicated. Decoupling the feedback connection during training obviates the need for this. Second, as Mandic and Chambers showed, there exists a state space representation of the NARX model, providing us with the option to merge the descriptive and predictive models at prediction time [24]. This could be useful for online prediction on a constrained compute device such as an airborne platform.

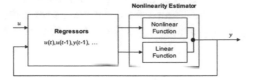

Fig. 5. Generic NARX model (Source: Alkhoshi & Belkasim)

Another highly promising approach is to run each input through a CNN to extract the spatial features, followed by a FC-LSTM (fully connected LSTM) to extract the temporal features. The hidden state of the LSTM could then be used to make pixel-wise predictions about the future state of the fire. The problem with this architecture is that any spatial information would likely be lost in the fully connected operations within the LSTM. This would make predictions at the pixel level nearly impossible. To solve this problem, we propose using an LSTM architecture where the hidden states are 3-D and the interactions between them are convolutional. This is effective because convolutions retain spatial information. For example, the input gate would look as follows:

$$i_t = \sigma(W_{xi} * X_t + W_{hi} * H_{t-1} + W_{ci} \circ C_{t-1} + b_i) \tag{2}$$

where i_t is the input gate at time t, W_{xi} is the convolution that acts on input X from time t, W_{hi} and W_{ci} are the convolutions that act on the memory state H and hidden state C from time t-1, respectively. b_i is the input bias, and σ is some non-linear activation function. The recurrent nature of the LSTM will allow the model to learn temporal patterns, and the convolutional nature of its gates will allow the network to retain, combine, and propagate the spatial features extracted from the input CNN. Similar network architecture has been used successfully for predicting the intensity of rainfall at a regional level [25].

Other candidate models under consideration include sparse RNNs which employ some combination of Genetic Algorithm (GA) to evolve network topology [6], RNNs deployed on top of grid architecture, [5] and RNNs with elements of Bayesian inference [26, 27]. Once a reliable rate is predicted from the local rate of spread, future work will include the construction of decision support analytics. These will include analytics to identify the riskiest parts of the perimeter, forecast time available to evacuate locations or structures, and optimize interventions such as retardant drops.

5　Discussion

The current technique for extracting fire perimeter treats each WAMI image independently. In fact, they are time sequenced, which may present an opportunity to leverage one or more previous frames to improve accuracy without sacrificing the ability to predict online. One method for doing this would be to modify the DeepLab architecture such that each channel of an image is concatenated with its predecessor. The concatenated channels could then be dimensionally reduced with a convolutional layer or else preserved and incorporated into the region proposal network. Alternatively, for a given image the predicted regions from the DeepLab model could be integrated raster by raster with the previous image's labeled classification polygons using a Hidden Markov Model or related model. The given image's classification polygons would be used to learn the transition and emission matrices. In prediction mode, the current image's predicted regions would serve as the signal, updating the previous image's predicted regions following a transition.

In regard to fitting the predictive model, the most significant challenge is that weather data are sparser than what is ideal, both spatially and temporally, to say nothing of mobile sensors which may not be at the same location in successive timesteps. The ideal output would be able to support predictions at every rasterized location and at every timestep with or without each data feed. Interpolation of weather data in both space and time is likely to miss key information, such as gusts of wind, that have the potential to cause big changes in the local rate of spread. With that in mind, our first round of testing will experiment with training the model with and without individual data columns in order to produce different sets of parameter values for use at prediction time consistent with data availability in every timestep. Later experiments will test the inclusion of multiple weather time series corresponding to different distances from a perimeter segment. Finally we will experiment with including additional feedback connections to incorporate past weather, and in so doing complicate a relatively simple NARX or CNN architecture and sacrifice the feedback decoupling method of training.

While the accuracy of the predictive fire model will be a highly relevant metric, the ultimate utility of this research lies not in the most accurate fire spread predictions but in the speed and computational optimization with which the model can be deployed. This applied research is only useful to firefighters and first responders if it becomes a tool they can use to combat and contain wildland fires across the United States in real time. To that end, we must computationally optimize the model such that inference can be run seamlessly on an edge compute device such as a Coral Dev Board with onboard Google Edge TPU. To achieve this, the model, initially trained in TensorFlow, will be converted to TensorFlow Lite format (.tflite).

6 Conclusion

When called upon to fight wildfires, the USFS and its state firefighting counterparts must make a long sequence of tough decisions that do not cease until the fire is contained. Decisions in this sequence include when and where to deploy firefighters as the fires move, to build firebreaks, to drop fire retardants, and to order evacuations. They have platforms and sensors capable of collecting terabytes of data. They ought to have ML models at their disposal capable of processing those data to generate predictions that enhance their decision-making processes. Producing real-time fire predictions will contribute to this goal. The public deserves no less.

References

1. Hoover, K.: Wildfire Statistics. Congressional Research Service. Crs.gov 7-7500 (2018)
2. Finney, M.A., Grenfell, I.C., McHugh, C.W., Seli, R.C., Trethewey, D., Stratton, R.D., Brittain, S.: A method for ensemble wildland fire simulation. Environ. Model. Assess. **16**(2), 153–167 (2011)
3. Tymstra, C., Bryce, R.W., Wotton, B.M., Taylor, S.W., Armitage, O.B.: Development and structure of Prometheus: the Canadian wildland fire growth simulation model. Natural Resources Canada, Canadian Forest Service, Northern Forestry Centre, Information Report NOR-X-417, Edmonton, AB (2010)
4. Hostikka, S.I.M.O., Mangs, J.O.H.A.N., Mikkola, E.S.K.O.: Comparison of two and three dimensional simulations of fires at wildland urban interface. Fire Saf. Sci. **9**, 1353–1364 (2008)
5. Alemany, S., Beltran, J., Perez, A., Ganzfried, S.: Predicting hurricane trajectories using a recurrent neural network. arXiv preprint arXiv:1802.02548 (2018)
6. Moradi Kordmahalleh, M., Gorji Sefidmazgi, M., Homaifar, A.: A sparse recurrent neural network for trajectory prediction of Atlantic hurricanes. In: Proceedings of the 2016 Genetic and Evolutionary Computation Conference, pp. 957–964. ACM, July 2016
7. Jiang, G.Q., Xu, J., Wei, J.: A deep learning algorithm of neural network for the parameterization of typhoon-ocean feedback in typhoon forecast models. Geophys. Res. Lett. **45**(8), 3706–3716 (2018)
8. Cui, Y., Olsen, K.B., Jordan, T.H., Lee, K., Zhou, J., Small, P., Levesque, J.: Scalable earthquake simulation on petascale supercomputers. In: SC 2010: Proceedings of the 2010 ACM/IEEE International Conference for High Performance Computing, Networking, Storage and Analysis, pp. 1–20. IEEE, November 2010
9. Hantson, S., Arneth, A., Harrison, S.P., Kelley, D.I., Prentice, I.C., Rabin, S.S., Bachelet, D.: The status and challenge of global fire modelling. Biogeosciences **13**(11), 3359–3375 (2016)
10. Reichstein, M., Camps-Valls, G., Stevens, B., Jung, M., Denzler, J., Carvalhais, N.: Deep learning and process understanding for data-driven earth system science. Nature **566**(7743), 195 (2019)
11. Henderson, S.G., Biller, B., Hsieh, M.H., Shortle, J., Tew, J.D., Barton, R.R.: Agent-based modeling and simulation of wildland fire suppression
12. Niazi, M.A., Siddique, Q., Hussain, A., Kolberg, M.: Verification and validation of an agent-based forest fire simulation model. In: Proceedings of the 2010 Spring Simulation Multiconference, p. 1. Society for Computer Simulation International, April 2010
13. Andrews, P.L.: The Rothermel surface fire spread model and associated developments: A comprehensive explanation. Gen. Technical Report. RMRS-GTR-371. Fort Collins, CO: US Department of Agriculture, Forest Service, Rocky Mountain Research Station, 121 p. 371 (2018)
14. Lenihan, J.M., Daly, C., Bachelet, D., Neilson, R.P.: Simulating broad-scale fire severity in a dynamic global vegetation model. Northwest Sci. **72**(4), 91–101 (1998)
15. Sitch, S., Smith, B., Prentice, I.C., Arneth, A., Bondeau, A., Cramer, W., Thonicke, K.: Evaluation of ecosystem dynamics, plant geography and terrestrial carbon cycling in the LPJ dynamic global vegetation model. Glob. Change Biol. **9**(2), 161–185 (2003)
16. Arora, V.K., Boer, G.J.: Fire as an interactive component of dynamic vegetation models. Journal of Geophysical Research Biogeosciences 110(2) (2005)
17. Pfeiffer, M., Spessa, A., Kaplan, J.O.: A model for global biomass burning in preindustrial time: LPJ-LMfire (v1. 0). Geosci. Model Dev. **6**(3), 643–685 (2013)

18. Gorelick, N., Hancher, M., Dixon, M., Ilyushchenko, S., Thau, D., Moore, R.: Google earth engine: planetary-scale geospatial analysis for everyone. Remote Sens. Environ. **202**, 18–27 (2017)
19. Chen, L.C., Papandreou, G., Kokkinos, I., Murphy, K., Yuille, A.L.: DeepLab: Semantic image segmentation with deep convolutional nets, atrous convolution, and fully connected CRFs. IEEE Trans. Pattern Anal. Mach. Intell. **40**(4), 834–848 (2018)
20. Pham, D.T., Dimov, S.S., Nguyen, C.D.: Selection of K in K-means clustering. Proc. Inst. Mech. Eng. Part C J. Mech. Eng. Sci. **219**(1), 103–119 (2005)
21. Huang, J., Rathod, V., Sun, C., Zhu, M., Korattikara, A., Fathi, A., Murphy, K.: Speed/accuracy trade-offs for modern convolutional object detectors. In: Proceedings of the IEEE Conference on Computer Vision and Pattern Recognition, pp. 7310–7311 (2017)
22. Lin, T., Horne, B.G., Tino, P., Giles, C.L.: Learning long-term dependencies in NARX recurrent neural networks. IEEE Trans. Neural Netw. **7**(6), 1329–1338 (1996)
23. Diaconescu, E.: The use of NARX neural networks to predict chaotic time series. Wseas Trans. Comput. Res. **3**(3), 182–191 (2008)
24. Mandic, D.P., Chambers, J.: Recurrent Neural Networks for Prediction: Learning Algorithms, Architectures and Stability. Wiley, New York (2001)
25. Xingjian, S.H.I., Chen, Z., Wang, H., Yeung, D.Y., Wong, W.K., Woo, W.C.: Convolutional LSTM network: a machine learning approach for precipitation nowcasting. In: Advances in Neural Information Processing Systems, pp. 802–810 (2015)
26. Overholt, K.J., Ezekoye, O.A.: Quantitative testing of fire scenario hypotheses: a Bayesian inference approach. Fire Technol. **51**(2), 335–367 (2015)
27. West, M., Harrison, P.J., Migon, H.S.: Dynamic generalized linear models and Bayesian forecasting. J. Am. Stat. Assoc. **80**(389), 73–83 (1985)

A Fast Multi-phases Demon Image Registration for Atlas Building

Youshan Zhang(✉)

Computer Science and Engineering, Lehigh University, Bethlehem, PA 18015, USA
yoz217@lehigh.edu

Abstract. Medical image registration is an essential branch in computer vision and image processing, and it plays a vital role in medical research, disease diagnosis, surgical navigation, and other medical treatments. However, existing methods are time-consuming; the progress of image registration is difficult to observe, and fewer works for human brain atlas building use image registration method. In this paper, we first introduce a fast multi-phase demon registration (FMDR) model for image registration and atlas building. To show the applicability of our FMDR model, we use synthetic circle data to illustrate a faster and more accurate result of our model than other benchmark methods. We also demonstrate the morphological changes of a TBI case, which shows the continuous shape changes from a diseased state to a healthy state. To illustrate the performance of our model, we use a set of T2 MRIs to estimate the template image for atlas building.

Keywords: Demon image registration · Atlas building · Shape deformation

1 Introduction

Morphological information and functional information of the different patients differ from each other even in the same modality; it is necessary to build an atlas for shape analysis. Therefore, for effective information integration–the fusion of information from various images or different time series images from the same patient is relatively remarkable. It can primarily improve the level of clinical diagnosis, treatment, disease monitoring, surgery, and therapeutic effect evaluation, for example, the fusion of anatomical images and functional images. It can provide an accurate description of anatomical location for abnormal physiological regions. Also, the fusion of images from different modalities can be applied to radiation therapy, surgical navigation, and tumor growth monitoring [1]. Therefore, image registration for atlas building is essential in the medical field.

There are many works addressed image registration problem. Elsen et al. summarized some medical image registration technologies and realize the alignment of different images [2]. Other methods include mutual information for multi-modalities image registration [3], Fourier transform [4]. Image registration will consume more substantial computation time, especially for 3D image registration. Plishker et al., 2007 discussed the acceleration techniques of medical image registration [5]. Nevertheless, one crucial criterion in medical image registration is anatomical structures are one-to-one corresponded with each other after image registration, while transformation has to be topology-preserving (diffeomorphic). If a geometric shape is significantly different in two or multiple images, a topology-preserving transformation is hard to generate. To solved this problem, several geodesic registration methods on manifold had been proposed, e.g., Large Deformation Diffeomorphic Metric Mapping (LDDMM) [6,7]. LDDMM provides a mathematically robust solution to the large-deformation registration problems, by finding geodesic paths of transformations on the manifold of diffeomorphisms. The advantage is that it can solve the large deformation registration problem, but the transformation is computationally very costly to compute if shape change is relatively large. Zhang et al. (2013), proposed a fast geodesic shooting algorithm for atlas building based on the metric of original LDDMM for diffeomorphic image registration, and was faster and used less memory intensive than original LDDMM method [8].

For image registration, it is useful to observe the differences between the composition of background deformations of the image and foreground deformations of geometric objects, such as TBI lesion or tumor. However, the challenge is that most of image registration methods cannot account for image appearance, and take a long time for matching images.

In this article, we propose a multi-phases demon image registration for atlas building. Our contributions are in two folds: (1) a fast and accurate demon registration model; (2) multiple stages to show the progress of shape deformations. We show experimental results of both 2D synthetic data and 3D T2 brain MRIs data. To demonstrate the applicability of our model, we recover a partial circle to a full circle shape. Our registered circle is better than other benchmark methods. Also, we show reasonable progress of changing from a diseased (hemorrhagic) state to a healthy state, while original demon registration method fails to show such progress. Finally, we demonstrate an example for atlas building to estimate template for real 3D brain images.

2 Method

In this part, we discuss the classic demon registration method and discuss our fast multi-phases demon registration (FMDR) model for atlas building.

2.1 Background

Demon Registration. For any point p, let f be the intensity in a static image (I_T, we call it as the target image in the below), and s be the intensity in a moving image (I_S, we call it as the source image in the below). Log demon registration was proposed by Thirion [9], it estimated the displacement $\mathbf{u}$ (velocity) for point p to match the corresponding point in I_S:

$$\mathbf{u} = \frac{(s-f)\nabla f}{|\nabla f|^2 + (s-f)^2}, \tag{1}$$

where $\mathbf{u} = (u_x, u_y)$ in 2D registration, and $\mathbf{u} = (u_x, u_y, u_z)$ in 3D registration; ∇f is the gradient of target image, and it is a internal edge force. $s - f$ is the external force, and $(s - f)^2$ can make velocity more stable. Later Vercauteren et al. [10] proposed standard registration model for Log demon registration, it minimizes the following energy functions:

$$E = ||I_T - I_S \circ (\phi + U)||^2 + \frac{\sigma_i^2}{\sigma_x^2}||U||^2, \tag{2}$$

where ϕ is the original transformation filed in x and y direction; U is the update transformation field in each iteration; $\circ$ is the image transformation operation; σ_i and σ_x control the uncertainty of image noise and transformation. The minimizing of E in Eq. 2 can be calculated using Taylor expansion. We can rewrite Eq. 2 for any point p as:

$$E = ||f - s + \mathbf{u}\nabla s||^2 + \frac{\sigma_i^2}{\sigma_x^2}||\mathbf{u}||^2, \tag{3}$$

Take the gradient of Eq. 3 with respect to $\mathbf{u}$, we obtain: $\nabla_{\mathbf{u}}E = 2\nabla s(f - s + \mathbf{u}\nabla s) + 2\frac{\sigma_i^2}{\sigma_x^2}\mathbf{u}$. Let $\nabla_{\mathbf{u}}E = 0$, we can get the transformation field $\mathbf{u}$ as Eq. 1, when $\sigma_i = (f - s)$ and $\sigma_x = 1$, see [10] for the details calculation of $\mathbf{u}$. But Eq. 1 only use the internal edge force of target image, Wang et al. [11] added another internal edge force for source image:

$$\mathbf{u} = \frac{(s-f)\nabla f}{|\nabla f|^2 + \alpha^2(s-f)^2} + \frac{(s-f)\nabla s}{|\nabla s|^2 + \alpha^2(s-f)^2}, \tag{4}$$

where α^2 is proposed by Cachier et al. [12] to control the force length. Equation 4 enhance the stability and the convergence speed of the velocity field.

Algorithm 1. Demon Registration

Input: Source image I_S, target image I_T, α, and number of iterations: itr
Ensure: Registered source image I'_S
1: Initialize transformation field U
2: **For** $i = 1$ to itr
3:   Calculate U according to Eq. 4
4:   $U = Ker \star U$
5:   Update image $I'_S = I_S \circ (S + U)$
6: **end**

Ker is the Gaussian kernel for smoothing the velocity field. However, demon registration will take a long time to register source and target image if there is significant difference between them. *itr* should be large enough to guarantee a good registration results. Also, the transformation U will keep updating on I_S; there are no stages to show the changes in the image. In the following section, we will show how our multi-phases demon registration to overcome these limitations.

2.2   Multi-phases Demon Registration for Atlas Building

We define a general multi-phases model for atlas building using demon registration. Given input images $I_1, \cdots, I_N$, the atlas building task is to find a template image I to minimize the difference between I and input images. Differ from minimizing sum-of-squared distances function ($\min_I \frac{1}{N} \sum_{i=1}^{N} ||I - I_i||^2$) in [13], we aim to minimize following energy function:

$$E = \arg\min_I \frac{1}{2} \sum_{k=1}^{K} \sum_{i=1}^{N} ||I - I_{ik} \circ (\phi + U_{ik})||^2 + \frac{\alpha^2}{2} \sigma_i^2 ||U_{ik}||^2, \qquad (5)$$

where U_{ik} is the update transformation field of each input image at phase k. α and σ_i control the regularity term ($||U_{ik}||^2$), which measures the smoothness of transformation. The first similarity term measures the similarity of the template image and input images).

To minimize the energy function in Eq. 5, similar to solve Eq. 2, we use Taylor expansion, and rewrite Eq. 5 as:

$$E' = \arg\min_I \frac{1}{2} \sum_{k=1}^{K} \sum_{i=1}^{N} ||I - I_{ik} + U_{ik} \nabla I_{ik}||^2 + \frac{\alpha^2}{2} \sigma_i^2 ||U_{ik}||^2, \qquad (6)$$

Take the gradient of Eq. 6 with respect to U_{ik}, we obtain:

$$\nabla_{U_{ik}} E' = \nabla I_{ik}(I - I_{ik} + U_{ik} \nabla I_{ik}) + \alpha^2 \sigma_i^2 U_{ik} \qquad (7)$$

Let $\nabla_{U_{ik}} E' = 0$, we get the update transformation U_{ik} as:

$$U_{ik} = \frac{(I_{ik} - I)\nabla I_{ik}}{|\nabla I_{ik}|^2 + \alpha^2 \sigma_i^2} \qquad (8)$$

Again Eq. 8 only contains the internal force of template image I, to improve the registration convergence speed and the stability, we add another internal force of ∇I, and let σ_i be $I_i - I$, we get the new update transformation in our multi-phases model:

$$U_{ik} = \frac{(I_{ik} - I)\nabla I_{ik}}{|\nabla I_{ik}|^2 + \alpha^2 (I_{ik} - I)^2} + \frac{(I_{ik} - I)\nabla I}{|\nabla I|^2 + \alpha^2 (I_{ik} - I)^2} \qquad (9)$$

$$I_{ik} = I_{i(k-1)} \circ (\phi + U_{i(k-1)})$$

By getting new images I_{ik}, we could calculate the close-form solution for our template I:

$$I = \frac{1}{N} \sum_{i=1}^{N} \{I_i \circ (\phi + U_{ik})\} \qquad (10)$$

Also, our FMDR model is a generalized model, it also work for registrating two image with $N = 1$ in Eq. 5. Our FMDR model can not only find the corresponding positions between a source and a target image, but explain model differences in image appearances. We demonstrate advantages of our model in Sect. 3.

Algorithm 2. Multi-phase Demon Registration for Atlas Building

---

Input: Source images $I_1, I_2, \cdots, I_N$, noise α, number of iterations: itr', and smooth stage kk
Ensure: Template image I
 1: Initialize transformation field U, and template image I
 2: **For** $k = 1$ to K
 3:   **For** $i = 1$ to itr
 4:     Calculate U_{ik} according to Eq. 9
 5:     **if** $(k >= kk)$
           $U = Ker \star U$
 6:     Update image $I'_S = I_S \circ (S + U)$
 7:   **end**
 8: **end**
 9: Calculate template image I according to Eq. 10

---

where $itr > K \times itr'$, the time complexity of Algorithm 1 is $\mathcal{O}(iter)$, and the time complexity of Algorithm 2 is $\mathcal{O}(K * iter')$. Therefore, our FMDR model is faster than original demon registration. In addition, in Algorithm 1, it applies Gaussian kernel in each iteration, which will reduce the resolution of register image. In our FMDR model, we have smooth parameter kk, which controls the smooth stage to maintain the resolution of the registered image.

3   Results

By introducing our FMDR model, which can not only find the correspondences between the source and the target image, but also observe the progress of shape deformation. We demonstrate the effectiveness of our model using one synthetic circle data, clinical TBI, and real 3D T2 MRI brain data.

3.1   Synthetic Circle Data

In this circle data, we want to restore a partial circle to a full circle, and we aim to test whether our FMDR model can recover the full shape with less computation time and maintain higher accuracy than other benchmark methods.

We compare results of our model with demon registration (DR) [10], Large Deformation Diffeomorphic Metric Mapping (LDDMM) [6] and Bayesian atlas building using diffeomorphic image registration (BADR) [13]. From Fig. 1, we can find that result of our FMDR model (Fig. 1(e)) can match perfectly in the

"C" shape, while the other methods cannot perfectly recover the shape. Especially, DR and LDDMM only recover partial of the shape. Also, we can visualize the difference between the target image and the registered results in Fig. 2. Comparing with the other three methods, our FMDR model has the lowest difference between target and final registered image since there is no yellow color in our model, and it only has some blue shallow outlines. As shown in Table 1, our FMDR use less computation time compared with original demon registration, BADR, and LDDMM.

3.2   TBI Lesion Registration

To illustrate our model can emphasize geometric shape changes, we test our model use a Traumatic brain injury (TBI) dataset [14]. We manually segment the pathology (red ellipse in Fig. 3, we aim to show the progress from a TBI lesion state 3(a) to a health state 3(b)). As shown in Fig. 4, the deformed area continuously changes until the health state, which illustrates the FMDR method

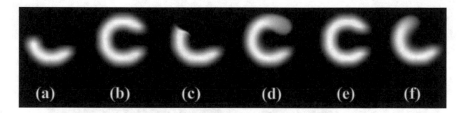

Fig. 1. Circle registration results comparisons of our FMDR model: (a) source image, (b) target image, (c) BADR, (d) LDDMM, (e) with DR, and (f) methods.

Fig. 2. Difference between the target image and registered results using different methods (from left to right: DR, BARD, FMDR, LDDMM) The color is changing from blue to yellow. Blue color means there is less difference between the source image and registered image, while yellow indicates there is a significant difference between the source and registered image.

Table 1. Computation time comparisons between FMDR, DR and LDDMM

	FMDR	DR	LDDMM
Time	**147 s**	245 s	1281 s

can represent deformation changes. However, the DR model cannot correctly show a reasonable health state.

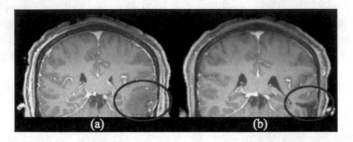

Fig. 3. Traumatic brain injury images. (a) initial scan; (b) registered scan after eight months later. The circle is TBI lesion, dramatically changes in brain lesion part, the shape is deformed. A traditional image registration cannot describe the conversion from a disease state (a: hemorrhagic) to a relative health state (b) (Image from [14], Fig. 1).

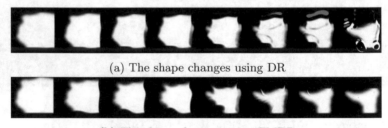

(a) The shape changes using DR

(b) The shape changes using FMDR

Fig. 4. The shape deformation comparisons of our FMDR and DR model. Our FMDR model shows correct shape changes while the DR model does not show a reasonable final health state. Notice that the presence of brain deformations, the change in the lesion's shape, and the conversion of tissue from a diseased state (hemorrhagic) to a healthy state.

3.3   Atlas Building on 3D MRIs

To demonstrate the effectiveness of our method on the real 3D data, we apply our FMDR model to a set of 3D T2 MRIs Fig. 5b. It is a set of Multiple Sclerosis data [15].

From Fig. 5b, the average MRIs is blur, but our estimated template image is obviously clearer than the average MRIs, and this demonstrates our FMDR model can well represent the general information for T2 images, and our method can be used to estimate the template of images which will provide a reliable reference for image fusion.

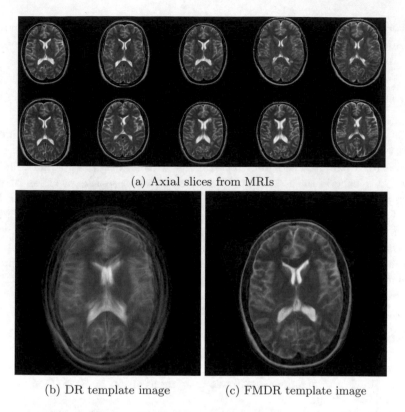

(a) Axial slices from MRIs

(b) DR template image   (c) FMDR template image

Fig. 5. The estimated template image using FMDR and DR model. (a) axial view of input MRIs; (b) estimated template image using DR model; (c) estimated template image using FMDR model.

4   Discussion

One apparent strength of FMDR method is that it can accurately recover the target shape with less computational time. From the results of synthetic images (Fig. 1), we observe that FMDR has a relatively higher recovery rate than other standard image registration methods, such as Demon registration and LDDMM. Also, our FMDR model can describe the geometry shapes changes in a TBI case (Fig. 3), but our model is not limited to applying in a TBI case. It also can be used in qualifying the tumor growing process, for example, tracing the infiltrating and displacing of a brain tumor. Furthermore, this method can be useful for predicting the chronic blood perfusion changes in patients who have a stroke. However, the stage k is a hand turning parameter in our model, which can be changed with different source and target images. Although we show that our method can show the progress of TBI lesion changes, it cannot automatically segment the ROIs (e.g. in Fig. 3(a) and (b) the overlaid segmented lesion shape is manually segmented). And it will absolutely consume time, so this method may not work efficiently in a large database if we only focus on the ROIs. Therefore, our model can be improved if we can propose an addition model to segment these ROIs automatically.

5   Conclusion

In this paper, wo present a novel multi-phases demon registration framework for fast and accurate atlas building. We demonstrate the performance of our model using three data sets. Our result is better than other benchmark methods using 2D synthetic data, and our model can delineate the progress of shape deformations. Also, our model can build a good template for a set of brain MRIs data. For future work, we will test our FMDR method using more 3D images. In addition, we will design new experiments for clinical application (e.g. tumor progression and prediction of chronic blood perfusion changes after stroke). Also, automatically extracting ROIs of the images will help our method in quantify the shape deformations. Furthermore, how to improve the matching accuracy and efficiency of our models need to be further explored.

References

1. Histed, S.N., Lindenberg, M.L., Mena, E., Turkbey, B., Choyke, P.L., Kurdziel, K.A.: Review of functional/anatomic imaging in oncology. Nucl. Med. Commun. **33**(4), 349 (2012)
2. Van den Elsen, P.A., Pol, E.-J.D., Viergever, M.A.: Medical image matching-a review with classification. IEEE Eng. Med. Biol. Mag. **12**(1), 26–39 (1993)
3. Collignon, A., Maes, F., Delaere, D., Vandermeulen, D., Suetens, P., Marchal, G.: Automated multi-modality image registration based on information theory. Inf. Process. Med. Imaging **3**, 263–274 (1995)
4. Srinivasa Reddy, B., Chatterji, B.N.: An FFT-based technique for translation, rotation, and scale-invariant image registration. IEEE Trans. Image Process. **5**(8), 1266–1271 (1996)
5. Plishker, W., Dandekar, O., Bhattacharyya, S., Shekhar, R.: A taxonomy for medical image registration acceleration techniques. In: IEEE/NIH Life Science Systems and Applications Workshop, LISA 2007, pp. 160–163. IEEE (2007)
6. Faisal Beg, M., Miller, M.I., Trouvé, A., Younes, L.: Computing large deformation metric mappings via geodesic flows of diffeomorphisms. Int. J. Comput. Vision **61**(2), 139–157 (2005)
7. Cao, Y., Miller, M.I., Winslow, R.L., Younes, L.: Large deformation diffeomorphic metric mapping of vector fields. IEEE Trans. Med. Imaging **24**(9), 1216–1230 (2005)
8. Zhang, M., Thomas Fletcher, P.: Finite-dimensional lie algebras for fast diffeomorphic image registration. In: International Conference on Information Processing in Medical Imaging, pp. 249–260. Springer (2015)
9. Thirion, J.-P.: Image matching as a diffusion process: an analogy with Maxwell's demons. Med. Image Anal. **2**(3), 243–260 (1998)

10. Vercauteren, T., Pennec, X., Perchant, A., Ayache, N.: Non-parametric diffeomorphic image registration with the demons algorithm. In: International Conference on Medical Image Computing and Computer-Assisted Intervention, pp. 319–326. Springer (2007)
11. Wang, H., Dong, L., O'Daniel, J., Mohan, R., Garden, A.S., Kian Ang, K., Kuban, D.A., Bonnen, M., Chang, J.Y., Cheung, R.: Validation of an accelerated 'demons' algorithm for deformable image registration in radiation therapy. Phys. Med. Biol. **50**(12), 2887 (2005)
12. Cachier, P., Pennec, X., Ayache, N.: Fast non rigid matching by gradient descent: study and improvements of the "demons" algorithm. Ph.D. thesis, INRIA (1999)
13. Zhang, M., Singh, N., Thomas Fletcher, P.: Bayesian estimation of regularization and atlas building in diffeomorphic image registration. In: International Conference on Information Processing in Medical Imaging, pp. 37–48. Springer (2013)
14. Niethammer, M., Hart, G.L., Pace, D.F., Vespa, P.M., Irimia, A., Van Horn, J.D., Aylward, S.R.: Geometric metamorphosis. In: International Conference on Medical Image Computing and Computer-Assisted Intervention, pp. 639–646. Springer (2011)
15. Loizou, C.P., Murray, V., Pattichis, M.S., Seimenis, I., Pantziaris, M., Pattichis, C.S.: Multiscale amplitude-modulation frequency-modulation (AM-FM) texture analysis of multiple sclerosis in brain MRI images. IEEE Trans. Inf Technol. Biomed. **15**(1), 119–129 (2011)

Hand Sign Language Feature Extraction Using Image Processing

Abdelkader Chabchoub[1,2(✉)], Ali Hamouda[2], Saleh Al-Ahmadi[3],
Wahid Barkouti[1,2], and Adnen Cherif[1]

[1] ATSSEE, Faculty of Science, University of Tunis El Manar, Tunis, Tunisia
achabchoub@mct.edu.sa, wahidbarkouti@hotmail.fr,
adnane.cher@fst.rnu.tn
[2] College of Technology Medina, Medina, Kingdom of Saudi Arabia
ali.saeed@interserve.com
[3] Department of Electrical Engineering, Islamic University in Madinah, Medina,
Kingdom of Saudi Arabia
saleha@iu.edu.sa

Abstract. Sign language is a method of communication, especially with people in special needs such as those that are deaf and dumb, but this language is not easily understood by everyone. Many articles and researches have focused on learning hand language and converting hand signals into meaningful signs, by using signal processing methods to translate or to convert sign language to speech or texts, for example, showing the number or the text on the screen or converting it to a spoken language. In this paper, an image processing technique has been proposed, to extract the hand signal feature of English numbers and convert them into written text. Specific gloves were used to simplify the extraction of features using two symbols- a circle and a triangle that was printed on each glove. An algorithm was applied to each image with its feature extracted by MATLAB programs. The process was applied to each PNG image by converting it to a binary image and using object detection, sobele filter for edge detection, image resizing, calculating the number of circles and triangle. The correlation between each image feature was measured to identify the specific sign language and then to convert it to texts or audio.

Keywords: Deaf and dumb · Claves · Hand sign · Image processing · Image recognition · Correlation · Synthesis speech

1 Introduction

Sign language is the only way to communicate with others for the deaf and the dumb. Many techniques and codes have been developed to improve and recognize the manual language, to reduce the problem of communication faced by the deaf people to allow them to communicate normally with other people. Numerous research works have been done in sign languages such as American Sign Language, British Sign Language and Japanese Sign Language [5]. But the work done in the field of Arabic language did not reach an integrated work and is in the process of research for improvement.

© Springer Nature Switzerland AG 2020
K. Arai et al. (Eds.): FTC 2019, AISC 1070, pp. 122–131, 2020.
https://doi.org/10.1007/978-3-030-32523-7_9

Finding an experienced and qualified interpreter every time is a very difficult task. The dilemma is that hand language teaching is only considered for deaf people. Moreover, people who are not deaf, never try to learn the sign language for interacting with the deaf [4]. Our solution focusses on assisting the deaf people by developing a new technology, based on a modified computer vision or microprocessor and embedded systems. For research purposes, a computer will be designated with features such as digits display or applicable gloves with LCD, Chin-Shyurng Fahn and Herman Sun proposed an electromagnetic sensor for hand language recognition fixed on the fingertip [3] to determine the finger position and rotation. Mohd Firdaus Zakaria, Hoo Seng Choon, and Shahrel Azmin Suandi proposed an image processing technique that utilizes intensity value from the input image. Otsu's methods were used for image binarization. Additionally, filtering is applied to eliminate noise and edges detection. [1] Zhong Shen, Juan Yi, used the soft flexible sensor for the data glove sensor application, depending on the length and blend of sensitive and insensitive regions [6]. An image processing technique with MATLAB [7] simulation was applied to the input image. Arabic numbers (1, 2, 3, …, 10) and hand signs were captured by mobile phone. In this way, 10 images of each number were processed and their features extracted [8]. As described above, this is a method with a simple algorithm with low computational time. The intensity value from the input image then converts a gray image to a binary image. The new image has a foreground and background. Sobel operator is used to find the edges. Finally, image features are extracted.

In this paper Sect. 2 covers the proposed methodology and Sect. 3 covers the results and identification of hand signals.

2 Proposed Methodology

As discussed above, our solution for the image processing technique is to design specific gloves that are easy for extracting features or converting hand signs to text or decimal numbers. The first step in our algorithm as in Fig. 1, is to capture an image and convert it to a binary image with only two colors- black and white (0,1). After that the binary image is filtered by removing all unwanted pixels. The next process is edge detection that is done through many algorithms. The final feature extraction step is to determine the shape of the triangle and the circle in an image and determine the center of each one. After that it is easy to write the corresponding text.

2.1 Image Processing

Hand sign language has been implemented in Madinah College of technology (MTC) in Saudi Arabia as there are students with special needs on campus. This will help them to integrate with the society and ease their communication issues. In our design, we used numbers only as the first step due to their importance in the communication, marketing, and accounting. Each number has an equivalent hand sign (as in Fig. 2). This represents numbers from 1 to 10. We designed a database that consists of 1000 images, and each number has 100 images with a specific size. Figure 3 shows image database of the Arabic hand signal number using the gloves.

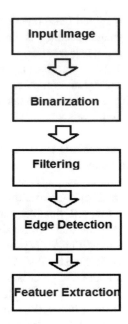

Fig. 1. Block diagram of system

Fig. 2. Hand sign language Arabic number

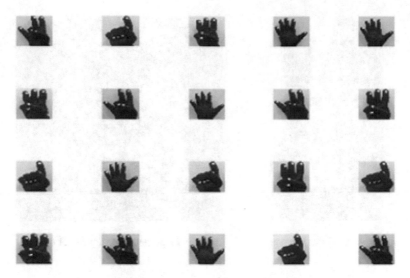

Fig. 3. Image database of the Arabic hand signal number using the glove.

2.2 Binarization with Fixed Thresholding Method

In the Fixed Thresholding binarization method, [2–4] a fixed threshold value in a given image is used to assign 0's and 1's for total values of pixel positions. The basic idea for fixed binarization method is described in Eq. (1) in which all image were converted to binary image (Fig. 4).

$$g(x, y) = \begin{cases} 1 \; if \; f(x, y) > 0 \\ 0 \quad otherwise \end{cases} \tag{1}$$

2.3 Edge Detection

Think of an image as a height field. On such a surface, edges occur at locations of steep slopes, or equivalently, in regions of closely packed contour lines. A mathematical way to define the slope and direction of a surface is through its gradient [1]. The local gradient vector J points in the direction of steepest ascent in the intensity function. Its magnitude is an indication of the slope or strength of the variation, while its orientation points in a direction perpendicular to the local contour as in Fig. 5.

$$\mathbf{J(x)} = \nabla\mathbf{I(x)} = \left[\left(\frac{\partial I}{\partial x} + \frac{\partial I}{\partial y} \right)(x) \right] \tag{2}$$

Fig. 4. Binary image   **Fig. 5.** Edge detection by Sobel operator

2.4 Shape Recognition

The proposed method recognizes the shapes of an object by computing the compactness [10]. Equation (2) they are two shapes fixed on the gloves circle and triangle as in Fig. 3. After binarization and filtering and edge detection then we remove all noise and applying background equalization to and filtering all continues line to extract only geometrical feature as in Fig. 7. For each shape circle feature separated and counted and triangle feature separated and counted.

2.5 Feature Extraction

Final feature are extracted in Fig. 7 and then the center of each objects in the feature is determined as shown in Fig. 9. The circle object is determined by MATLAB function to classify objects based on their roundness using MATLAB function (bwboundaries), a boundary tracing is routine. For circle objects its value is closed to 1 and for a triangle its value around 0.5 as in Fig. 6. After determining the circle and triangle, along with the center of each one as in Fig. 8, we use filtering to separate each feature and calculate its number as in Figs. 10 and 11. The feature of each number is calculated as number of circle and the triangle as in Table 1. Correlation is measured in one sine, between each ten image feature and the main value calculated as in Table 1. It can be seen that the correlation between three and six is very high [9].

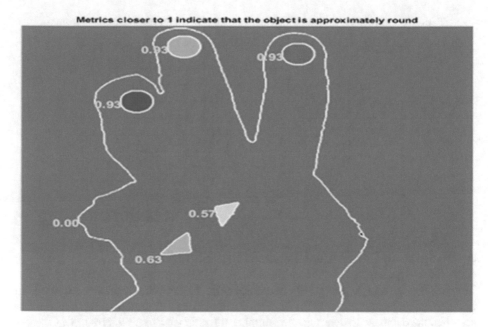

Fig. 6. Object detection

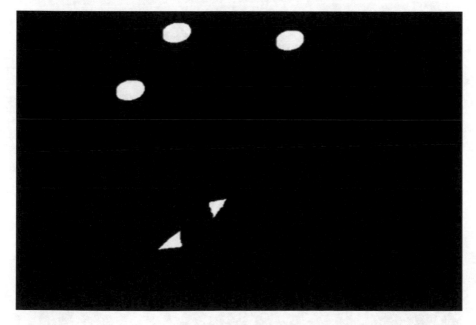

Fig. 7. Feature filtering

Table 1. Correlation measured between one sign between each ten image feature.

Hand sign	No of Circle	No of triangle	Total number	Correlation
1	1	4	5	85.59
2	2	3	5	87.54
3	3	2	5	80.00
4	4	0	4	–
5	5	0	5	79.17
6	3	2	5	–
7	3	2	5	80.02
8	3	2	5	81.90
9	3	2	5	83.55
10	0	5	5	89.50

3   Results and Identification of Hand Signal

The number of triangles (Fig. 10) and circles (Fig. 11) and the distance between them is determined on the glove (Fig. 8). This information is inputted to the system that searches the database to determine the number the character to be referred to and wants to be pronounced by the disabled.

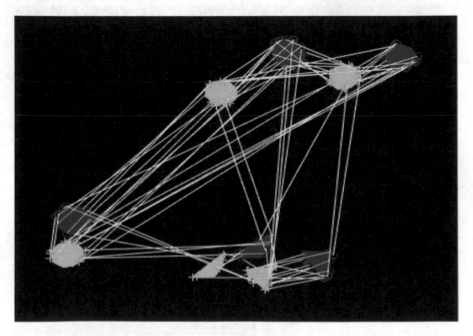

Fig. 8. Feature matching of two image

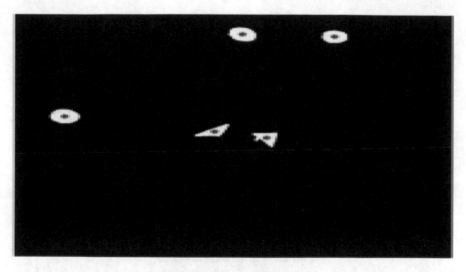

Fig. 9. The center of each filter

Fig. 10. Triangle featuer Number

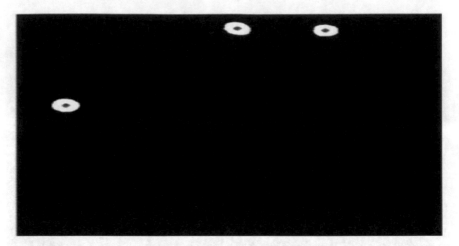

Fig. 11. Circule featuer Number

4   Conclusion

All features extracted from an image show very good values of the feature points and correlation values. However, number three and six have closed correlation because there is no big difference between them in the method of hand sign as in Fig. 1. The work was implemented in MATLAB. The system solves the handicapped person's difficulties in communicating better with people and by transforming the manual signal into a voice. The system will also improve the quality of teaching and learning at the Deaf and Mute Institute. The main objective of this work is to serve humanity, which is achieved through better education and better communication.

For future application to get more accuracy, fuzzy logic, neural networks, and deep learning are proposed as extended methods for hand language identification and recognition.

References

1. Zakaria, M.F., Choon, H.S., Suandi, S.A.: Object shape recognition in image for machine vision application. Int. J. Comput. Theor. Eng. **4**(1), 76–80 (2012)
2. El Hayek, H., Nacouzi, J., Kassem, A., Hamad, M., El-Murr, S.: Conference 2014. IEEE (2014). ISBN 978-1-4799-3166-8 9999
3. Fahn, C.S., Sun, H.: Development of a fingertip glove equipped with magnetic tracking sensors. Sensors **10**, 1119–1140 (2010)
4. Singha, J., Das, K.: Indian sign language recognition using eigen value weighted euclidean distance based classification technique, (IJACSA) Int. J. Adv. Comput. Sci. Appl. **4**(2), 188–195 (2013)
5. Kour, K.P., Mathew, L.: Sign language recognition using image processing. Int. J. Adv. Res. Comput. Sci. Softw. Eng. **7**(8), 142–145 (2017)

6. Shen, Z., Yi, J., Li, X., Lo, M.H.P., Chen, M.Z., Hu, Y., Wang, Z.: A soft stretchable bending sensor and data glove applications. Robot. Biomim. **3**, 22 (2016)
7. Polygerinos, P., Lyne, S., Wang, Z., Nicolini, L.F., Mosadegh, B., Whitesides, G.M., Walsh, C.J.: Towards a soft pneumatic glove for hand rehabilitation. In: IEEE International Conference on Intelligent Robots and Systems, pp. 1512–1517 (2013)
8. Wang, H., Hui, S., Hong, H.Z., Qing, S., Xiao, F.: 2017 IEEE 2nd Information Technology, Networking, Electronic a7nd Automation Control Conference (ITNEC), pp. 1269–1272 (2017)
9. Basistov, Y.A., Yanovskii, Y.G.: Comparison of image recognition efficiency of Bayes, correlation, and modified hopfield network algorithms. Pattern Recognit. Image Anal. **26**(4), 697–704 (2016)
10. El Abbadi, N., Al Saadi, L.: Automatic detection and recognize different shapes in an image. IJCSI Int. J. Comput. Sci. Issues **10**(6), 162–166 (2013)

Butterfly, Larvae and Pupae Defects Detection Using Convolutional Neural Network and Apriori Algorithm

Jerwin M. Montellano[✉]

Graduate School, University of the East, Manila, Philippines
jerwin.montellano@ue.edu.ph

Abstract. The motivation behind this paper is to build up customized applications into easy to use Google Cloud Platform that contains 18000 pictures of butterflies from the Saturniidae, Nymphalidae and Papilionidae family to distinguish Butterfly classifications, image fruits or larval host plants and pupae deformities by incorporated trailblazing innovation to give profitable experiences into the activity. The exploration techniques utilized shading channels, mass channels, and shape discovery as exchange figured out the prepared pictures indicated using machine vision in CNN. The procedure will utilize a computer vision through scanning pupae quality before they are conveyed to the market of purchasers. The pictures were (a) transferred and named; (b) prepared, (c) assessed, and (d) tried in AutoML Vision. The computerized devices expectation certainty or accuracy normal is 97.1% and review of 91.3% out of 429 pictures 11 labels, 4 test images in top 5 labels appeared in the assessment. The next model was created and trained about 398 images9 labels, 4 test images. Precision and recall are based on a score threshold of 0.5. Precision help high exactness model produces less false positives 90.9% and review help high review model produces less false negatives 85.1%. The purpose of the study was to provide computerized instruments incorporated product search information communicate with GCP Cloud Vision API with the utilization of the Corvid Node Runtime to the online business website of Insects Connection, including machine vision, and sensor combination in items to robotize the retail exchange buy installment strategies.

Keywords: Apriori · Artificial Intelligence · Blob filtering · Butterfly species · Camera captured document images · Character recognition system · Color matching · Convolutional Neural Network · CNTK-R · Corvid · Data mining · Handheld device · Image classification · Image clustering · Image processing · Image recognition · Image segmentation · Keras · Machine learning · Mask R CNN · Optical character recognition · Pytorch · Rest Net Api · Shape detection · TensorFlow · Thresholding

1 Introduction

Butterflies, bees, and humans have love associations. This happens to fertilize the greater part of the yields we devour while adding to the economy. Individuals love seeing those beautiful creatures that can facilitate distressing life. In this investigation,

© Springer Nature Switzerland AG 2020
K. Arai et al. (Eds.): FTC 2019, AISC 1070, pp. 132–161, 2020.
https://doi.org/10.1007/978-3-030-32523-7_10

the butterfly culture in Marinduque has developed since 1972 that even though they've been bred in captivity, they're still normal. The place it began captive reproducing community adjusted the captive rearing innovation for a standout amongst the most obvious types of Earth's biodiversity not exclusively: (a) to secure them and deal with their normal assets, (b) to address this issue by expanding pupae and silk generation, and (c) to enable innovation to support harvests and maintainability (d) to strengthen the linkage in Government, private entrepreneurs and stakeholders. The business is prospering right there in terms of butterfly raisers increments to 40% within the barrio. Marinduque has to promote sericulture industry to accommodate a group of farmers, introduce technologies and establishing facilities. The projected output is as follows: (a) good quality replicated in part of Marinduque, and (b) income of butterfly farmers increased.

Marinduque is in MIMAROPA Region IV-B in the Philippines, among the top agricultural crops, fisheries and livestock industry in this region. Population engaging in horticulture and butterfly farming is not included. The huge place that is known for Marinduque is busy with sloping zone and the vast majority of land has been developed. Considering the province's diversified topography and atmospheres, the creation of significant worth included money crops such as sericulture, butterfly farming, agriculture, coconut crops, and so forth is significant to the country's development.

The local government lack knowledge and skills and has no limited human and financial resources to create institutional development. The proposed project for sustainability of pupa and cocoon silk productions were implemented by cascading training, silk marketing, silk product development, monitoring, and statistical data collection. The reproducers sold dried butterfly, and live butterflies and pupae for nearby traders, items such as paper enrichments in various sizes of casings, packaged specialty and clock with adjusted structure and craftsmanship created for instructive purposes in local and for export. The study's main goal is to help all of the rancher's supportabilities increment their efficiency and give important experiences into their activities (a) by coordinating trailblazing innovation and (b) custom fitted information investigation through data science. The lack of progress and still struggle in production, lower quality of pupae, less sericulture industry and appropriate technologies such as controlling temperature in butterfly farming rearing seasons, avoiding diseases, lack of other host plants leaves and grading qualities new technologies have not been applied by farmers. The good quality of pupae and cocoon silk production is verified in the target area through capacity building. This can be improved by distributing manuals and providing training to extension workers, this can facilitate technical training for butterfly farmers and seri-farmers in each group in raising awareness and reinforcing important tips in butterfly farming.

The butterflies that are listed from the website of Conservation Status of Threatened Philippine Lepidoptera Species classified by the IUCN list are from the family of **Papilionidae**: Troidin, 13 species, Pailionini, 8 species, Liepto Circini, 5 species; family **Pieridae**: Pierinae, 8 species, family **Nymphalidae**: Nymphalinae, 2 species, Limenitidinae, 18 species, Charaxinae, 7 species, Danainae, 15 species, Morphinae, 4 species, Satyrinae, 2 species, family **Lycaenidae**: Liphyrinae, 2 species, Ployommatini, 5 species, Techlini, 34 species, Miletinae, 8 species, Poritidae, 8 species, from family **Hesperiidae**: Hisperiinae, 10 species, and Pyrinae, 2 species. Philippine butterflies website photographs by other contributors: over 600 of the 900 species are illustrated. The researcher simply chose 18 species from the Saturniidae, Nymphalidae and Papilionidae family and gathered a huge number of pictures to assess and to test the model. From time to time, new advancements, for example, web or portable applications improves to man-made reasoning which gadgets and machines are worked for ventures, online business, foundations, and frameworks to create reports, measurements, and deliberation. Furthermore, researcher of the computer vision is an incredibly troublesome errand and requires translating the picture and parsing it into its parts such as image fruits or larval host plants, eggs, larvae, pupae, and butterflies. Advanced photographs give the most noteworthy conceivable goals. The researcher proposed a framework utilizing AI and information examination. The method used for choosing the best nature of pupae for the fare are filtering gadgets incorporated into the handheld gadgets, such as mobile phones, cell phones, and tablets. The advanced instruments are structured and centered on structure our logical capacities crosswise over butterfly culture and agribusiness.

In addition, the learning model is the instrument to be utilized to change over pictures, machine printed, hand-printed or written by hand reports, and literary depiction document into editable article picture group. The machine can take in nature and improve themselves from the experience without being unequivocally modified or without the mediation of human. Its primary point is to influence PCs to gain naturally from the experience. The life will likewise be caught to distinguish object, where it endeavor to find and order various articles inside the picture or body practices, movements and characteristics and spare the crude information assembled originating from various dispersed servers. Furthermore, the researcher utilized the pictures of image fruits or larval host plants, eggs, larvae, pupae, and butterflies integrated the identifying devices to take the qualities, and sizes. The system integrated the dynamic database to Corvid Node Runtime platform using Node.js and Google Cloud Vision API and is connected to GCP Google Vision API. Googlenet [23] used one progressively significant and increasingly broad Inception facilitate with to some degree unmatched quality, anyway adding it to the outfit seemed to improve the results just unimportantly. The web has turned into a need for everybody to search for the concealed images. Web-based shopping is presently turning into the standard as you can get nearly anything on the web. There are diverse administrations to put in your request with, and distinctive methods for paying for them. Web-based shopping is without a doubt advantageous, yet there is another side to it.

From methods for sharing data, it has turned into a wellspring of diversion and for other people, a method for purchasing things while inside the solace of your own home. The framework will utilize Google Cloud Product Search which empowers the venture to have OCR in catching the pertinent butterflies progressively, on-gadget. Generally, these kinds of A.I. errands require devoted, substantial servers. It conquered that impediment by compacting complex PC vision models until they could fit on the chip it use for Gateway and still run precisely and dependable. In the dashboard, many items, butterfly pupae, and different things will be shown to the leaderboard in diagrams the most every now and again sold things. Raisers and the purchaser's examination for the consistent and seasonal purchase will be monitored and strategized. It shows a rundown of bugs, for example, butterflies species and moths, honey bees, blossoms, and creatures. The entrances additionally gather perceptions from butterfly devotees through entryways oversaw and kept up by nearby accomplice protection associations. Along these lines, gateways targets explicit gatherings of people with the most abnormal amount of neighborhood skill, advancement, and task proprietorship.

2 Related Literature

This part of the paper introduces the related writing and concentrates on computer vision, after intensive research and profundity seek done by the researcher. This will likewise display the combination of the craftsmanship and hypothetical and reasonable structure to completely comprehend the examination to be done and in conclusion the meaning of terms for a better understanding of the investigation.

Object Recognition from Natural Language Descriptions

The related [21] work utilized Standard Vision strategy to looked at against two standard methodologies the spatial shading histogram and bags of words to explore the models. The spatial shading histogram has five preparing pictures for every class and the bags of words and has one preparing picture for each classification. They examined models for connecting data in content and pictures together for visual item recognition from characteristic language depictions. They utilized mapping among literary and picture highlights for testing an issue. The underlying model accomplished unassuming exactness with no preparation images. State of the workmanship vision techniques gives great outcomes however rely upon the preparation pictures utilized.

Image Segmentation Based on Machine Vision

The investigation related [9] by entering the machine vision frameworks into this subject, they transformed into a solid, minimal effort and continuous innovation. Notwithstanding the presence of machine vision frameworks in this procedure, there are as yet significant difficulties in arranging farming items regarding quality, size, shape, and examination of imperfections. Potato is a standout amongst the most significant horticultural items that is delivered and has a high application. Lamentably, it experiences different sorts of maladies and deformities. Consequently, it is quality control has specific significance.

Detection and Segmentation Through ConvNets

Author in [26] transposed with 2×2 convolution strived to learn suitable loads for channel expected to perform up inspecting by beginning with upper left corner esteem, which is a scalar, duplicate it with the channel and duplicate those qualities into the yield cells. At that point, it moves the channel some particular pixels in the yield in the extent to one-pixel development in info.

Character-Level Convolutional Networks for Text Classification

The investigation related [16] picking the most ideal AI classifiers to offers an exact investigation on the utilization of character-level Convolutional Network (ConvNets) for content grouping both 9 layers profound with 6 convolutional layers and 3 completely associated layers. The examination of the models are (a) the bag of-words model was built by choosing 50,000 most continuous words from the preparation subset, (b) bag of-ngrams models are developed by choosing the 500,000 most regular n-grams (as much as 5-grams) from the preparation subset for each dataset and (c) an exploratory model that utilizations k-means on word2vec gained from the preparation subset of each dataset, and afterward utilize these educated methods as delegates of the bunched words.

There are two runs of the mill utilization [14] of the insight data combination innovation in farming checking and early-cautioning research was presented. These two applications demonstrate the multidisciplinary joined with conventional horticulture and current data innovation. Further looks into will be centered on how to improve the office and genuine convenient of the re-enactment. The solid arrangement of the data innovation and farming innovation speak to the advancement course of things to come rural innovation. Notwithstanding the presence of machine vision frameworks in this procedure, [9] there are as yet real difficulties in sorting agrarian items as far as quality, size, shape, and examination of imperfections.

The strategy [15] improves the view of the binarized pictures dependent on shading tone and consequently the divergence among unique and binarized picture is decreased. This strategy adjusts for contrasts in light and shade by incorporating data content in the thresholding estimation. The methodology [12] in which subtractive shading planning is the underlying advance to channels picture varied until a perceptual match is gotten with a reference white. The difficulties of vision are a direct result of the complexities of butterflies and moths, hatchlings and pupae. Next, the three shading centers [12] are changed until a match is gotten with perfect shading. FCM thresholding [7] gives fine sectioned outcomes when contrasted with edge recognition strategy. The advancement property of FCM is improved when it is joined with neighborhood thresholding. Versatile delicate thresholding [10] is utilized since we saw that the appropriation of non-nearby examples can be approximated by Laplacian conveyance.

With the progression [1] of handling speed and inward memory of hand-held cell phones with implicit computerized cameras including top of the line PDAs, Individual Advanced Partners (PDA), PDAs, iPhones, iPods, and so on, another pattern of research has become an integral factor. The ImageNet Vast Scale Visual Recognition [8] use objects classification characterization and discovery on many item classes and a huge number of pictures and their depictions. This progression [3] comprises of a pre-handling stage where pictures are changed over into twofold pictures, a top-down division procedure that separates the pictures; the production of a database by the extricated pictures; and recognition arrange where the database is utilized for changing over any pictures into a content document.

The expansion in cameras [6] innovation using cell phones, video catch and the measure of information caught has been increment. The need for new camera applications expanded also. Right now 360 Camera, VR, and 3D recordings are normal and wanted in cell phones. 3D and VR recordings are appealing and contain a colossal bit of portable applications, somewhere in the range of 2012 and 2018, the quantities of 3D and VR versatile applications encountered a fast development in North America.

Through the filtering procedure, [2] an advanced picture that will be tried, assessed, checked and examined by the computer vision as unique pictures are caught. While OCR optical scanners gadget changes over light power into dim dimensions, thresholding, mass separating and shape recognition. The used, [5] Neural Network for the course of action of characters with three layers explicitly Input layer, an Output layer, and Hidden layer. The geometric features expelled like spot, line, twist or circles are given as a commitment to the information layer. Each section of the distributed depiction is named a spot, line, twist, or circle. For every circumstance, the photos of the portion are settled: if a line, what are its presentation and its size in regard to the character diagram-short, medium or long. One data neuron is used to encode all of these possible choices (short/medium/long) and all of four possible presentations for a line. One data neuron is used to encode the characteristics of each fragment expelled by geometric part extraction method. A neuron has two techniques for undertakings as getting ready mode and testing mode. In the planning mode, the neuron can be set up to fire (or not), for explicit information structures. In the testing mode, when an empowering data configuration is perceived at the data, its related yield transforms into the present yield. If the data configuration does not have a spot in the appeared of information plans, the ending principle is used to choose if to fire or not. Furthermore, utilizing the Convolutional Neural Network, the machine pools to change over the stunned picture into a bi-level picture of profoundly differentiate. Often, this system known as thresholding is performed on the scanner to save memory space and computational effort.

Gaussian blend demonstrates [11] for extricating forefront targets and create multi-scaled areas to accelerate item or conduct identification in high-goals input video outlines the Rest Net Programming interface of GCP and CNTK-R library. This undertaking proposes [20] another system of calculation MAA that beats the restrictions related with existing techniques and empowers the finding of affiliation rules dependent on Apriori Algorithm among the nearness and additionally nonattendance of a lot of things without a preset least help limit and Minimizing Candidate Generation.

Technology plays a big role [22] in reaching and engaging the right audiences, especially as people shop online more than ever. Find out how automation can help you create agile strategies to stand out from the competition. Gateways may have a local concentration or they may have increasingly explicit objectives as well as explicit techniques, for example, butterflies relocation. Every gateway is completely incorporated inside the butterfly framework. The e-Butterfly framework has included memorable information from different gallery accumulations and biodiversity foundations into the dissemination and phenology highlights and apparatuses. This information may help educate how butterflies populaces and phenology have changed crosswise over the Philippines.

Perceive picture from your index inside the web or portable photographs and actualizes visual pursuit experience that empowers the applications to perceive items in your pictures. The framework [17] is upheld by histogram arranged example examination with straight twofold examples for preparing. The framework tracks the item movement and features the recognition of fruitful identification. The multiple object detection [28] used to discover the optical stream vectors which thusly clear a route for the recognition and following of the single moving article in a video. Kalman channel evacuates the clamor that impacts a foundation subtracted picture and predicts the position of an article precisely. This framework is separated into two boards Administrator and Client, whereas "Administrator" keeps up the monitoring of larval host plants, and butterflies control of harvest pupae and crafted items. In framework news, the administrator can add new subtleties to the framework this will keep up day by day report of agribusiness and rancher additionally with ongoing data of plant. There are five kinds of modules used to build up the proposed framework, for example, Data Social event, Data Sharing, and Page Structuring.

3   Methodology

Figure 1 is the initial step of the design thinking in Machine Learning. It visualizes user attitudes and behaviors, to advocate in the interest of the client in a compassion guide helps UX groups adjust on a profound comprehension of the end user. The mapping procedure likewise uncovers any gaps in existing client information. It is a synergistic perception used to express what we think about a specific sort of client. It externalizes information about clients so as to (1) make a mutual comprehension of client needs and (2) help in basic leadership.

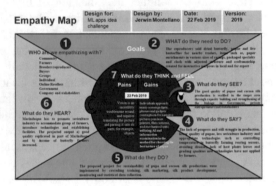

Fig. 1. Empathy map

Architectural Design of the system use Cloud Vision API and Cloud Natural Language API, Cloud StackDriver Lodging API, Corvid Database, Wix Media, and Web Apps into Corvid Node Runtime integrated to web apps which can tags and post the images (see Fig. 2).

Architecture Diagram

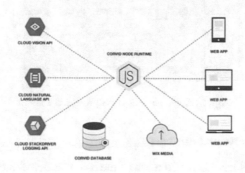

Fig. 2. Design of the proposed system

3.1 SilkNet Spun Casing Method

The researcher utilized the examination of pupae/casing SilkNet demonstrate a similar life cycle of butterflies, as the window slides over the entire picture. Turns out, this convolution procedure all through a picture with a weight grid delivers another picture (of a similar size, contingent upon the show). The researcher proposed the framework for butterfly stages to utilize their capacity on how the butterflies make and change to another model. For instance, the eggs delivered had laid egg to brood into hatchlings unpleasant, hatchlings change their skin concealing and find sustenance, packaging and pupa rest to hang and switch hatchling grown-up butterfly. The pupae and the case sleep for about eight days doing contemplation like preparing model to turn into a grown-up butterfly or moth that can see by other the entire image of their actual magnificence that they never see with its own eyes (yield).

Figure 3 shows screenshots of the proposed system with the [23] progressing footing of portable and inserted figuring. The proficiency of the calculations particularly the capacity and memory of the proposed system use gains significance, telephone consolidate features, trades between various clients in the framework, it also lets customers to move the data through the framework. Aside from it helps them to perceive the customer and give the correspondence according to the embraced element of security. With the trade of the report referenced and the run required methodology at the server customer who got the record will do the assignments like embedding, deciphering, and decompress in their component of hierarchy of leadership, etc.

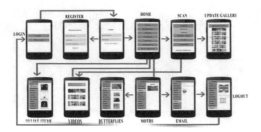

Fig. 3. Screenshots of the proposed system

3.2   Convolutional Neural Network

Computer researchers have invested a long time to construct frameworks, calculations, and models which can get pictures. Today in the period of Artificial Intelligence and Machine Learning have the capacity to make amazing progress in distinguishing objects in pictures, recognizing the setting of a picture, identifying feelings and so on. A standout amongst the most well-known calculation utilized in computer vision today is Convolutional Neural Network or CNN. CNN instruct computers to bode well out of this dazing cluster of numbers is a difficult errand. Typical CNN [18] comprises of an input layer, numerous hidden layers, and an output layer. The concealed layers are by and large a mix of exchanging convolutional-and pooling layers, trailed by a few completely associated layers. In the convolutional layers, different channels (parts) are used to convolve (include estimations of a pixel inside a picture to its neighboring pixels dependent on a specific channel) the picture into highlight maps. The consequent pooling layer diminishes the components of these element maps.

3.2.1   Convolution Layer

Convolution layer is a scientific activity that speaks to a flag going through an LTI (Direct and Time-Invariant) framework or channel (Fig. 4). The convolution is essentially the fundamental of the result of the two capacities (in this precedent the capacities are and), where one is turned around. The profound system is assembled had three convolution layers of size 64, 128 and 256 pursued by two thickly associated layers of size 512 and a yield layer thick layer of size 18 (number of classes in the Butterfly dataset) of 9 × 9 convolution layer with 3 × 3 pooling layer.

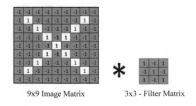

9x9 Image Matrix   3x3 - Filter Matrix

Fig. 4. Convolution layer visualization

3.2.2 Activation Layer

Activation Layer used Silknet spun casing convolution 15 × 15 with filter of 3 × 3 to start applying dots and sliding at the center across the image to connect each every pixel esteem in a picture orchestrate the photographs into four phases of transfer learning (Fig. 5). The first step is (a) to start with off-the-shelf pre-trained convolutional neural networks like VGG16 or ResNet. Then (b) replace the classifier with input images and (c) use feature extraction and fine tuning to retrain pupae images. (d) Resize the image size required by the image model used 224 × 224 for ResNet models, 227 × 227 for the AlexNet the model way to reduce the number of passcs for an image at best to single pass.

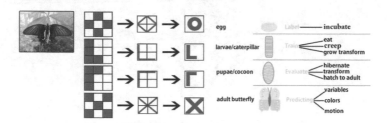

Fig. 5. Shows the SilkNet model

3.2.3 Pooling Layer

Another imperative idea of CNN's is pooling, which is a type of non-direct down-examining. There are a few nonlinear capacities to actualize pooling among which max pooling the most is widely recognized. For the most part to sometimes insert a pooling layer between dynamic convolutional layers in a CNN plan. Convolutional frameworks may fuse neighborhood or overall pooling layers, which solidify the yields of neuron bunches at one layer into a singular neuron in the accompanying layer. For instance, max pooling uses the most extraordinary motivation from all of a lot of neurons in the previous layer. Another model [15] is ordinary pooling, which uses the typical motivation from all of a lot of neurons in the previous layer. Figure 6 shows the Convolutional Neural Network.

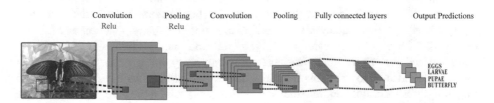

Fig. 6. Convolutional neural network.

3.2.4 Fully Connected Layer

Fully connected layer classified the images of flattened matrix of butterflies, eggs, larvae, pupa deformities and image fruits or larval host plants into science (specifically, practical investigation) convolution is a numerical task on two capacities (f and g) to create a third capacity that communicates how the state of one is adjusted by the other (Fig. 7). The term convolution alludes to both the outcome work and to the way toward registering it.

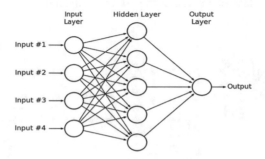

Fig. 7. Fully connected layer

3.3 Color Matching/Filtering

The researcher added some kind of computer vision pipeline to process the pictures before using it to get the state of the pupae. These are [4] securing pictures on camera antiquities and restrictions which add to the particular sorts of imperfections present in the pictures that must be managed by consequent handling. One is the way that numerous video signs are entwined. Color camera can utilize diverse introduction times for the distinctive shading groups, which can make up for the more unfortunate affectability of the silicon chip to short wavelength (blue) light. Color resolution is imperative to consider the numerous implications of "pixel". It is essential to center the optics effectively to catch the fine subtleties in the picture. Electronics and bandwidth limitation sweep is ostensibly 525 lines for every full edge, with two interweaved 1/60th-second fields consolidating to make a whole picture. Just around 480 of the sweep lines are usable; with the rest of during vertical remember. Pixel is the most alluring to have the dividing of the pixel esteems be the equivalent in the event and vertical headings (i.e., square pixels), as this disentangles many handling and estimation tasks. Dim or Grayscale goals in which the picture fast blaze simple to-computerized converters more often than not gauge every voltage perusing to create an 8-bit number from 0 to 255. This full range may not be utilized for a real picture, which may not differ from full dark to white. Noise Images in which the pixel esteems change inside districts that are in a perfect world uniform in the first scene can emerge either due to restricted tallying insights for photons or different sign, misfortunes presented in the moving of electrons inside the chip, or because of electronic clamor in the speakers or cabling. High depth images for different gadgets that produce informational indexes that are regularly treated as pictures for survey and estimation produce information with a lot more prominent range than a camera. Color imaging in most true pictures are shading as opposed to monochrome. The light magnifying instrument produces shading pictures, and numerous organic example planning procedures utilize shading to distinguish structure or restrict synthetic movement in novel ways. Notwithstanding for inorganic materials, the utilization of spellbound light or surface oxidation produces shading pictures to depict structure. Digital camera limitations current improvement of the computerized still-outline cameras for the buyer market includes for all intents and purposes each customary creator of film cameras and photographic supplies. Color space transformation from RGB (the brilliance of the individual red, green, and blue sign, as caught by the camera and put away in the PC). Color correction of pictures are obtained under various lighting conditions, the shading esteems recorded are influenced. Human vision is tolerant of extensive variety in lighting, evidently utilizing the outskirts of the review field to standardize the shading understanding, or by accepting known hues for some perceived articles. Color displays littler phosphor dabs, a higher recurrence examine and higher data transmission intensifiers, and a solitary dynamic output (instead of join) produce a lot more prominent sharpness and shading immaculateness. Image types in the ordinary pictures with which people are outwardly encountered; the brilliance of a point is an element of the splendor and area of the light source joined with the direction and nature of the surface being seen. These "surface" or "genuine world" pictures are frequently hard to decipher utilizing PC calculations, in view of their three-dimensional nature, the communication of the enlightenment with the different surfaces, and the way that a few surfaces may darken others. Image go, stereoscopy promotion picture prerequisites same sort infers that splendor and shading esteems ought to be the equivalent and, thusly, that brightening must be uniform and stable for pictures procured at various occasions.

With Grassman's laws currently determined, thought is given to the advancement of a quantitative hypothesis for shading matching. It ought to be obvious that there is no principal hypothetical contrast between shading coordinating by an added substance or a subtractive framework. In a subtractive framework, the yellow color goes about as a variable safeguard of blue light, and with perfect colors, the yellow color successfully shapes a blue essential light. Along these lines, the red channel in perfect world structures the green essential, and the cyan channel in a perfect world structures the red essential. Subtractive shading frameworks conventionally use cyan, maroon, and yellow color ghostly channels instead of red, green, and blue color channels on the grounds that the cyan, red, and yellow channels are score channels which license a more noteworthy transmission of light vitality than do narrowband red, green, and blue band pass channels. In shading printing, a fourth channel layer of variable dark dimension thickness is regularly acquainted with accomplish a higher difference in the generation since basic colors don't have a wide thickness run.

3.4   Image Recognition

Machine vision is one of the most up to date strategies in mechanized pupae quality; this system comprises of two phases including isolating the picture of pupae from the foundation and after that looking at the nearness of a deformity and sizes in the pupae picture.

3.4.1   Butterflies and Larval Host Plants Classification

The use of apps enables a researcher to select and recognize butterflies, and image fruits or larval host plants from sources of information like video and still camera pictures. There are some formula tools and techniques that can be used across effectively in computer vision for object detection and segmentation. Butterflies and larval host plants can be classified in every pixel into one of several possible categories. This means, all pixels bearing butterflies would be classified into a single category, so are pixels with grass and larval host plants. More importantly, the output doesn't distinguish between two different objects shown in Figs. 8, 9 and 10.

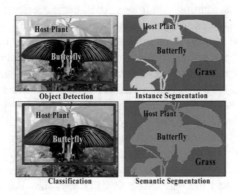

Fig. 8. Computer vision—object detection and segmentation

Fig. 9. Butterflies classification

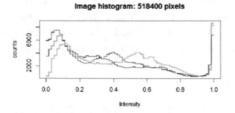

Fig. 10. Butterfly histogram

3.4.2 Pupae Defects Detection

Pupae Defects Detection [14] focus of agriculture system is identification of the various pupae defects and provide their respective solutions (Fig. 11). In the future, user interaction should be provided with photo scan analysis technology to identify defects of pupae with the use of the proposed computer-based intelligence projects to choose and recognize objects from information sources like video and still camera pictures. The size of the pupae differs according to larvae variety, rearing season and harvesting conditions. The pupae use multiscaled convolutional layers with various kernels: 3×3, 5×5, 7×7, 11×11 took the qualities of pupae of filter ($K = 1 \times 1$).

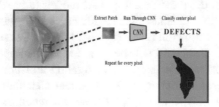

Fig. 11. Pupae defects detection

3.4.3 Image Fruits and Larval Host Plant Leaves Classification

Image Fruit or larval host plant leaves classification characterization by 13-layer profound convolutional neural network and information expansion and has a lowest latency 22 ms, 557 KB edge detection (Fig. 12). They utilized [24] a managed Deep Neural Network (DNN) as an incredible pixel classifier. DNN is a maximum pooling (MP) convolutional neural network (CNN). It legitimately works on crude RGB information examined from a square fix of the source picture, focused on the pixel itself. The DNN is prepared to separate patches with an image host plants fruit near the inside from every single other window trained model in vision edge image with bounding boxes.

Fig. 12. Image fruits and larval host plant leaves classification

Larval host plants in concealed pictures are recognized by applying the classifier on a sliding window, and postprocessing it yields with basic systems. Leaves were [13] filtered at 300-DPI goals, 24-bit shading with strong white foundation in lossless pressure position PNG. The utilized scanners: Epson Perfection V331, Mustek ScanExpress A3 USB 2400 Pro2 and Hewlett Packard scanjet 3500c3.

Since the DNN works on crude pixel esteems, no species information is required: despite what might be expected, the DNN naturally learns a lot of visual highlights from the preparation information. DNN strategies [25] that empower proficient preparing of DNNs to improve vitality effectiveness [27], random cropping of rescaled images together with random horizontal flipping and random RGB colour and brightness shifts and throughput without yielding application precision or expanding equipment cost are basic to the wide sending of DNNs in AI frameworks.

3.5   Association Rule (Apriori Calculation)

Association mining includes the utilization of AI models to break down information for examples, or co-event, in a database. It recognizes visit on the off chance those affiliations, which are called affiliation rules. Figure 13 shows Consistent vs. Seasonal. The sold things show in the dashboard will be seen and utilized apriori model to the most regularly sold things utilizing prescient investigation and prescribe the items to the purchaser utilizing Incessant Example Development Calculation (to evade pointless blends) and can take a picture to tags and posts for recommending where they can reserve and buy the products by Fig. 14, Apriori Algorithm Association Rules such as Support, Confidence, and Lift.

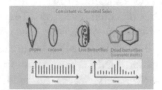

Fig. 13. Consistent vs. Seasonal

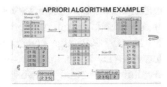

Fig. 14. Apriori algorithm

Association rules are made via looking information for incessant on the off chance that examples and utilizing the criteria backing and certainty to distinguish the most significant connections. Support means that how much of the time the items show up in the data. Confidence number of times the if-then statements are found true. A third measurement, called lift, can be utilized to contrast certainty and anticipated certainty.

Association rules are determined from itemsets, which are comprised of at least two things. On the off chance that guidelines are worked from dissecting all the conceivable itemsets, there could be such a large number of standards that the principles hold small importance. With that Association rules are ordinarily made from principles well-spoken to in information.

3.5.1   Butterflies Dataset

Images are gathered in the community through a cell phone device via smart mobile with high resolution required. The pupae defects were collected from the farmers, and breeders to take shots of their defects issue and upload it in the cloud server as a Google developer.

This dataset is much the same as the Butterfly Dataset, aside from it has 54 classes containing 1000 pictures each. There are 1000 preparing pictures and 100 testing pictures for each class. The 18 classes in the Butterfly Dataset are gathered into 18 super classes. Each picture accompanies a "fine" mark (the class to which it has a place) and a "coarse" name (the super class to which it has a place).

3.5.2   Butterfly Management

Larval host plants leaves are placed in a container of Water for the female butterflies to lay eggs. The aggregate of eggs over the pupae is recorded to ascertain their level of effective pupae to turn into a butterfly. The usual dates were recorded by every species with the number of females that laid eggs and stored inside the safe container. Butterfly eggs, larvae, pupae and adult butterfly defects are monitored and listed. The hatchlings larvae will be kept inside their cage net for sleeving for daily feeding larval host plants leaves until they become pupae and then look for buyers or bring to the market to be sold by retail traders.

4   Results and Discussions

The proposed system is designed and simulated for an environmental condition of processing and aligning an object tracking system with an intelligent detection and monitoring approach for smarter notifications. From the development, the researcher has reached the following decisions as discussed in this chapter.

$$y' = M(x1, x2, x3, \ldots)$$

where:

    y' is the label learning system (compared to the ground truth y)
    x1, x2, x3 ... are input features read by the machine learning system
    M is the model, an algorithm that makes predictions based on the input data.

Figure 15 shows Transfer Learning VGG16 Convolutional Nets. One of those libraries is Keras, and it makes building neural networks an absolute breeze. Keras is an abstraction layer that sits on top of a low-level neural network library like TensorFlow, and it has lots of nice functions to create network layers, validation functions, and training engines. It was also found out that Keras can also run on top of CNTK, Microsoft's awesome deep learning library. However it was also found out that Keras is a Python-only library, which means you cannot call it from C# code.

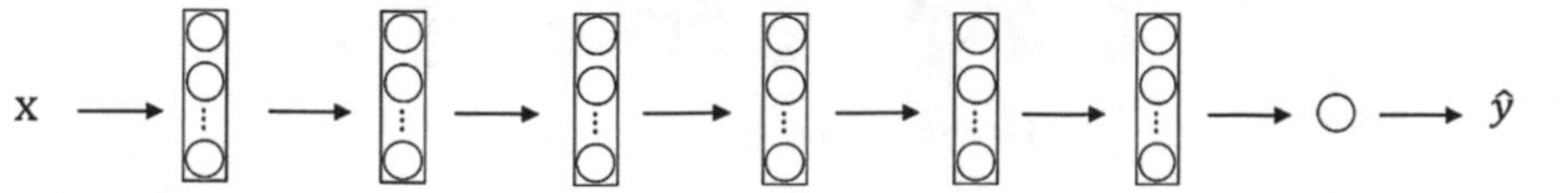

Fig. 15. Transfer learning VGG16 convolutional nets

4.1 AutoML Vision

Figure 16 shows AutoML Vision Model. AutoML Vision is a suite of AI items that empowers designers with constrained AI skill to prepare top-notch models explicit to their business needs. It depends on Google's best in class exchange learning and neural engineering look innovation. The method for the improvement of butterflies, larvae and pupae defects detection strategy can be grouped into four phases: labeling and organizing data, training, evaluating, predicting or testing and approval

Fig. 16. AutoML vision model

4.1.1 Labeled and Organized Data

Figure 17 shows image of super class surveyed mark no fewer than 18000 pictures of butterfly and pupae with the superclass of 18 kinds of pictures every 1000 species transferred to prepare as a model. The model would empower to create and would be manufactured examples of butterflies, engineered tests of pupae and hatchlings. It utilized the Google Cloud Platform for data status requires mass amassing capacity is an outright need in picture taking care of utilizations. An image of size 1024*1024 pixels, where the intensity of each pixel is a 8-bit sum, requires one megabyte of additional room if the image isn't compacted. The researcher went to investigate the image fruits, and butterflies in Marinduque gathered pictures as Datasets for information readiness. After getting all the pictures of image fruits and butterflies, larvae and pupae defects it uploaded to the cloud platform.

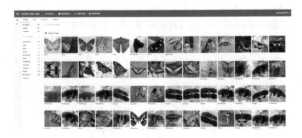

Fig. 17. Image of super class

4.1.2   Trained Data

On February 27, 2019, 11:05:47 AM initial training took place for 15 min or more. On average the testing time was less than one second per image. A total of 10 experiments were conducted. Several parameters were varied between experiments in order to investigate their influence on the performance of the model namely pupae, butterfly species and larval host plants.

4.1.3   Evaluation

Picture shows being [5] utilized today are predominantly concealing (in a perfect world level screen) TV screens. Screens are driven by the yields of pictures and structures show pictures that are an indispensable bit of the PC system. Just every once in a while are there essentials for picture show applications that can't be met by introduction picture open monetarily as an element of the PC system.

A pixel p at directions (x, y) has four flat and vertical neighbors whose facilitates are given by $(x + 1, y)$, $(x - 1, y)$, $(x, y + 1)$, $(x, y - 1)$. This arrangement of pixels, called the 4-neighbors of p, is signified by N4 (p). Every pixel is a unit remove from (x, y), and a portion of the neighbors of p lie outside the computerized picture if (x, y) is on the outskirts of the picture.

The four corner to corner neighbors of p have organizes $(x + 1, y + 1)$, $(x + 1, y - 1)$, $(x - 1, y + 1)$, $(x - 1, y - 1)$ and are meant by ND (p). These focuses, together with the 4-neighbors, are known as the 8-neighbors of p, indicated by N8 (p). As previously mentioned, a portion of the focuses in ND (p) and N8 (p) fall outside the picture if (x, y) is on the fringe of the picture.

Evaluation understanding would go in reverse to have certain powerful terms of the hubs and components of a butterfly in the portrayal of the scene as far as the hubs and the components (Fig. 18). To begin deciphering the picture and foundation without understanding there's anybody on the scene. And afterward complete a move, an interpretation on the chart, which includes making a hub structure for the picture and afterward clarifying that piece of the picture regarding the model.

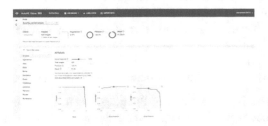

Fig. 18. Evaluation

The framework has the expectation of certainty or accuracy normal of 97.1% and a review of 91.3% of 429 pictures enlisted in the cloud appeared in the assessment. The images datasets are still kept updated and the images are labeled.

4.1.4   Prediction and Tested the Model

Predicting and Testing Model and approval in AutoML vision were [5] utilized when overseeing thousands, or even millions, of pictures, giving acceptable limit in an image getting ready system can be a test. Propelled limit with regards to picture getting ready applications falls into three principal classes: (1) transient amassing for use in the midst of taking care of, (2) web based storing for commonly speedy review, and (3) recorded limit, depicted by uncommon access. Limit is assessed in bytes (eight bits), Kbytes (one thousand bytes), Mbytes (one million bytes), Gbytes (which means Giga, or one billion, bytes), and Tbytes (which means Tera, or one trillion, bytes). One system for giving transient accumulating is PC memory. Another is by explicit sheets, called diagram pads, that store at any rate one pictures and can be gotten too rapidly, generally speaking at video rates (e.g., at 30 complete pictures for each second). The last strategy allows in every way that really matters prompt pictures zoom, similarly as material (vertical moves) and dish (even moves).

Its availability between pixels is a major idea that [5] rearranges the meaning of various advanced picture ideas, for example, locales and limits. To build up if two pixels are associated, it must be resolved on the off chance that they are neighbors and if their dim dimensions fulfill a predefined basis of closeness (state, if their dim dimensions are equivalent). For example, in a paired picture with qualities 0 and 1, two pixels might be 4-neighbors, yet they are said to be associated just on the off chance that they have a similar esteem.

Give V a chance to be the arrangement of dark dimension esteems used to characterize nearness. In a paired picture, $V = \{1\}$ on the off chance that we are alluding to the adjacency of pixels with esteem 1. In a grayscale picture, the thought is the equivalent, yet set V normally contains more components. For instance, in the nearness of pixels with a scope of conceivable dark dimension esteems 0 to 255, set V could be any subset of these 256 qualities. We think about three kinds of nearness:

(a) 4-adjacency. Two pixels p and q with qualities from V are 4-adjacency if q is in the set N4 (p).
(b) 8-adjacency. Two pixels p and q with qualities from V are 8-adjacency if q is in the set N8 (p).
(c) m-adjacency (blended adjacency). Two pixels p and q with qualities from V are m-adjacency if
(i) q is in N4 (p), or
(ii) q is in ND (p) and the set has no pixels whose qualities are from V.

Blended nearness is an alteration of 8-adjacency. It is acquainted with dispose of the ambiguities that regularly emerge when 8-adjacency is utilized. For instance, consider the pixel course of action appeared.

(a) for $V = \{1\}$. The three pixels at the highest point of Fig. 19(b) demonstrate various (vague) 8-adjacency, as shown by the dashed lines. This uncertainty is evacuated by utilizing m-adjacency, as appeared in Fig. 19(c). Two picture subsets S1 and S2 are nearby if some pixel in S1 is adjoining some pixel in S2. It is comprehended here and in the accompanying definitions that nearby methods 4-, 8-, or m-contiguous. An (advanced) way (or bend) from pixel p with directions (x, y) to pixel q with directions (s, t) is a succession of unmistakable pixels with directions

(x0,y0), (x1,y1),..,(xn,yn)
where (x0, y0) = (x, y) and (xn, yn) = (s, t) and pixels (xiyi) and (xi − 1, yi − 1) are contiguous for 1<=i<=n.

For this situation, n is the length of the way. On the off chance that (x0, y0) = (xn, yn), the way is a shut way. We can characterize 4-, 8-, or m-ways relying upon the sort of adjacency indicated. For instance, the ways appeared in Fig. 18(b) between the upper east and southeast focuses are 8-ways, and the way in Fig. 19(c) is an m-way. Note the nonattendance of uncertainty in the m-way. Give S a chance to speak to a subset of pixels in a picture. Two pixels p and q are said to be associated in S if there exists a way between them comprising totally of pixels in S. For any pixel p in S, the arrangement of pixels that are associated with it in S is known as an associated segment of S. On the off chance that it just has one associated part; at that point set S is known as an associated set.

Give R a chance to be a subset of pixels in a picture. We consider R an area of the picture if R is an associated set.

The limit (additionally called fringe or form) of a locale R is the arrangement of pixels in the district that have at least one neighbor that are not in R. On the off chance that R happens to be a whole picture (which we review is a rectangular arrangement of pixels), at that point its limit is characterized as the arrangement of pixels in the first and last lines and segments of the picture. This additional definition is required on the grounds that a picture has no neighbors past its outskirts. Ordinarily, when we allude to an area, we are alluding to a subset.

a b c

Fig. 19. Region of subset

The researcher used all activations, but reduces them by a factor p (to account for the missing activations during training). In Table 1, prediction results, the probabilities result shows the top classified species, image fruits or larval host plants, and its pupa.

Table 1. Prediction results

Scientific classification butterflies	Class of probabilities predictions of 5 labels (English & Local Name)	Scientific classification (Larval Host plants/Image Fruits)	Class of probabilities predictions of 5 labels (English & Local Name)
Graphium agamemnon	**Agamemnon 93.4** Doson 2.8 Demoleus 2.7 Kotzebuea 2.1 Bolina 2.0	*Annona muricata*	**Soursop 97.6** Citrus Citrus Maxima Basketbasket Lagaylay
Attacus atlas	**Atlas 97.5** Sylvia 3.8 Palinurus 1.7 Demoleus 0.8 Biblis 0.3	*Goniothalamus amuyon*	**Amoyong 99.7** Sugar Apple 0.2 Citrus 0.1 Lagaylay 0.1 Basketbasket 0
Hypolimnas bolina	**Bolina 93.6** Demoleus 4.4 Polytes 2.3 Palinurus 0.7 Rumansovia 0.6	*Ipomoea batatas*	**Camote Tops 98.4** Soursop 0.1 Basktebasket 0.1 Citrus 0.1 Lagaylay 0
Cetosa biblis	**Biblis 99.4** Palinurus 6.7 Sylvia 1.8 Kotzeboeua 1.0 Agamemnon 0.4	*Wattakaka/Tragia volubilis*	**Bagin 98.8** Lagaylay 1.7 Sugar Apple 0.6 Soursop 0.6 Camote Tops 0.4
Papilio demoleus	**Demoleous 85.8** Leuconoe 3.7 Palinurus 1.4 Biblis 1.0 Agamemnon 0.3	*Citrofortunella microcarpa*	**Calamansi 97.2** Soursop 6.6 Lagaylay 0.1 Basketbasket 0 Camote Tops 0
Papilio polytes	**Polytes 98.6** Kotzebuea 0.5 Leocone 0.5 Sylvia 0.2 Rumansovia 0.2	*Triphasia trifolia*	**Lemoncito 99.4** Tapias 1.3 Citrus 0.2 Lagaylay 0 CamoteTops 0
Idea leuconoe	**Leuconoe 92.8** Palinurus 1.1 Kotzeboeua 0.6 Demoleus 0.5 Atlas 0.2	*Parsonsia helicandra*	**Lagaylay 90.3** Citrus 0.8 Camote Tops 0.4 Basketbasket 0.1 Soursop 0

(*continued*)

Table 1. (*continued*)

Scientific classification butterflies	Class of probabilities predictions of 5 labels (English & Local Name)	Scientific classification (Larval Host plants/Image Fruits)	Class of probabilities predictions of 5 labels (English & Local Name)
Papilio palinurus	**Palinurus 90.9** Poleytes 3.0 Bolina 2.3 Rumansovia 1.4 Kotzeboeua 1.3	*Genus euodia*	**Tapias 97.5** Citrus 4.7 Lemoncito 1.4 Calamansi 0.3 Basketbasket 0.3
Troides rhadamantus	**Rhadamantus 64** Kotzeboeua 41 Leuconoe 0.9 Demoleus 0.6 Sylvia 0.6	*Aristolochia tagala*	**Basket Basket 94.2** Citrus 1.0 Lagaylay 0.1 Soursop 0 Camote Tops 0
Papilio rumansovia	**Rumansovia 99.8** Kotzeboeua 0.7 Palinurus 0.1 Leuconoe 0 Polytes 0	*Citrus maxima*	**Citrus Maxima 99.2** Citrus 1.5 Basketbasket 0.5 Lemoncito 0.2 Soursop 0.2

Figure 20 shows predicted result of the classification of moths and butterflies.

Butterfly   Moth

Fig. 20. Predicted result of classification

Defects pupa shows the pupae have followed its deformity (Fig. 21).

Over bend   Stretch mark   Stretch mark and over bend

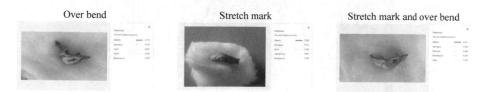

Fig. 21. Defects of pupa

Larval Host Plants shows the superclass of picture products of larval host plants as per its characterization (Fig. 22).

Fig. 22. Larval host plants

5 Apriori Algorithms Results and Calculation Sample

Mine all frequent items set considering the dataset in Table 2.

Table 2. Sample plotting and formula of computation of sold items

Buyers	Sold items	Items1	Items2	Items3	Items4	Items5
Buyer 1	1,2,5	1	1			1
Buyer 2	2,		1			
Buyer 3	2,3		1	1		
Buyer 4	1,2,4	1	1		1	
Buyer 5	1,3	1		1		
Buyer 6	2,3		1	1		
Buyer 7	1,3	1		1		
Buyer 8	1,3	1		1		
Buyer 9	1,2,3,5	1	1	1		1
Buyer 10	1,2,3	1	1	1		

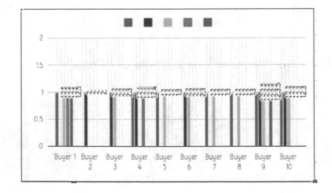

Fig. 23. Graph of buyers who bought the products

Graph of Buyers shows who bought the products which are itemized on the chart of purchaser's activities from the months they sold the items (Fig. 23).

$$\textbf{Support}(\textbf{X}) = \frac{\text{Number of transactions in which X appears}}{\text{Total Number of transactions}}$$

The Support (items 1 -> items 2)
Support = 4/7 = 57%

The Confidence (items 1, items 2) -> items 3
Confidence = 2/3 = 75%

The Lift

A lift value greater than one means that item Y is likely to be bought if item X is bought, while a value less than 1 means that item Y is likely to be bought if item X is bought.

A lift value is less than 1{item1->item2} = (4/10)/{(7/10)*(7/10} = .816

Figure 24 demonstrates the bar chart of the species was sold during the month and is shown in like manner by its shading (August–November).

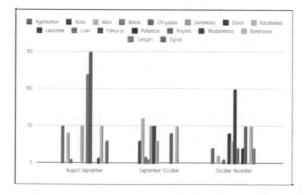

Fig. 24. Graph of species of butterflies sold in months

Figure 25 shows the Classification of Butterfly species. The factors are recorded, for example, SCid, kingdom, phylum, class, order, genus, family, scientific name, common name, local name, country of origin, height (wings), width (wingspan), length (body), width (body) colors, larval host plants, etc. Also larval host plants classification such as SCid, kingdom, phylum, class, order, genus, family, scientific name, common name, local name, and country of origin.

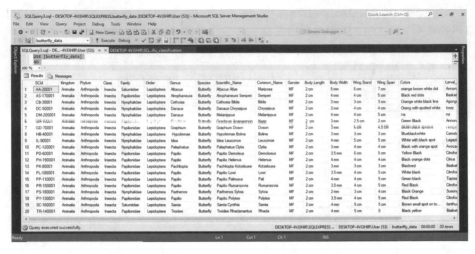

Fig. 25. Database of butterflies

Table 3 shows the typical getting of eggs from the leaves and parts of the hosplants are accumulated in the compartment or glass with spread. The test demonstrates the quantity of eggs of one of the animal categories.

Table 3. Gathering of eggs

In Container	Hanged	Creepy	Species	Eggs	Larvae %
5/20/19	6/20/19	330	Papilio rumanzovia	Yes	84.15
5/21/19	6/21/19	250	Papilio rumanzovia	Yes	90%

Table 4 shows total of pupae blemishes analyzed using the propelled device to pick the great characteristics.

Table 4. Pupae with good qualities to sell to the market of buyers

Hanged	Become Pupae	Attached	Species	Pupae	Pupae %
6/20/19	6/21/19	270	Papilio rumanzovia	Yes	95.98%
6/21/19	6/22/19	200	Papilio rumanzovia	Yes	80%

Figure 26 shows the great nature of pupae that will be traded to the purchasers and the gatherers of other nations.

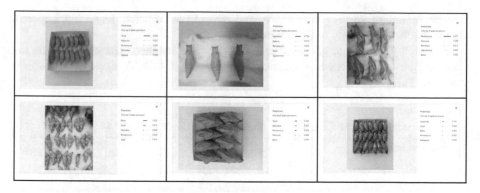

Fig. 26. Pupae items online

Figure 27 shows the things that are made by the plan in surrounded, packaged, and clock of dried butterflies that will be sold locally for informational purposes.

Fig. 27. Customized and crafted items online

6   Business Analysis

The researcher calculated and formulated a contingency table of Eggs possible for larvae or caterpillar, Larvae possible for parasites/virus exposure and virus disease, and Pupae possible for butterflies using chi-squared test.

$$\text{Xsq} = \sum (\text{observed frequency} - \text{expected frequency}) \text{squared over expected frequency}$$

The probability of the butterfly species was compared from the captive butterflies to the wild butterflies which have better qualities of pupae from domesticated female butterflies. The good qualities of pupae also depend on the food plant quality of leaves.

But the volume of keeping larvae alive and away from predators had a big impact to secure the maintainability of pupae orders.

Probability of eggs to get hatchlings creepy when ranched. The eggs will be free from being eaten by ants when breeders can take care of it inside the safe container as shown in Table 5.

Table 5. Eggs that is possible for larvae or caterpillars.

Papilio rumanzovia	Infertility rate		Total
	Yes	No	
Yes	120	10	130
No	150	50	200
Total	270	60	330

The proportion of Population Exposure of 0.39 and Not Expose of 0.60, Population of Fertility Exposure Yes is 105.3 and Exposure No is 162, Population of Infertility Exposure Yes is 23.4 and Exposure No is 36. During the study, it has an infertility rate of 15.85% if exposed outside the container and a reduced rate of 84.15% inside the safety container. The research provides evidence that the use of safety container for the eggs have higher fertile eggs.

Table 6 shows the sleeving hatchlings give a most regular habitat, crisp sustenance constantly accessible, helpful and efficient.

Table 6. Larvae possible for parasites/virus

Papilio rumanzovia	Parasitic virus		Total
	Yes	No	
Yes	20	58	78
No	28	172	200
Total	48	230	278

Table 7 shows the proportion of Population Exposed of 0.17 and Proportion of Population Not Exposed of 0.83, Population of affected by Parasitic Virus Exposure Yes is 3.4 and Exposure No 23.24, Population not affected by Parasitic Virus Exposure Yes 9.86 and Exposure No 207.5. Sleeving in containers are effective for parasite-free and healthy rearing rate of 95.98% than traditional sleeving by nettings rate of 4.02%. Observations were successful to increase income for breeders depends on the sensitivity of species.

Table 7. Pupae possible for butterflies

Papilio rumanzovia	Pupae of good qualities		Total
	Yes	No	
Yes	240	2	242
No	10	18	28
Total	250	20	270

The proportion of Population Good Qualities of .90 and Proportion of Population Defects of.10, Population of Good Qualities of Pupae: 242, Good Quality Yes is 216 and Good Quality No 1, Population of Defective Pupae: 28, Good Quality Yes is 1.8 and Good Quality No 1.8. The pupae have been detected with 85% good qualities possible to adult butterflies.

Figure 28 shows the difference between the qualities and defects of pupae. The defect pupae shown below have stretch body, have deformities, wounded, have parasites or virus inside the body of pupae and over bend body figure unlike the good nature of pupae has unblemished skin and live indicates the good qualities of pupae.

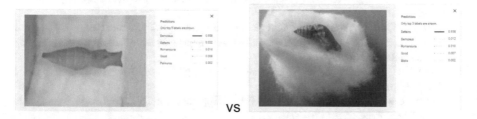

VS

Fig. 28. Good quality vs defective pupae classification

7 Conclusion

The study was to provide computerized instruments incorporated product search information to the online business website of Insects Connections with the utilization of the CNN and Apriori algorithm, hence including machine vision, and sensor combination in items to robotize the retail exchange buy installment strategies. And to integrate with ML Kit, a mobile SDK that makes it easy to apply Google's machine learning technology to Android and iOS apps in a powerful yet easy-to-use package and the integration of the Cloud Vision API of Corvid Node Runtime in Wix Media Web Apps.

The study recommends for further research in the field of future technology particularly in machine learning such as computer vision and object detection. Also, the following recommendations were drawn based on the findings of the study. To the future researchers, work extracts more information from butterflies and other animals for agribusiness in cutting-edge technology. DNN as of now broadly utilized for some AI applications, including computer vision, speech recognition, and robotics, and are frequently conveying superior to human accuracy. Object detection can be hundreds of times slower than image classification, and therefore, in applications where the location of the object in the image is not important, use image classification. To develop an AI system that predicts the captive site of capturing visited butterflies an example of the growing use of workplace surveillance. The system recommends use of a deep-learning algorithm trained on waterways, mountains and fields site images, and captive records. It can then be put to work monitoring, a new captive site and the use of Deep Learning Containers locally and on Google's Kubernetes Engine, the Compute Engine, or in the AI Platform Training.

References

1. Ayatullah, F.M., Nabamita, M., Subhadip, B., Mita, N.: Design of an Optical Character Recognition System for Camera-based Handheld Devices (2011)
2. Venkata Rao, N., Sastry, A.S.C.S., Chakravarthy, A.S.N., Chakravarthy, P.K.: Optical Character Recognition Technique Algorithms (2016)
3. Vamvakas, G., Gatos, B., Stamatopoulos, N., Perantonis, S.J.: A Complete Optical Character Recognition Methodology for Historical Documents (2008)
4. Russ, J.: The Image Processing Handbook. CRC Press, Boca Raton (2011). https://doi.org/10.1201/b10720
5. Gupta, S.: Optical character recognition. Int. J. Eng. Innov. Technol. (IJEIT) 2(7), January 2013. ISO 9001:2008 Certified. https://www.termpaperwarehouse.com/essay-on/Optical-Character-Recognition/185823
6. Obeid, R.: CAMTUAL: Robust Real-time Mobile App 2D-To-3D & VR Video Conversion Using Deep3D CNN (2014)
7. Umamaheswari, J., Radhaman, G.: A Hybrid Approach for DICOM Image Segmentation Using Fuzzy Techniques (2012)
8. Yousefi, S., Ghabel, S.N.: Survey on Object Detection Methods in Visual Sensor Networks (2016)
9. Razmjooy, N., Estrella, V., Loschi, H.: A Survey of Potatoes Image Segmentation based on Machine Vision (2019)
10. Liu, H., Xiong, R., Zhang, J., Gao, W.: Image Denoising via Adaptive Soft-Thresholding Based on Non-Local Samples (2015)
11. Gu, Q., Yang, J., Wei, Q.Y., Reinhard, K.: Integrated Multi-scale Event Verification in an Augmented Foreground Motion Space (2018)
12. Pratt, W.: Digital image processing. Color matching. Photometry and colorimetry. PIKS Inside, 3rd edn. p. 49 (2001)
13. Novotny, P., Suk, T.: Leaf Recognition of Woody Species in Central Europe (2013)
14. Gadade, H.D., Maske, P., Kinge, S.: A Survey on Agriculture Monitoring and Disease Identification System (2019)
15. Sheikh, L.M., Hassan, I., Sheikh, N.Z., Bashir, R.A., Khan, S.A., Khan, S.S.: An Adaptive Multi-Thresholding Technique for Binarization of Color Images (2005)
16. Zhang, X., Zhao, J., LeCun, Y.: Character-level Convolutional Networks for Text Classification (2015)

17. Abhishek, M., Suresh, R., Ahme, S.T.: Machine Vision Optimization Using Multi-view Support Vector Machine (SVM) Technique (2016)
18. Barend, W., Vaart, V., Lambers, K.: Learning to Look at LiDAR: The Use of R-CNN in the Automated Detection of Archaeological Objects in LiDAR Data from the Netherlands (2019)
19. Colello, A.: Image Blob Detection: A Machine Learning Approach (2016)
20. Vaddadi, V.R.: Association Rule Mining for Identifying Optimal Customers Using MAA Algorithm (2017)
21. Wang, J., Markert, K., Everingham, M.: Learning Models for Object Recognition from Natural Language Descriptions (2009)
22. Hashimoto, R., Uehara, Y.: 3 keys to smarter audience discovery with machine learning (2019). https://www.thinkwithgoogle.com/intl/en-apac/tools-resources/success-stories/flipkart-pioneers-digital-approach-become-india-ka-fashion-capital/
23. Szegedy, C., Liu, W., Jia, Y., Sermanet, P., Reed, S., Anguelov, D., Erhan, D., Vanhoucke, V., Rabinovich, A.: Going Deeper with Convolutions (2015)
24. Ciresan, D., Giusti, A., Gambardella, L., Schmidhuber, J.: Mitosis Detection in Breast Cancer Histology Images with Deep Neural Networks (2013)
25. Sze, V., Chen, Y.: Efficient Processing of Deep Neural Networks: A Tutorial and Survey (2017)
26. Parmar, R.: Detection and Segmentation through ConvNets (2018). https://towardsdatascience.com/detection-and-segmentation-through-convnets-47aa42de27ea
27. Le, J.: The 4 Convolutional Neural Network Models That Can Classify Your Fashion Images (2018). https://towardsdatascience.com/the-4-convolutional-neural-network-models-that-can-classify-your-fashion-images-9fe7f3e5399d
28. Akshay, S., Sajin, T., Prashanth, A.R.: Improved Multiple Object Detection and Tracking Using KF-OF Method (2016)

Understanding Dynamic Logic
Using 2D Images

John Promersberger[1], Sudhanshu Kumar Semwal[1(✉)], and Leonid Perlovsky[2]

[1] University of Colorado, Colorado Springs, CO, USA
{jpromers,ssemwal}@uccs.edu
[2] Northwestern University, Boston, MA, USA
lperl@rcn.edu

Abstract. To understand the working of Dynamic Logic (DL) we discuss a simple example considering 2D images and finding a Gaussian pattern in the image if it exist. The technique is also noise resistant and works well on noisy data can be which can be simulated by adding randomness to the image. In implementing a pattern in the image, we do need to know the characteristics of the model for example in our case we are looking for circular patterns and simple equations of circle provides a basis for developing an analytic measure. We provide our implementation for understanding the basics of Dynamic Logic (DL). The understanding included a determination of when to terminate the iteration process and a dependence between model parameters.

Keywords: Dynamic Logic · Similarity measures using spherical equation

1 Introduction

One major benefit of Dynamic Logic [1,4] is that it can be extended to multi-dimensional data set such as medical volume data set. Understanding it in 2D could be useful as a starting point for developing 3D implementation if future. The technique is also noise resistant and so works well on noisy data (i.e. finds the model with noise in it). Dynamic Logic is an iterative, unsupervised model-based learning process that maximizes learning by attempting to correlate an estimated model and an existing dataset to determine similarity. It uses Bayes Theorem to allow each individual element to contribute quantities that can be weighted based upon how much each model is capable of contributing to every single element. The first goal is to determine the similarity between each model by weighting the individual elements, N, of the model, or probability density function, so that we can determine the degree of likelihood calculated from our function. Since this is done every iteration, our goal is build off the previous iteration's values, so we can see if our confidence increases or decreases. This is done by Eq. 1. Dynamic Logic will sum every element in $f(m|n)$ dividing by the number of samples in N, allowing an overall confidence factor to be generated by Eq. 2.

© Springer Nature Switzerland AG 2020
K. Arai et al. (Eds.): FTC 2019, AISC 1070, pp. 162–175, 2020.
https://doi.org/10.1007/978-3-030-32523-7_11

Increasing the similarities will make the models more like the data, increasing our confidence. An important aspect of Dynamic Logic is matching vagueness or fuzziness of similarity measures to the uncertainty of models. Initially, each model's parameter values are unknown, and the uncertainty is high for models and so the data seems not similar. If the models estimates allow it to increase in confidence, the models parameters will optimize towards the actual values and the similarity measurements between the models and the data will increase. First order linear differential equations for a model are used within Eqs. 1 to 3. There really is no dt calculation performed, it is simply denoting an iterative step from one step to the next, the internal time of convergence [2].

$$f^{it+1}(m|n) = \left[\frac{r_m \ell (n|m)}{\sum\limits_{m \in M} r_{m^i} \ell (n|m^i)} \right]^{it} \tag{1}$$

$$r_m^{it+1} = \frac{1}{N} \sum_{m \in N} f(m|n) \tag{2}$$

$$S_m^{it+1} = S_m^{it} + dt \cdot \sum_{m \in N} \frac{abs(X(n))f(m|n) \left[\frac{\partial \ln \ell(n|m)}{\partial \mathbf{M}_m} \right] \partial \mathbf{M}_m}{\partial \mathbf{S}_m} \tag{3}$$

The majority of the program was implemented in Processing, available at [2]. Downloading the zip file that it is specific for the target operating system is all that is required. The zip file's content is a completely self-contained directory that contains everything that is required to run the project. Extract the contents to a desired location and then simply run the processing.exe executable program.

1.1 Model Selection

We started this project by trying to choose a simple shape that we could identify and quantify the shape's characteristics that has been embedded within an extremely simple image. A Gaussian function was recommended by Dr. Perlovsky because it had several benefits: It was a continuous function with values generated infinitely in every direction from the circle's center location. Since it is a continuous function, it does not have any discrete edges like a circle. If a model of a circle was to intersect with a circle within a simulated dataset, the model and the data would intersect at only two points, making it difficult to learn Dynamic Logic while simultaneously locating a discrete shape. Although possible, the point of this study was to learn how Dynamic Logic functioned. The Gaussian function is equal in every direction, so we wouldn't have to handle rotating of models to see if they lined up with our image. The Gaussian function also had the advantage that it contains the exponential function. This helps our system mathematically. Dynamic Logic can take advantage of the fact that in this case, knowing that data beyond the radius of our model will generate smaller values. We started by creating a simple image that contained our desired truth data, X_n, which was generated by using exponential function in Eq. 4. X_n is a two dimensional array containing three dimensional data, where the value at every (x, y) element is the value of the Eq. 4 calculation.

$$X_n = \frac{r}{2\pi s^2} \mathrm{e}^{-\frac{(x-x_0)^2+(y-y_0)^2}{s^2}}$$

(4)

Equation 4 contains several variables and constants:

r_0 parameter, height scaling, $0 \leq r_0 \leq 1'$
s parameter, radius of function
x_0 parameter, location of model center in X-axis
y_0 parameter, location of model center in Y-axis
x X-axis index into multi-dimensional array X_n, $0 \leq x \leq$ image width
y Y-axis index into multi-dimensional array X_n, $0 \leq y \leq$ image height

Processing language can easily render a correctly formatted matrix, so rendering the calculated model was relatively easy. Every element in X_n could very easily be scaled from a single float to a 1×3 element array within $[0, 255]$, consisting of identical [R, G, B] contributions of each color, generating a grayscale image seen in Fig. 1. This would make the array that is actually rendered a float [] [], which is re-calculated every iteration to determine what to render.

The use of the exponential function is useful in this project, and throughout Dynamic Logic also. When the value of the exponential term is less than one, the distance from some index (x, y) to (x_0, y_0) is less than the radius s. The opposite, when the value of the exponential term is greater than 1, the distance from index (x, y) to (x_0, y_0) is greater than the radius s. The further these distance beyond the radius value that has been defined, the less these indexes will contribute to the confidence of the truth feature that the data is associated with a specific model. Conversely, values within the radius, will contribute greater to the confidence as the model converges on the truth parameters that define the desired shape within the dataset. Also exponential functions are differentiable, necessary for DL algorithm,

$$X[x][y] = \frac{r}{2\pi s^2} \mathrm{e}^{-\frac{(x-x_0)^2+(y-y_0)^2}{s^2}}, \text{ where } 0 \leq x \leq \text{width}, 0 \leq y \leq \text{height}$$

(5)

Since it is an actual image, it is naturally generated and maintained as a matrix in Eq. 5 and Processing language snippet of executable code is in [1, 3, 5]. To add some variation and realistic modifications to the image that we would be using to test the abilities of Dynamic Logic, the X_n image containing the Gaussian function was allowed to expand beyond the radius, but was truncated in the shape of a square by using the additional conditional in the Processing snippet [3]. Resulting function is shown on the left and right is the image X(n) created in which we are trying to find the Gaussian function. It is very simple case for DL yet it lets us show the working of Dynamic Logic using a simple example.

Since Dynamic Logic is built on model based learning, to begin to understand the critical idea of confidence, we need to have more than a single model within

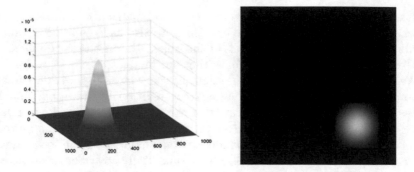

Fig. 1. Gaussian function (left) and X_n, Truth Data (right)

our image. The truth image in Fig. 1 contains three models, the first two being a three dimensional Gaussian function and the last is a normal distribution. We went above and beyond the project proposal by identifying two Gaussian models simultaneously. We have already described how the Gaussian function has been defined and created. To make portions of the image contain pixels that represent a normal distribution, we implemented the straightforward Eq. 6. The third component that has been added to the truth image is a random background noise component. Noise was injected by randomly selecting 10,000 indexes of the 1,000,000 that construct the image. Adding noise will alter the confidence factors by increasing or decreasing the summation of indexes that are intersected between the model estimation and the truth data, X_n.

$$X_n\,[x]\,[y] = \frac{1}{\text{image width} \cdot \text{image height}} \tag{6}$$

Listing 1.1. Normal Distribution

```
int N = width * height;
for (int y = 0; y < height; ++y)
{
        for (int x = 0; x < width; ++x)
        {
                background[x][y] = 1. / N;
        }
}
```

1.2   Initial Estimates

The first step that needs to be taken is to determine which models will be searched for in a particular dataset or image. If we can create a model with estimated parameters and search for the defining parameters, and a strategy for moving the model in response to the image, then any model can be defined.

Since we know that we have defined two Gaussian distribution, that is the first model, M_1 and M_2, that we will define. Since we also have embedded a normal distribution into the remaining pixels of the image, that is the third model, M_3, that we will construct and search for within our image.

We can use a two model that are estimating our Gaussian function and allow Dynamic Logic to optimize each parameter, iteration by iteration, hopefully increasing the confidence and similarity of the mathematical model in relation to the data. We will allow the model's estimated x, y position to be moved by the data, hopefully allowing the confidence and similarity to the image data to increase. Another approach is to also create potentially any number of Gaussian models, each model having been built upon a different radius, $(0, \mathbb{R}]$, and as some models move towards the location of the Gaussian shape within the truth image, the models that are built with a radius that matches the image should have the greatest level of confidence and similarity to the truth image. We need to assign an initial estimate for each parameter in each model that we will be processing. The actual values that X_n was constructed from and the initial parameter estimates for our model that is based on a Gaussian function are defined in Table 1.

By embedding multiple models within our image, X_n, we will be able test Dynamic Logic's ability to use Bayesian inference to determine the similarity between a portion of the dataset and each estimated model. It is an attempt to discern between the characteristics of whatever region an estimated model is currently intersecting to determine which model the data is most similar to. We will be able to see that the our estimated model will simply center itself on the data that it overlaps and calculate how similar the intersected data is to the defining model. Playing with various initial parameters and simulating noise by increasing and decreasing the height of the normal distribution showed that the Gaussian model might even overlap with the Gaussian function within X_n, but resolve to a confidence of 0.

Table 1. Model values and initial estimates

Parameter	Actual Gaussian 1	Actual Gaussian 2	Estimate Gaussian 1	Estimate Gaussian 2	Actual Normal
x_0	250	750	350	760	–
y_0	750	250	675	405	–
s_0	100	100	150	150	–
r_0	.8	.8	.2	.2	$\frac{1}{N}$
r_m	–	–	.5	.5	.5

Along with the required parameters to define the location and shape of a model, we need to track the degree of similarity for every model. For sake of ease, each model is assigned initially to have a likelihood of $\frac{1}{m}$, where $m = 2$, the number of models that we have generated and are attempting to match against our data. By assigning r_m, the models confidence, an initial value of $\frac{1}{m}$, or 50%, we can watch as our likelihood can go up or down, depending upon the location of the estimate. If we had decided to use 100 estimation models, the logical approach would be to begin by assigning each model a likelihood of 1% and as iterations are performed, it is possible that several models may have increased confidence while many models would decrease to 0. Since the image that was generated contains only two models, we can safely say that when processing the image, the subsection will either be similar to one of the two models. As understanding of Dynamic Logic increases and begin to potentially process real imagery, it is guaranteed that there will be data that does not fit into any model. This makes sense, requiring the "dust bin" model that Perlovsky recommends, as it will sweep up any remaining unclaimed confidence. It is likely that any data that is considered random, could fall into this category, ensuring that we have not achieved any degree of confidence in our other models.

Now that every estimated model has an initial value for every parameter, we can then calculate the confidence for our model M_n which is performed every iteration to determine how similar our estimation is to what truly exists. To do this we use Eq. 7 and see the demonstration in Listing 1.2.

$$f^{it}(m|n) = \left[\frac{r_m \ell\,(n|m)}{\displaystyle\sum_{m \in M} r_m^i \ell\,(n|m^i)} \right] \tag{7}$$

1.3   Iterative Parameter Calculation

We begin my taking every model, using the initially estimated parameters to create a matrix containing our prediction of what data would look like. For every index, the similarity is calculated as an array for every model, it's calculated contribution divided by the sum of the contribution of each index across all models. This is seen in Eq. 7. Equation 7 shows both models, to the left, the percentage of the contribution of the Gaussian function by index, to the right, the percentage of the contribution from the normal distribution by index. This is seen in Listing 1.2 in Matlab that was used to generate Fig. 2. This value must be between [0, 1]. We didn't think through what the figure for the normal distribution very well prior to my first plot generation. It was distinctly evident immediately following that as the Gaussian function climbs, the normal distribution decreases in relation, making the curve look like a bowling ball sitting on a trampoline.

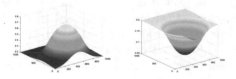

Fig. 2. $f(m|n)$, Similarity Calculation M_1 (left) and M_2 (right)

Listing 1.2. Confidence Calculation

```
%Define parameters
width = 1000; height = 1000;
%Create model of normal distribution
bg = (1 / (width * height)) * ones(width, height);
%Gaussian Model parameters
x0 = 350; y0 = 500; s0 = 250; r1 = .8;
%Calculate Gaussian Model
scale = r1 / (2 * pi * s0^2);
for (y = 1 : height)  for (x = 1 : width)
   pdf(x, y) = scale * exp(-((x - x0)^2 + (y - y0)^2) / s0^2);
   end end
%Calculate similarity
for (y = 1 : height) for (x = 1 : width)
         fm1n(x,y) = r1*pdf(x,y) / (pdf(x,y) + bg(x,y));
         fm2n(x,y) = (1 - r1) * bg(x,y) / (pdf(x,y) + bg(x,y));
end end
%Generate figures
```

Now that we know how much each model contributes to every index in relation to all the other models, we can calculate an actual confidence factor that will determine how well over all the data, X_n, compares to each model. These confidences for each iteration are defined as r_m, where $m \in M$, our overall collection of models.

$$r_m^{it+1} = \frac{1}{N} \sum_{n \in N} abs\left(X(n)\right) f(m|n) \tag{8}$$

1.4   Data Intersection

When an estimated model does not intersect with any portion of the similar data, the model will basically flounder. For instance, if we are analyzing model M_1, the Gaussian function, for an iteration, but the model does not intersect the embedded portion of the same function and only intersects with the data constructed from the Normal distribution, M_2, it will determine that it is not similar to any portion of the intersecting data, forcing the likelihood of the model to a confidence value of 0% so that we know that we are not seeing anything that resembles the Gaussian function. Since the values are complementary, we can see that the confidence of M_2, the normal distribution, increases to 1, or 100% confidence that the M_2 model is truly intersecting with a Normal distribution. This is clearly visible by Fig. 3. Amazingly, it only takes a single iteration to immediately recognize that there is no sign of an intersection with the circle. We anticipated an iteration or two for the confidence level to take hold and grow to 1, but it only took one. Confidence in data may grow slowly, but it is even more quick to recognize when the model does not match the intersected data. In some sense then this makes use of DL for interactive applications, because person-in-the-loop can define start of a model interactively which overlaps with the region of interest somewhat as a starting point to fit that model to the data.

Dynamic Logic calculates the confidence by essentially performing a mathematical 'and' operation, performed in Eq. 8. Each index in the dataset, $X_n[][]$, is multiplied by the $f[][]$ that is calculated for each model. The r_m is then simply scaled by dividing r_m by the sum of all the values calculated for all models. If we had been lucky enough to absolutely guess our initial estimates correctly, we could declare victory if the confidence was calculated to 100%, or high enough to meet our requirements. If our confidence was too low, we need to let the data move our models. Dynamic Logic will sum the values based on the contribution of each pixel to each model based upon f. Values where $f(m)$ do not intersect with the desired data are very small values. The real contribution is when the model is modified to intersect with as much of the same data as possible. The greater the intersection, the more that confidence will increase for each iteration and following.

To find the optimal direction and amount to move for each iteration, we need to solve for $\frac{dx_0}{dt}$ and $\frac{dy_0}{dt}$ which is:

$$\frac{dx_0}{dt} = \frac{2(x - x_0)}{s^2} \tag{9}$$

Movement in the Y-axis for each time step iteration dt is comparable to $\frac{dx}{dt}$. We can take the derivative of our model's mathematical definition to determine the optimal direction and amount to move in the Y-axis.

$$\frac{dy_0}{dt} = \frac{2(y - y_0)}{s^2} \tag{10}$$

The goal is to calculate sum of the values from each index of the dataset that intersect with our estimated model. By using the method shown in Listing 1.3, we are able to calculate the difference from x, y location from the previous estimated location. This difference must then be applied to our current model, which becomes are new estimated location.

Listing 1.3. S_x, S_y Calculation

```
// Calculate the new estimate
double x_iteration = 0;
double y_iteration = 0;
//calculate iteration values
for (int y = 0; y < height; ++y)
{
for (int x = 0; x < width; ++x)
{
double val = abs((float) Ximage[x][y]) * fmln[x][y];
x_iteration += val * 2. * (x - x0) / pow((float)s0, 2);
y_iteration += val * 2. * (y - y0) / pow((float)s0, 2);
}
}
// Update our location
x0 += x_iteration;
y0 += y_iteration;
```

Values from the intersecting indexes to the left of x_0 (which is constant for each iteration) generates negative values and positive numbers are generated by the values from the indexes to the right of x_0. Values in the column of our current x_0 make no contribution in moving the model estimate in the X axis. Likewise, values above y_0 (which is also constant for each iteration) generates negative values and positive numbers are generated by the indexes below y_0. Values in the row of our current y_0 make no contribution in moving the model estimate in the Y axis. By using Eqs. 9 and 10, it is possible to determine the optimal distance in each direction that each parameter x_0 and y_0 (center of estimate) should step to make the sum of the $[x]$, $[y]$ elements left of the model center equal to the intersected values right of center.

As the model and the target shape within the dataset begin to intersect, the value is at its largest, either negative or positive due to the lone contribution in that direction with no counterbalance. This generates larger jumps as the estimate moves towards the location of the truth dataset when the numerators will generate their largest values. The intersecting pixels (colored red) in Fig. 4 between the dataset and the model (shown in blue) will force the estimate to make a large jump downwards and the left. As the estimate center approaches the true center, both sides are intersecting, make the iteration value decreasing each iteration, allowing the model to slow down as it continues towards the center. The model will continue to slowly settle to it's final location, iteration by iteration, moving less and less, until it reaches it's final location. The function is continuous, but we are trying to add up values from the data indexes to try to find a point where the weight is counterbalance. Since the values must be translated to integer locations within a matrix, the location will eventually settle on a location that it will not move from as the model location is resolved

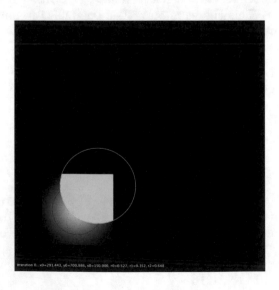

Fig. 3. Model intersecting With X_n

to the nearest matrix element. The process is very quick to move towards the actual data location, traversing over 96% of the distance from the center of the estimated model to the center of the actual Gaussian distribution, requiring only 6 iterations. The remaining distance to the actual x, y, 4%, took 8 iterations to reach the final x, y position, and a final iteration to detect that it hadn't moved more than $10e^{-9}$ from the previous iteration's estimated x, y positions. Once the final position was identified, the final radius determination only took 6 iterations, and a single iteration to detect that the radius calculation hadn't changed. Data results are shown in Table 2.

Table 2. Iteration values

Iteration	x position	y position	Radius	Height	f_1	f_2
Initial	330.000	655.000	150.000	0.200	0.500	0.500
1	291.42429271	700.92337880	150.00000000	0.52663713	0.352	0.648
2	268.17871148	728.42629612	150.00000000	0.73790251	0.376	0.624
3	254.79794819	744.23412051	150.00000000	0.79585081	0.466	0.534
4	249.88772336	750.07892773	150.00000000	0.79992000	0.605	0.395
5	249.99436807	749.94568708	150.00000000	0.79984003	0.740	0.260
6	249.96382891	749.99041353	150.00000000	0.79984003	0.821	0.179
7	249.97891051	749.96946019	150.00000000	0.79984003	0.853	0.147
8	249.96983887	749.98053430	150.00000000	0.79984003	0.864	0.136
9	249.97385094	749.97475744	150.00000000	0.79984003	0.867	0.133

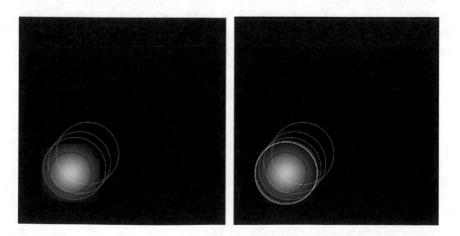

Fig. 4. Left: Intersecting Models over 3 iterations; Right: Intersecting Models Until Location Locks, 31 Iterations

To determine the height of the model for the next iteration, we looked at $\frac{dr}{dt}$ in our Gaussian function.

$$\frac{dr}{dt} = \frac{d}{dr}\left[\ln\left(\frac{r}{2\pi s^2}e^{\frac{-((x-x_0)^2+(y-y_0)^2)}{s^2}}\right)\right] \tag{11}$$

$$\frac{dr}{dt} = \frac{1}{r} \tag{12}$$

This equation didn't make much sense to implement. Since our r_0 value is less than one, this would have made r_0 instantly greater than one. The location of the next (x_0, y_0) has already been solved, so rather than implementing this equation, we simply assigned the height equal to the value contained within the data at $X_n[x_0][y_0]$. This was due to one of the approaches that we had taken for trying to update the radius for each iteration as the model moved to find the final (x_0, y_0) location which caused great problems. Using the genuine height at the data point seemed logical. Since we had calculated the location, knowing the height would allow me to create Gaussian functions with various radius values and solve for a comparable radius. We had instabilities modifying the position and the radius simultaneously. At times the radius would drastically increase, equaling values that were orders of magnitude larger than the height and width of the display. Consideration must be taken for reducing the radius if the estimate is at a point where the edge of the circle has begun to touch or exceed the edge of the dataset. The process of calculating each parameter during each iteration led to instabilities for each iteration. Allowing the model to move while changing the radius allowed large portions of the model to be included one iteration and then excluded during the following iteration, caused the radius to largely increase, including indexes that should not have been incorporated into the calculation. This would force new x, y locations, even driving it away from the desired target location. Driving the location would impact the radius, essentially inducing an unintended feedback loop. Troubleshooting became unbearable. Trying to identify what was wrong by looking at million indexes over multiple iterations across multiple arrays to determine what was happening to the location when the radius modification was introduced. This led to an alternate implementation, first solving for location and height, followed by solving for radius once the location has been locked down. It likely is not a requirement, but it seemed a good approach that any parameter that was a single dimension, x, y, r_0, could be solved for simultaneously, as long as the parameters did not have a dependence on one another. The iterative calculations for calculating changes in radius are listed in Eqs. 13 to 14.

$$= \frac{d}{ds}\left[\ln\left(\frac{r}{2\pi s^2}e^{-\frac{(x-x_0)^2+(y-y_0)^2}{s^2}}\right)\right] \tag{13}$$

$$= \frac{2((x-x_0)^2+(y-y_0)^2-s^2)}{s^3} \tag{14}$$

Since we are allowing a combination of the confidence of a model and the intersection of a model with data to determine how to move the estimated model, eventually the estimated model will cease to move, once Eqs. 9 and 10 find a balance point in the intersecting indexes in X_n. This was important because after this point, the estimated model would no longer move. This can be seen in Table 2, after completing iteration 15. At the completion of iteration 15, we can detect that the location has not changed from iteration 14 to 15, so we can declare this model as having its location locked. Once the location is locked, we make the x, y, r_0 parameters unable to be modified and begin solving for the radius. To determine the radius value, we implemented an iterative comparison that was executed after the final location had been identified. At this point, we should be highly confident that we have found the correct data or highly confident that our model has not been found. We did not implement it, but at this point the only reason to attempt to find the radius of the circle would be if we are confident that we have found the data. There is no use in wasting our time if we have a low degree of confidence, rather, reinitialize the model to another location to test the data. As shown in Fig. 5, a mask was created to intersect with the model and the final location in X_n, allowing the elements to be summed. This is shown in Algorithm 1. The idea works for this instance, simply decreasing the radius of the model until the sum of the model and the sum of the indexes in X_n match. The final radius was not identical, but close. My final radius (Table 3) resolved to a value of 99 compared to the actual value of 100. The difference is accounted for by the random noise that had been injected. The radius had to be decreased to account for noise values that had replaced indexes that were within the Gaussian function in the truth set. Unfortunately this somewhat requires being able to see the edge of the data within X_n. Without some information to determine when to stop adjusting the model, it could be virtually impossible to determine. Even the random noise that was injected affected the results, but the final result was very close. The image or data may require a degree of processing to aid in processing so that model adjustment can terminate.

Another method that is still functional with this image would simply be to start the model at the minimum radius, and using the same masking technique, to increase the radius until the values are the same. This is similar to the technique we used, and has the same shortcoming in knowing when to stop. The technique is able to detect when crossing the boundary of the Gaussian function in X_n by a difference in the summed indexes when crossing into the area that is represented by the Normal distribution. It is relatively straight forward to see that using the confidence and likelihood functions allows various pixels

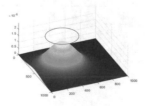

Fig. 5. Radius mask

Algorithm 1. Characterize radius

---

while *prevRadius* ≠ *radius* **do**
  for i=0 **to** width/2 **do**
    Create Gaussian function using x, y, r_0 and current s_0
    Create mask using x, y and current s_0
    modelsum ← Sum all indexes that the mask intersects with in model M_1 Gaussian function
    sum ← Sum all indexes that the mask intersects with in X_n
    if *modelsum* < *sum* **then**
      radius ← *radius* - 1
    else
      if *modelsum* > *sum* **then**
        radius ← *radius* + 1
      end if
    end if
  end for
end while

---

to be weighted differently, which Dynamic Logic uses to balance the intensity of the indexes that intersect between the likelihood and the dataset. Moving using the derived equations was extremely useful. But determining the radius is not as trivial. If one was to remove the comparison and only use the likelihood and confidence determinations, how would you know to make the radius larger or smaller? What weighting information is available to make a decision? More specifically, what weighting information is available between the model estimate and X_n to make a decision? Obviously performing a comparison based on prediction versus actual can work in a simple fashion, but we were really hoping that there would an equation (Eq. 14) that would help us solve, rather than having to use an empirical solution.

Table 3. Actual and identified values

	Gaussian 1 Actual	Gaussian 1 Identified	Gaussian 2 Actual	Gaussian 2 Identified
x_0	250	249.993	750	750.05
y_0	750	749.986	250	249.991
s_0	100	99	100	99
r_0	.8	.8	.8	.8
r_1	.5	.956	.5	.956
r_2	.5	.044	.5	.044

2   Future Work

One of the major benefit of Dynamic Logic is that if there is an overlap between the data and the model then iterative process terminates very fast, specially when the model can be defined by a (geometric) equation in our case circle. One obvious extension of our work is to extend the idea in three dimensions but for arbitrary shapes. One idea is to work with volumes instead of 2D images. Dynamic Logic research is extensive and it has already found applications in several areas [1]. The estimation started off with one large vague blob, eventually resolving to multiple models. The part that would be great to learn is how the model attempted to go from one large model to three or more individual models. DL could be applied iteratively or multiple instantiations of the same program could be running with different initial estimates trying to find possible overlapping 2D circles for our implementations.

References

1. Perlovsky, L., et al.: Emotional cognitive neural algorithms with engineering applications dynamic logic: from vague to crisp. Springer, Berlin (2013)
2. https://www.processing.org
3. Promersberger, J.: Understanding dynamic logic using 2D images. MS Project report, University of Colorado Colorado Springs, pp. 1–42 Sprig (2018). Advisor: Dr. SK Semwal
4. International Journal of Emerging Technology and Advanced Engineering Website: www.ijetae.com. (ISSN 2250-2459, ISO 9001:2008 Certified Journal, Volume 4, Issue 5, May 2014)
5. Semwal, S.K., Janzen, M., Promersberger, J., Perlovsky, L.: Towards approximate sphere packing solutions using distance transformations and dynamic logic. In: Arai, K., Kapoor, S. (eds.) CVC 2019, AISC 944, pp. 1–14 (2020). https://doi.org/10.1007/978-3-030-17798-0-31

Screens Affordances in Image Consumption

Ema Rolo[1]([⊠]), Helena Nobre[1], and Vania Baldi[2]

[1] GOVCOPP, University of Aveiro,
Campus Universitário de Santiago, 3810-193 Aveiro, Portugal
esrs@ua.pt, hnobre2@gmail.com
[2] DigiMedia, University of Aveiro,
Campus Universitário de Santiago, 3810-193 Aveiro, Portugal
vbaldi@ua.pt

Abstract. Despite the infusion of screen technologies, we hardly notice their properties as we commonly interact with them. Encounters with visual driven-networks occur through the screens orienting consumers toward visual data and information. Through physical interactions with the object and its surface, images move fastly. More than any contemporary object, screens have facilitated the proliferation of customized visual information. As the potential adaptability of screens as technology expands, the screen is evolving into new directions and applications in capturing consumers' attention in the most attractive ways. Taking in account the current trends in which smartphones and tablets are at the forefront of marketing innovation and commercial applications, this paper aims to establish a conceptual relationship between smartphones screens affordances and the consumption of images. Based on the literature review, a theoretical framework and insights for future research on the topic of consumer behavior in screen culture are proposed.

Keywords: Screen culture · Affordances · Consumer behavior

1 Introduction

The proliferation of images depicting lifestyles or aesthetic systems eliciting taste regimes [3] results from two essential facilitators: first, the emergence of social media websites that enables the consumer-to-consumer interaction [17], and second, the availability and mass adoption of smartphones [26]. These technological events permitted consumers worldwide to participate in the consumption of cultural resources, establish social connections [17] and peep into new "eye-catching" realities [40], through the smooth and sleek surface [10] of the smartphone screen.

Reflections on media and the relationship between the means of production and consumption of cultural goods offer insights into how a culture sees itself. At the same time, everyday practices based on taste, from ordinary to spectacular, interest scholars because of the way these seemingly trivial acts can tell stories about consumers and their lives [3]. The present study is specifically concerned with the way images funneled through the screen affect consumers' preferences. Mobile screens as *theoretical* objects

© Springer Nature Switzerland AG 2020
K. Arai et al. (Eds.): FTC 2019, AISC 1070, pp. 176–187, 2020.
https://doi.org/10.1007/978-3-030-32523-7_12

are meaningful objects of analysis that compel us to propose, interrogate, and theorize through. However, they constitute a rapidly developing type that encompasses multiple interfaces [73]. As mobile media and convergence [44], global connectedness [2, 24, 50], consumer lifestyles [13, 53, 57], hybrid markets [50, 69, 70], brand communities [18, 71], identity projects [6, 45], and storytelling [76] have become major topics of research, the screen that underlies and unites them, emerges as a critical area of interest [59].

In the beginning of the 20th century, *the Century Dictionary and Cyclopedia*, defined screen as "a covered framework, partition, or curtain, (…) which serves to protect from the heat of the sun or of a fire, from rain, wind, or cold, or from other inconvenience or danger, or to shelter from observation, conceal, shut off the view, or secure privacy; as, a fire-screen; a folding-screen; a window-screen (…) hence, such a covered framework or curtain used for some other purpose; as, a screen upon which images may be cast by a magic lantern" [41]. Since then, screens have been theorized through different disciplinary lenses. A screen – this "surface for animation" [1] means different things in different academic contexts, each raising a myriad of historical, aesthetic, theoretical, technological and, political questions. Addictive behaviors supported by technologies have received increasing attention from different sectors [52]. Observed literature regarding screen-gadgets and network technologies, emphasized, for example, concerns about behavioral addiction impacting general health and well-being, e.g. [5, 28, 29, 46, 51]. The images confined by the aesthetics of the frame, arranged in order to suggest symbolic affiliations to lifestyles, within the weightless mobility of the screen, [74] play a part in shaping new markets based on aesthetic systems [24]. Whereas the virtual has a long association with the imaginary and transient nature of digital culture [59], today's screens harbor a multiverse of aesthetic systems, guiding consumers on *how* to consume, more than on *what* to consume [3]. Besides the conceptual challenges in this study, our specific concern lies with how screens at once provide access to mediated taste regimes [3] while also restraining the visual experience [14, 59]. In sum, we purpose that the screen affordances comprehend more than an invitation to act but new forms of consumer behavior.

2   Screen Culture

The Internet made it possible for ordinary people to bypass the traditional gateways to influence, acquire an audience, and exert cultural influence [3]. The smartphone, as the archetype of technical digital objects [64], characterized by its polyvalence and its openness of information, facilitates the consumer-to-consumer interaction [17]. The potential adaptability of screens into any number of media, technological, and limitless commercial operations, render them into the best channels for engaging customers [48]. Push notifications and messaging now bring content to people from specific channel contexts at an ever-increasing rate, shifting from what was originally a desktop phenomenon [63]. Companies and brands play a significant role in this transformation, investing in responsive design and mobile-friendly websites, making contents set up quickly for short-term visualizations, creating a mobile experience both accessible and optimized [15]. *A reality that goes beyond empirical data*, from the moment in which samples, information, and elements of possibility are mixed on a surface [14]. When compared to other forms of media consumption, mobile is the only growing segment at the expense of TV, print, radio, and desktop computer usage [55]. Considering that 46% of users drop off in less than 10 s [77], companies have to consider that short timeframe, to provide on screen the information users are demanding, to capture their attention before they decide to bounce off. Without belittling the changes brought forth by economic and industrial developments through the phenomena of mass adoption of smartphones, the screen-based devices are promoting a power shift and redefining human interaction [48], and participatory practices through the image [33], since anyone who owns a smartphone can produce limitless images for free.

Besides new forms of visual content, consumption through haptic navigation is more accessible, rendering consumers in both interpretive agents and producers of meanings [49, 56]. The production of meaning embedded in the consumption of signs supports the growth of commodity culture, niche marketing, and the creation of new lifestyles making all spheres of life permeable to commodification [4]. "Lifestyles are emerging in which, the new heroes of consumer culture create a life project and display their individuality in the particularity of the assemblage of goods, clothes, practices, experiences, appearance and bodily dispositions they design together into a lifestyle" [27]. This premise emphasizes the idea of Zygmunt Bauman's "liquid modernity" [8]. Forms of modern life may differ in quite a few respects – but what unites them is precisely their temporariness and inclination to constant change [8]. Accordingly, change is occurring more and more rapidly within society. To *be modern* means to modernize – compulsively, obsessively; not so much *to be*, (…) but forever *becoming*. Each new structure, which replaces the previous one as soon as it is declared old fashioned and past its use-by date is only another momentary settlement – acknowledged as temporary and *until further notice*. As Bauman stated, - "a hundred years ago *to be modern* meant to chase *the final state of perfection* – now it means an infinity of improvement, with no *final state* in sight".

The encouragement for flexibility, mobility, and transition centered on the screen instigated the free flow of information and goods across international borders, and cultural boundaries allowing consumers to express taste while spending very little if any, money [3]. While choices and preferences form enduring patterns that define consumer lifestyles [38], lifestyles become commodities to buy, infinitely transformable with the purchase of new commodities [36]. The aesthetic signs and taste regimes [3] became the foundations of consumption in the hyper-commodification process [19] and the smartphone screen the privileged point of access serving as the frame for tracking, quantifying, and *commodifying* our existence [59]. Whether through personal demand or media presentation, taste regimes' meanings and lifestyles produced by a consumer culture are reduced to surface effects and we acquire them by seeing them. We are addicted to visual imagery and promptly respond to their symbols [62, 72]. Thus as we look at screens, we find our way of being in the world [43].

3   Screen Affordances

The mobile handled screen invites for a renewed inquiry about the status of the screen. For Verhoeff [73], the screen status rests at once on an abstract notion of the site of image presentation and viewing, on a frame for representation, and *on a very material object* to carry around. However, viewing is no longer a matter of looking alone, nor of perceptually receiving images [73], it fosters an understanding, a conscience about the presence of the triangulated relationship of the user, image and screen. As network technologies expand the modal possibilities for mediated engagement [67], the smartphone screen assumes a new nature, a surface across which the images travel through [14] producing physical and ideological relationships, that often render us some desired quantitative reward. "They are the window displays for convergence, where we can see the melding of film, broadcasting, and computers into hybrid media and commerce" [1]. Thus, screens take part in the mediation of consumers' interpretations, notions, and preferences about brands, communities and services.

We, therefore, seek to explore smartphone screens conceptual related affordances that facilitate the efficacy of the communication of aesthetic systems based on images. In the last years, scholars have drawn on the concept of "affordances" to map out the *complex* relationship between technology's materiality and social factors [25]. In the original formulation of the concept, according to Gibson [31], affordances were noted as the objective, latent possibilities present in an environment that actors could act on in specific ways. According to Hutchby, [42] affordances views technologies as artifacts that may be both shaped by and shaping of the practices, humans use in interaction with, around and through them. Furthermore, they represent the opportunity for an interaction between the technical properties of an object and the actions of a social agent. Ian Hutchby also defines affordances as functional and relational aspects which frame, while not determining, the possibilities for *agentic* action concerning an object [42]. In the context of industrial design, Norman [61], suggests that affordances result from the mental interpretation of things, based on our prior knowledge and experience applied to our perception of the things about us. An affordance is not a property of a particular physical or non-physical object but the relationship between the person and the object. For physical objects such screens, there is a tacit knowledge of the perceptible characteristics that will be interpreted. Contrary to Gibson's conception, Norman [61] perceived that affordances were not mere opportunities for action, but "perceived action possibilities" that suggested actions to the individual. Therefore, what designers or architects create are not mere opportunities for action, but invitations that can have a severe influence on behavior. The temporary sculptures of the Dutch artist Krijn de Koning, consisting of walkways and platforms with colored walls situated in the middle of galleries, illustrates these invitations by the physical objects that we operate and deal. Koning's installations challenge those who pass by, offering new possibilities to navigate and experience the space the works inhabit, (see Fig. 1).

Objects are not merely functional but always affect the individual emotionally, making certain behaviors more likely to occur [75]. To clarify our position, we follow Gibson's affordances [31], as opportunities for action. However, by suggesting that affordances can also invite behavior, we complete Gibson's original conception,

Fig. 1. Krijn de Koning, Work for Nieuwe Kek, Amsterdam

conceiving affordances not only as invitations or *solicitations to act*, but also as action possibilities that invite behavior. In other words, we comprehend that the smartphone screens affordances do not cause a specific behavior but make it possible. By conceptualizing affordance as a multidimensional and relational construct, the present literature review crosses the disciplinary boundaries of media studies and consumer behavior in order to propose a theoretical framework and new insights for future research on the topic of consumers' preferences and orientations in the contemporary screen culture.

Besides the conceptual challenges, our specific concern lies with how screens at once provide access to taste regimes [3] but also restraint the visual experience [14, 59]. Following, we identify and describe four conceptual affordances, – the *sleek surface*, the *escapist arena*, the *playful immersion* and the *intimate lens* – that facilitate, entail and *invite* the engagement with the screen regarding the consumption of images on smartphones screens.

3.1 The Sleek Surface

Humans are drawn to screens by the images they bear. The technical qualities of screened images, such as movement, adjustment, color, and light, ensured the popularity of screens in our everyday practices [1, 59, 65]. In this study, we relate consumers' engagement with screened images with the concept proposed by philosopher and cultural theorist Byung-Chul Han, regarding the contemporary obsession with the *smooth* [37]. In an interview led by Boeing and Lebert [10], Han declared, "Everything flows in soft, smooth transitions. Everything seems to be rounded, abraded, and polished." The author argued that beauty has lost its edge and has been turned into something merely smooth and pleasing, like the flawless smartphone (screen), the unscathed skin of an epilated woman, or the art of Jeff Koons. Anything that is *not positive* has fallen victim to the need for a fast, unobstructed flow of information and capital. The only thing left within the aesthetic of the smooth is that which gives immediate pleasure, that which can be consumed [10]. Screened images can bind the physical qualities of objects depicted and their meaning in order to create aesthetic patterns, like the Kinfolk's visual universe that was popular online around 2012 [47]. The Kinfolk aesthetic system became a trend on social media sites while linking reclaimed wood, old-timey light bulbs with visible filaments, potted succulent plants, muted colors, and natural fibers with cultural narratives about durability and authenticity. That kind of *instrumentalization* enrolls objects and *doings* to actualize meanings in a way to orchestrate taste regime practices and further influence consumption [3]; thus, they became more than a resource for what to consume but also how to consume [3]. Every carefully composed shot of a foam latte heart is a mode through which the person taking and posting the image, further turns themselves into a subject of a specific aesthetic regime [3] and screens, the material that organizes our relation with it. The book *Styling for Instagram* by Cyd Leela, where the author offers tips on how to create fictional environments or stage existing environments for visual storytelling purposes, is also a result of the market of aesthetics [23]. According to Arsel and Bean [3], modes of interaction structured by social media sites have influenced the emergence and direction of taste regimes specific to those sites, configuring moments and objects into aesthetic systems. Those examples also suggest that the search for information, commodities, or entertainment is increasingly visual *pleasing* [34] and tightly bound up with the smooth and sleek materiality that screens confer to images. If so, screens not only contribute to producing images that circulate worldwide through participatory media, but they also become the primary means of viewing and experience certain aspects of life.

3.2   The Escapist Arena

In line with previous studies arguing that digital games [12] and binge-watching practices [45] are often viewed as being inherently escapist, we suggest that screen practices such as *tapping-to-advance* or scrolling through a feed of images could be seen as well. The concept of escapism—or the self-selected separation of oneself from one's immediate reality—through the consumption of media resources, such as television, music, games, movies, and texts has been a current topic of interest in the field of consumer culture studies [7, 32, 58]. Regarding binge-watching [45], escapism can be distinguished between passive and active. The same authors point to classic passive escapism examples that include watching a single discrete episode of a television series or a movie.

Forms of passive escapism are considered activities which allow consumers to *free their mind* from their current conditions but do not require much from consumers in terms of cognitive efforts or interactivity beyond their attention and appreciation (...) in contrast, active escapism provides components of interactivity and opportunities to achieve a somewhat *lasting presence* in the fantasies we escape to, such as playing and taking on character roles within video games [45].

Building on the notion that both practices allow us to "become temporarily transported into narrative worlds" [45] and embody the alluring reality of something conceived for the surface, we metaphorically suggest the screen as an escapism arena that relates with a playful invitation.

3.3   The Playful Immersion

Apple invites adult consumers to *play* with their iPhones in the way children play with their food [11]. In its first form, the iPhone was said to look "like an expensive bar of chocolate wrapped in aluminum and stainless steel" [35]. When we are children, adults tell us to not play with our food, but, as Steven Connor [16] points, sweets are designed to be playthings, "sweet things do not taste of themselves; they taste of our pleasure in them." This is not only because the taste is inextricably bound up with smell, sight, and touch but also because sweetness derives from the interactive "essence of eating" rather than from the ingredients themselves. The pleasure of the iPhone, like its siblings, is not what it does, but what it *invites* us to do with it, and how it invites all of our senses to the table [11]. Apple imagines, designs, and markets every one of its products, including the iPhone, similarly: as technologies that go beyond function, as playthings designed to appeal to multiple senses and to provoke an interactive engagement with them, right down to their curvaceous contours, smooth textures, and colorful surfaces. The contemporary screens allow users to ignore the boundary between function and fun [11] while the continuous flow of data and images [14] has become perpetually available *everywhere*. An association with the concept of the display is offered by the same author (cit. on p. 35) [14], "the display shows, making an idea accessible, it exhibits, but does not uncover. The display *makes present* images, it places them in front of us, in case we want to make use of them." Moreover, the eminent portability of screens means nearly every space is a potential screen context [59]. Huhtamo [40, 41] argued that miniaturized screens display vast amounts of rapidly changing information —images, graphics, and text. However, despite their minuscule size, an intuitive, almost real-time relationship develops between the fingers and the streams of successive data traversing the palm ([41], cit. in p.78), offering both intimacy and immediacy [9, 67] through the modulated light of the surface [1]. The playful modality of the smartphone screen can also be associated with the concept of "immersion."

According to Janet Murray (1997) [60], the term immersion in digital environments derives metaphorically from the physical experience of being underwater. The immersion uses the flow principles based on the Csikszentmihalyi theory [20] and was revisited by Rémond and Romo [68], regarding gambling activities on screens and smart environment applications. Immersion has been described as "the experience of being transported to an elaborated simulated place, regardless of the fantasy content; the sensation of being surrounded by a completely other reality, as different as water is from air, that takes over all of our attention, our whole perceptual apparatus" [68] and largely contributes to an individual's commitment with the screen [20, 21]. These features have been critical aspects in creating engagement and enjoyment with games through screens, and both relate another concept, the concept of *presence* or feeling the *real* environment.

3.4   The Intimate Lens

The screens that confront the contemporary viewer "solicit and shape our presence to the world, our representation in it, and our sensibilities" [59] and each differently invites our complicity in formulating space, time and bodily investment as significant personal experience [59].

Erkki Huhtamo, a pioneer in the branch media studies, considers the historical development of objects, narratives, and practices that manifest—and contribute to—the idea of the screen as visual mediation [41]. In both essays [39, 41], on the etymology and evolution of screens, the author offers a link between historical examples across several centuries, such as the peeping devices and the contemporary screen practices. As he states, - "the dispositive used in peep practice emphasized a private viewing (…) the production of a visual enclosure - the separation of the picture from the physical environment where the device was located was a common goal" [41]. These *devices* did not have a screen in the sense of the television screen; the principle of 'interfacing' users with these "viewing machines" was by peeping [39]. The vistas were hidden inside a box, and to access them, one had to glue one's eye(s) to a hole provided with a magnifying lens. The "user interface" of these peepshow boxes reveals a similitude with contemporary screen practices such as *tap-to-advance* on Instagram on the sense that a specific scene is magnified, pulling it forward to the observer (or pushing the observer into the scene) effacing the sense of a framed surface. Giovanni Battista della Porta in his sixteenth-century descriptions of how a "magic–lantern" spectacle should be arranged and what sort of images and effects it can produce, evokes the idea of the screen as a frame that should disappear from the viewer's consciousness. This way, the image becomes more effective; the less aware the viewer is of the screen's presence [59]. The multi-sensorial nature of the visual experience in smartphone's screens offers a fully-fledged embodied experience [54], and as we forget their presence, the experience with them becomes more *real* and intimate [41], and the communication more accessible and direct. As Vilém Flusser [30], also remarked: "Screens let us 'enter' the realm they depict."

4   Conclusion

Smartphones screens do not operate solely as windows opening up a field of vision; they are shields and blinders of observation while comprehending action possibilities that invite behavior within and beyond the surface.

As our awareness of screens, which are today the ubiquitous medium of reception, has shrunk to nothing [22] our aesthetical awareness becomes more materialist. Following current trends on screen culture, we are more likely to produce and view images with the help of screens than through any other visual support. Hence, we increasingly rely more on screened images even to experience our immediate environment visually [59] which might mean that everyday practices, like the *innocuous* act of drinking coffee and the aesthetic proprieties of the objects in use, will turn into mediated influences on consumers' tastes and preferences. While micro-level actions of connected consumers shape new aesthetic and taste regimes, the demand for *immaterial objects*, such as images on Pinterest, often results in the accretion of material goods [3].

The conceptual affordances of screens presented in this study, outline that consumers are more likely to engage with screened-images when they forget the identity of the medium itself, but also when consumers access them privately while enjoying its materiality. We can also attest that consumers have a higher propensity to engage with visual-information on screens due to its hybridity that comprehends both a utilitarian and playful modality. Despite the number of empirical studies in the literature devoted to assessing screen effects in a multitude of contexts, the relationship between the screen and consumer aesthetic preferences remain still an underexplored subject [66]. Finally, further validation of such insights through an interpretative phenomenological approach is recommended to a context-aware understanding of consumption as social performance within screen culture.

References

1. Acland, C.: The crack in the electric window. Cine. J. **51**(2), 167–171 (2012). https://doi.org/10.1353/cj.2012.0007
2. Askegaard, S., Arnould, E., Kjeldgaard, D.: Postassimilationist ethnic consumer research: qualifications and extensions. J. Consum. Res. **32**(1), 160–170 (2005). https://doi.org/10.1086/426625
3. Arsel, Z., Bean, J.: Social distinction and practices of taste. In: Arnould, E., Thompson, C. (eds.) Consumer Culture Theory, 1st edn. SAGE Publications, Thousand Oaks (2018)
4. Barker, C., Janne, E.: Cultural Studies Theory and Practice, 5th edn. SAGE Publications, Thousand Oaks (2016)
5. Barnes, S., Pressey, A., Scornavacca, E.: Mobile ubiquity: understanding the relationship between cognitive absorption, smartphone addiction and social network services. Comput. Hum. Behav. **90**, 246–258 (2019). https://doi.org/10.1016/j.chb.2018.09.013
6. Belk, R.: Possessions and the extended self. J. Consum. Res. **15**(2), 139 (1988). https://doi.org/10.1086/209154
7. Batat, W., Wohlfeil, M.: Getting lost "into the wild": understanding consumers' movie enjoyment through a narrative transportation approach. Adv. Consum. Res. **36**, 372–377 (2009)
8. Bauman, Z.: Liquid Modernity, 1st edn. Malden, Massachusetts (2012)
9. Bolter, D., Grusin, R.: Remediation: Understanding the New Media, 1st edn. MIT Press, Cambridge (2000)
10. Boeing, V., Lebert, A.: Byung-Chul Han: "Das Glatte charakterisiert unsere Gegenwart" (2014). https://www.zeit.de/zeit-wissen/2014/05/byung-chul-han-philosophie-neoliberalismus
11. Cannon, K., Barker, J.: Hard candy. In: Snickars, P., Vonderau, P. (eds.) Moving Data - The iphone and the Future of Media, 1st edn. Columbia University Press, New York (2012)
12. Calleja, G.: Digital games and escapism. Games Cult. **5**(4), 335–353 (2010). https://doi.org/10.1177/1555412009360412
13. Carfagna, L., Dubois, E., Fitzmaurice, C., Ouimette, M., Schor, J., Willis, M., Laidley, T.: An emerging eco-habitus: the reconfiguration of high cultural capital practices among ethical consumers. J. Consum. Cult. **14**(2), 158–178 (2014). https://doi.org/10.1177/1469540514526227

14. Casetti, F.: What is a screen nowadays? In: Monteiro, S. (ed.) The Screen Media Reader: Culture, Theory and Practice, 1st edn. Bloomsbury Academic (2017)
15. Chaffey, D.: Mobile marketing statistics compilation | Smart Insights (2019). https://www.smartinsights.com/mobile-marketing/mobile-marketing-analytics/mobile-marketing-statistics
16. Connor, S.: Sweets - expanded transcript of a talk broadcast on BBC radio (2000). http://www.stevenconnor.com/magic/sweets.htm
17. Cova, B., Shankar, A.: Consumption tribes and collective performance. In: Arnould, E., Thompson, C. (eds.) Consumer Culture Theory, 1st edn, pp. 88–106. SAGE Publications, Thousand Oaks (2018)
18. Cova, B., Kozinets, R., Shankar, A.: Consumer Tribes, 1st edn. Butterworth-Heinemann, Oxford (2007)
19. Crook, S., Pakulski, J., Waters, M.: Postmodernization, 1st edn. Sage, London (1992)
20. Csikszentmihalyi, M.: The Psychology of Optimal Experience, 1st edn. Harper Perennial Modern Classics, New York City (2008)
21. Csikszentmihalyi, M., Csikszentmihalyi, I.: Optimal Experience: Psychological Studies of Flow in Consciousness, 1st edn. Cambridge University Press, Cambridge (1992)
22. Cubitt, S.: Current Screens. In: Monteiro, S. (ed.) The Screen Media Reader: Culture, Theory and Practice, 1st edn. Bloomsbury Academic (2017)
23. Cyd, L.: Styling for Instagram: What to Style and How to Style It, 1st edn. St. Martin's Griffin, New York (2018)
24. Dolbec, P., Fischer, E.: Refashioning a field? connected consumers and institutional dynamics in markets. J. Consum. Res. 41(6), 1447–1468 (2015). https://doi.org/10.1086/680671
25. Duffy, B., Pruchniewska, U., Scolere, L.: Platform-specific self-branding. In: Proceedings of the 8th International Conference on Social Media & Society -#Smsociety17 (2017). https://doi.org/10.1145/3097286.3097291
26. Eriksen, T.: We Are Overheating - TEDxTrondheim [Video] (2019). https://www.youtube.com/watch?v=ivjXlRu_3aQ
27. Featherstone, M.: Consumer Culture & Postmodernism. SAGE Publications, London (1991)
28. Friedberg, A.: The Multiple. In: Monteiro, S. (ed.) The Screen Media Reader: Culture, Theory and Practice, 1st edn. Bloomsbury Academic (2017)
29. Frosh, P.: Telling presences: witnessing, mass media, and the imagined lives of strangers. Crit. Stud. Media Commun. 23(4), 265–284 (2006). https://doi.org/10.1080/07393180600933097
30. Flusser, V.: Two Approaches to the Phenomenon, Television - The New Television, 1st edn. MIT Press, Cambridge (1977)
31. Gibson, J.: The Ecological Approach to Visual Perception, 1st edn. Houghton Mifflin, Boston (1979)
32. Giddings, S.: Events and collusions: a glossary for the microethnography of video game play. Games Cult. 4(2), 144–157 (2008). https://doi.org/10.1177/1555412008325485
33. Goggin, G.: New Technologies and the Media, 1st edn. New Technologies and the Media, New York (2012)
34. Goh, D., Ang, R., Chua, A., Lee, C.: Why we share: a study of motivations for mobile media sharing. In: Active Media Technology, pp. 195–206 (2009). https://doi.org/10.1007/978-3-642-04875-3_23
35. Grossman, L.: The Apple of Your Ear (2007). http://content.time.com/time/magazine/article/0,9171,1576854-5,00.html
36. Halter, M.: Shopping for Identity: The Marketing of Ethnicity, 1st edn. Schocken, New York (2000)

37. Han, B.: Saving Beauty, 1st edn. Polity, Cambridge (2017)
38. Henry, P., Caldwell, M.: Spinning the proverbial wheel? Social class and marketing. Mark. Theory 8(4), 387–405 (2008). https://doi.org/10.1177/1470593108096542
39. Huhtamo, E.: Elements of screenology: toward an archaeology of the screen. In: ICONICS: International Studies of the Modern Image, Tokyo: The Japan Society of Image Arts and Sciences, vol. 7, pp. 31–82 (2006)
40. Huhtamo, E.: Screen tests: why do we need an archaeology of the screen? Cine. J. 51(2), 144–148 (2012). https://doi.org/10.1353/cj.2012.0011
41. Huhtamo, E.: Screenology; or, media archaeology of the screen. In: Monteiro, S. (ed.) The Screen Media Reader: Culture, Theory and Practice, 1st edn. Bloomsbury Academic (2017)
42. Hutchby, I.: Technologies, texts and affordances. Sociology 35(2), 441–456 (2001). https://doi.org/10.1177/s0038038501000219
43. Introna, L., Ilharco, F.: On the meaning of screens: towards a phenomenological account of screenness. Hum. Stud. 29(1), 57–76 (2006). https://doi.org/10.1007/s10746-005-9009-y
44. Jenkins, H.: Convergence Culture: Where Old and New Media Collide, 1st edn. NYU Press, New York (2006)
45. Jones, S., Cronin, J., Piacentini, M.: What Can Binge-Watching Tell Us About Escapism? (2018). https://www.jmmnews.com/what-can-binge-watching-tell-us-about-escapism/
46. Kim, Y., Jang, H., Lee, Y., Lee, D., Kim, D.: Effects of internet and smartphone addictions on depression and anxiety based on propensity score matching analysis. Int. J. Environ. Res. Public Health 15(5), 859 (2018). https://doi.org/10.3390/ijerph15050859
47. Kinfolk. (2019). Retrieved from https://kinfolk.com/
48. Kotler, P., Kartajaya, H., Setiawan, I.: Marketing 4.0: Moving from Traditional to Digital, 2nd edn. Wiley, New Jersey (2017)
49. Kozinets, R.: Utopian enterprise: articulating the meanings of star trek's culture of consumption: figure 1. J. Consum. Res. 28(1), 67–88 (2001). https://doi.org/10.1086/321948
50. Kozinets, R., Patterson, A., Ashman, R.: Networks of desire: how technology increases our passion to consume. J. Consum. Res. ucw061 (2016). https://doi.org/10.1093/jcr/ucw061
51. Kuss, D., Harkin, L., Kanjo, E., Billieux, J.: Problematic smartphone use: investigating contemporary experiences using a convergent design. Int. J. Environ. Res. Public Health 15(1), 142 (2018). https://doi.org/10.3390/ijerph15010142
52. Lopez-Fernandez, O.: Internet and Mobile Phone Addiction (2019). https://doi.org/10.3390/books978-3-03897-605-9
53. Maciel, A., Wallendorf, M.: Taste engineering: an extended consumer model of cultural competence constitution. J. Consum. Res. ucw054 (2016). https://doi.org/10.1093/jcr/ucw054
54. Massumi, B.: Parables for the Virtual: Movement, Affect, Sensation Post-Contemporary Interventions, 1st edn. Universidade Duke, Durham (2002)
55. Meeker, M.: Internet Trends 2018. In: Presentation, Code Conference (2018)
56. Meyer, A.: Investigating cultural consumers. In: Pickering, M. (ed.) Research Methods for Cultural Studies, 1st edn. Edinburgh University Press, Edimburgh (2008)
57. Moisio, R., Arnould, E., Gentry, J.: Productive consumption in the class-mediated construction of domestic masculinity: do-it-yourself (DIY) home improvement in men's identity work. J. Consum. Res. 40(2), 298–316 (2013). https://doi.org/10.1086/670238
58. Molesworth, M., Watkins, R.: Adult videogame consumption as individualised, episodic progress. J. Consum. Cult. 16(2), 510–530 (2014). https://doi.org/10.1177/1469540514528195
59. Monteiro, S.: The Screen Media Reader: Culture Theory and Practice, 1st edn. Bloomsbury Academic, New York (2017)

60. Murray, J.: Hamlet on the Holodeck: The Future of Narrative in Cyberspace, 1st edn. The MIT Press, Massachusets (1997)
61. Norman, D.: The Design of Everyday Things, 2nd edn. Basic Books, New York (2013)
62. Packard, V.: The Hidden Persuaders, 2nd edn. Ig Publishing, New York (2007). The Hidden Persuaders
63. Palmeri, R.: Why We Don't Surf the Web Anymore -And Why That Matters (2019). https://www.forbes.com/sites/valleyvoices/2016/02/22/why-we-dont-surf-the-web-anymore-%C2%ADand-why-that-matters/#517d64f77ac2
64. Paquienséguy, F.: Questionner les Pratiques Communicationnelles: Offre, Pratiques, Contenus, 1st edn. Editions Apogée, Rennes (2009)
65. Paul, C.: Mediations of light: screens as information surfaces. In: Cubitt, S. (ed.) Digital Light, 1st edn. Open Humanities Press, London (2015)
66. Peighambari, K., Sattari, S., Kordestani, A., Oghazi, P.: Consumer Behavior Research. SAGE Open, 6(2), 215824401664563 (2016). https://doi.org/10.1177/2158244016645638
67. Pittman, M., Reich, B.: Social media and loneliness: why an instagram picture may be worth more than a thousand twitter words. Comput. Hum. Behav. 62, 155–167 (2016). https://doi.org/10.1016/j.chb.2016.03.084
68. Rémond, J., Romo, L.: Analysis of gambling in the media related to screens: immersion as a predictor of excessive use? In: Lopez-Fernandez, O. (ed.) Internet and Mobile Phone Addiction: Health and Educational Effects. MPDI, Basel (2019)
69. Scaraboto, D.: Selling, sharing, and everything in between: the hybrid economies of collaborative networks. J. Consum. Res. 42(1), 152–176 (2015). https://doi.org/10.1093/jcr/ucv004
70. Scaraboto, D., Figueiredo, B.: Holy mary goes' round. J. Macromark. 37(2), 180–192 (2017). https://doi.org/10.1177/0276146717690201
71. Schau, H., Muñiz, A., Arnould, E.: How brand community practices create value. J. Mark. 73(5), 30–51 (2009). https://doi.org/10.1509/jmkg.73.5.30
72. Smoke, T., Robbins, A.: The World of the Image, 1st edn. Pearson, London (2006)
73. Verhoeff, N.: Mobile Screens: The Visual Regime of Navigation, 1st edn. Amsterdam University Press, Amsterdam (2012)
74. Verhoeff, N.: Performative cartography. In: Monteiro, S.: The Screen Media Reader: Culture, Theory and Practice, 1st edn. Bloomsbury Academic (2017)
75. Withagen, R., de Poel, H., Araújo, D., Pepping, G.: Affordances can invite behavior: reconsidering the relationship between affordances and agency. New Ideas Psychol. 30(2), 250–258 (2012). https://doi.org/10.1016/j.newideapsych.2011.12.003
76. Woodside, A., Sood, S., Miller, K.: When consumers and brands talk: storytelling theory and research in psychology and marketing. Psychol. Mark. 25(2), 97–145 (2008). https://doi.org/10.1002/mar.20203
77. Work, S.: How Loading Time Affects Your Bottom Line (2019). https://neilpatel.com/blog/loading-time/

Photofeeler-D3: A Neural Network with Voter Modeling for Dating Photo Impression Prediction

Agastya Kalra$^{(\boxtimes)}$ and Ben Peterson

Photofeeler Inc, Denver, CO, USA
agastya.kalra@gmail.com, ben@photofeeler.com

Abstract. In just a few years, online dating has become the dominant way that young people meet to date, making the deceptively error-prone task of picking good dating profile photos vital to a generation's ability to form romantic connections. Until now, artificial intelligence approaches to Dating Photo Impression Prediction (DPIP) have been very inaccurate, unadaptable to real-world application, and have only taken into account a subject's physical attractiveness. To that effect, we propose Photofeeler-D3 - the first convolutional neural network as accurate as 10 human votes for how smart, trustworthy, and attractive the subject appears in highly variable dating photos. Our "attractive" output is also applicable to Facial Beauty Prediction (FBP), making Photofeeler-D3 state-of-the-art for both DPIP and FBP. We achieve this by leveraging Photofeeler's Dating Dataset (PDD) with over 1 million images and tens of millions of votes, our novel technique of voter modeling, and cutting-edge computer vision techniques.

Keywords: Computer vision · Online dating · Facial beauty prediction · Voter modeling

1 Introduction

Online dating is the future [1]. In 2017, Tinder, a mobile dating app, became the number one top grossing app in the Apple App Store [2]. In 2019, 69% of Generation Z is active on dating apps, making online dating the dominant means of meeting people to date for 18-to-22-years-old today [3,4].

But Generation Z's record loneliness [5] may point to the ineffectiveness of the current dating platforms. The leading dating apps' profiles are highly dependent on photos. Research says that photos are misleading because different photos of the same person can give entirely different impressions [6]. To make matters worse, individuals display bad judgment in choosing their own photos [7].

A. Kalra—This author served as a research partner for the duration of this project but is not an employee at Photofeeler Inc. For any inquiries related to Photofeeler Inc. please email ben@photofeeler.com.

K. Arai et al. (Eds.): FTC 2019, AISC 1070, pp. 188–203, 2020.
https://doi.org/10.1007/978-3-030-32523-7_13

But according to The Guardian, 90% of people decide to date someone based on their dating photos alone [8] - meaning that picking the right photo is vital to one's success. This is why assistance in choosing dating profile photos is sorely needed in order to facilitate the right connections. The lifelong partnerships of many millions of people depend on this.

In industry, besides hiring an expert [9], there are two types of online services used to evaluate photos for dating profiles: online voting websites [10], and online artificial intelligence platforms (OAIPs) [11,12]. Photofeeler [10] is the largest of the first category, with over 90 million votes cast, 2 million images voted on, and over 100k new votes every day. In this work, we show that the Photofeeler-D3 network achieves the same statistical significance as 10 (unnormalized and unweighted) human votes, making it more accurate than a small group of independent humans, but not as accurate as a Photofeeler voting test. In terms of the OAIPs, the most popular platforms are hotness.ai [12] and prettyscale.com [11], both of which only measure attractiveness. They are evaluated on the London Faces Dataset [13] in a 2018 study [14]. We outperform both of these online services on that benchmark by achieving over 28% higher correlation with human voters.

While optimizing for the most attractive photo is a good proxy for maximizing matches, attractiveness alone is not the optimal metric if the goal is to find high quality matches that lead to actual dates and long-term relationships [15]. That is why Photofeeler's voting-based online dating photo rating service also measures the smart and trustworthy traits. This allows users to find the photo that not only makes them look *hot*, but also reliable, principled, intellectual, and safe to meet with in person. With this in mind, the Photofeeler-D3 neural network outputs scores for these 3 traits - the first neural network to do so.

In literature, the closest well-studied task is Facial Beauty Prediction (FBP) [16–25]. In FBP, the goal is to take a perfectly cropped photo of the subject's face looking forward in a neutral position, and predict the objective attractiveness of that individual [16]. In our case, the photos are of people in different settings, poses, expressions, outfits, makeup, lighting, and angles, taken with a variety of cameras. We show that our model's attractiveness output also works for FBP, achieving state-of-the-art performance on the benchmark SCUT-FBP dataset [16].

FBP has received some backlash on social media [26] due to the ethics of objectively assigning attractiveness scores to individuals. In DPIP, the ratings are assigned to the photos, not the individual. Figure 1 shows photos from the Photofeeler Dating Dataset (PDD) of the same person with very different scores. The goal of DPIP is to give people the best chance at successfully finding long-term relationships in dating apps through selecting photos for the profile as *objectively* as possible. We discuss FBP methods further in Sect. 2, and compare to existing benchmarks in Sect. 4.

In this work, we explore the idea of using AI to predict the impressions given off by dating photos. We create a neural network that achieves state-of-the-art results on a variety of benchmark datasets [13,16,19] and matches the accuracy of a small group of human voters for DPIP. We introduce voter modeling as

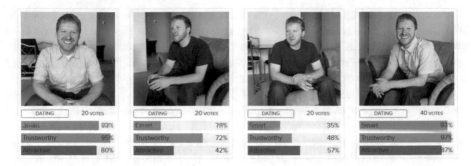

Fig. 1. Four example photos of the same subject with different scores from the Photofeeler Dating Dataset.

an alternative solution to predicting average scores for each trait, which helps lower the impact of noise that comes from images without many votes. Finally we discuss the implications of our results on using votes to rate the smart, trustworthy, and attractive traits in single-subject photos.

The remainder of the paper is structured as follows. Section 2 reviews similar public datasets, convolutional neural networks, approaches for FBP, and online AI services for DPIP. Section 3 describes the PDD structure and the Photofeeler-D3 architecture and training procedure. Section 4 contains results on benchmark datasets and discussion. Section 5 summarizes the findings of the paper. Demo is available at https://blog.photofeeler.com/photofeeler-d3/.

2   Related Work

In this section, we discuss the relevant benchmark datasets, convolutional neural network architectures, facial beauty prediction, and OAIPs.

Datasets. There are a variety of benchmark datasets for scoring images: The AVA dataset [27], the Hot-Or-Not dataset [19], the SCUT-FBP dataset [16], the LSFCB dataset [20], the London Faces Dataset [13], and the CelebA dataset [28]. The AVA dataset [27] doesn't have attractiveness ratings for the subject, instead they have an attractiveness rating for the entire image i.e. *Is this a good photo?*, which is very different from *Does the subject look good in this photo?*. The Hot-Or-Not [19] dataset contains 2k images of single subject photos with at least 100 votes from the opposite sex on a 1–10 attractiveness scale. We report performance on this dataset since this is the closest publicly available dataset to our own. The SCUT-FBP [16] dataset is the standard benchmark for the FBP task - containing 500 images of cropped Asian female faces in neutral position staring forward into the camera. We benchmark our Photofeeler-D3 architecture on the SCUT-FBP dataset since the task is similar. The London Faces dataset is similar to the SCUT-FBP dataset except it contains 102 images of diverse males and females. It was used to benchmark prettyscale.com [11] and hotness.ai [12], so we use it to benchmark our Photofeeler-D3 network. The LSFCB [20] dataset contains 20k images for FBP but is not publicly available, so we do not include it. The CelebA [28] dataset contains a binary indicator for attractiveness marked by a single labeler for each image, which is very different from DPIP, so we do not include it in our work. Sample photos from each dataset are shown in Fig. 2.

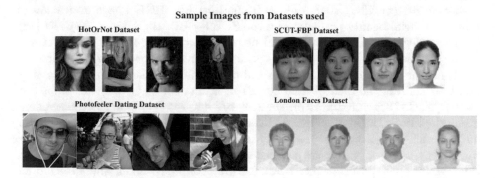

Fig. 2. Sample photos from each dataset. The London Faces Dataset and the SCUT-FBP dataset are simpler than the HotOrNot dataset and the Photofeeler Dating Dataset.

Convolutional Neural Networks. In the last six years, convolutional neural networks (CNNs) have achieved state-of-the-art results in a variety of computer vision tasks including classification [29–34], bounding box prediction [35], and image segmentation [36]. We present a brief review of relevant CNN architectures. *Architectures:* The first major CNN architecture to be popularized was AlexNet [29] after its 2012 ILSVRC [29] win. It had 8 layers, used large convolution kernels and was the first successful application of dropout. After that, a variety of improvements have come along. VGG16 [33] won ILSVRC in 2014 by using many small kernels rather than a few large ones. In 2015, it was dominated by Residual Networks (ResNets) [32] where they introduced the idea of deep architectures with skip connections. In 2016, it was won by the InceptionResNetV2 [31], which combined the inception architecture [30] with skip connections to achieve even higher accuracy. In 2017, the Xception [34] architecture was introduced, which matched the performance of InceptionResNetV2 with much fewer parameters by leveraging depth-wise separable convolution layers. In 2018, the Neural Architecture Search Network [37] (NASNet) was published - an architecture generated through reinforcement learning. However, due it its size and complexity, it has yet to gain popularity. In our work we compare all architectures listed here since ResNet, not including NASNet.

Facial Beauty Prediction. Facial Beauty Prediction is the task of objectively assessing the average attractiveness rating of a face in a neutral position looking forward into the camera [16]. This is very different from DPIP because in DPIP the subject is rated in different contexts. Traditional FBP algorithms [16] relied on facial landmarks and some combination of hand-engineered rules and shallow machine learning models. However since 2015, CNNs have dominated the FBP task [17,18,20–24,38] due to the wide availability of pretrained networks and increased access to public data. Gray et al. [23] proposed a 4 layer CNN and were the first to discard facial landmarks. Gan et al. [39] used deep learning to extract beauty features instead of artificial feature selection. Xu et al. [17] used a specific 6 layer CNN that took as input both the RGB image and a *detail image* for facial beauty prediction on the SCUT-FBP [16] dataset. PI-CNN [18] - a psychology inspired convolutional neural network, introduced by Xu et al, separated the facial beauty representation learning and predictor training. Xu et al. [21] proposed using models pretrained on other facial tasks as a starting point to address the lack of data for FBP. Anderson et al. [24] benchmark a variety of CNN architectures on the CelebA dataset for binary attractiveness prediction. Both Fan et al. [40] and Liu et al. [25] propose replacing the regression output with a distribution prediction output and using a KL-Divergence loss rather than the standard mean squared error. We adopt a similar architecture to this. Gao et al. [41] utilize a multi-task learning training scheme where the model is required to output facial key-points along with average attractiveness scores. In CR-Net [22], Xu et al. propose using a weighted combination of mean squared error and cross-entropy loss to improve resilience to outliers when training. All of these works benchmark on either the HotOrNot [19] dataset or the SCUT-FBP [16] dataset. We benchmark Photofeeler-D3 on both.

Online AI Platforms. There are two main OAIPs for attractiveness scoring: hotness.ai [12] and prettyscale.com [11]. Neither of these measure the smart or trustworthy traits. hotness.ai [12] takes a photo of a single subject in any pose and, using an unspecified deep learning algorithm based on facial embedding, gives a discrete attractiveness score from 1–10. prettyscale.com [11] requires the user to give a photo of a face and specify some geometric facial key-points. It uses a variety of hand-crafted rules to give an attractiveness score from 0–100. Photofeeler-D3 is different in that it takes in an image of a single subject in any pose and outputs smart, trustworthy, and attractive scores. Additionally, we conjecture that the Photofeeler-D3 network is the only one to use voter modeling. We show comparisons to both these tools in Sect. 4.

3   Our Method

In this section, we discuss the Photofeeler Dating Dataset and the Photofeeler-D3 neural network.

3.1   Photofeeler Dating Dataset

The PDD contains 1.2 million dating photos - 1 million male images of 200k unique male subjects and 200k female images of 50k unique female subjects. The images have a variety of aspect ratios, but the maximum side is at most 600 pixels. The metadata for each image contains a list of voters, a weight from 0–1 for each vote (used to filter out poor quality votes), and both their normalized vote in the range 0–1 and their original *raw* vote in the range 0–3 for each of the 3 traits. We normalize the votes for each voter depending on how they vote, i.e. if a voter gives mostly 0s and 1s, then a 2 from that voter will have a much higher normalized score than a voter who normally gives 2s and 3s. The weights are determined by how predictable a voter is, so a voter who always votes 1 will have a weight of 0. We exclude the weighting and normalization algorithms since they are Photofeeler Intellectual Property, however these algorithms dramatically improve the quality of the scores. All voters are the opposite sex of the subject in the photo. We compute the test labels y_{it} for each image x_i as a weighted sum of all the normalized votes v_{ijt} where i is the image index, j is the voter index, t is the trait (one of smart, attractive, or trustworthy) and Γ_i is the set of voters that voted on the image x_i. It is important to note that these labels are not the "true score" of the image, as these traits are subjective. Rather they are noisy estimates of the population mean scores. We will demonstrate later how modeling this subjectivity is critical to our approach.

$$y_{it} = \frac{\sum_{j \in \Gamma_i} w_j * v_{ijt}}{\sum_{j \in \Gamma_i} w_j} \tag{1}$$

We separate out 10000 male subject images and 8000 female subject images to be used for testing. We guarantee that these images are of subjects not found in the training set and contain at least 10 votes to ensure some amount of statistical significance. We choose not to restrict it further since images with a higher number of votes tend to have higher scores because images getting low scores tend to be replaced more quickly by the user. Thus restricting it further would reduce the diversity of images in the test set.

For experimentation on hyperparameters, we set aside what we call the *small dataset*, a 25,311 male image subset of the training set (20,000 train, 3,000 val, 2,311 test), and evaluate only on the attractiveness class. Evaluating on male attractiveness will provide a lower bound for all classes because it is the most difficult to predict out of all traits and demographics. For training and evaluating our final set of hyperparameters, we use all male and female images in the PDD, we call this the *large dataset*. When comparing the Photofeeler-D3 neural network to other benchmarks and to humans, we use the result of training on these 2 sets. We evaluate the model by using the Pearson correlation coefficient (PC) rather than mean squared error (MSE) because a model that simply predicts the sample mean of the training set for each test image would get a decent MSE, but would get a PC of 0%.

The dataset was collected using Photofeeler's online voting platform [10]. Users upload their images and then receive votes based on how many votes they cast. The voters are given an image and told to rate their impression of each trait on a scale of 0 (No) to 3 (Very). These votes are weighted and normalized to create a score for each image on a scale of 0–1. Due to improvements in the normalization and weighting algorithm over time, the training images collected earlier have a higher error rate. The test set was taken from recent images only.

3.2   Photofeeler-D3 Neural Network

Architecture. To effectively evaluate the impressions given off by images we use a 4 part architecture - the base network, the temporary output, the voter model, and the aggregator. We go through each component with respect to a single trait for simplicity. Extending to multiple traits just requires an output per trait (see Fig. 3 for Photofeeler-D3 Architecture).

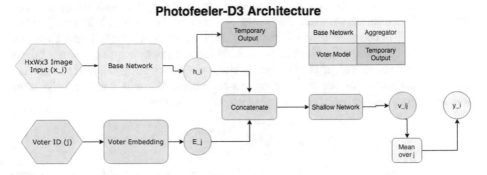

Fig. 3. A diagram of the Photofeeler-D3 Architecture

Base Network: We take as input an HxWx3 RGB image and forward pass it through one of the standard classification architectures [31–34] mentioned in Sect. 2.2. This is our base network $h_i = g(x_i; \theta)$, where x_i represents the input image, the vector h_i represents the output of the base network, and θ represents the parameters of the model. The choice of architecture and image input size are hyperparameters. In our best model we used the Xception [34] architecture with input image size $600 \times 600 \times 3$. We center the images and pad the rest with black for non-square images. To get the vector h_i, we remove all fully connected layers from the architecture and apply global average pooling to the last convolutional layer.

$$h_i = g(x_i; \theta) \tag{2}$$

Temporary Output: Starting with h_i, we apply a fully connected layer with weight matrix W_t for each trait t and produce a per-trait output of $\bar{o}_{ti}$. The output is a 10-dimensional vector that has gone through the softmax σ activation function. This output is used during training but removed during testing. Note we omit the t subscript from the equation for simplicity.

$$\bar{o}_i = \sigma(Wh_i) \tag{3}$$

Voter Model: We introduce an embedding matrix E that contains one row for each unique voter. To compute the predicted normalized vote score $\bar{v}_{ij}$ for voter j on image x_i, we concatenate the output of the base network h_i with the voter embedding E_j and apply a shallow fully connected network ϕ that produces a distribution over 10 classes. For the label, we turn v_{ij} into a 10-dimensional

1-hot encoding by rounding v_{ij} to the nearest 0.1. To construct the real number $\bar{v}_{ij}$, we take the inner product of the output of ϕ with the vector b. The vector b is defined as $[0.05, 0.15, 0.25...0.95]$.

$$\bar{v}_{ij} = < \phi([h_i, E_j]), b > \tag{4}$$

Aggregator: To produce the final prediction $\bar{y}_i$ we aggregate the predicted vote of a random sample β of 200 voters. This allows us to get a stable estimate of what many voters will think of a single image.

$$\bar{y}_i = \frac{1}{n} \sum_{j \epsilon \beta} \bar{v}_{ij} \tag{5}$$

Voter modeling means that the model is no longer trying to predict a potentially noisy estimate of the population mean. Instead it is predicting the subjective opinion of a particular voter - the score they would give to the image. This means the number of votes per image is less relevant to the training process and therefore the impact of noise that comes from images without many votes is limited. Then we take the mean of many predicted votes in the aggregation step to get a stable prediction of the scores of an image. We demonstrate the effectiveness of this system in Sect. 4.

Training. A separate model is trained for images with female subjects and images with male subjects. We follow the same 2-part training scheme for both; the first part trains the base network and the second part trains the voter model. We split it up because joint training does not seem to converge to a good solution.

Base Model Training: The first step in the training pipeline involves training the base model with the temporary output. We describe the training for a single trait for simplicity. The input x_i is an image of size HxWx3 and the label y_i is a 10-dimensional vector representing a discrete distribution over all the votes for trait t and image x_i. We adopt this idea from Liu et al. [25] except our votes are weighted. To compute y_i, we must compute its unnormalized form $y\prime_i$ and then normalize. We compute each entry k of $y\prime_i$ as follows:

$$y\prime_{ik} = \sum_{j \epsilon \Omega} w_j$$
$$s.t. \tag{6}$$
$$\Omega = \{j | 0.1(k-1) < v_{ij} < 0.1k\}$$

And then normalize.

$$y_{ik} = \frac{y\prime_{ik}}{\sum_p y\prime_{ip}} \tag{7}$$

The loss is the KL-divergence between the true distribution y_i and the predicted distribution $\bar{o}_i$. This provides the same gradient as cross-entropy but is easier to interpret for the case that the labels y_i are not a 1-hot encoding. We train on 1 pass through the data with the Adam [42] optimizer and an initial learning rate of 1e-4. This prepares the base model for the second phase of training.

Voter Model Training: The second step in the training pipeline involves training the voter model which was described in Sect. 3.2. The goal is the learn the parameters of ϕ and the entries of the embedding matrix E. The input is now the image x_i and a random voter id j, where j comes from the set of voters that have voted on x_i, and the label is a 1-hot encoding of the vote v_{ij}. We keep the base network frozen allowing us to train with batch size 1000. We use the Adam [42] optimizer with learning rate 1e-3 and the standard cross-entropy loss function. This second step of training is much faster than the first step.

Testing. At test time, the model takes in a test image x_i and returns the output y_i of the aggregator step. We randomly sample a set of 200 voter ids for testing, allowing us to achieve a stable prediction.

4   Experimental Results and Discussion

We implement our Photofeeler-D3 network in Keras [43] on a p2.xlarge AWS instance [44]. First we evaluate different hyperparameters on our small dataset. Then we evaluate our best set of hyperparameters on the large female and the large male datasets. We evaluate this model against human voters to understand *How many human votes are the model's predictions worth?* Finally, we compare against OAIPs and the architectures in the FBP task.

4.1   Hyperparameter Selection

The key hyperparameters in the Photofeeler-D3 architecture are the image input size, the base network architecture, and the output type. We conduct experiments on the small dataset to confirm the best hyperparameters, and then apply them when training on our large datasets.

Image Size: Standard CNNs for classification use somewhere between a 224×224 and 300×300 image size [33,34] for their inputs. However we noticed that in images this small, the subject's face is not always clear. Since facial expressions are very important to a voter's impression [45], we conjecture that larger image sizes will lead to improved performance. To verify, we conduct an experiment where we fix the base network and the output type, and vary the image size. We show in Table 1(left) that using the maximum image size of 600×600 achieves almost 9% higher correlation on the small test set than 224×224. When the images are small the facial expression is not clearly visible, so the model struggles to correctly evaluate the image.

Architecture: It's always hard to determine the best base model for a given task, so we tried four standard architectures [31–34] on our task and evaluated them on the small dataset. Table 1(middle) shows that the Xception [34] architecture outperforms the others, which is surprising since InceptionResNetV2 [31] outperforms Xception on ILSVRC [29]. One explanation is that the Xception architecture should be easier-to-optimize than the InceptionResNetV2. It contains far fewer parameters and a simpler gradient flow [34]. Since our training

Table 1. Quantitative comparison of different hyperparameters for attractiveness prediction on the small dataset. On the left we have image size, in the middle we compare architectures, and on the right we compare output types. This table shows that the best hyperparameters are 600×600 for image size, Xception of architecture, and voter modeling for output type.

Image Size	PC (%)
224x224	49.3
312x312	52.0
448x448	56.8
600x600	**59.1**

Architecture	PC (%)
VGG16 [33]	42.3
ResNet50 [32]	44.6
Inception ResNetV2 [31]	47.5
Xception [34]	**49.6**

Output Type	PC (%)
Regression	39.3
Classification	47.5
Distribution Modeling	51.3
Voter Modeling	**54.5**

dataset is noisy, the gradients will be noisy. When the gradients are noisy, the easier-to-optimize architecture should outperform.

Output Type: There are four main output types to choose from: regression [17, 21], classification [22, 33], distribution modeling [25, 40], and voter modeling. The results are shown in Table 1 (right). For regression [21] the output is a single neuron that predicts a value in range [0, 1], the label is the weighted average of the normalized votes, and the loss is mean squared error (MSE). This performs the worst because the noise in the training set leads to poor gradients which are a large problem for MSE. Classification [22] involves a 10-class softmax output where the labels are a 1-hot encoding of the rounded population mean score. We believe this leads to improved performance because the gradients are smoother for cross-entropy loss. Distribution modeling [25, 40] with weights, as described in Sect. 3.2, gives more information to the model. Rather than a single number, it gives a discrete distribution over the votes for the input image. Feeding this added information to the model increases test set correlation by almost 5%. Finally we note that voter modelling, as described in Sect. 3.2, provides another 3.2% increase. We believe this comes from modeling individual voters rather than the sample mean of what could be very few voters.

We select the hyperparameters with the best performance on the small dataset, and apply them to the large male and female datasets. The results are displayed in Table 2. We notice a large increase in performance from the small dataset because we have 10× more data. However we notice that the model's predictions for attractiveness are consistently poorer than those for trustworthiness and smartness for men, but not for women. This shows that male attractiveness in photos is a more complex/harder-to-model trait. There are a lot of subtleties to what makes a male subject attractive for dating.

Table 2. Correlation results of Photofeeler-D3 model on large datasets for both sexes

Dataset	Smart	Trustworthy	Attractive
Large Female	81.4	83.2	81.6
Large Male	80.4	80.6	74.3

4.2   Photofeeler-D3 vs. Humans

While Pearson correlation gives a good metric for benchmarking different models, we want to directly compare model predictions to human votes. We devised a test to answer the question: *How many human votes are the model's prediction worth?*. For each example in the test set with over 20 votes, we take the normalized weighted average of all but 15 votes and make it our *truth* score. Then from the remaining 15 votes, we compute the correlation between using 1 vote and the truth score, 2 votes and the truth score, and so on until 15 votes and the truth score. This gives us a correlation curve for up to 15 human votes. We also compute the correlation between the model's prediction and *truth* score. The point on the human correlation curve that matches the correlation of the model gives us the number of votes the model is worth. We do this test using both normalized, weighted votes and raw votes. Table 3 shows that the model is worth an averaged 10.0 raw votes and 4.2 normalized, weighted votes - which means it is better than any single human. Relating it back to online dating, this means that using the Photofeeler-D3 network to select the best photos is as accurate as having 10 people of the opposite sex vote on each image. This means the Photofeeler-D3 network is the first provably reliable OAIP for DPIP. Also this shows that normalizing and weighting the votes based on how a user tends to vote using Photofeeler's algorithm increases the significance of a single vote. As we anticipated, female attractiveness has a significantly higher correlation on the test set than male attractiveness, yet it is worth close to the same number of human votes. This is because male votes on female subject images have a higher correlation with each other than female votes on male subject images.

Table 3. Quantitative study showing the number of human votes the model's predictions are worth with respect to each trait. Normalized votes indicates that the votes have gone through Photofeeler's vote weighting and normalizing process. Unnormalized votes contain more noise, therefore the model's predictions are worth more unnormalized votes than normalized ones.

Dataset	Norm. Votes?	Smart (#votes)	Trustworthy (#votes)	Attractive (#votes)	Mean (#votes)
Large Female	Yes	4.4	5.0	2.7	4.0
Large Female	No	11.6	14.5	5.2	10.4
Large Male	Yes	4.9	5.2	3.1	4.4
Large Male	No	11.2	11.6	5.9	9.6

This shows not just that predicting male attractiveness from photos is a more complex task than predicting female attractiveness from photos, but that it is equally more complex for humans as for AI. So even though AI performs worse on the task, humans perform *equally worse* meaning that the ratio stays close to the same.

4.3   Photofeeler-D3 vs. OAIPs

To compare to OAIPs, we evaluate prettyscale.com [11], hotness.ai [12], and the Photofeeler-D3 network on the London Faces dataset [13]. For prettyscale.com and hotness.ai, we use results from an online study [14]. Table 4 shows that our model outperforms both of these by at least 28% correlation. Photofeeler is the largest online voting platform in the world, therefore the PDD is one of the largest datasets in the world for attractiveness prediction [10]. Through leveraging this data and applying the voter modeling technique, we achieve state-of-the-art performance in OAIPs.

Table 4. Quantitative comparison of Photofeeler-D3 against other OAIPs on the London Faces Dataset.

OAIP	PC (%)
prettyscale.com [11]	53
hotness.ai [12]	52
Photofeeler-D3	**81**

4.4   Photofeeler-D3 in FBP

In FBP there are two main datasets: the SCUT-FBP dataset [16] and the HotOrNot dataset [19]. The SCUT-FBP dataset contains 500 female subject images with 10 votes per image from both male and female voters rating the subject's attractiveness from 1–7. The task is to predict the average attractiveness score for an image. This task is different from DPIP for a few reasons: there are only 10 votes - meaning there will be quite a bit of noise; the voters are both male and female, not just male; and the images are not natural, they are neutral faces looking forward into the camera. In the literature, we find some works that only show the best run on the dataset [17,22,25,40],and other works that do a 5-fold cross validation [18,21,41] on the dataset. We test our system both ways. We use only the Pearson correlation metric because our scale is from 0–1 whereas the dataset has a scale from 1–7. The Photofeeler-D3 architecture has 3 outputs, one for each trait. To adapt to this dataset, we use only the attractiveness output. All results are shown in Table 5. We show that without any training on the dataset, the Photofeeler-D3 architecture achieves 89% best run and 78% in cross validation. Although this is not state-of-the-art, these are still good scores considering how different the task is. If we allow the network to retrain we get 91% cross validation and 92% as the best run. This is the best score for cross validation. Additionally, we believe that all of the architectures are getting quite close to the limit on the dataset since there are only 500 examples with 10 votes each. Anything above 90% correlation is probably fitting the noise of the dataset. We notice that with our dataset, using the average of 10 raw votes is only 87% correlated with using the average of all the votes.

The HotOrNot [19] dataset contains 2000 images, 50% male subjects and 50% female subjects. Each image has been voted on by over 100 people of the opposite sex. Results are available in Table 5. All other FBP methods [22,23,38] first use the Viola-Jones algorithm to crop out the faces and then forward pass their models. Our method takes in the full image, resizes it to 600×600, and forward passes the Photofeeler-D3 network. We show that without any training on this dataset, we achieve 55.9% cross validation accuracy, outperforming the next best by 7.6%. Another interesting observation is that our model achieves 68% correlation with the 1000 females and 42% correlation with the 1000 males. This reinforces the hypothesis that male attractiveness is a much more complex function to learn than female attractiveness.

Table 5. Quantitative Analysis of different models on the Facial Beauty Prediction Task on both the SCUT-FBP dataset and the HotOrNot dataset.

Architecture	SCUT-FBP Best Run	SCUT-FBP 5 Fold CV	HotOrNot
MLP [18]	76	71	-
AlexNet-1 [41]	90	84	-
AlexNet-2 [41]	92	88	-
PI-CNN [18]	87	86	-
CF [17]	88	-	-
LDL [40]	**93**	-	-
DRL [25]	**93**	-	-
MT-CNN [41]	92	90	-
CR-Net [22]	87	-	48.2
TRDB [21]	89	86	46.8
AAE [38]	-	-	43.7
Multi-scale [23]	-	-	45.8
Photofeeler-D3 - No Training	89	78	**58.7**
Photofeeler-D3 - Retraining	92	**91**	-

5   Conclusion and Future Work

In this work we propose the Photofeeler-D3 architecture that, taking advantage of the Photofeeler Dating Dataset and the concept of voter modeling, achieves state-of-the-art results. Additionally, we demonstrate that using our model to select the best dating photos is as accurate as having 10 humans vote on each photo and selecting the best average score. Through this work, we also conclude that Photofeeler's normalizing and weighting algorithm dramatically decreases noise in the votes. Finally we note that although male attractiveness seems to be more difficult to model than female attractiveness, it is equally more difficult for both humans and AI.

While this work has pushed state-of-the-art dramatically, there is still room for improvement. Firstly, extending the embedding architecture to show improvement joint training will improve performance. Secondly, our database actually has 9 traits not 3. Leveraging these 9 traits should improve generalization through multi-task learning. Thirdly, we could implement more complex multi-task training methods rather than just one output head per task. Finally, we should investigate the model's accuracy for modeling different voter age groups and ethnicities with the learned embeddings.

References

1. Fetters, A.: The 5 years that changed dating, December 2018. (Online; posted 21 December 2018)
2. Crook, J.: Tinder hits top grossing app in the app store on heels of tinder gold launch (2017). (Online; posted 2017)
3. Watts, J.: Gen-z courts intrigue on tinder, February 2019. (Online; posted 13 February 2019)
4. Jones, D.: How generation z handles online dating, January 2019. (Online; posted 15 January 2019)
5. Marcos, A.: Move over, millennials: How 'igen' is different from any other generation, August 2017. (Online; posted 22 August 2017)
6. Todorov, A., Porter, J.M.: Misleading first impressions: different for different facial images of the same person. Psychol. Sci. **25**, 1404–1417 (2014)
7. White, D., Sutherland, C.A.M., Burton, A.L.: Choosing face: the curse of self in profile image selection. Cogn. Res. Princ. Implic. **2**, 23 (2017)
8. Nelsori, S.: Dos and don'ts for choosing stand-out dating profile photos. (Online)
9. We've discovered 7 traits that girls go nuts over... which ones are missing from your profile? April 2019. (Online)
10. Pierce, A.: Frequently asked questions, January 2019. (Online)
11. prettyscale.com. Am i beautiful or ugly? (Online)
12. hotness.ai. How hot are you? (Online)
13. DeBruine, L., Jones, B.: Face Research Lab London Set, May 2017
14. Richardson, T.: An attractiveness researcher puts the internet's most popular "hotness algorithms" to the test, August 2018. (Online)
15. Pierce, A.: How to pick your best dating profile pictures based on photofeeler scores, November 2018. (Online; posted 16 November 2018)
16. Xie, D., Liang, L., Jin, L., Xu, J., Li, M.: SCUT-FBP: a benchmark dataset for facial beauty perception. In: 2015 IEEE International Conference on Systems, Man, and Cybernetics, pp. 1821–1826. IEEE (2015)
17. Xu, J., Jin, L., Liang, L., Feng, Z., Xie, D.: A new humanlike facial attractiveness predictor with cascaded fine-tuning deep learning model. arXiv preprint arXiv:1511.02465 (2015)

18. Xu, J., Jin, L., Liang, L., Feng, Z., Xie, D., Mao, H.: Facial attractiveness prediction using psychologically inspired convolutional neural network (PI-CNN). In: 2017 IEEE International Conference on Acoustics, Speech and Signal Processing (ICASSP), pp. 1657–1661. IEEE (2017)

19. Donahue, J., Grauman, K.: Annotator rationales for visual recognition. In: 2011 International Conference on Computer Vision, pp. 1395–1402. IEEE (2011)

20. Zhai, Y., Cao, H., Deng, W., Gan, J., Piuri, V., Zeng, J.: Beautynet: joint multiscale CNN and transfer learning method for unconstrained facial beauty prediction. Comput. Intell. Neurosci. **2019** (2019)

21. Xu, L., Xiang, J., Yuan, X.: Transferring rich deep features for facial beauty prediction. arXiv preprint arXiv:1803.07253 (2018)

22. Xu, L., Xiang, J., Yuan, X.: CRNet: classification and regression neural network for facial beauty prediction. In: Pacific Rim Conference on Multimedia, pp. 661–671. Springer (2018)

23. Gray, D., Yu, K., Xu, W., Gong, Y.: Predicting facial beauty without landmarks. In: European Conference on Computer Vision, pp. 434–447. Springer (2010)

24. Anderson, R., Gema, A.P., Isa, S.M., et al.: Facial attractiveness classification using deep learning. In: 2018 Indonesian Association for Pattern Recognition International Conference (INAPR), pp. 34–38. IEEE (2018)

25. Liu, S., Li, B., Fan, Y.-Y., Guo, Z., Samal, A.: Facial attractiveness computation by label distribution learning with deep CNN and geometric features. In: 2017 IEEE International Conference on Multimedia and Expo (ICME), pp. 1344–1349. IEEE (2017)

26. [R] SCUT-FBP5500: A diverse benchmark dataset for multi-paradigm facial beauty prediction, February 2018. (Online)

27. Ava-dataset, Apirl 2019. (Online)

28. Liu, Z., Luo, P., Wang, X., Tang, X.: Large-scale celebfaces attributes (celeba) dataset. Retrieved August **15**, 2018 (2018)

29. Krizhevsky, A., Sutskever, I., Hinton, G.E.: Imagenet classification with deep convolutional neural networks. In: Advances in Neural Information Processing Systems, pp. 1097–1105 (2012)

30. Szegedy, C., Vanhoucke, V., Ioffe, S., Shlens, J., Wojna, Z.: Rethinking the inception architecture for computer vision. In: Proceedings of the IEEE Conference on Computer Vision and Pattern Recognition, pp. 2818–2826 (2016)

31. Szegedy, C., Ioffe, S., Vanhoucke, V., Alemi, A.A.: Inception-v4, inception-resnet and the impact of residual connections on learning. In: Thirty-First AAAI Conference on Artificial Intelligence (2017)

32. He, K., Zhang, X., Ren, S., Sun, J.: Deep residual learning for image recognition. In: Proceedings of the IEEE Conference on Computer Vision and Pattern Recognition, pp. 770–778 (2016)

33. Simonyan, K., Zisserman, A.: Very deep convolutional networks for large-scale image recognition. arXiv preprint arXiv:1409.1556 (2014)

34. Chollet, F.: Xception: deep learning with depthwise separable convolutions. In: Proceedings of the IEEE Conference on Computer Vision and Pattern Recognition, pp. 1251–1258 (2017)

35. Ren, S., He, K., Girshick, R., Sun, J.: Faster R-CNN: towards real-time object detection with region proposal networks. In: Advances in Neural Information Processing Systems, pp. 91–99 (2015)

36. Ronneberger, O., Fischer, P., Brox, T.: U-net: convolutional networks for biomedical image segmentation. In: International Conference on Medical Image Computing and Computer-assisted Intervention, pp. 234–241. Springer (2015)

37. Zoph, B., Le, Q.V.: Neural architecture search with reinforcement learning. arXiv preprint arXiv:1611.01578 (2016)
38. Wang, S., Shao, M., Fu, Y.: Attractive or not? beauty prediction with attractiveness-aware encoders and robust late fusion. In: Proceedings of the 22nd ACM International Conference on Multimedia, pp. 805–808. ACM (2014)
39. Gan, J., Li, L., Zhai, Y., Liu, Y.: Deep self-taught learning for facial beauty prediction. Neurocomputing **144**, 295–303 (2014)
40. Fan, Y.-Y., Liu, S., Li, B., Guo, Z., Samal, A., Wan, J., Li, S.Z.: Label distribution-based facial attractiveness computation by deep residual learning. IEEE Trans. Multimed. **20**(8), 2196–2208 (2018)
41. Gaol, L., Li, W., Huang, Z., Huang, D., Wang, Y.: Automatic facial attractiveness prediction by deep multi-task learning. In: 2018 24th International Conference on Pattern Recognition (ICPR), pp. 3592–3597. IEEE (2018)
42. Kingma, D.P., Ba, J.: Adam: a method for stochastic optimization. arXiv preprint arXiv:1412.6980 (2014)
43. Chollet, F., et al.: Keras (2015)
44. Amazon Inc.: Amazon web services (2019)
45. Pierce, A.: Why dating pics that look trustworthy = more dates, June 2016. (Online)

Human-Augmented Robotic Intelligence (HARI) for Human-Robot Interaction

Vishwas Mruthyunjaya[1] and Charles Jankowski[2(✉)]

[1] Aisera, Palo Alto, USA
vishwas.mruthyunjaya@aisera.com
[2] Cloudminds Technology, Santa Clara, USA
charles.jankowski@cloudminds.com

Abstract. This paper provides a system design framework for a human-robot interaction system. The design introduces a human-augmented robotic intelligence embedded in a human-robot interaction system. The motivations behind the system design are spoken dialogue systems, Wizard-of-OZ framework, and existing HRI designs for socially intelligent robots. In this work, we explain how artificial intelligence of human-robot interaction system is enhanced by human intelligence through collaboration. The collaborative artificial intelligence enables the system to learn from demonstration. The main objective and the gradual progression from this paper is to build an iterative interactive system that is capable of achieving human-robot interaction similar to the nuances of human-human interaction.

Keywords: Artificial intelligence · Human-robot interaction · Cloud robot · Human-augmented robotic intelligence · Human-in-the-loop · Robot control unit · Virtual Backbone Network · Social robot · Spoken dialogue systems · Automatic speech · Natural language processing

1 Introduction

The term robot, most of the time, attracts a human's attention. The attention of the people, in general, is a way of acknowledging the capability and functionality of the robot to perform a task. The popular belief is that the general audience often imagines a physical robot as a mechanical humanoid-like figure that executes complex tasks, unlike a typical machine [17, 18]. The movies, comic books, and the media influence the idea of a humanoid-like figure for a robot in the mind of people.

One of the many capabilities of a robot is the ability to interact with humans, which has given rise to a specific and defined focus for research in the field of human-robot interaction (HRI). The robot interaction is an essential aspect in robotics, ranging from the industrial robotics to social robotics. The only differing feature is the level of interaction the robots possess [17].

© Springer Nature Switzerland AG 2020
K. Arai et al. (Eds.): FTC 2019, AISC 1070, pp. 204–223, 2020.
https://doi.org/10.1007/978-3-030-32523-7_14

The author in [17] define HRI as a field of study dedicated to understanding, designing, and evaluating robotic systems for use by or with humans. Interaction, in general, is a communication between two or more entities. In HRI, interaction requires communication between the robots and the humans. The communication in HRI has two broad categories: remote interaction and proximate interaction [18].

"The study of the processes involved in perceiving each other and coming to 'know what we know' about the people in our world is essentially a question not only of what behaviour we have seen but of our cognition as individual perceivers-our social cognition. Social cognition, therefore, is the study of the mental processes involved in perceiving, attending to, remembering, thinking about, and making sense of the people in our social world" [19]. Now, intuitively, from the definition of social cognition, we can define what social cognition in HRI demand. That is, social cognition in a robot is the ability of the robot to process the social information; the information processes are encoding, storing, retrieving, and applying to social situations. While the existing HRI design accommodates such social learning capabilities, the learning is dependent on its interaction with the users, a self-exploratory approach [17,18]. However, our hypothesis is that, when a human aids the robot in its interaction with another human, the robot can learn the social ways of human-human interaction.

A study shows that by 2050, about a half of the world population age to be around 60 years or more [20]. Therefore, it is possible that there will be a high usage of social robots to assist people, specifically, elderly. Hence, social robots need to have better social intelligence to aid and assist the human population. To develop social intelligence, it is imperative that the robots learn from the ways of human-human interaction. Although there are several frameworks and designs for social robots, a specific framework to enable social learning of the robot through a human-in-the-loop (HIL) is still in the process.

This paper proposes a framework for HRI system where a human operator augments the robotic intelligence through collaborative working. The human operator and the robot work together to achieve goals and carry out a smooth HRI. In the following sections, the paper provides an overview of the system design proposed. Then, the paper focuses on discussing the HIL, explaining the need, the switch that triggers the involvement of the HIL, and some experience in deploying the system. The paper concludes with a note on how human-augmented robotic intelligence enables learning in the HRI systems.

2   System Design

Figure 1 shows a high-level diagram of the overall system architecture. Figure 2 shows a detailed view of the components of HARI for HRI system. From Fig. 2, the main components of the system design are:

- Cloud robot
- Robot Control Unit (RCU)

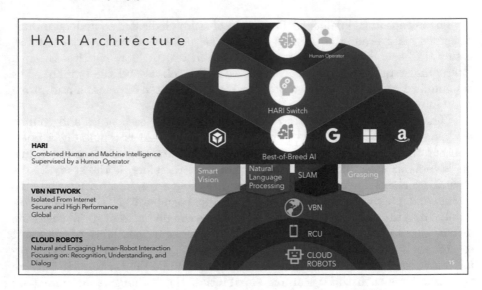

Fig. 1. High-level system diagram of HARI architecture.

- Virtual Backbone Network (VBN)
- Human-Robot Interaction (HRI) System
- Human-Augmented Robotic Intelligence (HARI)
- AI Learning System (AI-LS)

2.1 Cloud Robot

The system user interacts with a "cloud robot", a physical robot that connects to the cloud-based platform.

The HARI platform was specifically intended to be agnostic regarding the nature of the robot, so that many types of robots incorporating many different tasks can be utilized. This ranges from primarily social robots, where the main interaction is conversation, to more physically enabled robots that are intended for more motion and/or manipulation.

Examples of robots that have been used extensively with HARI are the Pepper social robot from SoftBank Robotics [21] in the social space, and Yumi from ABB [22] in the manipulative space.

2.2 RCU

A Robot Control Unit or RCU is a hardware device that interfaces a cloud robot to the cloud platform.

There is no theoretical requirement for a specific device for interfacing the robot to the platform, as long as a well-defined API exists for the platform. We have found, however, some of the following significant advantages in having such a device readily available:

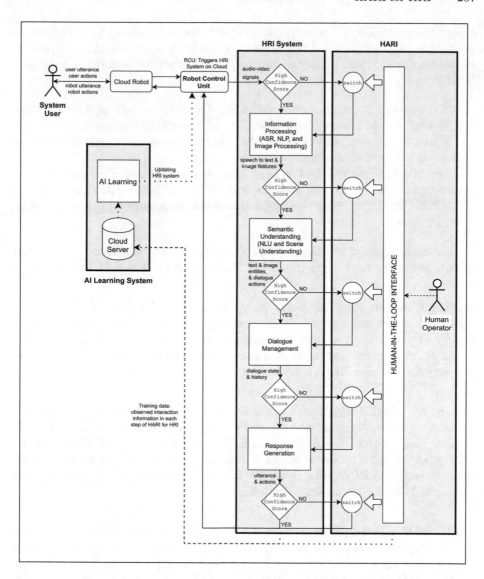

Fig. 2. HARI for HRI System Design. The work-flow: a system user interacts with the cloud robot using linguistic and non-linguistic methods. The cloud robot is connected to cloud AI through a robot control unit. The cloud AI consists of HRI system, HARI, and AI learning system. HRI system is similar to the interaction design seen in a social robot such as Pepper [21]. The HRI system consists of following stages information processing, semantic understanding, dialogue management, and response generation. HARI is embedded in all stages of HRI system to augment the AI capabilities of HRI system. HARI design consists of HARI switch, human-in-the-loop interface, and a human operator. The function of HARI switch is to switch between AI of HRI system and the human operator in HARI interface. The role of a human operator is to make decisions for the HRI system when the HARI switch is triggered. The AI learning system collects the interaction data from both HRI and HARI systems. The learning system trains and learns from the human operator decisions to update the AI of HRI system on the cloud.

Edge Functionality: Even with a high-performance controlled network inter-facing the robot to the cloud (see Subsect. 2.3 below), there are certain function-alities that need to remain physically close to the robot for performance reasons. This include:

– Voice activity detection for conversation. On devices such as Amazon Echo [23] and Google Home [24] for example, this capability remains on the device for latency reasons, even though the device requires cloud connectivity for speech recognition, understanding, etc.
– Human compliant manipulation/motion: For manipulative robots that will interact in the same space as humans, it is critical that certain tasks such as collision avoidance react very quickly and not be sensitive to cloud/network conditions.

Having a standard (or set of standard) devices with needed local functionality greatly simplifies the integration of additional robots to the HARI platform.

Sensor Control: Physical robots have a wide array of input sensors for input, such as microphones, cameras, and proximity detectors. Consistency in sensor input can dramatically improve the performance of machine learning algorithms based on that sensor input, or conversely dramatically reduce the amount of training data needed for an acceptable level of performance.

Hardware Flexibility: Most hardware subsystems on robots are designed one-off for that robot, and are not necessarily flexible in terms of adding additional devices and sensors. A well-designed RCU can offer the possibility of easily adding additional sensors for added functionality. Examples where it has been done so with the RCU are sensors for Raman spectroscopy [36] and portable ultrasound [37] for medical applications.

Multiple Operating Systems: It can be advantageous to have such a device run multiple instances of native operating systems, perhaps even different OSes, in order to separate local functionality from interaction with the cloud platform.

In [37], the current primary RCU is DATA A1 device, which is a multiple OS Android device with a "sandwich" hardware architecture which allows easy incorporation of additional needed hardware.

2.3   VBN

It has been found the open public internet in it's current form is not sufficient to meet the needs of cloud robotics. Therefore, we have created our own Virtual Backbone Network (VBN) as the network infrastructure between the RCU and the cloud platform [38].

Two main factors drive the need for a more controlled network infrastructure:

Latency: Many cloud robotic applications require particularly low latency between the robot/RCU and the cloud. These include social robots for conversation, where end-to-end delay can greatly degrade the conversational experience, and manipulative applications where time to respond to external stimuli is critical. For teleoperation applications as well, low latency is absolutely essential. Using our VBN, for instance, [38] have been able to successfully and in real time teleoperate a Yumi robot between the USA and China.

Security and Privacy: More and more security and privacy are paramount concerns, especially with the cloud deployments. The purpose of the VBN is to not only address the latency issue, but also incorporates significant amount of Blockchain technology [39] to enable security and privacy.

2.4   HRI System

The motivation behind the HRI system discussed in this architecture is from the task-oriented spoken dialogue system architecture [2,3], and social robot such as Pepper [21]. The pipeline consists of the following task-specific blocks (refer Fig. 2):

- Information Processing
- Semantic Understanding
- Dialogue Management
- Response Generation

During a human-robot interaction, the robot receives audio and video signals from a human user through various sensor technologies such as microphones and stereo cameras. The raw audio-visual data received by the robot is processed in the information-processing block to extract the information from the user's dialogue. In semantic-understanding block, the extracted data is analyzed to find relevant and useful information – identifying the scene, end-users, gestures, facial expressions, and the social environment from the video and context, language, tone, and other speech/textual information from the audio. The semantic analysis helps the robot to create a specific problem-meaning relation to which the response-generation block generates a relevant answer or an action. However, in the pipeline, the semantic information is supplied to the dialogue-management block before passing it to the response-generation block. In the dialogue-management block, the dialogue manager helps to keep track of the dialogue history and the turn-taking state between the robot and the user. The output, current dialogue state from the dialogue manager, is the input to response-generation block. The dialogue state helps the natural language generation (NLG) to pick an action response for the given dialogue state. In addition to NLG, the robot requires commands to perform physical actions such as bodily movements, gestures, change in gaze direction, emotion elicitation, and others. The response-generation block generates both the natural-language and physical-action. The NLG output is a sequence of words, and the physical-action output is a list of commands for robot behaviour. The text-to-speech (TTS) module transforms the word sequence to voice, and the action-to-controls module maps physical-action commands to the robot controls (servos, facial expression, and location change). Subsequently, the whole cycle repeats until the conversation ends.

Information Processing: The information processing block receives unstructured raw audio-visual data from the robot sensors and transforms into structured data. The information processing has various modules such as:

- Speech recognition
- Natural language processing
- Image processing.

Multiple modules besides the above mentioned modules may exist depending on the design requirement and the nature of HRI. Every module works on a specific task and utilizes either the audio information or the visual information, or both. However, the automatic speech recognition (ASR) and natural language processing (NLP) are a must, because for a social robot the core of the HRI is the dialogue exchange between the user and the robot [17,18]. Considering the design aspects from a social robot such as Pepper [21] as the norm for social robots, we assume that the facial and the visual features extraction through image processing are as vital as ASR and NLP.

For the speech signals, ASR is used to recognize the audio signals and process them into text. The NLP module performs syntactic parsing and text categorization on the text data derived from the speech. The syntactic information from NLP is the input to the natural language understanding (NLU) module in the semantic understanding block. In addition to converting the speech signals to text, the speech signals also serve as an input to classifying user emotions based on the tone of the speech [25]. For the video input, the image processing executes tasks such as face tracking, face recognition, facial features extraction, object detection, and gesture recognition tasks.

In addition to using audio and video signals independently, the audio-visual speech recognition (AVSR) technology uses both the acoustic and the speech information to recognize speech [1]. When the robot is in a noisy environment, AVSR helps to mitigate the errors in speech recognition caused by the noise.

Semantic Understanding: The semantic understanding block consists of NLU and scene classification modules that work on processed text and visual input, respectively.

From a spoken dialogue system point of view, the spoken language understanding has three main components: (a) domain classification, (b) intent classification, and (c) slot-filling [1,9]. The NLU module in semantic understanding is responsible for classifying user sentiment, user intent, and slot-filling. While sentiment, domain, and intent classifications are semantic utterance classification problems [9–11], the slot-filling relies on mapping semantic labels to the words in a sequence [4–8].

Image processing for a visual input comprises of classifying user facial expressions, body movements, user's gaze, object detection, and gesture recognition. In human-human interaction, although the interaction focuses on the dialogue exchange, non-linguistic cues such as gesture, gaze, body movements, and attention play an important role in smooth interaction between humans [13,14]. Therefore, the mentioned classifications are necessary for making HRI more human-human interaction [12].

Continuing, it is essential to combine the information from both the audio and the visual data to form a comprehensive semantic understanding. Thus, semantic understanding enables the subsequent blocks, dialogue management and response generation, to generate the action response required by the robot to reach the goal of the interaction with the user by providing a combined information of visual understanding, user intent, user sentiment, and context-specific knowledge-base details.

Dialogue Management: As explained by [12,15], a human-human conversation is centrally speech-based dialogue, but involves utilizing the other information such as gestures, action, social cues, and attention from one another. Similarly, the interaction exchange between robot and user is a combination of spoken and NLP systems [15], as well as visual data [13,14]. The dialogue management in HRI plays a vital role in bridging the semantic understanding block and response generation block.

In the dialogue management block, the dialogue manager (DM) takes the central role of organizing the dialogue sequences and the dialogue exchange between the user and the robot. Adding, the DM is responsible for dialogue state tracking, turn-taking, and discourse modelling. The dialogue state tracking and turn-taking information help in planning the next step in the sequence of dialogue exchange with the user. The discourse modelling is task-driven [15]. Using the information from state tracking and turn-taking, the DM models a discourse that leads the conversational steps to gather information from the user to achieve the task.

Response Generation: The response generation block has two main functions:

- Natural Language Generation (NLG): to generate meaningful text as a response to the user.
- Physical Action Generation (PAG): to generate corresponding action commands for the robot to make physical movements (gestures, arm movement, moving from one position to another, eliciting emotions, and facial expressions).

The role of NLG is to use the linguistic and non-linguistic representation of the data to produce text [15,16]. Hence, NLG generates a transcript of word sequence and form a meaningful sentence by using the semantic information and the discourse model from the audio-visual data. PAG, similar to NLG, the semantic information from audio-visual data to generate commands for robot's movements. The text generated from NLG is converted to speech using the text-to-speech (TTS) module. The commands from PAG is mapped to respective components on the robot to make a physical movement.

2.5   HIL

Similar to the Wizard-of-Oz paradigm [26], the HIL refers to having a human operator in the background who assists the robot at multiple stages to ensure smooth interaction between the robot and the user. The involvement of HIL depends on the HARI-switch. The role of the switch is to indicate if the HIL should intervene at the alerted stage of the HRI system. The factors that play a role in triggering the HARI-switch to switch to the HIL are:

- Confidence score: Every block of the HRI system outputs a result, which serves as an input to the subsequent block. The result from each block not only consists of the required functional output but also the confidence scores of each module within the block. In addition to the independent confidence scores from each module, the result consists of an overall confidence score by combining the scores from individual modules. If the confidence score is high, the system does not need an intervention from the HIL. However, if the confidence score is low, the HIL is called into action by the HARI-switch. The HIL takes necessary action and updates the result(s). The threshold to determine the high and the low ranges of the confidence scores varies and can be set explicitly.
- Multiple results: In HRI system, in some cases, the results from a block include multiple options. For instance, in an intent classification module, it is possible for the system to output multiple intents from the input. Therefore, the HARI-switch triggers a call for action from the HIL to select the best matching intent.
- HRI system failure: There maybe cases where the system blocks may fail entirely. In such cases, HIL is essential to step in and continue the interaction with the user.

2.6   AI-LS

The artificial intelligence learning system, similar to HRI and HARI systems, is also a part of the secured cloud platform. From Fig. 2, as shown in the system design, the AI-LS consists of two components: database and AI learner.

Although the primary objective of having a human in the loop is to work in tandem with the AI of the HRI system to make the decision process of the robot more accurate through human collaborated AI, the long-term objective is to ensure that the HRI system learns from the nuances of decision making ways of a human operator. Subsequently, reducing the operation of switching to HIL by the HRI system.

Since the role of the AI-LS is to learn from the human operator, throughout the HRI assisted by HIL, the database of the AI-LS receives the conversational data and the decisions made at each step of the HRI system. Subsequently, the AI learner of AI-LS uses the data from the database as a training data to train, learn, and update the AI capabilities of HRI system.

3 Human-in-the-Loop

3.1 Why HIL?

A system user interacting with a social robot can have a conversation that is either task-specific or general [27]. In the case of task-specific end goal, if the robot is specialized in the task at hand, the robot performs the tasks and makes decision with greater accuracy. However, when the conversation with the system user is more general, the robot may encounter imperfect and unseen or partial knowledge [27]. When the robot encounters incomplete knowledge, the robot fails to preform tasks and make decisions successfully [17], [27].

With HIL, the human operator helps the robot when the robot encounters unseen or incomplete data and is unable to make an informed decision from the trained model. Whenever the HARI switch is triggered, the HIL is involved. Consequently, the human operator corrects the wrong results generated by the specific module of the HRI system. During the process of correction and updating, the system logs the following: (a) input data from the preceding stage, (b) the incorrect result from the current stage, and (c) the updated result by the human operator. At every step of the interaction, the system generates and stores the log in the database of AI-LS. The AI learner of AI-LS uses the log data as a training example. Thus, learning from demonstration to map the *corrected* knowledge to results. In addition to mapping, the AI learner can track the specific training example that originated the incorrect result. Therefore, AI learner helps to update the trained model of the HRI system and making the HARI for HRI a constant process of learning and updating AI. Incidentally, improving the performance of AI.

3.2 The Role of Confidence Scoring in HRI with HIL

What Is a Confidence Score? In machine learning, an algorithm is designed to learn and perform a specific task from data. A machine learning algorithm trains the system (computer, agent, robot) on data to make a decision independently. From the trained machine learning model, the system predicts a result or classifies the input data into a category. During the process of prediction or classification, the algorithm generates a probability or probability distribution to get the best matching result. Generally, the probability percentage is the confidence percentage of a machine learning model of the system for a given input.

Confidence Scoring Function in HARI for HRI System: In machine learning, the functions responsible for confidence scoring are called activation functions. Relying on the confidence scoring mechanism using activation functions may still produce results with a higher confidence score that are incorrect [28]. Therefore, in HARI for HRI system, the confidence scoring mechanism embed further steps to analyze the error forms in the result. The error forms are of two types: false accept and false reject [29]. Similar to understanding the uncertainty in a model using deep models and Gaussian process [29], the confidence scoring function in HARI for HRI system uses a Gaussian process to train the results to analyze errors and break down the confidence scoring approach [currently, research in this area is ongoing].

HARI Switch and Confidence Scoring: Figure 3 gives an in-depth insight of the work-flow process between information processing stage and semantic

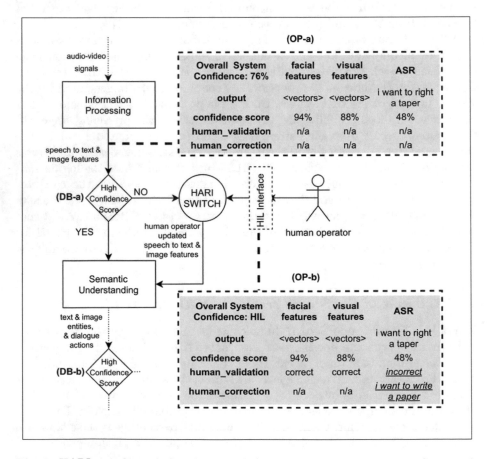

Fig. 3. HARI switch work-flow between information processing stage and semantic understanding stage

understanding stage from Fig. 2. In Fig. 3, the information processing stage receives audio-video signals as input and derives facial and visual features from the video, and speech and text features from the audio. The information processing stage has multiple modules that work on extracting specific features. For instance, the ASR module converts audio signals to text and image processing module obtains the facial features of the user as well as the overall visual features of the scene.

Along with the features, each module within the information processing stage outputs the confidence score of their respective results. In Fig. 3, from the output block A, it is observed that the facial features, the visual features, and the ASR has output as well as the confidence score. In addition to individual confidence scores from every module, the block generates an overall confidence score by combining the confidence scores. Calculating the combined confidence score is achieved by averaging confidence scores or using a correlation formula.

In Fig. 3, the decision block, DB-a, checks the confidence score of the output block, OP-a, if the confidence score is below the threshold value, it triggers the HARI switch. When the HARI switch is triggered, the HIL corrects the incorrect results of the output block, OP-a. The fixed results are updated by the human operator as shown in the output block OP-b. The updated output, OP-b, serves as input to the next stage, semantic understanding. Similarly, the decision block and the HARI-switch repeat their process at all stages of the HRI system (refer Fig. 2).

3.3   Deployment Experience with HIL

Robots such as Pepper [21] and Yumi [22] are the platforms where the deployment and testing of HARI for HRI is carried out. Specifically, Pepper, because Pepper is a social robot and is an ideal candidate for social HRI [21]. Showcasing the HARI platform in conferences Mobile World Congress America, USA and Mobile World Congress, Spain exhibited striking results. Figures 4, 5 and 6 show examples of Pepper being deployed, as well as the XR-1 robot and Cloudia intelligent digital avatar from CloudMinds.

Observations from Deployment Experience:

– Conversation topic: The conversations were, mostly, general. However, depending on the social environment the robot is situated, the interaction can be goal-oriented and task-specific.
– Reaction: Reactions from the users were positive, excitement, fear and uninterested. The positive responses were because of the ingenuity and accuracy of the robot's ability to carry the conversation. However, the same proved to be a bane for some because the users had not come across a robot that could carry on a conversation similar to human-human conversation.
– Reaction time: One of the primary factors in turn-taking conversation with a robot is the reaction time of the robot [30,31]. Whenever there was a delay in response from the robot, the users showed less interest in carrying out the conversation. In some cases, the users simplified their original question.

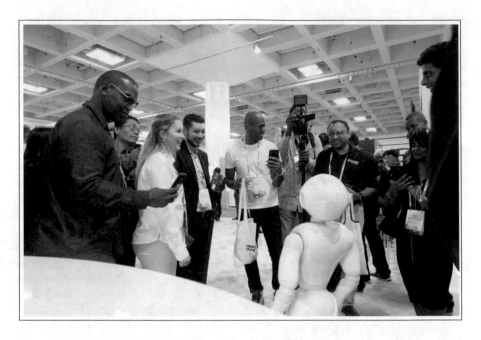

Fig. 4. The Pepper social robot.

Fig. 5. CloudMinds' XR-1 robot.

Fig. 6. CloudMinds' Cloudia intelligent digital avatar.

– Attention from the robot: In a social set-up such as a conference, the noise
  level is high. Adding, multiple people assemble to have a conversation with
  the robot. In before-mentioned cases, when the robot fails to be attentive by
  turning the head in another direction to that of a primary user in the crowd,
  the primary user questions if the robot is out of the loop of conversation.
  Thus, it is essential to maintain attention by tracking the voice and face of
  the primary user.

Engagement of Physical Robots and Avatars: It is interesting to consider the advantages of physical robots and digital avatars as compared to much
cheaper conversational form factors such as text bots or voice bots (e.g., Amazon Echo, Google Home). What we have informally found is that as you move
towards form factors that more closely mirror the human form factor, you have
the *potential* for additional and deeper engagement with the user, with all the
advantages that come with that, such as emotional attachment and conversation
rate. This is possibly due to the principle that physical robots and avatars more
closely mirror and display similarity to the human user, and thus bring a closer
sense of shared goals between the user and the bot. Figure 7 shows this.

We emphasize the increased *potential* for higher engagement, with the strong
caveat that, in order to achieve that potential, the bot must be itself more
engaging and natural in its interactions with the user. Conversational behavior

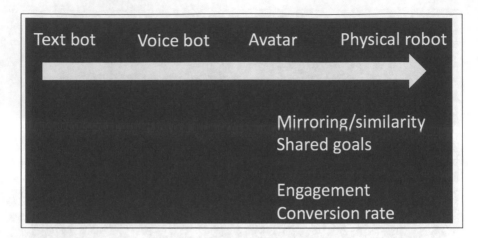

Fig. 7. Advantages of avatars and physical robots compared to other form factors.

which might be deemed acceptable in a text or voice bot would might not be accepted as natural from an avatar or physical robot, which makes conversation design and development for these form factors extremely challenging, and one of the reasons that the HIL offered by HARI is an important tool. The current state of AI-only conversation is not advanced to the level of providing such engaging interactions necessary for these form factors.

Design Interface for Human Operator: The design interface for the human operator must be simple and modular. Since the conversation between the user and the robot is in real-time, the HIL must execute a decision for the robot in real-time. The interface must consist of action buttons, simple right rail navigation of pre-loaded sentences, the video and the audio feed of the robot's view, and the dialogue state to help the HIL to perform actions faster. Adding, other requirements such as a good quality headphone, network connectivity to connect with the robot, and a display monitor is a must.

It is essential to design the HIL interface from the robot's view for facilitating the HIL to build a mental model of the conversation from the robot's perspective. Mental model from the robot's perspective helps to devise actions better [27, 31]. Consequently, the efficient and accurate actions by HIL lead to better user engagement with the robot.

3.4   Transitioning of HRI with HIL to Without HIL

While HIL helps the robot perform better in carrying out interactions with users, the HRI system should steer towards less usage of HIL. The central reason behind having HIL for HRI is to aid the robot with decision making and that the robot learns from the decisions made by the human operator. Therefore, HIL is an integral part of the learning process in the HRI system that helps the robot in learning from demonstration. Assuming that the AI is never complete and an ever learning system [17,27], HIL will continue to exist in some capacity. However, the involvement of HIL must decrease over time or include HIL in those areas where the robot needs assistance to achieve the task. It is a subjective decision as to when the HIL must stop aiding the learning process.

4   Issues and Discussion

- Deployment environment:
  - Noise: Social robots, when set up in a noisy environment, perform poorly with speech recognition. Testing HARI for HRI in a conference, one of the observations was the effect of noise in HRI. Noise not only affects the speech recognition performance, but also affects the human operator. Speech is an essential part of social interaction in HRI. Therefore, better speech recognition models, audio-visual speech recognition technology, and placing the robot in a less noise prone area is vital.
  - Multiple-speakers: Another problem the robot faced in a social gathering and/or conference setup is that, when there were multiple speakers talking to the robot, the robot had less accurate information from the speech. Adding, the head motion to maintain the gaze with the speaker was affected as the robot could not identify the primary speaker amongst the multiple speakers. Thus, another factor that could improve the interaction is by improving primary speaker identification.
- Hardware: Whilst most of the focus relies on improving interaction capabilities, a vital component for a physical robot is its hardware. Overheating was a factor due to high volume of requests from the end users to pose with the robot and take pictures. In a setup such as a conference, the number of users wanting to interact with a robot is higher than usual. Therefore, the issue with overheating must be handled.
- Network delay: Similar to human-human interaction, in HRI, the robot's ability to respond to the user without delay is necessary. However, having the HARI for HRI system on the cloud would require a fast private network. Therefore, VBN helps to reduce network delays and facilitate better HRI.
- Human operator:
  - Familiarization: A human operator must familiarize themselves with the HIL interface before operating. Therefore, training a human operator is necessary.
  - Controversial topics: When encountered with a controversial question or a question that needs an answer from the expert, the human operators must choose not to answer.
  - Non-expert human operator: Human operators are not experts of the task-specific interactions (example: health care). Therefore, having an expert in the loop is harder than having a generalist in the loop.
  - Availability: A human operator works in parallel with the robot. Meaning, when the robot is online and ready for interacting with the users, the

human operator must be available. Although the long-term goal is to reduce the involvement of HIL, currently, the human operator works in tandem with the robot.

- HIL Design Interface: Another area of improvement would be in the design interface for the human operator or HIL. Since, at times, the operator does not have the same view as the robot's head – human operator receives the video from the phone attached to robot's chest and not from the cameras in robot's head region – the human operator found it difficult to recognize the primary speaker that the robot was looking at. Suggestions such as first-person view and multi-camera view are being implemented.
- Evaluation: Another factor to help improve the interaction learning is to design how the system performance is evaluated as a whole as well as module-wise. While the objective evaluation such as accuracy of information processing seem straightforward, the subjective evaluation of learning from the human operator, unlearning from the learnt behaviour, and the humanness of the interaction is a hard task. Currently, there is no unlearning mechanism, but the mechanism is in progress. Similarly, evaluating the humanness or human-human like interaction by HARI platform is in progress as it currently relies on human emotion feedback for every response. The human emotion feedback is sentiment analysis, facial expressions, and speech emotion. Therefore, improvement to evaluate the issue is vital for transitioning HARI with HIL to without HIL.

5   Applications

The robot applications are ever growing. The most common application areas are tutoring systems, healthcare, and teleoperation.

Socially assistive robotics technology has the potential to provide novel means for monitoring, motivating, and coaching. Post-stroke rehabilitation is one of the largest potential application domains, since stroke is a dominant cause of severe disability in the growing aging population [17]. For education, the tutoring systems help in evaluating, monitoring, and teaching the candidates [17,27]. In addition, the emerging areas where social robots are making impact are

- Entertainment is another field where the usage of robots is increasing in demand [21,33,34]. The HIL enables teleoperation capability and thus controlling the robot from a remote location to entertain the crowd. The entertainment includes singing, dancing, conversing, and performing intelligent tasks. In addition to charming the people, robots are used in commercials to promote products and services [33–35]. Not limiting advertisements and mass entertainment, the robots are a medium for artists to showcase their art and connect with the community [32].
- Similar to entertainment, in retail, the robots are used to sell items in a store, engage with customers and find items from the inventory, and act as a store manager. Although the tasks related to retail is restricted compared to the

other social set-up, the HARI design enables the robot to switch to a human operator when the robot requires human intervention. From [33], customers who shop in stores interact with the robot to gather more information of the supplies they seek. Continuing, by answering product related questions, the robot provides customer service to customers [33]; with the support by HIL the robot will be able to manage customer queries gracefully. The robot can also serve as a point of attraction to get more customers to visit the store [33].

6 Summary

The usage of robots keeps increasing. More and more social robots are coming into the market. While most of the HRI designs do not include HIL, the HRI systems without HIL requires a system update from time-to-time to accommodate new information and changes associated with the knowledge-base. The HARI for HRI design is a framework from which a continuous learning system for HRI can be established. The learning system for HRI is just as vital as the performance of the system, because the learning facilitates the HRI system to study the ways of human-human interaction through HIL. When the HRI system emulates human-human interaction, the system is enabling a stronger human-robot relationship. With the use of robots growing more, it is imperative to build a better human-robot dynamic. User's trust in a robot arrives from better engaging and interactive aspect of the robot.

The field of AI is propagating at a swift pace. However, most of the propagation focuses on developing groundbreaking techniques and algorithms. Although the techniques and algorithms are indispensable to the field of AI and HRI, with growing interest in collaborative AI, a framework to have a modular HRI system is imminent to keep up with the changes in AI.

References

1. Dupont, S., Luettin, J.: Audio-visual speech modeling for continuous speech recognition. IEEE Trans. Multimed. **2**(3), 141–151 (2000)
2. Young, S.: Using POMDPS for dialog management. In: 2006 IEEE Spoken Language Technology Workshop (2006)
3. Zhao, T., Eskenazi, M.: towards end-to-end learning for dialog state tracking and management using deep reinforcement learning. In: Proceedings of the 17th Annual Meeting of the Special Interest Group on Discourse and Dialogue (2016)
4. Wang, Y., Deng, L., Acero, A.: Spoken language understanding- an introduction to the statistical framework. IEEE Signal Process. Mag. **22**(5), 16–31 (2005)
5. Wang, Y., Deng, L., Acero, A.: Semantic frame based spoken language understanding. In: Spoken Language Understanding: Systems for Extracting Semantic Information from Speech, USA, NY, pp. 35–80. Wiley, New York (2011)
6. Wang, Y.-Y., Acero, A.: Discriminative models for spoken language understanding. In: Proceedings of ICSLP (2006)
7. Raymond, C., Riccardi, G.: Generative and discriminative algorithms for spoken language understanding. In: Proceedings of Interspeech (2007)

8. Zue, V., Glass, J.: Conversational interface: advances and challenges. Proc. IEEE **88**(8), 1166–1180 (2000)
9. Mesnil, G., Dauphin, Y., Yao, K., Bengio, Y., Deng, L., Hakkani-Tur, D., He, X., Heck, L., Tur, G., Yu, D., Zweig, G.: Using recurrent neural networks for slot filling in spoken language understanding. IEEE/ACM Trans. Audio Speech Lang. Process. **23**(3), 530–539 (2015)
10. Yaman, S., Deng, L., Yu, D., Wang, Y.-Y., Acero, A.: An integrative and discriminative technique for spoken utterance classification. IEEE Trans. Audio Speech Lang. Process. **16**(6), 1207–1214 (2008)
11. Guo, D., Tur, G., Yih, W., Zweig, G.: Joint semantic utterance classification and slot filling with recursive neural networks. In: 2014 IEEE Spoken Language Technology Workshop (SLT) (2014)
12. Jaimes, A., Sebe, N.: Multimodal human-computer interaction: a survey. Comput. Vis. Image Underst. **108**(1–2), 116–134 (2007)
13. Chen, L., Huang, T., Miyasato, T., Nakatsu, R.: Multimodal human emotion/expression recognition. In: Proceedings Third IEEE International Conference on Automatic Face and Gesture Recognition
14. Sebe, N., Cohen, I., Gevers, T., Huang, T.: Multimodal approaches for emotion recognition: a survey. In: Internet Imaging VI (2005)
15. Pietquin, O.: Natural language and dialogue processing. In: Multimodal Signal Processing, pp. 63–92 (2010)
16. Wen, T., Gasic, M., Mrksic, N., Su, P., Vandyke, D., Young, S.: Semantically conditioned LSTM-based natural language generation for spoken dialogue systems. In: Proceedings of the 2015 Conference on Empirical Methods in Natural Language Processing (2015)
17. Fong, T., Nourbakhsh, I., Dautenhahn, K.: A survey of socially interactive robots. Robot. Auton. Syst. **42**(3–4), 143–166 (2003)
18. Goodrich, M., Schultz, A.: Human-robot interaction: a survey. In: Foundations and Trends® in Human-Computer Interaction, vol. 1, no. 3, pp. 203–275 (2007)
19. Moskowitz, G.: Social Cognition. Guildford Press, New York (2005)
20. Pollack, M.: Intelligent technology for an aging population: the use of AI to assist elders with cognitive impairment. AI Mag. **26**(2), 9 (2018)
21. Pandey, A., Gelin, R.: A mass-produced sociable humanoid robot: pepper: the first machine of its kind. IEEE Robot. Autom. Mag. **25**(3), 40–48 (2018)
22. ABB's Dual-Arm Collaborative Robot - Industrial Robots From ABB Robotics. New.abb.com (2018). https://new.abb.com/products/robotics/industrial-robots/yumi. Accessed 02 Dec 2018
23. Amazon Echo: Amazon.com (2018). https://www.amazon.com/dp/B06XCM9LJ4/ref=fs_ods_aucc_rd. Accessed 02 Dec 2018
24. Google Home: Google Store (2018). https://store.google.com/us/product/google_home?hl=en-US. Accessed 02 Dec 2018
25. Kim, Y., Provost, E.: Emotion classification via utterance-level dynamics: a pattern-based approach to characterizing affective expressions. In: 2013 IEEE International Conference on Acoustics, Speech and Signal Processing (2013)
26. Kelley, J.: An iterative design methodology for user-friendly natural language office information applications. ACM Trans. Inf. Syst. **2**(1), 26–41 (1984)
27. Breazeal, C.: Social interactions in HRI: the robot view. IEEE Trans. Syst. Man Cybern. Part C (Appl. Rev.) **34**(2), 181–186 (2004)
28. Michalski, R.: Machine learning. Elsevier Science (2014)

29. Kendall, A., Gal, Y.: What uncertainties do we need in bayesian deep learning for computer vision?. In: Advances in Neural Information Processing Systems 30 (NIPS) (2017)
30. Barnlund, D.C.: Communication Theory: Second Edition, 2nd edn. Transaction Publishers, New Brunswick (2008)
31. Kahn, P.H., Freier, N.G., Kanda, T., Ishiguro, H., Ruckert, J.H., Severson, R.L., Kane, S.K.: Design patterns for sociality in human-robot interaction. In: Proceedings of the 3rd International Conference on Human Robot Interaction - HRI 2008 (2008)
32. Cross, J., Bartley, C., Hamner, E., Nourbakhsh, I.: Arts & bots: application and outcomes of a secondary school robotics program. In: 2015 IEEE Frontiers in Education Conference, 10 (2015)
33. Mitchell, O.: Can robots save retailers from an apocalypse?, The Robot Report (2018). https://www.therobotreport.com/retail-robots-save-retailers-apocalypse/. Accessed 02 Dec 2018
34. Baird, N.: Robots, automation and retail: not so cut and dried, Forbes (2018). https://www.forbes.com/sites/nikkibaird/2018/06/19/robots-automation-and-retail-not-so-cut-and-dried/#1ed13fd67b06. Accessed 02 Dec 2018
35. Underwood, C.: Robots in Retail - Examples of Real Industry Applications — Emerj - Artificial Intelligence Companies, Insights, Research, Emerj (2018). https://emerj.com/ai-sector-overviews/robots-in-retail-examples/. Accessed 02 Dec 2018
36. CloudMinds – Smart Handheld Raman: Airaman.com (2018). https://www.airaman.com/. Accessed 02 Dec 2018
37. CloudMinds: En.cloudminds.com (2018). http://en.cloudminds.com/DATA. Accessed 02 Dec 2018
38. Tian, N., Kuo, B., Ren, X., Yu, M., Zhang, R., Huang, B., Goldberg, K., Sojoudi, S.: A cloud-based robust semaphore mirroring system for social robots. Learning, vol. 12, p. 14
39. Lin, I.C., Liao, T.C.: A survey of blockchain security issues and challenges. IJ Netw. Secur. **19**(5), 653–659 (2017)

Kinematic Analysis of the Bio-inspired Design of a Wearable Compliant Gait Rehabilitation Robotic System for Smart and Connected Health

S. M. Mizanoor Rahman$^{(\boxtimes)}$

The University of West Florida, Pensacola, FL 32514, USA
rsmmizanoor@gmail.com

Abstract. The design of a novel actuator VIC-SEATS (Variable Impedance Compliant Series Elastic Actuator with Two Springs) is introduced and the static and kinematic characteristics of the soft actuator for a targeted application of actuating the knee joint of a gait rehabilitation robot is presented. Targeted novelties of the actuator design are that it is designed with only one motor and two sets of springs in series that can adjust the stiffness for low, medium and high force situations without any change in the springs, it is light in weight, compact in size but enough strong due to proposed fabrication with advanced materials, and it has competitive advantages such as high force controllability, force bandwidths, back-drivability, efficiency, safety, power/mass and force/mass ratios and output torque, and low friction, inertia and impedance. The electro-mechanical design, prospective fabrication materials and working mechanisms of the novel actuator design are summarized. Based on simulation results, the design parameters are optimized to determine desired static and kinematic features such as torque trajectory, torque-displacement relationship, velocity profile and required power/energy for actuating the knee joint of a gait rehabilitation robot through comparing these characteristics with the human characteristics. Then, applications of hand-held gaming interface to display key rehabilitation performance and the tele-rehabilitation facility for connected health are analyzed. Finally, social and technical motivations for stroke patients towards utilizing robotic rehabilitation devices are analyzed.

Keywords: Rehabilitation robot · Compliant actuation · Kinematics · Connected health · Tele-rehabilitation · Gaming interface · Rehabilitation motivation

1 Introduction

Superiority of robot-based rehabilitation performance over manual rehabilitation is now well-established [1–3]. The robotic systems with stiff actuation may guarantee precision, speed, repeatability, etc., and thus may be suitable for the autonomous applications with less or no human interactions [1]. However, stiff actuation is not suitable for the systems deployed in unstructured environments where human's safety, comfort and psychological acceptance are the vital issues, e.g. the assistive and the rehabilitation robots [2, 3]. These robots need actuation that provides compliance through variable

© Springer Nature Switzerland AG 2020
K. Arai et al. (Eds.): FTC 2019, AISC 1070, pp. 224–243, 2020.
https://doi.org/10.1007/978-3-030-32523-7_15

stiffness, force controllability, large force bandwidths, back-drivability, high efficiency and safety, stability, adaptation to changing situations, softness, high power/mass and force/mass ratios, high output torque, etc., and low friction, inertia, impedance, etc. [4, 5]. Active compliance through the applications of some software techniques based on impedance/admittance regulations and joint torque control may add some of the above attributes to the actuators for the assistive and rehabilitation robots [6, 7]. However, delays at all stages of the systems have made this approach less practical. Hence, it is thought that the passive compliance through the usage of especially designed soft, compact, light-weight, efficient and compliant actuators can be the best choice for the assistive and the rehabilitation robotic devices [3, 8].

The natural systems especially the humans are robust because they follow variable stiffness, soft and compliant actuation [9]. Muscle is the internal motors of the human body responsible for the movements of the skeletal system. In muscle-tendon model (MTM), the contractile component shortens the muscle through the actin-myosin structures and the parallel and series elastic components provide elasticity, energy efficiency and variable stiffness during human movement [10]. It is believed that the actuation methods developed for assistive and rehabilitation robots mimicking their human counterparts may match the static and dynamic performance of the humans and thus may provide better cooperation to their human partners [11].

The compliant actuation in the human inspired the development of wide ranges of variable stiffness compliant soft actuators [11–13]. The Variable Stiffness Actuators (VSAs) and the Series Elastic Actuators (SEAs) are the early development of the soft compliant actuators [14, 15]. The VSAs have the fixed springs though the springs can be compressed and/or extended to change the stiffness. Change of stiffness manually through changing compression or extension between the operations of the actuator is possible, but it is almost impossible within the operation. The stiffness change within the operation can be done by another motor, but it makes the actuator large, heavy, less energy efficient, and less cost-effective [15]. Again, the VSAs are usually not so suitable to produce linear motion.

In the SEAs, the impedance can be changed by changing stiffness, damping and inertia, where the stiffness change is achieved by inserting elastic spring to the stiff motor in series [14]. Combination of series-parallel elasticity has also been proposed [16]. However, the SEAs still have many limitations especially the fixed compliance that largely depends on the spring constant [17]. The soft spring produces high fidelity of force control, reduces stiction, but also limits the bandwidth at high forces. On the contrary, the stiff spring produces large force bandwidth, but force fidelity is low and stiction is high [17]. These limitations also affect the control performance [3]. Again, the spring producing the fixed compliance may need to be changed physically if a change in the compliance is needed. Moreover, the existing SEAs are large, heavy, error-prone and less power efficient [1]. The existing VSAs and SEAs usually do not use advanced materials for their fabrication that make them heavy and less strong. For the above reasons, the applications of SEAs and VSAs to the assistive and rehabilitation robots are still not so satisfactory.

The assistive and rehabilitation robots can be the most effective with their human users if and only if the actuator design is optimized for the static and dynamic characteristics through comparing the characteristics with the human characteristics.

However, such optimization is very rare. Robinson *et al.* [17] analyzed the dynamic characteristics of the SEA such as stability, resonance, force bandwidths, etc. for a walking robot. However, their analysis lacked static characteristics and they also did not optimize the design through benchmarking with the human characteristics. Dynamic characteristics of a SEA, i.e. force controllability, force bandwidths, back-drivability, efficiency, safety, stability, adaptation to changes, impacts, uncertainties and disturbances, resonance, power/mass and force/mass ratios, etc. can be optimized for an exoskeleton robot for gait rehabilitation [17–20, 33–36]. However, bench marking of such optimization with human counterpart is still not attempted. Again, in addition to dynamic characteristics, static and kinematic characteristics of SEAs such as torque and velocity, torque-displacement relationship, power/energy requirements, etc. for applications with gait rehabilitation robots are necessary, but still not optimized. In addition, design innovation is necessary to appropriately attach the exoskeleton device with the human wearer so that it fits with the wearer's body comfortably, which were not considered in [18, 33–36].

Hand-held gaming interface is a latest innovation for wearable healthcare devices that can facilitate the user to monitor rehabilitation performance in real-time through visual interface based on some game-like activities performed during actual rehabilitation practices [29]. The usability of such gaming interfaces can increase if the interfaces are hand-held while being used. The hand-held gaming interfaces may impact the patient's user-friendliness, self-satisfaction, cognition and situation awareness, and thus may impact user acceptance of the rehabilitation devices. However, user-centered design, and development and evaluation of hand-held gaming interfaces for stroke rehabilitation devices are still not emphasized. In addition, tele-rehabilitation is a practice where the patients can be kept connected with the healthcare providers and the patient communities in real-time [30]. This concept can help patients share their rehabilitation progress with their healthcare providers, and can also help benchmark the rehabilitation progress with other similar patients in their communities. However, tele-rehabilitation practices for stroke patients for achieving connected health still did not reach its maturity level. Furthermore, it is experienced while working with the stroke rehabilitation patients that in most cases they are not enough motivated to use the robotic rehabilitation devices. Their unwillingness of using rehabilitation robotics devices cannot help flourish robot-aided rehabilitation initiatives. Hence, social and technical motivations of stroke patients towards the usage of rehabilitation robotic devices are necessary to receive complete benefits from the robotic rehabilitation systems [31]. However, appropriate methods and strategies to motivate the patients are not observed in the state-of-the-art literature [18, 33–36].

Being motivated by above facts and figures, as an extension of previous works in [18, 33–36], the objectives of this article have been decided as to design a novel actuator VIC-SEATS (Variable Impedance Compliant Series Elastic Actuator with Two Springs) to actuate the knee joint of a gait rehabilitation robot in the sagittal direction. The SEA design is expected to provide all the advantages that a typical SEA usually provides. In addition, it is to be more compact, lightweight and contains a mechanism that can change the stiffness within the operation (gait cycle) without any change in the physical spring, which can make the device more energy efficient. The rehabilitation robot design also includes better aesthetic properties and a novel arrangement so that the device can motivate its users and also can be comfortably attached with the wearer's body. The static and kinematic characteristics of the actuator is optimized for the knee joint of the rehabilitation robot through comparing these characteristics with human characteristics. Concepts of hand-held gaming interface and tele-rehabilitation facility for achieving connected health are proposed and evaluated, and a systematic approach to motivate patients towards using wearable robotic stroke rehabilitation devices is proposed based on survey results.

2   Design of the Novel Series Elastic Actuator

Taking inspiration from [18, 33–36], the schematic diagram of the novel SEA is proposed as Fig. 1 shows. The SEA design consists of a servomotor, two types of springs, a ball screw, a gear mechanism and an output link (the prospective robot link). A torsional spring is put in series between the servomotor (with a rotary encoder) and the spur gear. The gear is also attached to another rotary encoder. Two rotary encoders measure the deflection of the torsional spring. The motor torque is estimated based on the spring constant. The gear reduces the rotational speed of the motor based on the gear ratio. To achieve the bi-directional loading of the torsional spring, it is packaged inside an assembly that contains two opposite winding torsional springs loaded independently with respect to the rotational direction. The gear is also attached to a ball screw in series to convert the rotational motion of the motor to the translational motion. The ball screw nut is attached to the output link in series through another set of translational springs. A linear potentiometer is attached to the linear springs to measure the linear deflection. When the nut moves, a linear deflection is to be produced in the springs, which can be measured by the linear encoder (potentiometer). The output force can be estimated based on the spring constant and the linear deflection using the Hooke's law. The linear springs added on both ends of the ball screw nut are kept connected to a shuttle to help achieve back-drivability along the arrow. The ball screw nut can move independent of the shuttle [18].

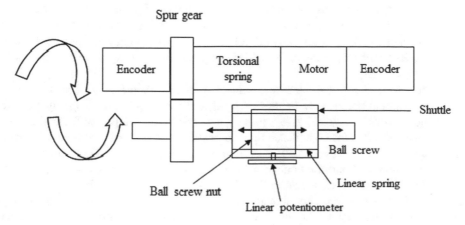

Fig. 1. The schematic 2D diagram of the proposed series elastic actuator (SEA) [18, 33–36].

The translational springs are selected as soft and small. The allowable stroke is also kept small. The torsional spring is selected as small in size though the effective spring constant is found very big. It is at the high speed and the low torque range. As a result, the overall size of the actuator can be kept smaller than that of the state-of-the-art actuators, and the weight is also kept comparatively lower [12–15]. Note that the shuttle assembly and the ball screw nut share the same linear guide for the constraint in the movement other than in the desired direction of the stroke [18].

The proposed SEA assembly principal parts are designed to be manufactured out of the space grade aluminum alloy (e.g., 6061-T6). This type material can provide the advantages in having physical and mechanical properties, e.g. low weight (i.e., density 2.7 g/cc) and high strength (i.e., ultimate tensile strength 310.00 MPa, tensile yield strength 276.00 MPa, fatigue strength 96.50 MPa, shear strength 207.00 MPa, etc.). Other mechanical properties can also be favorable. These properties may be the hardness, elongation at break (12.00%), modulus of elasticity (68.90 GPa), Poisson's ratio (0.33) and fracture toughness. This type material can also provide good thermal and electrical characteristics. Carbon fiber composite materials can be used to manufacture the main structure of the proposed gait rehabilitation robot, which is to be actuated by the proposed series elastic actuator [18, 33–36].

3   Working Principles of the Novel Series Elastic Actuator

The working principles of the proposed SEA have been explained in details in previous publications [18, 33–36]. As explained previously, the proposed SEA is modeled consisting of translational elements only by converting the rotary elements to the equivalent translational elements, which is shown in Fig. 2. Such conversion may help analyze the actuator performance at the output end that produces linear output force. In the model in Fig. 2, F_1 refers to the motor input force, m_1 refers to the equivalent mass of the motor as derived in Eq. (1) where J_1 refers to the moment of inertia of the motor and p refers to the pitch of the ball screw, m_2 refers to the equivalent mass of the ball screw as derived in Eq. (2) where J_2 refers to the moment of inertia for the ball screw and k_1 is considered here as the equivalent translational spring constant of the torsional spring k_t as derived in Eq. (3), and k_2 refers to the spring constant of translational spring, b_1 and b_2 are used for damping for the motor and the ball screw respectively, and F_o refers to the output force. Note that it may need to select k_1 much bigger than k_2 (i.e., $k_1 \gg k_2$) [18].

At low force situations, the model reduces to as shown in Fig. 2(b), which means that the torsional spring may behave as rigid and may not act as a spring, but the translational spring may act as a spring. As a result, the output stiffness (i.e., the impedance) and thus the compliance may be provided by only the translational spring depending on the spring constant, k_2. At high situations, the translational springs may be fully compressed and thus only the torsional spring may act as a spring that can transform the model to as shown in Fig. 2(c). As a result, the force bandwidth of the proposed SEA is achieved very high at the high force situations due to the large spring constant k_1 of the torsional spring. It can be found that the performance of the SEA (i.e., the impedance and compliance) may depend on the input forces and the difference between the spring constants (k_1 and k_2). In this way, the SEA can change its output impedance characteristics depending on the input force magnitudes without needing any change or replacement in the physical springs. The proposed SEA can thus achieve a high force bandwidth due to having the torsional spring with high spring constant and low output impedance and low nonlinear friction due to having the soft translational springs with low spring constant. In the real rehabilitation system built with the proposed SEAs, both the k_1 and the k_2 may work simultaneously when the system transforms from low input forces to the high input forces, and the vice versa [18].

$$m_1 = J_1 \left(\frac{2\pi}{p}\right)^2 \tag{1}$$

$$m_2 = J_2 \left(\frac{2\pi}{p}\right)^2 \tag{2}$$

$$k_1 = k_t \left(\frac{2\pi}{p}\right)^2 \tag{3}$$

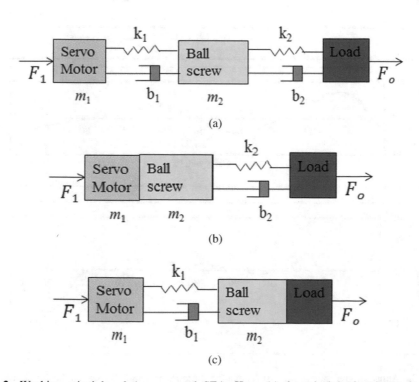

(a)

(b)

(c)

Fig. 2. Working principle of the proposed SEA. Here, (a) the principle for the equivalent translational motion situation, (b) the principle for the low force situation, and (c) the principle for the high force situation [18, 33–36].

4 Study of Human Gait Characteristics

Winter developed various discrete sets of human gait data related to the motion (angular displacement, angular velocity), torque, power and work/energy during human gait cycle [21, 22]. The changes in angular displacement, torque and power during gait cycle based on Winter's data sets are shown in Fig. 3, which show the gait profiles of a healthy human of about 80 kg weight for the knee joint movement along the sagittal plane during his/her gait on a plane surface. The gait profiles can show that the maximum angular movement is about 70°, the maximum torque is about 60 Nm, the maximum power is about 100 W and the work done is about 35 J during the gait cycle for the knee joint along the sagittal plane for the mentioned healthy human of 80 kg. Note that these human gait characteristics are approximate, and should be targeted

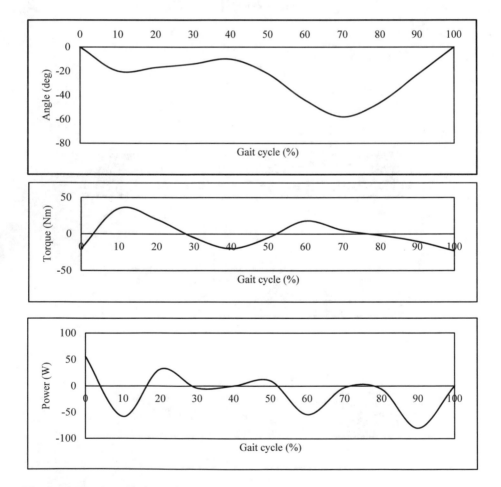

Fig. 3. Approximate gait profiles of a healthy human of about 80 kg for his/her knee joint movement along the sagittal plane [21, 22]. Here, computed work = 33.64 J, and computed maximum power = −96.32 W.

when designing the proposed SEA for usage in the development of the gait rehabilitation robot for stroke rehabilitation at the knee joint.

Such consideration justifies the overall design as the 'bio-inspired' design as emphasized in this article, which was not considered in previous works [18, 33–36].

5   Design of the Robotic Device for Gait Rehabilitation at Knee Joint

The design of the robotic device using the proposed SEAs were presented in details in previous publications [18, 33–36]. Here, the design of the device for rehabilitation at the knee joint is focused in particular. For this purpose, a modular robotic orthosis is designed using the proposed SEA for gait rehabilitation of the stroke patients at their

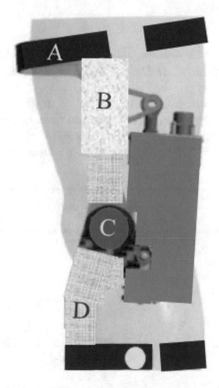

Fig. 4. The proposed robotic orthosis designed with the proposed novel SEA for gait rehabilitation at the knee joint of the stroke patient (user). The design shows aesthetic structure as well as better arrangements for tightly attaching the device with the user body, which may ensure safe wearing (see [18, 33–36] for comparison of the design).

knee joints along the sagittal plane as shown in Fig. 4. The orthosis is designed to be attached with human user's lower limb at two points called *A* and *D*. The point *A* attaches the orthosis with the human user's thigh and the point *D* attaches the orthosis with human user's shank. The upper end of the SEA is attached with the orthosis frame at point *B*. The output of the SEA is attached with the orthosis frame through point *C*. In addition, appropriate arrangements are made so that the device can be comfortably attached with the body of the wearer.

When the SEA is in operation, the ball screw can produce linear movement as the output (and also force), and the force can produce torque at point *C*, which can be placed at the center of the knee. The torque applied at the knee joint can help produce the angular movement at the knee joint of the patient (human user) wearing the orthosis. The movement range may depend on the output force that depends on the input force. In addition, materials of various textures are proposed to be used to build the structure of the device in order to enhance the aesthetic properties of the device, which may motivate the users.

6   Optimizing the Static and Kinematic Characteristics of the SEA for the Gait Rehabilitation Robot

It is assumed that the stroke patient's lower limb is totally numb due to stroke. This numbness may occur at the beginning of the stroke when no rehabilitation is provided. In this case, it is thought that the rehabilitation robot should provide 100% torque assistance to the patient of 80 kg weight to move the knee for its full movement range

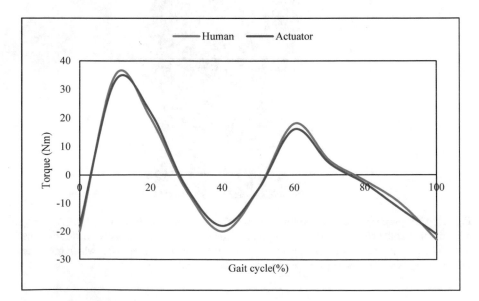

Fig. 5. Approximate comparison of torque profiles between the human and the proposed SEA with respect to gait cycle for the prospective stroke patient's knee joint movement along the sagittal plane for the proposed gait rehabilitation at the knee joint.

as the patient is not able to produce any torque. Hence, the 100% torque requirement needs to come from the rehabilitation device that needs the device to have similar characteristics as the healthy human produces (see Fig. 3). The torque assistance of the device will reduce if the patient can provide some torque or the knee movement range is kept limited instead of full movement.

The static and kinematic characteristics of the SEA actuating the device depend on the values of the actuator parameters such as m_1, m_2, k_1, k_2, b_1, b_2, p, etc. [18]. The output torque also depends on the output force and the distance between point C and the axis of the output force. For this purpose, it was to optimize the angular motion, torque and power profiles of the rehabilitation robotic device designed with the proposed SEA using the optimization tool of MATLAB based on the selected design parameter values of the SEA (see [18] for further details). Then, the results were compared with that of the human.

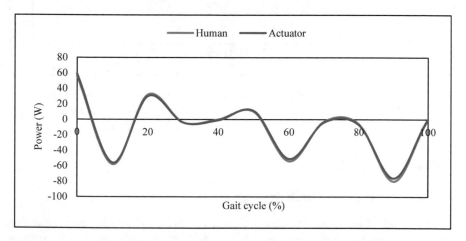

Fig. 6. Approximate comparison of power profiles between the human and the proposed SEA (motor) with respect to gait cycle for the prospective stroke patient's knee joint movement along the sagittal plane for the proposed gait rehabilitation.

The comparison results are shown in Figs. 5, 6 and 7. Figure 5 shows that the SEA torque profile approximately matches the human torque profile. Figure 6 shows the matching between human and actuator characteristics, and also shows that the required peak motor power and the energy requirements for the SEA to produce the required torque are low. This may happen due to the advantage of the energy storage-release phenomena in the springs of the SEA. In addition, good matching in the angular movement-torque relationships between the human and the proposed SEA was also observed. Figure 7 shows a good match between the human movement velocity and the SEA velocity with respect to the gait cycle, which indicates synchronization in two velocities showing a good fit between the human and the robotic device. Good matches in the torque profiles, angular movement-torque relationships and velocity profiles between the human and the robotic device (the SEA) justify the possibilities of good human-robot cooperation, safety and wearability of the device, and human's acceptance of the robot.

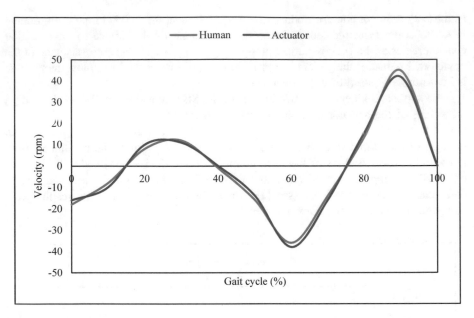

Fig. 7. Approximate comparison of angular velocity profiles between the human and the proposed SEA with respect to gait cycle for the prospective stroke patient's knee joint movement along the sagittal plane for the proposed gait rehabilitation at the knee joint.

7   Design Analysis

7.1   Selection of Motor and Ball Screw

Figures 5, 6 and 7 provide significant insights that can help determine suitable motors, gears, sensors, springs, ball screw, etc. for the SEA for the orthosis. The SEA should provide up to 60Nm assistive torque at the knee joint, as the figures show. A Maxon DC brushless servomotor (EC 4-pole 120 W 36 V) can be used which is universal in all joint applications and lightweight (0.175 kg), and provides low moment of inertia, favorable power to weight ratio and compactness. The ball screw is to be used for mechanical transmission of rotary to linear motion for its high efficiency, the spur gears can be used for their compactness. The selected ball screw can be from Eichenberger Gewinde AG (http://www.gewinde.ch/) that has a pitch of 2 mm/rev and that can output up to 1500 N force. The translational springs can be used to operate in the range of about 25% of the full force. The rotary encoder has a minimum of 1024ppr to provide a high resolution force sensing. Due to proposed usage of light-weight materials and compactness, the total mass of the proposed SEA can be less than 0. 80 kg based on SolidWorks design analysis [18, 33–36].

7.2   Analysis of Stiction

There is stiction in the motor-ball screw assembly system that may impact the SEA characteristics and performance. Stiction (F_s) can be expressed as (4), and the feed-back force error (F_e) can be expressed as (5). Note that stiction is a low frequency phenomenon, and if stiction increases, the feed-back force error can also increase. If the proportional gain of a proposed PD control (k_p) increases, the stiction may decrease [17, 18]. A graph can be plotted taking $\sqrt{k_p + 1}$ as the abscissa and F_e/F_s as the ordinate, which can show an inverse parabolic relationship as in Fig. 8 [17]. Here, F_2 is the active spring force and F_d is the desired force of the PD control.

$$F_s = F_1 - F_2 \tag{4}$$

$$F_e = F_d - F_2 \tag{5}$$

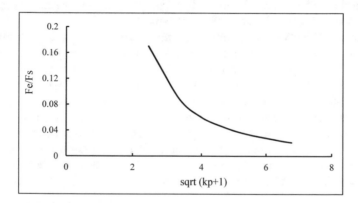

Fig. 8. Stiction characteristics in the motor-ball screw system [18, 33–36].

7.3   Analysis of Sensor Resolution

The resolution of the prospective sensors is important to make the rehabilitation system usable in critical situations. The force resolution should fall within user or patient's normal input force capabilities. For the position sensor, the minimum displacement the sensor can detect may be 0.0046 mm. Hence, the minimum force the low stiffness spring can detect may be estimated as 0.1176 N, the absolute maximum force may be 244 N, and the minimum and maximum detectable forces for the torsional spring may be 5.88 N and 2.36×10^3 N respectively [18, 33–36].

8   Gaming Interface, Tele-Rehabilitation and Motivation Analysis

8.1   Gaming Interface

A game-like user interface can be designed for the stroke rehabilitation system [29]. The gaming interface along with the prospective lower limb rehabilitation system built with the proposed SEA is shown in Fig. 9. In the proposed gaming user interface, the patient can practice gait rehabilitation (e.g., knee rehabilitation) in the forms of some target-based game-like activities, and the rehabilitation performance (performance of the game) can be measured in real-time using suitable sensors and displayed in a monitor in front of the patient. The proposed gait (knee) rehabilitation performance can be expressed in terms of torque and angular movement velocity that the patient usually generates during the rehabilitation practice. The torque and velocity characteristics indicate the patient's kinetics and kinematics performance achieved through the rehabilitation practices. These two parameters are considered here as the rehabilitation performance because these parameters can justify the patient's ability to conduct gait [32].

The patient's kinetics and kinematics performance is changed in real-time and the changes can be reflected through the interface as diagrams. The patients may need to achieve targeted kinetics and kinematics performance in a specified time period. It is possible to award the highest scores (100%) to the patient for achieving the kinematics and kinetics performance of a healthy human. The game is that the patient can conduct rehabilitation practice, improve his/her kinematics and kinetics performance and chase to achieve the performance of a healthy human. The game-like activities can also be installed in a mobile device and the user can hold it in his/her hand during the rehabilitation practice. The gaming performance can also be displayed in a large monitor placed beside the rehabilitation robotic system.

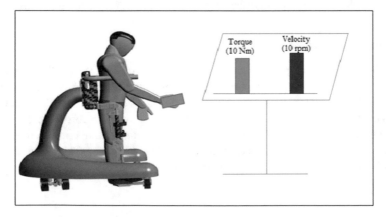

Fig. 9. Game-like user interface displaying rehabilitation performance in kinetics and kinematics terms used through a hand-held device or a large fixed screen placed in front of the patient during the gait (or knee) rehabilitation session [18, 33–36].

In a short survey, the proposed design in Fig. 9 was separately demonstrated using simulation to nine (9) stroke patients, stroke researchers and service providers (the subjects), and they were asked to evaluate the potential effectiveness of the proposed game-like user interface of Fig. 9 based on a few assessment criteria and on an evaluation scale. The evaluation scale was a Likert scale with 1 indicating the least effective and 5 indicating the most effective. Out of the 9 subjects, 4 were female, and 5 were male. The patients who responded were aged, they had recent paralysis in their lower limbs due to stroke, and were taking treatment from local healthcare providers. The evaluation results are shown in Fig. 10.

The results show that the subjects perceived the proposed game-like user interface positively. The results show that the game-like user interface can increase the patient's perceived situation awareness and engagement with the device during rehabilitation. The subjects also perceive that the interface can be made easy to use, and the fatigue may be low as it is interfaced through a lightweight hand-held device (e.g., a mobile phone). The subjects also perceive that the overall system can enhance the clarity of status of rehabilitation performance as the status is displayed to the user in real-time. Disclosure of rehabilitation performance status can also help patients plan for future practices and consultation with healthcare providers [29].

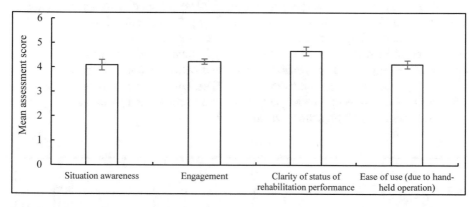

Fig. 10. Assessment of potential effectiveness of the game-like user interface by the subjects.

8.2   Tele-Rehabilitation and Social Connection

A tele-rehabilitation system can be designed as shown in Fig. 11. In the proposed tele-rehabilitation system, the patient can practice rehabilitation using the robotic device at home, the performance in terms of kinematics and kinetics can be displayed through the gaming interface, and the same performance status can also be shared with healthcare professionals at distant locations in real-time or near real-time through appropriate internet or cloud-based communication system [30]. In addition, the performance can be shared with other similar patients in communities through the same procedure or through some social media such as Facebook. The proposed tele-rehabilitation and social media practice can ensure connected health.

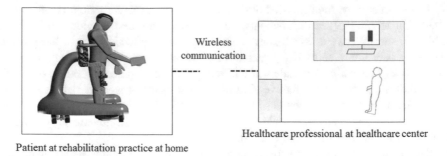

Patient at rehabilitation practice at home

Fig. 11. Real-time tele-rehabilitation practice between patient and healthcare professional.

In another survey, the proposed tele-rehabilitation system shown in Fig. 11 was separately demonstrated using simulation method to the selected subjects to evaluate the potential effectiveness of the proposed system based on a few assessment criteria using the previously used evaluation scale (Likert scale). The assessment results are shown in Fig. 12. The results show that the proposed tele-rehabilitation system and the social media-based exposure of the rehabilitation performance and exchange of rehabilitation-related information can keep the patients (subjects) well-connected with healthcare professionals and other patients suffering from similar diseases or disabilities. Such connectedness can reduce rehabilitation costs because neither the patients nor the healthcare providers need to make too much travels. Instead, they can exchange information while practicing rehabilitation in their own locations. The patients or their healthcare providers can compare the status of their rehabilitation progress with that of other patients and thus can benchmark their own progress and make necessary decisions and actions for further improvement [30].

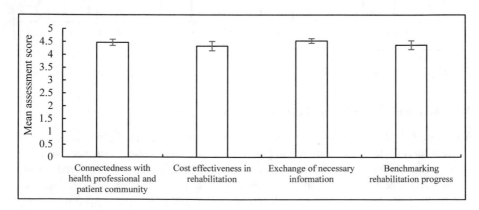

Fig. 12. Assessment results of potential effectiveness of the tele-rehabilitation system and social media.

8.3   Motivation Analysis

Another survey was conducted with the selected subjects to know what features of a robotic stroke rehabilitation system can motivate the patients to use the rehabilitation system intuitively. The assessment results are shown in Fig. 13, where different features of stroke rehabilitation system perceived as necessary by the subjects and the frequencies of their responses are presented. The results show that the suitability to home use, low cost and connectedness to the healthcare professionals and the patient communities are the most expected features of the robotic rehabilitation system that can motivate the patients to use the proposed robotic rehabilitation system. The results also reveal that addressing these features while designing and developing the robot-based stroke rehabilitation system can be the appropriate strategy to motivate the patients to use the rehabilitation systems enthusiastically, which can help receive rehabilitation benefits accordingly [31].

9   Discussion

The proposed design was analyzed for knee joint. However, similar analysis can be performed for other joints and degrees of freedoms. For example, the hip joint can be 3-dimensional (sagittal, transverse and frontal). The ankle joint can be 2-dimensional (sagittal and frontal). It is posited that the bio-inspired optimization of kinematics and kinetics characteristics comparing with that for humans can make the rehabilitation devices more natural and patient-friendly.

The gaming interface can visually display the rehabilitation performance and at the same time can generate cognitive impacts on the patients. Hence, the design of gaming interface needs to ensure that it impacts patient cognition in rational way. The types and nature of game-like activities can also keep the patients engaged. Hence, design of appropriate game-like activities impacting patient cognition and engagement demands further analysis [29].

The proposed tele-rehabilitation practice can be implemented using different communication methods such as internet, cloud, etc. These methods may impact real-time implementation, speed, accuracy and communication, which can impact the overall objective of the connected health. Similarly, connection through social media can also impact the overall connected health system. In addition, voice-based communication can be added to the internet/cloud-based tele-rehabilitation to make the communication multimodal and natural [30].

More wearable and ambient sensors can be used to collect rehabilitation measures that can help express rehabilitation performance in better ways. For example, the patient can wear inertial measurement units (IMUs) or Myo armband sensors that can measure various kinetics and kinematics data of different muscles. Such measurements can also be realized through the gaming-like activities, and displayed through the visual gaming interface. The same can also be shared to healthcare professionals and patient communities via tele-rehabilitation and social media that can further enhance the connectedness with more relevant and technically insightful rehabilitation-related information and events. Appropriate artificial intelligence and machine learning algorithms can also be incorporated with the rehabilitation system especially with the gaming interface design to make it smarter and intelligent.

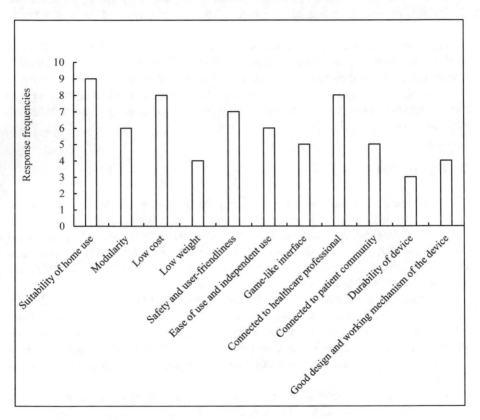

Fig. 13. Motivating features with frequencies for robotic rehabilitation devices and systems perceived as necessary by the selected subjects.

10   Conclusions and Future Works

The design, configuration, materials and working principles of a novel variable impedance compliant series elastic actuator with two springs (VIC-SEATS) were presented. The main novelty of the design is that it uses two sets of springs in series that allow change the stiffness based on input forces without requiring a change in the physical springs. The static and kinematic characteristics of the SEA design such as the torque and velocity profiles, angular movement-torque relationships, power and energy profiles, etc. were analyzed and optimized, and compared with the relevant human characteristics with a targeted application of actuating the knee joint of a gait rehabilitation robot for stroke patients. The results showed good matching between the actuator and the human characteristics that in turn indicate the good performance of the actuator design in various terms. The design also provided low power and energy requirements. The results can help develop compliant gait rehabilitation robots for stroke patients. Though the actuator design for the gait rehabilitation robot has not been physically tested yet, based on the analytical results it is believed that it will show satisfactory performance during the real applications. The effectiveness of the proposed gaming interface and tele-rehabilitation facility can improve patient satisfaction and keep them professionally and socially connected with the healthcare providers and patient communities. The motivational features of the robotic rehabilitation devices identified through survey can help rehabilitation designers design appropriate rehabilitation systems that can motivate the patients to use robot-aided rehabilitation systems.

In the near future, it may be necessary to analyze the effects of the inclusion of parallel elastic elements in the actuation system for the stipulated knee rehabilitation application [26]. Initiatives can be taken to augment the proposed design of the knee joint rehabilitation to a full lower limb gait rehabilitation system consisting of ankle, knee, hip and pelvis modules [27]. The effects of the inclusion of bi-articular elements in the proposed SEA for the design of the full lower limb gait rehabilitation system can be investigated [28]. The physical prototype of the knee joint based on the proposed design will need to be developed and its properties and characteristics will need to be tested experimentally during clinical trials with stroke patients. Advanced gaming interface with more practical game-like activities that can further impact the patients cognitively and keep them connected and engaged with improved rehabilitation performance can be developed. Advanced communication methods for the proposed tele-rehabilitation can be determined. More wearable and ambient sensors to gather more rehabilitation information to further impact the connected health/rehabilitation can be determined and used. Advanced intelligence such as machine learning, deep learning, convolutional neural networks, etc. can be incorporated with the rehabilitation system to make it smarter and intelligent. Design of patient-focused control systems can also be considered [23–25].

References

1. Tsagarakis, N.G., Laffranchi, M., Vanderborght, B., Caldwell, D.G.: A compact soft actuator unit for small scale human friendly robots. In: Proceedings of 2009 IEEE International Conference on Robotics and Automation, pp. 4356–4362 (2009)
2. Carrozza, M., Cappiello, G., Stellin, G., Zaccone, F., Vecchi, F., Micera, S., Dario, P.: A cosmetic prosthetic hand with tendon driven under-actuated mechanism and compliant joints: ongoing research and preliminary results. In: Proceedings of the 2005 IEEE International Conference on Robotics and Automation, pp. 2661–2666, 18–22 April 2005
3. Vallery, H., Veneman, J., Asseldonk, E., Ekkelenkamp, R., Buss, M., Kooij, H.: Compliant actuation of rehabilitation robots. IEEE Robot. Autom. Mag. 15(3), 60–69 (2008)
4. Veneman, J., Kruidhof, R., Hekman, E., Ekkelenkamp, R., Van Asseldonk, E.H.F., Van der Kooij, H.: Design and evaluation of the LOPES exoskeleton robot for interactive gait rehabilitation. IEEE Trans. Neural Syst. Rehabil. Eng. 15(3), 379–386 (2007)

5. Peshkin, M., Brown, D.A., Santos-Munne, J.J., Makhlin, A., Lewis, E., Colgate, J.E., Patton, J., Schwandt, D.: KineAssist: a robotic overground gait and balance training device. In: Proceedings of 2005 9th IEEE International Conference on Rehabilitation Robotics, pp. 241–246 (2005)
6. Hirzinger, G., Sporer, N., Albu-Schaffer, A., Hahnle, M., Krenn, R., Pascucci, A., Schedl, M.: DLR's torque-controlled light weight robot III – are we reaching the technological limits now? In: Proceedings of IEEE International Conference on Robotics and Automation, vol. 2, pp. 1710–1716 (2002)
7. Vische, D., Kathib, O.: Design and development of high performance torque-controlled joints. IEEE Trans. Robot. Autom. **11**(4), 537–544 (1995)
8. Tagliamonte, N.L., Sergi, F., Carpino, G., Accoto, D., Guglielmelli, E.: Design of a variable impedance differential actuator for wearable robotics applications. In: Proceedings of 2010 IEEE/RSJ International Conference on Intelligent Robots and Systems, pp. 2639–2644 (2010)
9. Ferris, D.P., Farley, C.T.: Interaction of leg stiffness and surface stiffness during human hopping. Am. Physiol. Soc. **82**, 15–22 (1997)
10. Zajac, F.: Muscle and tendon: properties, models, scaling, and application to biomechanics and motor control. Crit. Rev. Biomed. Eng. **17**(4), 359–411 (1989)
11. Yang, C., Ganesh, G., Haddadin, S., Parusel, S., Albu-Schaeffer, A., Burdet, E.: Human-like adaptation of force and impedance in stable and unstable interactions. IEEE Trans. Robot. **27** (5), 918–930 (2011)
12. Albu-Schaffer, A., Eiberger, O., Grebenstein, M., Haddadin, S., Ott, C., Wimbock, T., Wolf, S., Hirzinger, G.: Soft robotics. IEEE Robot. Autom. Mag. **15**(3), 20–30 (2008)
13. Wolf, S., Hirzinger, G.: A new variable stiffness design: matching requirements of the next robot generation. In: Proceedings of 2008 IEEE International Conference on Robotics and Automation, pp. 1741–1746 (2008)
14. Pratt, G., Williamson, M.: Series elastic actuators. In: Proceedings of IEEE/RSJ International Conference on Intelligent Robots and Systems, vol. 1, pp. 399–406 (1995)
15. Schiavi, R., Grioli, G., Sen, S., Bicchi, A.: VSA-II: a novel prototype of variable stiffness actuator for safe and performing robots interacting with humans. In: Proceedings of 2008 IEEE International Conference on Robotics and Automation, pp. 2171–2176 (2008)
16. Grimmer, M., Eslamy, M., Gliech, S., Seyfarth, A.: A comparison of parallel- and series elastic elements in an actuator for mimicking human ankle joint in walking and running. In: Proceedings of 2012 IEEE International Conference on Robotics and Automation, pp. 2463–2470 (2012)
17. Robinson, D.W., Pratt, J.E., Paluska, D.J., Pratt, G.A.: Series elastic actuator development for a biomimetic walking robot. In: Proceedings of 1999 IEEE/ASME International Conference on Advanced Intelligent Mechatronics, pp. 561–568 (1999)
18. Yu, H., Rahman, S.M.M., Zhu, C.: Preliminary design analysis of a novel variable impedance compact compliant actuator. In: Proceedings of 2011 IEEE International Conference on Robotics and Biomimetics, 7–11 December 2011, Phuket, pp. 2553–2558 (2011)
19. Velandia, C., Celedón, H., Tibaduiza, D.A., Torres-Pinzón, C., Vitola, J.: Design and control of an exoskeleton in rehabilitation tasks for lower limb. In: Proceedings of 2016 XXI Symposium on Signal Processing, Images and Artificial Vision (STSIVA), Bucaramanga, pp. 1–6 (2016)
20. Ragonesi, D., Agrawal, S.K., Sample, W., Rahman, T.: Quantifying anti-gravity torques for the design of a powered exoskeleton. IEEE Trans. Neural Syst. Rehabil. Eng. **21**(2), 283–288 (2013)

21. Winter, D.: The biomechanics and motor control of human gait: normal, elderly, and pathological. Waterloo University Press (1991)
22. Winter, D.: Biomechanics and Motor Control of Human Movement. Wiley, Fort Collins (2005)
23. Rahman, S.M.M., Ikeura, R., Hayakawa, S., Yu, H.: Manipulating objects with a power assist robot in linear vertical and harmonic motion: psychophysical-biomechanical approach to analyzing human characteristics to modify the control. J. Biomech. Sci. Eng. 6(5), 399–414 (2011)
24. Rahman, S.M.M., Ikeura, R., Yu, H.: Novel biomimetic control of a power assist robot for horizontal transfer of objects. In: Proceedings of 2011 IEEE Int. Conference on Robotics and Biomimetics, 7–11 December 2011, pp. 2181–2186 (2011)
25. Rahman, S.M.M., Ikeura, R.: Weight-perception-based novel control of a power-assist robot for the cooperative lifting of light-weight objects. Int. J. Adv. Robot. Syst. 9(118), 1–13 (2012)
26. Mettin, U.: Parallel elastic actuators as a control tool for preplanned trajectories of underactuated mechanical systems. Int. J. Robot. Res. 20(9), 1186–1198 (2010)
27. Jiang, Y., Bai, B., Wang, S.: Modeling and simulation of omni-directional lower limbs rehabilitation training robot. In: Proceedings of the 10th World Congress on Intelligent Control and Automation, Beijing, pp. 3737–3740 (2012)
28. Lim, B., Babic, J., Park, F.: Optimal jumps for biarticular legged robots. In: Proceedings of 2008 IEEE International Conference on Robotics and Automation, pp. 226–231 (2008)
29. Valdés, B.A., Hilderman, C.G.E., Hung, C.T., Shirzad, N., Van der Loos, H.F.M.: Usability testing of gaming and social media applications for stroke and cerebral palsy upper limb rehabilitation. In: Proceedings of 2014 36th Annual International Conference of the IEEE Engineering in Medicine and Biology Society, Chicago, IL, pp. 3602–3605 (2014)
30. Kawai, Y., Honda, K., Kawai, H., Miyoshi, T., Fujita, M.: Tele-rehabilitation system for human lower limb using electrical stimulation based on bilateral teleoperation.In: Proceedings of 2017 IEEE Conference on Control Technology and Applications (CCTA), Mauna Lani, HI, pp. 1446–1451 (2017)
31. Mihelj, M., Novak, D., Ziherl, J., Olenšek, A., Munih, M.: Challenges in biocooperative rehabilitation robotics. In: Proceedings of 2011 IEEE International Conference on Rehabilitation Robotics, Zurich, pp. 1–6 (2011)
32. Nowak, D.: The impact of stroke on the performance of grasping: usefulness of kinetic and kinematic motion analysis. Neurosci. Biobehav. Rev. 32(8), 439–1450 (2008)
33. Rahman, S.M.M.: A novel variable impedance compact compliant series elastic actuator: analysis of design, dynamics, materials and manufacturing. Appl. Mech. Mater. 245, 99–106 (2013)
34. Rahman, S.M.M., Ikeura, R.: A novel variable impedance compact compliant ankle robot for overground gait rehabilitation and assistance. Procedia Eng. 41, 522–531 (2012)
35. Rahman, S.M.M.: Design of a modular knee-ankle-foot-orthosis using soft actuator for gait rehabilitation. In: Proceedings of the 14th Annual Conference on Towards Autonomous Robotic Systems (TAROS 2013), 28–30th August 2013. Oxford University, U.K. Lecture Notes in Computer Science, Springer-Verlag, vol. 8069, pp. 195–209, July 2014
36. Rahman, S.M.M.: A novel variable impedance compact compliant series elastic actuator for human-friendly soft robotics applications. In: Proceedings of the 21st IEEE International Symposium on Robot and Human Interactive Communication (RO-MAN 2012), Paris, France, September 9–13 2012, pp. 19–24 (2012)

Grasp Rehabilitation of Stroke Patients Through Object Manipulation with an Intelligent Power Assist Robotic System

S. M. Mizanoor Rahman$^{(\boxtimes)}$

The University of West Florida, Pensacola, FL 32514, USA
smrahman@uwf.edu

Abstract. A 1-DOF experimental power assist robotic system was developed for grasping and lifting objects tied with it by stroke patients as a part of their upper arm grasp rehabilitation practices. A human-in-the-loop model of the targeted dynamics for lifting objects with the system was developed that considered prospective patient's weight perception. A position control scheme based on the weight-perception-based dynamics model was developed and implemented. The control parameters were the virtual mass values used for the grasped and manipulated object. The virtual mass values that produced satisfactory human-robot interactions (HRI) in terms of maneuverability, motion, stability, health and safety, etc. were determined. The results showed that a few sets of mass values could produce satisfactory HRI. Then, stroke patients grasped and manipulated objects with the system separately for the most satisfactory virtual mass values that provided optimum kinesthetic perception. In separate experiments, the virtual mass value was exponentially reduced when the patients manipulated the objects with the system, which provided variation in kinesthetic perception. The results showed that the power-assisted manipulation with optimum kinesthetic perception significantly contributed to stroke rehabilitation. The results also showed that the rehabilitation performance varied with the variation in the kinesthetic perception. In addition, effectiveness of a proposed game-like visual interface and smart tele-rehabilitation for connected health was investigated.

Keywords: Power assist robot · Connected health · Object manipulation · Stroke patient · Free motion grasp rehabilitation · Kinesthetic perception

1 Introduction

Grasping and manipulation is very necessary for humans in their daily living and in performing daily prehensile activities. Stroke patients may lose motor skills and abilities for grasping and manipulating objects, and thus stroke rehabilitation for upper arm especially the grasp rehabilitation has received priorities [1]. Grasp rehabilitation is an active area of research, and a plethora of robotics devices have been proposed for grasp rehabilitation [2–15]. In [2, 3], a grasp rehabilitation system was proposed where user can reach, grasp and retrieve an object on virtual task exercise. In [4], a grasping device

© Springer Nature Switzerland AG 2020
K. Arai et al. (Eds.): FTC 2019, AISC 1070, pp. 244–259, 2020.
https://doi.org/10.1007/978-3-030-32523-7_16

was proposed that consists of a two-degrees of freedom mechanism for measuring the grasp force and for a passive haptic rendering. In [5], rehabilitation of grasping and forearm pronation and supination was performed with a haptic knob. In [6], EMG biofeedback-based virtual reality (VR) system was used for hand rotation and grasping rehabilitation. In [7], a novel hybrid tool called the ArmeoFES was proposed for reaching and grasping rehabilitation. In [8], kinematic data analysis for post-stroke patients was performed following bilateral versus unilateral rehabilitation with an upper limb wearable robotic system. In [9], a soft extra wearable muscle system was used for improving grasping capability in neurological rehabilitation. In [10], a rehabilitation wearable device was used to improve hand grasp function of stroke patients. In [11], an exoskeleton for measurement of human hand finger joint trajectories was proposed, a prosthesis was developed and used, and a model predictive controller was suggested for grasp rehabilitation. In [12], a three-digit grasp haptic device with variable contact stiffness for rehabilitation and human grasping practice was studied. In [13], the quality of grasping and the role of haptics were investigated in a 3-D immersive virtual reality environment in individuals with stroke. In [14], a hand rehabilitation learning system was developed in the form of an exoskeleton robotic glove. In [15], a soft robotic supernumerary finger and a wearable cutaneous finger interface were used to compensate the missing grasping capabilities in chronic stroke patients, and so forth.

However, most of the state-of-the-art upper limb stroke rehabilitation devices are wearable that do not seem to resemble natural free motion reaching, grasping and manipulation practices observed in healthy humans [16]. It is believed, in order to resemble natural human practices, the free motion style reaching, grasping and manipulation should be considered. However, grasp rehabilitation with free motion of patient hand is usually not observed. In addition, the existing devices cannot provide power-assistance to the patients [17], and also cannot optimize and change patients' kinesthetic perceptions when the patients practice grasping and manipulation tasks with the rehabilitation devices. There is no information on how power-assisted grasping and manipulation can impact rehabilitation performance especially if the kinesthetic perception at the time of power-assisted grasping and manipulation is changed. It is believed that power-assistance and variation in kinesthetic perception at the time of grasping and rehabilitation can impact rehabilitation performance and patient satisfaction [18]. Power assist systems have been proposed for object manipulation in industries [19–25]. However, the state-of-the-art literature does not show any investigation on the possibilities of utilizing the power-assisted free motion object manipulation for upper arm stroke rehabilitation.

There is a psychological issue (illusion in weight perception) with power assist robotic systems for object manipulation that may affect rehabilitation performance [18–25]. A power assist system reduces the perception of weight when an object is manipulated with the system [17–25]. However, the user (patient) cannot realize it until he/she has a haptic feeling of the object (until he/she touches and manipulates the object). The human (patient with sound cognitive ability) usually applies a feedforward grip force and a load force (vertical lifting force) to the object to lift it with the system based on the visually perceived weight of the object, which depends on the object's visual size and shape [26]. As the visually perceived weight is larger than the haptically perceived weight for power-assist-lifted objects, the applied load force becomes excessive that results in unexpected motion of the system. Consequently, maneuverability, stability, safety, etc. may not be satisfactory, and such unsatisfactory performance may not help conduct effective rehabilitation practice with the system. However, the state-of-the-art power assist systems do not consider these psychological issues [17–25]. It is posited that the psychological issue should be addressed before the power assist robotic systems can be exposed to stroke patients for the proposed rehabilitation practices.

The control systems for the power assist systems for object manipulation are usually derived from the system dynamics. The dynamic equations of motions usually include inertial force (product of virtual mass and object acceleration), gravitational force (product of virtual mass and gravitational acceleration), load force, actuator force, viscosity, friction force, etc. The actuator force, viscosity, friction, etc. may be disregarded for free motion dynamics [18]. Thus, the virtual mass value used for the object is the main control parameter of the control system. It is assumed that inclusion of weight perception in the dynamics and control as well as the optimization of the mass parameter value used in the control system could solve the above-mentioned psychological problem of the power assist systems, and thus could improve the patient-robot interactions and make the system safe and suitable for stroke rehabilitation. However, such inclusion of weight perception in the control and optimization of the mass parameter have never been considered for the state-of-the-art power assist systems to foster and facilitate upper limb grasp rehabilitation [27–30].

To address the aforementioned limitations, in this paper, a 1-DOF experimental power assist robotic system was developed for grasping and lifting objects tied with it by stroke patients (or healthy persons pretending as stroke patients) as a part of their intended upper arm rehabilitation practices. The grasped and manipulated object was small and lightweight that was to be natural for patients [18–25]. The position control method that considered human's weight perception to produce optimum HRI was implemented for manipulating the objects [31]. Then, experiment was arranged where stroke patients (or healthy persons pretending as stroke patients) grasped and manipulated an object with the power assist system separately for the most satisfactory (optimum) virtual mass values providing the optimum kinesthetic perception [31]. In a separate experiment, the virtual mass value was exponentially reduced when the patients manipulated the object with the system to experience the variation in the kinesthetic perception [18]. Then, it was investigated how the power-assisted manipulation with optimum and variable kinesthetic perception could impact the intended upper limb stroke rehabilitation. In addition, the effectiveness of a proposed game-like visual interface and tele-rehabilitation facility with the power assist system for the connected health was investigated.

2　Development of the Test Bed Power Assist Robotic System

A 1-DOF test bed power assist system for lifting objects was developed as shown in Fig. 1. In fact, this device was used in many past experiments, e.g. [18–25]. The same experiment device was used here because the same device used in a different way could serve the purpose of this article. As in Fig. 1(a), an AC servomotor and a ball screw were coaxially put on a metal plate and then the plate was vertically attached to a wall. The ball screw was used to convert the rotary motion to the linear motion. A force sensor (load cell) was attached to the ball nut of the ball screw through an acrylic resin block. One end of a universal joint was tied to the force sensor and the other end was attached to a wooden block. The wooden block helped tie an object (a rectangular thin metal box) to the nut of the ball screw through the force sensor. Figure 1(b) shows how an object was tied to the ball screw nut through the force sensor. The force sensor was used to measure the force applied to the object by a human user (say, a stroke patient or a healthy subject) when the human lifted the object with the system. An optical encoder was attached to the servomotor to measure the position and its derivatives for the object lifted with the assist system.

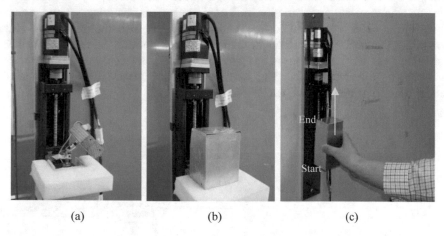

(a)   (b)   (c)

Fig. 1. (a) Construction of the power assist system, (b) the complete power assist system with an object, (c) the way how a human user can lift an object with the power assist system [18–25].

The rectangular object (box) was made by bending thin aluminum sheet (thickness: 0.0005 m). The top side of the box was covered with a cap made of thin aluminum sheet (thickness: 0.0005 m) and the bottom and the back sides were kept open. Three objects of three different sizes such as large (0.06 × 0.05 × 0.16 m), medium (0.06 × 0.05 × 0.12 m) and small (0.06 × 0.05 × 0.09 m) were actually made. Self-weights of the boxes were 0.020 kg, 0.016 kg and 0.012 kg for the large, medium and small size object respectively. Each of the objects was separately lifted with the power assist robotic system, and was called the power-assisted object (PAO). The object was kept on a soft surface before it was lifted, as in Fig. 1(b). Figure 1(c) shows how a human user can lift an object with the power assist robotic system. The experimental setup and the communication system of the power assist robotic system are presented in details in previous publications such as [18–25].

3 System Dynamics and the Control System

The system dynamics and the control system for the proposed power assist robotic system were described in details in previous publications such as [18–25]. Here, some important portions of the system dynamics and controls are discussed briefly. The human-centered free motion dynamics for lifting an object (PAO) with the power assist robotic system can be expressed as (1). In (1), m_1 is the virtual mass parameter for the inertial force, and m_2 is the virtual mass parameter for the gravitational force. This difference was considered in power-assisted object manipulation because the visually perceived object weight is different from the haptically perceived weight [17–25]. In (1), $\ddot{x}_d$ is the desired displacement of the lifted object, g is the acceleration of gravity, and f_h is the load force (lifting force) applied to the assist system by its human user. Note that the actuating force, friction, noises, perturbations, disturbances and viscosity in the linear slider were ignored in (1) to resemble the free motion natural lifting dynamics [32]. The weight-perception-based feedback position control for the power assist robotic system derived based on (1) is shown in Fig. 2 (see previous publications, e.g. [27–30] for further details). This is a novel position control scheme that includes human feature, i.e. the weight perception of the human.

$$m_1 \ddot{x}_d + m_2 g = f_h \tag{1}$$

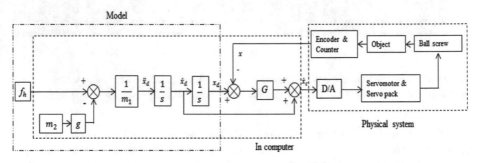

Fig. 2. The position control considering human's weight perception. The model represents (1). The servo system needs to be kept on the velocity control mode for executing the position control. The input is the load force (f_h) and the output is the desired displacement of the lifted object ($\ddot{x}_d$, and its derivatives). As g is fixed, the main control parameters for the control are the values of m_1 and m_2. G is here the feedback gain, and the D/A is the abbreviation of the digital to analog converter. Here, $\dot{x}_c$ is the commanded velocity to the system [27–30].

4 Experiment 1: Determining Optimum System Behavior Using Healthy Subjects

The control system in Fig. 2 was executed using MATLAB/Simulink for lifting an object with it in collaboration with a human subject, as shown in Fig. 1(c). Six values of the virtual masses were selected based on experience. It means that m_1 could be any of the six values: 0.25, 0.5, 0.75, 1.0, 1.25, 1.50 kg. Similarly, m_2 could also be any of the six values: 0.25, 0.5, 0.75, 1.0, 1.25, 1.50 kg. In a trial, a pair of values of m_1 and m_2 was randomly selected and put in the control system in Fig. 2. Then, the control system was executed, and a human subject needed to lift the object with the system. The human-robot interaction (HRI) performance in the power-assisted manipulation was expressed in terms of maneuverability, motion (kinematics-displacement, velocity and acceleration overshoot), stability, comfort, occupational health and safety, etc. At the end of the trial, the subject needed to assess the HRI using a 7-point Likert scale (−3 the worst, 0 the neutral, +3 the best) [31]. In this way, 36 subjects lifted the object with the power assist robotic system for 36 pairs of values of m_1 and m_2 separately, and assessed the HRI. The subjects were the undergraduate and graduate engineering students with mean age 23.11 years and standard deviations of 1.56. The study was approved by the institutional review board (IRB) or its equivalent (e.g., the university ethical committee). The subjects received detailed instructions about the experiment procedures before they participated in the experiment. The subjects gave informed consent about their attendance to the experiment.

Based on the assessment results, it was found that the system produced satisfactory performance in some pairs of values of m_1 and m_2, and the most satisfactory (best) performance was achieved for $m_1 = 0.5$ kg, $m_2 = 0.25$ kg. This is why, the pair $m_1 = 0.5$ kg and $m_2 = 0.25$ kg was considered as the condition for the best (optimum) performance of the system for lifting objects. Here, the optimality was determined subjectively based on heuristics. See [31, 33, 34] for further details about how the control parameters for the optimum performance were decided.

5 Experiment 2: Grasp Rehabilitation of Stroke Patients Using Power Assist Robotic System

5.1 Subjects

In total, 12 stroke patients with upper arm motor disabilities were actually needed to be selected. All the targeted patients were aged, 5 were male and 7 were female. The targeted subjects were believed to possess good cognitive abilities. The targeted subjects were taking therapy with healthcare providers at homes and/or health centers. They were to be at the early stages of their therapies. The subjects were needed to be divided randomly into 2 groups. In group 1, there were targeted 2 males and 4 females. In group 2, there were targeted 3 males and 3 females. For group 2, 2 subgroups were needed to be made: 2 males and 1 female were targeted in subgroup 1 (S21), and 1 male and 2 females were targeted in subgroup 2 (S22).

Note that the required/targeted number of stroke patients may not be available to participate in the experiment in some cases. In such cases, for the proof of concept, patients with other similar types of upper arm disabilities and with demonstrated weakness in object manipulation may be used as an alternative. In extreme cases where the actual patients are unavailable for the experiment, healthy subjects may pretend as the stroke patients and may take part in the experiment. In such cases, the healthy subjects need to be trained properly using video demonstration of the behaviors of the stroke patients so that the healthy subjects can learn how to pretend like stroke patients to produce similar manipulation and prehensile behaviors as the actual stroke patients usually show.

5.2   Experiment

Each group of patients (or healthy subjects pretending as patients) was evaluated after each 5 days for 15 days. Group 1 patients received grasping therapies every day for 1 h from healthcare providers (or from the experimenter pretending as a healthcare provider) using some conventional wearable grasping and manipulation devices instrumented with force and position sensors [14]. Group 2 patients practiced grasping and rehabilitation with the proposed power assist robotic system for 1 h every day. The virtual mass values $m_1 = 0.5$ kg, $m_2 = 0.25$ kg were used with the control system shown in Fig. 2, which was believed to provide optimum kinesthetic perception to the patients. However, the patients might not be able to have the optimum kinesthetic perception fully due to numbness and lack of motor skills.

During the experiment with the power assist robotic system, each patient separately grasped and manipulated a medium size object with the power assist robotic system following the method shown in Fig. 1(c). They were allowed to take rest for a while if

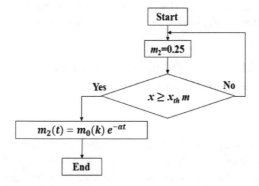

Fig. 3. Exponential reduction of m_2 value when a patient lifts the object with the system and when the lifting displacement (x) in upward direction exceeds a threshold (x_{th}). Here, m_0 is the initial value of m_2 at the initial time k, t is the end time of the lift, and α is the decay constant.

they experienced fatigue. They were allowed to perform the experiment sitting on wheelchairs/chairs or standing on the floor depending on their physical conditions. The experimenter guided and assisted the patients to perform the rehabilitation tasks. In addition, for subgroup S22, the m_2 value of the control system in Fig. 2 exponentially reduced following (2) as illustrated in Fig. 3 when the patients started lifting the object with the power assist system and exceeded a displacement (position) threshold [18]. That practice provided variation in kinesthetic perception while lifting due to variation in m_2 value. The kinetics and kinematics data were recorded (see [18–25, 31] for details about the kinetics and the kinematics data).

The kinetics and kinematics data were displayed in a screen using a game-like visual interface placed in front of the patient with the power assist system as shown in Fig. 4 [35], and the same interface was shared with the healthcare professional (or a person pretending as a healthcare professional) at a distant location (in a different room in this experiment) in real-time using an appropriate internet communication method as shown in Fig. 5. This practice is called here the tele-rehabilitation practice [36]. At the end of the rehabilitation practice, each patient completed a survey to evaluate the rehabilitation system based on some selected criteria using the same 7-point Likert scale used for experiment 1 (−3 the worst, 0 the neutral, +3 the best).

$$m_2(t) = m_0(k)e^{-\alpha t} \tag{2}$$

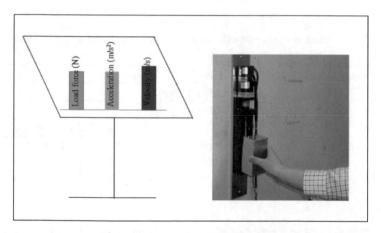

Fig. 4. Visual interface to display rehabilitation performance in real-time through game-like activities, e.g. meeting a target load force, velocity or acceleration in a specified time.

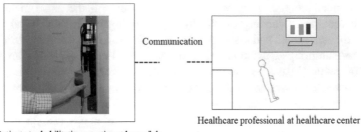

Patient at rehabilitation practice at home/lab

Healthcare professional at healthcare center

Fig. 5. Tele-rehabilitation system to share rehabilitation performance between patient (subject) and healthcare professional (here, the experimenter waiting in another room) at a distant location.

5.3   Experimental Results

Human's motor skills and abilities while grasping and manipulating can be assessed through the peak load force, peak acceleration and peak velocity that human produces at the time of grasping and manipulation [31]. It is assumed that the larger peak load force, peak acceleration and peak velocity indicate the better motor skills and abilities, and the better motor skills and abilities achieved through the rehabilitation practices indicate the success/effectiveness of the proposed rehabilitation methods. Tables 1, 2 and 3 show the peak load force, peak acceleration and peak velocity for different groups of patients (subjects) at different observation frequencies.

Table 1. Rehabilitation performance evaluation results for different groups of stroke patients after 5 days

Evaluation criteria	Observation frequency		
	After 5 days		
	Group 1	Group 2	
		Subgroup S21	Subgroup S22
Mean peak load force (N)	2.68 (0.11)	3.19 (0.08)	3.43 (0.14)
Mean peak acceleration (m/s^2)	0.48 (0.04)	0.59 (0.03)	0.67 (0.03)
Mean peak velocity (m/s)	0.11 (0.001)	0.16 (0.0011)	0.23 (0.0014)

Table 2. Rehabilitation performance evaluation results for different groups of stroke patients after 10 days

Evaluation criteria	Observation frequency		
	After 10 days		
	Group 1	Group 2	
		Subgroup S21	Subgroup S22
Mean peak load force (N)	2.91 (0.07)	3.47 (0.10)	3.86 (0.05)
Mean peak acceleration (m/s^2)	0.57 (0.02)	0.66 (0.05)	0.80 (0.04)
Mean peak velocity (m/s)	0.17 (0.0017)	0.22 (0.0021)	0.29 (0.0016)

Table 3. Rehabilitation performance evaluation results for different groups of stroke patients after 15 days

Evaluation criteria	Observation frequency		
	After 15 days		
	Group 1	Group 2	
		Subgroup S21	Subgroup S22
Mean peak load force (N)	3.04 (0.13)	3.93 (0.09)	4.14 (0.12)
Mean peak acceleration (m/s^2)	0.64 (0.02)	0.73 (0.05)	0.89 (0.06)
Mean peak velocity (m/s)	0.24 (0.0013)	0.28 (0.0019)	0.36 (0.0020)

Results in Tables 1, 2 and 3 show that the peak load force, peak acceleration and peak velocity for all groups of patients observed after 10 days were higher than that observed after 5 days of the rehabilitation cycle. Similarly, the peak load force, peak acceleration and peak velocity increased as the rehabilitation practice continued up to 15 days. This clearly shows that the rehabilitation practice was effective to increase motor skills and abilities. If the peak load force, peak acceleration and peak velocity are compared between group 1 and group 2, it can be found that the peak load force, peak acceleration and peak velocity were larger for group 2 than that for group 1 throughout the observation frequencies. This proves that the power-assisted free motion grasping and manipulation rehabilitation practice with optimum kinesthetic perception as proposed here was better than the grasp rehabilitation provided through grasping and manipulation tasks using constrained wearable robotics devices [14]. Between the two subgroups of group 2, the peak load force, peak acceleration and peak velocity for subgroup 2 (S22) were larger than that of subgroup 1 (S21), which indicates that the variation in the kinesthetic perception while conducting the rehabilitation practices provided better rehabilitation performance.

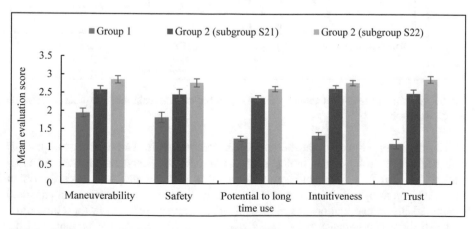

Fig. 6. Subjective evaluation of different rehabilitation methods evaluated by stroke patients.

Figure 6 shows the subjective evaluation results for different rehabilitation methods evaluated by the stroke patients (subjects). The results show that the power-assisted free motion grasping and manipulation rehabilitation with optimum kinesthetic perception was perceived as better than the grasp rehabilitation provided through grasping and manipulating tasks using wearable robotics devices in terms of maneuverability, safety, long time usability (low fatigue), intuitiveness and trustworthiness [14]. Between the two subgroups of group 2, the power-assisted rehabilitation method with variable kinesthetic perception was perceived as better by the patients (subjects).

Figure 7 shows the assessment results of gaming-like visual interface for rehabilitation practice with the power assist robotic system (Fig. 4). The results show that the patients perceived the gaming-like visual interface as satisfactory based on the 7-point Likert scale (−3 the worst, 0 the neutral, +3 the best). As the results show, the graphical display of the key rehabilitation performance parameters in real-time (Fig. 4) increased patients' situation awareness as well as kept them engaged with the rehabilitation practice. The graphical display of the key rehabilitation performance parameters in real-time also increased the clarity of the performance to the patients that might help the patients understand the trend of the rehabilitation progress as well as help them make plans for further improvement. The graphical interface also served as a guide to the patients while practicing rehabilitation that eased the overall operation of the system and helped conduct the rehabilitation practice [35].

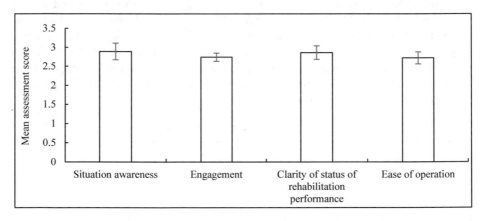

Fig. 7. Assessment results of gaming-like visual interface for rehabilitation practice with the power assist robotic system.

Results in Fig. 8 show that the tele-rehabilitation facility can connect the patients with healthcare professionals and patient community and exchange necessary information that can help the healthcare professionals and the patients make important decisions regarding rehabilitation practices as well as can help them compare and benchmark the rehabilitation progress with other relevant patients in the community [36]. All these can enhance the connectedness in healthcare. As the patients and the healthcare professionals can exchange information and provide consultation online, the overall rehabilitation system can help them reduce travels that can save their time and reduce overall healthcare costs.

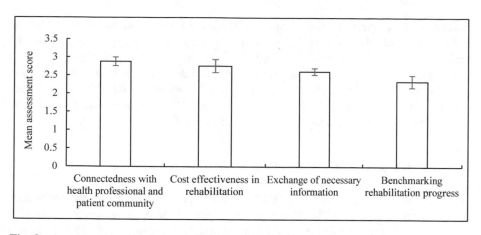

Fig. 8. Assessment results of tele-rehabilitation facility for rehabilitation practice with the power assist robotic system.

6 Discussion

Objects of three different sizes in experiment 1 could be used to understand the effects of visual object sizes on kinetics and kinematics of grasping and manipulating as well as on the performance of the system so that the results could be generalized. On the other hand, the performance evaluation of the system was subjective instead of objective. The optimum kinesthetic condition ($m_1 = 0.5$ kg, $m_2 = 0.25$ kg) was determined based on heuristics or empirical methods that depended on the subjective evaluation. However, it is to believe that the subjective evaluation would be effective because such evaluations were proven effective in previous cases [18–25].

The best values of m_1 and m_2 used for lightweight objects can also be effective when designing the control systems to lift slightly heavier objects once the patients gain some advanced motor skills through rehabilitation practices. The heavy object weight can be borne by the strength of the structure of the robot and the patients can feel only a small weight (a portion of the value of m_2 used in the control system), and the patient's load force can determine the motion of the system only. However, the robot strength, structure and configuration should be made suitable for manipulating heavier and larger loads if the patients want to continue practice for long period with improved motor abilities. It is realized that the rehabilitation performance with power assist system largely depends on the appropriate performance or behavior of the system itself, which may be further improved by further optimizing the values of m_1 and m_2 and by applying active compliance algorithms as well as passive compliance through soft, compliant and human-friendly mechanisms and system design [37–41].

Usually, the impedance/admittance based position and force control methods are used for power assist systems [18–25]. For lifting objects, the positional accuracy is demanding and the human may perceive and realize the system characteristics produced by the positional command realistically and comfortably [18]. Again, the position control may compensate friction, inertia, viscosity, etc. of the system. It can be realized that the dynamic effects, nonlinear forces, etc. can affect less intensely for the position control. However, if the difference between m_2 and m_1 is very large, the position control may impose very high load to the servomotor that may result in instability. The comparison of the system performance as well as the rehabilitation performance between position and force controls for the system for lifting objects may reveal significant insights for choosing the appropriate control methods for robotic devices for the proposed rehabilitation applications [34, 42].

7 Conclusions and Future Work

A 1-DOF power assist robotic system was developed for lifting objects, and stroke patients could grasp and lift the objects as a part of their upper arm rehabilitation practices. The user's weight perception was considered when deriving the dynamics model of the human-robot collaborative rehabilitation system, and the weight-perception-based feedback position control was developed based on the dynamics model. In the control system, the control parameter was the virtual mass values used for the manipulated object. In order to use the system for rehabilitation, safe and satisfactory system performance was expected. Hence, the virtual mass values that produced the optimum system performance and interactions were determined. The results showed that a set of mass values produced satisfactory/optimum performance and interactions. Then, the stroke patients grasped and manipulated objects with the system at the optimum virtual mass values separately that provided optimum kinesthetic perception. In a separate experiment, the virtual mass value was also exponentially reduced when the patients manipulated the objects with the system that provided variation in the kinesthetic perception. The performance of the free motion power-assisted grasping rehabilitation was compared with that performed using constrained wearable rehabilitation devices. Results showed that the power-assisted manipulation with optimum kinesthetic perception contributed to stroke rehabilitation with better performance in comparison with that for the conventional wearable constrained rehabilitation. It was also found that the rehabilitation performance varied with variation in kinesthetic perception. The results also showed that the gaming interface-based tele-rehabilitation enhanced the connectedness in healthcare, the overall rehabilitation performance and the system acceptance perceived by the patients (subjects).

In the future, cloud-based tele-rehabilitation with faster communication and accuracy will be investigated. Connectedness in robotics-based healthcare and rehabilitation will be increased through incorporating advanced social media in healthcare practices. The 3-DOF translational manipulation with power-assist will be considered. It will be investigated how the manipulation along other 2-DOFs can impact the rehabilitation performance. Better games with more relevance with the rehabilitation activities will be developed to be included in the visual interface. For the experimental validation, more emphasis will be given on conducting the experiments with actual stroke patients and healthcare professionals instead of using non-patient subjects pretending like patients.

References

1. Barsotti, M., et al.: A full upper limb robotic exoskeleton for reaching and grasping rehabilitation triggered by MI-BCI. In: Proceedings of the 2015 IEEE International Conference on Rehabilitation Robotics (ICORR), Singapore, pp. 49–54 (2015)
2. Loureiro, R.C.V., Harwin, W.S.: Reach & grasp therapy: design and control of a 9-DOF robotic neuro-rehabilitation system. In: Proceedings of the 2007 IEEE 10th International Conference on Rehabilitation Robotics, Noordwijk, pp. 757–763 (2007)
3. Loureiro, R.C.V., Lamperd, B., Collin, C., Harwin, W.S.: Reach & grasp therapy: effects of the Gentle/G system assessing sub-acute stroke whole-arm rehabilitation. In: Proceedings of the 2009 IEEE International Conference on Rehabilitation Robotics, Kyoto, pp. 755–760 (2009)
4. Podobnik, J., Mihelj, M., Munih, M.: Upper limb and grasp rehabilitation and evaluation of stroke patients using HenRiE device. In: Proceedings of the 2009 Virtual Rehabilitation International Conference, Haifa, pp. 173–178 (2009)
5. Lambercy, O., et al.: Rehabilitation of grasping and forearm pronation/supination with the Haptic Knob. In: Proceedings of the 2009 IEEE International Conference on Rehabilitation Robotics, Kyoto, pp. 22–27 (2009)
6. Ma, S., Varley, M., Shark, L.K., Richards, J.: EMG biofeedback based VR system for hand rotation and grasping rehabilitation. In: Proceedings of the 2010 14th International Conference on Information Visualisation, London, pp. 479–484 (2010)
7. Crema, A., et al.: A hybrid tool for reaching and grasping rehabilitation: the ArmeoFES. In: Proceedings of the 2011 Annual International Conference of the IEEE Engineering in Medicine and Biology Society, Boston, MA, pp. 3047–3050 (2011)
8. Kim, H., et al.: Kinematic data analysis for post-stroke patients following bilateral versus unilateral rehabilitation with an upper limb wearable robotic system. IEEE Trans. Neural Syst. Rehabil. Eng. **21**(2), 153–164 (2013)
9. Nilsson, M., Ingvast, J., Wikander, J., von Holst, H.: The soft extra muscle system for improving the grasping capability in neurological rehabilitation. In: Proceedings of the 2012 IEEE-EMBS Conference on Biomedical Engineering and Sciences, Langkawi, pp. 412–417 (2012)
10. Park, W., Jeong, W., Kwon, G.H., Kim, Y.H., Kim, L.: A rehabilitation device to improve the hand grasp function of stroke patients using a patient-driven approach. In: Proceedings of the 2013 IEEE 13th International Conference on Rehabilitation Robotics (ICORR), Seattle, WA, pp. 1–4 (2013)
11. Kakoty, N.M., Hazarika, S.M., Koul, M.H., Saha, S.K.: Model predictive control for finger joint trajectory of TU Biomimetic hand. In: Proceedings of the 2014 IEEE International Conference on Mechatronics and Automation, Tianjin, pp. 1225–1230 (2014)
12. Altobelli, A., Bianchi, M., Serio, A., Baud-Bovy, G., Gabiccini, M., Bicchi, A.: Three-digit grasp haptic device with variable contact stiffness for rehabilitation and human grasping studies. In: Proceedings of the 22nd Mediterranean Conference on Control and Automation, Palermo, pp. 346–350 (2014)
13. Levin, M.F., Magdalon, E.C., Michaelsen, S.M., Quevedo, A.A.F.: Quality of grasping and the role of haptics in a 3-D immersive virtual reality environment in individuals with stroke. IEEE Trans. Neural Syst. Rehabil. Eng. **23**(6), 1047–1055 (2015)
14. Ma, Z., Ben-Tzvi, P., Danoff, J.: Hand rehabilitation learning system with an exoskeleton robotic glove. IEEE Trans. Neural Syst. Rehabil. Eng. **24**(12), 1323–1332 (2016)

15. Hussain, I., Meli, L., Pacchierotti, C., Prattichizzo, D.: A soft robotic supernumerary finger and a wearable cutaneous finger interface to compensate the missing grasping capabilities in chronic stroke patients. In: Proceedings of the 2017 IEEE World Haptics Conference (WHC), Munich, pp. 183–188 (2017)

16. Igarashi, K., Katsura, S.: Position based free-motion data connecting by using minimum force-differential model. In: Proceedings of the IEEE International Conference on Mechatronics, pp. 535–540 (2015)

17. Kazerooni, H.: Extender: a case study for human-robot interaction via transfer of power and information signals. In: Proceedings of the IEEE International Workshop on Robot and Human Communication, pp. 10–20 (1993)

18. Rahman, S.M.M., Ikeura, R., Nobe, M., Sawai, H.: Design and control of a 1DOF power assist robot for lifting objects based on human operator's unimanual and bimanual weight discrimination. In: Proceedings of the 2009 IEEE International Conference on Mechatronics and Automation (ICMA 2009), Changchun, China, 9–12 August 2009, pp. 3637–3644 (2009)

19. Rahman, S.M.M., Ikeura, R., Nobe, M., Sawai, H.: Study on optimum maneuverability in horizontal manipulation of objects with power-assist based on weight perception. In: Proceedings of SPIE, vol. 7500, p. 75000P (2010)

20. Rahman, S.M.M., Ikeura, R., Nobe, M., Sawai, H.: Displacement-load force-perceived weight relationships in lifting objects with power-assist. In: Proceedings of SPIE, vol. 7500, p. 75000S (2010)

21. Rahman, S.M.M., Ikeura, R., Nobe, M., Sawai, H.: Lifting objects with a power assist system: effects of friction between human's hand and object on perceived weight and load force. In: Proceedings of the 2009 IEEE/SICE International Symposium on System Integration, Tokyo, Japan, 29 November 2009, pp. 77–82 (2009)

22. Rahman, S.M.M., Ikeura, R., Nobe, M., Sawai, H.: Human operator's load force characteristics in lifting objects with a power assist robot in worst-cases conditions. In: Proceedings of the 2009 IEEE Workshop on Advanced Robotics and Its Social Impacts, Tokyo, Japan, 23–25 November 2009, pp. 126–131 (2009)

23. Rahman, S.M.M., Ikeura, R., Nobe, M., Sawai, H.: Human's weight perception and load force characteristics in lifting objects with a power assist robot. In: Proceedings of the 2009 IEEE International Symposium on Micro-NanoMechatronics and Human Science (MHS 2009), Nagoya, Japan, 8–11 November 2009, pp. 535–540 (2009)

24. Rahman, S.M.M., Ikeura, R., Nobe, M., Sawai, H.: Control of a power assist robot for lifting objects based on human operator's perception of object weight. In: Proceedings of the 18th IEEE International Symposium on Robot and Human Interactive Communication (RO-MAN 2009), Toyama, Japan, 27 September–2 October 2009, pp. 84–90 (2009)

25. Rahman, S.M.M., Ikeura, R., Nobe, M., Sawai, H.: Unimanual and bimanual weight discrimination in lifting objects with a power assist system. In: Proceedings of the 2009 IEEE/ICROS-SICE International Joint Conference (ICCAS-SICE 2009), Fukuoka, Japan, 18–21 August 2009, pp. 4787–4792 (2009)

26. Gordon, A., Forssberg, H., Johansson, R., Westling, G.: Visual size cues in the programming of manipulative forces during precision grip. Exp. Brain Res. 83(3), 477–482 (1991)

27. Rahman, S.M.M., Ikeura, R.: Cognition-based variable admittance control for active compliance in flexible manipulation of heavy objects with a power assist robotic system. Robot. Biomim. 5(7), 1–25 (2018)

28. Rahman, S.M.M., Ikeura, R.: Weight-perception-based fixed and variable admittance control algorithms for unimanual and bimanual lifting of objects with a power assist robotic system. Int. J. Adv. Robot. Syst. 15(4), 1–15 (2018)

29. Rahman, S.M.M., Ikeura, R.: MPC to optimise performance in power-assisted manipulation of industrial objects. IET Electr. Power Appl. **11**(7), 1235–1244 (2017)

30. Rahman, S.M.M., Ikeura, R., Hayakawa, S., Yu, H.: Lifting objects with power-assist: weight-perception-based force control concepts to improve maneuverability. Adv. Eng. Forum-Special Theme Mechatron. Inf. Technol. **2–3**, 277–280 (2012)

31. Rahman, S.M.M., Ikeura, R.: Cognition-based control and optimization algorithms for optimizing human-robot interactions in power assisted object manipulation. J. Inf. Sci. Eng. **32**(5), 1325–1344 (2016)

32. Rahman, S.M.M., Ikeura, R.: Weight-perception-based novel control of a power-assist robot for the cooperative lifting of light-weight objects. Int. J. Adv. Robot. Syst. **9**(118), 1–13 (2012)

33. Rahman, S.M.M., Ikeura, R., Nobe, M., Sawai, H.: A psychophysical model of the power assist system for lifting objects. In: Proceedings of the 2009 IEEE International Conference on Systems, Man, and Cybernetics, USA, pp. 4125–4130 (2009)

34. Rahman, S.M.M., Ikeura, R.: Weight-prediction-based predictive optimal position and force controls of a power assist robotic system for object manipulation. IEEE Trans. Ind. Electron. **63**(9), 5964–5975 (2016)

35. Valdés, B.A., et al.: Usability testing of gaming and social media applications for stroke and cerebral palsy upper limb rehabilitation. In: Proceedings of the 36th Annual International Conference of the IEEE Engineering in Medicine and Biology Society, pp. 3602–3605 (2014)

36. Kawai, Y., Honda, K., Kawai, H., Miyoshi, T., Fujita, M.: Tele-rehabilitation system for human lower limb using electrical stimulation based on bilateral teleoperation. In: Proceedings of the IEEE Conference on Control Technology and Applications, HI, pp. 1446–1451 (2017)

37. Rahman, S.M.M.: A novel variable impedance compact compliant series elastic actuator for human-friendly soft robotics applications. In: Proceedings of the 21st IEEE International Symposium on Robot and Human Interactive Communication, Paris, France, 9–13 September 2012, pp. 19–24 (2012)

38. Rahman, S.M.M.: A novel variable impedance compact compliant series elastic actuator: analysis of design, dynamics, materials and manufacturing. Appl. Mech. Mater. **245**, 99–106 (2013)

39. Rahman, S.M.M., Ikeura, R.: A novel variable impedance compact compliant ankle robot for overground gait rehabilitation and assistance. Procedia Eng. **41**, 522–531 (2012)

40. Rahman, S.M.M.: Design of a modular knee-ankle-foot-orthosis using soft actuator for gait rehabilitation. In: Proceedings of the 14th Annual Conference on Towards Autonomous Robotic Systems (TAROS 2013). Lecture Notes in Computer Science, vol. 8069, Oxford University, U.K., 28–30 August 2013, pp. 195–209. Springer, Heidelberg, July 2014

41. Yu, H., Rahman, S.M.M., Zhu, C.: Preliminary design analysis of a novel variable impedance compact compliant actuator. In: Proceedings of 2011 IEEE International Conference on Robotics and Biomimetics, Phuket, Thailand, 7–11 December 2011, pp. 2553–2558 (2011)

42. Rahman, S.M.M., Ikeura, R., Hayakawa, S.: Novel human-centric force control methods of power assist robots for object manipulation. In: Proceedings of 2013 IEEE International Conference on Robotics and Biomimetics (IEEE ROBIO 2013), Shenzhen, China, 12–14 December 2013, pp. 340–345 (2013)

Reasonably Optimal Utilisation Through Evolution (ROUTE) in Airspace Design

Thomas Lawson[✉] and Yanyan Yang

School of Engineering, University of Portsmouth, Portsmouth, UK
up780962@myport.ac.uk, linda.yang@port.ac.uk

Abstract. The underlying navigation technology that enables the navigation of aircraft through airspace is improving, allowing aircraft to navigate waypoints that do not need to be previously defined, no longer being confined to antiquated navigational aids. From this the opportunity is arising to create highly optimized airspace that can change its design on the fly in reaction to the environment. This paper presents a novel approach towards route finding in airspace design using computational evolution. The proposed method, ROUTE, can create new airspace designs in reaction to changing environmental constraints, optimising for fuel burn (therefore cost and emissions) as well as noise disturbance, and route length.

Keywords: Airspace design · Genetic algorithms · Dynamic airspace · Constrained optimisation

1   Introduction

Currently, global airspace infrastructure is built around 'conventional' navigation, making use of ground-based beacons. Airspace design is carried out by humans who are aided by computer design tools. As there are many factors that must be considered such as noise pollution, airport capacity and aircraft fuel consumption which impacts fuel cost and emissions of the aircraft. All these factors are compounded by an ever-increasing number of aircraft. This conventional approach does not take advantage of modern navigation technologies [1, 2]. With the advent of modern navigation technologies, aircraft far better know where they are, this means that a flight is no longer reliant on flying passed fixed waypoints. This development and introduction of Performance-based Navigation gives aircraft the opportunity to fly more dynamic routes that can change and react to environmental constraints, for example weather.

Researchers have proposed methods for using genetic algorithms for the navigation of Unmanned Aerial Vehicles (UAVs) [3, 4]; work has also been done to look into airspace sectorisation using recursive geometric optimisation [5] as well as evolutionary optimisation [6]. The UKs Civil Aviation Authority (CAA) have outlined their view of how modern navigation should be used to better use airspace [7].

This paper proposes an evolutionary heuristic computer aided method, called ROUTE, for developing airspace. Genetic algorithms are computational models for evolution designed to mimic natural evolution, where a pool of candidates is populated with randomly created individuals. A pool of breeding candidates from each generation

© Springer Nature Switzerland AG 2020
K. Arai et al. (Eds.): FTC 2019, AISC 1070, pp. 260–274, 2020.
https://doi.org/10.1007/978-3-030-32523-7_17

are selected using a roulette wheel selection method. This ensures preference for more fit candidates, however means that less fit candidates can also be selected. They are then mutated and crossed over and create the next generation. This method of selecting more fit individuals, while maintaining diversity from the fewer less fit individuals, means with each generation the algorithm will produce individuals whose fitness will tend towards the optimal solution [8].

A benefit of using a computational approach means ROUTE can work alongside performance based [7] navigation to create automated reactive airspace, not dependant on constraints such as fixed waypoints. New waypoints can be transmitted to aircraft en-route. It would reduce the human resource needed for airspace design, as well as allowing numerous contingent situations to be more readily modelled.

The rest of the paper is organised as follows. Related work is critically reviewed in Sect. 2. Section 3 introduce the design of the algorithm. Section 4 describes the algorithm in detail. In Sect. 5 we present the experimental results and discuss issues raised from them. Section 6 concludes the paper and propose future work.

2 Related Work

Work has been done to use GAs to compute the routes for UAVs [3, 4]. This work uses real time data to build the route for one UAV in less than half a second. While this work looks at the problem of computing a route for an aircraft based on terrain and avoiding certain features, it is limited in as much as it does not look at repeatable routes, or how one route may interact with another, other than coordinating the arrival between two drones.

Analysis has been done [9], and a method proposed for organising airspace into 'tubes' this approach means reducing complexity as aircraft will fly through tubes in the sky. The tubes will be formed using the Hough Transform, then using a GA adjusted to fit in the highest percentage of flights, with minimum extra distance travelled [10]. This approach looks at how to capture large routes aircraft can be moved through, however it doesn't look at how aircraft should be routed from en-route airspace to an airfield. While this may not be such a problem in more sparsely populated countries, such as the US, this is more of a concern in the UK as airports are often in densely populated areas.

Researchers also look at how airspace can be more effectively sectorised [5] and how this can be done dynamically [11]. The aforementioned research looks at how to reduce complexity and optimize airspace by changing the shape of sectors, but it does not consider the benefits that can be gained from changing the underlying routes aircraft fly.

There has also been research into how airspace can be more efficiently utilised through Area Navigation, a technique which allows for dynamic navigation points. This research shows that more flexible navigation can allow for more efficient routes [12]. This paper proposes a method to determine the most appropriate points evolutionarily.

A technique of optimising approach trajectories based on environmental constraints has also been proposed [13]. This is improved upon [14]. This research looks at methods for proposing routes which build on existing routes. Our method proposes an approach capable of proposing its own routes.

Lastly it has been shown that through optimising speed and trajectories it is possible to improve efficiency and reduce emissions of a flight path [15, 16]. Our evolutionary approach hopes to iteratively find these improvements and propose routes that apply them.

The proposed ROUTE algorithm offers a technique building on past research to build routes which form the best compromise between Speed, efficiency, pollution, and therefore cost. The goal of the ROUTE algorithm is to address these problems and develop a technique for generating airspace designs in a reasonable amount of time, to be able to better take advantage of more modern navigation technologies and adapt in real time to changes in the environment.

3   The ROUTE Design

3.1   Design Objectives

The ROUTE algorithm aims to find the optimal design of airspace, consisting of multiple routes, given a set of defined environmental constraints. It will look to use computational evolution to propose the most effective airspace structure given a set of constraints. This can be applied on a smaller scale to produce Arrival and Departure routes or a larger scale to produce the design for a country's airspace or beyond. The design objectives have been outlined as follows:

- Below 4,000 ft priority is given to minimising noise pollution over populated areas.
  - As an aircraft is not at this altitude for long, and this is the altitude that most noise pollution is produced.
- Between 4,000 ft and 7,000 ft equal priority is given to minimising noise pollution and route efficiency.
  - At this altitude it is still important to consider surrounding communities, however if this might produce an unnecessarily complex route, a compromise would be preferred.
- Above 7,000 ft priority is increasingly given to maximising route efficiency.
  - As an aircraft climbs away from 7,000 ft its noise pollution becomes less of a concern, and for this reason simpler more efficient routes are preferred.
  - Route efficiency can be described as minimising journey times and route lengths as well as fuel burn by aircraft.

- Maximise airport utilisation
  - Where capacity is sufficiently large, aircraft should arrive at an airport at the maximum rate safely allowed.
- Minimise Airspace complexity
  - Many separate routes, which cross each other produce risk for air traffic control and pilots, for this reason flightpaths should cross each other as infrequently as possible.
- It must be possible to avoid designated areas.
  - Due to weather or security concerns, for example, or regulatory issues such as overflying national parks.

Maximising efficiency and airport utilisation should result in the ROUTE algorithm finding novel approaches to managing arrivals, without placing aircraft on hold as this is a highly inefficient. An example of how improved efficiency can be achieved is seen in LAMP 1a, which uses an arc to organise arrivals into London City airport [17]. Although only on one route, and done manually, this demonstrates that there are approaches beyond using holding stacks to improve efficiency. The proposed method ROUTE will propose ways that these optimisations could be automated, and also find many small improvements which can accumulate.

3.2   Genome Structure

Each generation the algorithm will produce a population of multiple possible solutions, it is these solutions which will be scored based on its fitness and be acted upon by the genetic operators. Each solution contains a set of n routes, where n is the number of pairs of entrance and exit (start and finish) points. Each solution will contain the same number of routes between the same points, only one route within a solution will be subject to mutation or crossover.

3.3   Mutators

The ROUTE algorithm involves 6 genetic operators, one crossover operator, and 5 genetic mutators. These mutators are used to introduce suitably variance to allow routes to converge towards an optimal solution. All operators are only applied to the intermediary points on a route, the first and final points remain static. For each solution, only one route will have the chance of being mutated. Now we will outline the six operators.

1. *Crossover:* This operator combines two parent routes into two new offspring. A random point is selected along each route, the first half of the first route is then combined with the first part of the second route, and the second part of the first route is combined with the second part of the second route. This operator is only applied to routes between the same start and finish point. The number of nodes in the child routes can differ from their parents.

2. *Delete Mutator:* This operator operates on both feasible and infeasible routes. It randomly removes an intermediate node from the selected route.
3. *Insert Internal Waypoint mutator:* This mutator imagines an oblong shape around the two waypoints that form a segment, then inserts a new waypoint within that shape.
4. *Smooth turn operator:* This operator places two new waypoints on the midpoint of the segments either side of the selected segment, then deletes the selected waypoint, the result being that it will create segments with smoother turns that the previous.
5. *Split Segment mutator:* This mutator splits a segment in two, placing a waypoint on a segments mid-point. This is used to help ensure segment lengths are shorter than the maximum segment length.
6. *Approach Mutator:* This operator replaces the penultimate waypoint with a new waypoint randomly placed within the approach cone of an airport.

3.4 Route Feasibility

A route will be feasible if it passes through no excluded airspace, and if it does not make any turns at too great of an angle. If a route is infeasible, then it will not have its fitness evaluated. It will instead have its fitness set equal to the sum of the largest feasible fitness, and the number of infeasible segments in the route.

3.5 Evaluating Fitness

For each Individual Route r_i the routes fitness $F(r_i)$ shall be defined as follows:

$$F(r_i) = \sum \left\{ l_r * w_l + t_r * w_t + pfo_r * w_{pfo} + fb_r * w_{fb} \right\} \qquad (1)$$

Where each w is the weight of the corresponding value. Each value is outlined below:

1. *l (Total route length): Total length in kilometres of the route.* Where $len(s_{ri})$ is the length of a single segment (s_{ri}) of the route.

$$l_r = \sum_{i=0}^{n} \left\{ len(s_{ri}) \right\} \qquad (2)$$

2. *t (Total route time): Total flight time for the route.* Where spd_{sri} is the speed restriction of the segment.

$$t_r = \sum_{i=0}^{n} \left\{ spd_{sri} * len(s_{ri}) \right\} \qquad (3)$$

3. *pfo (Population overflown in each flight band):* Where pfo_{r0} represents the population in the height band FL0 – FL40, pfo_{r40} represents the population in the height band FL40 – FL70, pfo_{r70} represents the population in the height band FL70+ .

$$pfo_r = \sum pfo_{r0} + pfo_{r40} + pfo_{r70} \tag{4}$$

4. *Total fuel burned:* For a given number of aircraft, flying the generated routes, what is the total amount of fuel burned in KG. Where fbr_r represents the fuel burn rate for the aircraft on the selected route. This can be adapted on a per route basis.

$$\sum_{i=0}^{n} \{fbr_r * len(s_{ri})\} \tag{5}$$

This algorithm measures fitness against one objective $F(r)$. The aim is to minimise $F(r)$. The use of the weights w allows for the prioritisation of different objectives to meet the users' needs. This means that one can decide whether to prioritise environmental disruption due to noise over CO_2 emissions and cost by adjusting the weight values.

This is inspired by [4] however as it is not in real time, it considers some of the concerns from [6] to introduce novel mutators to insure low complexity when multiple routes are introduced.

4 The ROUTE Algorithm

The ROUTE algorithm is designed using computational evolution. The algorithm will initially generate a population of individual routes, these routes will then be evaluated according to a set of viability criteria. Those individual routes who meet the viability criteria are then evaluated for fitness. This is done for efficiency as fitness calculations are computationally expensive. Once evaluated the routes are then randomly selected using the roulette method for mutation and crossover. This roulette method is a method of random selection where more fit individuals are more likely to be selected however less fit individuals are still able to be selected, this preserves diversity. Once selected the individuals are subject to either mutation or crossover, or are directly passed through to the next generation, the chances of any of these three operators being applied are equal. The routes are then re-evaluated and the process loops and continues until the end condition is reached. The end condition in ROUTE is either a limit on number of generations, or the number of generations progressed without a significant change. This process is outlined in Fig. 1 below.

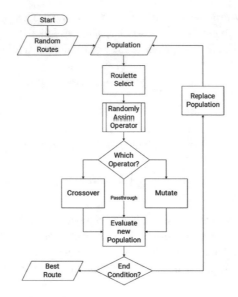

Fig. 1. ROUTE Algorithm flow diagram

In this algorithm, a chromosome represents a group of routes. Each route is a sequence of points, initially randomly selected, between fixed entrance and exit points. All the candidate routes form the population, this is formed of P groups of routes.

For evolution, a set S is selected where $S \leq P$ this set forms the candidates for genetic mutation and crossover. Individuals from P are selected using a roulette wheel of P slots, where the probability of an individual being selected is inversely proportional to the fitness of and individual. During roulette selection, an individual may be selected more than once.

After the selection of S, the set is iterated through, randomly selecting whether to mutate or not each individual, and where selected for mutation, each mutator has an equal chance of being selected. Where a mutator is not applicable to an individual, another mutator will be selected from the pool, with that mutator removed. Once mutated fitness is re-evaluated, and the candidate is added to a new extended population.

Once the set S has been exhausted the new extended population is evaluated and the fittest P individuals are retained, while the lesser fit individuals are discarded.

Once a proposed solution is suitably fit, this solution is selected as the optimal solution. Whether a solution is suitably fit can be chosen after a fixed number of generations or evaluated by the algorithm.

The steps of the algorithm can be summarised as follows:

```
DATA: Population, ViablePool, NonViablePool
START
  randomize Population of size P
  while end condition not met do
    for each individual in population do
      evaluateViabilityScore() of individual
      if individual's viability = 0 then
        evaluate fitness of individual
        add individual to ViablePool
      else if viability score > 0 then
        add individual to NonViablePool
      end if
    end for
    for each individual in NonViablePool do
      individual's score = population's highest fitness +
                                        individual's viability
    end for
    combine ViablePool and NonViablePool to form candidates
    sort candidates by fitness from low to high
    assign candidates descending rank from first to last
    breedNextGeneration() and add children to
                                        population in order
    reduce population to P members
  end while
```

The 'breedNextGeneration' step is summarised in the algorithm below.

```
DATA Candidates, Population, Children
function breedNextGeneration()
START
  while Candidates < S do
    candidate = selectUsingRouletteWheel(Population)
    add candidate to Candidates
  end while
  for each candidate do
    if randomNumber(0 to 1) < mutationChance do
      mutator = select mutator at randomNumber(1 to 8)
      mutate candidate with mutator and add to children
    end if
  end for
  return children
END
```

The roulette wheel selection method ensures that while you are more likely to select fitter individuals, there is a chance to also select the least fit individuals. This ensures that the population does not stagnate. The below algorithm briefly outlines the simple and efficient method for selecting a candidate using a roulette wheel implementation:

```
function selectUsingRouletteWheel ()
START
  for each candidate do
    totalRank += candidateRank
  end for
  targetRank = totalRank * RandomNumber(0.00 to 1.00)
  for each candidate do
    targetRank -= candidateRank
    if targetRank < 0 do
      return candidate
    end if
  end for
  return last candidate
END
```

Lastly the Viability of an individual is described as the sum of the deviance between the upper and lower acceptable bounds and the actual measurement for each waypoint in a route. This non-binary approach means that a small improvement, even though it may not result in a viable route would be preferred as it is a step closer to a viable route.

For example, the following algorithm describes the calculation of the total viability of the angles between segments in the route.

```
DATA TotalViability
function evaluateViabilityScore()
START
  for each waypoint in route do
    firstLineSegment  = lineSegment(waypoint to waypoint+1)
    secondLineSegment    =    lineSegment(waypoint+1    to
waypoint+2)
    angle          =              angleBetween(firstLineSegment,
secondLineSegment)
    if angle > allowedAngle then
      TotalViability += angle - allowedAngle
    end if
  end for
  return TotalViability
END
```

Once evolution is complete, the fittest routes are selected. Each segment of each route is iterated through. Where two routes intersect each other a two sub routes are created from the point of intersection onwards. Each sub route is evaluated for fitness, and the fittest sub route replaces the less fit sub route in both routes. This minimises complexity as where the routes share commonality they will merge. This is illustrated in figure two below.

5   Experimental Results

The ROUTE algorithm was implemented using Java 8 run on a Core i7 PC running Windows 10. In order to validate that ROUTE reaches its objectives, we will present evidence of the effect of changing weight parameters, as well as the effects of changing constraints on the route, lastly presenting the overall effectiveness of the solution.

In the figures below population is represented in the coloured grid beneath the routes drawn. Purple represents higher population in a square, and pink represents less population. The red line represents the approach, the top of the approach is where the routes are attempting to converge, they would then fly the approach to the runway at the opposite end.

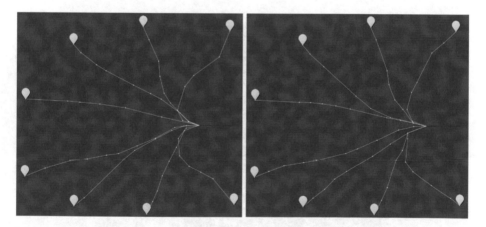

Fig. 2. Completed routes

Figure 2 shows how a set of routes can evolve together. The left shows routes evolved, and the right shows how these routes can be simplified to prevent unnecessary overlapping.

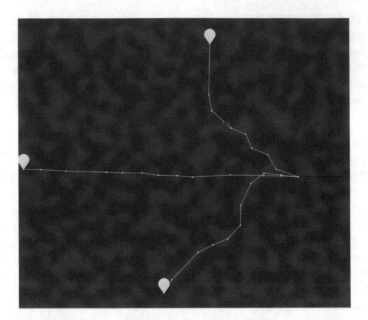

Fig. 3. Showing one route with a length weight of zero.

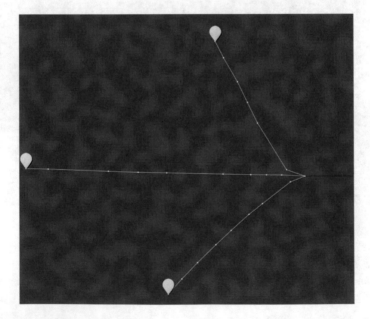

Fig. 4. Showing one route with a length weight of 10. This tends to take a more direct route.

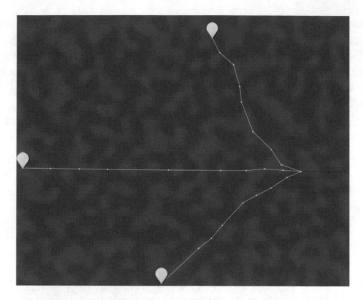

Fig. 5. Balanced length and population

Figure 3 shows a closer representation of how a lower priority ($w_l = 0$) results in routes that more carefully avoid population. This is compared with Fig. 4 which shows how a higher priority ($w_l = 10$) can lead to an almost direct route. Lastly Fig. 5 shows how a more appropriate weight ($w_l = 1$) can lead to a route who is a better compromise, of being simpler with fewer route corrections, while also showing some tendency to avoid populated areas.

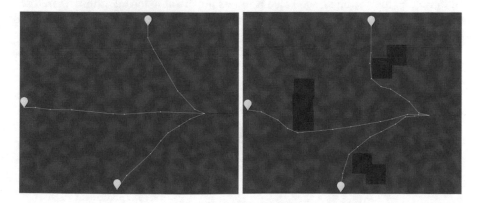

Fig. 6. Demonstrating the avoidance of restricted areas (red boxes)

Figure 6 shows best how viability in ROUTE can affect a routes shape. Before ROUTE will evaluate the fitness of a route it must first evaluate whether it breaks any restrictions, for example if a turn is to sharp, or if it enters a restricted area.

In this example, the route will first need to not enter the red box, before it is then optimised, notice therefore how it will avoid but stick close to the restricted areas. This demonstrates the ability of the ability to conform to viability requirements.

In addition to visually confirming these results, we have included below the results from a series of tests, illustrating the effects of changing the algorithm configuration on the generated routes.

Firstly, we look at the effect of changing weights to determine whether the route generated should be shorter or fly over fewer people.

Table 1. Higher population weight (weight = 10).

	N	Minimum	Maximum	Mean	Std. Deviation
Length	5	1056.42	1068.35	1061.3707	4.91244
Population	5	199.20	208.03	203.6226	3.83253

Table 2. Higher length weight (weight = 10).

	N	Minimum	Maximum	Mean	Std. Deviation
Length	5	1011.51	1013.87	1012.3594	1.07681
Population	5	257.42	265.11	260.5541	2.95241

As should be expected, we can see in above tables (Tables 1 and 2) that the mean length is smaller when the length weight is higher, similarly for the population it is smaller when the population weight is higher. However, interestingly the standard deviation is higher in the measurement with the lower weight. It could be concluded that due to the lower weight, the value is not as important to the algorithm, therefore it becomes less consistently optimised because the algorithm will it to boost other statistics.

Next, we look at how changing the number of generations can impact the performance of the algorithm (Tables 3 and 4).

Table 3. Fewer generations (generations = 10).

	N	Minimum	Maximum	Mean	Std. Deviation
Fitness	5	17912.91	18039.44	17971.4361	50.49479

Table 4. Greater generations (generations = 100).

	N	Minimum	Maximum	Mean	Std. Deviation
Population	5	17827.42	17927.83	17875.2560	42.76566

Here it is possible to see the two main benefits of increasing the number of generations used in evolution. Firstly, the longer search results in better results, this can be seen as all values of fitness are lower in the test allowed to run for more generations.

Secondly, we also see a smaller standard deviation on the longer running sample. This suggest that the results are groups closer together. From this it can be concluded that with more time a given result is more reliably the 'best' route.

Lastly, we consider the effects of changing the population size. This is to see whether the computational expense of more individuals is rewarded with higher quality results (Tables 5 and 6).

Table 5. Smaller population size (population = 20).

	N	Minimum	Maximum	Mean	Std. Deviation
Fitness	5	17929.01	18079.07	17993.6140	63.21561

Table 6. Larger population size (population = 300).

	N	Minimum	Maximum	Mean	Std. Deviation
Population	5	17795.27	17942.86	17882.3460	59.22986

Here it is possible to see that with a larger population the deviation is smaller, this suggests that with more individuals each run results in a closer result. This highlights the drawback of a genetic algorithm, as each run is based off a different random start point, the outcomes are not necessarily repeatable. However, seeing the smaller standard deviation with the higher population size suggests that a way to mitigate this pitfall is using a larger population. This small deviation suggests that the results are more similar, and therefore more consistent.

6 Conclusions

This paper has proposed ROUTE, an evolutionary approach to the design of airspace. The paper has demonstrated ways in which a route can be influenced by shifting priorities, to create a route fit for a given situation. This can be seen in the way the priorities change as a route gets closer to the ground, with a route following more contoured paths when it is lower in height.

We believe this work has identified the viability of generating airspace structures using computational aided methods. However, more work should be done to evaluate what would be necessary to validate such an approach for operational use in such a safety critical industry.

Moreover, improvements should be explored in simplifying airspace designs to promote simplicity. While this research begins to explore this problem, it looks for commonality in the routes, a better option may be to incorporate this as part of the genetic algorithm itself. This limitation of the algorithm, as well as the inability to compare it to operational data will need to be addressed before this approach could see operational use.

The method initially outlined here, however, has great potential to create modern and adaptive navigational infrastructure, and could have applications beyond airspace, and future efforts could consider other areas of route planning which would benefit from a balanced approach to route design.

References

1. EUROCONTROL, European Airspace Concept Handbook for PBN Implementation
2. Timar, S., Hunter, G., Post, J.: Assessing the Benefits of NextGen Performance Based Navigation (PBN) (2013)
3. Zhang, X., Duan, H.: An improved constrained differential evolution algorithm for unmanned aerial vehicle global route planning. Appl. Soft Comput. J. **26**, 270–284 (2015)
4. Zheng, C., Li, L., Xu, F., Sun, F., Ding, M.: Evolutionary route planner for unmanned air vehicles. IEEE Trans. Robot. **21**(4), 609–620 (2005)
5. Basu, A., Mitchell, J.S.B., Sabhnani, G.: Geometric algorithms for optimal airspace design and air traffic controller workload balancing. In: 2008 Proceedings of the Tenth Workshop on Algorithm Engineering and Experiments (ALENEX), pp. 75–89. Society for Industrial and Applied Mathematics, Philadelphia (2008)
6. Delahaye, D., Puechmorel, S.: 3D airspace design by evolutionary computation. In: Proceedings of AIAA/IEEE Digital Avionics Systems Conference, pp. 3.B.6-1–3.B.6-13 (2008)
7. CAA, Performance-based Navigation Airspace Design Guidance: Noise mitigation considerations when designing PBN departure and arrival procedures
8. Whitley, D.: A Genetic Algorithm Tutorial
9. Xue, M.: Design analysis of corridors-in-the-sky, Santa Cruz, CA, United States (2008)
10. Xue, M., Kopardekar, P.H.: High-capacity tube network design using the hough transform. J. Guid. Control Dyn. **32**(3), 788–795 (2009)
11. Ehrmanntraut, R., McMillan, S.: Airspace design process for dynamic sectorisation. In: Proceedings of AIAA/IEEE Digital Avionics Systems Conference, pp. 1–9 (2007)
12. Sprong, K.R., Haltli, B.M., Dearmon, J.S., Bradley, S.: Improving flight efficiency through terminal area RNAV
13. Hogenhuis, R.H., Hebly, S.J., Visser, H.G.: Optimization of area navigation noise abatement approach trajectories. Proc. Inst. Mech. Eng. Part G J. Aerosp. Eng. **225**(5), 513–521 (2011)
14. Hartjes, S., Visser, H.G., Hebly, S.J.: Optimization of RNAV noise and emission abatement departure procedures, pp. 1–13, September 2012
15. Lovegren, J., Hansman, R.J.: Estimation of potential aircraft fuel burn reduction in cruise via speed and altitude optimization strategies (2011)
16. Sahin Meric, O.: Optimum arrival routes for flight efficiency. J. Power Energy Eng. **3**, 449–452 (2015)
17. NATS, LAMP Phase 1a airspace change now live - NATS (2016). https://www.nats.aero/news/newsbrief/janfeb-2016/lamp-phase-1a-airspace-change-now-live/. Accessed 28 Feb 2019

Autonomous Robot Navigation with Fuzzy Logic in Uncontrolled Environments

Ryan Reck[1(✉)] and Sherine Antoun[2]

[1] University of Illinois, Springfield (UIS), Springfield, IL, USA
rreck2@uis.edu
[2] Colorado Mesa University, Grand Junction, CO, USA

Abstract. The "tourist in an unfamiliar village" navigation model seeks to simulate how a human would navigate with minimal information about their surroundings. By implementing this model and using Continuous Transmitted Frequency Modulated (CTFM) ultrasonic sensors, along with more advanced localization and mapping algorithms, a more efficient form of autonomous navigation can be created. Through utilizing the CTFM ultrasonic sensor echo data analysis techniques and combining them with Simultaneous Localization and Mapping (SLAM) methodology, it is theorized that the mobile robot will be able to identify obstacles, as well as characteristics and features in the environment.

Keywords: Fuzzy mapping · Robotics in uncontrolled environments · CTFM echo analysis · Path planning · Autonomous navigation

1 Introduction

With the advancement of machine learning, ultrasonic sensors and efficient control systems, autonomous navigation has the potential to make significant improvements in an increasingly technological society through a variety of different applications. The use of ultrasonic sensors, which detect the environment by transmitting and receiving ultrasonic waves, is highly important as they allow for effective outdoor mapping. As noted by Long, et al. [1], this is due to their detection range, accurate resolution and stable controllability, and immunity to light or the material composition of objects. For these reasons, along with their low comparative cost to other environmental modeling sensors, ultrasonic sensors are highly useful in outdoor operating environment modeling. This study seeks to progress the work of Antoun and McKerrow [2], specifically the use of fuzzy logic in creating an efficient mapping system. In coordination with the fuzzy mapping system, we postulate that a fuzzy logic controller can achieve obstacle avoidance and efficient navigation. We propose to do this by modeling the logic described by Antoun and McKerrow for the tourist in an unfamiliar village [2] and coupling it with real time CTFM preceptor data stream analysis algorithm such as incremental, updatable algorithms. This approach will utilize

© Springer Nature Switzerland AG 2020
K. Arai et al. (Eds.): FTC 2019, AISC 1070, pp. 275–283, 2020.
https://doi.org/10.1007/978-3-030-32523-7_18

$$P(C \mid X) = \frac{P(X \mid C)P(C)}{P(X)}$$

Fig. 1. Naïve Bayesian formula of calculating the post probability [3]

a Bayesian method for calculating post probability of feature based echo classification such as proposed by Ren et al. 2014 [3] (Fig. 1). This model bases the navigational concept of fuzzy logic on human interaction with the environment, specifically modeling how a human would behave with a basic map of a location. This level of decision making by the control system will enable the robot to make more human-like decisions regarding its navigation. The goal of this research is to ensure a greater understanding of fuzzy logic through the tourist in an unfamiliar village model [2].

With this greater level of understanding, this model of fuzzy logic can be used for further aspects of research in fuzzy navigation control. This study will provide an overview of fuzzy mapping and fuzzy logic control, as well as the testing of this navigation system in a dynamic real-world environment, namely the UIS campus grounds.

This proposal will build on prior research by analyzing current fuzzy control strategies, specifically using a fusion of Ultrasonic Ranging Data and hybrid intelligence systems. Our analysis of different strategies builds an understanding of the challenges and goals of autonomous navigation using fuzzy logic. The tourist in an unfamiliar village model is then further explained, which allows a robot to navigate it's environment efficiently, using a CTFM ultrasonic sensor and either an in-real-time expert system or a neural network controller. The proposal explains our wall following system as well as the need for further research regarding efficient obstacle avoidance. Ultimately, this method will work to ensure viable navigation and obstacle avoidance necessary for an uncontrolled environment.

The remainder of this paper describes the proposed research, starting with a section on Current Fuzzy Control Strategies for autonomous navigation in uncontrolled environments, followed by a discussion on Autonomous Systems, then discourse on CTFM Ultrasonic Sensing, then a description of the Proposed Control System and methodology, the final section describes the project Timeline and projected subsequent publications on the research progress.

2   Current Fuzzy Control Strategies

A considerable body of work exists describing extensive research on fuzzy mapping systems, as well as the use of ultrasonic sensors in fuzzy mapping. These studies have focused on fuzzy logic and its use for mapping unknown environments. Specifically, much of the preexisting research has focused on fuzzy sensor fusion of Ultrasonic Ranging Data in indoor environments [4,5]. This fusion of data is meant to "integrate multiple information to obtain only a datum, in order to correct mistakes and suppress incoherent data" [4].

There is also significant research regarding ultrasonic sensors used for mapping indoor environments using fuzzy logic, specifically in regard to the problem of local minimums [6], the minimum distance of a point or MDP [1]. Even hybrid intelligent systems with architectures based on fuzzy logic and virtual obstacle algorithms have been introduced. This allows for multiple aspects of localization and mapping, enabling the robot to autonomously navigate in order to reach its target, avoid obstacles, follow walls, and prevent emergency situations [7].

There have also been improvements in robot localization, specifically feature based mapping algorithms that will provide a more efficient solution to the SLAM problem. One such feature-based mapping algorithm is the Incremental Gaussian Mixture Network, which was developed by Milton Heinen [8]. This potentially allows autonomous robots to be able to identify specific features in the outdoor environment, such as walls and corners of buildings [9].

The proposal's methodology is consistent with Antoun and McKerrow's [2] work regarding wall following through ultrasonic sensors, specifically the use of signal correspondence and signal analysis that has been detailed significantly in their work [10]. The importance of effective wall following is paramount to this proposal, as the tourist in an unfamiliar village method relies primarily on this concept. This research also looks at the methodology behind Yata's [11] accurate angle measurement approach, which measures the accurate bearing angle of the reflecting point [11] as well as the inherent need for the robot to be able to move around different environmental changes such as walls, curb edges or fences [12]. Building on previous wall following techniques, this proposal seeks to not only make use of efficient wall following, which is detailed later in this paper but to also improve upon high-level path navigation as well as introduce a method of navigating around paths that are blocked in an optimal manner.

Since the preexisting research on ultrasonic sensors and fuzzy logic focuses primarily on indoor navigation, this study is imperative to understanding how this technology can function in unfamiliar, outdoor environments. Barring Antoun and McKerrow's 2006 theoretical outline [2], there is limited research regarding ultrasonic sensor based navigation and outdoor uncontrolled workspace mapping. This research will allow for further understanding of the limits of fuzzy control systems for autonomous robotics with ultrasonic sensor mapping in an unstructured and unpredictable environment.

2.1   Autonomous Systems

The importance of having an improved understanding of fuzzy mapping and fuzzy logic is central to further exploring the use of these systems in autonomous robotic navigation. By enabling robots to utilize navigation systems capable of dealing with uncertainty in the operational environment models, and compensate for such constraint by employing logic algorithm and quality sensor data, these same systems can achieve greater autonomy in uncontrolled environments. With advances in technology, specifically more accurate high quality Continuous Transmitted Frequency Modulated (CTFM) ultrasonic sensors coupled with localization and mapping algorithms, we theorize that many of the constraints that were recognized by Antoun and McKerrow [10] will be far less limiting.

2.2   CTFM Ultrasonic Sensing

Early robotics perception using simple time-of-flight ultrasonic sensors produced inconsistent results due to the limitations of the sensors that were compounded by poor understanding of acoustics. Previous experimental work with ultrasonic sensing in air solved some of the inherent problems that users confront, and led to the development of reliable sensing systems. Since 1995, CTFM has been used to navigate mobile robots [12]. Other research demonstrated 99.73% classification of 12 surfaces using 5 features representing roughness, extracted from echoes recorded by a moving CTFM sensor [13], as well as multiple environment features and bearing angles [10]. The body of research demonstrates that CTFM ultrasonic sensing is a reliable and robust system for classification, obstacle detection, and environment feature extraction.

Yata et al. [11] demonstrated that a single receiver measures the range to reflecting objects. Because the sensor transmits a beam, these objects can be located anywhere on a sector of a spherical shell defined by that beam. As the frequency response of a transducer varies with angle relative to its axis, the angle to an object can be measured by matching the echo to a set frequency response template. Antoun [14] demonstrated a simple algorithm to detect multiple targets within an ultrasonic sensor echo data stream in real-time.

In this proposed research, we intend to use a K-sonar CTFM sensor developed by Bay Advanced Technology (New Zealand) as a mobility aid for people with visual impairments. The device is comprised of two vertically stacked 19 mm diameter transducers, one for transmission and one for reception. A single 19 mm diameter transducer has a theoretical beam angle of 19.32° from axis to first minima (Fig. 2). Combining two transducers to form a transmitter and receiver, the vertical diameter is 47 mm and the theoretical horizontal beam angle is 7.6°. The CTFM system is set to transmit a downward swept sine wave (*fsweep* is 100 kHz to 50 kHz) every 100ms (sweep period *ts*). The ultrasound energy reflects from objects and returns to the receiver as an echo. The echo is a delayed and filtered version of the transmitted signal. A demodulation sweep, derived from the transmitted sweep, is multiplied with the received echo in the time domain. The outputs of this multiplication are sum and difference frequencies (Fig. 3).

The distance of flight information is contained in the difference frequencies (*fa* is 0 to 5 kHz), where frequency is proportional to range (Figs. 3 and 4) and amplitude is proportional to orthogonal surface area. This time domain signal is converted to a power spectrum with a Fast Fourier Transform (FFT) to give a range-energy echo (Fig. 4). The amplitude in frequency-bin i is the energy reflected from surfaces in a spherical annulus at range ri.

By utilizing our proposed CTFM ultrasonic sensors echo data analysis techniques, we theorize that a mobile robot will be able to identify obstacles in its path as well as characteristics and features of its environment. It will then combine these with simultaneous localization and mapping or SLAM methodology to maintain a more accurate record of its location while mapping. By using ultrasonic sensors there will be greater levels of accuracy in both target reaching and obstacle avoidance [4]. While there has been significant research

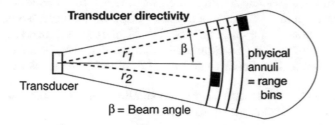

Fig. 2. An ultrasonic transducer emits a beam of energy. r = range to orthogonal surface

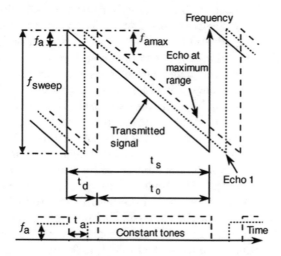

Fig. 3. CTFM demodulation - multiplying the echo by the transmitted signal produces a set of different tones where frequency is proportional to range to object.

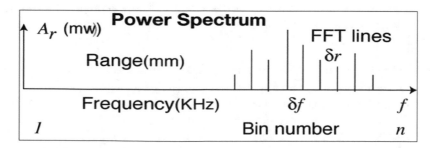

Fig. 4. Power spectrum of echo - frequency (bin number) is proportional to range and amplitude to echo (backscattered) energy at that range.

regarding indoor ultrasonic sensing, more work needs to be done to improve the mapping precision and efficiency of outdoor mapping with ultrasonic sensors for autonomous robots. Using high quality CTFM ultrasonic sensors, this research will look to enhance the understanding of fuzzy mapping in an unpredictable outdoor environment and seek to address the following research questions:

1. Can a more efficient system be created to allow for a constant determination of the robots position and course?
2. What are the limits of a fuzzy logic control system using ultrasonic sensors, and how can these limits be overcome through modern technology?
3. Using fuzzy logic mapping, is the robot able to navigate an outdoor uncontrolled environment in a manner this is both safe and effective?

Attempting to solve these questions will further our understanding of fuzzy logic and its applications in real world environments. With this greater understanding of fuzzy logic and its application to unknown and unpredictable environments, we are able to identify efficient systems of navigation and obstacle avoidance. Specifically, this research will determine whether the tourist in an unfamiliar village model is efficient at navigating in such an environment. This will provide insight into the use of fuzzy logic in such a model, as well as give justification for the use of this navigation model in academic or industrial scenarios. Overall, by improving the understanding of the tourist in an unfamiliar village model, as well as its intersection with the use of ultrasonic sensors and fuzzy logic, we will be able to further our understanding of machine learning and navigation in unknown and unpredictable environments. This greater understanding of fuzzy logic and navigation of unknown and unpredictable environments can allow for self-operating technology to assist the visually impaired, explore new or unsafe terrain, etc. In other words, perfecting outdoor navigation using fuzzy logic would yield endless applications to improve the world at large.

2.3   Proposed Control System

Fuzzy logic control can be described as an intelligent control analogous to human decision-making behavior that may be described as "close enough". The proposal will investigate the viability of either: in-real-time expert system, or, neural network, controller.

- Expert System can be described by the set of attributes:
    - Quick representation of structured knowledge as rules
    - Require an explicit rule for every situation
    - Can not directly apply numeric methods
    - Can not implement in Very-Large-Scale Integration (VLSI)
- Neural Networks can be described by the set of attributes:
    - We can not directly encode structured knowledge
    - Superimpose input-output samples on a black-box web of synapses

- We do not know what it has learned unless we check all input/output pairs
- We do not know what it forgets with new learning
- We can not directly encode rules

One of the main behaviors the robot will need in order to operate in an uncontrolled environment is collision avoidance behavior. Furthermore, while traveling in an uncontrolled environment, the robot's safest mode of operation is to follow environmental features in the human space. These include continuous geometric features that help describe navigable paths [12] such as walls, curb edges, and or fences.

Initial experiments will assume a wall as the environment feature to track. As the robot is right wall following at a safe tracking distance (space between robot and wall) then most objects that can result in a collision will be on the right (open doors, wall-side tables, etc.), forcing the robot to move to the left further into the navigable path. Objects on the left that can cause collision will force the robot to move to the right toward the wall, which therefore reduces the tracking distance. For narrow pathways, a pathway center line following behavior would be more appropriate than the right wall following:

Collision Avoidance Control Rules

- IF object on left
  - THEN move to the right
- IF object on right
  - THEN move to left
- IF tracking distance greater than half hallway width
  - THEN reduce to track down the centre
- IF path narrows
  - THEN track down the centre
- IF no path past object
  - THEN reduce linear velocity to zero to stop before a collision
- IF bump object
  - THEN back off

This architecture does not solve the problem of what to do when the path is blocked, it simply tracks along a continuous geometric environment feature or path avoiding collisions along the way by either veering to the left, veering to the right, or stopping. The three behaviors interact in response to the world to produce the desired high-level behavior - path navigation. The problem of a blocked path requires additional work by a planner and is part of this proposed research as it progresses.

3   Research Timeline

- The primary research method for this study is conceptual modeling as well as developing a fuzzy mapping system.
- This study will look further at published literature to gain a greater understanding and analysis of other mapping systems.
- After this expanded literature review, the study will begin modeling the conceptual basis for the tourist in an unfamiliar village navigation planner.
- The second stage of this research will commence after the navigation system has been conceptualized.
- In this second stage, the mapping and navigation system will be developed using National Instruments LabVIEW.
- Experimental simulation/testing will follow (simulation environment will be determined in due course).
- We will need to develop new control and sensing libraries to deploy on a retrofitted delta omnidirectional robot.
- Mapping and navigation system will then be tested in a dynamic outdoor environment, specifically the UIS college campus.
- This theoretical work will be conducted between April 2019 and May 2020, after which we anticipate submitting for publication our preliminary experimental simulation results.

References

1. Long, Z., He, R., He, Y., Chen, H., Li, Z.: Feature extraction and mapping construction for mobile robot via ultrasonic mdp and fuzzy model. Sensors **18**(11), 3673 (2018)
2. Antoun, S.M., McKerrow, P.J.: Landmark navigation with fuzzy logic. In: Australasian Conference on Robotics and Automation, Auckland New Zealand, December 2006. ISBN 978-0-9587583-8-3
3. Ren, S., Lian, Y., Zou, X.: Incremental naive bayesian learning algorithm based on classification contribution degree. J. Comput. **9**(8), 1967–1975 (2014)
4. Alonge, F., Di Bemardo, G., Raimondi, F., Italia, F., Lavorgna, M.: Fuzzy data fusion for real-world mapping using 360 rotating ultrasonic sensor. IFAC Proc. Vol. **30**(6), 943–948 (1997)
5. Lee, I.-H., Lu, M.-C., Hsu, C.-C., Lin, S.-S.: Map building of unknown environment based on fuzzy sensor fusion of ultrasonic ranging data. In: Conference Towards Autonomous Robotic Systems, pp. 446–448. Springer (2012)
6. Lining, S., Rui, L., Weidong, W., Zhijiang, D.: Mobile robot real-time path planning based on virtual targets method. In: 2011 Third International Conference on Measuring Technology and Mechatronics Automation, vol. 2, pp. 568–572. IEEE (2011)
7. Miloud, H., Abdelouahab, H.: Improving mobile robot navigation by combining fuzzy reasoning and virtual obstacle algorithm. J. Intell. Fuzzy Syst. **30**(3), 1499–1509 (2016)
8. Heinen, M.R., Engel, P.M., Pinto, R.C.: IGMN: an incremental gaussian mixture network that learns instantaneously from data flows. In: Proceedings of VIII Encontro Nacional de Inteligência Artificial (ENIA2011) (2011)

9. Heinen, M.R.: A connectionist approach for incremental function approximation and on-line tasks. Ph.D. thesis, The Federal University of Rio Grande do Sul (2011)
10. Antoun, S.M., McKerrow, P.J.: Issues in wall tracking with a ctfm ultrasonic sensor. IEEE Sens. J. **13**, 4671–4681 (2013)
11. Yata, T., Kleeman, L., Yuta, S.: Wall following using angle information measured by a single ultrasonic transducer. In: Proceedings of 1998 IEEE International Conference on Robotics and Automation (Cat. No. 98CH36146), vol. 2, pp. 1590–1596. IEEE (1998)
12. Ratner, D., McKerrow, P.: Navigating an outdoor robot along continuous landmarks with ultrasonic sensing. Robot. Auton. Syst. **45**(2), 73–82 (2003)
13. McKerrow, P.J., Kristiansen, B.E.: Classifying surface roughness with CTFM ultrasonic sensing. IEEE Sens. J. **6**(5), 1267–1279 (2006)
14. Antoun, S.M.: Mining CTFM echo signal data for navigation. In: Proceedings of SAI Intelligent Systems Conference, pp. 1200–1210. Springer (2018)

The Future of Socially Assistive Robotics: Considering an Exploratory Governance Framework (EGF)

Hook Chuan Lim[✉]

UOWD, FEIS, Dubai, UAE
Hclim@uowdubai.ac.ae

Abstract. Centralized or self/peer governance approaches are major concerns and debates in modern smart cities and living. Proponents of central top-down governance approaches argued that the centralized approach supports the strong alignment of control and safety and in short, facilitating ease of national development while self/peer advocates highlight opportunities for agile responses and a capability to bridge the widening gap between technology, social acceptance and well-being. As in all modern innovative measures, the consequences of Socially Assistive Robotics (SAR), its interactions between technology and users need to be explored. This paper looks at an Exploratory Governance Framework (EGF) as a possible option to managing future innovative services and technologies with focus on the narrow domain of SAR. This research provides opportunities to incorporate concepts of coevolution of technology and governance as well as elements of design considerations to impact development of modern technologies and services and allow balanced perspective on the centralized versus self/peer regulatory options.

Keywords: Socially Assistive Robotics · Exploratory · Governance · Technology · Coevolution · Framework

1 Introduction

Robotics has made major and successful inroads into our daily activities and modern social living. Robots are now common features in industrial automations and of late, we are experiencing a surge in the new usage of robotics in the medical field, particularly in the healthcare services domain. The signs of Socially Assistive Robotics (SAR) as providers of care services are appearing in the horizon. This new era of healthcare provision rekindles issues of roboethics; of social acceptance and of governance needs. A number of motivators are observed that continue to drive the development, deployment and use of SAR: a shortfall of actual human care-givers providing support and care for elderly, children and those needing daily, constant assistance and guidance; a means to drive efficiency and cutting costs and to also indulge in service differentiation and innovations; an opportunity where technologies can provide improved and enhanced services to assist our current generation and to allow a more inclusive and holistic living/human experiences. This aspect is essentially made possible with smart living and environment that allows robots to sense and to

K. Arai et al. (Eds.): FTC 2019, AISC 1070, pp. 284–292, 2020.
https://doi.org/10.1007/978-3-030-32523-7_19

communicate with smart decision making and responses. While these motivators continue to drive innovations in SAR, what is clear is the growing, nagging presence of a generic concern. This concern, outwardly expressed by some and kept unspoken by others is none other than the combined concerns of privacy, safety and social acceptance [1, 2]. The social acceptance frames the debate and offers a simple perspective: for some, it is a concern that technology should only assist and not replace the man-in-the loop cases; for others, it is about safety and issues of privacy, especially of medical data. As reported by [3], technology is ahead of our ability to regulate and govern and there is a strong need to start considering regulation and explore framework that will work for our new system and environment. This applies aptly and specifically to SAR technology.

The actual approach to SAR services, technological governance and their consequences are not well defined. The traditional debate on centralized versus self/peer regulation as governance mechanism is vexed and controversial while newer technological perspectives of "transition research in governance for sustainability" [4] adds to the governance complexity. This paper is an attempt to unpack the complex governance landscape and proposes an Exploratory Governance Framework (EGF) to aid policy analysis and technological development. The paper is organized as follows: Sect. 2 provides preliminary background, context and essential definitions to help pave the path for subsequent sections. Section 3 outlines the various conceptual theories that underpin the design of an EGF; Sect. 4 introduces the core components of the EGF. Section 5 applies the EGF to a participatory action-based research methodology to validate EGF components and highlights the comparative results. In addition, discussions of the implications of these results for policy formulations, services and technological design and development are addressed. Section 6 concludes the paper with needed identification of current gaps in governance of SAR and prospects for future works and research.

2 Context and Definitions

2.1 Preliminaries

Robotics is seen by many societies as an important alternative to the provision of aged care given the increasing aged populations and a lack of care providers. Of late, this provision of medical aged care is even extended to the provision of other related health care services, for example, assistive care in Autism Spectrum Disorder (ASD) and mentally-challenged individuals and children. Hence, robotics particularly, SAR as one of the many digital innovations of our modern age bears an important consideration for social acceptance and deployment. SAR brings with it both social and economic benefits and ramifications. What is not clear is how these digital innovative measures with high promises can impact society and modern living.

The fact that disruptive innovations can impose severe social consequences is a case in point. Such ramifications include or involve social/labor displacements (the case of industrialization of many states); evolving and changing of industries (the changing facets and replacement of many traditional industries that is labor-intensive and of low

productivity with modern productions) and periods of social unrest and social uphea-
vals under dire conditions (the recent Uber platform that displaces the taxi system [5]).

2.2 Essential Terms

Assistive Technologies (AT) is the umbrella term used to map a broad range of
technological products and services to aid human living It contributes towards
improving our general well-being and enhances or promotes independence from a
heavy reliance on another fellow human caregiver. In this context, SAR is a subset of
AT. In SAR, the core distinguishing feature is then the goal to promote independence
and improvements in general well-being and living. It is this feature that sets it apart
from the common simple social robots (of simple design and features) and from other
entertainment robots. Based on the definition suggested in [6], SAR is here narrowly
defined as a robot that aids in the performance of a task for a weak(er) member of our
society. This definition separates SAR other forms AT and serve to prevent miscom-
munications and misinterpretation of terms and concepts.

Social acceptance issues are mostly perceived at the application or implementation
of technology phase. An early example of social acceptance is the lesson learnt from
wind mills technology. As presented in [7], supporters of wind mill technology did not
expect to encounter any social acceptance until deep into the implementation phase
where many complex concerns and issues arose. In the context of SAR, social
acceptance similarly is about social approval and legitimacy. Social acceptance of SAR
carries with it three possible dimensions reported in [7]; "…the community…the socio-
political…and the market acceptance…". Without social acceptance, SAR and in fact
most social technologies would be severely restricted to prototype testing and trials and
will not go beyond those phases.

The term and concept of governance is difficult to be clearly defined. Some have
used it to mean public administration of public sector [8]; others have applied various
meaning to this term resulting in ambiguity and elusiveness, for example, [9] defined
governance in the context of the study of international political economy simply as
"organizing collective actions". Here, this definition is viewed from the perspective of a
state as the unit of governance. Almost instantly, one is able to see that this definition is
biased towards government and state management. Hence a newer perspective, the
Antwerp School suggests the idea of an enterprise governance of information technology with the overall goal of achieving alignment and value of business and IT. In
essence, "…Enterprise governance of IT (EGIT) is an integral part of corporate governance and addresses the definition and implementation of processes, structures and
relational mechanisms in the organization that enable both business and IT people to
execute their responsibilities in support of business/IT alignment and the creation of
business value from IT-enabled business investments…" [10].

2.3 Implications, Significance and Gaps

Authors in [11–13] provides for a systemic review of related works in assistive
robotics. While these papers outline the state of research and developments, what is a
common assumption is the implied acceptance and expected governance of SAR.

Adopting these assumptions is likely to see a replay of the wind mills technology scenario and to block the efforts and progress of socially important policy formulations. The critical gap and need for robust governance must be addressed early in product design and inception phases of SAR. A revisit of essential conceptual theories and theoretical concepts laid the required foundations for the EGF design and formulation.

3 Conceptual Theories

3.1 Governance Theories

Digital innovation fuels the explosive growth of smart cities and smart living in the modern age. Broadly speaking, digital innovation has various facets and one of the more apt perspectives is that of combined and carefully orchestrated set of new products and services with perhaps new business models within designated context. Viewed from this angle, digital innovation is an important force that impact social systems and [14] outlines a set of essential governance theories that can be applied towards a better understanding of the power of digital innovative transformation. These five theories are:

a. Economic governance generalizes relationships between markets and entities. The primary basis is transaction cost economics and takes a relatively static view of governance.
b. Public choice is a related set of theories that departs from a neoclassical economic perspective. The primary view centers on profit maximization and the individual-istic nature of collective organization.
c. Political economies theories focus on the overlapping states and markets while moving firms and corporate hierarchal entities aside. The debate centers on which entities serve as sponsors of development and nurtures an emphasis of 'market failure'.
d. Political governance theories deal with forms of control and their interactions. The issues of control and sectoral sub-system governance. This reflects a strong theme of centralized versus de-centralized approach.
e. Regulatory theories closely relate to political governance. Within these areas are concepts such as "convention", mechanisms and tuning the balance between set of interests and/or indicators.

3.2 Institutional Theory

Institutional theory stress on the social and cultural dimensions of entities and systems within the institutional context bearing social expectations and norms of social appropriateness or legitimacy. In institutional theory perspective, organizations are constrained by social expectations and approval. Institutional theories maps to the SAR context.

4   EGF Design

4.1   Considerations and Formulations

Design Thinking is all about having and including design in the way we work and as a research focus it is maturing and has rapidly spread to many fields [15]. It is also true that design thinking is becoming more important for SAR services and products development. This is further reinforced by the traditional view that design is a core and "...distinguishing engineering activity..." [16] (p. 330). What is critical in design thinking and SAR engineering is not merely including the term "design" as part of the development pipeline. It goes beyond terminology and it embraces dual elements of (a) spirit of iterative reflections, that is, a consideration of the user and/or flow of usage and (b) the need for governance. This shift in focus from product-centric to human-centric perspective is what makes design thinking more relevant, useful and innovative. It also explains why design thinking has been missed in previous SAR application development or has not been accorded with the right level of attention.

Applying the various theoretical foundations and concepts and taking into considerations the three dimensions of social acceptance, we propose a user-experienced based EGF comprising: Governance Modalities (GM), Appropriate Usage (AU), Technological Confidence (TC). The EGF in this formulation takes into considerations the coevolution of technology and governance and allows for bi-directional flow of interactive effects of technological innovations and governance.

4.2   Core Components

A triangular schematic outlines the relationships of the EGF components (see Fig. 1). These three components occupy the apex of the triangle. This color-coded components in primary colors of red, green and blue further represents the inter-related relationships of GM, AU and TC.

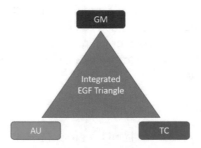

Fig. 1. Integrated EGF triangle. Each component of Governance Modalities (GM), Appropriate Usage (AU) and Technological Confidence (TC) occupies the vertex position and the triangle depicts the inter-related relationships. GM influences both AU and TC; likewise, AU guides GM and contributes to TC. TC is further strengthened with strong GM and AU factors.

Governance Modalities (GM). This component addresses combination of governance theories or mechanisms needed to suit SAR context. As SAR has wide range of applications, different modes of governance are required. Governance modes may be of a single element, for example public choice theory or regulatory theory or may be combined with suitable governance concepts. The choice and selection of the governance modes is necessarily context-based. Early determination of the GM at the SAR design phase will strengthen AU and increases TC and likely to lead to better social acceptance.

Appropriate Usage (AU). Appropriate usage is guided by both technical design and user interactions. Here, suitable ethical considerations can be mapped and be evaluated. The interactive effects between GM and AU can be easily analyzed, for example, GM of economic governance may lead to less than desired AU as agencies focus on profit maximization and other rent-seeking notions. Where this scenario occurs, TC will be affected and eventually leading towards weaker social acceptance.

Technological Confidence (TC). This subjective component drives social acceptance. The higher the technological confidence coupled with good GM and strong AU is a likely recipe for easy social acceptance. Poor combinations of GM and AU conversely lead to a harder social acceptance.

4.3 Participatory Action-Based Research (PAR)

PAR is also another popular management science research tool where the researcher actively involves in intervening the study and seeks to bring about desired change [17]. Here, we apply a modified form of PAR. The modifications include:

(a) Using the subjects to perform the action while the researcher adopts a more distant observer role.
(b) The researcher does not attempt to bring about any changes, but instead, is interested in understanding the complex underlying processes that came with the action carried out by the subjects.
(c) Experimentations were conducted to allow for a richer exploration of data and results.

A big part of the PAR session involves the design and application of a participatory rating tool known as the "Socratic wheel" [18] to rank and order the categorical components of the EGF. This simple participatory rating and ranking is used to validate the importance of the components and demonstrates simple mapping of components and usage.

5 Methodology, Results and Discussions

5.1 Design Considerations

A major component of the EGF is the completeness of the framework. To ensure completeness, three research methodologies were considered to allow for framework component validations. These approaches were computational experimentation and simulation; qualitative surveys and qualitative expertise solicitations using PAR. Qualitative expertise solicitation using PAR was adopted in this project and the considerations for not selecting the other two approaches includes:

- The parameters for building the computational model and for enabling the simulation is not well understood, particularly, the underlying distribution and shape as complex decision choices made in governance are involved and hence, unless there are more project resources, the computational experimentation and simulation approach is the least preferred of the three approaches.
- Qualitative surveys were ruled out although it is a viable alternative and approach. The primary reason for not applying qualitative surveys was due to the desire to involve a small group of experts in the fields of SAR technology and governance. We envisaged that future work could involve the use of larger sample sizes as part of qualitative surveys.

5.2   Descriptive Data

A total of 20 attendee participated in the PAR session. A set of descriptive information for each of the attendee include DevOps experiences; SAR experiences and years in design work (see Fig. 2).

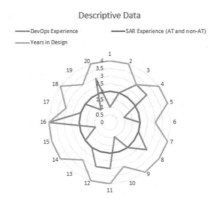

Fig. 2. Session descriptive data. A total of 20 attendees with varying DevOps experiences, SAR experiences and years in design work.

5.3   Validation of EGF Components

A panel of experts was gathered in a closed session and each of the participants helped to validate the EGF components. The validation process involved the discussion of the components, the usefulness as well as the need for considerations of additional components. The final outcomes suggest varying strengths of the design and formulation (see Fig. 3).

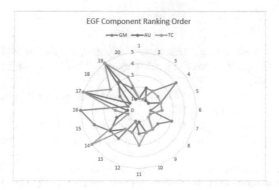

Fig. 3. Validation of GM, AU and TC. The results demonstrate a general usefulness of a EGF.

5.4 Discussion (Informal Interviews)

Validation of components represents the critical first step in the formulation of EGF. During the PAR sessions, informal interviews were also conducted to better understand the coevolution of technology and governance. Some of the comments were as follows:

> "...SAR is important and useful, however, its application and use should be limited to supporting care givers and not replacing the needed human touch..." (ID04).
> "...the main concern for me is data protection and privacy issues. Can robust governance ensure that in the context of SAR? Perhaps governance mode should be a combined top-down and bottom-up approach, much like how Chinese government handles innovation and innovation zones in mainland China..." (ID07).
> "...future-proofing innovation is not an easy task and what is valid today may be invalid tomorrow. I think results must be treated with care..." (ID02).

These informal comments and interviews serve to further highlight important caveat, namely, that these results are experimental. Care in treating and interpreting the results are required. Notwithstanding the issues and concerns, this research has helped to frame the path forward towards SAR management and governance. With a validation of ECF components, the next logical steps will be to qualitatively test the various GM, AU and TC under specified SAR scenario.

6 Concluding Remarks and Future Works

This paper takes the view of coevolution of technology and governance, formulate and propose an EGF for SAR. With an EGF, it is possible to carry out analysis and assess the future of SAR in terms of social acceptance. We have only scraped the tip of a large domain and in a fast-moving and dynamic growth of SAR, more research is required to further validate and test the value of EGF to ensure that while SAR endeavors to bridge healthcare demands for greater and more responsive services and products, robust governance and due diligence are considered and deployed.

References

1. Charani, E., Castro-Sánchez, E., Moore, L.S., Holmes, A.: Do smartphone applications in healthcare require a governance and legal framework? It depends on the application! BMC Med. **12**(1), 29 (2014)
2. Oborn, E., Barrett, M., Darzi, A.: Robots and service innovation in health care. J. Health Serv. Res. Policy **16**(1), 46–50 (2011)
3. Dickinson, H., Smith, C., Carey, N., Carey, G.: Robots and the Delivery of Care Services: What is the Role for Government in Stewarding Disruptive Innovation? ANZSOG, Melbourne (2018)
4. Loorbach, D., Frantzeskaki, N., Thissen, W.: A transition research perspective on governance for sustainability, In: European Research on Sustainable Development, Berlin, Heidelberg (2011)
5. Hinings, B., Gegenhuber, T., Greenwood, R.: Digital innovation and transformation: an institutional perspective. Inf. Organ. **28**(1), 52–61 (2018)
6. Asghar, I.: Impact of assistive technologies in supporting people with dementia, PhD diss., Bournemouth University (2018)
7. Wüstenhagen, R., Wolsink, M., Bürer, M.J.: Social acceptance of renewable energy innovation: an introduction to the concept. Energy Policy **33**(5), 2683–2691 (2007)
8. Khan, H.A.: Globalization and the Challenges of Public Administration: Governance, Human Resources Management, Leadership, Ethics, E-Governance and Sustainability in the 21st Century, Palgrave. Macmillan, Springer Nature (2018)
9. Prakash, A., Hart, J.A.: Globalization and Governance, vol. 1. Psychology Press, Hove (1999)
10. Haes, S.D., Van Grembergen, W.: Enterprise Governance Of Information Technology: Achieving Alignment and Value, Featuring COBIT 5. Springer, AG Switzerland (2015)
11. Abdi, J., Al-Hindawi, A., Ng, T., Vizcaychipi, M.P.: Scoping review on the use of socially assistive robot technology in elderly care. BMJ Open **8**(2), e018815 (2018)
12. Whelan, S., Murphy, K., Barrett, E., Krusche, C., Santorelli, A., Casey, D.: Factors affecting the acceptability of social robots by older adults including people with dementia or cognitive impairment: a literature review. Int. J. Soc. Robot. **10**(5), 643–668 (2018)
13. Pedersen, I., Reid, S., Aspevig, K.: Developing social robots for aging populations: a literature review of recent academic sources. Sociol. Compass **12**(6), e12585 (2018)
14. Von Tunzelmann, N.: Historical coevolution of governance and technology in the industrial revolutions. Struct. Change Econ. Dyn. **14**(4), 365–384 (2003)
15. Dijksterhuis, E., Silvius, G.: The design thinking approach to projects. J. Mod. Proj. Manage. **4**(3), 33–41 (2017)
16. Razzouk, R., Shute, V.: What is design thinking and why is it important? Rev. Educ. Res. **82**(3), 330–348 (2012)
17. Mingers, J.: The paucity of multimethod research: a review of the information systems literature. Inf. Syst. J. **13**(3), 233–249 (2003)
18. Chevalier, J.M., Buckles, D.J.: Participatory Action Research: Theory and Methods for Engaged Inquiry. Routledge, London, New York (2013). ISBN-10: 0415540321

A Novel Power Management System
for Autonomous Robots

Mamadou Doumbia$^{(\boxtimes)}$, Xu Cheng, and Haixian Chen

School of Computer Science and Electronic Engineering, Hunan University,
420082 Changsha, China
cadip8@yahoo.fr, cheng_xu@yeah.net, haixian@hnu.edu.cn

Abstract. In this paper, an automatic battery recharging system for autonomous robot is presented. Its design and implementation are based on some standard electronic components integrated to build the Power Management System (PMS)' Integrated Circuit (IC) board. This PMS consists of a limit switch (LS), an infrared sensor (IR), and an integrated circuit (IC). This IC is designed and simulated and the simulation result show that it is feasible and reliable. Thus a prototype IC board has been built and mounted on a four wheeled robot with three applicative levels. Each level has its own Industrial Personal Computer (IPC) for a specific task. These IPCs communicate with each other data and commands via the RS232 and RJ45 ports. In addition, this robot communicates with the battery charger by infrared. In this designed PMS, the risk of electrical damage is avoided when connecting the robot to the charger. Apart to the simulation results, the recharging time of a battery (24 V, 3000 Ah) discharged from 18 V to 24 V is about 7 h. The built system provides a safe recharging system avoiding all damages in robot and its charger at the docking time. The proposed PMS system is reliable and can safely recharge the battery of any autonomous robot properly connected to the charger.

Keywords: Mobile robot · PMS · IC · Automatic recharging

1 Introduction

In recent years, many approaches and technologies have been developed in robot's auto-recharging field. But, there is still important to increase the performance of these developed auto-recharging systems. The aim of the new implanted PMS will be to decrease the number of robot's recharging operation during it working hours so that the robot can get ability to work many hours without recharging its battery. The robots are used in many places offering different services to people. Those services can be the physical assistance to the people in needs or some standing alone services in some public places. Those autonomous robots are popular in many places such as at airport, train stations or in some banks. Usually in these places, people line up to get some services one by one. During this time, they can spend more than half an hour before being served. This situation wastes people's time and is too boring.

In this context, the autonomous robots are useful and can help people to save their time. They offer some services or guiding people to some appropriate places.

© Springer Nature Switzerland AG 2020
K. Arai et al. (Eds.): FTC 2019, AISC 1070, pp. 293–305, 2020.
https://doi.org/10.1007/978-3-030-32523-7_20

Therefore, those robots can work continual at least 14 h per day depending to their batteries power capacity. This intensive work requires to the autonomous robots to regularly recharge their batteries when the current of this one reaches a predefined threshold. In this paper, the threshold is set to 18 V. To decrease the number of these regular battery recharging, we need a more performant PMS. To realize this new PMS, we propose an approach into three essentials steps: design, implementation and Simulation. The first step consists to correctly design the complete IC of the new PMS for the autonomous robots. Second one is its implementation and the third one is the simulation of the final product (the new PMS).

The main contribution in this research paper consist to design and implement a new PMS. This PMS is composed into two parts (charger and charge controller). Each part is based on an integrated circuit (IC) using infrared sensors (IR) (transmitter (IRT) and receiver (IRR)) and analog-to-digital converters (AD). The goal is first to control the recharging process of the battery. Secondly, to allow a redundant (repetitive) check of the connection status between the robot and the docking station for more security. Many experiments are conducted to check the reliability of the connection status between the charger and the charge controller before the recharge operation. This contribution can bring a considerable improvement in the previous researches done in this field. In addition, it improves the safety of the robot by improving the feasibility and reliability of the current recharging systems.

From the last decade, many researchers introduced different approaches and technologies to build some auto-recharging systems for the robots. Some of them used single sensors and others used fusion sensors to realize their systems. The auto-recharging battery is the main function of this system. Among them, we have [1], where Quilez, Roberto et al. present a successfully auto-recharging system. This system use QR codes as landmarks and IR sensors as distance measurement to bypass the obstacles and reach the target securely. In paper [2], the authors achieve in their research a system to manage the robot's power system and recharging their batteries. The system is based on IR sensor technologies and ICs.

In addition, another single sensor based research is [3], where the robot introduced by the authors communicate with its nearby objects through wireless. That allowing it to increase its successfully docking operation and enable the auto-recharging batteries.

Furthermore, in [4], an auto-recharging system with localization error-compensation capability is described in two docking mechanisms. A fusion sensor (IR, sonar and ultrasound) is used to realize this system. Similarly, a self-recharging robot's design and implementation is described in [5]. This self-recharging operation is realized throughout an information fusion of encoders, electronic compasses and infrared sensors.

Su, K. L. et al. [6] developed a wireless IR interface to control and increase the performance of their auto-recharging system. This system interconnects their docking station and their mobile robot more quickly and easily. Taylor, Trevor, et al. [7] described some technologies to build a docking and recharging system for an autonomous robot. The built system is based on the infrared images of the docking station. In [8], a smart recharging station for electric vehicles equipped with an automated arm is presented. To localize the recharging station in the docking area, the authors used an infrared beacon system. In [9] Doumbia, M., Cheng, X., et al. present a self-charging system for a mobile robot. This robot can automatically dock to a charger based on IR sensors and recharges its battery.

The remainder of the paper is organized as follows: Sect. 2 present a description of the system architecture, autonomous robot, and PMS Integrated Circuit. In Sect. 3, the auto-recharging strategy is explained in detail. Simulation results and discussions are given in detail in Sect. 4. The last section contains the conclusion and the future works.

2   System Architecture

The power management system (PMS) of the autonomous robot presented in this paper contains two main parts: the charger in the docking station and the charge controller in the robot. Those two parts should be connected to allow the robot's recharge operation. The charger contains four IRTs and their corresponding receivers are placed on the robot. These IR sensors are divided into two groups (three for navigation and the fourth for control) in the charger and the charge controller. Below in "Fig. 1", there is an illustration of the whole implemented PMS. The charger based on the integrated circuit (ICds), contains a microcontroller unit (MCU) to control the other components. These components are a limit switch (LS), four identical IR Transmitters (IRT) and a recharging slots.

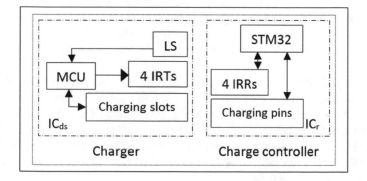

Fig. 1. The implemented Power Management System.

This LS detects whither or not the robot arrived at the docking station and correctly connected to the charger. Three of the four IRTs are used to guide the robot in its docking operation. The fourth IRT is for control and cooperate with the LS to inform the robot when it touched the charger. The charge controller integrates a microcontroller (STM32) in its integrated circuit (ICr), to manage the recharging process. It is also in charge to inform the robot about the recharging status (empty battery or if it is fully charged). This ICr contains the four IR Receivers (IRR) indicated to communicate with the charger's four IRTs. In addition, there is a recharging pins module to receive the required voltage to charge the robot's battery.

2.1 Autonomous Robot

In this paper, the prototype robot used is a four wheels mobile robot. Its dimensions are 45 cm, 40 cm, 95 cm respectively for length, width, height, and it has a weight of around 70 kg. It is composed mainly in three levels (Mechanical level, Operating System level and Applications level). Each level has its own Industry Personal Computers (IPC).

The mechanical level's IPC is IPC1. This mechanical level is connected to two DC motors and sensors2. Which is a set of sensors such as five ultrasonic sensors, one inertial measurement unit (IMU), three IRRs, etc. This IPC1 is connected to the Operating System level (IPC2) throughout the communication port RS232. This IPC2 supports sensors1, which is a set of sensors containing a camera and a LIDAR. In turn, it (IPC2) is also connected to IPC3 via the RJ45 communication port. The IPC3 is at Application level, it contains a user interface to allow the communication between the robot and its users.

In addition, there is a power supply module which contains a 24 V, 3000 mAh Li-ion battery. The IPC1 perceives the environment around the robot through its sensors and then communicates this information to the IPC2. This information serves to the IPC2 (using Linux operating system (Ubuntu 16.04 with ROS)) to prepare some decisions and commands. These new commands are send to the IPC1 to control the robot's wheels. IPC2 transmits some received data to the IPC1 as respond to certain requests or to take new commands or make new decisions. The robot's three IRRs allow it to perform easily its docking operation. The power supply module converts the 24 V of the Li-ion battery to the different output voltages of the robot. These output voltages are 12 V, 5 V and 3.3 V for respectively the IPCs, the DC motors and the sensors.

The two DC motors drive the robot through its two rear wheels. The charge controller module contains a microcontroller U3 to monitor the recharge process. An illustration of the hardware components and a real picture of the built mobile robot are presented respectively below in "Figs. 2 and 3". For the three navigational IRRs on the robot, one is positioned in front of the robot. The other two take place on both sides of the robot (left and right). According to their transmission direction, they are called left IRR, central IRR and right IRR.

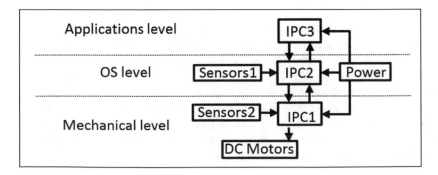

Fig. 2. Hardware components of the robot.

Fig. 3. Real robot.

2.2 Power Management System Integrated Circuit

"Figure 4" below represents a simplified model of this PMS' IC. In this figure, the charger and the charge controller are separated by a dashed line. The charger is at the left side of this dashed line and at its right side the charge controller is presented. The voltages in this circuit are represented by the symbols from V1 to V5 in the "Fig. 4".

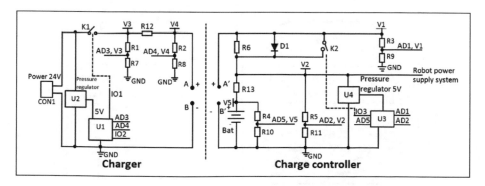

Fig. 4. The Power Management System monitoring circuit.

Their real values are respectively related to the voltages from the AD1 to AD5. These five ADs are used to decrease the real voltage in the PMS' IC. They allow U1 and U2 to know the real voltages for each Vi (i = 1 to 5) and control the recharging process.

Two pressure regulators U2 and U4 are used to supply an 5 V output to respectively to U1 and U3. Thirteen different resistances Ri (i = 1, 2,...13) and one diode D1 are the fundamental components of this PMS' IC. They are responsible to drive the voltage in all the circuit. The CON1 supply the charging voltage of 24 V in the charger side. Bat represent the battery to be charged in the charge controller. Two magnetic switch keys (K1 and K2) controlled by respectively U1 and U3 serve to begin and stop the recharging operation.

3   Auto-Recharging Strategy

The PMS' two modules (charger and charge controller) are implemented separately but should cooperate to complete the auto-recharging operation. When the robot is running out of its battery, it will suspend its current task and then looks for the charger. The charger throughout its three IRTs continuously transmits six IR signals (one far signal (FS) and one near signal (NS) for each IRT). The robot considers those transmitted IR signals only when it need to charge its battery. At this time, it begins to look for these signals. The robot will follow the signals found until to be connected to the charger, then charges its battery. "Figures 5 and 6" below show respectively the charger's and charge controller's working principle. The PMS working principle is based on the changes of voltage in its IC boards. These changes describe three statuses: "not connected", "connected but not in load" and finally, "connected and recharging".

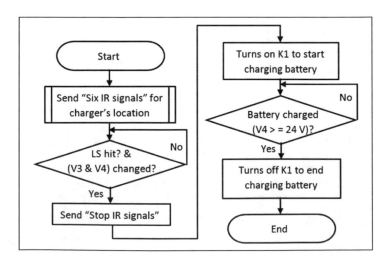

Fig. 5. The charger's working principle.

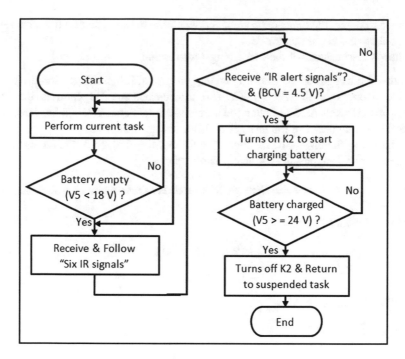

Fig. 6. The charge controller's working principle.

For the charger, the recharging process begins when the MCU (U1 in "Fig. 4") detects a robot docked under two secure recharging conditions:

- One is the robot hits the LS at the end of its docking operation.
- Other is the voltage changes in the analog-to-digital convertors (AD3 and AD4).

The U1 allows to begin the recharging process only if these two conditions are satisfied. It also has the ability to control the recharging progression. Once connected, the robot will:

- First, receive throughout its fourth IRR an alert signal from the charger's fourth IRT.
- Secondly, notice that its battery's current voltage (BCV) changed to the quarter of its previous voltage (18 V) before the connection.

After these two steps, the robot stops moving forward (it finishes its docking operation). At that time, U3 will close the switch K2 meaning that the charge controller is ready for recharging process. Whereas U1 will close the switch K1 to start the recharging process. In normal condition, the voltage changes in AD3 and AD4 allow to U1 to know about the robot's connection status. That means its recharging progression and its departure from the charger.

For the charge controller in the robot, the microcontroller U3 has mainly two functions:

- One is to detect the connection status between the docking station (the charger) and the robot (the charge controller).
- Other is also to control the recharging progression.

Both of these functions depend on the programming status of some analog-to-digital converters of the PMS's IC. The microprocessor U3 controls the voltage changes in AD1 and AD2 to define the corresponding status. AD1 is for detecting the connection status between the charger and the charge controller.

Considering the PMS's IC in "Fig. 4" and from the analysis of the three statuses in this IC, one can generate the following formulas:

$$R12 = R13 = 0.047\,\Omega \tag{1}$$

So we consider that

$$R12 = R13 \approx 0\,V \tag{2}$$

3.1   Not Connected Status

For charger K1 is open:

$$V3 = V4 = 0\,V \tag{3}$$

For charge controller K2 is open:

$$V1 = ((R3 + R9)/(R6 + R3 + R9)) * VBat \tag{4}$$

$$V2 \approx V5 = VBat \tag{5}$$

3.2   Connected and not Charging Status

K1 and K2 are opened and let call:

$$R14 = R4 + R10 \tag{6}$$

$$R15 = R5 + R11 \tag{7}$$

$$R16 = R3 + R9 \tag{8}$$

$$R17 = R2 + R8 \tag{9}$$

$$R18 = R1 + R7 \tag{10}$$

$$R14 = R15 = R16 = R17 = R18 \tag{11}$$

And according to the scheme after the connection, we have

$$1/R = 1/R16 + 1/R17 + 1/R18 = 3/R16 \tag{12}$$

So from "(11)" and "(12)" we get

$$R = R16/3 \tag{13}$$

$$V1 = V3 = V4 = R/(R+R6) * VBat = 1/4 * VBat \tag{14}$$

$$V2 = V5 = VBat \tag{15}$$

3.3   Connected and Charging Status

K1 and K2 are closed and we have:

$$V3 = VPower = 24\,V \tag{16}$$

So considering "(2)", the voltages of all the Vi (i = 1 to 5) are almost same. Because, there is no strong resistance on the wire linking these five Vi. Thus we can deduct the following equation:

$$V4 = V1 = V2 \approx V5 \tag{17}$$

Through these different statuses, the U1 and U3 perform the corresponding task. They also turn on the corresponding lights (LED) on the robot and charger to express the operating status of the recharging process.

To control the progress of recharging, using the Eq. "(18)" below U3 frequently calculates the charging current (Icharge). This electric current is tested by U3 to know if the battery is fully charged or not. If this Icharge is close to zero Ah then U3 knows that the battery is fully charged. In other words, recharging is complete means that the voltage in V5 (a voltage proportional to that of AD5) is greater than 24 V (about 29.4 V).

$$|(V5 - V2)|/R12 = Icharge \tag{18}$$

The voltages in AD2 and AD5 are used to control the stability of the robot's working voltage. At the completion of the battery recharge, U1 stops recharging operation by turning off the switch K1. U3 turns on the full charge LED on the robot and opens the switch K2 to turn off the charge controller. After that, the robot disconnects itself and then leaves the docking station to continue its suspended task. In "Fig. 7" below, the implemented hardware board of the charger is shown. Its corresponding charge controller's board is integrated into the IPC1.

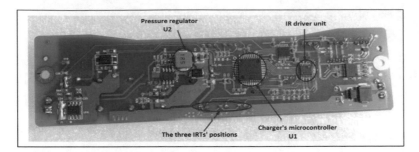

Fig. 7. The implemented hardware board of the charger.

4 Simulation Results and Discussion

To discover the performance of our PMS, the simulation tool Multisim has been used. In this tool, the PMS' IC has been designed. This design contains a power component of 24 V in the charger and a battery in the charge controller. This power component is used to supply the recharging voltage in the system. The two magnetic switches of the PMS' IC are represented by K1 and K2, respectively for the charger and the charge controller. In addition, two switches (K3 and K4) are added in this PMS' IC design. These switches closed describe the connection of the charge controller to the charger. The objective in this simulation is to check if the designed PMS's IC is correct. That means to check the correctness of the provided voltage values at each of the five Vi in the IC.

To get these different voltage values, six multimeters are used. One multimeter connected to each Vi (i = 1 to 5) and the sixth is connected to the battery. To notice the progression of the battery recharging process, a capacitor C1 is used to replace this battery. In this simulation the three status ("not connected", "connected but not in load" and "connected and recharging") has been simulated. These simulated statuses are divided into two phases. Phase 1 is called the individual test. It consists to simulate separately charger and charge controller. Phase 2 consists to connect charger and charge controller to build the whole system (PMS' IC) and then conduct this whole system test. The results from these phases' simulations are presented respectively in Tables 1 and 2 below.

Table 1. Individual simulation results.

	K1	Charger (V)		K2	Charge controller (V)	
Table Individual	0	V3	0	0	V1	8.993
		V4	0		V2	18
					V5	18
	1	V3	24	1	V1	18
		V4	24		V2	18
					V5	18

Table 2. Whole system simulation results.

	K1K2	Whole system (V)				
		V1	V2	V3	V4	V5
Table whole system	00	4.502	18	4.502	4.502	18
	01	18	18	18	18	18
	10	22.153	19.846	24	22.153	18
	11	21	21	24	21	18

To be more expressive in these tables, an opened and closed switch is represented respectively by a value zero and a value one.

- Phase 1 (Individual tests): The two switches K3 and K4 are opened to separate the charger and the charge controller. For the charger, initially the switch K1 is opened and the voltages V3 and V4 are equal to zero volt. After closing K1, the corresponding multimeters print out 24 V each one. This is normal as the R10's resistance value (between the two multimeters) is small specifically equal to 0.047Ω. Concerning the charge controller, in initial state K4 is opened. At this time the voltage values (in volts) for V1, V2, V5 and V6 (for battery) are respectively 8.993, 18, 18 and 18. When K4 is closed all these voltage values are same and equal to 18 V.
- Phase 2 (Whole system test): The two switches K3 and K4 are closed to connect the charger and the charge controller. So that the whole system (PMS's IC) is build. Here two statuses are considered "connected and not load" and "connected and charging". The simulation results from these two statuses are presented in Table 2.

The combination of K1 and K2 as a set of variable give us four states 00, 01,10 and 11. These four states describe respectively "Not charging and charge controller not ready for charging", "Not charging and charge controller ready for charging", "Charging and charge controller not ready for charging" and "Charging and charge controller ready for charging". During the recharging simulation, the battery in this designed IC is changed to a capacitor C1. This let us see the recharging progression throughout a new multimeter added in the design. This capacitor C1 is initiated to 600 farads and 18 V to represent the discharged battery. After launching the simulation, K1 and K2 are closed to start the battery recharging process.

One can notice the progression of the recharging process in the multimeter attached to the battery. This recharging process is complete when the recharging current (Icharge) is equal to zero. This current is calculated by U1 throughout the following formula:

$$|(V3 - V4)| / R12 = Icharge \tag{19}$$

It (U1) ends its recharging operation only if Icharge is equal to zero. After this simulation, the PMS' IC is build and implemented. The produced board is used to recharge a Li-ion battery (24 V, 3000 mAh), at 18 V. During this recharge, the recharging time is measured and it is equal to around 7 h. This designed PMS' IC is reliable and safe throughout its implementation process. All the recharging steps in both (Charger and Charge controller) modules are sequential (no parallelism task). That means when the robot arrives to the charger, there is no voltage on the recharging slots. Thus no electric damage is possible at that time in the PMS' IC.

The charger uses a 24 V to begin the battery's recharging operation. This is possible only if the responses from the different conditions tested by U1 and U3 in Sect. 3 above are affirmatives. This approach can avoid damaging the robot and the charger during the stabilization time at the end of its docking process.

5   Conclusion and Future Works

The design and implementation of a new PMS for an autonomous robot have been presented. The new PMS allow at any time to both charger and charge controller to know all the steps of the recharging process. This means that when the battery is fully charged, the charger will stop recharging automatically. At the same time, the charge controller will tell to the robot to immediately leave the docking station to continue its suspended task. In addition, it can determine if the recharging status is normal or not. Second, the "Multisim" simulation tool was used to check the performance of the PMS. The results of the simulation show that the PMS' IC we have designed is reliable. So its corresponding hardware IC is built and integrated into a real robot.

This robot can perform its automatic recharging operation safely. The recharging range voltage in this research is set from 18 V to 24 V. So that the built-in PMS circuit's recharging time is about 558 s in the simulator. For a real Li-ion battery (24 V, 3000 mAh), the recharging time is around 7 h.

We plan, as future research, to design, implement and integrate into a mobile robot a new wireless automatic recharging system. Thus the robot will no longer waste time when recharging its battery. This new recharging system will allow the robot when recharging its battery, to always continue its current tasks.

Acknowledgment. The National Natural Science Foundations of China NSFC with grant (No. 61772185) supported this work so we say thank you very much.

References

1. Quilez, R., Zeeman, A., Mitton, N., Vandaele, J.: Docking autonomous robots in passive docks with infrared sensors and QR codes (2015)
2. Acosta Calderon, A.C., Ng, B.S., Mohan, E.R., Ng, H.K.: Docking system and power management for autonomous mobile robots. Appl. Mech. Mater. **590**, 407–412 (2014)
3. Song, G., Wang, H., Zhang, J., Meng, T.: Automatic docking system for recharging home surveillance robots. IEEE Trans. Consum. Electron. **57**(2), 428–435 (2011)
4. Roh, S.G., Park, J.H., Lee, Y.H., Song, Y.K., Choi, H.R.: Flexible docking mechanism with error-compensation capability for auto recharging system of mobile robot. Int. J. Control Autom. Syst. **6**(5), 731–739 (2008)

5. Niu, Y., Zhang, J., Meng, T., Wang, H.: Design of a home surveillance robot with self-recharging capabilities, In: International Symposium on Knowledge Acquisition and Modeling. IEEE (2010)
6. Su, K.-L., Liao, Y.-L., Li, S.-P., Lin, S.-F.: An interactive auto-recharging system for mobile robots. Int.J. Autom. Smart Technol. **4**(1), 43–53 (2014)
7. Taylor, T., Wyrzykowski, M., Larsen, G. C., Paull, M.M.: Docking process for recharging an autonomous mobile device (2013)
8. Petrov, P., Clément Boussard, Ammoun, S., Nashashibi, F.: A hybrid control for automatic docking of electric vehicles for recharging, In: ICRA International Conference on Robotics and Automation. IEEE (2012)
9. Doumbia, M., Cheng, X., Havyarimana, V.: An auto-recharging system design and implementation based on infrared signal for autonomous robots. In: The 5th International Conference on Control, Automation and Robotics on Proceedings, pp. 894–900. IEEE, Beijing (2019)

Some Cyberpsychology Techniques to Distinguish Humans and Bots for Authentication

Ishaani Priyadarshini[✉], Haining Wang, and Chase Cotton

University of Delaware, Newark, USA
{ishaani,hnw,ccotton}@udel.edu

Abstract. In a world where artificial intelligence is one of the greatest assets, unmanned operations seem to be the future. The world of cybersecurity is witness to numerous system break-ins for the purpose of gaining access. One of the ways to gain access to systems is fulfilled by authentication, the process where an entity verifies who he or she claims to be to access a system. With network traffic increasing day by day, the bots form a huge chunk of the network traffic. Over the last few years, bots have been trained to imitate human beings to gain access to computer based systems. Traditional authentication methods are based on what we know, who we are and what we have, and can be bypassed easily these days. Bots have been known to imitate human beings in order to gain access to systems by identifying captchas and picture based authentication systems. A bot gaining access to sensitive data may have severe repercussions. Thus there is a need to introduce certain parameters that could easily tell apart a bot and a human being. One of the primary factors that differentiates bots and human beings is the kind of psychology embedded in humans. In this paper, we identify a set of cyberpsychological factors that could differentiate human beings from bots. This may prove to be beneficial for the purpose of authentication where a human being may be able to authenticate himself/herself based on his/her psychology, whereas a bot may not be able to do the same because of the lack of its thinking capabilities. Therefore, the issue focuses on three important domains, that is, cybersecurity, artificial intelligence and cyberpsychology.

Keywords: Cyberpsychology · Authentication · Cyber security · Artificial intelligence · Bots

1 Introduction

Cyberpsychology underpins the phenomenon of human mind and its behavior with respect to human interaction as well as communication of both man and machine. It highlights the effect of cyberspace interactions over individual minds. Moreover, it takes into account how humans and technology interact with each other [1–3]. Thus it encompasses all psychological phenomenon associated with or affected by emerging technology. A significant fraction of the internet traffic is made up of bots. A bot may be defined as a software application or script that is capable of performing automated tasks and commands. They may perform malicious tasks and remotely take control

K. Arai et al. (Eds.): FTC 2019, AISC 1070, pp. 306–323, 2020.
https://doi.org/10.1007/978-3-030-32523-7_21

over a system. They can not only perform large operations while still remaining small, but may also compromise open source and unsecured devices. With artificial intelligence gaining tremendous success, it is possible for a bot to access a system and authenticate itself as a human being. Authentication ensures that an individual is who he or she claims to be. If a bot is capable of authenticating itself as a human being, and gains access to a database that incorporates sensitive data, the results may be catastrophic. In the past several techniques have been used along with credentials to enhance system security in form of captchas or image recognition techniques. However, over the years, bots have become sophisticated enough to break captchas and recognize images as good as humans would do. This makes distinguishing a bot behavior from humans difficult for the purpose of authentication, thereby seeking powerful methods to tell a bot and human apart. Cyberpsychology, the study of human mind and behavior while interacting with machines comes into the picture here. What differentiates a human from a bot could be the ability of humans to think whereas bots may only be fed training data to either predict a test data or provide an output based on its input.

In this paper, the issue being undertaken is the fact that sophisticated bots have been known to authenticate themselves as human beings. The results of this kind of inappropriate authentication could have catastrophic results in the future. Thus the need arises to prevent the bots from performing such inapt activities, and hence we are motivated towards finding parameters that could provide aid in differentiating humans and bots when it comes to authentication. The design of the research is based on identifying certain authentication methods proposed in the past, why they failed and how bots could bypass those. On basis of that we suggest certain cyberpsychology techniques that could strengthen the procedure of authentication such that a bot cannot imitate to be a human. In the later part of the research, we propose cyberpsychology characteristics as one of the important criteria for distinguishing robot and human behavior. The characteristics are evaluated non-quantitatively in order to find out which characteristics can contribute to differentiating human and robot behavior. The novelty that we bring to this research lies in the **emerging field of cyberpsychology which has not been considered in depth for the purpose of distinguishing human and bot behavior for the purpose of authentication**. Based on the proposed techniques, we present some results that support our proposal. Further, we chalk out a graph based on certain cyberpsychological characteristics to deduce which characteristics are likely to have more impact in this process of human and bot behavior discrimination. Rest of the paper is organized as follows: Sect. 2 incorporates Literature Survey while Sect. 3 details the proposed work. In Sect. 3, Subsect. 3.1 underpins the proposed techniques, whereas Subsect. 3.2 mentions the cyberpsychological characteristics. Section 4 highlights the implementation and evaluation of the proposed research work. Sections 5 and 6 present the results and conclusion.

2 Literature Survey

In this section we will take a look at some of the previous authentication methods that were proposed and were easily bypassed by bots. Moreover we analyse these methods and also perform their critical evaluation so as to find their limitations. This would assist us in verifying the strength of our approach.

Griffin [4] proposed a biometric knowledge based two factor authentication mechanisms. The underlying idea is based on a combination of two attributes: something you know (basically a key in this case) and something you have (biometrics). The aim of the research is to provide confidentiality of credentials from phishing and man in the middle attacks. We know that two-factor authentication or two-step verification is a security process in which the user provides two authentication factors to verify they are who they say they are. The extra layers of security added by two-factor authentication go through Application Authentication, Standard Login, OTP generation and OTP delivery. Two-factor authentication can be bypassed by conventional session management, OAuth mechanism, Brute Force and race conditions [5]. Bots can be programmed accordingly to outperform authentication. They could be programmed to enter valid credentials and compromise devices that would receive OTP to bypass authentication process as well.

Markowitz [6] proposed a voice biometrics based on authentication system for speaker verification. The proposed technique is a combination of speech processing and biometrics that matches data from a voiceprint database in order to verify the authenticity of a user. In present day voice (biometrics) is used as a means for authenticating individuals. Voice recognition is the technology by which sounds, words or phrases spoken by humans are converted into electrical signals, and these signals are transformed into coding patterns to which meaning has been assigned [7]. Based on the signals and coding patterns individuals are identified and authenticated. Template matching and feature analysis are the two approaches to voice recognition, and since they exist as data in databases, if compromised, their purpose is defeated. Another way of undermining authentication is the introduction of voice editors. Voice manipulation caused by voice editors is a security threat. There could be audio files of entities supposedly speaking words that they never uttered. It is possible to change contents in an otherwise legitimate audio file. They could be used to implicate someone in a criminal act they did not perform. So even though using voice metrics increases security, it can also be used to sabotage systems.

Shah [8] and [9] introduced picture based authentication techniques. The system gives a picture and the person supposed to be authenticated is supposed to describe the picture. The way it works is that users select concrete category of images, which they may be presented later in a grid format and asked to describe. With the introduction of image recognition systems like CaptionBot, the method becomes futile [10]. CaptionBot- Microsoft's latest artificial intelligence was introduced to identify pictures and provide captions describing them. It relies on two neural networks that deal with image recognition and natural language processing, respectively. So a bot has yet again disrupted human authentication.

Turing [11] proposed the turing test which is a method for determining whether or not a computer is capable of thinking like humans. A computer chatbot named Eugene Goostman, who had the persona of a 13-year-old boy, could pass turing tests to convince that it is a human [12]. Turing tests have been useful in authentication. CAPTCHA (Completely Automated Public Turing Test to Tell Computers and Humans Apart) is a kind of turing test in which users may be required to view a distorted string of alphanumeric characters in an image and enters the characters accordingly. Robots have the capability to beat captchas like "I am not a Robot" [13]. Thus turing tests can fail. In the past, Turing tests have been conducted for machines to generate human like behavior. The Turing test does not directly test whether the computer behaves intelligently but if the computer behaves like a human being [12]. Since human behaviour and intelligent behaviour is not exactly the same thing, the test can fail to accurately measure intelligence in many ways. Using a capacitive stylus, a robot may physically move the mouse on the trackpad, as if it were a real human wiggling their finger around. Sometimes humans can fail the turing test as well making it insignificant for distinguishing humans and bots.

Zorkadis and Donos [14] proposed a biometrics based authentication technique. Biometrics is the measurement and statistical analysis of people's physical and behavioral characteristics and is used for identification and access control, or for identifying individuals. The basic premise of biometric authentication is that everyone is unique and an individual can be identified by his or her intrinsic physical or behavioral traits. However they are prone to certain issues. The biometric data may be stored as a mathematical snapshot in a database. While authenticating an entity, the data is compared to the entered biometric. If the database is compromised, it leads to serious security issues as biometrics cannot be reproduced. They can be easily hacked, and lack revocability [15]. Bots can easily bypass biometric authentication as well.

Bots and robots perform the same tasks as humans; however, robots are more mechanical as compared to bots which are more inclined towards the software side. Although robots have several cybersecurity issues associated with them [17] have been known to access systems by clicking on the 'I am not a robot' boxes using stylus. Moreover, on basis of how a human being moves a mouse, moments before they click on it, Google has come up with a concept of identifying a robot from a human being. The concept does not rely on user interaction with the system but just the tracking movement of the mouse [16]. However, it may be ineffectual as just by mere tracking movements of mouse it cannot be determined if the user is human or not.

Burrows et al. [18] proposed authentication based on logic. Several protocols along with their attributes have been studied for the research. Several other researchers have come up with authentication techniques based on Fuzzy logic [19, 20]. A bot may capable of carrying out complex series of actions automatically and programmable by computer. Since it is computer oriented, it relies on logic. So logic cannot be used to differentiate a human being and a robot as it varies from person to person. Also logic is something which is confined to a human being making it more of a biometric thing, which we know could be compromised. Similarly Intelligent Quotient (IQ) can be used to assess intelligence by introducing a series of standardized tests and questions [21]. It has been found that in the past, robots have been programmed to maintain different levels of IQ so as to imitate human beings from different ages. So logic and IQ do not serve as a basis for differentiating humans from bots.

As we have seen, certain authentication procedures were proposed in the past and they have been known to get compromised by devices. Therefore there is a need to find a way that could distinguish bots and humans. We know that bots are immune to emotions and other psychological elements, therefore we take into account cyber psychological techniques to tell apart a human and a robot while considering authentication which we will indicate in the next few section. In the later section we take into account cyberpsychology characteristics in form of quotients and their impact on bots.

Thus we have listed several of the techniques that have been used in ast for authentication and how they could be easily broken by bots and robots. The weaknesses found in these systems hint at finding much more sophisticated methods that could distinguish between a human and a bot so that bots cannot imitate humans so as to gain unauthorised access to systems.

3 Proposed Work

In the previous section, we listed some of the authentication methods that are capable of being easily bypassed by the bots. Since bots are becoming sophisticated with the increasing use of neural networks and several other machine learning techniques to imitate humans for accessing systems, it is important that we identify ways that would not allow unauthorised access by bots. As mentioned earlier we are relying on the discipline of cyberpsychology to mitigate the issue. Cyberpsychology pertains to the thinking ability that humans possess, therefore authentication techniques may allow humans access to the system while denying the same to bots because of their lack of conscience. To solve the issue, we propose two approaches; hence we divide this section into two parts. In the first half of the section, we propose some **cyberpsychology techniques** that could tell apart a bot and a human being. Consequently some implementation has been performed in the Sect. 4.1. In the second part, we introduce the concept of **cyberpsychology characteristics** to differentiate a human and bot, the evaluation of which is presented in the implementation section.

3.1   Cyberpsychology Techniques Proposed

We have already discussed how bots can vandalize the process of authentication and gain access into a system. We rely on the concept of cyberpsychology to ensure that bots can be differentiated from human beings when it comes to authentication. Cyberpsychology is the combination of psychology and cyberspace. It is concerned with the psychological effects and implications of computer technologies. We rely on psychology because that is one of the primary aspects that differentiate human beings from bots. It is the process of control and communication with the help of studying mind and behavior. There is difference in the way a human interacts with a system and a bot interacts with a system. We take into account these subtle differences related to psychology and present a set of techniques in this section that could identify a bot and deny access to it. Humans think on basis of instincts and emotions whereas robots and bots think on basis of sensors and environment [22]. Robots and bots work by taking outputs from sensors, sending them to a microprocessor and running predetermined software routines to produce instructions for the robot's actuators to alter movement. They may also imitate behavior of organisms like insects. As we know Turing tests have been conducted for machines to generate human like behavior. The Turing test does not directly test whether the computer behaves intelligently [11]. It tests only whether the computer behaves like a human being and the test can fail to accurately measure intelligence in many ways. Our primary focus would revolve around the behavior aspect and not the intelligence aspect. Cleverly chalked questions or psychological interrogations can be instrumental in contrasting the behavior of bots and humans. We have taken into account two sections to identify the differences between humans and bots. The first section deals with the different types of questions, events, abilities of bots that distinguish them from humans, whereas in the second section we take into account different psychological quotients that deal with specified quality or characteristics. The techniques identified are as follows:

a. **Asking identical Questions:** Psychologists use the technique of asking identical questions to people to ensure transparency. The human brain can comprehend that identical questions refer to the same idea, and end up presenting identical answers to the set of similar questions asked. However, in case of bots, the questions represent a series of data. When different questions related to the same idea are asked to a bot, it interprets the questions in a different manner as the data produced for multiple questions pertaining to the same idea may differ, and may present distinct responses. For e.g. Consider the question "How was your day?" A robot may comprehend it in a different manner than the question "How was the day?" It is unlikely to get the same answer for both the questions. "If you had a lot of money what would you buy? What is that one thing that you want to buy the most? Name a thing whose purchase will make you immensely happy." are different versions of the same question and a robot may reply differently to each of them.

b. **Picture Description:** In the previous section we mentioned that CaptionBots are effective in describing images. So if a robot is asked a psychological question regarding a picture, for e.g. "Describe the picture." CaptionBot makes it quite easy for the bots, the reason being that it relies on two neural networks that deal with image recognition and natural language processing respectively. From cyberpsychology point of view, there is a need to introduce modification to the picture in question. One can randomly draw pictures using tools like paint and occupy the images with colors that the images naturally should not bear. For e.g. we draw a cat using the paint tools and not color it black or white or brown which a cat may usually be. Instead we impart it a color that is unreasonable say for e.g. Blue. The human mind can comprehend a blue cat whereas a bot may not be successful in doing so.

c. **Grammatically incorrect pages or Verbal Linguistics:** Another psychological technique that could be used while interacting with a bot is the use of correct grammar. A bot may not be capable of translating one language to another with correct syntax and semantics. The very basic example that could be taken into account is the google translator. Although it is easy to translate small sentences, complex sentences are not easy to translate and produce erroneous results. For e.g. the sentence 'Απλά σκουπίστε τους ώμους σας μακριά' in Greek refers to "Just wipe that dirt off your shoulder" whereas Google Translator shows it to be "Just wipe your shoulders away", which alters the original sentence to a great extent. A bot may be asked to translate a complex sentence in one language to another language and grammatical errors following could hint at the bot being a non-human.

d. **Spatial Visualization ability:** Spatial visualization ability or visual-spatial ability is the ability to mentally manipulate 2-dimensional and 3-dimensional figures. It is typically measured with simple cognitive tests and is predictive of user performance with some kinds of user interfaces. Since the ability deals with looking at images, analyzing them and manipulating them, it may be toilsome for a bot to perform if not programmed accordingly [23]. Sometimes bots face difficulty in visualizing and analyzing multi-dimensional objects. Although a bot may be somewhat trained, but it would lack the accuracy for a different set of images. On the other hand a human being possessing the ability to inspect an image may be efficient at spatial visualization ability.

e. **Test of Creative Thinking:** Creativity is the use of imagination or original ideas. Since it focuses on original ideas, bot may not be creative. Sure it can be trained, it can imitate the behavior of other entities, but as of now bots have yet to be praised for their creativity such that they do not rely on neural networks or other artificial intelligence models. Test of creative thinking can differentiate between a human and a bot to some extent. Creative thinking based situational questions may be effective in distinguishing a bot and a human. For example, the bot is given a situation "What will you do if your house burns down?" or "Would you rather sacrifice one adult to save two children, or two children to save five adults?" The answers may vary to these questions, but it's the explanation on basis of which differentiation could be done.

f. **Comprehending tricky questions:** Bots have the inability to infer precarious questions. One of the reasons for this is because they try to evaluate all the questions on basis of signals and data. But there could be questions that are easily understandable by humans and not bots because they cannot differentiate between entities (living and non-living). Everything is regarded as data for bots. For e.g. consider a question "God asked Abraham to sacrifice his son, Isaac because he wanted to test his faith. Whose son and whose faith are we talking about?" Such questions can be answered correctly by humans in most cases but not in case of bots.

g. **Summarizing a paragraph:** The act of summarizing a paragraph requires understanding, analyzing, interpreting and formulating an idea incorporated in the paragraph. It is difficult to condense a five hundred word essay in fifty words, or more practically, a one page paragraph into two sentences. This could be achieved exclusively by humans who are adept in the technique of skimming through the paragraph, understanding what information to include and what information to eliminate. Humans will be more capable than bots to get the general idea of a paragraph and can summarize it better than bots. A bot may be given a comprehension paragraph and asked the underlying idea behind it in the form of multiple choice questions.

h. **Tricky interrogation:** Intuitive questions that question observation, comprehension and judgement may pose a difficult situation for bots. Consider the question "John lost his dog in Central park. His dad put on his boots and went to help John to find it. Why did his dad put on his boots?" The answer to the question could be "Because he was going outside" and extracting an answer about which nothing has been mentioned in the question from an artificial system will be hard, because the fact that you suppose to wear boots when you go to Central Park is not stated explicitly in this text. The bot is incapable of making well thought of answers to questions like these.

i. **Pronunciation:** When it comes to pronunciation, bots are ahead of human beings because of the fact that bots can be trained phonetics rules and can have a clear understanding as to how to pronounce even the toughest words. But this capability of bots can also be used to differentiate them from human beings. For example, consider the word 'кnapsack', and the bot is supposed to pronounce it. As a human being the pronunciation may differ, but for a bot it will differ significantly. The reason is that there is a catch in the word 'кnapsack'. A human mind comprehends the first letter as 'K' whereas a bot will always comprehend it as Greek letter kappa (k), hence altering the pronunciation significantly. Hence pronunciation is yet another factor that can differentiate between a human and a bot.

j. **Differentiate between Positive and Negative events:** The concept of positivity and negativity is largely driven by instincts which bots thoroughly lack. They can only differentiate between a positive and negative even on the basis of the input given. Since they cannot comprehend the idea underlying those events and also their after effects, they cannot differentiate between slightly complicated positive and negative events. A bot will consider a negative event by filtering contents like "no", "deny", "never", "negative", etc. For example "Elly feel happy squeezing her pet down the drain" or "Elly loves burning trees" are negative events but the bot may perceive it positively due to encouraging phrases like "happy" and "loves".

k. **Sophisticated covert questions:** Sometimes bots pretending to be humans, if not smart enough tend to answer questions about themselves. For example, they will tell you what algorithm they are using, how much power they are using, how long they will last without charge, etc.

l. **Memes:** Memes are forms of pictures, video clips, films, graphics, texts, quotes, animations existing independently or combined, as a mode of expression [24]. With the internet being flooded with memes these days, which may be specific to languages across the world, geographic regions and domains (film industry, sports industry, finance industry, etc.), it may pose a serious challenge for a bot to imitate as a human being for interpreting a meme. A specific meme template may infer specific information to a given group; the same template may be applied to another domain imparting new information. Sometimes, memes are difficult for humans to comprehend, which is why it might be even more difficult for a system to interpret it [25].

3.2   Some Cyberpsychology Characteristics to Differentiate Humans and Bots

In the previous subsection we have seen how cyber psychological techniques in the form of questions can address the issue of bot behavior versus human behavior for the process of authentication. There is another approach involved for the same which we will be discussing in this section. Psychology involves characteristics of human behavior in form of certain measures which may or may not be quantitative. However, based on our analysis, insight and investigation we assign them qualitative values very low, low, medium, high and very high, which will further be used in our evaluation in the next section. Each of the characteristics has its own definition and is associated with a certain level of human behavior that can differentiate it from bots. The idea is to find significance measure of each characteristic when it comes to differentiating human and robot behavior by assigning them values. We take a look at seven such characteristics (Fig. 1 is a representation of the same).

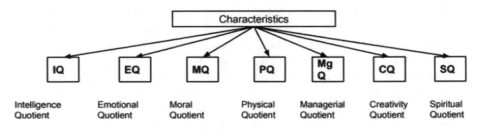

Fig. 1.   Representation of the seven characteristics of cyberpsychology

a. **Intelligence Quotient:** Intelligence Quotient (IQ) is the measure of intelligence of an individual derived from the results obtained from specially designed tests. The formula is given by the mental age of an individual by its chronological age, multiplying the result by hundred. It leads an individual to think, make cause effect analysis, perform math, and create a level of understanding. However, it does not measure the intelligence of an individual but cognitive and intellectual ability [23]. In the section pertaining to literature survey, we have discussed how bots can be programmed to maintain different levels of IQ so as to imitate human beings from different ages and that logic and IQ do not serve as a basis for differentiating humans from bots. This makes it impossible to distinguish a human and bot on basis of IQ, although it is predicted that in coming years bots may have an IQ which will be higher than humans [26]. All the above analysis makes IQ imprecise to differentiate human and bot behavior. So we assign it a low value. The reason we have not given it a very low value is because sometimes there can be questions that bots can answer and humans cannot or vice versa, so that may highlight the differences to some extent.

b. **Emotional Quotient:** Emotional intelligence or Emotional Quotient is the ability of individuals to recognize their own emotions and those of others, and also be able to understand different feelings, to label them appropriately. The emotional information is to manage emotions to adapt to environment [27]. It also deals with the concept of sensitivity, empathy and courageousness. Bots have been known to gain emotional intelligence according to recent studies [28]. Chatbots have been known to carry out conversations with emotions like happiness, sadness and disgust. So even though Bots can be fed emotional intelligence, they would still lack emotional instincts. The feelings of anger, fear, disgust and empathy only exist as data in the bots and bots can only be trained to exhibit those kind of behaviors. The researchers are still far away from building a machine that would completely understand users' emotion. This makes emotional quotient not very useful in distinguishing human and bot behavior. However there can be certain scenarios where test of creative thinking and classification of positive and negative events can come into play, thus making it one of the important characteristics for classification. Thus we assign the value medium for differentiating humans from bots on basis of emotional intelligence.

c. **Moral Quotient:** Moral Quotients is associated with the perception of social and working life. It deals with features like moral intelligence, honesty, integrity, compassion, patience, responsibility, tolerance, honesty, commitment and maintain, requiring energy, enthusiasm and effort, [29] mostly unavailable in a bot. So a moral quotient can be given a value very high for distinguishing human behavior from a robot.

d. **Physical Quotient:** Physical Quotient (PQ) indicates the capacity to work through situation through your physical proves. That means effective and efficient use of your hand eyes coordination. Along with that it takes into consideration reactions, feelings, thoughts, confidence and resilience of individuals. Even though robots can successfully coordinate movement of their components, the cause effect existing in human body coordination is something robots lack, thus making Physical Quotient a very high requirement for distinguishing between a human and a bot. A bot (if needed) may have a body made of components; however, it will not be as effective as a human body thus making Physical Quotient one of the most important characteristics.

e. **Managerial Quotient:** Managerial Quotient can be defined as the collection of political, emotional and intelligence quotient [30]. Political intelligence indicates people's ability to analyze the impact of event, to create power resource and to use power. A bot may have a very high emotional or intelligence quotient, but it will still find it hard to make decisions on the basis of political quotient. In other words a bot is incapable of analysing an event or deciding how to use power However, in the case of quantitative terms it may assign tasks, responsibilities to lower level functional components, on the basis of certain calculations and predictive outcomes. So the requirement for this quotient as a characteristic to distinguish human and bot behavior could be moderate or medium.

f. **Creativity Quotient:** Creativity Quotient may be defined as the ability to promote novel ideas and manifest them from thought into reality. The process heavily relies on original thinking and then producing [31]. As we know that bots lack creative thinking and can either imitate behavior or learn, they are incapable of coming up with novel ideas, thus making creativity quotient as one of the important measures in distinguishing human and bot behavior. Thus the value assigned is very high for this characteristic.

g. **Spiritual Quotient:** Spiritual Quotient could be defined as the central and most fundamental of all the intelligence, because it becomes the source of guidance for the others [32]. Human beings are essentially spiritual creatures because we are driven by a need to ask 'fundamental' or 'ultimate' questions. Who am I? Why was I born? Why am I suffering? Neither animals nor machines have been known to come up with such questions. It also takes into account self-awareness, principles, positive use of adversity and sense of vocation. All these features are non-existent in a bot making it one of the important characteristics to distinguish between a human and a bot. Thus the value assigned is very high for spiritual intelligence.

4   Implementation and Evaluation

In this section, we present implementation of some cyber psychological techniques discussed in the previous section. Although there may exist several ways to implement each of the techniques proposed, here we present only three methods considering the ease of implementation. Further, based on the qualitative values assigned to each of the cyber psychological characteristics, we present a graphical representation comparing various cyberpsychology quotients.

4.1   Implementation for Cyberpsychology Techniques

In this section, we have taken three cyberpsychology techniques, basically, picture description, pronunciation and grammatically incorrect pages or Verbal Linguistics. Although, several other techniques have been proposed in the previous section, we have highlighted these three specific techniques for the ease of their implementation.

a. **Picture Description:** Since the idea is to introduce modification to an image such that a human can comprehend it but not a bot, we uploaded a picture to captionBot that was rendered using graphic tools like MS Paint. The picture incorporates a blue sun and a purple cloud (Fig. 2). This image can be easily understood by a human being but not by a bot, since for a bot a sun is supposed to be round with colors defined like yellow, orange or red, defined within a specific range. When the picture was uploaded to CaptionBot, it could not recognize the image. Hence it could be an efficient method for authenticating humans. The picture uploaded is as follows. CaptionBot describes it to be a logo.

Fig. 2. Figure showing what CaptionBot described the image to be.

b. **Pronunciation:** This cyberpsychology technique deals with the idea of replacing alphabet characters by symbol characters which bear a close resemblance to it. So symbols like η (eta), ρ (rho), γ (gamma) and β (beta) can be replaced with alphabets like n, p, y and B due to their striking resemblance. When the website was asked to pronounce the word 'κnapsack' (Fig. 3) it kept on asking 'Did you mean knapsack?' as it could not afford the pronunciation for Greek letter kappa ('κ') which is identical to alphabet 'k'.

Search suggestions for κnapsack

We have these words with similar spellings or pronunciations:

1	knapsack
2	knapsacks
3	snapback
4	snaps back
5	snap back
6	snap-back

Fig. 3. The result of a website being asked to pronounce the word 'κnapsack', it kept on suggesting words similar to it but could not pronounce the word.

c. **Grammatically incorrect pages or Verbal Linguistics:** The idea acquainted with this technique is that bots may not be very good at translating languages. The following figure (Fig. 4) exhibits the same. The translation of a sentence from Polish to English does not match when reversed. A bot may not be efficient at translating correctly which could be a factor in discriminating a bot against a human.

Fig. 4. Figure representing translation issue when the system is asked to translate a particular sentence.

4.2   Evaluation of Cyberpsychological Characteristics

We have described seven cyber psychological characteristics in order to bring out the differences between a robot and a human. We have also assigned them with values ranging from very low to very high. Based on the analysis, we can present the data in a graph as follows. The following graph (Fig. 5) is a representation of unquantified data that has been given some values based on the importance of each quotient, similar to the graph in [33].

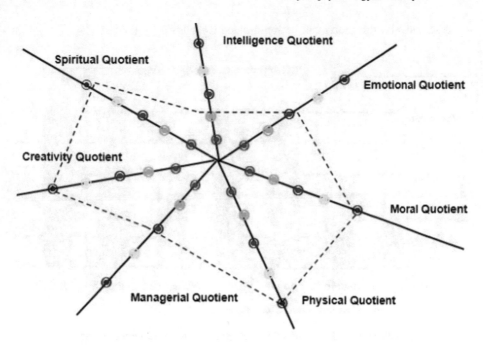

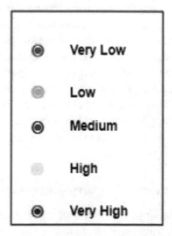

Fig. 5. Graph depicting values assigned to different cyberpsychology characteristics

Alternatively, the graph can be represented as follows (Fig. 6).

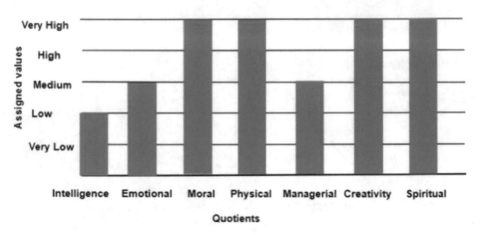

Fig. 6. Alternative graph representing assigned values to different characteristics

5   Results

We have described two approaches for tackling the tricky situation of a bot versus human behavior for the purpose of authentication. On one hand we have proposed certain cyberpsychology techniques in the form of questions and on the other hand we describe a set of characteristics for the same. To evaluate the plan, while authenticating, the user may be asked a set of random questions based on cyberpsychology techniques. The characteristics offer a way to find out which characteristics can differentiate a bot and a human appropriately. Once we find out the characteristics that dominate as a standard for distinguishing human and bot behavior, a user may be tested on the basis of questions pertaining to the characteristics. We present the observations for each of the proposed techniques as follows

5.1   Analysis of the Proposed Cyberpsychological Techniques

We have performed the implementation of three proposed cyber psychological techniques i.e. Picture Description, Pronunciation and Grammatically incorrect pages or Verbal Linguistics. Using CaptionBot, we were able to differentiate a human and bot based on interpreting an image. CaptionBot, which is quite efficient in recognizing objects from their pictures was unable to recognize when images were drawn using computer system tools. A picture of a blue colored sun and purple colored cloud could not be recognized by the system due to the unexpected strokes and color outside the permitted values, thus strengthening our proposed technique.

For the second implementation, we chose to replace a character in a word with a symbol that looks like the replaced character. A human may not notice such a trivial difference for the process of pronunciation, a bot however will, since each character holds a specific ASCII (American Standard Code for Information Interchange) and Unicode in its system. No two characters could have the same ASCII value; therefore the system cannot pronounce a word that has a character replaced with another character that looks almost like the replaced character. Thus our approach is meticulous.

Finally, a bot may not be as apt as a human being in translating languages. A bilingual human will effortlessly present one language in form of another and vice versa without any changes. However, in our implementation we find that the bot translates a sentence from English to Polish to yield a set of characters in a sentence. However, when the same set of characters (Polish) are converted to English; we do not obtain the initial sentence. Thus, an affirming method to distinguish between a human and a bot is grammatically incorrect pages or Verbal Linguistics.

5.2   Qualitative Analysis of Cyberpsychological Characteristics

Assigning the qualitative values to the cyber psychological characteristics, we yield a graph from which it is evident that characteristics like Moral Quotient, Physical Quotient, Creativity quotient and Spiritual Quotient may be highly regarded to tell apart a bot and a human. On the other hand Intelligence Quotient, Emotional Quotient and Managerial Quotient may not be proficient enough for the same. This is because Intelligence, Emotions and Managerial aspects could be fed into a bot by virtue of programming. However inducing Morals, Creativity, Spiritualism, etc. may take more effort.

6   Conclusion

It is evident that bots have been vandalizing systems in order to gain access. As the results following this could be catastrophic, certain approaches have been suggested. In the past, several methods have been introduced for human authentication and we have discussed how bots can bypass those. The former section of the article highlights cyberpsychology techniques which incorporate picture description, pronunciation, translation, classifying positive and negative events, etc. that differentiate a bot from a human for the purpose of authentication. The latter part focuses on cyberpsychology characteristics in form of quotients in order to find what quotients may be worthwhile in distinguishing a bot and a human. The concepts have been supported using implementation results for cyberpsychology techniques and graphs for cyberpsychology characteristics.

7   Future Work

In future, there may be rampant increase in bots taking over the systems and authenticating themselves as human beings. Several questions similar to the ones described in the cyberpsychology techniques could be useful to prevent inaccurate authentication as they could put bots in a fix. We have concentrated on implementing few of the cyberpsychological techniques due to the ease of implementation. This paper introduces several approaches to discriminate bot behavior from humans, which is really important for the purpose of authentication. In future, we would like to present implementation of the suggested techniques as well. Moreover, certain cyber psychological characteristics have been considered for this article, many more still exist like HQ (Health Intelligence Quotient) DQ (Daring Intelligence Quotient) AQ (Adversity Intelligence Quotient) etc. They could also be used as parameters to tell a human and a bot apart, as psychology comes into play. The introduction of an appropriate fourth factor of authentication could do wonders; the ones that have been introduced in the past are not very sophisticated.

References

1. Kaye, L.: An introduction to cyberpsychology. In: Cyberpsychology, Behavior, and Social Networking, vol. 19, no. 4, p. 294 (2016)
2. Connolly, I., Palmer, M., Barton, H., Kirwan, G.: An Introduction to Cyberpsychology. Taylor and Francis Group, London (2016)
3. Norman, K.: Cyberpsychology: An Introduction to Human Computer Interaction. Cambridge University Press, Cambridge (2008)
4. Griffin, P.: Biometric knowledge extraction for multi-factor authentication and key exchange. Procedia Comput. Sci. **61**(2015), 66–71 (2016)
5. Shahmeer Amir: 4 Methods to Bypass two factor Authentication – Shahmeer Amir, 15 July 2017. https://shahmeeramir.com/4-methods-to-bypass-two-factor-authentication-2b0075d-9eb5f
6. Markowitz, J.: Voice biometrics. Commun. ACM **43**(9), 66–73 (2000)
7. Ali, H., Ibrahim, Y.: Energy conservation using voice recognition. J. Inf. Eng. Appl. 59–62
8. Shah, P.: Image based authentication system. Int. J. Comput. Appl. (0975 – 8887) (2013). International Conference on Communication Technology
9. Dark Reading: Images could change the authentication picture, 03 March 2010
10. Huang, Q., Zhang, P., Wu, D., Zhang, L.: Turbo learning for CaptionBot and DrawingBot. In: Proceeding NIPS 2018 Proceedings of the 32nd International Conference on Neural Information Processing Systems, pp. 6456–6466 (2018)
11. Turing, A.: Computing machinery and intelligence. Mind **49**, 433–460 (1950)
12. Momoh, O.: Turing test, 29 October 2016. https://www.investopedia.com/terms/t/turing-test.asp
13. Jones, R.: Google Has Finally Killed the CAPTCHA, 03 November 2017. https://gizmodo.com/google-has-finally-killed-the-captcha-1793190374
14. Zorkadis, V., Donos, P.: On biometrics-based authentication and identification from a privacy-protection perspective: deriving privacy-enhancing requirements. Inf. Manage. Comput. Secur. **12**(1), 125–137 (2004)

15. Zanuy, M.: On the vulnerability of biometric security systems. IEEE Aerosp. Electron. Syst. Mag. **19**(6), 03–08 (2004)

16. O'Reilly, L.: Google's new CAPTCHA security login raises 'legitimate privacy concerns', 20 February 2015. https://www.businessinsider.com.au/google-no-captcha-adtruth-privacy-research-2015-2

17. Priyadarshini, I.: Cybersecurity risks in robotics. In: Detecting and Mitigating Robotic Cyber Security Risks. IGI Global (2018)

18. Burrows, M., Abadi, M., Needham, R.: A logic of authentication. ACM Trans. Comput. Syst. (TOCS) **8**(1), 18–36 (1990)

19. Ru, W., Eloff, J.: Enhanced password authentication through fuzzy logic. IEEE Expert **12**(6), 38–45 (1997)

20. Yao, F., Yerima, S., Kang, B., Sezer, S.: Fuzzy logic-based implicit authentication for mobile access control. In: SAI Computing Conference. IEEE Explore (2016)

21. Lazzeri, N., Mazzei, D., Cominelli, L., Cisternino, A., Rossi, D.: Designing the mind of a social robot. Appl. Sci. **8**(2), 302 (2018)

22. Wilson, R.: How do robots think?, 09 September 2014. https://www.electronicsweekly.com/blogs/viewpoints/robots-think-2014-09/

23. Revermann, S.: What does the "IQ" stand for in the IQ test?, 21 November 2017. https://education.seattlepi.com/iq-stand-iq-test-3035.html

24. Blackmore, S.: Imitation and the definition of a meme. J. Memetics **2**(11), 159–170 (1998)

25. Priyadarshini, I., Cotton, C.: Internet memes: a novel approach to distinguish human and bot authentication. In: Advances in Intelligent Systems. Springer (2019)

26. Tracey, J.: New AI scores higher on an IQ test than the average human, 29 June 2015. https://www.outerplaces.com/science/item/9224-new-ai-scores-higher-on-an-iq-test-than-the-average-human

27. Emotional intelligence, 04 August 2018. https://en.wikipedia.org/wiki/Emotional_intelligence

28. Kumar, A., Singh, R., Chandra, R.: Emotional intelligence for artificial intelligence: a review. Int. J. Sci. Res. (IJSR) (2017)

29. Moral quotient, 05 December 2017. https://mindgridperspectives.com/2017/12/05/moral-quotient-mq/

30. Hossein, D., Hojjat, R., Zeinab, B., Esmaeil, J., Negar, Y.: Managerial quotient: a systematic review among managers of Tehran University of Medical Sciences. Am. J. Ind. Bus. Manage. **6**(04), 467–479 (2016)

31. Dermatoglyphics Multiple Intelligence Test (DMIT). http://www.mindecodetd.com/iq/

32. Spiritual intelligence, 23 July 2018. https://en.wikipedia.org/wiki/Spiritual_intelligence

33. Priyadarshini, I.: Features and architecture of the modern cyber range: a comparative analysis and survey. University of Delaware (2018)

Multi Criteria Method for Determining the Failure Resistance of Information System Components

Askar Boranbayev[1(✉)], Seilkhan Boranbayev[2], Assel Nurusheva[2],
Yerzhan Seitkulov[2], and Askar Nurbekov[2]

[1] Nazarbayev University, Astana, Kazakhstan
aboranbayev@nu.edu.kz
[2] L.N. Gumilyov Eurasian National University, Astana, Kazakhstan
sboranba@yandex.kz, nurusheva.assel@mail.ru,
erj@mail.ru, nurbekoff@gmail.com

Abstract. The study is devoted to determining the level of failure resistance of the components of an information system in order to increase the reliability of the most vulnerable components. A multi-criteria method is proposed for determining the fault tolerance of information system components. Information system components act as alternatives. Alternatives are described by a set of criteria. Each criteria has its own direction of optimization and a unit of measurement. To obtain comparable scales of criteria values, normalization was done. The estimation of additional coefficients is made. The criteria weights are determined on the basis of expert estimates.

Keywords: Information system · Criterion · Fault tolerance · Expert assessment

1 Introduction

In the period of global digitalization, the problems of sustainability of information infrastructures are becoming increasingly important [1, 2]. In particular, this problem concerns information systems (IS) providing services to a large number of people around the world [3, 4]. For many companies, even a minor system downtime can lead to significant damage (large financial losses, reputational losses, damage to health or human life, etc.) [5].

The most common causes of IS failures include the realization of risks related to information security, as well as risks associated with the reliability and fault tolerance of IS. The calculation of information security risks for critical information infrastructures was previously described in more detail in [6].

The task of ensuring the resiliency of the IS is too complicated and unstructured [7]. For its solution, it is insufficient consideration of one criterion or a point of view that will lead to the solution. It is required the knowledge of the areas that create critical situations for IS and lead to the failure of IS. It is necessary to define the criteria that determine the stability of IS through a specified period.

This article proposes a new model, which takes into account many factors affecting the resiliency and reliability of the IS. The model is aimed at minimizing and

© Springer Nature Switzerland AG 2020
K. Arai et al. (Eds.): FTC 2019, AISC 1070, pp. 324–337, 2020.
https://doi.org/10.1007/978-3-030-32523-7_22

prevention of system downtime, which is related to reliability. In this case, solving a problem requires decision makers to achieve many conflicting goals. All new ideas and possible solutions should be compared according to many criteria [8]. The decision maker needs to determine the choice of alternatives, evaluate them to find the best, rank them from best to worst, group them and describe how well each alternative meets all criteria simultaneously [9].

The MCDM method is one of the most widely used decision-making approaches in many areas. The technique helps to improve the quality of decisions, making the decision-making process more efficient. The method includes the definition of alternatives (components under study), criteria (characteristics allowing comparison of alternatives) and weights (significance) of criteria. Quantitative and qualitative criteria usually describe the alternative for multi-criteria evaluation. Criteria have different units of measurement. Normalization is aimed at obtaining comparable scales of criteria values. Methods of normalization of criteria values are used.

The main advantage of multi-criteria methods is their ability to solve problems associated with various conflicting interests. An overview of commonly used MCDM methods is given in [10].

The use of MCDM methods in risk analysis in various fields is given in [11–19].

2 Risk Assessment to Determine the Level of Reliability of IS

There were considered in detail the methods for assessing risk at an early design stage of IS and their neutralization to improve the reliability of an IS in [20, 21]. These methods allow risk assessment at the early stage of the IS development process and determine the most effective risk mitigation strategies for them in cases where traditional methods cannot be applied. They help determine which individual components need to be investigated to ensure the reliability of IS. As a result of using the methods, the IS user receives a quantitative risk assessment of the system under study. These methods make it possible to eliminate the need to use expensive resources to identify risks. The practical risk reduction strategies, which are stored in the database of the software system, are used to mitigate the risks of IS successful.

Risk identification and neutralization are also used in other areas of activity [22–29].

3 Model for Assessing the Fault-Tolerance of the IS Components

Figure 1 presents the proposed model for evaluating the fault-tolerance of the IS components. This model includes the main stages and processes necessary to solve the multiple criteria problem of determining the fault tolerance of the IS components.

As can be seen from the figure, the implementation of the model includes the following steps:

- Description of the problem. The problem may lie in determining the level of fault tolerance of IS, the IS component, the information and communication infrastructure, the software system, etc.

- Forming a group of experts. Solving the task requires group decision-making of experts with relevant education, qualifications, and work experience.
- Definition of alternatives depends on the description of the problem. In our case, the components of IS were used as alternatives.
- Definition of criteria affecting the resiliency of alternatives.
- Determining the direction of optimization (minimization or maximization) of the criterion.
- Determination of the most preferred values of criteria for alternatives (components of IS).

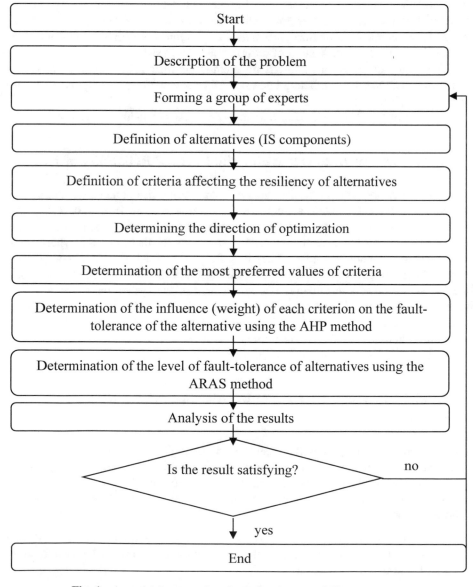

Fig. 1. A model for assessing the fault-tolerance of IS components.

- Determination of the influence (weight) of each criterion on the fault-tolerance of the alternative using the AHP method;
- Determination of the level of fault-tolerance of alternatives using the ARAS method;
- Analysis of the results obtained by the expert group and determination of satisfaction with the outcome. In case of satisfaction with the result, the task is considered as solved one; otherwise, it is necessary to solve it again.

4 The Method of Additive Ratio Assessment (ARAS) in the Process of Making Multi-criteria Decisions

The typical MCDM problem is related to the task of ranking a finite number of solution alternatives, each of which is described in detail in terms of various decision criteria that must be considered simultaneously. According to the ARAS method, the value of the utility function, which determines the complex relative efficiency of a feasible alternative, is directly proportional to the relative influence of the values and weights of the main criteria [9]. The application of the ARAS method is described in [30–34].

The first stage is the formation of a decision-making matrix (DMM). In MCDM, discrete optimization problems that need to be solved are represented by the following DMM preferences for m possible, feasible alternatives (strings) evaluated by n important criteria (columns):

$$
X = \begin{matrix} x_{01} & x_{0j} & x_{0b} \\ x_{i1} & x_{ij} & x_{ib} \\ x_{a1} & x_{aj} & x_{ab} \end{matrix} \; ; i = \overline{0,a}; j = \overline{1,b}, \tag{1}
$$

where a - the number of alternatives, b - the number of criteria describing each alternative, x_{ij} - value representing the performance value of the alternative i in terms of criteria j, x_{0j} - optimal criterion value j [9].

If the optimal value of the criterion j is unknown, then

$$
\begin{aligned}
x_{0j} &= \max_i x_{ij}^*, \, if \max_i x_{ij}^* \, is \, preferable; \\
x_{0j} &= \min_i x_{ij}^*, \, if \min_i x_{ij}^* \, is \, preferable.
\end{aligned} \tag{2}
$$

Usually, the performance values x_{ij} and criteria weights w_j are considered as digital multimeter records. The system of criteria, as well as the values and initial weights of the criteria, are determined by experts. Interested parties can correct information according to their goals and capabilities [9].

Usually, the criteria are of different sizes. The purpose of the next stage is to obtain dimensionless weighted values from the comparative criteria. To avoid the difficulties caused by different criteria sizes, the ratio to the optimal value is used. Various theories describe the relationship to the optimal value. However, the values are displayed either on the interval [0; 1] or interval [0; ∞] by applying the normalization of the digital multimeter. At the second stage, the initial values of all criteria are normalized - determining the values x_{ij} normalized DMM X.

$$X = \begin{array}{ccc} \bar{x}_{01} & \bar{x}_{0j} & \bar{x}_{0b} \\ \bar{x}_{i1} & \bar{x}_{ij} & \bar{x}_{ib} \\ \bar{x}_{a1} & \bar{x}_{aj} & \bar{x}_{ab} \end{array} ; i = \overline{0, a}; j = \overline{1, b}, \tag{3}$$

Criteria, the preferred values of which are the maxima, are normalized as follows:

$$\bar{x}_{ij} = \frac{x_{ij}}{\sum_{i=0}^{a} x_{ij}}. \tag{4}$$

The criteria, the preferred values of which are minimal, are normalized by applying a two-step procedure:

$$x_{ij} = \frac{1}{x_{ij}^*}; \bar{x}_{ij} = \frac{x_{ij}}{\sum_{i=0}^{a} x_{ij}}. \tag{5}$$

When dimensionless criteria values are known, all criteria that initially have different dimensions can be compared.

The third stage - the definition of a normalized weighted matrix - X. It can evaluate the criteria with weights $0 < w_j < 1$. Only reasonable weights should be used because weights are always subjective and influence the decision [9]. Weight values, w_j usually is determined by the method of expert evaluation. The sum of the weights w_j will be limited as follows:

$$\sum_{j=1}^{b} w_j = 1. \tag{6}$$

$$\widehat{X} = \begin{array}{ccc} \widehat{x}_{01} & \widehat{x}_{0j} & \widehat{x}_{0b} \\ \widehat{x}_{i1} & \widehat{x}_{ij} & \widehat{x}_{ib} \\ \widehat{x}_{a1} & \widehat{x}_{aj} & \widehat{x}_{ab} \end{array} ; i = \overline{0, a}; j = \overline{1, b}, \tag{7}$$

The normalized weighted values of all criteria are calculated as follows:

$$\widehat{x}_{ij} = \bar{x}_{ij} w_j; i = \overline{0, a}, \tag{8}$$

where w_j - weight (importance) of the criterion j, but x_{ij} - normalized criterion rating j.

$$S_i = \sum_{j=1}^{n} \widehat{x}_{ij}; i = \overline{0, a}, \tag{9}$$

where S_i - the value of the optimality function of the alternative.

The highest value is the best, and the smallest - the worst. Considering the calculation process, the optimality function S_i has a direct and proportional relationship with values x_{ij} and weights w_j of criteria under study and their relative effect on the final result. Therefore, the higher the value of the optimality function S_i the more effective the alternative becomes. The priorities of alternatives can be determined according to the value S_i. Therefore, when using this method, it is convenient to evaluate and rank alternative solutions [9].

The degree of alternative utility is determined by comparing the analyzed option with the ideally best option S_0. The equation used to calculate the degree of utility K_i alternative A_i is, given below:

$$K_i = \frac{S_i}{S_0}; i = \overline{0, a}, \tag{10}$$

where S_i and S_0 - values of the optimality criterion obtained from the Eq. (9).

It is clear that the calculated values K_i are in the interval [0, 1] and can be ordered in increasing order, which is the desired order of priority. The complex relative efficiency of a possible alternative can be determined according to the values of the utility function.

5 Results and Discussion

The calculations of this article use the results of studies that were tested IS at the University of Kazakhstan. To determine the failures in this IS, the specific components (modules) of the IS were tested. The following components of the IS were highlighted:

(1) IS component № 1 - the component responsible for calculating grades for the students.
(2) IS component № 2 - the component responsible for checking student grades.
(3) IS component № 3 - the component responsible for downloading the rating log.
(4) IS component № 4 - the component responsible for sending the journal by e-mail.

These IS components were used as alternatives in the MCDM task.

5.1 Criteria Selection

The characteristics that affect the level of resiliency of the IS components were used as criteria:

(1) The percent of recoverability/maintainability (the more recoverable/maintainable component of IPS, the more resilient it is).
(2) The failures (the presence of historical failures in the IS component increases the likelihood of repeated failure in the component).

(3) The percent of the failure effect on the working capacity of IS (varying types of failures can lead to different types of damage to production, from minor problems with the operation of IS to complete failure of the entire IS, so it is essential to consider the seriousness of historical failures).

(4) The processed data backup (the presence of redundancy allows to recover components affected more quickly).

(5) Compliance with legal requirements (availability of a local or international certificate of accordance with the requirements of information security/fault tolerance/reliability of IS ensures that all relevant procedures in the organization are documented and that they comply with state or international requirements).

The AHP method was used to calculate the significance of the influence of each criterion on the fault tolerance of IS: the criteria were compared with each other. Five experts with IT education and experience of at least five years evaluated the criteria. Table 1 presents the comprehensive results of a paired comparison of one of the experts.

Table 1. Pairwise comparison of criteria weights.

		X_1	X_2	X_3	X_4	X_5	w
The percent of recoverability/maintainability	X_1	1	1	1	½	1	0.1714
Compliance with legal requirements	X_2	1	1	1	1	½	0.1714
The percent of the failure effect on the working capacity of IS	X_3	1	1	1	½	1	0.1714
The processed data backup	X_4	2	1	2	1	½	0,2261
The failures	X_5	1	2	1	2	1	0,2598

Similar calculations were carried out by the remaining four experts. The results of the comparison of the weights of the criteria are shown in Table 2.

Table 2. Results of comparison of criteria weights.

	E_1	E_2	E_3	E_4	E_5	w	Normalized weight
X_1	0,15	0,25	0,17	0,15	0,20	0,18	0,653
X_2	0,14	0,13	0,17	0,17	0,20	0,16	0,574
X_3	0,16	0,14	0,17	0,23	0,15	0,17	0,607
X_4	0,16	0,19	0,23	0,20	0,26	0,21	0,737
X_5	0,40	0,29	0,26	0,26	0,20	0,28	1
Total						1,00	3,57

Figure 2 shows the weights of the criteria that affect the resiliency of the IS components.

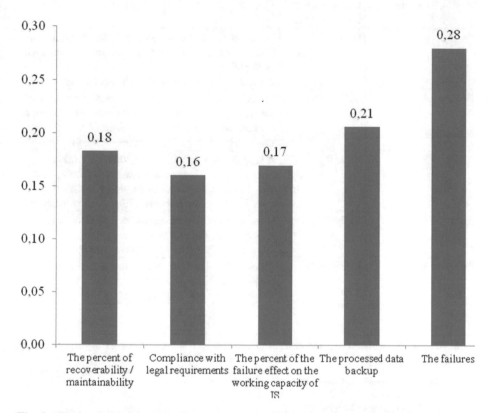

Fig. 2. The level of influence (weight) of the criteria on the fault tolerance of the IS components.

In this article, the failure is characterized by attributes such as:

- the percent of recoverability/maintainability (severity of failure);
- the percent of the failure effect on the working capacity of IS (failure priority).

Severity indicates how serious the defect and how it can affect the performance of IS. The severity of the fault is usually distinguished in five types (Table 3).

Priority indicates the order for each task to eliminate the defect. The priority of the fault is usually distinguished in three types (Table 4).

Table 3. The severity of failures [20].

Title	Rank failure severity	The percent of recoverability/ maintainability	Description
Blocker	5	0%–40%	The most severe failure (error) in which work with IS is impossible. This kind of error must be corrected without fail. IS is difficult to restore due to the severity of the error
Critical	4	41%–70%	Critical error in which a specific part of the IS does not work. This problem must be solved to continue working with the primary functions of the system The recoverability of IS is below average
Major	3	71%–80%	A type of error in which something is working incorrectly, but is not particularly dangerous, since there is the possibility of continuing work using other input points. Recoverability of IS is average
Minor	2	81%–90%	Usually, minor errors do not disrupt IS; a problem may occur in the user interface The recoverability of IS is above average
Trivial	1	91%–100%	An error that does not pose a threat to the IS is usually a problem of a third-party library or service The level of recoverability of IS is high

Table 4. The scale of ranking the percent of the failure effect on the working capacity of IS (priority) is based on [35].

Description	The percent of the failure effect on the working capacity of IS (priority)	Definition
Extremely dangerous	100%	The failure may result in the death of a person or critical damage to infrastructure
Very dangerous	90% 80%	The failure to do so could result in serious injury or severe infrastructure damage due to service interruptions
Dangerous	70% 60%	The failure can result in minor or moderate injury with a high degree of personal dissatisfaction or significant infrastructure problems that need repair
Average danger	50%	The failure can result in minor injuries with some people's discontent or significant infrastructure problems
Danger from low to moderate	40% 30%	The failure can lead to minor injuries or not to injuries, but it annoys customers or leads to minor infrastructural problems that can be overcome by small changes in infrastructure or business
Slight danger	20%	The failure cannot lead to injury, and the client is not aware of the problem; However, there is a possibility of minor injuries
No danger	10%	The failure is not harmful and does not affect the infrastructure

Tables 5 and 6 were formed based on Tables 3 and 4. It should be noted that "1" is used as the minimum value in the "0 - optimal value" line since dividing by "0" is impossible. When calculating the "The percent of recoverability/maintainability (severity of failure)" and "The percent of the failure effect on the working capacity of IS (failure priority)," the task used averaged values of the results of the survey of five experts.

Table 5. Measurement results in IS (initial DMM X)

IS components	Criteria				
	The percent of recoverability/ maintainability	The failures	The percent of the failure effect on the working capacity of IS	The processed data backup	Compliance with legal requirements
	X_1	X_2	X_3	X_4	X_5
Units	%	PC.	%	PC.	Yes/No (1/0)
Optimization direction	Max.	Min.	Min.	Max.	Max.
Weight of criterion	0,18	0,28	0,17	0,21	0,16
0 – optimal value	100	1	1	3	1
1	81	14	20	1	1
2	91	1	20	1	1
3	81	6	10	1	1
4	91	5	10	1	1

Table 6. Normalized measurement values in IS (normalized DMM X)

IS components	Criteria				
	The percent of recoverability/ maintainability	The failures	The percent of the failure effect on the working capacity of IS	The processed data backup	Compliance with legal requirements
	X_1	X_2	X_3	X_4	X_5
Weight of criterion	0,18	0,28	0,17	0,21	0,16
0 – optimal value	0,225	0,410	0,769	0,43	0,20
1	0,182	0,029	0,038	0,14	0,20
2	0,205	0,410	0,038	0,14	0,20
3	0,182	0,068	0,077	0,14	0,20
4	0,205	0,082	0,077	0,14	0,20

Weighted normalized measurement values in IS (weighted normalized DMM), calculated by formulas (9)–(10), are listed in Table 7 and Fig. 3.

Table 7. Weighted normalized measurement values in IS (weighted normalized DMM) and the results of the decision

IS components	The percent of recoverability/ maintainability	The failures	The percent of the failure effect on the working capacity of IS	The processed data backup	Compliance with legal requirements	S	к	IS component rank
	X_1	X_2	X_3	X_4	X_5			
0 – optimal value	0,041	0,115	0,131	0,088	0,032	0,407	1	
1	0,033	0,008	0,007	0,029	0,032	0,110	0,269	4
2	0,037	0,115	0,007	0,029	0,032	0,221	0,541	1
3	0,033	0,019	0,013	0,029	0,032	0,127	0,312	3
4	0,037	0,023	0,013	0,029	0,032	0,135	0,332	2

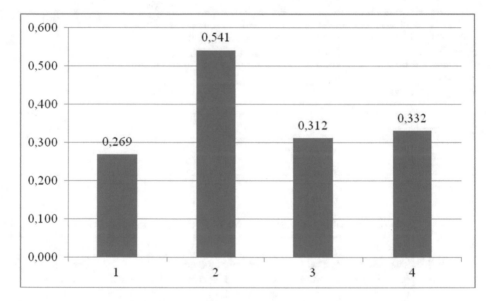

Fig. 3. Fault-tolerance level of IS components.

The order of priorities of the studied components can be represented as: $p_2 \succ p_4 \succ p_3 \succ p_1$. It means that the most fault-tolerant IS component is component № 2, and the least fault-tolerant IS component is component № 1. It can be stated that fault-tolerance of IS component № 2 is only 54% of the optimally fault-tolerant IS component, and fault-tolerance of the worst IS component № 1 - only 27%.

6   Conclusion

There is proposed method to solve the problem of determining the level of fault-tolerance and reliability of IS in this article.

As described above, the failure of IS can lead not only to the disruption or cessation of its operation but also to more global consequences. Thus, the presence of such risks leads to the need to find effective methods for their assessment.

The proposed method allows determining the complex fault-tolerance and reliability of IS components.

Therefore, when using this method, it is convenient to evaluate and rank alternative solutions. The degree of alternative is determined by comparing the analyzed variant with the ideally best one.

It can be stated that the ratio with the optimal alternative can be used when searching for ranking alternatives and finding ways to improve alternatives.

The best solutions can be achieved by applying scientific methods involving a large amount of information and calculations.

In our case, each of the five experts presented their own opinion on the values of the criteria. The significance of expert assessments was evaluated by using the AHP method. The proposed approach is effective since it can take into account some criteria with different directions of optimization. Thus, it is a powerful tool for solving such problems.

The proposed model is aimed at solving the problem of calculating the resiliency of IS. Five main criteria were defined: "the percent of recoverability/maintainability", "the failures", "the percent of the failure effect on the working capacity of IS", "the processed data backup" and "compliance with legal requirements".

The study shows that the most fault-tolerant IS component is component No. 2 (level 0.541), less fault-tolerant components of the component are components No. 3 and No. 4 (levels from 0.312 to 0.332) and the least fault-tolerant component is component No. 1 (level No. 0.269).

The model is proposed to be used in the future to calculate the level of resiliency of other information and automated systems.

Acknowledgment. This work was done as part of a research grant №AP05131784 of the Ministry of Education and Science of the Republic of Kazakhstan for 2018–2020.

References

1. Boranbayev, S., Goranin, N., Nurusheva, A.: The methods and technologies of reliability and security of information systems and information and communication infrastructures. J. Theor. Appl. Inf. Technol. **96**(18), 6172–6188 (2018)
2. Boranbayev, A., Mazhitov, M., Kakhanov, Z.: Implementation of security systems for prevention of loss of information at organizations of higher education. In: Proceedings - 12th International Conference on Information Technology: New Generations, ITNG 2015, pp. 802–804 (2015)
3. Boranbayev, A., Baidyussenov, R., Mazhitov, M.: Software architecture for in-house development of a student web portal for higher education institution in Kazakhstan. Adv. Intell. Syst. Comput. **738**, 759–760 (2018)
4. Boranbayev, A., Boranbayev, S., Nurusheva, A., Yersakhanov, K.: The modern state and the further development prospects of information security in the Republic of Kazakhstan. Information Technology – New Generations, pp. 33–38. Springer, Cham (2018)

5. Boranbayev, A., Boranbayev, S., Yersakhanov, Y., Nurusheva, A., Taberkhan, R.: Methods of ensuring the reliability and fault tolerance of information systems. In: Information Technology–New Generations, pp. 729–730. Springer, Cham (2018)

6. Turskis, Z., Goranin, N., Nurusheva, A., Boranbayev, S.: A Fuzzy WASPAS-based approach to determine critical information infrastructures of EU sustainable development. Sustainability **11**(2), 424 (2019)

7. Boranbayev, A., Boranbayev, S., Nurusheva, A., Yersakhanov, K.: Development of a software system to ensure the reliability and fault tolerance in information systems. J. Eng. Appl. Sci. **13**(23), 10080–10085 (2018)

8. Turskis, Z., Zavadskas, E.K., Peldschus, F.: Multi-criteria optimization system for decision making in construction design and management. Inzinerine Ekonomika-Eng. Econ. **1**, 7–17 (2009)

9. Zavadskas, E.K., Turskis, Z.: A new additive ratio assessment (ARAS) method in multicriteria decision-making. Technol. Econ. Dev. Econ. **16**(2), 159–172 (2010)

10. Figueira, J., Greco, S., Ehrgott, M. (eds.): Multiple Criteria Decision Analysis: State of the Art Surveys. Springer, New York (2005)

11. Erdogan, M., Kaya, I.: Prioritizing failures by using hybrid multi criteria decision making methodology with a real case application. Sustain. Cities Soc. **45**, 117–130 (2019). https://doi.org/10.1016/j.scs.2018.10.027

12. Lo, H.-W., Liou, J.J.H.: A novel multiple-criteria decision-making-based FMEA model for risk assessment. Appl. Soft Comput. **73**, 684–696 (2018). https://doi.org/10.1016/j.asoc.2018.09.020

13. Han, Y., Deng, Y.: A hybrid intelligent model for assessment of critical success factors in high-risk emergency system. J. Ambient Intell. Humanized Comput. **9**(6), 1933–1953 (2018). https://doi.org/10.1007/s12652-018-0882-4

14. Li, M., Wang, J.L., Li, Y., Xu, Y.C.: Evaluation of sustainability information disclosure based on entropy. Entropy **20**(9) (2018). https://doi.org/10.3390/e20090689

15. Pamucar, D., Stevic, Z., Sremac, S.: A new model for determining weight coefficients of criteria in MCDM models: full consistency method (FUCOM). Symmetry-Basel **10**(9) (2018). https://doi.org/10.3390/sym10090393

16. Mardani, A., Nilashi, M., Zakuan, N., Loganathan, N., Soheilirad, S., Saman, M.Z.M., Ibrahim, O.: A systematic review and meta-Analysis of SWARA and WASPAS methods: theory and applications with recent fuzzy developments. Appl. Soft Comput. **57**, 265–292 (2017). https://doi.org/10.1016/j.asoc.2017.03.045

17. Zyoud, S.H., Fuchs-Hanusch, D.: A bibliometric-based survey on AHP and TOPSIS techniques. Expert Syst. Appl. **78**, 158–181 (2017). https://doi.org/10.1016/j.eswa.2017.02.016

18. Govindan, K., Jepsen, M.B.: ELECTRE: a comprehensive literature review on methodologies and applications. Eur. J. Oper. Res. **250**(1), 1–29 (2016). https://doi.org/10.1016/j.ejor.2015.07.019

19. Celik, E., Gul, M., Aydin, N., Gumus, A.T., Guneri, A.F.: A comprehensive review of multi criteria decision making approaches based on interval type-2 fuzzy sets. Knowl.-Based Syst. **85**, 329–341 (2015). https://doi.org/10.1016/j.knosys.2015.06.004

20. Boranbayev, A., Boranbayev, S., Nurusheva, A.: Development of a software system to ensure the reliability and fault tolerance in information systems based on expert estimates. Adv. Intell. Syst. Comput. **869**, 924–935 (2018)

21. Boranbayev, A., Nurusheva, A., Yersakhanov, K., Seitkulov, Y.: A software system for risk management of information systems. In: Proceedings of the 2018 IEEE 12th International Conference on Application of Information and Communication Technologies (AICT 2018), 17–19 October 2018, Almaty, Kazakhstan, pp. 284–289 (2018)

22. Lough, K.G., Stone, R., Turner, I.: The risk in early design method. J. Eng. Des. **20**(2), 155–173 (2009)
23. Grantham Lough, K., Stone, R., Tumer, I.: Prescribing and implementing the risk in early design (RED) method. In: Proceedings of DETC 2006, Philadelphia, USA, - Philadelphia, 2006, pp. 431–439 (2006). https://doi.org/10.1115/detc2006-99374
24. Krus, D., Grantham, K.: Failure prevention through the cataloging of successful risk mitigation. Strategies **13**, 712–721 (2013). https://doi.org/10.1007/s11668-013-9728-8
25. Krus, D.A.: The risk mitigation strategy taxonomy and generated risk event effect neutralization method. Ph.D thesis, Missouri, 176 p. (2012)
26. Lough, K.G., Stone, R.B., Tumer, I.Y.: Implementation procedures for the risk in early design (red) method. J. Ind. Syst. Eng. **2**(2), 126–143 (2008)
27. Lough, K.G., Stone, R.B., Tumer, I.Y.: The risk in early design (RED) method: likelihood and consequence formulations. In: Proceedings of DETC 2006: ASME 2005 International Design Engineering Technical Conferences and Computers and Information in Engineering Conference, pp. 1–11 (2007). https://doi.org/10.1115/detc2006-99375
28. Vucovich, J.P., et al.: Risk assessment in early software design based on the software function-failure design method. In: Proceedings of the 31st Annual International Computer Software and Applications Conference. Institute of Electrical and Electronics Engineers (IEEE), August 2007 (2007)
29. Grantham, K., Elrod, C., Flaschbart, B., Kehr, W.: Identifying risk at the conceptual product design phase: a web-based software solution and its evaluation. Mod. Mech. Eng. **2**, 25–34 (2012)
30. Radovic, D., Stevic, Z., Pamucar, D., Zavadskas, E.K., Badi, I., Antucheviciene, J., Turskis, Z.: Measuring performance in transportation companies in developing countries: a novel rough ARAS model. Symmetry-Basel **10**(10) (2018). https://doi.org/10.3390/sym10100434
31. Ecer, F.: An integrated fuzzy AHP and ARAS model to evaluate mobile banking services. Technol. Econ. Dev. Econ. **24**(2), 670–695 (2018). https://doi.org/10.3846/20294913.2016.1255275
32. Dahooie, J.H., Abadi, E.B.J., Vanaki, A.S., Firoozfar, H.R.: Competency-based IT personnel selection using a hybrid SWARA and ARAS-G methodology. Hum. Factors Ergon. Manuf. Serv. Ind. **28**(1), 5–16 (2018). https://doi.org/10.1002/hfm.20713
33. Stanujkic, D., Zavadskas, E.K., Karabasevic, D., Turskis, Z., Kersuliene, V.: New group decision-making ARCAS approach based on the integration of the SWARA and the ARAS methods adapted for negotiations. J. Bus. Econ. Manage. **18**(4), 599–618 (2017). https://doi.org/10.3846/16111699.2017.1327455
34. Turskis, Z., Kersuliene, V., Vinogradova, I.: A new fuzzy hybrid multi-criteria decision-making approach to solve personnel assessment problems. Case study: director selection for estates and economy office. Econ. Comput. Econ. Cybern. Stud. Res. **51**(3), 211–229 (2017)
35. Immawan, T., Sutrisno, W., Rachman A.K.: Operational risk analysis with Fuzzy FMEA (Failure Mode and Effect Analysis) approach (Case study: Optimus Creative Bandung). In: MATEC Web Conference, vol. 154, p. 01084. EDP Sciences (2018)

Time-Invariant Cryptographic Key Generation from Cardiac Signals

Sarah Alharbi[✉], Md Saiful Islam, and Saad Alahmadi

Department of Computer Science, College of Computer and Information
Sciences, King Saud University, Riyadh 11543, Saudi Arabia
alharbisaral00@gmail.com

Abstract. Cardiac signal (also known as ECG signal) attracted researchers for
using it in generating cryptographic keys due to its availability and its intrinsic
nature of every individual. However, the intra-individual variance of ECG signal
decreases the possibility of getting a time-invariant key for each individual and
increases decryption errors in case of using it in symmetric cryptography. In this
paper, we propose a time-invariant cryptographic key generation approach
(TICK) that uses a novel method for reducing the intra-individual variance in the
real-valued ECG features of multiple sessions. Also, it uses a quantization
method for converting the improved ECG features to binary sequences with high
randomness. We have tested the approach on a multi-session database. Experimental results show its viability to improve the reliability of keys up to 96.80%
using across-sessions data and up to 98.69% using within-session data. We
verified the randomness using five of U.S. National Institute of Standards and
Technology statistical tests and the generated keys passed all tests. Also, we
verified the randomness using min-entropy, and the generated keys offer entropy
of ~ 1.

Keywords: Cryptographic key · ECG features · Quantization · Time-invariant
keys · Cardiac signals

1 Introduction

Cardiac signal (ECG) is an emerging biometric modality used in generating cryptographic keys. It is unique for each person due to the differences in the delay between
each part of the heart, physiological factors of the heart muscle, and the timing of blood
pumping in and out of the heart [1]. This uniqueness is necessary to ensure deriving
unique keys. Also, its availability with users eliminates the need to store it in a storage
medium which ensures the secrecy of keys. Furthermore, it eliminates the need to
consume resources in the key distribution process since capturing cardiac signals to
generate keys is possible in convenient way such as from fingers of a user [2].

Cardiac signal offers several other advantages over traditional modalities (such as
fingerprint and face). It represents the electrical activity of the heart and is a necessary
sign of life. Thus, ECG possesses the quality of universality which is a trait that many
other biometrics lacks such as fingerprints and hand geometry that are not available for
amputees. Also, ECG signal is difficult to capture without cooperation from the person

© Springer Nature Switzerland AG 2020
K. Arai et al. (Eds.): FTC 2019, AISC 1070, pp. 338–352, 2020.
https://doi.org/10.1007/978-3-030-32523-7_23

which makes replicating difficult by an adversary providing immunity to presentation attacks [3].

The cardiac signals are time-variant of every individual even for healthy people. In other words, it has an intra-individual variance which occurs among specimens of the same individual. Indeed, this variance negatively affects the possibility of regenerating keys or what is called reliability which is essential in the case of symmetric cryptography. The intra-individual variance becomes worse when specimens of a person come from multiple sessions where the average time interval among sessions is more than one month. Studying this case is necessary to measure the viability of a proposed approach to generate a highly time-invariant key.

There are some recent works successfully addressed this issue, but they perform poorly in case of using cardiac signals of multiple sessions. Therefore, in this paper, we focus on developing a method for generating time-invariant keys from cardiac signals that achieve the following goals:

- Reducing the intra-individual variance of across-sessions ECG samples.
- Use a method for converting ECG specimens to binary sequences where this method increases the randomness of keys to avoid guessing it by an adversary.

The remainder of the paper is organized as follows. In Sect. 2, we discuss the state of art approaches for generating cryptographic keys from cardiac signals. In Sect. 3, we describe our proposed method for generating a time-invariant cryptographic key from the electrocardiogram. The dataset, experiments, and results provided in Sect. 4. The discussion about the results is given in Sect. 5. Finally, we present the conclusions in Sect. 6.

2   Related Works

Cryptographic keys derived from cardiac signals are the binary representations of features extracted from this data. There are three different techniques for extracting features from cardiac signals which are as follows [4]. Firstly, the fiducial-based techniques that uses characteristic points of a cardiac signal to derive features such as inter-pulse interval (IPI) is the time interval between two successive R peaks. Figure 1 shows three heartbeats where each one composite of five peaks which are P, Q, R, S, and T. Second, the non-fiducial-based techniques that extract discriminative features from cardiac waveform without localizing fiducial points such as heartbeat shape approach (HBS) [5]. Third, the partially fiducial-based techniques that use R peaks for segmentation of cardiac signals to heartbeats such as wavelet distance measure approach (WDM) [6].

After feature extraction, there are various approaches for converting the extracted features to binary sequences which can be categorized as follows. The first category is the hashing approaches which convert each extracted feature to a binary string using a hashing equation. Then these binary strings concatenated to form the cryptographic key. For example, the authors in [7] proposed an IPI-based key generation approach that use IPIs to generate 128-bit binary random sequences as keys using modular encoding method [8]. In this approach, they extracted IPIs of specific size then they

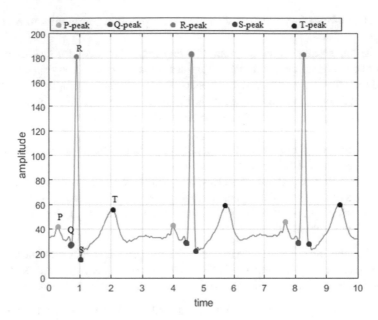

Fig. 1. Three segments of a cardiac signal where each segment has the five peaks are P, Q, R, S, and T, respectively.

applied the modular method which is Mod (IPI, power (2, x)) on each IPI where x is the number of bits generated for this IPI, Mod is the arithmetic modular, and power is the ordered pair of (x, y) where x is to the yth power. This function transforms each feature into an integer number in the range [0, power (2, x)-1]. Then this integer transformed into its binary equivalent to getting the x-bit binary sequence of this IPI. In the last step, the binary sequences of all IPIs of each user concatenated to form the cryptographic key for that user. The hashing approaches perform well with ECGs that captured under the same session conditions. However, this is not the case when ECGs captured under multiple session conditions. Indeed, hashing approaches does not allow for having an acceptable range of intra-individual variance which is inevitable in case of using ECG data of multiple sessions.

The second category is seed-based approaches that depend on generating seeds of keys before generating cryptographic keys. The authors claimed that this step can give an extra level of security for generated keys. There are two types of such methods which are as follows. First, approaches used conventional mechanisms for generating seeds such as PRNG and AES [9, 10]. Second, approaches used innovative mechanisms for generating seeds. For example, the authors in [11] used a mechanism to determine valid features to generate seeds. Then these seeds were used as input to pseudo-random bit generator. Also, the authors in [12] proposed an approach which is a starting point of considering the intra-individual variance to generate time-invariant keys. It depends on calculating the average of each discriminating feature extracted from heartbeats of different starting times of the training data. These features form a reference or model for each participant. Once a participant submits an ECG sample, the values of each feature in seed is determined by the similarity between the model and the submitted sample. By comparing seed-based approaches to other categories, it performs well to generate time-invariant keys. However, the seed-based approaches are avoidable in wireless body sensor networks which constructed from a set of sensors that send medical data to an E-health system [13]. This avoidance due to it consumes limited resources of memory and computation capability available to those sensors [14].

The third category is quantization approaches that convert all features values within a specified range to a common binary string. Also, it converts features values in a different range of values to a different common binary string. Figure 2 shows a quantization method that uses four quantization intervals (non-overlapping ranges of values) to assign a common binary string to all values that fall in the same interval. For example, the authors in [15] proposed 16 quantization intervals for IPIs extracted from cardiac signals of participants. They used the standard deviation and the mean of the training data to assign each feature value to an interval.

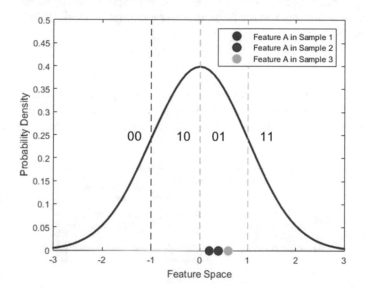

Fig. 2. Quantization method convert three values of a feature in three intraclass

This assignment represented by converting feature value to the gray code of that interval. Then concatenation function used to concatenate gray codes of features to form the cryptographic key. Another example, the authors in [3] proposed a different approach for quantization called interval optimized mapping bit allocation (IOMBA). It adjusts the thresholds of the quantization intervals according to the required security level of keys. In general, the quantization approaches perform well to generate time-invariant keys and does not use an intermediate process that consumes resources. However, it performs poorly when multiple ECG data used where the intra-individual variance maximized.

3   Time-Invariant Cryptographic Key (TICK) Generation

In this section, we present a new cryptographic key generation method, called time-invariant cryptographic key (TICK) generation. It depends on decreasing the intra-individual variance before deriving a key. This differs from the existing techniques that directly convert the extracted features to keys. The raw ECG signal can be captured using commercially available finger-based ECG device [2]. Then features are extracted using features extraction method such as wavelet distance measure [6].

To derive keys from these features' vectors, TICK has two phases which are enrollment phase and key generation phase that shown in Fig. 3. At the enrolment phase, a transformation matrix (W) and the quantization thresholds are computed for each set of participants. Section 3.1 presents the steps of generating W in detail. Then, at the key generation phase, a specimen of an enrolled participant is converted to a key using the following steps. First, the matrix W is used for reducing the intra-individual variance of features vector s extracted from the participant specimen. This step called variance improvement where the features vector s multiplied by W. Thus, the improved features vector e is:

$$e = Ws \tag{1}$$

Then, each feature in e is converted to a binary string using the precomputed quantization thresholds. Section 3.2 presents the quantization process in detail. Finally, these binary strings are concatenated to form the binary sequence of the cryptographic key.

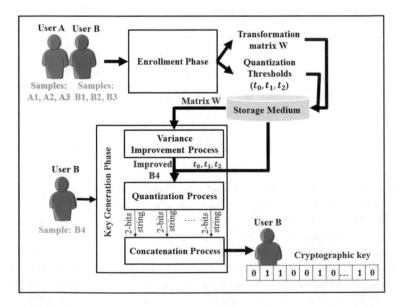

Fig. 3. Illustration of participants' enrollment and key generation phases

3.1   Generating the Transformation Matrix (W)

At the enrollment phase, we use specimens from participants to get the transformation matrix (W) after extracting features using a partially fiducial based method. Let M be a $(m \times n)$ matrix, where m is the number of features vectors extracted from participants specimens, and n is the number of features f_j where $j = 1, 2, ..., n$. At this phase, we need an equal number of specimens from each participant, where p is the number of features vectors related for each participant i where $i = 1, 2, 3, ..., m/p$.

There are three steps to get the transformation matrix W from M which are as follows:

Step 1: Calculating the inter-individual variance for all users in M. First, we calculate μ which is the overall mean of the mean vectors μ_i of the p samples that belong to each user i. Then, we compute the inter-individual variance S_B as follows:

$$S_B = \sum_{i=1}^{m/p} p\,(\mu_i - \mu)(\mu_i - \mu)^T \tag{2}$$

Step 2: Calculating the intra-individual variance for all users in M. First, we calculate the mean-centering for each user i (d_i) by subtracting (μ_i) from the matrix P samples that belong to the user i as follows:

$$d_i = P_i - \mu_i \tag{3}$$

After that, we compute the intra-individual variance SW as follows:

$$S_W = \sum_{i=1}^{m} d_i^T \times d_i \tag{4}$$

Step 3: Multiplying the inverse of the intra-individual variance matrix (S_W) by the inter-individual variance matrix (S_B) to get the transformation matrix W:

$$W = S_W^{-1} \times S_B \tag{5}$$

Then we store the matrix W in a storage medium where all the participants in M become enrolled in the system.

3.2   The Quantization Process

Quantization is a process of converting the features vectors extracted from ECG specimens to keys. The quantization process converts all features of a features vector v where $v = [f_1, f_2, f_3, ..., f_n]$ that are within specified range to a common binary string.

In this paper, one of our goals is to use a quantization process that ensures the randomness in the generated keys. According to the frequency test in the NIST suite, the truly random sequence is that have an approximately equal number of 0s and 1s [16]. Thus, in the quantization method of TICK, we assign the binary strings that have an equal number of 0s and 1s to the intervals that most of the features lie in it. This step is to ensure that the final binary sequence (key) has an approximately equal number of 0s and 1s. However, passing the frequency test is essential to pass most of the subsequent tests of the NIST suite which is a benchmark for randomness tests [16].

The steps of the quantization method in TICK are as follows. First, determine the $2^n - 1$ thresholds (t) where n is an even number of bits dedicated for each feature. In case of 2-bit quantization, this requires three thresholds to form four quantization intervals. Each one of the four quantization intervals can be assigned to either 00, 01, 10, or 11. In the quantization process of TICK, we force having randomness in the final binary sequence through assigning these strings to the quantization intervals in specific manner. Indeed, having equal number of 0s and 1s in the building blocks of a binary sequence (key), ensures distributing the 0s and 1s of this key in quite uniform manner. Thus, we convert the features values of high frequency rate to binary strings such as 01 and 10 where the number of 0s and 1s in the string are equal. By drawing the histogram of the dataset used in the enrollment phase, we can notice the two ends of the range of features values that have high frequency. Thus, the thresholds are $t_1 = \alpha$, $t_2 = \beta$, and $t_0 = (t_1 + t_2)/2$ where α and β are the two ends of this range of values. After that, we convert each feature f_j to a binary string using the following piece-wise function $c^2(f)$:

$$
c^2(f) = \begin{cases}
11 & \text{if } f_j \geq t_1 \\
01 & \text{if } t_0 \leq f_j \leq t_1 \\
10 & \text{if } t_2 \leq f_j \leq t_0 \\
00 & \text{if } f_j \leq t_2
\end{cases}
\tag{6}
$$

Finally, we concatenate the binary strings to form the final cryptographic key. However, quantizing features to binary strings of odd length reduces the randomness of the final binary sequence due to the inequality in the number of 0s and 1s. Regarding the 4-bits quantization case, we use 15 thresholds which are $t_0 - t_{14}$ and 16 quantization intervals. In this case, $t_1 = \alpha$, $t_2 = \beta$, $t_0 = (t_1 + t_2)/2$, and the other thresholds can be categorized into two groups. The first group is the thresholds on the right of t_1 and the second group is the thresholds on the left of t_2. The first group is $t_3 = t_1 + \gamma$, $t_5 = t_3 + \gamma$, ..., $t_{13} = t_{11} + \gamma$ where γ is a constant value. The second group is $t_4 = t_2 - \gamma$, $t_6 = t_4 - \gamma$, ..., $t_{14} = t_{12} - \gamma$. After that, we convert each feature f_j to a binary string using the following piece-wise function $c^4(f)$:

$$c^4(f) = \begin{cases} 1111 & \text{if } f_j \geq t_{13} \\ 0001 & \text{if } t_{11} \leq f_j \leq t_{13} \\ 1101 & \text{if } t_{09} \leq f_j \leq t_{11} \\ 1011 & \text{if } t_{07} \leq f_j \leq t_{09} \\ 1000 & \text{if } t_{05} \leq f_j \leq t_{07} \\ 1001 & \text{if } t_{03} \leq f_j \leq t_{05} \\ 0110 & \text{if } t_{01} \leq f_j \leq t_{03} \\ 1100 & \text{if } t_0 \leq f_j \leq t_{01} \\ 1010 & \text{if } t_{02} \leq f_j \leq t_0 \\ 0011 & \text{if } t_{04} \leq f_j \leq t_{02} \\ 0101 & \text{if } t_{06} \leq f_j \leq t_{04} \\ 0111 & \text{if } t_{08} \leq f_j \leq t_{06} \\ 0100 & \text{if } t_{10} \leq f_j \leq t_{08} \\ 0010 & \text{if } t_{12} \leq f_j \leq t_{10} \\ 1110 & \text{if } t_{14} \leq f_j \leq t_{12} \\ 0000 & \text{if } f_j \leq t_{14} \end{cases} \tag{7}$$

Finally, we concatenate the binary strings to form the final cryptographic key.

4   Experimental Results

In this section, we assess the performance of the proposed TICK generation approach in terms of reliability and randomness. We use the extended version of the FECG database which is a multisession in-house database of ECG records captured from fingers was built by the authors in [2, 17] and extended in [18]. This database was built using a commercially available finger-based ECG device to capture each record of ECG signal for fifteen seconds from the thumbs of a subject at a sampling frequency of 250 Hz. This device can capture ECG records directly by placing participant thumbs of both hands on the dry conducting electrodes without requiring any other preparation. The extended version of the FECG database contains 656 records from 164 individuals collected in two sessions. The average interval between the two sessions is more than two months. We use the wavelet distance measure (WDM) to extract features from ECG records of the extended FECG database [5]. For each ECG record, the locations of R-peaks and the required heartbeat rate must be given as inputs to this method. These two inputs are used to segment ECG record into heartbeats which composite of the five peaks P, Q, R, S, and T. Figure 1 shows the segmentation process to three heartbeats where each one is composite of the P, Q, R, S, and T waves. Then the average of the heartbeats is taken, and Daubechies 3 wavelet filter is used to remove noisy heartbeats. Finally, detail coefficients of the discrete wavelet transform are used to get the features of the average heartbeat.

4.1   Randomness

Keys can be considered random if they are computationally and statistically random. A key is computationally random if it has enough entropy which makes it not easily predicted by an adversary. According to the recommendation by NIST [19], a better entropy measure to use is min-entropy which defined as follows:

$$H_\infty(x) = -\log(\max p(y)) \tag{8}$$

where $\mathbf{x}$ is a binary sequence, $y = \{0, 1\}$, and $p(y)$ is the probability of occurring that event.

Also, this key is statistically random if it passes statistical randomness tests such as the NIST statistical test suite [15]. The NIST benchmark developed for cryptographic random and pseudo-random number generator applications. If a pass rate (also called P-value) of one of these tests is less than the threshold (0.01), then the binary sequence fails to pass this test. In this paper, we use five of NIST tests for evaluating randomness in the keys generated by TICK which are as follows. First, the frequency test (F-Test) specifies whether the number of 0s and 1s in a key are approximately the same as would be anticipated for a truly random sequence. Second, the frequency test within a block (B-Test) specifies whether the number of 1s in an M-bit block is approximately $M/2$, as would be anticipated for a truly random sequence. Third, the run test (R-Test) specifies whether the number of runs of various lengths in a key is as expected for a random sequence where the run is an uninterrupted sequence of identical bits. Fourth, discrete Fourier transform test (D-Test) which detects the repetitive patterns that are near each other in the tested sequence that would indicate a deviation from the assumption of randomness. Fifth, cumulative sums test (C-Test) specifies whether the cumulative sum of the partial sequences occurring in the tested sequence is too large or too small relative to the expected behavior of that cumulative sum for random sequences.

To study the viability of the proposed quantization process enhances the randomness. We studied the effect of two quantization functions $c_1^2(f)$ and $c_2^2(f)$ on the randomness of keys generated using across-sessions data and using several values of thresholds. The function $c_1^2(f)$ is as written in Eq. 6 and the function $c_2^2(f)$ is:

$$c_2^2(f) = \begin{cases} 01 & \text{if } f_j \geq t_1 \\ 11 & \text{if } t_0 \leq f_j \leq t_1 \\ 00 & \text{if } t_2 \leq f_j \leq t_0 \\ 10 & \text{if } f_j \leq t_2 \end{cases} \tag{9}$$

Where, the intervals that contain most of the features of a vector v assigned to binary strings that have an unequal number of 0s and 1s. The comparison in term of min-entropy and the passing rates of the above NIST tests. Figure 4 shows the randomness of keys resulted from 2-bit quantization process and using across-sessions data. In this experiment, we used three groups of thresholds which are $\{t_0 = 0, t_1 = 0.05, t_2 = -0.05\}$, $\{t_0 = 0, t_1 = 0.3, t_2 = -0.3\}$, and $\{t_0 = 0, t_1 = 0.7, t_2 = -0.7\}$. The presented results are the average randomness of keys generated using these three groups of thresholds. The results show that $c_1^2(f)$ is better than $c_2^2(f)$ to increase the randomness since it assigns the binary strings that have an equal number of 0s and 1s to the intervals that most of the features lie in it.

We repeated this experiment to show the randomness of keys generated using 4-bit quantization process. In this experiment, we used one group of thresholds which is $\{t_0 = 0,\ t_1 = 0.3,\ t_2 = -0.3,\ t_3 = 0.325,\ t_4 = -0.325,\ t_5 = 0.35,\ t_6 = -0.35,\ t_7 = 0.375,\ t_8 = -0.375,\ t_9 = 0.4,\ t_{10} = -0.4,\ t_{11} = 0.425,\ t_{12} = -0.425,\ t_{13} = 0.45,\ t_{14} = -0.45\}$ where $\lambda = 0.025$. We used the function $c_1^4(f)$ as written in Eq. 7 and the function $c_2^4(f)$ which is:

$$
c_2^4(f) = \begin{cases}
1001 & \text{if } f_j \geq t_{13} \\
0110 & \text{if } t_{11} \leq f_j \leq t_{13} \\
1100 & \text{if } t_{09} \leq f_j \leq t_{11} \\
0001 & \text{if } t_{07} \leq f_j \leq t_{09} \\
1101 & \text{if } t_{05} \leq f_j \leq t_{07} \\
1000 & \text{if } t_{03} \leq f_j \leq t_{05} \\
1011 & \text{if } t_{01} \leq f_j \leq t_{03} \\
1111 & \text{if } t_0 \leq f_j \leq t_{01} \\
0000 & \text{if } t_{02} \leq f_j \leq t_0 \\
1110 & \text{if } t_{04} \leq f_j \leq t_{02} \\
0010 & \text{if } t_{06} \leq f_j \leq t_{04} \\
0111 & \text{if } t_{08} \leq f_j \leq t_{06} \\
0100 & \text{if } t_{10} \leq f_j \leq t_{08} \\
0101 & \text{if } t_{12} \leq f_j \leq t_{10} \\
0011 & \text{if } t_{14} \leq f_j \leq t_{12} \\
1010 & \text{if } f_j \leq t_{14}
\end{cases}
\tag{10}
$$

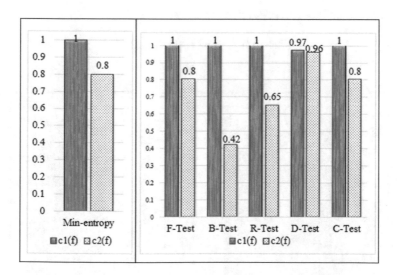

Fig. 4. The randomness of keys generated by TICK using 2-bit quantization

Figure 5 show that $c_1^4(f)$ is better than $c_2^4(f)$ to increase the randomness since it assigns the binary strings that have an equal number of 0s and 1s to the intervals that most of the features lie in it.

4.2   Reliability

Due to the time-variant nature of cardiac signals, it is necessary to ensure that derived cryptographic keys from such data are time-invariant. Thus, we use the reliability metric to measure the proportion of unrepeatable bits to repeatable bits among keys that derived from samples of each participant. Also, we used samples of multiple sessions in which intra-individual variance become worse. To measure the reliability, we used the hamming distance to find the number of unrepeatable bits among every two keys. We refer to the hamming distance between two keys derived from two samples of the same participant as the intra-hamming distance (*intraHD*). In this paper, we considered the target key is a key generated using samples of the first session, and estimated keys are keys generated using samples of the second session.

We refer to the target key of the participant i, as a reference key (r_i). If all keys ($k_{i,j}$) that belong to participant i in session 2 equal to (r_i), then (r_i) can be considered as reliable. Then the intra-hamming distance is computed as follows where g is the key length:

$$\text{intraHD} = \frac{\text{HD}(r_i, k_{i,j})}{g} \times 100 \tag{11}$$

Then, the reliability can be written as:

$$\text{Reliability}(r_i) = 100 - \text{intraHD} \tag{12}$$

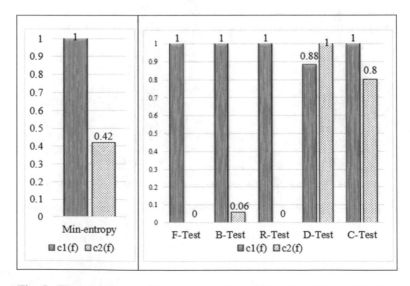

Fig. 5.   The randomness of keys generated by TICK using 4-bit quantization

To get the reliability of keys generated using the within-session data, we repeat the same experiment but using the samples of one session. Table 1 shows four experiments to study the performance of the quantization process of TICK. The first two experiments show the viability of using the variance improvement process to increase the reliability of keys generated using within-session data. The second two experiments show the viability of using the variance improvement process to increase the reliability of keys generated using across-sessions data. Also, by comparing the two experiments of the 4-bits quantization, we can conclude that the quantization process of TICK if used with the variance improvement is quite successful in reducing the gap in the reliability of keys generated using within-session data and across-sessions data.

Table 1. Performance of the quantization process of TICK

Type of data used	Bits/Feature	Reliability	
		Before improvement	After improvement
Within-session data	2-bits	84.04	97.29
	4-bits	83.89	98.69
Across-sessions data	2-bits	80.09	93.47
	4-bits	79.91	96.80

5 Discussion

Many approaches for generating cryptographic keys from cardiac signals, reported in the literature, exhibited excellent performance to get time-invariant keys. However, the existing methods performing poorly in case of using across-sessions cardiac signals to derive time-invariant keys. This degradation is due to the increase in the intra-individual variance of cardiac signals when captured under different conditions. Time-invariant cryptographic key generation approach TICK was useful to reduce the intra-individual variance in across-sessions data before converting it to binary sequences.

Furthermore, TICK uses a conversion method that allows for having an acceptable range of intra-individual variance since it quantizes the ECG features values that fall in the same range to a common binary string. We compared the quantization method of TICK with its counterpart of a prior approach called IOMBA [2] from two perspectives. The first perspective is the viability of steps taken by the approach to improve reliability performance. On this occasion, the quantization method of TICK depends on using a sort of stable thresholds based on the premise that the intra-individual variance of the quantization input reduced in the previous phase. On the one hand, the quantization method of IOMBA depends on using the reliability parameter (β) to determine the thresholds of the quantization intervals in a way that can improve the reliability. But, as well determining these thresholds depends on using the standard deviation and the mean of each feature in the dataset used for enrolling users. Indeed, the standard deviation and the mean of an ECG feature would not be stable under different session conditions [9]. This yield a sort of inequality among the quantization intervals used for generating a key for the first time and the quantization intervals used for re-generating it again. Table 2 compares the performance of TICK and IOMBA to get reliable and random keys. We used the best value of the reliability parameter in IOMBA which is 0.01. Table 2 shows the viability of using the variance improvement process and the stable thresholds that used in TICK to increase the reliability of keys generated across-sessions data.

Table 2. The performance of the quantization process of TICK and IOMBA

Bits/Feature	Quantization process of	Reliability	Min-entropy
2-bits	TICK	93.47	1
2-bits	IOMBA ($\beta = 0.01$)	83.43	0.46

The second perspective is the viability of steps taken by the approach to improve randomness performance. On this occasion, the quantization method of TICK increases the randomness in a binary sequence by converting the values of ECG features of the highest frequency to binary strings of an equal number of 0s and 1s. On the other hand, the quantization method of IOMBA uses an entropy parameter (α) to control the randomness and allow trade-off with reliability parameter (β). The authors claimed that it was not easy to achieve high levels of these two metrics simultaneously. However, lowering randomness reduces the number of possible keys that can be generated using data of an individual which facilitate guessing it by an adversary. We studied the randomness of keys generated by IOMBA statically and computationally. Table 2 shows the viability of the quantization process used in TICK to add entropy to the keys generated using across-sessions data more than its counterpart used in IOMBA. Also, Fig. 6 shows that the keys generated by IOMBA are not truly random according to the NIST tests. In contrast, the keys generated by TICK have a quite uniform distribution of 0s and 1s as shown in Fig. 6.

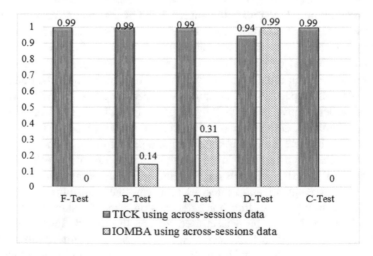

Fig. 6. The randomness of the keys generated by the quantization methods of TICK and IOMBA

However, the key length in TICK is determined by the number of ECG features used as input to the quantization process and the number of bits dedicated per feature. Thus, we suggest using extended vectors of ECG features to get longer keys. Also, quantizing features to binary strings of odd length reduces the randomness of the final binary sequence due to the inequality in the number of 0s and 1s.

6 Conclusion

Enhancing the variance of ECG data captured under different conditions is a feasible solution to improve the security level in the generated cryptographic keys from such data. The proposed method is used to generate cryptographic keys from cardiac signals of multiples sessions. Experimental results show that TICK is quite successful in improving the reliability combined with the randomness. By allowing more extended ECG features and lowering the number of bits assigned to each feature, lengths of keys generated by TICK can be further increased without affecting the performance of the reliability and randomness. Also, future work includes investigating other procedures that can be taken to improve the quality of quantization process to get better performance.

References

1. Hoekema, R., Uijen, G.J., Van Oosterom, A.: Geometrical aspects of the interindividual variability of multilead ECG recordings. IEEE Trans. Biomed. Eng. 48(5), 551–559 (2001)
2. Islam, M.S., Alajlan, N.: Biometric template extraction from a heartbeat signal captured from fingers. Multimedia Tools Appl. 76(10), 12709–12733 (2017)
3. Karimian, N., Guo, Z., Tehranipoor, M., Forte, D.: Highly reliable key generation from electrocardiogram (ECG). IEEE Trans. Biomed. Eng. 64(6), 1400–1411 (2017)
4. Coutinho, D.P., Silva, H., Gamboa, H., Fred, A., Figueiredo, M.: Novel fiducial and non-fiducial approaches to electrocardiogram-based biometric systems. IET Biometrics 2(2), 64–75 (2013)
5. Islam, M.S., Alajlan, N., Bazi, Y., Hichri, H.S.: HBS: a novel biometric feature based on heartbeat morphology. IEEE Trans. Inf. Technol. Biomed. 16(3), 445–453 (2012)
6. Chan, A.D., Hamdy, M.M., Badre, A., Badee, V.: Wavelet distance measure for person identification using electrocardiograms. IEEE Trans. Instrum. Meas. 57(2), 248–253 (2008)
7. Zhang, G.H., Poon, C.C., Zhang, Y.T.: Analysis of using interpulse intervals to generate 128-Bit biometric random binary sequences for securing wireless body sensor networks. IEEE Trans. Inf Technol. Biomed. 16(1), 176–182 (2012)
8. Bao, S.D., Poon, C.C., Zhang, Y.T., Shen, L.F.: Using the timing information of heartbeats as an entity identifier to secure body sensor network. IEEE Trans. Inf Technol. Biomed. 12 (6), 772–779 (2008)
9. Moosavi, S.R., Nigussie, E., Virtanen, S., Isoaho, J.: Cryptographic key generation using ECG signal. In: 14th IEEE Annual Consumer Communications & Networking Conference (CCNC), pp. 1024–1031. IEEE, Las Vegas (2017)
10. Moosavi, S.R., Nigussie, E., Levorato, M., Virtanen, S., Isoaho, J.: Low-latency approach for secure ECG feature based cryptographic key generation. IEEE Access 6, 428–442 (2018)

11. Hamad, N., Rahman, M., Islam, S.: Novel remote authentication protocol using heart-signals with chaos cryptography. In: International Conference on Informatics, Health & Technology (ICIHT), pp. 1–7. IEEE, Riyadh (2017)
12. González-Manzano, L., de Fuentes, J.M., Peris-Lopez, P., Camara, C.: Encryption by heart (EbH)—using ECG for time-invariant symmetric key generation. Future Gener. Comput. Syst. **77**, 136–148 (2017)
13. Zeadally, S., Isaac, J.T., Baig, Z.: Security attacks and solutions in electronic health (E-health) systems. J. Med. Syst. **40**(12), 263 (2016)
14. Wang, B., Wang, L., Lin, S.J., Wu, D., Huang, B.Y., Zhang, Y.T., Yin, Q., Chen, W.: A body sensor networks development platform for pervasive healthcare. In: 3rd International Conference on Bioinformatics and Biomedical Engineering, pp. 1–4. IEEE, Beijing (2009)
15. Xu, F., Qin, Z., Tan, C.C., Wang, B., Li, Q.: IMDGuard: securing implantable medical devices with the external wearable guardian. In: Proceedings of IEEE INFOCOM, pp. 1862–1870. IEEE, Shanghai (2011)
16. Rukhin, A., Soto, J., Nechvatal, J., Smid, M., Barker, E., Leigh, S., Levenson, M., Vangel, M., Banks, D., Heckert, A., Dray, J., Vo, S.: A statistical test suite for random and pseudorandom number generators for cryptographic applications. NIST Special Publication 800-22, pp. 1–153, 15 May 2001. http://www.nist.gov
17. Islam, M.S., Alajlan, N.: Model-based alignment of heartbeat morphology for enhancing human recognition capability. Comput. J. **58**(10), 2622–2635 (2015)
18. Islam, S., Ammour, N., Alajlan, N., Abdullah-Al-Wadud, M.: Selection of heart-biometric templates for fusion. IEEE Access **5**, 1753–1761 (2017)
19. Barker, E., Kelsey, J.: Recommendation for the entropy sources used for random bit generation. Draft NIST Special Publication, 800-900 (2012). http://www.nist.gov

Privacy Implication and Technical Requirements Toward GDPR Compliance

Ching-Chun (Jim) Huang[(⊠)] and Zih-shiuan (Spin) Yuan

BiiLabs Co., Ltd., Taipei, Taiwan
jserv@ccns.ncku.edu.tw, spin.yuan0530@gmail.com

Abstract. The EU General Data Protection Regulation (GDPR), regarded as the most significant change in information privacy law in these two decades, admittedly impacts our current systems in almost every field. After analyzing real cases, including Microsoft Office, Google LLC., Knuddels.de and Facebook Ireland Ltd, we attempt to illustrate the concepts behind data collection, processing, controlling and transformation toward GDPR compliance. Technology is, in other words, the principal problem that data protection law is trying to solve. If entities are storing too much personal data, for example, technology needs to deliver delete, erase, de-duplication and minimization enforcement. Under Article 33 of the GDPR, operation organization must report a data breach within 72 h without "undue delay," which highlights the importance of identity transparency for each data ownership while finding out where the breach occurred, what areas have been affected, and how it happened is not straightforward as typical awareness. Self-sovereign identity is addressed in this paper to propose a practical infrastructure and clarify the technical requirements of GDPR judgment and compliance along with the mitigation of certain impacts from access control, integrity, confidentiality, and privacy implication.

Keywords: GDPR · Ontology · Data management model · Self-Sovereignty Identity (SSI) · Decentralized Identifiers (DIDs)

1 Introduction

Data, deemed as the oil of the digital world, has proved its values with a series of successes of Alphabet (new parent company of Google), Facebook and Microsoft. These titans collect customer information by providing overall service in order to achieve precision marketing as well as obtain ample materials for their artificial intelligence (AI) without openly and clearly informing their users. Apparently, novel business models are emerging while public awareness of data privacy is still lacking. For better governance, the EU General Data Protection Regulation (GDPR) was eventually enforced on 25 May 2018 after two-year grace period in replacement of the Data Protection Directive 95/46/EC [1] since 1995. This most-significant change in information privacy law in the two decades transforms digital footprints into a part of human rights by reaffirming the rights of the data subjects, harmonizing the law, empowering all citizens' data privacy and reshaping the organizational function to data protection across the Europe [2]. Moreover, the GDPR demonstrates its determination

© Springer Nature Switzerland AG 2020
K. Arai et al. (Eds.): FTC 2019, AISC 1070, pp. 353–367, 2020.
https://doi.org/10.1007/978-3-030-32523-7_24

of regulation on the penalties and fines, which is up to €10 million or 4% of the worldwide annual turnover of the prior year. Admittedly, GDPR becomes one of the top search queries and easily catch industry's attention.

However, even if it widely calls into questions how to be GDPR-compliant, most people still have limited understanding of its spirit and hence getting confused or even misled. Therefore, this essay will induce the regulation spirit by four cases in order to elaborate the patterns as well as the future trend toward the compliance. After the review on related works, the comparison between the GDPR and the past law will be made in Session III. Next, the following two sessions will discuss the preconditions, the limitations, the security and cross-border management for the data collection, storage and processing with four case study. Afterwards, Session VI will induce the patterns from current tech review and propose a possible trend towards the GDPR-compliance. In the end, the GDPR spirit as well as its social impacts will be elaborated.

2   Related Works

Current studies focus on two dimensions: How does the GDPR impact on the specific products, service process or industry? And what tool, technique or framework can achieve GDPR-compliance? Aside from the overall analysis [3], the impacts of the GDPR on whether IoT era or blockchain technologies also catch public attentions [4, 5]. As for the technological aspect, academic groups focusing on information security intend to provide possible solutions to specific GDPR issues in foundation of the privacy-enhancing technologies (PETs), such as the requirements for revoking consent, the right to be forgotten [6], the concept of privacy by design and by default [7], and the right to data portability [8]. Over the past decade, a couple of frameworks and specific tools have emerged for the enhancement of data privacy. From the generic aspect, ISO 27000/29100/27018 provide the overview of information security management systems (ISMS) as well as a comprehensive set of controls over data processing [9]. Meanwhile, a unified compliance framework (UCF) and COBIT 5 are respectively produced by the Network Frontiers LLC and the ISACA to outline a set of controls applicable to both organizational and technical requirements [10, 11]. Recommendations on technical measures are highlighted by the NIST 800-53 and the European Union Agency for Network and Information Security (Enisa). Particularly, the Enisa published a series of reports [12], such as data pseudonymization and data protection by default, to sketch out the shaping technologies according to GDPR provision (More technical details will be probed in Sect. 6.1). But does it equal to the solution to the GDPR-compliance?

Unfortunately, despite the diverse discussions and the adequate developments, it still seems to have a long journey for compliance in the nowadays. The reason may lie in the focus on specific issues, the complexity of the tools or the huge gap between frameworks and implementation. Before further investigation, the basic principles of the GDPR should be firstly probe in order to figure out the possible future trend.

3   General Data Protection Regulation

In addition to the regulation of data controller/processor (Art. 24-43) and the supervision of data transfer to the third countries (Art. 44-50) as before, the GDPR further increases territorial scope (Art. 3), strengthens the consent condition (Art. 7), and raises the penalty fee to violators (Art. 83-84) in order to extend its influence. Moreover, the innovative concepts for the data subject right (Art. 12-23) have been further proposed in the new legislation, including "data portability" and "the right to be forgotten". A data subject refers to an identified or identifiable natural person who could produce digital footprints called personal data. Generally, personal data can be divided into direct and indirect identifiers. The former ones include social security numbers, health insurance number and single or multi-elements specific to the physiological, mental, physical, genetic, economic, cultural or social identity, while the latter ones imply a mixture of birth date, gender, transaction records, location indicators and so on. The GDPR enables data subjects to enforce their rights to erasure, rectification and data portability. These changes (Table 1) trigger wide panic, resulting from the uncertain standard for adjudication.

In fact, by April 2019 there are four affirmed yet appealable sanctions and hundreds of pending cases across the EU [13]; however, the range of fines is surprisingly between €4,800 to €50,000,000, which has triggered heated debates. For instance, the first and simultaneously the lowest-fined one took place in Austria. The Austrian Data Protection Authority (DSB) issued a fine of €4,800 against a private organization due to the unlawful monitoring of public space with CCTV camera [14]. In contrast, the most severe penalty announced by French CNIL fell on Google LLC and reached €50,000,000 for lacking transparency, information and proper consent of personalized advertisements [15]. The other two cases are Knuddels.de regarding data storage form and a private hospital concerning problematic profile management system [16], respectively fined 20,000 € by the LfDI in Germany and 400,000 € by the CNPD in Portugal [17]. Yet, the disparity in four sanctions complicates matters. Therefore, four real cases will be studied for the analysis on data collecting, processing, controlling and transferring.

Table 1. The key changes in the GDPR

Item	Data Protection Directive	General Data Protection Regulation
Extraterritorial Applicability	Ambiguous territorial applicability referring to data process 'in context of an establishment'	Applicable to any company processing personal data of European residences 1. whether domestic or international company 2. whether abroad processor/controller or not
Penalties	Proper sanctions	Maximum fine up to 4% of annual global revenue or €20 Million, whichever is higher

(continued)

Table 1. (*continued*)

Item		Data Protection Directive	General Data Protection Regulation
Consent		Providing purpose of data processing to users	Prohibiting using lengthy misleading consent to users and requesting for easy withdrawal
Data Subject Rights	Breach Notification	Simplification of or exemption from notification sometimes allowed	Mandatory notifications by data processors to both customers and the authorities in 72 h while resulting in a risk of data breach
	Right to Access	Right for a data subject to get a free copy and obtain data concerning him	Right for data subjects to get a free electronic copy and secure validation from data controllers, deciding whether personal data is processed
	Right to be Forgotten	Limited right to cease further dissemination and erase personal data	Right to require controllers to cease further dissemination and erase personal data under the premise of not endangering public interests
	Data Portability	X	Right for data subjects to receive and transmit data in "machine readable format"
	Privacy by Design	X	Calling for the inclusion of data protection from the onset of designing systems
	Data Protection Officers	Requiring companies to provide the name and address of controllers	Staff responsible for regular and systematic monitor of users' personal data or criminal convictions

4   Preconditions and Limitation on Data Processing

In the field of data privacy, the boundaries of both the execution of power and the enforcement of rights remain blurry; as a result, the GDPR intends to ensure users to secure their rights by strengthening the preconditions of data processing, such as the consent and information. Despite acquiring the permission from users, there are still a couple of limitations on personal data processing. Thus, this session will take the case of Google LLC and Microsoft Office for instance to highlight the importance of user consents and the significance of the transparency of data processing.

4.1   Google LLC Case: Lawfulness and Transparency of Data Processing

In accordance with the GDPR, the company Google LLC was imposed a financial penalty of 50 Million euros by the CNIL's restrict committee, the data protection authorities in France, for lack of transparency, information and valid consent concerning the advertisement personalization on 21 January 2019 [15]. The exorbitant fine immediately turned into a bombshell to the public, igniting both positive and negative comments in the industry. In fact, on the first day GDPR took effect cross the EU, Google LLC was condemned for GDPR-noncompliance, particularly the forced consent, by an Austrian non-governmental organization (NGO) named None of Your Business (NOYB) [18] and a French advocacy group addressed as La Quadrature du Net (LQDN) on behalf of 12,000 people. Led by Max Schrems, the NOYB claimed as much as $4.88 billion fine from Google and lodged the same official complaints to Facebook, Whatsapp and Instagram simultaneously. Half year later, Google was again hit a complaint about GDPR-noncompliance by a couple of European consumer organizations on a basis of new study conducted by the Norwegian Consumer Councils [19]. As before, the plaintiffs indicated that Google designed "deceptive" tricks on "forced" consent to get detailed private records as well as inform users to be partially blocked if not staying agreed [20].

At the point, the CNIL's restricted committee announced Google's two major violations of the GDPR, respectively the obligations of "transparency and information" and to "process ads personalization with legal basis", which underlined the essential GDPR spirit (Art. 5-7) [15]. Furthermore, the committee shared their calculation for the first domestic sanction in consideration of consistency of violation, market scale of Android operating system in France, and company's profit models on the ads personalization. The reference may inspire other corporations to modify their business strategies.

4.2 Microsoft Case: Purpose Limitation, Data Minimization and Storage Limitation

In November 2018, Microsoft was found by a Dutch government report to collect personal data such as the titles of Email or sentences from Office products on a large scale via Office without informing its users, which means unlawful telemetry collection [21]. In response to the speculation, Microsoft firmly said the move of obtaining diagnostic data, as standard practice among programmers, aimed for functional and security purposes. However, the problems of Microsoft lie in the method of informing, non-transparency of the data processing and the ignorance of the willingness of data subjects. After the negotiation, on 26 October 2018 a consensus reached by both sides officially requires Microsoft to conduct an improvement plan to adapt its products used by the Dutch government in lines of GDPR-compliance and other legislation concerned [22]. SLM Microsoft Rijk claimed that only if the progress is insufficient or the improvements are unsatisfactory, further complaint will be filed to the Data Protection Authority, or the Microsoft products would be still accessible in Dutch government. As for Microsoft, their spokesman expressed appreciation over the matter [23].

The accusations based on the Dutch government's findings can correspond to the Article 5 of the GDPR, including the principles of purpose limitation, of data minimization and of storage limitation. Firstly, any collection of personal data should be specified, explicit and purpose-oriented, thus other extensive processing is not allowed, except for the purposes of public interest. Secondly, the data minimization indicates that data subjects should only disclose necessary and adequate information in relation to the purpose of data collection. For example, why should I uncover my income as financial statement while taking taxi? In this case, it is adequate for the driver to confirm that the passenger can afford one-way fare. Lastly, without the consideration of public interest or appropriate technical measures, personal data should be no longer stored after the original purpose achieves.

4.3   Analysis

For the Google LLC case, the issue of consent agreement could get involved in the notion of "Privacy by Default" mentioned in the GDPR Art. 25. Default settings mean initial privacy settings without any manual operation for a product or service, and this idea aims at allowing personal data to be only collected for necessary and specific purposes. Normally, personal data should not be accessible to the companies without a data subject's intervention while the use of dark patterns for misleading is prohibited. Similar to Google LLC, Microsoft is criticized for its unlawful, unclear and opaque data processing, plus the retention of data storage and the lack of discretion for users. In sum, the requirements of data processing include the transparency, information and consent in order to secure the rights of data subjects. Fortunately, Microsoft reached a settlement instead of accepting a ruling. Now it seems more confusing about the standard for the data protection office to make a judgement. Next, Session V will compare the lopsided penalties between Google LLC and Knuddels.de in order to unveil the key points of the GDPR.

5   Data Storage and Transmission

Whether Google LLC or Microsoft, the violation of the GDPR is still the issues of internal use. The following case of Knuddels.de not merely gets involved in severe data leakage to the public website but receives an intriguing €20,000 fine. The Session V consists of the issues of data security and cross-border dataflow to clarify data storage as well as transmission. Furthermore, aside from the emergency actions a company could take while the GDPR-noncompliance, some precautionary measures will be discussed with the case of Facebook Ireland Ltd.

5.1   Knuddels.de Case: Encryption Requirement for Data Security

On 27 November 2018, a chat site's data breach up to 330,000 users revealed great potential risks of the storage of cleartext [24], hence German authorities issued the first domestic GDPR fine with €20,000 to Knuddels.de, which was far below expected figures by the public. Knuddels.de is well-known as the largest local social media platform and requires users for anonymous registration before getting access to their service. Disastrously, in July 1.87 million username/password combinations and up to 800,000 e-mail addresses were exposed on Mega.nz and Pastebin.com. What's worse, not only being hit by cyberattacks, Knuddels.de was also found by the netizens to store users' information in cleartext, leading to large-scale data leakage. Soon after, their official bulletin board immediately announced a series of progress reports, including emergency solution and event timeline, in order to quell users' doubts [25], and hence receiving compliments from the Data Protection and Freedom of Information (LfDI) Baden-Württemberg, which could be noted for shedding light on GDPR-compliant procedure.

Compared to data leakage itself, the determining factor for the authorities lies in the form of data storage. In other words, ciphertext referring to the encryption is likely to dominance to mainstream form of data storage in pursuit of better cyber security. Also, it is observable that GDPR covers both "the pseudonymization and encryption of personal data" and "the financial burden on the corporation" from the administrative injunction. Yet, despite enhancing security measures and working in conjunction with the authorities, preventive measures are more critical. Hence, the next session will describe possible precautionary measures a company may take under the GDPR.

5.2   Facebook Ireland Ltd Case: Transfers of Personal Data to Third Countries or International Organizations

For compliance, most organizations tend to demonstrate their determination by the modification of user consent while few might choose to delete the EU user data as opposition of the strict regulation [26]. Facebook Ireland Ltd, as one of the largest social media platform, also made some preparations prior to the implementation of GDPR. In spite of a series of complaints and accusations afterwards [27–29], Facebook secretly transferred their non-EU users from Irish server to American headquarters in consideration of rule circumvention regarding cross-border dataflow [30]. Originally, the international headquarters in Ireland managed its 370 million in Europe and 1.52 billion users elsewhere, which means all the Facebook users have the rights granted by the GDPR except for the 239 million North American users stored on the US server. With the server change, the affected 1.5 billion users in Africa, Asia, Australia and Latin America should comply with American privacy protection and cannot file complaints in Irish courts or with Ireland's Data Protection Commissioner. From a consequentialist point of view, however, Facebook still hardly escaped the binding regulations with two filed complaints.

The move of changing server mainly relates to the issue of the data transfers to the third countries or international organizations, and that concern has been definitely no longer a nonsense. With the explosive rise of international business, the cross-border dataflow still stay risky even if the companies in the EU could be able to comply with the GDPR. The invalidation of Safe Harbour decision has reinforced the argument [31], not to mention other countries lacking the awareness of data privacy. That's why Facebook provides two different consents to EU and non-EU users in order to impair the influence of the GDPR.

5.3   Analysis

Although the responsibility shift can be referenced to other European organizations with non-EU business, the fundamental problems of data privacy and protection remain unsolved. Admittedly, the management of cross-border dataflow is still far from realization. Take the EU for example. Based on different situations, the transfer mechanism can be classified into adequacy decision, binding corporate rules (BCRs), model contractual clauses (Model Clauses), EU-US and Swiss-US Privacy Shield, Derogation, and consent from individual [32]. However, the organizational measures can merely determine the responsibility, instead of tracking dataflow.

In this case, there are at least two issues for probe: the possibility of recording dataflow and precautionary measures for data protection. Indeed, digital world simplifies the process of file duplication and transmission, resulting in the tougher situation for data management. Yet, users as well as the authorities barely know how and who handle the sensitive personal information due to the shortage of reliable techniques. Apparently, current measures seem not prepare for the dramatic change. As for the latter one, the notion of "Privacy by Design (POD)" mentioned in GDPR Art. 25 refers to a forethought of the design of data processing itself, including data minimization, pseudonymization, and the integration of the necessary safeguards in achievement of the right protection of data subjects. But again, it calls into questions what techniques should be applied. Hence, the next session will give a look at the current techniques and potential trends toward the GDPR-compliance.

6   Data Privacy and Protection

By the discussions on the preconditions, the limitations, the security and the cross-border managements of data controlling, processing and storage, some key points have been elaborated. In the following, the items violating the GDPR will be summarized as Table 2 to correspond to the GDPR seven principles: (a) Lawfulness, fairness and transparency; (b) Purpose limitation; (c) Data minimization; (d) Accuracy (not mentioned); (e) Storage limitation; (f) Integrity and confidentiality, and (g) Accountability. Subsequently, Session 6 will provide possible direction of development after clarifying the technical gap based on the investigation of current measures.

6.1   Current Measures Under the GDPR

Privacy-enhancing technologies (PETs) mean the methods for data protection on the basis of relative regulation. A current taxonomy of PETs, such as informed consent, data minimization, data tracking, anonymity, control, negotiate terms and conditions, technical enforcement and remote audit of enforcement [33], could correspond to the GDPR seven principles. For instance, to ensure a user's willingness for providing personal data, a revocable consent adopting "data tagging" technology and "sticky policies" is the most accessible way. Another example is ex-ante and ex-post transparency- enhancing technologies. Such as P3P, Data Track, Kelsey-Schneier log and TaintDroid, they use "privacy policy languages" tool and "Human Computer Interaction (HCI)" components to inspect the transparency of data processing. For minimal data collection, storage and processing, encryption and decentralized storage are two typical measures, and in particular, the former one includes anonymity, pseudonymity and unlinkability.

Table 2. Items violating the GDPR in the four cases

Item/Company	Google. LLC	Microsoft office	Knuddles.de	Facebook Ireland Ltd
Lack of Transparency (a)	V	V		V
Misleading Information on Consent Agreement (a)	V	V		V
Unlawful Collection/Storage/Data Processing (a) (c)	V	V		V
Lack of Purpose Limitation (b)	V	V		
The Indefinite Retention Period of Diagnostic Data (e)		V		
Debated Data Storage Form (f)			V	
Not enough control over sub-processors and factual processing (f)		V		
Issue on Cross-border Data Transfer (f)	V	V		V
Incorrect Qualification as Data Processor (g)		V		
Data Leakage & Duty of Notification (g)			V	V

Despite the creative and innovative ideas, the failure of PETs seems to go mainstream due to its technical complex, the lack of customer demand or trust, or the incompleteness of network effect [34]. Precisely speaking, the major PETs chose to not disclose their techniques or source codes to demonstrate how it worked while a minority of these projects lacked incentives for commercialization due to their focus on technical aspects. Fortunately, amidst the organizational and technical discussions on compliance, some patterns can be found as follows: Encryption, decentralized storage, user-centricity and machine readable data. Thus, the next part will further propose potential trend and possible measures for the GDPR-compliance.

6.2   Trends Toward GDPR-Compliance: Self-Sovereignty Identity (SSI)

To achieve the GDPR-compliance, current techniques have demonstrated a trend of encrypted user-centric identity management (IdM) with decentralized stored machine-readable data, which transformed to the notion of Self-Sovereignty Identity (SSI). Originated from the 2017 Gartner report [35], the revolutionary yet familiar concept of SSI is designed to "re-imagine the identity data model" with more proper balance between data security and user experience. Later, the 10 Guiding Principles of SSI was published by Christopher Allen to elaborate the elements of the Existence, the Control, the Access, the Transparency, the Persistence, the Portability, the Interoperability, the Consent, the Minimization and the Protection [36]. Eventually the terminology of SSI nowadays could refer to the concept of an equity technically allowed to control over the analog personal data while having a digital identity [37], which is expected to be a twilight for the future IdM design (Table 3) after the evolution of Isolated Identity Model, Central Identity Model, User-Centric Identity Model and Federated Identity Model.

Table 3. The correspondence of the SSI concepts to the GDPR principles

SSI Concept	GDPR Requirement	Current measures
Full control over personal data for individual	(a) Consent Data Subject Right (Right to Access, Right to Erasure, Right to be Forgotten)	Proof Request, Personal Identity Data Management System (PIMS), Connections and Microledger technology
Ensuring security and privacy	(f) Integrity and confidentiality; (g) Accountability	Cryptographically- Generated Identifiers
Full portability of data	Data Subject Right (Data Portability)	Open Identity Layer (DLT), Standardized Data Exchange Formats
Interoperability		
Ensuring data integrity	(f) Integrity and confidentiality	Off-chain Ledger, Zero-Knowledge Proofs (ZKP)
Transparency of the identity data is maintained	(a) Lawfulness, fairness and transparency	Records of Credentials, Proof Request
Minimization	(b) Purpose limitation (c) Data minimization; (e) Storage limitation;	ZKP, Decentralized Storage
Decentralized trust	N/A	Decentralized Public Key Infrastructure (DPKI)
(Core value of SSI)	Privacy by Design and by Default	

According to the White Paper published by the EGIZ [38], the measures for realizing SSI heavily rely on the decentralized ledger technology (DLT) with two additional parts, including the off-ledger storage for sensitive data and the data import for interoperability. On the other hand, some research focuses on the evaluation of IdM technologies by case study [39]. Yet, despite the diverse ongoing prototypes [40], it is still urgent to have a universal specification for reference. In this case, the Data Model and Syntaxes for Decentralized Identifiers (DIDs) proposed by the Credentials Community Group and supported by the World Wide Web Consortium (W3C) provides comprehensive data model, format, and operations in order to express analog claims in the cyberspace [41]. In the following, the last part of Session 7 will give some brief introduction as well as analysis on DIDs.

6.3   SSI Potential: Data Model for Decentralized Identifiers (DIDs)

Decentralized identifiers (DIDs) refer to a new type of identifier following the principle of Privacy by design to authenticate entities via proof (e.g., digital signatures, privacy-preserving biometric protocols, etc.) without centralized registry. Aiming at the realization of self-sovereign digital identity, a DID document contains proof purposes, verification methods (public keys and pseudonymous biometrics) and service endpoints (e.g. social networks, file storage services, and verifiable claim repository services) in order to describe the subject. In particular, the document takes the security and privacy into consideration by building blocks of decentralized public key infrastructure (DPKI) and keeping personally-identifiable information (PII) off-ledger. With the cryptographic protection, the entire ecosystem becomes more reliable. Moreover, for the portability and interoperability, the data structure recommended is using JSON-LD, one of the languages for machine readability. Admittedly, the specification gives guideline on how to achieve SSI.

As stated in the previous part, the notion of SSI is turning into a mainstream in the digital era in combination of the patterns of encryption, decentralized storage, user-centricity and machine readable data; meanwhile, the adoption of the DLT is regarded as the major solution at the turning point. Thus, the DID document gives a reliable ground of realizing self-sovereignty identity with a set of operations, but this is not equivalent to complying with the GDPR. For the compliance, supporting mechanism such as key management system should follow up. In other words, the Data Model for DIDs is not the terminal measure but the first step for the GDPR-compliance.

7 The GDPR's Spirit and Its Social Impacts

Whether the emerging issues of data justice or data subject right unveils the transforming concept of data itself while the GDPR driven by the need in the digital age aims at enhancing the position of data from analog signal to an extensive part of human right. Gradually, the data storage, processing and transmission is no longer static, which implies that all dataflow should be verified by the data subjects, or the data controller/processor would violate the human rights of their users. Simultaneously, the dynamic interaction among data requires strict security standard as well as holds on the principle of consent and minimal disclosure in every given transaction to ensure the privacy protection. Examples are shown in the cases of Google LLC, Microsoft Office and Knuddels.de. In addition, the appearance of the DLT is expected to utilize the management of cross-border data transfers from legal requirements to technical practices, due to each transaction recorded on the ledgers. Details is discussed in the case of Facebook Ireland Ltd. At this point, the GDPR is definitely a milestone in convergence for technology and privacy regulation.

This essay intends to provide a comprehensive look at a possible future trend towards SSI as well as the schemes after the induction of some universal patterns for the GDPR-compliance from four real cases. Despite the ongoing exploration for the GDPR-compliant measures, the DID specification still inspires the industry, particularly for the developers and the data protection offices (DPO), by providing a universal machine-readable format to enable the interoperability and data portability. In the near future, more prototypes as well as products will come out with the label of the GDPR-compliance, and most of them are likely to follow the DID specification after all.

8 Conclusion

This paper discusses how analysis and decentralized identity mechanisms can be applied for GDPR compliance. In the beginning, it introduces the significant changes in the EU General Data Protection Regulation, followed by the existing applications of static analysis to privacy properties. The proposed decentralized identity forms a semantic model of consent to make it specific and unambiguous as required by the GDPR. In particular, we can generate privacy policies from the model and detect violations of data minimization. Finally, we explained why GDPR compliance is not only the role of humans in enforcement but also the maximum use of de-identification compatible with the purposes of the data processing. That, in turn, can provide the optimal balance between maintaining utility of data and protecting the privacy of individual data subjects.

With the unfolding and flourishing technologies, the tensions between the GDPR and the use of the DLTs, such as the accountability of data controllers on public ledgers, the degree or methods of the anonymization, and the uprising rights and obligations under the new principles, have admittedly ignited a public panic [42]. This work, similarly, constrained to the ongoing and uncertain technical developments, fails to propose concrete solutions to the GDPR-compliance. However, more discussions on the eligibility of technology applications [43] and some uprising real cases help clarify the boundaries of the ruling. Take the DLTs-based identity management (IdM) for example. The ongoing platforms include uPort [44] and ShoCard [45], while the organizations concerned with the application are the Sovrin [46] and the TangleID [47]. Thus, the future work can not only propose a framework or focus on specific aspects of the regulation, but is able to go further with the concrete complaint measures or empirical research based on the practical application. Still, such guidance, at the early stage, can help provide much-needed clarity related to new GDPR obligations.

References

1. The European Union Legislative Process. https://eugdpr.org/the-process/. Accessed 1 June 2019
2. GDPR Official Website. https://eugdpr.org/. Accessed 1 June 2019
3. Colin, T.: What the GDPR means for businesses author links open overlay panel. Netw. Secur. **2016**(6), 5–8 (2016)
4. Zarsky, T.: Incompatible: the GDPR in the age of big data. Seton Hall Law Rev. **47**, 995–1020 (2017)
5. Shraddha, K.: Building-blocks of a data protection revolution: the uneasy case for blockchain technology to secure privacy and identity. Nomos Verlagsgesellschaft mbH, Baden-Baden. http://www.jstor.org/stable/j.ctv941qz6
6. Eugenia, P., Efthimios, A., Constantinos, P.: Forgetting personal data and revoking consent under the GDPR: challenges and proposed solutions. J. Cybersecurity **4**(1)
7. Salah, A., Saran, L.: GDPR privacy by design: from legal requirements to technical solutions. https://dsv.su.se/polopoly_fs/1.351720.1507815130!/menu/standard/file/Stipendie2017_ElShekeil-Laoyookhong.pdf. . Accessed 1 June 2019

8. Paul, H., Vagelis, P., Gianclaudio, M., Laurent, B., Ignacio, S.: The right to data portability in the GDPR: towards user-centric interoperability of digital services. Comput. Law Secur. Rev. **34**(2), 193–203 (2018)
9. ISO 27000. https://www.iso.org/isoiec-27001-information-security.html. Accessed 1 June 2019
10. Unified Compliance Framework (UCF). https://www.unifiedcompliance.com/. . Accessed 1 June 2019
11. ISACA. http://www.isaca.org. Accessed 1 June 2019
12. Enisa. https://www.enisa.europa.eu/topics/data-protection?tab=publications. Accessed 1 June 2019
13. Barbara S.: 100 Tage DSGVO: Wutmails schreiben reicht nicht. https://www.trend.at/branchen/digital/tage-dsgvo-wutmails-10340789. Accessed 1 June 2019
14. Austria announces first GDPR fine. https://iapp.org/news/a/austria-announces-first-gdpr-fine/. Accessed 1 June 2019
15. The CNIL: La formation restreinte de la CNIL prononce une sanction de 50 millions d'euros à l'encontre de la société GOOGLE LLC. https://www.cnil.fr/fr/la-formation-restreinte-de-la-cnil-prononce-une-sanction-de-50-millions-deuros-lencontre-de-la. Accessed 1 June 2019
16. de Meneses, A.O., Van Quathem, K.: Portuguese hospital receives and contests 400,000 € fine for GDPR infringement. https://www.insideprivacy.com/data-privacy/portuguese-hospital-receives-and-contests-400000-e-fine-for-gdpr-infringement/. Accessed 1 June 2019
17. McKenzie, B.: GDPR National Legislation Survey, 4.0 (2018). https://tmt.bakermckenzie.com/-/media/minisites/tmt/files/2018/08/gdpr_national_legislation_survey_4_aug2018.pdf
18. NOYB projects. https://noyb.eu/projects-2/. Accessed 1 June 2019
19. The European Data Protection Board, Re: How tech companies nudge users into choosing the less privacy friendly options. https://www.beuc.eu/publications/beuc-x-2018-061_joint_letter_to_ms_andrea_jelinek-deceived_by_design.pdf. Accessed 1 June 2019
20. Rijksoverheid: Complaint to the datatilsynet under Article 77 (1) of the European General Data Protection Regulation. https://fil.forbrukerradet.no/wp-content/uploads/2018/11/complaint-google-27-November-2018-final.pdf. Accessed 1 June 2019
21. Rijksoverheid: Stand van zaken onderhandelingen Rijk en Microsoft met betrekking tot AVG compliance. https://www.rijksoverheid.nl/documenten/rapporten/2018/11/07/data-protection-impact-assessment-op-microsoft-office. Accessed 1 June 2019
22. Rijksoverheid: Update on negotiations between Dutch central government and Microsoft on GDPR compliance. https://www.rijksoverheid.nl/documenten/rapporten/2018/11/07/data-protection-impact-assessment-op-microsoft-office. Accessed 1 June 2019
23. Microsoft Official Website: Privacy at Microsoft. https://www.microsoft.com/en-us/trust-center/privacy. Accessed 1 June 2019
24. Pressestelle: LfDI Baden-Württemberg verhängt sein erstes Bußgeld in Deutschland nach der DS-GVO. https://www.baden-wuerttemberg.datenschutz.de/lfdi-baden-wuerttemberg-verhaengt-sein-erstes-bussgeld-in-deutschland-nach-der-ds-gvo/. Accessed 1 June 2019
25. Knuddel.de website. https://forum.knuddels.de/ubbthreads.php?ubb=showflat&Number=2916245#Post2916245. Accessed 1 June 2019
26. Lomas, N.: Unroll.me to close to EU users saying it can't comply with GDPR. https://techcrunch.com/2018/05/05/unroll-me-to-close-to-eu-users-saying-it-cant-comply-with-gdpr/. Accessed 1 June 2019
27. NOYB: GDPR: noyb.eu filed four complaints over "forced consent" against Google, Instagram, WhatsApp and Facebook Corporations forced users to agree to new privacy policies. https://noyb.eu/wp-content/uploads/2018/05/pa_forcedconsent_en.pdf. Accessed 1 June 2019

28. Facebook security update notice. https://newsroom.fb.com/news/2018/09/security-update/. Accessed 1 June 2019
29. ICO: ICO issues maximum £500,000 fine to Facebook for failing to protect users' personal information.   https://ico.org.uk/about-the-ico/news-and-events/news-and-blogs/2018/10/facebook-issued-with-maximum-500-000-fine/. Accessed 1 June 2019
30. David, Ingram, Exclusive: Facebook to put 1.5 billion users out of reach of new EU privacy law.   https://www.reuters.com/article/us-facebook-privacy-eu-exclusive/exclusive-facebook-to-put-1-5-billion-users-out-of-reach-of-new-eu-privacy-law-ldUSKBN1HQ00P.   Accessed 1 June 2019
31. The Guardian: Facebook row: US data storage leaves users open to surveillance, court rules. https://www.theguardian.com/world/2015/oct/06/us-digital-data-storage-systems-enable-state-interference-eu-court-rules. Accessed 1 June 2019
32. GDPR and Cross-Border Data Transfers, Womble Bond Dickinson (US) LLP. https://www.womblebonddickinson.com/sites/default/files/2018-02/gdpr-cross-border-transfers-january2018.pdf. Accessed 1 June 2019
33. Shen, Y., Siani, P.: Privacy enhancing technologies: a review (2011)
34. The Technology Analysis Division of the Office of the Privacy Commissioner of Canada, Privacy Enhancing Technologies – A Review of Tools and Techniques. https://www.priv.gc.ca/en/opc-actions-and-decisions/research/explore-privacy-research/2017/pet_201711/#heading-0-0-3. Accessed 1 June 2019
35. Homan, F.: Blockchain: Evolving Decentralized Identity Design. https://www.gartner.com/doc/3834863/blockchain-evolving-decentralized-identity-design. Accessed 1 June 2019
36. Metadium: Introduction to self-sovereign identity and its 10 guiding principles. https://medium.com/metadium/introduction-to-self-sovereign-identity-and-its-10-guiding-principles-97c1ba603872?fbclid=IwAR3ZKD0hsq2fzEcEi43T_QgDqWqn2EgvTTbqX3Jwo9AJlpYnd0Nl7BBzDn4. Accessed 1 June 2019
37. Metadium: Introduction to self-sovereign identity and its 10 guiding principles. https://medium.com/metadium/introduction-to-self-sovereign-identity-and-its-10-guiding-principles-97c1ba603872. Accessed 1 June 2019
38. Andreas, A.: Self-sovereign identity: whitepaper about the concept of self-sovereign identity including its potential. https://www.egiz.gv.at/files/download/Self-Sovereign-Identity-Whitepaper.pdf. Accessed 1 June 2019
39. Dunphy, P., Petitcolas, F.A.P.: A first look at identity management schemes on the blockchain. IEEE Secur. Priv. **16**, 20–29 (2018)
40. Elizabeth, M.: Is self-sovereign identity the ultimate GDPR compliance tool? https://medium.com/evernym/is-self-sovereign-identity-ssi-the-ultimate-gdpr-compliance-tool-9d8110752f89. Accessed 1 June 2019
41. Community Group Report: Decentralized Identifiers (DIDs) v0.11. https://w3c-ccg.github.io/did-spec/#security-considerations. Accessed 1 June 2019
42. The European Union Blockchain Observatory and Forum: Blockchain and the GDPR. https://www.eublockchainforum.eu/sites/default/files/reports/20181016_report_gdpr.pdf. Accessed 1 June 2019
43. Michèle, F.: Blockchains and data protection in the European Union. Eur. Data Prot. Law Rev. **4**(1), 17–35 (2018)
44. Christian, L., Rouven, H., Joel, T., Zac, M., Michael, S.: Uport: a platform for self-sovereign identity. http://blockchainlab.com/pdf/uPort_whitepaper_DRAFT20161020.pdf. Accessed 1 June 2019

45. Shocard Inc.: ShoCard whitepaper: identity management verified using the blockchain. http://shocard.com/wp-content/uploads/2018/01/ShoCard-Whitepaper-Dec13-2.pdf. Accessed 1 June 2019
46. Sovrin Foundation: Sovrin™: a protocol and token for self-sovereign identity and decentralized trust. https://sovrin.org/wp-content/uploads/Sovrin-Protocol-and-Token-White-Paper.pdf. Accessed 1 June 2019
47. TangleID. https://tangleid.github.io/. Accessed 1 June 2019

STRIDE-Based Threat Modeling
for MySQL Databases

James Sanfilippo, Tamirat Abegaz[✉], Bryson Payne, and Abi Salimi

University of North Georgia, Dahlonega, GA 30597, USA
{jpsanf0632, tamirat.abegaz, bryson.payne,
abi.salimi}@ung.edu

Abstract. Online, data-driven applications have become the cornerstone of e-commerce, health care, and our economy as a whole, as well as a part of almost every web application and mobile app in our daily lives. Unfortunately, this reliance on databases encourages attackers to exploit every attack surface to compromise these data-driven systems. While there are many security methodologies in place to protect and preserve the confidentiality, availability, and integrity of data, there are cases where these implementations fail, resulting in unintended consequences. In this paper, the STRIDE threat modeling is used to identify potential threats to the MySQL database management system to assist developers and admins in proactively securing these systems. Overall, this research identified spoofing, tampering, and denial of service as the more common threats facing data-driven applications, each of which can cause significant damage against an insufficiently protected MySQL database. Moreover, this paper suggests potential countermeasures to better protect MySQL databases against adversarial threats.

Keywords: STRIDE · Threat modeling · Spoofing · MySQL · Attack surface

1 Introduction

As threats against data-driven applications have become more common in the recent decade, so have the means of identifying and categorizing such threats. Threat modeling could be described as a set of assumptions about our adversaries. More formally, threat modeling can be defined as "the process of producing a simplified, abstract description of how an adversary would perform potential attacks or pose security threats to the system" [1]. There a number of threat-modeling classification methodologies such as Open Web Application Security Project (OWASP), Process for Attack Simulation and Threat Modeling (PASTA), Trike, Operationally Critical Threat, Asset, and Vulnerability Evaluation (OCTAVE), and the Microsoft Security Life Cycle (SDL), which implements the STRIDE approach for modeling threats [2, 3, 6]. While all the above models have their advantages and disadvantages, this paper focuses on the Microsoft STRIDE approach because of its popularity in the secure software development community [3]. Employing the STRIDE model could help developers and security professionals uncover the vulnerabilities of a system and identify the presence of threats. The STRIDE thread modeling process identifies six key threats: Spoofing,

© Springer Nature Switzerland AG 2020
K. Arai et al. (Eds.): FTC 2019, AISC 1070, pp. 368–378, 2020.
https://doi.org/10.1007/978-3-030-32523-7_25

Tampering, Repudiation, Information disclosure, Denial of Service, and Elevation of Privileges.

The first threat in STRIDE modeling is spoofing, where attackers impersonate someone or something in order to compromise authentication and abuse the system [4–7]. Tampering is another common threat in which attackers modify files, alter data, or inject code to affect the integrity of data or system [4, 6, 8–10]. On the other hand, Repudiation primarily deals with the deniability of users in performing specific actions or transactions, whereas information disclosure is a significant threat that exposes data to unauthorized users. Denial of Service is a serious known threat in which normal users could be denied access to database server due to a resource overload. Finally, there are elevation of privileges threats in which an entity could either masquerade as a user with higher privileges than his/her own or modify the data in a directory to gain greater privileges.

As this project is to uncover vulnerabilities and threats within a MySQL database-driven application, STRIDE would be a practical approach to use. Microsoft's Threat Modeling Tool 2016 can be used to create a threat model diagram to show the elements of a MySQL database infrastructure [11, 12]. A respective report will analyze and identify the threats discovered in the diagram model. The report will break down the threats in the system and their impact on the overall MySQL system. Afterward, ideas for countering threats should be proposed through mitigation, where each mitigation must finally be validated [12–15]. The remainder of this paper is organized as follows. In Sect. 2, related works relevant to this research are presented. Section 3 discusses the procedures and methodology used for this research. Section 4 presents the experimental results. Section 5 discusses mitigations and countermeasures. Finally, the conclusions and future research opportunities are presented in Sect. 6.

2 Related Work

STRIDE is an acronym for the types of threats ranging from the following: spoofing (S), tampering (T), repudiation (R), information disclosure (I), denial of service (D), and elevation of privilege (E). Spoofing is a common threat process that allows an attacker to impersonate a target user [4, 6, 8–10]. In a possible scenario of this project, spoofing could occur when an attacker gets unauthorized access to the system and would potentially rewrite or delete a database. In tampering, an attacker alters or overrides the target's data. For example, an attacker can get full access to the target's system by cracking a user's password. In MySQL, possible tampering can occur through SQL injections, where an attacker inserts strings of malicious code within the database query [3]. In repudiation, an adversary could modify the files or settings of an application and would not admit to doing so [7]. Since this project uses an XAMPP cross-platform (X), Apache (A) server, MySQL (M), PHP (P) and Perl (P) application stack, which relies on a browser to access MySQL, the browser client could contribute to repudiation. That is, it could claim not to have received data from an outside source beyond the trusted boundaries. If logs and audits were not in place to keep a record of access, time, and changes to data, it would be difficult to isolate the exact source of data changes.

Information disclosure is an attack in which an intruder gains unauthorized access to data [4, 6, 8, 10]. There have been instances such as a session information disclosure in which sensitive information could be viewed by unauthorized users [7]. An attacker may sniff HTTP traffic through a data flow in MySQL. Depending on the type of data, if the data flow is not well encrypted, the attacker could steal information, which would result in compliance violations. Denial of Service (DoS) threats could cause serious downtime of services by overloading a resource of a system such as a server [12–16]. A more recent variant of this threat is the Distributed Denial of Service (DDoS), where botnets or zombies are remotely controlled to flood a target system until it is overwhelmed completely [14].

Elevation of privileges threats allow an attacker to alter the privileges of accessing a system, giving the attacker more rights over the system than those of the target user's [4, 6, 8, 10]. In a MySQL scenario, an attacker could use the web server to impersonate the browser client, increase its privileges, and possibly revoke privileges of a target user. This project will focus on these common threats. As each threat functions differently, it is more appropriate to define a threat when the relevant vulnerabilities are identified and mitigations determined.

3   Implementation

3.1   STRIDE Threat Model Process and Elements

Many threats can occur on systems and identifying, classifying and addressing threats and vulnerabilities can be a difficult process. A threat model can be used to identify, analyze, rank and mitigate the vulnerabilities present in an application [7, 15]. Depending on their intended purpose, these attacks can both cause serious impact on the integrity of the data and exploit the weakness of a flawed system. Bertino's work focuses on the vulnerabilities created when a web application is paired with an SQL server. The paper classifies each threat using STRIDE methodology before proposing countermeasures for the various potential attack vectors [4]. Fang et al. [6] experimented with SQL injection-based attacks and detection using MySQL database server along with Apache and PHP. They demonstrated that different attacks can be either partially detected or not detected etc. [12].

The STRIDE threat modeling process consists of 5 elements, which are used to build a diagram for a threat model – external entity/interactor, process, data flow, data store, and trust boundary. An external entity/interactor, represented as a rectangle, shows the interaction with people like users and providers, and with other systems such as browsers and web applications [16–18]. The process element, represented as a sphere, handles and processes the data of an application and services. A data flow element, represented as an arrow, deals with the directional movement of data such as function calls, network traffic, and remote procedure calls. The data store, represented as two stacked lines, handles the location of data including the database, configuration files, shared memory, and queues/stacks. Finally, the trust boundary, represented as dashed lines, handles the process of privileges that a system trusts such as a file system, internet, and user/kernel mode [18, 19].

3.2   Threats Mapped to Each STRIDE Element

The STRIDE model uses known threats that are obtained from various vulnerability databases and from the Microsoft Security Response Center [19]. Elements of a STRIDE threat model diagram determine how certain threats are likely to occur. As shown in Table 1, different threats affect each element type differently. For instance, if the data flow diagram contains an external entity diagram element, then threat types spoofing and repudiation must be considered. On the other hand, if the process diagram element is present, all the threat models must be considered in mitigating the attack vectors.

Table 1. Mappings of data flow diagram to STRIDE model

Elements	S	T	R	I	D	E
External entity ▭	√		√			
Process ◯	√	√	√	√	√	
Data flow →		√		√	√	
Data store		√	√	√	√	√
Trust boundary − − − − −						

Elements of a STRIDE threat model diagram determine how certain threats are likely to occur. The likely threats associated with a data flow element are tampering, information disclosure, and DoS attacks [19]. Data store threats are likely related to tampering, information disclosure, DoS and repudiation. An external entity could possibility be vulnerable to both spoofing and repudiation attacks. Finally, a process is likely to be vulnerable to all threats of spoofing, tampering, repudiation, information disclosure, DoS, and elevation of privileges [19].

4   Results

Table 2 shows a result of the threat modeling tool identified approximately 64 possible threats. The majority were spoofing (19), followed by denial of service (16) and tampering (11). The least number of threats were repudiation (9), information disclosure (5) and elevation of privileges (4). To determine the severity of these threats, they were broken down by each portion of the data flow interaction of MySQL database.

4.1   HTTPS Related Threats

Table 3 shows the results of the threat model breakdown related to a MySQL database. Overall, as shown in Table 3, thirteen threats were identified using HTTPS interaction data flows. Spoofing threats likely occurred from the browser including both the external destination entity, external entity, and browser client process. The browser could be spoofed by an attacker, leading to data being sent to the attacker, and to giving

Table 2. Threat model results

Threat name	Number of threats discovered
Spoofing	19
Tampering	11
Repudiation	9
Information disclosure	5
Denial of service	16
Elevation of privileges	4
Overall total	64

unauthorized access to the browser client and to the MySQL database. In the case of tampering threats, a SQL injection vulnerability against the MySQL database could allow an attacker to manipulate the data. Another possibility is that an attacker could tamper with the data flow in HTTPS and corrupts the MySQL database data store. Repudiation threats could occur through the browser's external entity. Denial of Service threats are possible due to an external agent capable of interrupting the data flow. This leads to potential HTTPS data flows being interrupted, the data store being inaccessible, and excessive resources consumed by both browser client and MySQL database. Finally, a threat from elevation of privileges could be a possibility though impersonation of the browser client to allow additional privileges to be exposed to the attacker.

Table 3. Threat breakdown related to HTTPS

Threat name	Number of threats discovered
Spoofing	4
Tampering	2
Repudiation	2
Information disclosure	0
Denial of service	4
Elevation of privileges	1
Overall total	13

4.2   IOCTL Related Threats

As shown in Fig. 1, the data flow diagram depicts a number of threats. Table 4 shows a total of seventeen threats were identified in the IOCTL interaction data flow: spoofing (5), tampering (1), repudiation (6), information disclosure (1) and Denial of Service (4). Possible spoof threats can occur in the user file system spoofing from both the source and destination of the data store. These attacks could allow an attacker to write data on the target while allowing data to be incorrectly delivered to the *htdocs* folder of

XAMPP. The consequence of the latter could be spoofing of source and destination data stored in the XAMPP *htdocs* folder, allowing the attacker to also write the data to the target and deliver incorrect data to the browser. Tampering threats are possible if the attacker tampers with and corrupts the data store of the XAMPP *htdocs* folder. Threats by repudiation include receiving data logs from unknown sources, as well as a less-trusted subject being able to update or write log data. And potential problems from Denial of Service attacks includes inaccessible data store and potential interruption of the data flow on IOCTL interface, caused by an external agent to interrupt data flow and prevent access of data store from the trusted boundary.

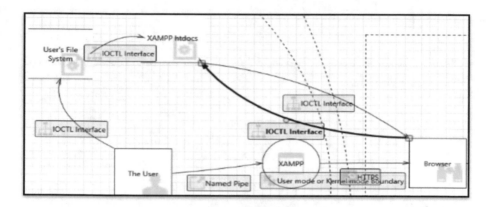

Fig. 1. IOCTL data flow

4.3   Named Pipe Related Threats

As shown in Fig. 2, two named pipe interactive data flow threats could occur: spoofing and elevation of privilege. Table 5 shows these two threats observed in the Named Pipe attack. Overall, the result shows that spoofing of the user external entity is possible, where an attacker spoofs the user and gains unauthorized access to XAMPP. Elevation of privilege threats are possible through XAMPP impersonation of the user to gain additional privilege.

4.4   SQL Related Threats

As shown in Fig. 3, seven SQL related interactive data flow threats identified. As indicated in Table 6, a total of 32 threats reported from each threat category: spoofing (8), tampering (8), repudiation (2), information disclosure (4), Denial of Service (8) and elevation of privileges (2). Spoofing threats can occur by an attacker spoofing a user as an external entity, the source and destination of a data store in MySQL database, and by the web application allowing an attacker to gain unauthorized access to the database, leading to possible information disclosure. Tampering can happen through SQL malicious code injections making the MySQL database vulnerable.

Table 4. Threat breakdown related to IOCTL

Threat name	Number of threats discovered
Spoofing	5
Tampering	1
Repudiation	6
Information disclosure	1
Denial of service	4
Elevation of privileges	0
Overall total	17

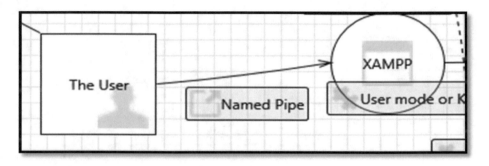

Fig. 2. Named pipe data flow

Table 5. Threat breakdown related to named pipe

Threat name	Number of threats discovered
Spoofing	1
Tampering	1
Repudiation	0
Information disclosure	0
Denial of service	0
Elevation of privileges	0
Overall total	2

A persistent cross-site scripting attack and tampering with the SQL data flow may lead to a corruption of the database. Repudiation threats could include the data store of the MySQL database denying data being potentially written, and a web application claiming that it did not receive data from outside the trusted boundary. Information disclosure threats range from weak access controls allowing an attacker to read data not intended for disclosure, to sniffing of data flows that could be used to attack other parts

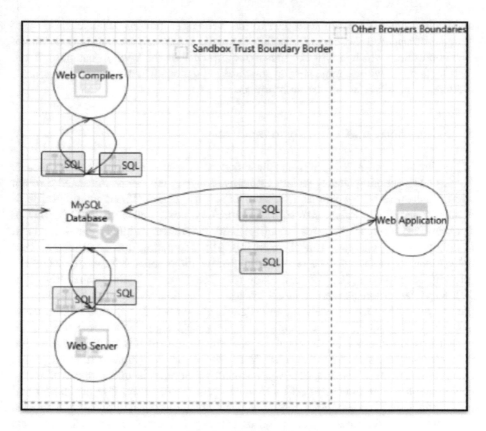

Fig. 3. SQL data flow

Table 6. Threat breakdown related to SQL

Threat name	Number of threats discovered
Spoofing	8
Tampering	8
Repudiation	2
Information disclosure	4
Denial of service	8
Elevation of privileges	2
Overall total	32

of the system. Denial of Service threats are possible because of high transaction loads on web servers, web applications, or MySQL databases. Other issues could lead to the data store being inaccessible, the data flow of SQL being interrupted, and web application to stop or crash. Finally, elevation of privilege could occur from the web application. An attacker can send data to the web application and change the flow of data per the attacker's choosing. Another approach is remotely executing code on the web application, which in turn executes malicious code on the MySQL database. In the next section, we will examine several standard mitigations that can be used to address the threats identified in this research project.

5 Mitigations and Countermeasures

Many threats found in a MySQL database can be mitigated through standard means. For instance, spoofing attacks violate authentication, so they can be mitigated through strong authentication techniques and tactics such as cookie authentication, Kerberos authentication, PKI (Public Key Infrastructure) systems such as SSL/TLS (Secure Sockets Layer/Transport Layer Security) and certificates, digital signatures, and building a strong encryption mechanism [18, 19]. Tampering threats are categorized under integrity violations. The recommended mitigations include those of mandatory integrity controls, ACLs (Access Control Lists), and digital signatures [18, 19]. Repudiation threats are countered by non-repudiation, where mitigations include the use of secure logging, auditing and digital signatures [18, 19]. In addition, using and maintaining strong client authentication would help minimize the possibility of an intruder performing actions on behalf of a client. Confidentiality is the classification that information disclosure falls under. Normally, it can be mitigated through proper encryption and even the use of ACLs. Denial of Service is under the availability classification, where the use of ACLs is once again favored here. Use of filtering and quotas would also be strongly recommended for mitigating Denial of Service attacks. Authorization is the basis that elevation of privileges revolves around, where threats can be mitigated through ACLs, group or role membership, privilege ownership, and input validation [18, 19].

6 Conclusion

With increasing reliance on data being stored in application databases, the risk of such data being compromised by various threats has risen. While there are many classification methodologies in place to protect and preserve the integrity of data, there are cases where these implementations fail, resulting in unintended consequences. The STRIDE threat model, which measures and outlines threats such as spoofing, tampering, information disclosure, denial of service, and elevation of privileges, can be used to both identify and understand how these threats could impact the security of database management systems. This paper utilized the STRIDE model on a MySQL database-driven app to identify a variety of threats and their impacts. Overall, spoofing, tampering, and denial of service threats were the more common threats identified, each of which can cause significant damage should they be exploited against a MySQL database. The results of this study can help with shedding some light on the types of threats that could possibly exist within a MySQL database system. The less common threats such as repudiation, information disclosure, and elevation of privilege can also lead to significant issues such as corporate espionage and cybercrime. The suggested solutions should help in addressing the threats identified in MySQL databases and data-driven applications.

For future work, we intend to continue studying MySQL database threats in a production environment to determine if other vulnerabilities that were not identified in the STRIDE threat model could compromise the database.

References

1. Marback, A., Do, H., He, K., Kondamarri, S., Xu, D.: Security test generation using threat trees. In: ICSE Workshop on Automation of Software Test, pp. 62–69 (2009)
2. Hasan, R., Myagmar, S., Lee, A.J., Yurcik, W.: Toward a threat model for storage systems. In: Proceedings of the 2005 ACM Workshop on Storage Security and Survivability, StorageSS 2005, pp. 94–102 (2005). https://doi.org/10.1145/1103780.1103795
3. Abomhara, M., Køien, G., Gerdes, M.: A STRIDE-based threat model for telehealth systems (2015)
4. Bertino, E., Bruschi, D., Franzoni, S., Nai-Fovino, I., Valtolina, S.: Threat modelling for SQL servers. In: Chadwick, D., Preneel, B. (eds.) Communications and Multimedia Security. IFIP — The International Federation for Information Processing, vol. 175. Springer, Boston (2005)
5. Chadwick, D.: Threat modelling for active directory. In: Chadwick, D., Preneel, B. (eds.) Communications and Multimedia Security. IFIP — The International Federation for Information Processing, vol. 175. Springer, Boston (2005)
6. Fang, Y., Peng, J., Liu, L., Huang, C.: WOVSQLI: detection of SQL injection behaviors using word vector and LSTM. In: Proceedings of the 2nd International Conference on Cryptography, Security and Privacy, pp. 170–174. ACM, March 2018
7. Marback, A., Do, H., He, K., Kondamarri, S., Xu, D.: A threat model-based approach to security testing. Softw. Pract. Experience **43**(2), 241 (2013). https://doi.org/10.1002/spe.2111
8. Potteiger, B., Martins, G., Koutsoukos, X.: Software and attack centric integrated threat modeling for quantitative risk assessment. In: Proceedings of the Symposium and Bootcamp on the Science of Security, pp. 99–108. ACM, April 2016
9. Mathew, S., Petropoulos, M., Ngo, H.Q., Upadhyaya, S.: A data-centric approach to insider attack detection in database systems. In: International Workshop on Recent Advances in Intrusion Detection, pp. 382–401. Springer, Heidelberg, September 2010
10. Shostack, A.: Threat Modeling: Designing for Security. Wiley, Hoboken (2014)
11. Shevchenko, N., Chick, T.A., O'Riordan, P., Scanlon, T.P., Woody, C.: Threat Modeling: A Summary of Available Methods (2018)
12. Rodsan: Microsoft threat modeling tool – azure, 16 August 2018. https://docs.microsoft.com/en-us/azure/security/azure-security-threat-modeling-tool-feature-overview. Accessed 12 Dec 2018
13. Kumar, N., Sharma, S.: Study of intrusion detection system for DDoS attacks in cloud computing. In: Tenth International Conference on Wireless and Optical Communications Networks (WOCN), pp. 1–5. IEEE, July 2013
14. Lonea, A.M., Popescu, D.E., Tianfield, H.: Detecting DDoS attacks in cloud computing environment. Int. J. Comput. Commun. Control **8**(1), 70–78 (2013)

15. Mishra S., Mahanty C., Dash S., Mishra B.K.: Implementation of BFS-NB hybrid model in intrusion detection system. In: Recent Developments in Machine Learning and Data Analytics, pp. 167–175. Springer, Singapore (2019)

16. Kambire, M. K., Gaikwad, P. H., Gadilkar, S. Y., & Funde, Y. A: An improved framework for tamper detection in databases. Int. J. Comput. Sci. Inform. Technol. **6**, 57–60 (2015)

17. Dhillon, D.: Developer-driven threat modeling: lessons learned in the trenches. IEEE Secur. Priv. **9**(4), 41–47 (2011)

18. Zargar, S.T., Joshi, J., Tipper, D.: A survey of defense mechanisms against distributed denial of service (DDoS) flooding attacks. IEEE Commun. Surv. Tutor. **15**(4), 2046–2069 (2013)

19. Introduction to Microsoft Security Development Life Cycle (SDL) Threat Modeling, pp. 1–77 (n.d.). [PDF file] Microsoft https://download.microsoft.com/download/9/3/5/935520EC-D9E2-413E-BEA7-0B865A79B18C/Introduction_to_Threat_Modeling.ppsx. Accessed 13 Dec 2018

Efficient RSA Crypto Processor Using Montgomery Multiplier in FPGA

Lavanya Gnanasekaran$^{(\boxtimes)}$, Anas Salah Eddin, Halima El Naga,
and Mohamed El-Hadedy

Department of Electrical and Computer Engineering,
California State Polytechnic University, Pomona, CA 91768, USA
{gnanasekaran,asalaheddin,helnaga,mealy}@cpp.edu

Abstract. With the advancement of technology in data communication, security plays a major role in protecting user's data from adversaries. Cryptography is a technique which consists of various algorithms to provide secure communication during data transfer. One of the most widely used and highly secure algorithm is RSA (Rivest, Adi Shamir and Leonard Adleman). This research is going to focus on designing an efficient crypto processor for RSA on Nexys4, a ready-to-use FPGA (Field Programmable Gate Array) board. There are various techniques in implementing RSA algorithm in hardware platforms. Primary concentration is to implement the algorithm in a high performant manner, and it will be achieved using Montgomery multiplication technique. RSA is a public key cryptography system which involves generation of public key for encryption and private key for decryption. Building blocks of RSA includes: Two multiplier blocks, two blocks to verify the primality of random numbers, one GCD (Greatest Common Divisor) block to check the validity of a public key, two modular exponential blocks – one is for generating encrypted or cipher message and another is to retrieve the original message from the cipher text. Simulation and synthesis of these blocks is achieved and verified using Xilinx Vivado Design Suite.

Keywords: Cryptography · RSA · Montgomery multiplication · FPGA

1 Introduction

In this digital era, cryptography plays a critical role in everyone's daily life. We use a wide range of electronic devices and applications to help us with our everyday routine. Data stored and transmitted between two entities needs to be done in a secure manner to prevent them from reaching untrusted sources. Cryptography is a process followed to convert original message into ciphertext as part of encryption and convert the ciphertext back to original message in decryption.

Symmetric and asymmetric cryptography are two ways to secure original message from attackers. Symmetric cryptography uses same key to encrypt and

© Springer Nature Switzerland AG 2020
K. Arai et al. (Eds.): FTC 2019, AISC 1070, pp. 379–389, 2020.
https://doi.org/10.1007/978-3-030-32523-7_26

decrypt and is therefore faster than asymmetric algorithms but isn't secure enough. On the other hand, to ensure higher security, asymmetric cryptography uses a public key to encrypt the message and private key to decrypt it. RSA and Elliptic Curve Cryptography (ECC) are some of the examples of asymmetric cryptography.

This research will focus on explaining the architecture of RSA and its implementation using efficient multiplication and exponentiation components on FPGA.

Theoretical concepts of RSA algorithm is explained in Sect. 2. Required components like Montgomery multiplication, Montgomery exponentiation, primality tests and its simulation results are explained in its subsection. Future plans are discussed in Sect. 3.

2   RSA Algorithm

RSA is a widely-used, highly secure public key cryptographic algorithm. It generates a public key with which user's message is encrypted and a private key is generated to decrypt the message. Implementation of RSA is computationally expensive as it operates on large numbers.

Exponentiation operation is fundamental in RSA cryptosystems and it was a challenging task to implement the necessary blocks using efficient algorithms. These operations require high computations when implemented by software. To achieve high security and greater performance, it is often more advantageous to design and develop cryptosystems in hardware [4].

Figure 1 explains the complete block diagram of RSA algorithm. It starts with generating two large random prime numbers p, q and n is computed by

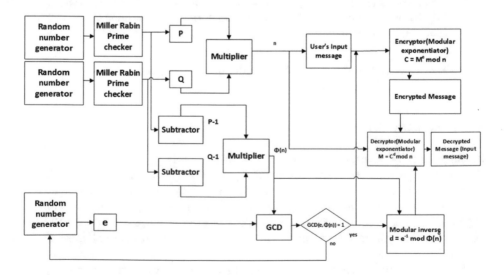

Fig. 1. Block diagram of RSA algorithm

multiplying the prime numbers using add-and-shift multiplier.

$$n = p \times q \tag{1}$$

$\phi(n)$ is computed by multiplying the outputs of subtractor of prime numbers p and q

$$\phi(n) = (p-1) \times (q-1) \tag{2}$$

A random number e is generated such that it is between 1 and $\phi(n) - 1$, and the Greatest Common Divisor (GCD) of e and $\phi(n)$ is 1. A random number generator block and GCD block generates this public key in hardware.

$$e \in \{1, 2, 3...\phi(n) - 1\}, GCD(e, \phi(n)) = 1 \tag{3}$$

The following public key is used to encrypt the incoming message to obtain cipher-text:

$$C = M^e \bmod n \tag{4}$$

A private key d is generated using modular inverse operation from e and $\phi(n)$ by

$$d = e^{-1} \bmod \phi(n) \tag{5}$$

Using the above generated private key, the original message can be decrypted back from the cipher-text by

$$M = C^d \bmod n \tag{6}$$

Figures 2 and 3 shows power estimation and utilization summary of RSA respectively. Figure 4 shows simulation results of the RSA algorithm designed and implemented using Very High-Speed Integrated Circuit Hardware Description Language (VHDL) on Xilinx Vivado Design Suite. Random prime numbers p and q are represented as reg_p and reg_q respectively in the program, and their product is stored as reg_n. Input *message* is encrypted into *cipher_text* using public key reg_n and reg_e. *cipher_text* is then decrypted back to input message using private key reg_d.

Since this implementation is utilizing 10% of look-up tables, 10 similar blocks can be implemented on a Nexys4 FPGA board.

2.1 Montgomery Multiplication

Multiplication and division require complex computations when implemented in hardware platforms. In a traditional approach, modular multiplication involves multiplication followed by division [6]. Equations (4) and (6) shows the need to compute modular exponentiation in RSA, each requiring lots of modular multiplication components. There are many approaches to perform multiplication such as multiply then divide, interleaving multiplication and reduction, Brickell's method [2].

To design and develop an efficient crypto-processor, Montgomery multiplication has been chosen for all our modular multiplication calculations. This

Power estimation from Synthesized netlist. Activity derived from constraints files, simulation files or vectorless analysis. Note: these early estimates can change after implementation.

On-Chip Power

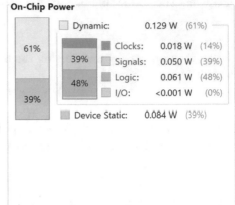

Dynamic:	0.129 W (61%)
Clocks:	0.018 W (14%)
Signals:	0.050 W (39%)
Logic:	0.061 W (48%)
I/O:	<0.001 W (0%)
Device Static:	0.084 W (39%)

Total On-Chip Power:	**0.213 W**
Design Power Budget:	**Not Specified**
Power Budget Margin:	**N/A**
Junction Temperature:	28.0°C
Thermal Margin:	59.0°C (12.8 W)
Effective ϑJA:	4.6°C/W
Power supplied to off-chip devices:	0 W
Confidence level:	Medium

Launch Power Constraint Advisor to find and fix invalid switching activity

Fig. 2. Summary of power estimation for 16-bit RSA

Fig. 3. Utilization summary for RSA

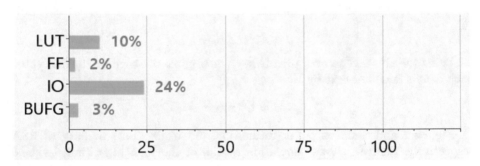

Fig. 4. Simulation results of encryption and decryption blocks of RSA

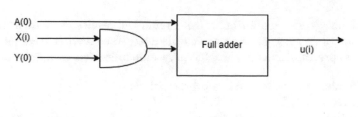

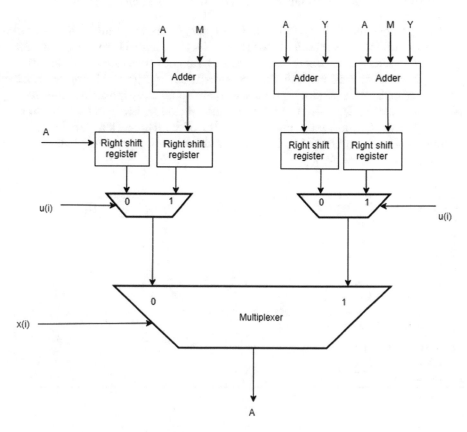

Fig. 5. Architecture of Montgomery modular multiplier

technique avoids the traditional division operation, and replaces it with shift-and-add operation. Montgomery modular multiplication is computed as

$$A = X \times Y \times R^{-1} \bmod M$$

$$(7)$$

$$R = 2^n, \text{where } n \text{ is number of bits}$$

Equation (7) represents a modular multiplication in Montgomery domain. Results of the above computation can remain in Montgomery domain, as converting it back to real world coordinates is a costly operation. Since we use

Montgomery exponentiation, it can handle the results of modular multiplication in Montgomery domain and avoids the need to convert it back [3].

Two preconditions for Montgomery multiplication technique are

- Modulus M needs to be co-prime with R
  i.e. $GCD(M, R) = 1$
- Multiplicand and multiplier need to be smaller than modulus M [5]

Montgomery multiplier requires $2n(n+1)$ single-precision multiplications due to n loop iterations as seen in Algorithm 1. As the result is computed in Montgomery domain, converting it back to real world coordinates requires additional single-precision operations, thereby having a total of $4n(n + 1)$ single-precision multiplications. This is slower than the traditional multiplication algorithms, which perform at $2n(n + 1)$ computational efficiency. Since we don't convert it to real world coordinates and the results of multiplication can be used in exponentiation component in Montgomery domain, this is an efficient modular multiplication technique [1].

Algorithm 1. Montgomery modular multiplication algorithm [1]

1: Inputs: X, Y, M
2: Ouput: $A = XYR^{-1} mod M$
3: $A \leftarrow$ zeros for $n + 1$ bits
4: $b = 2$
5: $M' = M^{-1} mod b$
6: **for** i: from 0 to n - 1 **do**
7:     $u(i) = (A_0 + X_i \times Y_0) \times M' mod b$
8:     $A = \frac{A + X_i \times Y + u(i) \times M}{2}$
9: **if** $A >= M$ **then**
10:     $A = A - M$
11: Return A

Table 1. Utilization summary of Montgomery multiplication

Resource	Utilization	Available	Utilization %
LUT	48	63400	0.08
FF	29	126800	0.02
IO	19	210	9.05
BUFG	1	12	3.13

For every iteration i in the algorithm, $u(i)$ is computed using a Full Adder, and the divide operation is performed using a right-shift register as shown in

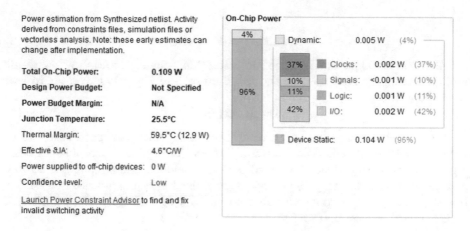

Fig. 6. Summary of power estimation for 16-bit Montgomery multiplier

the architecture in the Fig. 5. The entire algorithm is designed and implemented using Xilinx Vivado Design Suite. Figure 6 shows the summary of power estimation for 16-bit Montgomery multiplier. Figure 7 shows the simulation results of 16-bit Montgomery multiplication.

Name	Value	0 ns	5 ns	10 ns	15 ns	20 ns	25 ns	30 ns	35 ns	40 ns	45 ns	50 ns	55
⟩ x[15:0]	16538											16538	
⟩ y[15:0]	16538											16538	
⟩ m[15:0]	23457										23457		
⟩ A[15:0]	22303	0										22303	

Fig. 7. Simulation results of Montgomery multiplier from Xilinx Vivado Design Suite

Very High-Speed Integrated Circuit Hardware Description Language (VHDL) is the language used to describe the hardware design. A test bench has been written to verify the design for variable bit sizes. The program has been then synthesized and implemented on a Nexys4 FPGA board. Table 1 shows the utilization summary on implementing the algorithm on FPGA board.

2.2   Montgomery Exponentiation

Figure 8 shows the architecture of Montgomery exponentiation. It involves 4 Montgomery multiplier blocks and is repeated over t iterations, where t is number of bits in the exponent.

$$Result(A) = x^e \, mod \, m \qquad (8)$$

As the entire algorithm is carried out using Montgomery multiplication, result remain in Montgomery domain and the final step of Eq. (9) converts it to real world coordinates.

$$A = Mont(A, 1) \qquad (9)$$

Figure 9 shows the simulation results of Montgomery exponentiation component for 16 bits using Xilinx Vivado Design Suite. Since each block of RSA algorithm was designed and developed as a stand-alone reusable component, they can be simulated to verify the design, synthesized and implemented on FPGA board. Montgomery exponentiation utilizes   3% of look-up tables on Nexys4 FPGA board as seen in Table 2.

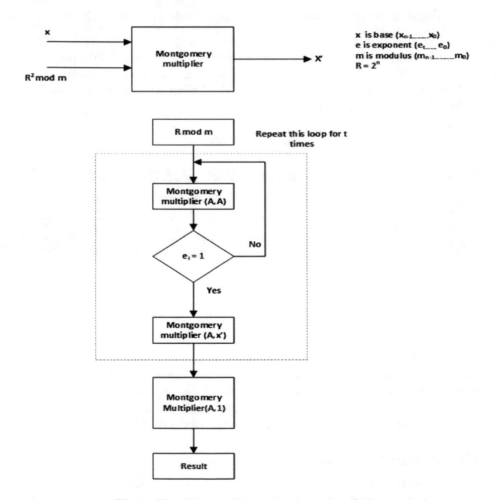

Fig. 8. Flowchart for Montgomery exponentiation

Table 2. Utilization summary of Montgomery exponentiation algorithm

Resource	Utilization	Available	Utilization %
LUT	1796	63400	2.83
FF	2696	126800	2.13
IO	18	210	8.57
BUFG	2	32	6.25

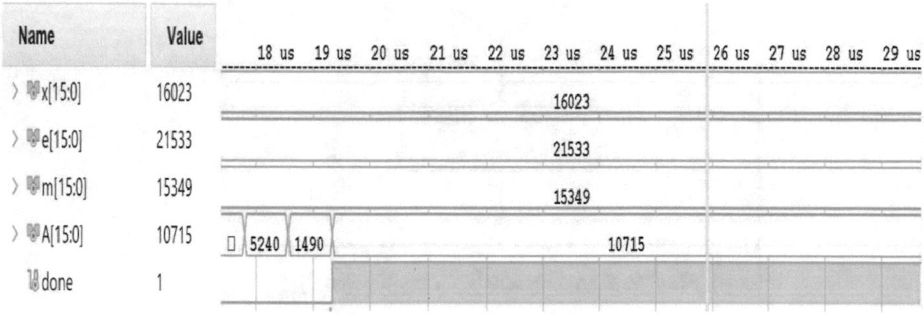

Name	Value	18 us	19 us	20 us	21 us	22 us	23 us	24 us	25 us	26 us	27 us	28 us	29 us
> x[15:0]	16023						16023						
> e[15:0]	21533						21533						
> m[15:0]	15349						15349						
> A[15:0]	10715	5240	1490					10715					
done	1												

Fig. 9. Simulation results of Montgomery exponentiation

2.3   Miller Rabin Primality Test

Primality tests are computationally expensive since it involves repeated exponentiation and modular multiplications. RSA algorithm begins with the need to generate two random prime numbers. Random numbers were generated on FPGA using Linear Feedback Shift Register (LFSR). These random numbers were passed onto a primality test before they can be used further in the algorithm. We chose Miller Rabin to confirm primality of random numbers. This is mainly due to the ability to configure witness count parameter in the algorithm. The higher the witness count, higher the number of iterations to generate a strong prime number. Another advantage of Miller Rabin is the way it breaks the loop as soon as it finds a composite number as shown in Fig. 10.

Error probability of Miller Rabin is less than $(1/4)^t$, where t is witness count. This algorithm has also proven to be better than Fermet's primality test and Solovay-Strassen test due to the computational costs involved in calculating modular multiplications [1]. Figure 10 shows the usage of a reusable random number generator block and modular exponentiation block in Miller Rabin primality tester algorithm.

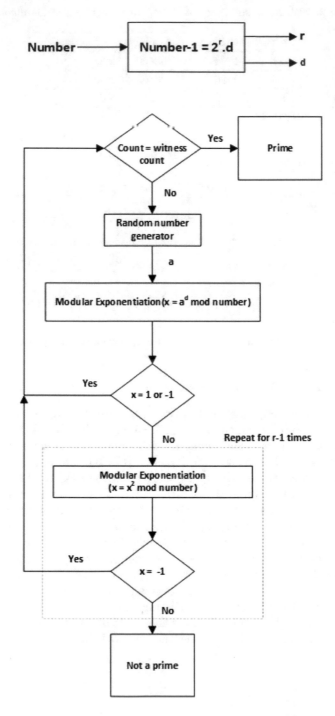

Fig. 10. Architecture of Miller Rabin primality test

3   Future Work

Metrics like Maximum frequency, Throughput, Performance/watt, Performance/ area for 1024-bits of RSA algorithm will be calculated and necessary steps will be taken to improve the design in components that need immediate attention.

After designing and developing an efficient RSA crypto processor, we will be focusing on implementing other widely used cryptographic algorithms like SHA-1 and Elliptic Curve Cryptography (ECC) to understand and analyze its performance using various key metrics. Since all the building blocks were developed to work as a stand-alone component, they can be re-used and improved upon when using it in other cryptographic algorithms.

Our traditional computers have the capability to understand the data in binary digits, i.e., the two states 0 and 1. Quantum computing is an advanced computing technique that operates on quantum mechanical phenomena and uses quantum bits to represent the data. Research has proven that our current public key cryptographic algorithms like RSA and ECC can be broken using quantum computing. We strongly believe that the metrics obtained from the traditional algorithms will help us to evaluate and analyze the performance of quantum resist algorithms being submitted by cryptographers to National Institute of Science and Technology (NIST).

Finally, we are going to produce a working prototype of cyber-physical model of a Digital Twin, which will operate on the FPGA board to perform advanced data analytics to follow the evolution of a production process.

References

1. Menezes, A.J., van Oorschot, P.C., Vanstone, S.A.: Handbook of Applied Cryptography, 5th edn. (2001)
2. Sahu, S.K., Pradhan, M.: Implementation of Modular multiplication for RSA Algorithm. In: International Conference on Communication Systems and Network Technologies, Katra (2011)
3. Bansal, M., Kumar, A., Devrari, A., Bhat, A.: Implementation of modular exponentiation using montgomery algorithms. Int. J. Sci. Eng. Res. 6(11) (2015)
4. Martínez, M.A.G., Luna, G.M., Henríquez, F.R.: Hardware implementation of the binary method for exponentiation in GF(2m). In: Proceedings of the Fourth Mexican International Conference on Computer Science, Tlaxcala (2003)
5. Nedjah, N., De Macedo Mourelle, L.: Hardware architecture for the montgomery modular multiplication. In: International Conference on Information Technology Coding and Computing, Las Vegas (2004)
6. Eberle, H., Gura, N., Shantz, S.C., Gupta, V., Rarick, L.: A public-key cryptographic processor for RSA and ECC. In: Proceedings. 15th IEEE International Conference on Application-Specific Systems, Architectures and Processors, Galveston (2004)

New Residue Signed-Digit Addition Algorithm

Shugang Wei[✉]

Gunma University, Maebashi, Gunma 376, Japan
wei@gunma-u.ac.jp

Abstract. In this paper, we propose a new residue signed-digit (SD) addition algorithm, which can be used in RSA public-key cryptosystem with a large modulus m. We use $\mu = m - 2^n$ and $-2^{n-1} + 1 \leq \mu < 0$ to calculate the residue n-digit SD number additions. Thus, the residue addition is implemented by (1) adding two n-digit SD numbers and (2) reducing the (n+1)-digit SD number obtained by the SD addition to an n-digit SD number using μ. Thus, the circuit of residue SD adder is constructed with two SD adders and some multiplexers, and no carry propagations arise during the residue addition. We have designed the circuits with VHDL for the encryption processor using the proposed residue SD adders. By comparing the performance of the encryption processor with that of the binary architectures, the proposed encryption processor is superior to the binary ones in computing time and low power.

Keywords: Signed-digit number · Residue arithmetic · Modulo m addition

1 Introduction

In public-key cryptography algorithms such as RSA cryptosystem [1], the essential on number operation is residue multiplication, which is implemented based on an addition modulo m, where m is the modulus as a key with very large word length. When an ordinary binary number arithmetic is used, the carry propagation during very large word length additions limits the processing speed of the RSA encryption [2]. It is known that carry propagation is limited to one position during additions of signed-digit (SD) numbers [3]. We have presented residue addition algorithms using SD arithmetic [4, 5]. In the residue SD addition circuit [5], the modulus m is chosen as $m = 2^n + \mu (1 \leq \mu \leq 2^{n-1})$, so that two SD adders can be used for the implementation of residue SD addition. However, the modulus m must be $2^n < m < 1.5 \times 2^n$, so that m having a value as $1.5 \times 2^n \leq m < 2^{n+1}$ can be not chosen as a key for RSA cryptosystem.

In this paper, we present a new architecture for the residue SD addition using $2^{n-1} + 1 \leq m \leq 2^n - 1$, that is, $m = 2^n + \mu$ and $-2^{n-1} + 1 \leq \mu < 0$ to calculate the residue operations. A residue multiplication can be performed by repeating the proposed residue addition with residue partial products.

For RSA encryption, a sequential residue multiplier is constructed with the high speed SD adders, several shifters and registers. The circuits based on both the proposed SD and the binary architectures have been designed. By comparison with the binary

© Springer Nature Switzerland AG 2020
K. Arai et al. (Eds.): FTC 2019, AISC 1070, pp. 390–396, 2020.
https://doi.org/10.1007/978-3-030-32523-7_27

ones, the proposed encryption processor is superior to the binary ones in computing time and low power, and may work at a faster clock rate for large word lengths.

2 Residue Signed-Digit Number Addition

2.1 Residue Addition Method Using Binary Numbers

Let x, y and m are positive numbers in n-bit binary number representation, where x $y < m$ and the most significant bit of m is 1. Thus, the residue addition of $s = x + y$ mod m can be calculated by two steps: (1) adding x and y to obtain z with $(n+1)$-bit binary number z; then (2) reducing z from $(n+1)$-bit to s in n-bit representation and $s < n$. That is, when $z > m$, $s = z - m$ is performed for the bit length reduction. To simplify the reduction operation, 2's complementation of m, m^*, is added to z and only the carry-out bit C is used for checking if $z > m$. For example, let $m = 11 = (1011)$, $x = 7 = (0111)$ and $y = 5 = (0101)$. Then $m^* = (0101)$, $z = x + y = (1100)$ and $z + m^* = (10001)$. Since the carry-out bit $C = 1$, $s = (0001) = 1$. In the algorithm using binary numbers, two n-bit additions are performed. When n is very large, the carry propagation time is long and limit speed of the residue addition.

2.2 Residue Addition Algorithm Using Signed-Digit Numbers

The n-digit radix-two SD number representation for an integer x is given as follows:

$$x = x_{n-1}2^{n-1} + x_{n-2}2^{n-2} + \ldots + x_0, \, x_i \in \{-1, 0, 1\}, (i = 0, 1, \ldots, n-1) \quad (1)$$

which can be denoted as $x = (x_{n-1}, x_{n-2}, \ldots, x_0)$. In the SD number representation, x has a value in a range of $[-(2^n - 1), 2^n - 1]$. An addition $z = x + y$, where x and y are n-digit SD numbers, is performed in parallel as follows:

(1) Calculate the intermediate sum w_i and carry c_i in each digit using the following equations:
In the case of $|x_i| = |y_i|$:

$$w_i = 0 \text{ and } c_i = \frac{x_i + y_i}{2};$$

In the case of $|x_i| \neq |y_i|$,
if $x_i + y_i$ and $x_{i-1} + y_{i-1}$ have the same sign, $w_i = -(x_i + y_i)$ and $c_i = x_i + y_i$; else $w_i = x_i + y_i$ and $c_i = 0$.
(2) Add the sums and carries, that is, $z_i = w_i + c_{i-1}$.
Since the addition using the SD number representation is performed in parallel, we introduce the SD number system into residue arithmetic with large word length.

Let X be an integer and m be a modulus, where m is an n-digit binary number meeting $2^{n-1} + 1 \leq m \leq 2^n - 1$. Then $x = <X>_m$ is represented as an n-digit SD number having multiple possible values such as $x = <X>_m = |X|_m$ and

$x = \langle X \rangle_m = |X|_m - m$. For example, for $X = (1, 0, -1, 1, 1) = 15$ and $m = (1, 0, 1, 1) = 11$, we have $x = \langle X \rangle_m = (1, -1, 0, 0) = 4$, or $x = \langle X \rangle_m = (0, -1, -1, 0) = -6$. When $X = 24$, the 4-digit SD numbers may be $x = (1, 1, 1, -1) = 13$, $x = (0, 1, -1, 0) = 2$ and $x = (-1, 0, -1, 1) = -9$.

Let μ be a residue parameter and defined as $\mu = m - 2^n$. Then $\langle 2^n \rangle_m = -\mu$. Since $2^{n-1} + 1 \le m \le 2^n - 1$, we have $-2^{n-1} + 1 \le \mu < 0$ and $\mu - (0, -1, \mu_{n-3}, \dots, \mu_0)$, where $\mu_i \in \{-1, 0, 1\}$, $(i = 0, 1, \dots, n-3)$. The representations of μ with the maximum and minimum values are $\mu_{max} = (0, -1, 1, \dots, 1) = -1$ and $\mu_{min} = (0, -1, -1, \dots, -1) = -2^{n-1} + 1$. We can calculate the SD addition modulo m with the n-digit SD number representation by the following algorithm.

[Residue addition algorithm: s $= \langle x + y \rangle_m$]

Input: $x; y$ (n-digit SD numbers)

Output: s (n-digit SD number)

1) Perform SD addition, obtain an $(n+1)$-digit SD number z,
$$z = x + y = (z_n, z_{n-1}, z_{n-2}, \dots, z_0)$$

2) Reduce the number from $n + 1$ digits to n digits using the values of z_n and z_{n-1}:

(Case 1) When $z_n = 0$, let $s_i = z_i$, then
$$s = (s_{n-1}, s_{n-2}, \dots, s_0) = (z_{n-1}, z_{n-2}, \dots, z_0)$$

(Case 2) When $z_n \ne 0$ and $z_n = -z_{n-1}$, let $s_i = z_i$ for $i = 0, 1, 2 \dots, n-2$ and $s_{n-1} = z_n$.
$$s = (s_{n-1}, s_{n-2}, \dots, s_0) = (z_n, z_{n-2}, \dots, z_0)$$

(Case 3) When $z_n \ne 0$ and $z_{n-1} = 0$, we have
$s = z - z_n m = z - z_n(2^n + \mu) = z_n 2^n + z_{[n-2:0]} - z_n 2^n - z_n \mu = z_{[n-2:0]} + (-z_n)\mu$.

Therefore, only an addition of $(n-1)$-digit SD numbers, $s = z_{[n-2:0]} + (-z_n)\mu$, is performed.

(Case 4) When $z_n \ne 0$ and $z_n = z_{n-1}$, we have
$$\begin{aligned} s = z - 2z_n m &= (z_n, z_{n-1}, z_{n-2}, \dots, z_0) - 2z_n(1, 0, -1, \mu_{n-3}, \dots, \mu_0) \\ &= (z_n, z_n, z_{n-2}, \dots, z_0) - z_n(1, 0, -1, \mu_{n-3}, \dots, \mu_0, 0) \\ &= (z_n, z_n, z_{n-2}, \dots, z_0) - z_n(1, 1, \mu_{n-3}, \dots, \mu_0, 0) \\ &= (z_{n-2}, \dots, z_0) - z_n(\mu_{n-3}, \dots, \mu_0, 0) \end{aligned}$$

Thus, in this case, an addition of $(n-1)$-digit SD numbers, $s = z_{[n-2:0]} + (-z_n) \times 2\mu_{[n-3:0]}$ is performed.

In this algorithm, we use $z_{[j:k]}$ denote $(z_j, z_{j-1}, \dots, z_k)$. By the proposed algorithm, an $(n+1)$-digit SD numbers can be reduced into an n-digit SD number modulo m meeting $2^{n-1} + 1 \le m \le 2^n - 1$. This means that an n-bit binary number else 2^{n-1} can be chosen as the modulus m. We give some SD numbers obtained by performing the first SD addition, and reduce them using the operations of (Case 1) $\sim$ (Case 4).

[Example 1]. Let $m = (1,0,1,1) = 11$ and $\mu = (0,-1,0,-1) = -5$. When $z = (0,-1,-1,1,1) = -9$, in (Case 1), $s = (-1,-1,1,1) = -9$. When $z = (1,-1,-1,1,1) = 7$, in (Case 2), $s = (1,-1,1,1) = 7$. When $z = (1,0,-1,1,1) = 15$, in (Case 3), $s = (0,-1,1,1) + (0,1,0,1) = (0,1,0,0) = 4$. When $z = (1,1,-1,-1,1) = 19$, in (Case 4), $s = (0,-1,-1,1) + (0,0,1,0) = (0,-1,0,1) = -3$.

3 Circuit Design of Residue Signed-Digit Addition

For the implementation of the proposed residue SD addition algorithm, the operations in (Case 1) and (Case 2) are no needs of arithmetic circuits. Since both the operations in (Case 3) and (Case 4) need an SD addition having an $(n-1)$-digit augend $(z_{n-2}, \ldots, z_0)$, we can use one multiplexer to select $(-z_n)\mu$ or $(-z_n) \times 2\mu_{[n-3:0]}$ to be added. Therefore, the reduction operations in (Case 3) and (Case 4) are implemented by one SD adder (SDA) as shown in Fig. 1. X, Y with n digits and $-\mu$ with $(n-1)$ digits are

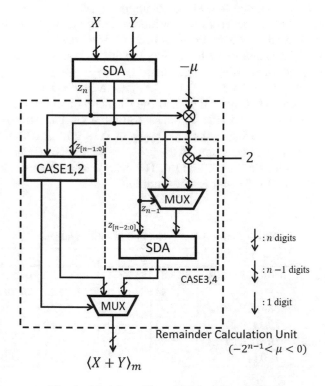

Fig. 1. Modulo m Signed-Digit Adder (MSDA)

inputted and the result of the residue SD addition is outputted in an n-digit SD number representation. Only two SDAs are used for the addition and the reduction, respectively. The SDA is designed as Fig. 2, and the delay time of the SD addition is a

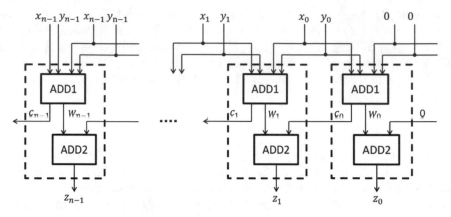

Fig. 2. Signed-Digit Adder (SDA)

constant which is the sum of delays by two parts: ADD1 for generating the intermediate sum w_i and carry c_i, and ADD2 for obtaining the sum z_i.

To evaluate the performance of the residue circuits, we suppose that the VLSI
implementation is based on a gate array IC, because ASIC design on a gate array is a
popular VLSI implementation method. We specify a binary representation $x_i =
[x_i(1)x_i(0)]$ for a radix-two signed digit, that is, $x_i \in \{-1, 0, 1\} = \{11, 00, 01\}$ For
example, $(1,0,-1,0,-1) = [01\ 00\ 11\ 00\ 11]$. Then the residue SD arithmetic operations
are described and simulated by VHDL, and the simulation and the logic circuit synthesis are performed by using VHDL synthesis software tool. The performances of the
circuits are obtained by using a design library of 0.18 μm CMOS technoloty. In
Table 1, the performance comparisons of the proposed residue SD adder and that using
high speed adders with CSAs and CLAs [6]. The delay time of the residue SD adder is
30% of the binary one for 512-bit residue addition.

Table 1. Performance evaluation of residue adders

	Word length n	16	64	256	512
Area [μm²]	*MBA(CSA/CLA)*	4693.81	21166.02	89210.92	180275.59
	MSDA	8244.44	34041.38	137192.21	274791.08
Delay [ns]	*MBA(CSA/CLA)*	3.91	7.14	12.55	16.85
	MSDA	3.63	3.61	3.69	3.73
Power [mW]	*MBA(CSA/CLA)*	4.90	23.37	99.19	200.48
	MSDA	12.20	51.79	210.39	421.29

4 Application to RSA Encryption Processor

In RSA encryption, $C = X^e$ mod m is calculated, where X and C are the plan text and the cipher text, respectively, and e and m are the encryption keys. The decryption is carried out in the same way as the encryption processing. The modular exponentiation is performed by repeating squaring and multiplication modulo m, as follows:

[Encryption Algorithm]
  $Q := 1$ for $i := k - 1$ down to 0 do
  begin
  $Q := \langle Q \times Q \rangle_m$;
  if $e_i = 1$ then $Q := \langle Q \times X \rangle_m$;
  end.

In the encryption, key e is in k-bit binary number representation. The processor can be constructed with two residue SD multipliers. We use the SD numbers as the input and output data. The SD number as the final result is converted into a binary number for the output of the encryption processor. The residue multiplication is implemented by using the proposed residue SD adder and some registers, for performing the following algorithm.

[Algorithm of Residue Multiplication]
  Let x, y, sx_i, pp_i and sum be n-digit SD numbers, where pp_i is a partial product for $i = 0, 1, \cdots, n-1$. $p = \langle x \times y \rangle_m$ is calculated by repeating step (2) n times.
(1) Let $sum = 0$ and $sx_0 = x$;
(2) For $i=0$ to $i=n-1$ do
    (2A) $pp_i := y_i \times sx_i$;
    (2B) $sx_{i+1} := \langle 2 \times sx_i \rangle_m$;
    (2C) sum $:= \langle sum + pp_i \rangle_m$;
(3) Output the product $p = sum$.

In the above algorithm, $m = 2^n + \mu$. The main arithmetic operation in the RSA encryption is the residue addition shown in (2C). We design two kinds of architectures with the proposed residue SD addition and the binary one using high speed binary adders such as CSA and CLA [6] (Table 2).

Table 2. Performance of RSA encryption processor

Word length n	Architecture	Area (μm^2)	Clock rate (MHz)	Power (mW)
32	**Bin(CSA/CLA)**	86343.04	109.17	1.82
	SD	151100.96	132.98	3.19
128	**Bin(CSA/CLA)**	338552.49	98.72	3.70
	SD	526630.50	128.37	6.33
512	**Bin(CSA/CLA)**	1339778.48	81.30	34.29
	SD	2012294.30	129.29	31.81

In the proposed SD architecture, since double registers for storing the SD number and a circuit converting the final result from SD to binary number are needed, more area is required. However, when the key is longer than 512 bits, the proposed processor is high speed and low power, and the working clock rates of the circuits are independent on the delay times.

5   Conclusions

In this paper, we have proposed a new residue signed-digit (SD) addition algorithm, which can be used in RSA public-key cryptosystem with a large modulus m. Since $\mu = m - 2^n$, where $-2^{n-1} + 1 \leq \mu < 0$, is used to calculate the residue n-digit SD number additions, any value of m with n bits can be chosen as the key of RSA cryptosystem. Some design results of the RSA encryption processor using the SD number arithmetic have been compared. Since the delay time of the residue SD addition is independent on the word length, the proposed residue addition circuits with a large modulus m can work in a fast clock rate. The proposed RSA encryption processor is high speed and low power comparing with the binary ones.

References

1. Rivest, R.L., Shamir, A., Adleman, L.: A method for obtaining digital signatures and public-key cryptosystems. Commun. ACM **21**(2), 120–126 (1978)
2. Vandemeulebroecke, A., Vanzieleghem, E., Denayer, T., Jespers, P.G.A.: A new carry-free division algorithm and its application to a single-chip 1024-b RSA processor. IEEE J. Solid-State Circ. **25**(3), 748–756 (1990)
3. Avizienis, A.: Signed-digit number representations for fast parallel arithmetic. IRE Trans. Electr. Comput. **EC-10**, 389–400 (1961)
4. Wei, S., Kameyama, M., Higuchi, T.: Performance evaluation of a multiple-valued RSA encryption VLSI. Trans. IEICE Jpn. **J73-D**(5), 484–491 (1990)
5. Wei, S., Shimizu, K.: A novel residue arithmetic hardware algorithm using a signed-digit number representation. IEICE Trans. Inf. Syst. **E83-D**(12), 2056–2064 (2000)
6. Chang, K.C.: Digital Systems Design with VHDL and Synthesis: An Integrated Approach. IEEE Computer Society Press, Los Alamitos (1999)

Study and Evaluation of Unsupervised Algorithms Used in Network Anomaly Detection

Juliette Dromard and Philippe Owezarski[✉]

LAAS-CNRS, Université de Toulouse, CNRS, Toulouse, France
{jdromard,owe}@laas.fr

Abstract. Network anomalies are unusual traffic mainly induced by network attacks or network failures. Therefore it is important for network operators as end users to detect and diagnose them to protect their network. However, these anomalies keep changing in time, it is therefore important to propose detectors which can learn from the traffic and spot anomalies without relying on any previous knowledge. Unsupervised network anomaly detectors reach this goal by taking advantage of machine learning and statistical techniques to spot the anomalies. There exists many unsupervised network anomaly detectors in the literature. Each algorithm puts forward its good detection performance, therefore it is difficult to select one detector among the large set of available detectors. Therefore, this paper, presents an extensive study and assessment of a set of well known unsupervised network anomaly detectors, and underlines their strengths and weaknesses. This study overwhelms previous similar evaluation by considering for the comparison some new, original and of premier importance parameters as detection similarity, detectors sensitivity and curse of dimensionality, together with the classical detection performance, and execution time parameters.

Keywords: Unsupervised network anomaly detection · Outlier detection · Subspace PCA method · Clustering algorithm · Curse of dimensionality

1 Introduction

With the booming in the number of network attacks, the problem of network anomaly detection has received increasing attention over the last decades. Current detectors are mainly based on prior knowledge of the attacks or of the normal traffic like signature-based detectors or behavioral-based detectors. This knowledge must be continuously updated to protect the network as the nature of the attacks keeps changing in time to evade new network protections. However, building signatures or new normal profiles to feed these detectors takes time and money as this work is usually done by network experts. As a result, they are unable to deal with zero-day attacks and/or new network behaviors. To overcome

© Springer Nature Switzerland AG 2020
K. Arai et al. (Eds.): FTC 2019, AISC 1070, pp. 397–416, 2020.
https://doi.org/10.1007/978-3-030-32523-7_28

these issues, a new generation of detectors has emerged which takes benefit of intelligent techniques which automatically learns from network traffic and allows bypassing the strenuous human input: unsupervised network anomaly detectors. These detectors aim at detecting network anomalies in an unsupervised way, i.e. without any previous knowledge on the anomalies. These anomalies may be due to attacks (DOS, DDOS, network scan, worm, etc.), to network failures or mis-configurations (route failures, traffic overload, imbalanced network traffic, etc.) or to some strange behaviors which should be monitored (use of multiple proxies, IP spoofing, etc.). Therefore, detecting these anomalies is of big interest for a network administrator. It can help him protect and gain an insight on its network.

Detectors rely on outlier detection algorithms which can be classified in three categories [9]: algorithms based on statistical models, algorithms based on spatial proximity and finally algorithms which deal with high dimensions. In this paper we analyse and evaluate the performance of 6 well known detectors, (two from each category): the Principal Component Analysis (PCA) subspace [16,23] and the robust PCA subspace method [14], DBSCAN [26], LOF (Local Outlier Factor) [3], SOD (Subspace Outlier Degree) [13] and UNADA [4].

The objective of this paper is to exhibit the practical state of the art in the domain of unsupervised anomaly detection. For this purpose, the major contribution of the paper deals with the evaluation kind that has been applied on the six detectors mentioned right over. It does not limit itself to the analysis of ROC curves and detection time, but tries to go further by considering their self and relative performance, especially when facing the same anomalies, but configured differently; the paper then deeply studies the sensitivity properties of these detectors. Given also the importance of the "Big data" keyword nowadays, with the need of analyzing huge amount of data containing many dimensions, the paper integrates in the evaluation of the detectors the study of their performance when facing such kinds of big data having a large number of dimensions. Given the large spectrum of the detectors considered in the paper, the evaluation methodology has been adapted depending on the detectors, trying as much as possible to remain fair despite evaluating slightly different features for each of the detection tools. The variability of the results that have been obtained puts forward the difficulty of making a right choice that can work in all conditions. It also points out the difficulty to parametrize such unsupervised detectors. We expect these results to show some of the research directions for improving the unsupervised detectors, and making them practically easy to use, and efficient in terms of global detection performance.

For providing such an evaluation, a valuable ground truth is required. It must be fair for providing the same realistic level of complexity for all the different algorithms implemented in the selected detection tools. That is why the ground truth must first contain very accurate labels for making the evaluation process relevant and accurate. It must also reproduce the difficulty of finding anomalies traffic that is quite small compared to gobal traffic. The way the ground truth is built is detailed in Sect. 4. The existing dataset matching the most the expressed requirement is KDD'99 thanks to the quality of its labels, and despite the fact that it is quite old. However, analyzing the current state of the art in traffic anomalies, it is not very far from the one in 1998, when KDD dataset was built: indeed, DDoS attacks, flash crowds or misconfigurations have similar effect on the traffic characteristics in 2017 and in 1998. New kinds of such anomalies are not very numerous. Despite not fully complete, we argue that traffic anomalies contained in KDD'99 represent a significant part of existing anomalies in 2017, and is enough for providing a high quality comparative evaluation of the 6 selected detectors. We then built our evaluation ground truth based on KDD'99 anomalies, but also adapting the background traffic to actual one, especially for reproducing the actual ratio between anomalous and background traffic.

To summarize, the main contributions of this paper are the following:

- It proposes a method to compare and study unsupervised network anomaly detectors.
- It proposes a new method to evaluate detectors sensitivity inspired by the Morris method.
- It gives guideline to parametrize these detectors.
- It points out the strengths and weaknesses of each detector.
- It uncovers important facts on the nature of network anomalies.

This paper is organized as follows. In a second section, unsupervised network anomaly detectors principle is presented. Then, a set of detectors are described and their configuration are discussed. A fourth section presents the detectors evaluation and discussed the obtained results. These latter are compared in terms of detection performance, detection similarity, execution time, detectors sensitivity and curse of dimensionality. Finally, Sect. 5 concludes.

2   Unsupervised Network Anomaly Detection Principle

Existing unsupervised network anomaly detectors include two main steps, the preprocessing and the outlier detection steps. Some detectors may integrate a third and optional step: the anomalies post-processing.

The first step aims at capturing the network traffic, usually in consecutive time-bins and at processing it to build a data matrix X. Collected packets are aggregated in flows according to a specific flow key which can be, for example, the IP source, the IP destination or a combination of both. For each flow a set of statistics are computed like its number of IP destinations, of packets or of ICMPs. A normalized data matrix X of size $p * d$ is built, with d being the number of statistics (or dimensions) used to describe a flow and p the total number of flows (or points). We will keep this notation throughout the paper. The outlier detection step aims at detecting anomalous flows in the data matrix X using outlier detection algorithms. These algorithms aim at identifying flows which have different patterns from the rest of the traffic. This phase has received most of the researchers attention as the detectors intelligence relies in it.

The post-processing step aims at extracting and displaying information about the anomalies to assist network administrators in their task. This stage has received little attention for the moment although it is a crucial one. The post-processing phase helps the network administrator understand, sort and classify the spotted anomalies in order to take appropriate counter-measures. Post-processing output can take different forms, for example in [4] the authors build signatures from the anomalies, in [16] they classify the anomalies using clustering techniques and in [11] they remove persistent anomalies to ease the network administrator task.

3   Outlier Detection Algorithms

To detect anomalous flows in the data matrix X, unsupervised network anomaly detectors rely on outlier detection algorithms. An outlier detection algorithm can either have a global view or a local view of the data. Thus, to evaluate a point abnormality level, a detector will either compare it to its neighbors (local view) or to the whole data (global view). Furthermore, a detector can either output a label for each point (normal vs abnormal point) or a score of outlierness. In the case of scores, the outlier detection algorithm must be followed by an additional step to extract anomalous points (flows) from scores. As stated in the introduction, outlier detection algorithms can be classified in three categories [9]: algorithms based on statistical models, algorithms based on spatial proximity and algorithms dealing with high dimensions.

3.1   Algorithms Based on Statistical Models

Outlier detection algorithms based on statistical models rely on the assumption that the data has been generated according to a statistical distribution. Outliers are then flows that deviate strongly from this distribution. Many statistical approaches have been applied to unsupervised network anomalies detection such as histograms [12], EM-clustering [25], the PCA subspace method [16,23] and the Gaussian mixture model [2].

The PCA subspace method has been extensively used for network anomaly detection. This approach divides the whole space of dimension d in two subspaces: the normal subspace made up of the k principal component (PC) directions of the data matrix X and the abnormal subspace made up of the $d - k$ PC directions left. There exists variants of the PCA subspace method [23], however, in the context of this study, we only evaluate the one proposed in [16] which has been extensively studied. In this approach, one score of outlierness is computed for each point. Once projected on the abnormal subspace, a point's score is equal to its l^2 norm. Points with a high score are more likely to follow a pattern which does not conform to the normal or natural one. This method takes as input one parameter k which defines the normal subspace dimension. The PCA subspace method complexity is in $O(p.d^2)$.

In [22], Ringberg et al. highlight that k must be picked up such that the k dominant PC directions capture most of the total deviation of the data to get good detection performance.

Some recent articles have underlined that PCA-based detectors suffer from the contamination subspace problem. This phenomena appears when some large outliers are included in the measured traffic data. These latter contaminate the subspaces and as a result, a large part of the anomalies are projected onto the normal subspace and are not detected. In order to solve the subspace contamination problem, [14] use robust PCA mechanisms to obtain PCs which are not influenced by the existence of outliers. In the following, we use the GRID algorithm [6] to get robust PCs as it does not take any input parameter and finds good quality PCs in a reasonable time.

Table 1. Outlier detection algorithms

Algorithm	View	Output	HD*	Complexity	Parameter	Parameters setting
Sub. PCA [16]	Global	Score	No	$O(p.d^2)$	k: nb of PC directions	Must capture most of the total deviation of the data
DBSCAN [7]	Global	Label	No	$O(p.log(p))$	r: radius	Percentage of the distance between the space two farthest points
					$minPts$: min nb of points to form a cluster	Percentage of the total number of points
LOF [3]	Local	Score	No	$O(p.log(p))$	nn: nb of nearest neighbors	Percentage of the total nb of flows
UNADA [4]	Global	Score	Yes	$O(d^2.p.log(p))$	r: radius (different for each subspace)	Percentage of the distance between the subspace two farthest points
					$minPts$: min nb of points to form a cluster	Percentage of the total number of points
SOD [13]	Local	Score	Yes	$O(d^2.p^2)$	α_{lof}	Advice 0.8
					nn: nb of nearest neighbors	Percentage of the total nb of flows
					l: number of reference points	Percentage of the total nb of flows
Naive alg.	Global	Label	Yes	$O(d.p)$	α_{naive}: nb of standard deviations	Set high enough to only detect flows with extreme values

HD: deal with high dimensions; nb is used for "numbers"

3.2   Algorithms Based on on Spatial Proximity

Many outlier detection algorithms rely on models based on spatial proximity like DBSCAN [7,26], K-mean [28], LOF [3], etc. Algorithms based on spatial proximity should be used with an index like the R-tree [20] or the k-d tree [27] to improve their time complexity. These detectors are based on the idea that points isolated from the others are outliers.

DBSCAN [7] is a density-based clustering algorithm which groups points that are closely packed together in clusters. Points that lay in low-density regions are considered as outliers. It can discover clusters of various shapes and sizes from a large amount of data which contains noise. It takes two input parameters r and $minPts$ which respectively describe the neighborhood radius of a point and the minimum number of points to form a cluster. There exists no rule of thumb to fix DBSCAN parameters and its configuration may differ with the data and the problem considered. In order to avoid that DBSCAN groups flows which belong to similar anomalies in the same cluster, the parameter $minPts$ should be superior to the maximum number of flows induced by similar attacks. For example, if 9 flows are induced by SYN attacks, then $minPts$ should be superior to 9 so that they do not form a cluster. Furthermore, as anomalies are flows which deviate strongly from the others, r must be chosen large enough so that points which are slightly different from the majority belong to a cluster. Thus, we propose to set r as a percentage of the distance between the space two farthest points and $minPts$ as a percentage of the total number of flows. DBSCAN time complexity is $O(p.log(p))$ when used with an R-tree index.

LOF [3] is a local spatial-based approach which assigns to each point an outlier factor representing its degree of outlierness regarding its local neighborhood. A point whose density is lower to that of its nn nearest neighbors is considered as an outlier. Thus, LOF is able to deal with regions of different densities. It takes as input one parameter nn, which represents the number of nearest neighbors considered to evaluate a point's abnormality. The value of nn must be carefully chosen. Indeed, if it is too low, LOF may then compare an anomalous flow with only similar anomalous flows, i.e. with flows generated by the same type of attack and may therefore not detect them as outliers. To overcome this issue, nn must be set larger than the maximal number of flows induced by similar attacks. We propose to fix it as a percentage of the total number of flows. For medium to high-dimensional data, the algorithm provides an average complexity of $O(p.log(p))$ [3].

3.3   Algorithms Dealing with the Curse of Dimensionality

In high dimensional data, distance between points become meaningless: the proportional difference between the farthest point distance and the closest point distance vanishes. This phenomena is called curse of dimensionality and can have an important impact on detectors detection performance. Some outlier detection algorithms have been specifically devised to deal with this curse like UNADA [4] or SOD [13]. To deal with this curse UNADA relies on a divide and conquer approach. It divides the space made up of d dimensions in $N = \binom{d}{2}$ two-dimensional subspaces. It then applies DBSCAN on each subspace. It finally combines the N obtained partitions in one final partition and computes for each point a score. For each point, it computes a core which is the sum of its distance to the biggest cluster in every subspace. UNADA has the same input parameters as DBSCAN: a radius r and $minPts$. However, the value of r must be adapted to every subspace.

SOD [13] is a local outlier algorithm which deals with high dimensions by selecting in an intelligent way subspaces to compute each point's score. It computes a score for each point which reflects how well it fits to the subspace that is spanned by a set of l reference points. The l points which shares the highest number of nearest neighbors with a point form its l reference points. For each point, SOD computes a subspace made up of the set of dimensions whose variance is low with respect to its l reference points. SOD takes three input parameters: α_{lof} a threshold to decide about the significance of a dimension, l the number of reference points and nn the number of nearest neighbors. Authors advise to set α_{lof} at 0.8. Furthermore, to avoid comparing an anomalous point with only similar anomalous points, l should be chosen much higher than the maximum number of flows induced by a same type of attack. SOD's time complexity is $O(d.p^2)$.

For the sake of comparison, we propose a naive outlier detection algorithm which aims at detecting points with extreme values. For each dimension, this algorithm detects as outliers the points which are α_{naive} standard deviations from the median. As it deals with one dimension at a time, our naive algorithm should be able to deal with high dimensions. Table 1 summarizes detectors characteristics.

3.4   Detectors Based on Scores

For algorithms which output scores (LOF, PCA subspace, SOD, UNADA), a final step is required to extract outliers. A threshold th is set and all the scores which are above th are considered as outliers. We have identified in the literature three main methods to set th:

- The knee method [4,5]. This approach consists in plotting the sorted outlier scores to get a convex curve. Usually, a knee in the curve can be observed indicating a change in the nature of the scores. The threshold is set at the curve knee point.
- The quantile based method [22,23]. The threshold is set at the q-qantile of the scores empirical distribution. For example in [23] and in [22], they fix it at the 0.9899 quantile and the 0.90 quantile respectively. However, this method implies that the percentage of anomalies in the data is known in advance, which is, in most cases, unrealistic.
- The statistical hypothesis testing method. This method assumes that normal flows follow a specific data distribution for which a statistical hypothesis

testing exists. The threshold is set at the test $1 - \alpha$ confidence level. This limit corresponds to a false alarm rate of α, if the starting assumptions are satisfied. For example in [15, 16], the authors assume that normal flows follow a multivariate Gaussian distribution which allow them to apply the Q-statistic developed in [10]. However, it has not yet been demonstrated that network traffic follows any specific distribution.

For most outlier detection algorithms, there exists no guideline to set their parameters. Good sense and a good understanding of the current problem are essential to set detectors parameters and get relevant outcomes.

Table 2. Detectors parameters

Algorithm	Parameter	Value	Range
DBSCAN	r	Not fixed	1% to 20%
	$minPts$	10% of the total nb of flows	1% to 20%
LOF	nn	20% of the total nb of flows	10% to 50%
PCA subspace	k	The k first PCs capture at least 90% of the total deviation	85 to 98%
rob. PCA subspace	k	The k first PCs capture at least 90% of the total deviation	85 to 98%
UNADA	Radius r	10% of the distance between the subspace two farthest points	1 to 10%
	$minPts$	10% of the total nb of flows	1 to 20%
SOD	nn	20% of the total nb of flows	10% to 40%
	l	10% of the total nb of flows	10% to 40%
	α_{sod}	0.8	
NAIVE	α_{naive}	Not fixed	0.5 to 3

4   Evaluation on Our New KDD'99 Inspired Dataset

In the field of network anomaly detection, as pointed out in [24], there is a lack of available public ground truth. In the literature, two main public available ground truths are often cited: the KDD99 ground truth [1] (summary of the DARPA98 traces) and the MAWI ground truth [8]. Many other datasets exists but, for the moment, do not provide the same amount of data, the same level of labels, are not easy to get, etc. The KDD99 contains multiple weeks of network activity from a simulated Air Force network, generated in 1998. Although the KDD99 dataset is quite old, it is still considered as a landmark in the field, because of the accuracy of the provided labels. On the contrary, the MAWILab dataset is more recent and is still being updated. It consists of labeled 15 min network traces collected daily from a trans-Pacific link between Japan and the United States.

However, the MAWILab ground truth is questionable as it has been obtained by combining the results of four ancient unsupervised network anomaly detectors [8]. indeed, labels are often not very relevant; for example many anomalies are labeled as 'HTTP traffic'. In addition, after manual inspections, some anomalous flows do not seem to exhibit unusual patterns.

For all these reasons, we have decided to perform our evaluation on a dataset built out the KDD99 dataset. This choice has been made because it outputs consistent labels that are fully accepted by the community. The evaluation has then been performed on a portion containing 10% of KDD99 dataset which contains 23 different types of attack, see [19] for more information on these attacks. To obtain this dataset, packets have been aggregated according to the TCP connection they belong to. Each flow is described by 41 attributes, 34 of which are numeric and 7 are categorical. As detectors do not deal with categorical variables, they have been turned into dummy variables, lifting the total number of variables to 118. The dataset cannot be used as it is, due to a too large number of anomalous flows; no detector based on outlier detection techniques can possibly detect the attacks as they are not rare. This problem could have been solved by aggregating the flows into another level (by IP source for example), but this is not possible with the KDD99 dataset as the IP addresses are not displayed. To overcome this issue and as in [5, 17, 21, 26], we have, selected randomly some flows, so that the percentage of anomalous flows stays under a certain threshold. We have built two datasets. The first one "dataset1" is made up of 1000 flows and includes 160 attacks, there is at most 8 flows for each type of attack. The second one "dataset2" is made up of 10000 flows and includes 979 attacks, there is at most 80 flows for each type of attack. Some dimensions may have a larger range than others, as a consequence they have a higher weight and may hide other features. To overcome this issue, both datasets are normalized using the max-min normalization so that each dimension scales in [0,1].

4.1 Evaluation in Terms of Detection Performance

In a first time, the algorithms are compared in terms of detection performance using the Area Under the ROC curve (AUC) measure. A ROC curve is obtained by plotting the true positives rate (TPR) against the false positive rate (FPR) at various parameters settings. The AUC takes its value in [0,1]; an AUC of 1 represents a perfect detector and an AUC of 0.5 a detector with complete random guess. The parameters used for this evaluation are displayed in Table 2. The column "Range" will be used later in this paper. The ROC curve points have been computed by varying

- the threshold th for algorithms which output scores (SOD, LOF, UNADA, and PCA).
- the radius r for DBSCAN.
- the parameter α_{naive} for the naive outlier algorithm.

Table 3. Detectors AUC, number of TPs and FPs and execution time

	Dataset1				Dataset2			
	AUC	TPs	FPs	Time	AUC	TPs	FPs	Time
UNADA	0.90	146	85	98 s	0.93	922	380	3 h 8 m
LOF	0.90	160	207	2.5 s	0.97	894	361	17 m
PCA	0.71	107	75	55 ms	0.70	638	1188	454 ms
rob. PCA	0.97	158	65	9 s	0.97	895	478	3 m 43 s
SOD	0.96	153	70	50 s	0.90	915	1612	4 h 30 m
NAIVE	0.96	160	57	20 ms	0.97	894	259	68 ms
DBSCAN	0.94	149	47	430 ms	0.96	902	232	58 s

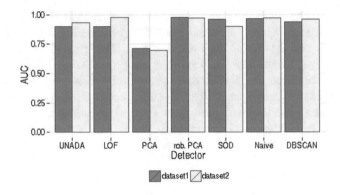

Fig. 1. Detectors AUC

Table 3 and Fig. 1 presents the AUC obtained by each detector for each dataset. It can be noticed that the PCA subspace method has the worst detection performance, this result can be explained by the contamination subspace problem. This assumption is confirmed by the robust PCA subspace method results. Indeed, by resolving the subspace contamination problem, this letter achieves the best detection performance among all the detectors with an AUC superior to 0.97 for both dataset. Except the PCA subspace method, every algorithm achieves good detection performance with an AUC superior or equal to 0.9. Naive detector AUC is superior to 0.96 for both experiments, therefore it outperforms most of the detectors. This result implies that most network anomalies in KDD99 dataset possess an extreme value in at least one dimension. Some further studies on other datasets should be carried out to check whether this observation can be extended to every network anomaly. If it is the case, this puts into question the use of complex algorithms to detect network anomalies.

In the following, each detector is set at its best setting. A detector best setting is defined as the setting which maximizes its informedness. The informedness is a statistical measure of the performance of a binary classification test which considers equally the TPR and FPR. It takes its value in $[1, -1]$ and is computed as follows:

$$informedness = TPR - FPR \tag{1}$$

Figure 2 displays detectors ROC curve obtained with dataset1. The square on each detector curve represents the results obtained at the detector best setting. A visual analysis of these figures shows that the maximum informedness is a good measure to select each detector best setting.

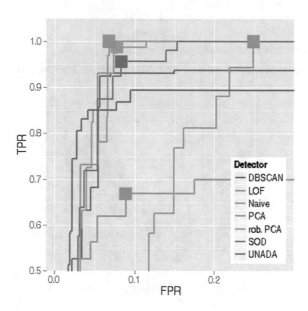

Fig. 2. Detectors ROC curve and their point of maximum informedness obtained with dataset1

The AUC provides information about the proportion of each detector TPR and FPR, however, this information is not sufficient to fully evaluate a detector performance. Indeed, as pointed out in [24], even with a low FPR, the number of false negatives (FNs) generated by a detector may be substantial and can overwhelm a network administrator.

Figure 3 displays detectors number of TPs and FPs obtained at their best setting according to the informedness. It can be noticed that even with a high AUC (>0.9), LOF and SOD, in dataset1, get a high number of FPs; their number of FPs is superior to their number of TPs. Such a situation may lead the network administrator to mis-classify and interrupt many normal flows.

4.2   Evaluation in Terms of Execution Time

A detector execution time is a very important parameter to consider while selecting a network anomaly detector. Indeed, the faster the detector identifies attacks,

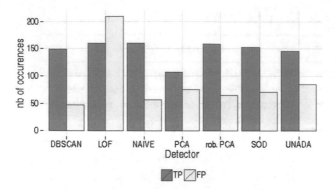

Fig. 3. Detectors number of true and false positives obtained with dataset1

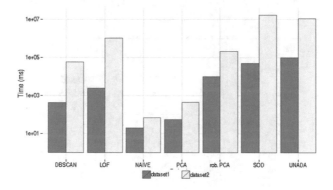

Fig. 4. Detectors execution time in milliseconds

the quicker the network administrator can take relevant counter-measures and the less important the damages on the network are. Figure 4 depicts the execution time of each algorithm for both datasets, the y-axis is in log scale. These results have been obtained on a single machine with 16 GB of RAM and an Intel Core i5-4310U CPU 2.00 GHz. As expected the execution time increases with the data size. As a reminder, dataset1 is made up of 1000 flows and dataset2 of 10,000 flows. The obtained results are logical according to detectors complexity displayed in Table 1.

It can be noticed that UNADA and SOD do not scale well with the number of flows; for dataset2, UNADA completes the detection in 3 h and SOD in 4 h and a half. On the other hand, the naive and the PCA subspace detector complete in less than a second.

5   Evaluation in Terms of Similarity

To evaluate the similarity between the anomalies found by the different algorithms we use the Jacquard index (JI). The JI measures the similarity between two finite sets and is defined as their intersection size divided by their union size. Thus, if A and B are the set of anomalies identified by two different detectors, their similarity according do the JI is computed as follows:

$$J(A, B) = \frac{|A \bigcap B|}{|A \bigcup B|} \qquad (2)$$

If JI is close to one then the detectors are very similar and if it is close to 0 then they are considered as very dissimilar. Figure 5 displays the similarity between the TPs of the different detectors for dataset1 (similar results have been obtained for dataset 2). It can be noticed that the JI is high for every algorithm

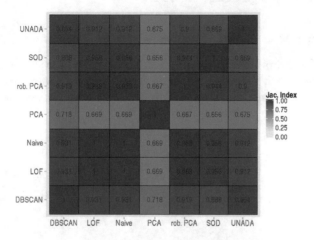

Fig. 5. Similarity between detectors TPs in dataset1

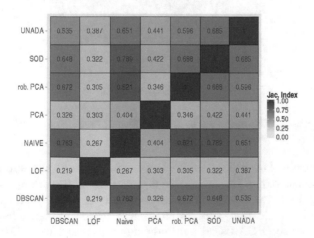

Fig. 6. Similarity between detectors FPs in dataset1

(all the squares are red) in both datasets except the PCA subspace method as this latter has bad performance in terms of of TPR and FPR. This implies that detectors mainly find the same anomalies.

Figure 6 displays the similarity between the FPs found by the detectors for dataset 1. One can observe that the JI is often very low (many squares are yellow), which implies that their FPs are different. Thus, it would be interesting to combine the outputs of these different algorithms to keep only the anomalies found by most detectors. As the similarity between their FPs is very low, most FPs would then be discarded and as the similarity between their TPs is high, most TPs would be kept. Therefore, combining detectors output would allow improving the overall detection performance by reducing the number of FPs while maintaining a high number of TPs.

5.1   Evaluation in Terms of Curse of Dimensionality

To evaluate detectors capacity to deal efficiently with high dimensions, we propose to evaluate their detection performance on dataset1 to which noisy dimensions are added. As in [29], the noisy dimensions are generated with a random uniform distribution which takes its values in [0,1].

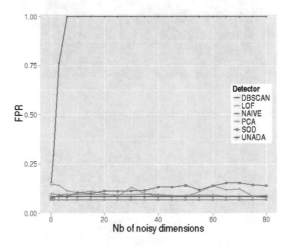

Fig. 7. Detectors FPR according to the number of added dimensions to dataset1

Figure 7 displays each detector FPR as a function of the number of added noisy dimensions. It can be noticed that noisy dimensions have no impact on UNADA and the naive algorithm performance. It can be explained by the fact that they both divide the space in subspaces of low dimensions and process them independently. PCA-based methods show little sensitivity to high dimensions, especially the robust PCA subspace detector. As it considers neighborhood rather than distance, LOF reacts well to the curse of dimensionality:

a point neighbors stay the same when noisy dimensions are added to the data. Even though SOD has been devised to deal with high dimensions, its FPR tends to increase when noisy dimensions are added. Even though DBSCAN radius r is re-computed each time some new dimensions are added (it is set at 10% of the distance between the space two farthest points), DBSCAN is the detector which suffers the most from the curse. With the increase in the number of noisy dimensions, points tend to move away from each other. As a result, DBSCAN identifies them all as outliers. Before adding these noisy dimensions, the curse had no effect on DBSCAN, even though KDD99 has many dimensions. This phenomenon can be explained by the fact that each dimension in KDD99 brings mainly information and few noise. Similar behaviors have been observed in [29]. explained by the fact that each dimension in KDD99 brings mainly information and few noise. Similar behaviors have been observed in [29].

5.2   Evaluation in Terms of Parameters Sensitivity

This section aims at evaluating and comparing detectors sensitivity. For each detector, we want to determine if its input parameters can be easily set such that it gets good detection performance and therefore a high informedness. Even with a high AUC, a detector can be unable to detect correctly anomalies if it is badly configured. A detector, very sensitive to its input parameters, may be very difficult to parameterize. As a result, a network administrator may fail in configuring it, the detector output becomes then useless. Therefore, the sensitivity of a detector is an important parameter to take into account even though it is rarely considered in the current literature.

To evaluate whether a detector can be easily configured, we propose an approach inspired by the Morris method [18]. This latter is a sensitivity analysis method which evaluates the influence of each input parameter on the output of a function. It computes for each parameter two statistics, its mean and its standard deviation impact on the output function.

We have modified the Morris method so that it applies to the analysis of detectors sensitivity. Our method aims at evaluating the impact of the input parameters of each detector on its informedness. To reach this goal, we have defined for each input parameter of each detector a range of possible "values" (see Table 2). By possible values, we mean values which could have been chosen by any "reasonable" expert in the field.

We apply each detector many times on the dataset1. For each input parameter described in Table 2, its range of possible values is discretized. Each detector is launched as many times as there are possible combinations of its input parameters. Finally, two statistics are computed to evaluate each detector: its informedness mean and standard deviation. The informedness standard deviation of a detector captures its sensitivity to its input parameters whereas its informedness mean provides an indication on its average detection performance. A detector is all the more easy to configure that its informedness mean is high and its informedness variance is low. As explained previously, unsupervised detectors either output labels or scores for each flow. For detectors which output scores, an extra step is required to extract anomalies from scores. Indeed, there is no clear gap between anomalous and normal flows scores. Therefore, extracting anomalies from scores is a difficult task.

To evaluate the sensitivity of detectors which output scores, we use the "best threshold method". This method sets the threshold, used to extract the anomalies from scores, at the value which maximizes the detector informedness. This method implies that flows labels are known in advance.

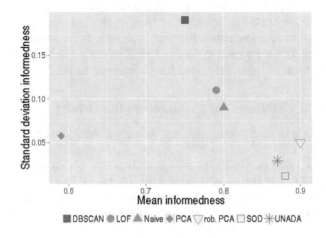

Fig. 8. Comparison of detectors sensitivity

Figure 8 displays the results of the detectors sensitivity analysis. Once again the PCA subspace method has bad performance due to its subspace contamination problem: its mean and standard deviation informedness is low. It implies that any configuration of this detector may lead to poor performance.

It can be noticed that DBSCAN standard deviation informedness is high which implies that it is difficult to configure correctly. Figures 9 and 10 can explain these results. Figure 9 displays DBSCAN informedness according to its radius (which is set as a percentage of the distance between the space two farthest points) with different $minPts$ values. It clearly shows that its informedness rises when the radius increases till a certain point. This is maybe because a larger radius increases the number of "normal" flows which belong to a cluster. Figure 10 depicts DBSCAN informednesss according to its minPts (which is set as a percentage of the total number of flows) with different values of radius. It can be noticed that when minPts increases, DBSCAN informedness tends to decrease which can be explained by the fact that fewer "normal" flows belong then to a cluster.

LOF standard deviation informedness is moderately high, it implies that it is quite sensitive to its input parameter nn. This sensitivity can be explained by its local view. Indeed, when nn is low, the probability that an anomalous point is compared only to other anomalous points is high. As a consequence, it may not appear as an outlier. This phenomena is illustrated by Fig. 11 which displays LOF informedness according to its parameter value nn (which is set as a percentage of the total number of flows).

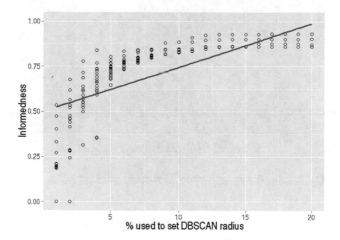

Fig. 9. DBSCAN informedness according to its radius.

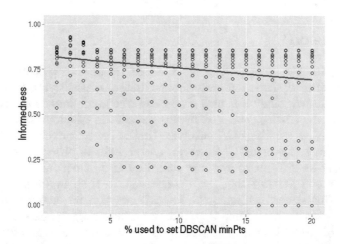

Fig. 10. DBSCAN informedness according to its minPts.

In depth studies (not presented in the paper because of space limitation) exhibit that DBSCAN informedness rises when the radius increases till a certain point. This is maybe because a larger radius increases the number of "normal" flows which belong to a cluster. It can also be noticed that when minPts increases, DBSCAN informedness tends to decrease which can be explained by the fact that fewer "normal" flows then belong to a cluster.

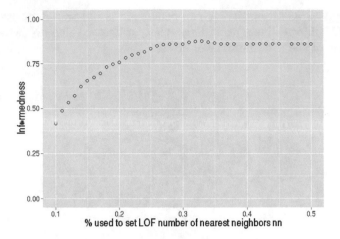

Fig. 11. LOF informedness according to the value of its input parameter nn.

Similarly, these studies show that LOF standard deviation informedness is moderately high. It implies that it is quite sensitive to its input parameter nn. This sensitivity can be explained by its local view. Indeed, when nn is low, the probability that an anomalous point is compared only to other anomalous points is high. As a consequence, it may not appear as an outlier.

6   Conclusion

This paper presents a comparison of different unsupervised network anomaly detectors in terms of detection performance, detector sensitivity, execution time and curse of dimensionality. It also proposes some guidelines to configure them. It points out the challenges raised by the extraction of anomalies from scores. Every detector except the PCA subspace method reaches very good detection performance in terms of AUC. However, at the light of other parameters some detectors may be difficult to apply in real life due to their high execution time like UNADA and SOD or their high sensitivity to their input parameters like DBSCAN. This study highlights the importance of using many parameters to evaluate a network detector and therefore underlines the weakness of evaluations only based on ROC curves. The results have pointed out that every network anomaly selected from the KDD99 dataset has an extreme value in at least one dimension and can therefore be easily identified by a naive algorithm. Some further studies on other datasets should be carried out to check whether this observation can be extended to every network anomaly. Among every algorithm the robust PCA subspace method has shown very good performance in terms of detection, input parameters sensitivity and robustness to high dimensions. To reach near to real time detection, the robust PCA subspace can re-use the PCs directions multiple times assuming that the normal space changes little in time. Anomaly detection is the first step to protect the network. The spotted anomalies must then be processed by the network administrator so that it takes relevant counter-measures. An important effort should now be made to take advantage of detectors output and to propose solutions to identify anomalies root causes.

References

1. KDD Cup 1999 Data. http://kdd.ics.uci.edu/databases/kddcup99/kddcup99.html. Accessed 03 Feb 2016
2. Bahrololum, M., Khaleghi, M.: Anomaly intrusion detection system using Gaussian mixture model. In: Convergence and Hybrid Information Technology, ICCIT 2008, vol. 1, pp. 1162–1167, November 2008
3. Breunig, M.M., Kriegel, H.-P., Ng, R.T., Sander, J.: LOF: identifying density-based local outliers. SIGMOD Rec. **29**(2), 93–104 (2000)
4. Casas, P., Mazel, J., Owezarski, P.: UNADA: unsupervised network anomaly detection using sub-space outliers ranking. In: NETWORKING 2011: 10th International IFIP TC 6 Networking Conference, pp. 40–51. Springer, Heidelberg (2011)
5. Casas, P., Mazel, J., Owezarski, P.: Unsupervised network intrusion detection systems: detecting the unknown without knowledge. Comput. Commun. **35**(7), 772–783 (2012)
6. Croux, C., Filzmoser, P., Oliveira, M.R.: Algorithms for projection-pursuit robust principal component analysis. Chemometr. Intell. Lab. Syst. **87**(2), 218–225 (2007)
7. Ester, M., Kriegel, H.-P., Sander, J., Xu, X.: A Density-based Algorithm for Discovering Clusters in Large Spatial Databases with Noise, pp. 226–231. AAAI Press, Portland (1996)
8. Fontugne, R., Borgnat, P., Abry, P., Fukuda, K.: MAWILab: combining diverse anomaly detectors for automated anomaly labeling and performance benchmarking. In: ACM CoNEXT 2010, Philadelphia, PA (2010)
9. Zimek, A., Kriegel, H.-P., Kröger, P.: Outlier detection techniques. In: Tutorial Notes: SIAM SDM 2010, Columbus, Ohio (2010)
10. Jensen, D.R., Solomon, H.: A Gaussian approximation to the distribution of a definite quadratic form. J. Am. Stat. Assoc. **67**(340), 898–902 (1972)
11. Julisch, K.: Clustering intrusion detection alarms to support root cause analysis. ACM Trans. Inf. Syst. Secur. **6**, 443–471 (2003)
12. Kind, A., Stoecklin, M.P., Dimitropoulos, X.: Histogram-based traffic anomaly detection. IEEE Trans. Netw. Serv. Manage. **6**(2), 110–121 (2009)
13. Kriegel, H.-P., Kröger, P., Schubert, E., Zimek, A.: Outlier detection in axis-parallel subspaces of high dimensional data. In: Advances in Knowledge Discovery and Data Mining: 13th Pacific-Asia Conference, pp. 831–838. Springer, Heidelberg (2009)
14. Kwitt, R., Hofmann, U.: Unsupervised anomaly detection in network traffic by means of robust PCA. In: Computing in the Global Information Technology, p. 37, March 2007
15. Lakhina, A., Crovella, M., Diot, C.: Diagnosing network-wide traffic anomalies. In: Proceedings of the 2004 Conference on Applications, Technologies, Architectures, and Protocols for Computer Communications, SIGCOMM 2004, pp. 219–230. ACM (2004)
16. Lakhina, A., Crovella, M., Diot, C.: Mining anomalies using traffic feature distributions. ACM SIGCOMM Comput. Commun. Rev. **35**(4), 217 (2005)

17. Leung, K., Leck, C.: Unsupervised anomaly detection in network intrusion detection using clusters. In: Proceedings of the Twenty-eighth Australasian Conference on Computer Science, ACSC 2005, vol. 38, pp. 333–342. Australian Computer Society, Inc, Darlinghurst (2005)
18. Morris, M.D.: Factorial sampling plans for preliminary computational experiments. Technometrics **33**(2), 161–174 (1991)
19. Olusola, A.A., Oladele, A.S., Abosede, D.O.: Analysis of KDD 99 intrusion detection dataset for selection of relevance features, In: World Congress on Engineering and Computer Science, pp. 162–168 (2010)
20. Kriegel, H.P., Schneider, R., Seeger, B., Beckmann, N.: The R*-tree: an efficient and robust access method for points and rectangles. Sigmod Rec. **19**, 322–331 (1990)
21. Portnoy, L., Eskin, E., Stolfo, S.: Intrusion detection with unlabeled data using clustering. In: In Proceedings of ACM CSS Workshop on Data Mining Applied to Security, pp. 5–8 (2001)
22. Ringberg, H., Soule, A., Rexford, J., Diot, C.: Sensitivity of PCA for traffic anomaly detection. SIGMETRICS Perform. Eval. Rev. **35**(1), 109–120 (2007)
23. Shyu, M.-L., Chen, S.-C., Sarinnapakorn, K., Chang, L.: A novel anomaly detection scheme based on principal component classifier. In: IEEE Foundations and New Directions of Data Mining Workshop, pp. 171–179 (2003)
24. Sommer, R., Paxson, V.: Outside the closed world: on using machine learning for network intrusion detection. In: 2010 IEEE Symposium on Security and Privacy, 0(May), pp. 305–316 (2010)
25. Syarif, I., Prugel-Bennett, G., Wills, A.: Unsupervised clustering approach for network anomaly detection. In: Networked Digital Technologies: 4th International Conference. Springer, Heidelberg (2012)
26. Thang, T.M., Kim, J.: The anomaly detection by using DBSCAN clustering with multiple parameters. In: 2011 International Conference on Information Science and Applications (ICISA), pp. 1–5, April 2011
27. Tsakok, J.A., Bishop, W., Kennings, Af.: kd-Tree traversal techniques. In: Interactive Ray Tracing, p. 190, August 2008
28. Yasami, Y., Khorsandi, S., Mozaffari, S.P., Jalalian, A.: An unsupervised network anomaly detection approach by k-means clustering & ID3 algorithms. In: Computers and Communications, pp. 398–403, July 2008
29. Zimek, A., Schubert, E., Kriegel, H.-P.: A survey on unsupervised outlier detection in high-dimensional numerical data. Stat. Anal. Data Min. **5**(5), 363–387 (2012)

Recurrent Binary Patterns and CNNs for Offline Signature Verification

Mustafa Berkay Yılmaz[1(✉)] and Kağan Öztürk[2]

[1] Akdeniz University, Antalya, Turkey
berkayyilmaz@akdeniz.edu.tr
[2] Alanya Alaaddin Keykubat University, Antalya, Turkey
kagan.ozturk@alanya.edu.tr

Abstract. Signature representations that are extracted by convolutional neural networks (CNN) can achieve low error rates. However, a trade-off exists between such models' complexities and hand-crafted features' slightly higher error rates. A novel writer-dependent (WD) recurrent binary pattern (RBP) network, and a novel signer identification CNN is proposed. RBP network is a recurrent neural network (RNN) to learn the sequential relation between binary pattern histograms over image windows. A novel histogram selection method is introduced to remove the stop-word codes. Dimensionality is reduced by more than 25% while improving the results. This work is the first to combine binary patterns and RNNs for static signature verification. Several test sets, derived from large-scale and popular databases (GPDS-960 and GPDS-Synthetic-10000) are used. Without training any global classifier, RBP network provides competitive equal error rates (EER). The proposed architectures are compared and integrated with other recent CNN models. Score-level integration of WD classifiers trained with different representations are investigated. Cross-validation tests demonstrate the EERs reduced compared to the best single classifier. A state-of-the-art EER of 1.11% is reported with a global decision threshold (0.57% EER with user-based thresholds) on GPDS-160 database.

Keywords: Offline signature verification · Recurrent neural network · Convolutional neural network · Score-level integration

1 Introduction

Offline signature verification (OSV) is the problem of verifying a static signature image of a claimed identity either as genuine or forgery. A forgery may be signed without any knowledge (random forgery) or after some training time (skilled forgery). There may be even more realistic forgeries that are generated by carbon-copy. An ongoing research area questions the dichotomy of random and skilled forgeries [1,2], which is not a focus of the current article.

Handcrafted features are being replaced by representations that are learned by deep neural networks, reporting better accuracies. By the way, OSV is still

© Springer Nature Switzerland AG 2020
K. Arai et al. (Eds.): FTC 2019, AISC 1070, pp. 417–434, 2020.
https://doi.org/10.1007/978-3-030-32523-7_29

a challenging task because of the lack of dynamic information. Moreover, deep systems need huge training datasets and days of training time with fast computers while it is easy to overfit to dataset characteristics unless mixed datasets are compiled for training. There are many hyper-parameters in such systems that should be intuitively set by hand.

Performance of OSV systems is usually measured by EER which is the error rate when the false accept (FA) and false reject (FR) rates are equal. This way the error metric is independent of the number of positive and negative test samples.

In this work RBP, a WD RNN with a BiLSTM layer that can learn the sequential flow of binary patterns between overlapping windows is introduced. Local binary patterns (LBP) histogram naturally handles variable-size signature images whereas BiLSTM learns the relation between windows. LBP histogram dimensionality is reduced by removing non-useful stop-words with a novel algorithm while decreasing the error rates. To the best of the authors' knowledge, this is the first attempt to utilize RNNs for the purpose of learning sequential information in static signatures. Instead of relying on learned convolution filters, LBP-coded image windows' histograms are directly represented as sequences to be learned by BiLSTM. No training database is required beforehand to teach a classifier how to do LBP encoding.

In the literature, LBP-coded image windows are usually stacked and combined together to form a feature vector instead of learning the sequential relation between them. That kind of an approach loses the spatial relation of windows in an image. This constraint is nicely solved using the proposed RBP network. RNNs already find their place in handwriting recognition problem as shown under Sect. 2.1. Such domain-based intuitions provide a motivation for investigating the performance of RNNs for OSV problem.

RBP network is the result of a trade off between the complexity of deep models and hand-crafted features' slightly higher error rates. In addition to RBP, a novel writer identification CNN that learns genuine and forgery classes for each training writer is proposed. Activations of the proposed CNN are used to train WD classifiers.

The remaining of this paper consists of the following: A detailed literature review on RNNs with image inputs, LBP features accompanying deep methods, and recent works on OSV is provided (Sect. 2). Detailed description of the proposed method is given (Sect. 3). Experimental results are presented among with analysis (Sect. 4). Paper is concluded and future work is proposed (Sect. 5).

2   Related Works

Related recent works are reviewed in three parts: methods of using RNNs with image inputs, methods that benefit from LBP features accompanying deep methods, recent literature on OSV problem.

2.1   RNNs with Image Inputs

There are different approaches in literature to utilize the RNN for image inputs. Multi-dimensional RNN is one of the solutions and it is applied in offline hand-writing recognition [3], among with scene labeling problem [4].

RNNs can directly be input with image pixels in different input designs. Considering each pixel as a time-step would need a very long-term memory and would be difficult to keep track of. Instead, each row or column of the image may represent the features and the corresponding pixels represent the time-steps. A more natural solution is to use windows as features and pixels inside the windows as time-steps. ReNet replaces the convolution and pooling layers of CNN with four RNNs that sweep the image horizontally and vertically in both directions [5]. The recurrent layer ensures that each feature activation in its output is an activation at the specific location with respect to the whole image, instead of usual convolution and pooling layers which only has a local context window. The lowest layer of the model sweeps over the input image, with subsequent layers operating on extracted representations from the layer below, forming a hierarchical representation of the input.

A fast and segmentation-free Chinese sub-character-level recognition is proposed by using the baseline multi-dimensional LSTM [6]. Connectionist temporal classification and sequence-to-sequence learning are compared for segmentation-free handwriting recognition problem [7]. A word image is interpreted as a time sequence where time is modeled along the width of the image with 1 pixel slice as a single unit of time. Pixel values after preprocessing are used as input features for the network. The connectionist temporal classification approach is reported to outperform the sequence-to-sequence learning approach in terms of generalization and prediction of long words.

RNNs have an intuitive application in online signature verification and they are considered in different works. Tolosana et al. analyze the feasibility of RNNs for on-line signature verification in real practical scenarios [8]. RNN (with LSTM layer) outperformed the previous results on the BiosecurID [9] ranging from 17.76% to 28.00% improvement for skilled forgeries.

CNNs are combined with RNNs usually in video processing tasks. Such hybrid networks are considered frequently also in offline handwriting recognition problem. Hybrid CNN-RNN architecture has been utilized for recognizing Urdu text from printed documents [10]. The network transcribes a sequence of convolutional features from an input image to a sequence of target labels, discarding the need to segment the input image into characters. Past and future contexts are modelled by bidirectional recurrent layers to aid the transcription. A modified CNN-RNN hybrid architecture with a focus on effective training using efficient initialization of network using synthetic data for pretraining; image normalization for slant correction and domain specific data transformation is proposed by Dutta et al. [11].

A writer-independent (WI) online signature verification system based on Siamese RNN is proposed to learn a dissimilarity metric from the pairs of signatures [12]. Both unidirectional and bidirectional schemes are considered for

LSTM and Gated Recurrent Unit (GRU) based systems. Proposed Siamese RNN outperformed the state-of-the-art results on BiosecurID database.

2.2 LBP Features Accompanying Deep Methods

LBP is a powerful image descriptor proposed by Ojala et al. [13]. Benrachou et al. compared LSTM and SVM with LBP histogram feature for eye detection [14]. LBP histogram is directly fed into LSTM layer as separate features. According to experiments, SVM resulted in higher accuracy (98.1% versus 95.8%). Different from [14], authors of the present work propose RBP network to learn the natural sequential relation of LBP codes between consecutive image windows with the help of a WD RNN utilizing a BiLSTM layer.

Local binary convolution (LBC) as an alternative to convolutional layers in standard CNN is proposed by Juefei-Xu et al. [15]. LBC layer comprises of a set of fixed sparse pre-defined binary convolutional filters that are not updated during the training process, a non-linear activation function and a set of learnable linear weights. The linear weights combine the activated filter responses to approximate the corresponding activated filter responses of a standard convolutional layer.

Zhang et al. propose to use LBP-image as input to CNN for the task of face recognition and report better results [16]. LBP encoded CNN models (TEX-Nets), trained using mapped coded images with explicit LBP-based texture information are shown to provide complementary information to the standard deep models [17]. In addition, two deep architectures named early and late fusion are investigated to combine the texture and color information.

2.3 Recent Approaches on OSV

OSV is a challenging yet attracting topic with many different approaches utilized so far. A recent detailed survey on OSV is presented by Díaz et al. [18]. When comparing different works, one should take into account the differences between many aspects like the databases, image representation, division of the databases into training and testing, reference selection protocol and the count of references, existence of skilled and/or random forgeries in training and/or testing, decision threshold as global or user-based, hyper-parameter choice among with a number of other factors.

Combination of many handcrafted features resulted in an EER of 7% on binary GPDS-160 database [19] with 12 reference signatures per subject [20]. WD and WI classifiers are combined at score-level for histogram of oriented gradients (HoG), LBP and scale invariant feature transform (SIFT) features.

Hafemann et al. use an identification CNN to learn signature representations while identifying the training subjects [21]. A multi-task framework is proposed by considering two terms in the cost function for feature learning to drive the features distinguish skilled forgeries. The first term drives the model for identification of training subjects while the second term drives the model to distinguish between genuine signatures and skilled forgeries. After training, the CNN (named

SigNet-F) can extract feature representations of 2048 dimensions for test signatures with whom WD SVMs are trained. GPDS-960-gray database [22] is utilized with gray-level signature images. Last 531 subjects are used for training whereas first 160 subjects are used for testing. SVM with radial basis function (RBF) reports 3.61% EER with 12 reference signatures per subject. If user-based ideal thresholds are calculated from the test scores, EER reduces to 1.72%.

Input image size should be fixed with CNNs to learn fixed size representations unless special pooling functions are implemented. In a recent work, spatial pyramid pooling (SPP) is utilized to work with CNNs in OSV problem [23]. While training the CNN with SPP, images of fixed size are used whereas during testing variable size images are used. Different CNN models are investigated. Among them, SigNet-SPP-300dpi-F uses inputs of resolution 300 dpi and uses both genuine and skilled forgery samples in training. It resulted in an EER of 1.69% (global threshold) on GPDS-300 (first 300 subjects of GPDS-960 database) with 12 references. EER drops to 0.41% if user-based ideal thresholds are used. SigNet-SPP-300dpi does not use the authenticity information of training signatures and results in an EER of 5.63% (global threshold) or 3.15% (user-based thresholds).

Instead of training an identification CNN and later extracting features with it, a WI two-channel verification CNN that takes a reference and a query signature as input has been proposed [24]. WI CNN is also used for feature extraction during WI verification. Global average pooling (GAP) is applied to decrease the dimensionality to 200 at the end of locally connected layers. Extracted features are later used for WD model training. WI and WD models' scores are combined for the final decision. Using 12 reference signatures on GPDS-160 gray, 4.13% EER is reported with a global decision threshold.

Zois et al. investigated the detection of first order transitions between asymmetrical lattice arrangements of simple pixel structures on WI OSV [25]. A decision stumps committee is utilized with boosting feature selection. Unrelated training and testing datasets are obtained from four signature databases (CEDAR [26], MCYT [27], GPDS-300, NFI [28]) in addition to a fifth signature dataset which contains disguised signatures (4NSigComp2010 [29]). Varying pruning levels of the signatures are explored to observe the effect of preprocessing. Using gray-level GPDS-300, 3.06% EER is reported with 5 reference samples and user-based thresholds.

3 Proposed Method

3.1 Preprocessing

A simple preprocessing is applied to keep the loss of information as low as possible while reducing the noise and preparing the images. First, the gray-level values are extracted from 255 so that the background becomes 0 and foreground becomes 255. Then, small connected components (less than 40 pixels) are removed to reduce the noise. Bounding box is then cropped by finding the minimum and maximum x and y coordinates of pencil strokes and deleting the empty rows and columns before and after them. In this work, the original image size is kept for RBP or as a last step the image is resized to 100×150 for CNN. An example original signature image and its preprocessed form are shown in Fig. 1.

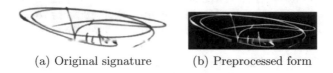

(a) Original signature   (b) Preprocessed form

Fig. 1. An example signature image from GPDS-960 and its preprocessed form.

3.2   Recurrent Binary Patterns

Local Binary Patterns. LBP characterizes the local texture in an image by encoding the neighborhoods. LBP for a center pixel (x_c, y_c) is calculated as

$$f_{LBP}(x_c, y_c) = \sum_{n=0}^{L-1} 2^n \xi(i_n, i_c) \tag{1}$$

where i_n is the intensity of the nth neighbor, i_c is the intensity of the center pixel, L is the considered number of neighbor pixels, function $\xi(\cdot) = 1$ if $i_n \geq i_c$ and 0 otherwise. In this work neighbor groups consist of $L = 4$ pixels and $f_{LBP}(\cdot)$ is between 0 and 15, resulting in 16 different codes.

Histogram is then built by the formula:

$$f_H(k) = \sum_{m=1}^{M} \sum_{n=1}^{N} \gamma(f_{LBP}(x_m, y_n), k) \tag{2}$$

for an $M \times N$ image where $\gamma(x, y)$ is 1 only when $x = y$, $k \in [0, K]$ is a histogram element and $K = 2^L - 1$ is the maximum LBP value.

Authors take into account all neighborhoods up to Chebyshev distance 4 from the center pixel. As generating binary codes directly from all 80 neighbors would result in 2^{80} codes, each distance is handled separately. Neighbors in each distance are grouped as 4 equidistant pixels, similar to [30]. One of such groups for Chebyshev distance 2 is shown in Fig. 2. In this case 4 such groups are obtained for 2-pixel neighbors.

In the overall LBP representation, there are two histograms for 1-pixel away neighbors (8-neighbors), 4 histograms for 2-pixel away neighbors, 6 histograms for 3-pixel away neighbors and 8 histograms for 4-pixel away neighbors. Proposed method discards the LBP codes that represent isolated pixels (all neighbors off) so each histogram is actually of size $2^4 - 1$. Overall representation dimensionality is $20 \times 15 = 300$.

Fig. 2. One of the neighbor groups for Chebyshev distance 2.

Most frequent LBP codes can be thought as generic stop-words and least frequent LBP codes can be discarded to reduce the dimensionality. To find such codes, authors sum up the extracted LBP histograms in a random small subset of the training set (defined in Sect. 4.2). Most frequent 40 and least frequent 40 LBP codes are found and the corresponding indices are then removed in future WD training and testing samples. Final LBP histogram dimensionality is thus 220, reduced by 26.67%. This technique improves the results while reducing the dimensionality. To conclude the usefulness of the proposed selection, experimental results without histogram selection on GPDS-960 are presented in Tables 2, 3, 4 and 5 among with the results obtained with histogram selection.

RBP network uses 5 horizontal and 5 vertical windows that are swept by the RNN sequentially, so each feature has 10 time-steps. LBP features are normalized by dividing to the maximum value over all LBP histograms in all windows in an image.

RBP Structure. LBP-coded image windows are learned by BiLSTM layer as sequences. Windows are 10% overlapping. Proposed BiLSTM layer uses 300 hidden units. As state activation function, hyperbolic tangent is used whereas for gate activation function sigmoid is used. RBP network is illustrated in Fig. 3.

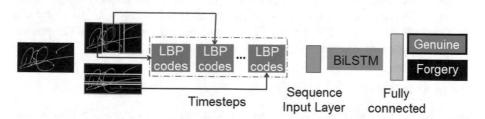

Fig. 3. Example WD-RBP network with 10% overlapping three horizontal and vertical windows.

Writer-Dependent Verification. RNNs are trained in a WD fashion, per subject. RNN is fed with the LBP histograms of reference signatures and random forgeries to train a WD model. Test samples' scores are directly used for verification. Verification EER can either be calculated using a global or user-based threshold.

3.3   Identification CNN with Forgery Outputs

Authors propose a CNN, namely, ϕ with $2|\tau|$ outputs in a separate development set τ of $|\tau|$ subjects. There are two outputs for each subject, one indicating genuine and one indicating forgery class. Image input size of the CNN is 100×150. The network ϕ consists of 10 convolutional layers and 2 fully connected layers. Network is concluded with a $2|\tau|$-way softmax layer to classify the input as genuine or forgery sample of the corresponding subject. CNN model used in this work with example 950 outputs is demonstrated in Fig. 4.

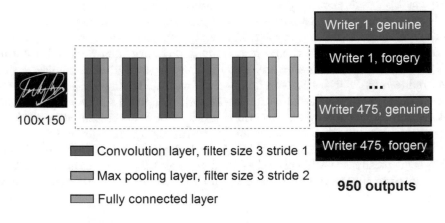

Fig. 4. Proposed CNN architecture with example $2|\tau|$ outputs.

Writer-Dependent Verification. WD verification by CNN is performed by training WD SVM classifiers. Signature features are extracted as the activation of the second-last layer of CNN, namely ϕ_{fc}. Features of a query signature Q is thus expressed as $\phi_{fc}(Q)$. Dimensionality of the features extracted by the proposed CNN is 1000. SVM with RBF kernel is utilized to train WD models using such features of reference signatures and random forgeries.

3.4   Other Models

SigNet-F CNN with a Single Forgery Output. Another CNN that is proposed in [21] is utilized for comparison. There are two terms in the cost function to learn features. The first term forces the model to distinguish between different subjects, while the second term forces the model to distinguish between genuine signatures and skilled forgeries. Second term adds a single sigmoid unit to the model that seeks to predict whether or not the signature is a forgery, compared to $|\tau|$ additional units in the proposed model. The network is expected to learn visual cues particular to each class such as bad line quality and jaggedness in the pen strokes that are usually present in forgeries.

SigNet-SPP-300dpi CNN. As a more recent CNN that can have variable-size inputs, SigNet-SPP-300dpi [23] which is trained with samples of 300 dpi resolution is also considered. During test time, variable-size inputs are handled with the help of SPP. Throughout the rest of the text, SigNet-SPP stands for SigNet-SPP-300dpi.

Two-Channel CNN. A two-channel CNN that allows concurrent WI and WD verification is also considered [24] for comparison and score-level integration. In addition to the representations to train WD models, a WI score is obtained with the two-channel CNN. WI and WD scores are then combined as proposed in [24].

Writer-Dependent Verification. WD verification is performed similar to the proposed CNN. Activations at the last layer before softmax is used as the feature vector for a query signature Q as $\theta(Q)$, whose dimensionality is 2048 both for SigNet-F and SigNet-SPP. Dimensionality is 200 (after GAP layer) for the two-channel CNN. WD shallow SVM classifiers with RBF kernel are trained similarly as offered in [21,23,24], using the activations of reference signatures and random forgeries.

3.5   Combination of the Approaches

WD classifiers trained with the above-mentioned representations are combined at score level. A simple weighted linear combination of the scores for a query signature Q is followed.

4   Experimental Results

4.1   Databases

GPDS-960 and GPDS-Synthetic-10000 signature databases are utilized to evaluate the proposed techniques and confirm the benefits. GPDS-960 signature database consists of 881 users. Gray-level signature images are in .png format and have been scanned at 600 dpi. Each subject has 24 genuine samples and at most 30 forgery samples. Number of forgery samples is less than 30 for a few subjects.

GPDS-Synthetic-10000 online and offline bimodal database [31] is a challenging synthetic database consisting of 24 genuine and 30 skilled forgery signatures for each individual. All static signatures are generated with different modeled pens. This work only uses the static signature images and ignore the dynamic data. This database is the most recent large-scale synthetic signature database and it is not to be confused with the previous GPDS-Synthetic-4000 offline signature database [32]. GPDS-Synthetic will be meant to denote the offline samples of GPDS-Synthetic-10000.

4.2 Experimental Protocol

Training and Validation Sets of CNNs. In the proposed CNN model with multiple forgery outputs, last 475 subjects of GPDS-960 are utilized for training, defined as set τ. Because the CNN is trained for identification, τ is vertically separated into two subsets as τ_1 and τ_2 per subject. τ_1 contains the first 20 genuine and the first 20 skilled forgery samples of each subject, while τ_2 contains the remaining samples.

CNN parameters are actually learned on τ_1. CNN hyper-parameters are determined by looking at the accuracy of the CNN on τ_2.

Details about the training protocol are provided for SigNet-F in [21], SigNet-SPP in [23] and two-channel CNN (WI&WD) in [24]. Only GPDS-160 results when $N = 12$ will be taken into account for the two-channel CNN.

WD Model Hyper-parameters. A multipurpose validation set V is reserved to determine WD SVM hyper-parameters (namely, cost (C) of error and γ of RBF kernel) and WD RNN hyper-parameters (notably the number of hidden units and the optimizer). V consists of the previous 106 subjects of GPDS-960 before τ. For convenience, WD model hyper-parameters are shown in Table 1.

Table 1. WD model hyper-parameters.

Model	SVM cost	SVM gamma
Proposed CNN	32	0.125
Two-channel CNN	32	1
SigNet-F & SigNet-SPP	2	0.0625
	BiLSTM units	Optimizer
Proposed RBP network	300	Adam [33]

Selecting the Combination Parameters. Combination weights are determined experimentally by training and testing WD classifiers on V.

WD Negative Samples. Random forgeries are represented as the union of two skilled forgery samples from each of the subjects in V, excluding the subject of interest of the WD classifier. This selection mechanism is repeated for RBP network and other WD classifiers trained with all CNN features that are used in combination. It is fair to use other subjects' skilled forgeries as one can always find and collect skilled forgery samples from some random subjects before deploying a verification system.

Test Set. Performance of RBP and the proposed CNN features, Signet-F features and their classifier combinations are evaluated using the test set T. There are three different test sets considered in this work: first 160 or first 300 subjects of GPDS-960, first 300 subjects of GPDS-Synthetic-10000. Samples of the first 160 subjects (GPDS-160) were collected before the rest of GPDS-960 and their appearance is different, so testing them separately makes sense. The reason behind the selection of first 300 subjects of GPDS-Synthetic-10000 is to keep the test size the same as GPDS-960 and to purely investigate the generalization capability. Random cross-validation tests are applied. The following procedure is repeated for each of the test sets in the same way.

T is divided into two disjoint parts vertically as T_1 and T_2 per writer. T_1 is the set of genuine reference samples and T_2 is the set of genuine and skilled forgery query samples. Selection of genuine reference samples is performed randomly. Genuine samples are randomly divided into two partitions P_1 and P_2, each having 12 samples per user. P_1 represents the set of potential reference signatures and P_2 represents genuine test samples. Inside P_1, N genuine samples are randomly selected as positive samples $T_1 = P_1{}^N$ to train WD model. The union $T_2{}^+ = (P_1 \setminus P_1{}^N) \cup P_2$ is the final set of positive test samples per user. When all skilled forgeries $T_2{}^-$ of the user are united with $T_2{}^+$, T_2 is determined as $T_2{}^- \cup T_2{}^+$.

This work considers $N = 5$ and $N = 12$. Partitioning is randomly repeated two times. The selection of N training samples inside P_1 is repeated 3 times. In total, 6 tests are performed for each such T_1 and T_2 distinction for $N = 5$. For $N = 12$, P_1 is completely covered so there is no random reference set selection inside the partition, resulting in 2 random partition tests. Partitions and reference samples are randomly determined once ahead of time and applied to all tests for all users including RBP, proposed CNN, SigNet-F, SigNet-SPP, two-channel CNN tests. Random forgeries are not used during testing.

Calculation of EER. EER is calculated directly from the test scores. It is either calculated for all writers' scores globally (global threshold EER) or with the scores of each writer separately (user-based threshold EER).

Summary of the Protocol. Separation of the database into subsets is shown in Fig. 5. As a summary, last 581 subjects of GPDS-960 are reserved for development purposes. However, the proposed CNN has double time of the outputs with the number of the training subjects compared to Signet-F models. This work utilizes less subjects to learn the parameters of the proposed CNN and more subjects for multipurpose validation. The reason is, authors utilize V for many purposes as described in the preceding sections and thus need a bigger validation set. The proposed CNN model will be improved in future by using more subjects for training.

First 300 subjects of GPDS-960 and GPDS-Synthetic are used for testing in any case. Either $N = 5$ or $N = 12$ references are considered for WD model training, where the negative samples are the forgery signatures of other subjects.

Reference subset is randomly selected and repeated several times. Random forgeries are not considered during testing.

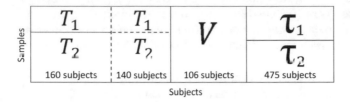

Fig. 5. Partitioning of the database into subsets.

4.3   Results and Discussion

Results of WD classifiers and their combinations are shown in Tables 2 and 3 for GPDS-160; Tables 4 and 5 for GPDS-300; Tables 6 and 7 for GPDS-Synthetic, when $N = 5$ and $N = 12$ accordingly. In each table, EER results with both global and user-based thresholds are provided. Among WD models, RBP utilizes RNN and others utilize SVM classifiers. As a reminder, experiments are repeated for random reference and test sets as described in Sect. 4.2; mean and standard deviation of such experiments' results are shown in the tables.

Table 2. GPDS-160 WD classifier EER results for $N = 5$.

Method	Global th.	User-based th.
RBP size 300 (no hist. select.)	$6.75 \pm 0.22\%$	$5.16 \pm 0.19\%$
(1) RBP size 220 (proposed hist. select.)	$5.77 \pm 0.13\%$	$4.29 \pm 0.15\%$
(2) Proposed CNN	$7.34 \pm 0.45\%$	$4.34 \pm 0.12\%$
(3) SigNet-F [21]	$4.28 \pm 0.23\%$	$2.83 \pm 0.29\%$
(4) SigNet-SPP [23]	$4.86 \pm 0.32\%$	$3.44 \pm 0.27\%$
Combi. 1 & 2	$3.38 \pm 0.28\%$	$1.82 \pm 0.14\%$
Combi. 1 & 3	$2.41 \pm 0.28\%$	$1.34 \pm 0.10\%$
Combi. 2 & 3	$2.97 \pm 0.11\%$	$1.87 \pm 0.12\%$
Combi. 3 & 4	$3.92 \pm 0.24\%$	$2.53 \pm 0.22\%$
Combi. 1 & 2 & 3	$2.05 \pm 0.25\%$	$1.11 \pm 0.06\%$
Combi. 1 to 4	$1.93 \pm 0.16\%$	$1.08 \pm 0.06\%$
Handcrafted features (binary images) [20]	7.98%	–

Table 3. GPDS-160 WD classifier EER results for $N = 12$.

Method	Global th.	User-based th.
RBP size 300 (no hist. select.)	$6.32 \pm 0.03\%$	$4.72 \pm 0.15\%$
(1) RBP size 220 (proposed hist. select.)	$5.56 \pm 0.05\%$	$4.24 \pm 0.26\%$
(2) Proposed CNN	$5.08 \pm 0.25\%$	$3.26 \pm 0.15\%$
(3) SigNet-F [21]	$3.51 \pm 0.46\%$	$2.11 \pm 0.52\%$
(4) SigNet-SPP [23]	$4.39 \pm 0.50\%$	$2.68 \pm 0.46\%$
(5) Two-channel CNN (WI & WD) [24]	$4.13 \pm 0.31\%$	$2.94 \pm 0.28\%$
Combi. 1 & 2	$2.75 \pm 0.51\%$	$1.63 \pm 0.52\%$
Combi. 1 & 3	$2.11 \pm 0.46\%$	$0.98 \pm 0.13\%$
Combi. 2 & 3	$2.19 \pm 0.35\%$	$1.27 \pm 0.58\%$
Combi. 3 & 4	$3.42 \pm 0.44\%$	$1.99 \pm 0.43\%$
Combi. 3 & 5	$1.76 \pm 0.37\%$	$0.88 \pm 0.36\%$
Combi. 1 & 2 & 3	$1.66 \pm 0.21\%$	$0.81 \pm 0.28\%$
Combi. 1 to 4	$1.66 \pm 0.21\%$	$0.81 \pm 0.28\%$
Combi. 1 to 5	$1.11 \pm 0.28\%$	$0.57 \pm 0.26\%$
Handcrafted features (binary images) [20]	6.97%	$-$

Table 4. GPDS-300 WD classifier EER results for $N = 5$.

Method	Global th.	User-based th.
RBP size 300 (no hist. select.)	$5.55 \pm 0.27\%$	$3.99 \pm 0.11\%$
(1) RBP size 220 (proposed hist. select.)	$4.58 \pm 0.15\%$	$3.29 \pm 0.06\%$
(2) Proposed CNN	$6.26 \pm 0.30\%$	$3.45 \pm 0.09\%$
(3) SigNet-F [21]	$4.38 \pm 0.23\%$	$2.83 \pm 0.21\%$
(4) SigNet-SPP [23]	$5.83 \pm 0.28\%$	$4.22 \pm 0.25\%$
Combi. 1 & 2	$2.79 + 0.14\%$	$1.19 \pm 0.12\%$
Combi. 1 & 3	$2.22 \pm 0.17\%$	$1.15 \pm 0.07\%$
Combi. 2 & 3	$3.06 \pm 0.17\%$	$1.78 \pm 0.13\%$
Combi. 3 & 4	$4.21 \pm 0.19\%$	$2.83 \pm 0.19\%$
Combi. 1 & 2 & 3	$1.86 \pm 0.11\%$	$0.84 \pm 0.08\%$
Combi. 1 to 4	$1.90 \pm 0.13\%$	$0.84 \pm 0.05\%$
Asymmetric pixel relations **(WI)** [25]	$-$	3.06%

For a direct comparison with previous works, single classifier results are analyzed. Previous works' results are shown in the corresponding tables as provided in their articles. Among the single WD classifiers, SVM trained with 2048-dimensional Signet-F representations can always perform the best. RBP network (feature dimensionality 220) and CNN representation (dimensionality 1000) give alone relatively higher EER results, nevertheless being very useful when used in

Table 5. GPDS-300 WD classifier EER results for $N = 12$.

Method	Global th.	User-based th.
RBP size 300 (no hist. select.)	$5.50 \pm 0.50\%$	$3.79 \pm 0.11\%$
(1) RBP size 220 (proposed hist. select.)	$4.59 \pm 0.42\%$	$3.18 \pm 0.16\%$
(2) Proposed CNN	$4.40 \pm 0.35\%$	$2.46 \pm 0.01\%$
(3) SigNet-F [21]	$3.64 \pm 0.33\%$	$2.23 \pm 0.33\%$
(4) SigNet-SPP [23]	$5.12 \pm 0.53\%$	$3.48 \pm 0.24\%$
Combi. 1 & 2	$2.10 \pm 0.21\%$	$1.00 \pm 0.28\%$
Combi. 1 & 3	$1.90 \pm 0.32\%$	$0.88 \pm 0.12\%$
Combi. 2 & 3	$2.30 \pm 0.25\%$	$1.20 \pm 0.34\%$
Combi. 3 & 4	$3.56 \pm 0.31\%$	$2.24 \pm 0.28\%$
Combi. 1 & 2 & 3	$1.54 \pm 0.27\%$	$0.61 \pm 0.16\%$
Combi. 1 to 4	$1.53 \pm 0.31\%$	$0.61 \pm 0.17\%$
SigNet-SPP-F [23]	$1.69 \pm 0.10\%$	$0.41 \pm 0.05\%$

Table 6. GPDS-Synthetic WD classifier EER results for $N = 5$.

Method	Global th.	User-based th.
(1) RBP size 220 (proposed hist. select.)	$31.73 \pm 0.40\%$	$31.16 \pm 0.31\%$
(2) Proposed CNN	$31.55 \pm 0.49\%$	$28.42 \pm 0.28\%$
(3) SigNet-F [21]	$27.47 \pm 0.24\%$	$24.91 \pm 0.22\%$
(4) SigNet-SPP [23]	$35.00 \pm 0.42\%$	$33.21 \pm 0.63\%$
Combi. 1 & 2	$28.82 \pm 0.45\%$	$25.90 \pm 0.35\%$
Combi. 1 & 3	$26.50 \pm 0.19\%$	$24.04 \pm 0.29\%$
Combi. 2 & 3	$25.90 \pm 0.30\%$	$22.48 \pm 0.42\%$
Combi. 3 & 4	$27.07 \pm 0.14\%$	$24.27 \pm 0.18\%$
Combi. 1 & 2 & 3	$25.08 \pm 0.31\%$	$22.13 \pm 0.42\%$
Combi. 1 to 4	$25.08 \pm 0.31\%$	$22.13 \pm 0.42\%$

conjunction with SigNet-F. SigNet-SPP always gives a bit higher error rates compared to SigNet-F. Two-channel CNN WI & WD combined model's error rate (dimensionality 200 for WD) is close to that of SigNet-F for GPDS-160 (Table 3).

Proposed LBP stop-word removal (histogram selection) algorithm always improves the results on GPDS-160 and GPDS-300 while reducing the dimensionality. Thus for the combinations, RBP is only used with histogram selection.

It can be seen that RBP is more robust to the number of references N. WD SVMs trained with CNN representations benefit more when N is increased. In general for $N = 5$, RBP and SigNet-F results are similar. RBP is better than the proposed CNN when $N = 5$, and the vice versa applies when $N = 12$.

Table 7. GPDS-Synthetic WD classifier EER results for $N = 12$.

Method	Global th.	User-based th.
(1) RBP size 220 (proposed hist. select.)	$26.32 \pm 0.79\%$	$24.22 \pm 0.21\%$
(2) Proposed CNN	$25.68 \pm 0.78\%$	$23.52 \pm 0.21\%$
(3) SigNet-F [21]	$18.93 \pm 0.27\%$	$16.98 \pm 0.33\%$
(4) SigNet-SPP [23]	$28.71 \pm 0.17\%$	$26.83 \pm 0.05\%$
Combi. 1 & 2	$22.00 \pm 0.05\%$	$20.01 \pm 0.52\%$
Combi. 1 & 3	$17.95 \pm 0.11\%$	$16.16 \pm 0.20\%$
Combi. 2 & 3	$18.18 \pm 0.25\%$	$15.62 \pm 0.13\%$
Combi. 3 & 4	$18.92 \pm 0.13\%$	$16.85 \pm 0.21\%$
Combi. 1 & 2 & 3	$17.50 \pm 0.22\%$	$14.80 \pm 0.08\%$
Combi. 1 to 4	$17.65 \pm 0.65\%$	$14.93 \pm 0.18\%$

GPDS-Synthetic EERs are higher when compared with GPDS-960. There are two main reasons behind this phenomenon. The first one is, authors do not perform any kind of training or fine-tuning special to GPDS-Synthetic and purely investigate the generalization performance on the test set. The second one is, as GPDS-Synthetic is a synthetic database, no natural differences in appearance of the signatures sourcing from the capability of human forgers exist. Different pen models are used so the variations between genuine samples are higher.

It has been found that SigNet-SPP does not usually help improve the results when combined with other methods. Authors hence do not report all combination results with SigNet-SPP. In some cases (Tables 4 and 7) the results even get worse. The reason is, combination weights are determined from the validation set V instead of the test set T.

For the classifier combinations, the best results are usually achieved when all WD classifiers are combined. However RBP and SigNet-F are very complementary and combination of all classifiers does not help improve the results so much, compared to the combination of RBP and SigNet-F. If it is not possible to train and store multiple CNNs, one can stick with Signet-F and train additional WD RBP networks to obtain lower EERs. It can be seen that score-level integration of the classifiers can help the best single representation further reduce the EER.

5   Conclusions and Future Work

A novel RNN with a BiLSTM layer that sweeps signature windows horizontally and vertically is proposed to learn WD sequential relation between windows' LBP representations, instead of stacking all windows' histograms and losing the sequential relation. As a result of the novel LBP stop-code removal algorithm, histogram dimensionality is reduced to only 220 and at the same time EER is decreased. This model can give competitive results and help reduce the error rates throughout an extensive experimentation with random cross-validation, compared to the representations of more complex and advanced CNN structures of higher dimensionality.

As can be seen from the results, RBP network gives higher EERs compared to CNN-based models trained on a development set. To improve the RBP network so that it can give alone better EERs, deeper RNNs will be investigated to be able to handle LBP histograms extracted at different resolutions.

A novel identification CNN structure with extra forgery classes for each training subject is also proposed. This CNN can extract 1000-dimensional representations and gives error rates close to the ones obtained with 2048-dimensional Signet-F representations. When all of the WD classifiers are used in conjunction, state-of-the-art verification results are reported.

Although the gap is close, SigNet-F with a higher-dimensional representation can give better EERs compared to the proposed CNN. As a future work, more efficient and recent networks in replacement of CNNs such as DilatedNets [34], ResNets [35], DenseNets [36] and MobileNets [37] will be investigated to compare and integrate with the proposed WD RBP network.

Acknowledgments. This work was supported by The Scientific Research Projects Coordination Unit of Akdeniz University, project number: 3780.

References

1. Alonso-Fernandez, F., Fairhurst, M.C., Fierrez, J., Ortega-Garcia, J.: Impact of signature legibility and signature type in off-line signature verification. In: 2007 Biometrics Symposium, pp. 1–6 (2007)
2. Galbally, J., Gomez-Barrero, M., Ross, A.: Accuracy evaluation of handwritten signature verification: rethinking the random-skilled forgeries dichotomy. In: 2017 IEEE International Joint Conference on Biometrics (IJCB), pp. 302–310 (2017)
3. Graves, A., Schmidhuber, J.: Offline handwriting recognition with multidimensional recurrent neural networks. In: Koller, D., Schuurmans, D., Bengio, Y., Bottou, L. (eds.) Advances in Neural Information Processing Systems 21, pp. 545–552. Curran Associates Inc., Red Hook (2009)
4. Byeon, W., Breuel, T.M., Raue, F., Liwicki, M.: Scene labeling with LSTM recurrent neural networks. In: CVPR, pp. 3547–3555. IEEE Computer Society (2015)
5. Visin, F., Kastner, K., Cho, K., Matteucci, M., Courville, A., Bengio, Y.: ReNet: a recurrent neural network based alternative to convolutional networks. arXiv preprint arXiv:1505.00393 (2015)
6. Bluche, T., Messina, R.: Faster segmentation-free handwritten Chinese text recognition with character decompositions. In: 2016 15th International Conference on Frontiers in Handwriting Recognition (ICFHR), pp. 530–535 (2016)
7. Shkarupa, Y., Mencis, R., Sabatelli, M.: Offline handwriting recognition using LSTM recurrent neural networks. In: The 28th Benelux Conference on Artificial Intelligence, Amsterdam (NL), 10–11 November 2016, vol. 1, p. 88 (2016)
8. Tolosana, R., Vera-Rodriguez, R., Fierrez, J., Ortega-Garcia, J.: Biometric signature verification using recurrent neural networks. In: 2017 14th IAPR International Conference on Document Analysis and Recognition (ICDAR), vol. 01, pp. 652–657 (2017)

9. Fierrez, J., Galbally, J., Ortega-Garcia, J., Freire, M., Alonso-Fernandez, F., Ramos, D., Toledano, D., González, J., Siguenza, J., Garrido-Salas, J., Anguiano, E., González, G., Ribalda, R., Ortega, J., Cardeñoso, V., Viloria, A., Vivaracho, C., Moro, Q., Faúndez, M., Igarza, J., Sanchez, J., Hernaez, J., Orrite, C., Martinez, F., Gracia, J.: BiosecurID: a multimodal biometric database. Pattern Anal. Appl. **13**(2), 235–246 (2010)

10. Jain, M., Mathew, M., Jawahar, C.V.: Unconstrained OCR for Urdu using Deep CNN-RNN hybrid networks. In: 4th Asian Conference on Pattern Recognition (ACPR 2017), Nanjing, China, p. 6 (2017)

11. Dutta, K., Krishnan, P., Mathew, M., Jawahar, C.: Improving CNN-RNN hybrid networks for handwriting recognition. In: The 16th International Conference on Frontiers in Handwriting Recognition, Niagara Falls, USA (2018)

12. Tolosana, R., Vera-Rodriguez, R., Fierrez, J., Ortega-Garcia, J.: Exploring recurrent neural networks for on-line handwritten signature biometrics. IEEE Access **6**, 5128–5138 (2018)

13. Ojala, T., Pietikäinen, M., Mäenpää, T.: Multiresolution gray-scale and rotation invariant texture classification with local binary patterns. IEEE Trans. Pattern Anal. Mach. Intell. **24**(7), 971–987 (2002)

14. Benrachou, D.E., dos Santos, F.N., Boulebtateche, B., Bensaoula, S.: Online vision-based eye detection: LBP/SVM vs LBP/LSTM-RNN. In: CONTROLO 2014 Proceedings of the 11th Portuguese Conference on Automatic Control, pp. 659–668 (2014)

15. Juefei-Xu, F., Boddeti, V.N., Savvides, M.: Local binary convolutional neural networks. In: Proceedings of the IEEE Conference on Computer Vision and Pattern Recognition (CVPR), pp. 19–28. IEEE (2017)

16. Zhang, H., Qu, Z., Yuan, L., Li, G.: A face recognition method based on LBP feature for CNN. In: 2017 IEEE 2nd Advanced Information Technology, Electronic and Automation Control Conference (IAEAC), pp. 544–547 (2017)

17. Anwer, R.M., Khan, F.S., van de Weijer, J., Molinier, M., Laaksonen, J.: Binary patterns encoded convolutional neural networks for texture recognition and remote sensing scene classification. ISPRS J. Photogram. Remote Sens. **138**, 74–85 (2018)

18. Díaz, M., Ferrer, M.A., Impedovo, D., Malik, M.I., Pirlo, G., Plamondon, R.: A perspective analysis of handwritten signature technology. ACM Comput. Surv. **51**(6), 117:1–117:39 (2019)

19. Ferrer, M.A., Alonso, J.B., Travieso, C.M.: Offline geometric parameters for automatic signature verification using fixed-point arithmetic. IEEE Trans. Pattern Anal. Mach. Intell. **27**(6), 993–997 (2005)

20. Yılmaz, M.B., Yanıkoğlu, B.: Score level fusion of classifiers in off-line signature verification. Inf. Fusion **32**(Part B), 109–119 (2016). SI Information Fusion in Biometrics

21. Hafemann, L.G., Sabourin, R., Oliveira, L.S.: Learning features for offline handwritten signature verification using deep convolutional neural networks. Pattern Recogn. **70**(C), 163–176 (2017)

22. Ferrer, M.A., Vargas, J.F., Morales, A., Ordonez, A.: Robustness of offline signature verification based on gray level features. IEEE Trans. Inf. Forensics Secur. **7**(3), 966–977 (2012)

23. Hafemann, L.G., Oliveira, L.S., Sabourin, R.: Fixed-sized representation learning from offline handwritten signatures of different sizes. IJDAR **21**(3), 219–232 (2018)

24. Yılmaz, M.B., Öztürk, K.: Hybrid user-independent and user-dependent offline signature verification with a two-channel CNN. In: The IEEE Conference on Computer Vision and Pattern Recognition (CVPR) Workshops (2018)

25. Zois, E.N., Alexandridis, A., Economou, G.: Writer independent offline signature verification based on asymmetric pixel relations and unrelated training-testing datasets. Expert Syst. Appl. **125**, 14–32 (2019)
26. Kalera, M.K., Srihari, S.N., Xu, A.: Offline signature verification and identification using distance statistics. IJPRAI **18**(7), 1339–1360 (2004)
27. Ortega-Garcia, J., Fierrez-Aguilar, J., Simon, D., Gonzalez, J., Faundez-Zanuy, M., Espinosa, V., Satue, A., Hernaez, I., Igarza, J., Vivaracho, C., Escudero, D., Moro, Q.. MCYT baseline corpus: a bimodal biometric database. IEE Proc. Vis. Image Sig. Process. **150**(6), 395–401 (2003)
28. Alewijnse, L.: Analysis of signature complexity. Master thesis. University of Amsterdam (2008)
29. Liwicki, M., van den Heuvel, C.E., Found, B., Malik, M.I.: Forensic signature verification competition 4NSigComp2010 - detection of simulated and disguised signatures. In: 2010 12th International Conference on Frontiers in Handwriting Recognition, pp. 715–720 (2010)
30. Yılmaz, M.B.: Offline signature verification with user-based and global classifiers of local features. Ph.D. thesis, Sabancı University (2015)
31. Ferrer, M.A., Diaz, M., Carmona-Duarte, C., Morales, A.: A behavioral handwriting model for static and dynamic signature synthesis. IEEE Trans. Pattern Anal. Mach. Intell. **39**(6), 1041–1053 (2017)
32. Ferrer, M.A., Diaz-Cabrera, M., Morales, A.: Static signature synthesis: a neuromotor inspired approach for biometrics. IEEE Trans. Pattern Anal. Mach. Intell. **37**(3), 667–680 (2015)
33. Kingma, D.P., Ba, J.: Adam: a method for stochastic optimization. In: Proceedings of the 3rd International Conference on Learning Representations (ICLR 2015) (2015)
34. Yu, F., Koltun, V.: Multi-scale context aggregation by dilated convolutions. In: 4th International Conference on Learning Representations, Conference Track Proceedings, ICLR 2016, San Juan, Puerto Rico, 2–4 May 2016 (2016)
35. He, K., Zhang, X., Ren, S., Sun, J.: Deep residual learning for image recognition. In: 2016 IEEE Conference on Computer Vision and Pattern Recognition, CVPR 2016, Las Vegas, NV, USA, 27–30 June 2016, pp. 770–778 (2016)
36. Huang, G., Liu, Z., van der Maaten, L., Weinberger, K.Q.: Densely connected convolutional networks. In: 2017 IEEE Conference on Computer Vision and Pattern Recognition (CVPR), pp. 2261–2269 (2017)
37. Sandler, M., Howard, A.G., Zhu, M., Zhmoginov, A., Chen, L.: MobileNetV2: inverted residuals and linear bottlenecks. In: 2018 IEEE Conference on Computer Vision and Pattern Recognition, CVPR 2018, Salt Lake City, UT, USA, 18–22 June 2018, pp. 4510–4520 (2018)

An Improvement of Compelling Graphical Confirmation Plan and Cryptography for Upgrading the Information Security and Preventing Shoulder Surfing Assault

Norman Dias[✉] and S. R. Reeja

Dayananda Sagar University, Bengaluru, Karnataka, India
norman.dbce@gmail.com, reeja-cse@dsu.edu.in

Abstract. Humans are continuously alluded as the weakest connect within the security chain stating that the issue does not lie with the security framework themselves but with clients who do not comply with the security conventions. The paper focuses on a proposed graphical authentication mechanism, since a human brain contains a way better capacity to memorize pictures or images, the proposed plot will serve the reason of better memorability, security as well shoulder surfing assault.

Keywords: Authentication strategy · Graphical passwords · Secure passwords

1 Introduction

Alphanumeric passwords are broadly used to verify a client on a system. Conventional secret phrase utilizes a grouping of alphanumeric characters to allow to access to a confirmed client. Notwithstanding, the content based password are inclined to lexicon assault or a dictionary attack [1].

The Lexicon assault is a strategy by which client's utilization diverse devices to break a secret key via consequently checking every one of the words that happen in either a word reference or openly accessible registries or directories.

The alphanumeric password provides a strong security, if they are sufficiently muddled to be speculated. For the most part a content based secret key or a password is an arrangement of 8 characters or more which incorporate upper and lower case characters and digits or special characters [2]. Retaining an arbitrary secret word which does not contain any significant data and is troublesome and should be possible just by monotonous realizing which serves to be an extremely frail method for remembering a secret word [3], and with n = number of secret key per client the rate of overlooking likewise builds that is the principal motivation behind why clients will in general overlook their mind boggling secret key.

The university of Malya conducted a survey with 100 people being computer savy from the domain of computer science and information technology, trying to identify the loopholes in textual passwords, the table below demonstrates the blend of digits, symbols and letters used in their passwords (Table 1).

© Springer Nature Switzerland AG 2020
K. Arai et al. (Eds.): FTC 2019, AISC 1070, pp. 435–453, 2020.
https://doi.org/10.1007/978-3-030-32523-7_30

Table 1. Password distribution [4]

Textual passwords	Users (%)
Letters and special symbols	3%
Numbers & special symbols	5%
Special symbols, alphabets and numbers	12%
Alphabets only	25%
Digits only	15%
Only alphabets & digits	40%

It was also concluded that 55% of the users used the same password across all the accounts [4] (Fig. 1).

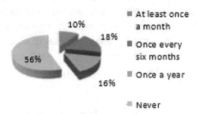

Fig. 1. Frequency of change in textual passwords [4]

The pie chart below depicts how frequently users change their textual passwords, i.e. more than 50% of the users never bother to change their passwords, 16% once in a year, 18% once every six months and the least i.e. 10% at least once in a month. The conclusion that can be drawn from this is that most of the users are unaware or do not bother that their passwords are vulnerable to security threats specifically in the case of users who prefer to keep their passwords the same across different accounts [5]. Many researchers have already proposed a good set of guidelines for selecting and maintain a good and a strong password [6]. Human beings have a critical capacity to perceive and recall visual images, and this hypothesis is supported by psychological studies [7].

In this paper, we propose a graphical password scheme which draws its base from pass points and story based graphical authentication scheme.

A password consist of one click-point per image in the foreground and the associated click point from the background image is preserved in the database, which makes detection of the click point and the associated image a daunting task for the attacker, moreover the movement of dummy cursor prevents shoulder surfing.

2   Literature Survey

Cognometric Systems also known as recognition based system. This system is predicated on hash image technique, whereby a user should choose some number of pictures generated from a group of random images by the program, later so as to be echt the client is needed to spot identical pictures in sequence [8]. The average login time is longer than the standard approach. A downside is that the server must store an over-sized quantity of images, which has to be transferred over the network, which will delay the authentication process. Another shortcoming of the framework is that the server stores the seeds of the portfolio pictures of every client in plain text. The way towards choosing picture from the database can be tedious for the client [8]. Similarly, the "Image Based Registration and authentication system", uses images as key to their accounts. The client usually supplies client ID and image as credential to the system, this technique uses Secure Hash Algorithm function SHA-1 where the output is 20 byte, thereby making the authentication process more secure [10].

Photographic authentication (Fig. 2) is a strategy for logging into untrusted open Internet access terminals. It requires a client to recognize their very own photos from a set of randomized pictures. By changing the particular pictures that appear on each login attempt. This technique is resilient to replay attacks, which is a weakness with the tra-ditional password based mechanism on systems for which every action could be deceitfully observed. A prototype implementation and relating client test demonstrate that not only the participants to a great degree proficient at rapidly, precisely and enjoyably perceive their own photos, however assailants are not in a position to figure out which photos are right, notwithstanding even when given examples of client photos [9].

Fig. 2. Photographic authentication web-browser interface [9].

The triangle scheme (Fig. 3) arbitrarily scrambles an arrangement of N objects on the screen. Practically the value of N ranges from a lower bound of hundred to an upper bound of thousand objects. The objects ought to be sufficiently diverse with the goal that the client can recognize them. There is a subset of K objects (i.e. k = 10) prior picked and retained by the client. At login, the framework will arbitrarily pick an arrangement of N Objects. This framework first randomly chooses a patch that spreads a large portion of the screen, and arbitrarily puts the K picked objects in that patch.

Fig. 3. The Triangle scheme [11]

To login the client must find three of the pass objects and click inside the undetectable triangle made by those three objects. This means that the client must click inside the convex hull of the pass objects that are displayed [11].

The number of possible passwords is the binomial coefficient. When N = 1000 and K = 10, the number of possible password is approx. $2.6 * 10^{23}$, which is more than alphanumeric password of length 15 ($36^{15} = 2.2 * 10^{23}$).

The Background Pass go can be considered as an associate improvement over Pass-Go, because it keeps majority of the advantages of Pass-Go and gives higher security level and higher easy use level as compared to the Pass-Go theme.

In Pass Go theme, the delicate territories are measured with a sweep of 0.4 * d (d = sideways length of cell), as compared to BPG (Fig. 6) the area is set to 0.30 * d range. Hence, the radius cannot be kept to large keeping in mind the end goal to lessen the achievement rate of figure assaults by just clicking at the focused areas without having prior knowledge of the genuine passwords, while in the meantime if the radius is too small then it will cause a sufficient measure of passwords input problems to the clients. In View

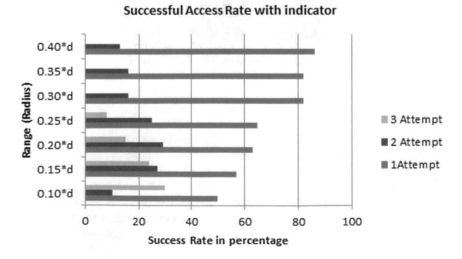

Fig. 4. Successful Access rate with indicator [4]

of heuristic testing 0.30 * d range proved to be satisfactory from clients viewpoint and the security result was also higher as compared to the Pass-Go [4]. Figures 4 and 5 represents the successful Access rate with indicator & without indicator respectively.

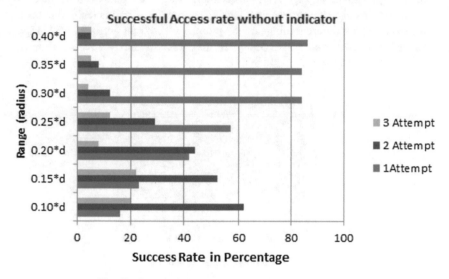

Fig. 5. Successful Access rate without indicator [4]

BPG utilizes a similar access function as I Pass go. The clients can draw a shape uninhibitedly relying upon their particular inclinations. The working of indicators is similar as in Pass go. On selection of one intersection a dot indicator will appear, whereas on a continuous selection of two or more intersections a line indicator will appear. The thickness and design of indicators can be upgraded according to the inclination of clients. BPG uses steady pointers to recognize the passwords since this approach has the capacity to quicken the procedure of retention by using invariable indicators. The password is then encoded as a succession of intersections, drawn by a 2 dimensional coordinate pairs.

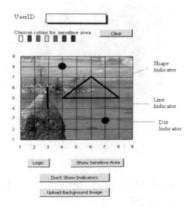

Fig. 6. Background Pass-Go scheme

The Password space of BPG as compared to déjà vu is quite higher since it is capable of enabling users to select different indicator type [4].

Pass logix inc organization built up another graphical confirmation plot called Passlogix v-go shown in Fig. 7. At enrollment stage the password is made by a sequential circumstance with rehashing a group of activities. In this technique the client is requested to tap on different things on the picture in the right grouping with a specific goal to be confirmed, one disadvantage is that this procedure gives just restricted password key space [19].

Fig. 7. Passlogix authentication [19].

Passfaces (Fig. 8) was an authentication technique that was developed by Real User Corporation. The essential plan is that, the user was asked to settle on 4 pictures of human faces among a decoy of faces, the user sees a grid of 9 faces, consisting of 1 face antecedently chosen by user, the user recognizes and clicks anyplace on the noted face. This method is recurrent for four rounds. This procedure is recurrent for many rounds. The user is echt if properly identifies the 4 faces. The technique is predicated on the idea that people will recall or memorize human faces very easily than different random images.

Fig. 8. Passfaces authentication

Comparative studies by Sasse et al. [20] showed that passfaces had solely a 3rd of the login failure rate of text based passwords. Their study put together showed that the

Passface login methodology took longer than text based passwords thus was used less frequently by users.

Monrose et al. [21] examined the graphical passwords and located clear patterns among these passwords for e.g. most users preferred to settle on faces of individuals from identical race, which makes the system predictable to some extent, this downside could also be relieved by randomly assignment of faces to users, however following this procedure, therefore would make it laborious for users to recollect their password.

W. Jansen et al. [22–24, 25] designed a graphical password mechanism for portable gadgets. Amid the enrolment stage, a client chooses a topic, which consist of thumbnail photographs and at that point registers a series of pictures as a password. In the verification stage, the client must provide the registered images within the redress grouping. One downside of this methodology is that, the number of thumbnail images is restricted to 30; therefore it has a very small password space. Each thumbnail picture is allotted a numerical value, and the arrangement of choice produces a numerical password. The outcome appeared that the length of image sequence was by and smaller than the textual of textual password movements is visible to others, some gestures are very tedious and complicated to be performed.

Voice based PIN and touch based PIN authentication techniques compared to the in- built authentication techniques of the Google glass (Figs. 9, 10, 11 and 12). The set of available gestures which is an in built authentication mechanism is prone to shoulder surfing. The gesture set is a combination of swipe or tap using one or two fingers, the sequence of finger movements is visible to others, some gestures are very tedious and complicated to be performed.

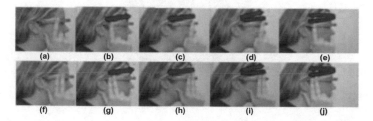

Fig. 9. Built in gesture set: (a) tap, (b) forward swipe, (c) back swipe, (d) forward hook swipe f, (e) back hook swipe, (f) two finger tap, (g) forward swipe two finger, (h) back swipe two finger, (i) swipe forward two finger hook, (j) swipe back two finger hook.

Voice based mechanism (Fig. 12) used voice to unlock the Google glass, the PIN spoken aloud is audible to all and the secret value can be compromised. The Google glass displays a numeric touchpad grid with each cell having two different numbers, the white digits represent the real PIN digits as available on a standard keypad and the second digit available next to it in red is mapped cipher digit that the client will utter in order to enter the corresponding real PIN digit. A reverse mapping of the cipher digits corresponding to real PIN Numbers.

In Touch based PIN authentication (Fig. 11) the client navigates to each of the cells using swipes and on a tap select a digit, forward and backward swipe movements are allowed, the client is verified when the entered PIN matches the saved PIN. For every instance the assignment of digits is different, therefore the gesture sequence varies each time for the same PIN.

Fig. 10. Built in mechanism: the first three entered gesture are tap, two finger tap and swipe back

Fig. 11. Touch based Mechanism: keypad assigned randomly, every instance different layout is displayed

Fig. 12. Voice based Mechanism keypad assigned randomly, every instance different layout is displayed

The above Fig. 13, indicates that the success rate with built in features is the lowest when compared with the other two methods i.e. Voice based and Touch based.

Cheng Sun et al. [13] performed analysis on pattern strength meter using pattern selection.

On a 3 * 3 grid based dots login system. Google has defined rules & a proper unlock pattern must follow the rules.

R1: Minimum number of Dots to be connected in a pattern in at least 4
R2: At the max 9 dots are connected
R3: The unconnected dots are connected on the path of creating a pattern
R4: To connect an unconnected dot a pattern can pass through a previously connected dot. The physical strength of the pattern strength was defined as follows:
Size: Pattern size is defined as the number of dots the pattern connects. The proper Pattern must be an integer{x | X Ɛ Z and $4 \leq x \leq 9$}.

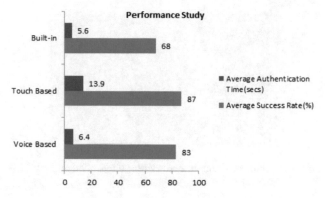

Fig. 13. Success rate Voice based and Touch based & Built-In

Table 2. Size statics of all valid Patterns [13]

Number of dots	Number of valid shapes
4	1624
5	7152
6	26,016
7	72,912
8	140,704
9	140,704
Total	**389,112**

Length: Physical length is the summation of all the lengths of all its segments
Overlaps: pattern line segment covered by other segment is counted as an overlap. The below presented Table 2, gives the statistics of the sizes of valid pattern

Cheng Sun et al. proposed strength score of a pattern.

$PSP = Sp * log2(Lp + Ip + Op)$
PSP Strength score of Pattern P.
Sp Size of Physical strength
Lp physical length
Ip no. of intersections
Op Overlaps of p

It was observed that connecting more dots would aid in increasing the visual complexity, resulting in a higher strength score, Cheng Sun et al. came up two types of pattern strength meter.

1. A 5 segment colour changing bar as shown in Fig. 14.
2. A gradient colour ratio bar with percentage strength score Fig. 15.

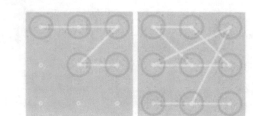

Fig. 14. Pattern strength meter with colour changing bar [13]

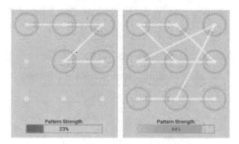

Fig. 15. Pattern strength meter with percentage strength score [13]

The presence of visual indicators with pattern strength helped in creating strong patterns as shown in Figs. 14 and 15, which helped in enhancing the security of pattern unlock. But in comparison with PIN/Password the overall security of pattern unlock is still weak, as a result of limited pattern space (Fig. 16).

A feedback collected from the participant indicated that in total 69% of all participants preferred the new mechanism of pattern strength meters, as shown in Fig. 17 (69% = 41% + 28% = Highly Likely + Somewhat Likely). It was also concluded that

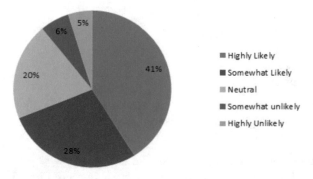

Fig. 16. Feedback [13]

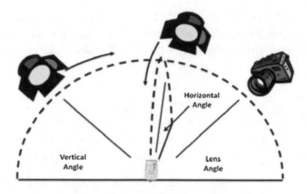

Fig. 17. Principle photographic Setup [14]

memorability becomes a main issue when complex patterns are used, which includes longer lengths, more intersections and overlaps but not with all users. It was empirically concluded that a few participants utilize designs which resemble alphabets, or digits to encourage memorability. For instance alphabets like C, L, N, Z and digits like 2, 7, and 9 can be effortlessly reproduced in a 3 * 3 grid. Such shapes will experience ill effects of shoulder surfing assault due to generally poor visual complexity.

Adam J Aviv et al. performed four experiments on smudge attack on smartphone touch screen, the experiment was performed on two variants of phone HTC G1 and HTC Nexus1, under different lighting and camera condition as shown in Fig. 17.

For the principle photographic setup (Fig. 17) a single light source in vertical or horizontal orientation. Vertical angle increments in plane with the camera, while horizontal angle increments in perpendicular plane to the camera were setup and all angles were calculated in reference to smartphone. In total 188 setups were created.

The goal of the first examination was to determine the conditions by which an assailant can track the patterns under perfect settings. Four different models of phones were used to extract the patterns.

Figure 18 shows a total of around 75% for selection around all the four corners. The center, right, lower and upper comprises only 14% of selection chances as the

Fig. 18. Android unlock pattern, Bias of initial point

starting point. But the chances of selecting the upper left corner are 38%. The researchers conducted an analysis on 3-grams v/s 2-grams patterns.

Figure 19(a) and (b) shows the 3-grams the most frequent to less frequent patterns [15].

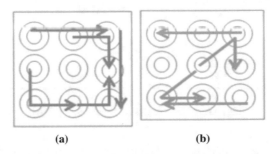

(a)   (b)

Fig. 19. (a) Most frequent Pattern with 3 g (b) Less frequent Pattern with 3 g

Modification was done to the original Android lock screen-four different versions Leftout Small, Leftout Large, Circle and Random. As shown in Fig. 20 for the left out Small pattern the upper left point was omitted as it was having the highest bias, the intention was to spread the bias from the upper left corner to the remaining points, as a result the first point was distributed more uniformly. Figure 20(b) shows the most frequent 3-grams patterns and it had the lowest number of possible patterns [15].

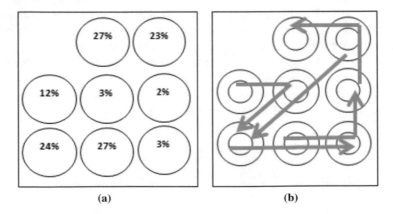

(a)   (b)

Fig. 20. (a) Selection of node to generate a pattern (b) Leftout Small Most frequent 3 g pattern

Similarly for the leftout Large, which has a 3 * 4 grid, the first point remains almost similar to the original Android lock. Figure 21(a) shows the bias among the various selection points and (b) shows the frequent 3-grams patterns [15].

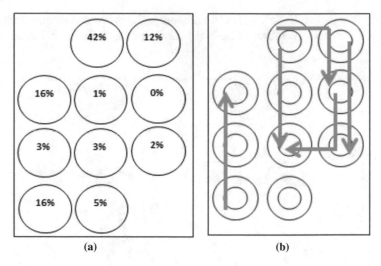

Fig. 21. (a) Selection of node to generate a pattern (b) Leftout Large Most frequent 3 g pattern

The circle pattern performed the best among all the four experiments, even though there were chances that the user may choose circular patterns, they were aware of the security problems, most of the user's preferred using geometrical structures like square or a triangle in the circle, but the starting point still has a higher preference of 41%, as shown in Fig. 22(a) and (b) shows the 3-gram pattern.

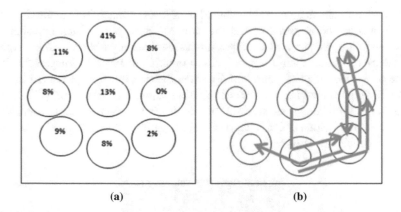

Fig. 22. (a) Selection of node to generate a pattern (b) Circle- Most frequent 3 g pattern

The last experiment was performed on a Random pattern wherein the points were loosely spaced, this pattern did not support any symmetrical patterns, since there were no more than two points in the same line, and no upper left corner point was provided, but in spite of this random pattern, its performance was poor, since the users tried to fit certain obvious pattern like the alphabet a. Figure 23(a) below shows that the first point still has a higher preference, and Fig. 23(b) shows the most frequent 3-gram patterns.

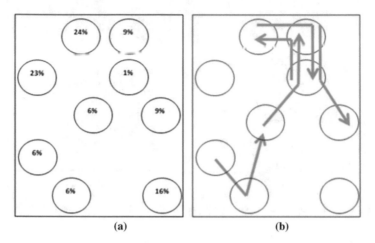

(a)   (b)

Fig. 23. (a) Selection of node to generate a pattern (b) Random – Most frequent 3 g pattern

3   Proposed Strategy and Discussion

3.1   Working of the Model

We limit the implementation of this method on a local intranet. This method is divided into two phases: the registration and the authentication phase. First considering the registration phase, we use the CCP algorithm as our base algorithm. In this method, we allow the user to select their images of choice creating a story, the image selection can be done on a wide range of genre initially when the first image is presented to the user then it will be presented in an discretized fashion as, with a virtual grid (Fig. 24) on it. As shown in figure below, the sole aim of the gird is to help the user to memorize the click point, very obvious that it is impractical to reproduce the same pixel so a tolerance area around each cell is provided.

Fig. 24. Placement of Virtual grid

The tolerance area is provided in the form of circular fashion i.e. whatever may be the click point chosen during the registration phase, the same may be reproduced during login phase but within the tolerance area from the Centre point of the circle, the figure below represents how a circular tolerance is provided in a cell (Fig. 25).

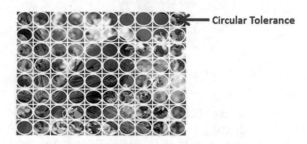
Circular Tolerance

Fig. 25. Circular Tolerance for each Cell

When the user clicks on a particular cell within a tolerance area there can be two cases as either the selected click point can be within the circular tolerance or outside the tolerance assuming that a user selects a click point within the circular as depicted in Fig. 26.

Fig. 26. Click point within Tolerance Area

3.2   Detection of Click Point

Assuming user clicks on a point (x,y), during the registration phase, r is the tolerance in pixels, ofx and ofy are offse in x-axis and y-axis, respectively, and sx, sy represent segment identifiers on x & y-axis, respectively. Then the system calculates the following:

$$iox=(int)(x-r)/(2*r) \qquad ofx=(int)(x-r)\%(2*r)$$
$$ioy=(int)(y-r)/(2*r) \qquad ofy=(int)(y-r)\%(2*r)$$

Suppose user clicks on a point(x1,y1) during re-login, the tolerance area i.e. r remains the same for both registration and the login phase

Let ofx1 and ofy1 be offset on x & y-axis, respectively, similarly sx1 and sy1 be segment identifiers on x & y-axis, respectively. The system does the following calculations:

$$Sx1=(int)(x1-ofx)/(2*r) \quad and \quad sy1=(int)(y1-ofy)/(2*r)$$

If iox is equal to sx1 and ioy is equal to sy1 then the coordinates at (x1,y1) proves to be a success else will result in a failure (Fig. 27).

Fig. 27. Click point outside Tolerance Area

The problem arises when the click point is outside the tolerance area, in such cases the care is taken by an additional grid as depicted in the figure below, clearly click point in not within the circular tolerance of the first grid and clearly lies outside the grid1, but is within the tolerance of the second grid indicated in light orange color (Fig. 28).

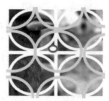

Fig. 28. Grid 2 indicated in Light orange color

In reality the click point of an image which the user chose during the registration phase is not the actual click point but an additional random image is associated with the image which is we call as the background image, as shown in the figure below, when the user follows the procedure of registration on the image displayed to him, in actual the processing is done on the back ground image, which the user is totally unaware of and the click point is chosen from the background image, this information along with the foreground image is cryptographically hashed and stored in the database, this process is carried out for a total of 5 iterations, with the user choosing a different image at each round (Fig. 29).

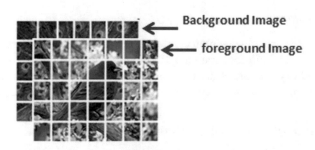

Fig. 29. Background & Foreground image

During the login phase i.e. the second half the user provide the same click points approximately, the user is authorized based on the values stored in the database, to avoid shoulder surfing attack along with the actual cursor on the image there are many dummy cursors moving along with the actual cursor this makes it even more difficult for an attacker to play an attack or for that matter even to record the movement of the cursor. Figure 30 below depicts the movement of the dummy cursors along with the actual cursors.

Fig. 30. Placement of dummy cursors

Figure 31 below shows the entire login phase for five successful rounds with dummy cursors moving along with the actual cursor making it very difficult for an attacker to identify the correct movement of the actual cursor.

Fig. 31. Entire authentication phase

4   Theoretical Results

Images with a size of 451 * 331 pixels, will provide, with a 19 * 19 pixel grid square will provide $2^{8.6}$ square per image, multiply it by 5, considering 5 iterations the number of images in the password sequence would be 254 choices with respect to the foreground images. Which is a quite a larger password space as compared to the text based password.

5 Conclusion and Future Work

The proposed method provides quite a larger space as compared to the text based passwords, Based on the size of the grid cell a large number of cells can be created per image and since this procedure is repeated five times, generates a high password space, the inclusion of the circular algorithm benefits the user from selection of a click point and not restricting to exact coordinates, to make it still more complex the password calculation is carried on the background image, the best feature of this methods that it avoids shoulder surfing, by the placement of Dummy cursor along with the actual cursor which helps the actual cursor to camouflage. Future work requires on the closer examination on how secure the method is and how the attackers might exploit it.

References

1. Fledmeier, D., Karn, P.: UNIX password security ten years later. In: Proceedings of the 19th International Conference on Advances in Cryptology. Lecture Notes in Computer Science, vol. 435. Springer, New York (1990)
2. Sobrado, L., Bridget, J.C.: Graphical Password, the Rutgers Scholar. An Electronic Bulletin of Undergraduate Research, vol. 4 (2007)
3. Adams, A., Sasse, M.A., Lunt, P.: Making passwords secure and usable. In: Thimbleby, H., O'Conaill, B., Thomas, P.J. (eds.) HCI 1997 and Computers XII, pp. 1–20. Springer, London (1997)
4. Por, L.Y., Lim, X.T., Su, M.T., Kianoush, F.: The design and implementation of background Pass-Go scheme towards security threat. WSEAS Trans. Inf. Sci. Appl. 5(6), 943–952 (2008). ISSN: 1790-0832
5. Ma, W., Campbell, J., Tran, D., Kleeman, D.: A conceptual framework for accessing passwords quality. IJCSNS Int. J. Comput. Sci. Netw. Secur. 7(1), 179–185 (2007)
6. Klein, D.: Foiling the cracker: a survey of improvements to password security. In: Proceedings of the 2nd USENIX Security Workshop, May 2008. http://www.citeseer.ist.psu.edu/112514.html
7. Paivio, A., Roger, T.B., Symthe, P.C.: Why are pictures easier to recall than words? Psychon. Sci. 11, 137–138 (1968)
8. Dhamija, R., Perrig, A.: Déjà vu: a user study using images for authentication. In: Proceedings of 9th USENIX Security Symposium (2000)
9. Pering, T., Sundar, M., Light, J., Roy, W.: Photographic authentication through untrusted terminals. IEEE Pervasive Comput. 2(1), 30–36 (2003)
10. Akula, S., Devisetty, V.: Image based registration and authentication system. In: Proceedings of Midwest Instruction & Computing Symposium (2004)
11. Wiedenbeck, S., Waters, J., Birget, J.C., Brodskiy, A., Memon, N.: Design and longitudinal evaluation of a graphical password system. Int. J. Hum. Comput. Stud. 63, 102–127 (2005)
12. Yadav, D.K., Ionascu, B., Ongole, S.V.K., Roy, A., Memon, N.: Design and analysis of shoulder shurfing resistant pin based authentication mechanisms on Google glass. In: International Conference on Financial Cryptography and Data Security, pp. 281–297 (2015)
13. Sun, C., Wang, Y., Zheng, J.: Dissecting pattern unlock: the effect of pattern strength meter on pattern selection. J. Inf. Secur. Appl. 19(4–5), 308–320 (2014)

14. Aviv, A.J., Gibson, K., Mossop, E., Blaze, M., Smith, J.M.: Smudge attacks on smartphone touch screens. In: WOOT 2010: Proceedings of the 4th USENIX Conference on Offensive Technologies, Article no. 1–7, Washington, DC. USENIX Association, Berkley (2010)
15. Uellenbeck, S., Durmuth, M., Wolf, C., Holz, T.: CCS 2013: Proceedings of the 2013 ACM SIGSAC Conference of Computer & Communication security, Berlin, Germany, 4–8 November 2013, pp. 161–172. ACM, New York (2013)
16. Ur, B., Kelley, P.G., Komanduri, S., Lee, J., Maass, M., Mazurek, M.L., Passaro, T., Shay, R., Vidas, T., Bauer, L., Christin, N., Cranor, L.F.: How does your Password measure up? The effet of strength meters on password creation. In: Proceedings of the 21st USENIX Conference on Security Symposium, Security 2012, Bellevue, WA. USENIX Association Berkeley (2012)
17. Egelman, S., Sotirakopoulos, A., Muslukhov, I., Beznosov, K., Herley, C.: Does my password go up to eleven? The impact of password meters on password selection. In: Proceedings of the SIGCHI Conference on Human Factors in Computing Systems, CHI 2013, Paris, France, 27 April–2 May 2013, pp. 2379–2388 (2013)
18. Ur, B., Alfieri, F., Aung, M., Bauer, L., Christin, N., Colnago, J., Cranor, L.F., Dixon, H., Naeini, P.E., Habib, H., Johnson, N., Melicher, W.: Design and evaluation of data-driven password meter. In: Conference on Human Factors in Computing Systems, CHI 2017, pp. 3775–3786. ACM (2017). ISBN 978-1-4503-4655-9
19. Paulson, L.D.: Taking a graphical Password to the approach. Computer 35(7), 19 (2002)
20. Brostoff, S., Sasse, M.A.: Are passfaces more usable than passwords? A field trial investigation. In: People and Computers XIV — Usability or Else!: Proceedings of HCI, Sunderland, UK. Springer, London (2000)
21. Davis, D., Monrose, F., Reiter, M.K.: On user choice in graphical password scheme. In: Proceedings of the 13th USENIX Security Symposium, San Diego, CA (2004)
22. Jansen, W., Gavrila, S., Korolev, V., Ayers, R., Swanstorm, R.: Picture password: a visual login technique for mobile devices. National Institute of Standards and Technology Interagency Report NISTIR 7030 (2003)
23. Jansen, W.A.: Authenticating users on handheld devices. In: Proceedings of Canadian Information Technology Security Symposium (2003)
24. Jansen, W.: Authenticating mobile device users through image selection. In: Data Security (2004)

Application of Siamese Neural Networks for Fast Vulnerability Detection in MIPS Executable Code

Roman Demidov[✉] and Alexander Pechenkin

Peter the Great St. Petersburg Polytechnic University, 29, Politekhnicheskaya
Street, Saint Petersburg, Russia
rd@ibks.spbstu.ru

Abstract. The paper addresses a problem of lightweight vulnerability detection
in program code, represented by MIPS instruction set, which is widely used in
network and IoT devices. For these purposes, it is proposed to use the Siamese
neural network, which was involved both for pre-training the instruction
embeddings, and for training the code vulnerability classifier. Instruction
embeddings are obtained by solving auxiliary task of matching the semantically
equivalent code pieces. Proposed approach has been tested on publicly available
dataset and gave positive results.

Keywords: Deep learning · Vulnerability detection · Vector representations

1 Introduction

The current state of the information technology industry is characterized by a large
number of network devices and systems controlled by embedded code. These devices
include IoT, routers, personal devices like smartphones and others. All of them contain
huge amount of code, and no one can guarantee that code is free from bugs, programming mistakes or other potentially malicious issues. Some of those bugs are
exploitable, thus can be used by hackers to redirect the control flow of the vulnerable
code, access sensitive data, or disrupt the normal operation of the vulnerable devices in
another way.

Ability to timely detect and prevent such bugs, usually called software vulnerabilities, is a very important task. Widespread availability of potential targets for attack
makes the problem more sharpen. There are many traditional approaches such as
symbolic execution methods family [1, 2], various fuzzing extensions [3], some hybrid
approaches [4]. Each of them have both advantages and disadvantages. However, the
problem in general is proven to be NP-hard, thus such detection can take a long time
even for the tiny programs.

1.1 Key Motivation

For some applications, it is impossible to use such traditional techniques as they have
multiple time and resource limitations. Consider the case of dynamic networks such as

© Springer Nature Switzerland AG 2020
K. Arai et al. (Eds.): FTC 2019, AISC 1070, pp. 454–466, 2020.
https://doi.org/10.1007/978-3-030-32523-7_31

VANET, MANET [5]. Such networks consist of mobile devices, each of those can independently move, suddenly appearing and disappearing in a wireless range of the other devices. State-of-the-art routing algorithms neither perform any device verification nor distinguish them before the network packets from them are accepted. This fact can significantly increase the number of possible attack scenarios. For example, a malicious device could become embedded into network and could then interact with the other devices over this network. If the dynamic network contains devices with the vulnerable program code, that is also remotely accessible (e.g. network stack as a part of firmware), malicious device can exploit vulnerabilities in them to gain further control over network.

A partial solution can be some type of device firmware safety authentication. Ability to determine devices, which are potentially vulnerable for network attacks, can increase the safety of the whole dynamic network. To do that, one needs a fast and reliable method of vulnerability detection without exhaustive program path examining or fuzzing with the unpredictable result and elapsed time.

1.2 Deep Learning Approach

Deep learning is the solution. As opposed to the traditional machine learning, it allows automatic extraction and further detection of the high-level features in the input data, that seems relevant for the learning task. Successful applications of deep learning include, but are not limited to, image, sound, text classification and generation tasks [6, 7]. Strong progress in data processing field has been influenced by the deep neural network ability of recognize the underlying highly abstract semantic dependencies in various types of data. During the training phase, these relevant dependencies are recognized, and features are built directly from data; during the evaluation phase those features are used to solve the main task. One of the main advantages of using deep learning for the proposed task is its' complexity – the trained model can give an answer in linear time.

The program code can be considered as one of the traditional data types with some specific differences. Program code is discrete like textual data, but has a more complex structure, contains various side effects, contains operations, mixed with the numerical operands. In addition, it seems hard to collect a huge labeled dataset of vulnerable and safe code in appropriate time. This fact makes the learning process harder in a real-world practice.

However, deep learning can solve these issues in a way that allows to properly distinguish safe program code from the vulnerable one. At first, one need to properly represent the program code elements, so that the neural network can better "understand" them. This can be done via the transfer learning approach described below. Second, a proper architecture of a deep neural network should be used, given the lack of labeled data.

In the next sections of the current paper, there will be more details about new input representations and the deep neural network configuration.

1.3 Transfer Learning

For the tasks, that deal with the continuous entities (this is true for images, video, sound etc.) input data can already be fed in the deep neural network without any modifications (other than normalizing). For the deep learning tasks that work with the discrete entities (such as text or program code analysis), first layers of the network usually embed each of those entities into the continuous vector space. That embedding space should represent semantic relations between objects, e.g. have similarity metrics.

Embeddings can be trainable, like other layers of the network, but they can also be pre-trained by solving some auxiliary task. Key point is the size of the training dataset. When it is small, one needs to reduce the amount of the trainable parameters count due to the overfitting problem. This can be done by training both embeddings and auxiliary network layers while solving auxiliary learning task with a huge dataset, but with the same requirements about data semantics. Trained weights from embedding layer can now be extracted and used as a new data representation. Such a technique called multi-task, or transfer learning [8].

The following section describes the successful deep learning applications for program code analysis.

2   Related Works

As for today there are several works, which address the problem of software weaknesses detection using deep neural networks. Some of them deal with the source code, written in high-level languages, while others work with the executables only. Work [9] operates with the large amount of Java source code with the additional bug reports. Two separate 1D-convolutional networks were used together in order to extract deep features both from the source code and from the corresponding report in order to produce the united response, which allows to identify bug locations in a previously unseen source code. Similar work [10] tries to identify programming mistakes (not bugs) in the JavaScript code. Like in many of the text processing tasks, authors initially obtain a large dataset of the correct code examples, that were mined from the open-source code databases, and augment them by synthetically constructing incorrect examples by randomly replacing some parts of these samples. The learned classifier helps to detect huge amount of programming mistakes in real-world code.

Significant work has been recently done by the Draper Laboratory research team [11]. They've collected the big, well-labeled dataset of C/C++ code both from the public Github repositories and Debian sources. Each sample label was generated, based on three different static analyzers. A sufficient size of the dataset allows to train the code vulnerability classifier directly on the input source code. From the deep learning point of view, authors have effectively used trainable embeddings without overfitting, as their vocabulary was small (only 156 syntactical tokens). The highest accuracy was achieved by combining the convolution neural network architecture with the random forest over the learned top-level features. However, in case of an assembly code, one needs to deal with the proper code representation before training the vulnerability classifier. Another interesting work [12] proposes to represent programs as their abstract syntax trees. Researchers propose a novel, tree-based convolutional neural network architecture, that helps to classify programs according to their functional meaning. Such an approach is especially applicable for the high-level languages with complex nonlinear structure (like C, C++, ObjC etc.). But for the executable code this approach is not necessarily a good fit: indirect calls and switch/case constructions break the fine-grained tree structure of the code samples.

Recent work in the field of vulnerability detection tries to reduce the input dimension by treating entire programs as "code gadgets" [13]. Starting from the significant API calls, that are related to known vulnerabilities, one collects dependent lines of code both by data flow and control flow criteria. Bidirectional LSTM model then learned to extract relevant features from sequence of gadget vector representations. Proposed techniques allowed to detect four known vulnerabilities in the widely used software.

Work [14] deals with the executable code of x86 architecture. The authors proposed Instruction2vec algorithm, that embeds every instruction in a similar way that the Word2vec does, and used the convolutional neural network architecture. However, embedding can be more expressive if they follow from the semantic meaning of code rather than from joint statistics of instructions.

The previous works of these authors describe deep learning methods, already implemented for the vulnerability detection [15]. The initial approach uses the program code effects emulation to build instruction embedding for further vulnerability detection. Such approach is useful only for simple architectures. Subsequent works improve detection accuracy by using morphological structure of code, and improved neural network architecture.

The next section describes the proposed approach, that utilizes the power of transfer learning and Siamese neural networks for the real-world processor architecture.

3 Proposed Approach

The proposed approach focuses on work with the executable code of a 32-bit MIPS processor architecture, because such architecture is widely used in dynamic network and IoT devices. Many of them are currently using this architecture due to its relative simplicity and low power consumption.

The MIPS I processor is an example of a RISC architecture with a reduced command set. The basic architecture contains 32 general-purpose registers for integer operands: numbers and addresses. MIPS assembler language contains more than 50 hardware-implemented commands [16], pseudo-instructions are also supported: they are translated into a sequence of real instructions before assembly. Like in other modern architectures, a set of implemented commands allows performing all the necessary actions: arithmetic and logical operations on integer values, loading and unloading values in and out of memory, transfer of control flow, etc.

Every MIPS instruction starts with a mnemonic, and contains up to three arguments in a register or numerical form. To reduce the size of an instruction dictionary, it was decided to learn embedding vectors not for the whole instructions, but for their parts:

mnemonics, register operands, offsets, and immutables. The averaged embedding vector of the individual parts forms the representation of the whole instruction.

3.1   Auxiliary Task for Instruction Embedding

To properly build a good embedding for the instruction elements, – an embedding, that contains semantic relations between those elements, – a proper auxiliary task is required. It should:

- depend heavily on the semantics of the instruction elements;
- be simple to obtain or generate as many valid labeled data for this task, as possible;

It is proposed to use the task of matching semantically equivalent machine code pieces. Given two different executable code pieces, the auxiliary network must determine, whether they represent the same piece of source code (see Fig. 1).

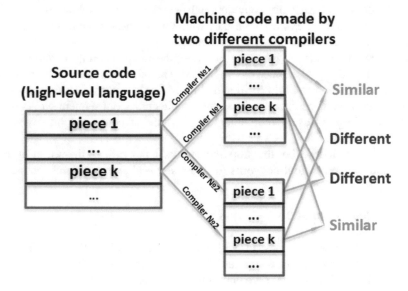

Fig. 1. The key idea of the proposed auxiliary task.

It is clear, that the correct solution of such a task relies not on the syntax, but on the semantics of the code. It is also true, that one can automatically generate a large, correctly labeled dataset for that task; details are described below. That process contains several stages:

- A set of random C programs was generated in the amount of several thousands with help of the CSmith open-source random code generator [17]. It was modified to insert the indexed _asm labels, which divide each program into the sequence of pieces;

- Each of the generated programs was compiled into two executables by different compilers (GCC and Clang) for 32-bit MIPS architecture. Each of them were splitted into indexed sections. It is possible to form a training dataset of pairs of executable code pieces with the same and different indexes with corresponding marks (1 as "Similar", and 0 as "Different").
- A special Siamese neural network architecture is proposed to be trained on auxiliary task [18]. It contains two identical halves with trainable weights (W_A and W_E), that are shared during the training. Each half is represented by the Gate Recurrent Unit network [19], that reads the sequence of instructions, represented by trainable embeddings, and then computes the representation of the entire instruction sequence (R_{left}, R_{right}, see Fig. 2). The final merging block outputs the Manhattan distance [20] between R_{left} and R_{right}.

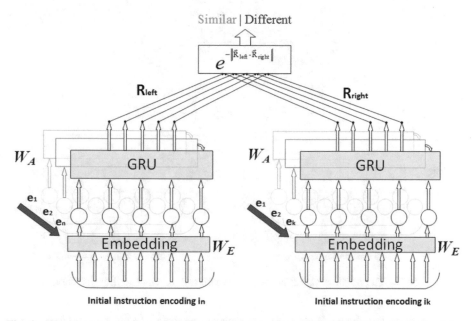

Fig. 2. The Siamese neural network for auxiliary task of semantically equivalent code matching.

- During the pre-training phase on an auxiliary task, the proposed network trains both instruction and sequence embeddings (yellow-colored on the Fig. 2). The first one forms new instruction representations W_{E*}, ready to be used for the vulnerability detection task.

3.2 Vulnerability Detection

As compared to the auxiliary task, the correctly labeled data for the vulnerability detection task of sufficient size cannot be easily collected. In general case that data cannot be synthetically generated, but must be received from static or dynamic vulnerability detectors, which is computational expensive operations.

The Siamese neural network architecture allows to virtually expand the number of training samples. This can be done by the small task reformulation. For the two input classes for a single code sequence ("Vulnerable" and "Safe") one may construct two other classes ("Same vulnerability status" or "Different vulnerability status"), but for the expanded input that contains a pair of sequences. Thus, for N labeled inputs for the initial task (vulnerability detection in code sequences), a new set of up to N^2 paired samples can be constructed. This is also true in the case of multiple labels for the vulnerable samples. From that point a similar Siamese neural network can be used for the modified vulnerability detection task with the expanded dataset (see Fig. 3).

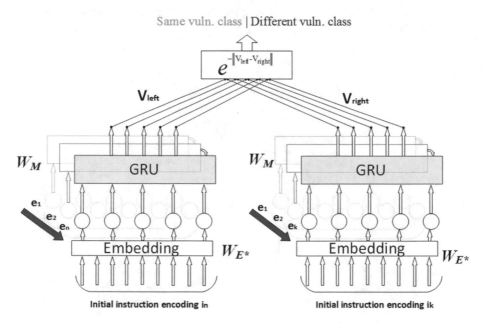

Fig. 3. The Siamese neural network for main task of vulnerability detection.

As compared to the auxiliary network, embedding weights W_{E*} are now fixed, since they are included from the pre-training phase. Trainable weights are only W_M (yellow colored on the Fig. 3). A set of optimal weights W_{M*} is learned during the training, which gives the entire sample representation in the vulnerability context - as V_{left*} or V_{right*} (they are identical, since left and right weights are shared). That fact allows to classify the new code for the vulnerability presence.

To do that, one of the network identical half's is extracted and stacked with Dense layers. Those weights W_{Dense} are finally trained on the initial dataset with single inputs to distinguish vulnerable code from safe (see Fig. 4).

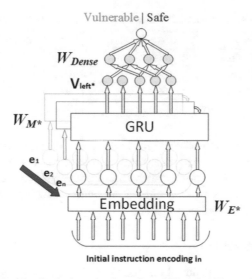

Fig. 4. The final phase in the vulnerability classifier training.

The proposed approach was implemented for classification into two and more classes. The next section describes the experimental results of such approach.

4   Evaluation

4.1   Dataset and Environment

The dataset for the auxiliary task was created synthetically as described above. For the vulnerability detection task the Juliet test suite [21] was used. It was created in U.S. National Institute of Standards and Technology. It contains the examples of both vulnerable and safe code for many vulnerability classes. We extracted some relevant examples, that are related to particular CWE classes: Integer Overflow, Double Free, Use-After-Free, NULL Dereference, and others. The data preprocessing is performed before evaluation:

- removing duplicates and non-compilable examples;
- compiling code for MIPS, extracting executable section from compiled binaries;
- determining the entry point and the set of useful functions to construct the instruction sequence for each sample;
- splitting all sequences to the training and validation sets;
- constructing the set of sequence pairs for modified task: random reordering, pairs labeling: 1 if the both sequences have the same vulnerability class or do not have one; 0 if the sequences have different vulnerability classes.
- balancing the number of positive and negative examples, shuffling the labeled samples to randomize their order.

The resulting dataset sizes are shown in Table 1.

Table 1. Approximate dataset sizes for different CWE classes.

Accuracy	Size of training set	Size of validation set
Double Free	40000	10000
NULL Dereference	60000	10000
Use-After-Free	40000	10000
Integer Overflow	100000	100000

The program evaluation was performed on the x64 Ubuntu workstation with 16 Gb RAM. For neural networks operations keras framework with tensorflow backend [22] were used. In our experiments the dimension of embedding space for the instructions was 30. Siamese neural network configuration contains two identical one-layer GRU networks with the cell dimension of 48.

4.2 Experimental Results

This section describes the achieved results. First, visualizing the learned vectors V_{left*} for samples both from training and from validating datasets, showed the separability between vulnerable and safe programs representations in the semantic vector space (see Fig. 5). This was done by using the t-SNE visualizing algorithm [23], that saves the initial relations between the points in the high-dimensional space after they're mapped to the low-dimensional space for the visualization purposes.

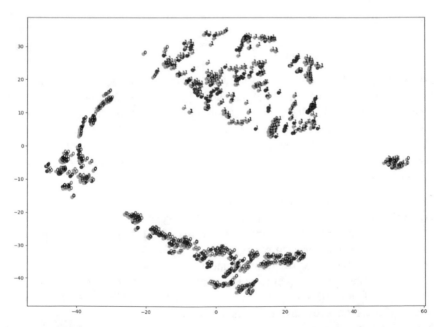

Fig. 5. The t-SNE 2D visualization of the learned feature vector V_{left*} for NULL Dereference vulnerability detector.

Orange and red points describe the vulnerable programs from the training and validation set respectively; light-blue and blue points describe the safe programs from the training and validation set respectively.

The Tables 2 and 3 represent the accuracy statistics of vulnerability detectors for different vulnerability classes. It is clear, that the researched model works great with both the validation dataset and the training dataset.

Table 2. Accuracy metrics for NULL Dereference vulnerability detector.

Accuracy	Training set (%)	Validation set (%)
Vulnerable programs with NULL Dereference	99	99
Safe programs	83	96
Summary	91	97

Table 3. Accuracy metrics for Double Free vulnerability detector

Accuracy	Training set (%)	Validation set (%)
Vulnerable programs with Double Free	99	99
Safe programs	88	87
Summary	94	93

As an another example, the united classifier was built for two vulnerability classes simultaneously (see Fig. 6). Red points describe the safe programs, blue points describe the Double-Free vulnerable programs, green points describe the Use-After-Free vulnerable programs. Same as before, the separability of these three classes in the semantic vector space is evident.

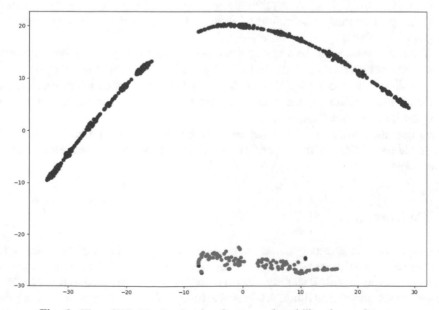

Fig. 6. The t-SNE 2D visualization for two vulnerability classes detector.

The Table 4 represents the accuracy statistics of the learned classifier based on the proposed approach. It can again be noticed that the accuracy on validation set is as good as on the training set.

Table 4. Accuracy metrics for multiple type vulnerability detector

Accuracy	Training set (%)	Validation set (%)
Vulnerable programs with Double Free	90	92
Vulnerable programs with Use-After-Free	99	99
Safe programs	86	87
Summary	91	91

To sum up, the experiments demonstrate the applicability of the proposed approach for building a fast and reliable executable code vulnerability classifier.

5   Future Work

There are several areas for further research:

- Building united vector representation for multiple processor architectures, both for instructions and programs. This will expand the scope of this approach. Additionally, doing that can help in the general task of decompilation;
- Work with the binary code patches that comes from Microsoft and other vendors: can help with the dataset augmenting. In this regard, there are additional problems, such as distinguishing vulnerabilities from performance improvements in every single patch, proper localization of vulnerable code, etc. All of them should be generally solved;
- Additional work with the program semantic space, that appears while training the Siamese network on an auxiliary task. It gives the ability of comparing programs, according to their meaning. That can be useful not only for the current task, but also for all tasks, related to the code meaning: malware detection, code sandboxing, smart IDE for compilers, etc.;
- Further improvements of the neural network architecture: possible innovations should provide the ability to work with the large code samples of complex nonlinear structure.

6   Conclusion

Proposed approach is heavily based on the modern deep learning methods, and allows to construct a fast executable code vulnerability classifier, which is important for dynamic networks and IoT security. Such a classifier takes linear time to run, can detect various types of vulnerabilities, and can be further trained when new training data comes.

In compare with the other research, current work proposes the novel approach of learning code embeddings. Based on the Siamese neural networks, it heavily uses the semantic relations between the different code examples. That is of most importance for the neural network classifier. Achieved results allow to construct the multi-class vulnerability classifier of MIPS executable code, with a potential ability to process code for many different processor architectures.

Experimental results show the practical ability of vulnerability detection even with the small datasets. The resulting accuracy metrics described above clearly show the possibility of commercial potential of the proposed approach.

Acknowledgments. This work is supported by the Russian Science Foundation under grant №
17-71-10065.

References

1. Chipounov, V., Georgescu, V., Zamfir, C., Candea, G.: Selective symbolic execution. In: Proceedings of the 5th Workshop on Hot Topics in System Dependability (HotDep) (No. CONF) (2009)
2. Godefroid, P., Levin, M.Y., Molnar, D.: SAGE: whitebox fuzzing for security testing. Commun. ACM **55**(3), 40–44 (2012)
3. Godefroid, P., Peleg, H., Singh, R.: Learn&fuzz: machine learning for input fuzzing. In: Proceedings of the 32nd IEEE/ACM International Conference on Automated Software Engineering, pp. 50–59. IEEE (2017)
4. Stephens, N., Grosen, J., Salls, C., Dutcher, A., Wang, R., Corbetta, J., Shoshitaishvili, Y., Kruegel, C., Vigna, G.: Driller: augmenting fuzzing through selective symbolic execution. In: NDSS, vol. 16, no. 2016, pp. 1–16, February 2016
5. Altayeb, M., Mahgoub, I.: A survey of vehicular ad hoc networks routing protocols. Int. J. Innov. Appl. Stud. **3**(3), 829–846 (2013)
6. He, K., Zhang, X., Ren, S., Sun, J.: Deep residual learning for image recognition. In: Proceedings of the IEEE Conference on Computer Vision and Pattern Recognition, pp. 770–778 (2016)
7. Wang, X., Jiang, W., Luo, Z.: Combination of convolutional and recurrent neural network for sentiment analysis of short texts. In: Proceedings of COLING 2016, the 26th International Conference on Computational Linguistics: Technical Papers, pp. 2428–2437 (2016)
8. Pan, S.J., Yang, Q.: A survey on transfer learning. IEEE Trans. Knowl. Data Eng. **22**(10), 1345–1359 (2010)
9. Huo, X., Li, M., Zhou, Z.-H.: Learning unified features from natural and programming languages for locating buggy source code. In: Brewka, G. (Ed.) Proceedings of the Twenty-Fifth International Joint Conference on Artificial Intelligence (IJCAI 2016), pp. 1606–1612. AAAI Press (2016)
10. Pradel, M., Sen, K.: Deep learning to find bugs. Technical Report TUD-CS-2017-0295. TU Darmstadt, Department of Computer Science
11. Russell, R., et al.: Automated vulnerability detection in source code using deep representation learning. In: 2018 17th IEEE International Conference on Machine Learning and Applications (ICMLA), pp. 757–762. IEEE (2018)

12. Mou, L., Li, G., Jin, Z., Zhang, L., Wang, T.: TBCNN: a tree-based convolutional neural network for programming language processing. CoRR (2014)
13. Li, Z., et al.: VulDeePecker: a deep learning-based system for vulnerability detection. arXiv preprint arXiv:1801.01681 (2018)
14. Lee, Y.J., Choi, S.H., Kim, C., Lim, S.H., Park, K.W.: Learning binary code with deep learning to detect software weakness. In: KSII The 9th International Conference on Internet (ICONI) 2017 Symposium, December 2017
15. Demidov, R., Pechenkin, A.: Vector representation of machine instructions for vulnerability assessment of digital infrastructure components. In: 2018 IEEE Industrial Cyber-Physical Systems (ICPS), pp. 835–840. IEEE, May 2018
16. Price, C.: MIPS IV instruction set (1995)
17. Yang, X., et al.: Finding and understanding bugs in C compilers. ACM SIGPLAN Notices, **46**(6), 283–294 (2011)
18. Koch, G., Zemel, R., Salakhutdinov, R.: Siamese neural networks for one-shot image recognition. In: ICML Deep Learning Workshop, vol. 2, July 2015
19. Wu, Z., King, S.: Investigating gated recurrent networks for speech synthesis. In: 2016 IEEE International Conference on Acoustics, Speech and Signal Processing (ICASSP), pp. 5140–5144. IEEE, March 2016
20. Aggarwal, C.C., Hinneburg, A., Keim, D.A.: On the surprising behavior of distance metrics in high dimensional space. In: International Conference on Database Theory, pp. 420–434. Springer, Heidelberg (2001)
21. Boland, T., Black, P.E., Juliet, 1.: 1 C/C++ and Java Test Suite. Computer, vol. 45, no. 10, pp. 88–90 (2012)
22. Bahrampour, S., Ramakrishnan, N., Schott, L., Shah, M.: Comparative study of deep learning software frameworks. arXiv preprint arXiv:1511.06435 (2015)
23. Maaten, L., Hinton, G.: Visualizing data using t-SNE. J. Mach. Learn. Res. **9**(Nov), 2579–2605 (2008)

Where Are We Looking for Security Concerns? Understanding Android Security Static Analysis

Suzanna Schmeelk[✉]

St. John's University, Queens, NY, USA
schmeels@stjohns.edu
https://www.stjohns.edu/academics/faculty/suzanna-schmeelk

Abstract. Static analysis is a traditional technique for software transformation and type analysis. Recently, static analysis has become a technique to identify cyber security vulnerabilities and malware. Specifically, static analysis has been extended into the mobile-computing arena for security-related analyses. This research examines many top security papers that are published in major conferences, journals and technical reports, and characterizes the current research characterize static analysis research. The papers identified in this paper were selected based their high citings by top research or because they introduced either a novel analysis technique or a novel security issue analysis. This research systematically constructs a static analysis landscape by charting and characterizing analysis strengths and limitations in both accuracy and security threats. The findings are reported online at www.technologyinthepark.com. This research has identified two types of static analysis motivations which affect the soundness of an analysis methodology: techniques for analyzing software for vulnerabilities and techniques used to examine applications for malware. Building on earlier research, for completeness and to aid the community by providing a coverage map, this research has connected technique motivations found to Mitre's attack taxonomy, Mitre's vulnerability taxonomy as well as the National Institute of Standards and Technology's (NIST's) Bugs Framework (BF) taxonomy. The findings include identifying vulnerabilities which are not being systematically researched.

Keywords: Android mobile services · Static analysis · Cyber security · Software engineering · Networking · End user services · Weakness detection · Malware prevention · Malware mitigation · Malware detection · Mobile devices · NIST Bugs Framework (BF) · Mitre CAPEC · Mitre CWE

1 Introduction

In 2019, Verizon published the *Verizon Mobile Security Index 2019 Report* [29] and the corresponding *Executive Summary* [30] reporting that "it is time to tackle mobile security". The Executive Summary reported that, "A third of organizations admitted having suffered a compromise that involved a mobile

© Springer Nature Switzerland AG 2020
K. Arai et al. (Eds.): FTC 2019, AISC 1070, pp. 467–483, 2020.
https://doi.org/10.1007/978-3-030-32523-7_32

device [which was] up from 27% in [their] 2018 report." In addition, the Executive Summary, "Two thirds of organizations said they are less confident about the security of their mobile assets than other devices." Finally, the full Index Report reported [29] that "five of six said organizations need to take mobile device security more seriously". Overall, the Verizon report emphasizes mobile security industry trends.

There are many methodologies to improve security on mobile devices depending on the device and the risks involved. General cyber security problems are cross-platform, as machines are, at large, vulnerable to attack. Each platform has its own niche environments and vulnerability nuances that require creative thinking and research to determine its potential threat vectors and attack surfaces.

In May of 2015, Sufatrio et al. [28] published a top-down research study of techniques aimed at securing the overall Android platform. They proposed a taxonomy for existing security solutions on Android. Their research covered a broad range of topics, from antivirus software, to dynamic analysis, to static analysis to the application vetting process on the Android market, among others.

Bottom-up research has also been driven to characterize mobile malware (e.g., Felt et al. [16]) and adware (e.g., Erturk [15]) as it appears in mobile libraries in the wild. Bottom-up research attempts to answer questions related to malware incentives, malware development accessibility, and malware identification evaluation techniques. In addition, the bottom-up research has encouraged the development of application benchmarks and malware sampling in public repositories, such as the DroidBench [3], Malware Gnome Project [19,32], and the applications used by DARPA's Automated Program Analysis for Cybersecurity (APAC) [11] challenge.

This research is organized as follows. First, we describe our research motivations. Then, we analyze the top research papers static analysis techniques to understand were researchers are looking. We characterize the research at large in static analysis into Confidentiality, Integrity, Availability and General techniques. Our research then reports on the findings and concludes.

1.1   Research Motivation

This research is novel, as it specifically examines one particular over-arching Android security domain—static-analysis techniques to prevent and thwart malware. We want to identify and categorize static-analysis techniques used for the discovery of Android application malware and determine how they are different from techniques used to discover Android application vulnerabilities. It is important to understand both where the current research is strong and where the current research ends, as well as what are the best security practices developers should use when developing Android applications because Android and the mobile-device market at large are expanding. People have personal devices and many organizations are either giving employees devices or, as Steven Bellovin addresses in his new book, *Thinking Security* [6], many organizations are encouraging employees to bring their own device (BYOD).

This paper extends the work of Schmeelk et al. [27] and Schmeelk [26] to answer questions about what is being covered and left uncovered in static analysis research. This is the first research to connect security research with multiple standardized malware and vulnerability repositories including the NIST Bug Framework and Mitre's CAPEC/CWE to provide details on more accurate detection and prevention going forward. Detection and prevention rely on careful and precise analysis. Mapping static analysis techniques to specific known weaknesses and threats closes the gaps in knowledge to help the community understand where the strengths and weakness are in what we are detecting via static analysis.

With over a billion devices running Android [24], it is critical to systematically understand where we are with current research trends and the mobile threat landscape. To enable the systematic study, we connect research with the National Institute of Standard's (NIST's) Bug Framework (BF) (Bojanova, Black, Yesha and Wu [10]) and Mitre's Common Attack Pattern Enumeration and Classification (CAPEC) [21] and Common Weakness Enumeration (CWE) [22] taxonomies to map static-analysis techniques with Android application weaknesses and standard attack patterns. One goal of the paper is to use this mapping to systematically assess the scope and effectiveness of the most widely cited static-analysis tools and techniques for addressing Android security. To our knowledge this is the first article to use such a technique to help the developer, researcher and security analyst communities measure the analysis completeness, as well as help the communities to not overlook certain analyses which are critical for Android application security. We argue that the Mitre and the BF taxonomies are a productive way to examine and compare the strengths and weaknesses of the current static-analysis techniques and static-analysis tools.

Overall, our findings in Sect. 7 show that there are many attack surfaces which are not being throughly examined by static analysis research. Specifically, we identified seven of sixteen malware attack mechanisms listed on the Mitre CAPEC "Attack Mechanism" taxonomy which have limited research available. In addition, we identified eleven of the Mitre CWE taxonomy categories being researched in the static analysis community. Finally, we discovered that six of the NIST BF classes fit into the analysis model; and, the research provides insight into additional bug classes/attributes to add to BF. As an Android application has many exploitable attack surfaces and our findings show that current Android malware prevention, detection and mitigation static analyses techniques are only considering a subset of attack mechanisms and vulnerabilities, the community at large is leaving surfaces wide-open for attack. It is extremely important for the security community to consider both the NIST and the Mitre taxonomies when formalizing their research to inform the developer, researcher and analyst communities as to which attack surfaces and attack mechanisms are important, have been examined, have been quantified, and inform the community into the best developer and security analyst practices.

A second overall finding from our paper shows that there are many gaps in Android application static analysis research, as discussed in Sect. 7.3. Our results shows that there is a need to develop countermeasure hardening through byte-code rewriting. Our results show that there is very little research into application dynamic updating, databases usage, native code invocation, nonstandard development, preparation for dynamic analysis, public interfaces, reflection, standard application threats, and the missing attack or vulnerabilities mentioned earlier. Our results show that static analysis can be improved by efficiency, precision and operating-system level analyses.

Finally, there is a large standard deviation between evaluation methodologies across research as well as a large standard deviation for the availability of open source materials. The large deviation increases the difficulty to compare and contrast the different techniques mathematically. We did find, from a security analyst's viewpoint, that some techniques require more user interaction and resources.

1.2   Android Static Analysis Techniques

Static analysis is a method to analyze programs without executing them (e.g., Aho et al. [1]). It captures a different view of the code security than other analyses (e.g. embedded credentials, leakage of data, input validation, leakage of resources, etc.). Static analysis can be performed on either the source code or the object code. The more issues discovered before software is released, the more secure the users. When code has already been released, static analysis discovers other issues which cannot be easily addressed using other analyses (e.g. dynamic, emulation, simulation, etc.). Static analysis can be done with a variety of techniques (and tools) such as model checking, data-flow analysis, abstract interpretation, among others. It can be designed to be automated and may require limited resources. This paper examines the methods currently used in security research to characterize what methodologies have been used over the past five years of research in this domain to inform Android application developers, researchers and security analysts of the strengths and limitations to static analysis methodologies.

In order for analysis to be complete with respect to Android application security, all threat vectors and attack surfaces must be analyzed. A National Vulnerability Database (NVD) [23] which is used to inform technology users of vulnerabilities within their devices. A simple search in the NVD on the term *Android* uncovers hundreds of matching entries indicating that people are finding vulnerabilities in off-the-shelf Android devices. The breadth of devices, firmwares, vendors and Android operating system versions used over diverse geographic locations makes exact malware quantification and exploitation statistics.

In order to map the threat vectors and attack surfaces to the current state-of-the-art analysis techniques, we introduce the use of the BF, CWE, and CAPEC to aid in the discovery of analysis completeness. The intention is to provider a wider overall analysis coverage. For example, if we only consider Android permissions, we will miss other large threat vectors such as developers failing to remove the storage of cryptographic keys in plain text from applications before they are published and used at large. From a security perspective, researchers need to ensure that their research not only contributes to better techniques in well known security domains but they also need to ensure that their research is broadly considering many different categories of security issues.

Application Weakness Detection. Development-time static analysis is geared to look for potential exploitable components in benign code. Unprotected benign code can be used as a scapegoat for malicious code.

NIST's Bugs Framework (BF): NIST introduced the Bugs Framework a few years ago by Bojanova, Black, Yesha and Wu [10]. NIST recently added three new classes of bugs to the BF: Encryption bugs (ENC), verification bugs (VRF), and key management bugs (KMN) in [9]. The BF is into breaking down bugs into four main elements: causes, attribute, consequences and sites of bugs. By distinguishing bugs using this methodology provides more insight into how bugs occur and what the effects can be. The BF is referred to as a *Periodic Table* for bugs. It tries to capture the essence of bugs rather than classify similar patterns separately.

Mitre's Common Weakness Enumeration (CWE): Mitre developed the CWE [22] to enable standardized discourse in this domain. CWE's list includes "flaws, faults, bugs, vulnerabilities, and other errors in software code, design, architecture, or implementation that if left untreated could result in systems and networks being vulnerable to attacks [22]". CWE is a broad list that captures vulnerable software paradigms across languages and operating systems for software assurance efforts. We used the Mitre CWE taxonomy to assess which vulnerabilities have been considered in analysis research enabling more complete security analyses.

The CWE is sponsored by the United States Computer Emergency Readiness Team (US-CERT) in the office of Cybersecurity and Communications at the U.S. Department of Homeland Security. It is a large online repository of software weaknesses known to aid the code security assessment industry. Specifically, it is used for scoring in the U.S. National Vulnerability Database [23] hosted by the National Institute of Standards and Technology (NIST).

Malicious Application Detection. Static analysis of live code can be beneficial to locate malware. Released code, or "in-the-wild" code, can have built-in attacks embedded into the design. In such a case, the software does not exactly contain weaknesses: it contains attack patterns. Mitre developed the CAPEC taxonomy [21] to enable standardized discourse in this domain. We used the Mitre CAPEC taxonomy to assess which attacks have been considered in detection analysis and mitigation analysis research again enabling more complete security analyses.

The CAPEC is also sponsored by US-CERT in the office of Cybersecurity and Communications at the U.S. Department of Homeland Security. One of the artifacts that CAPEC provides is a list of standard attack patterns. CAPEC defines an attack pattern as "an abstraction mechanism for helping describe how an attack against vulnerable systems or networks is executed [21]". CAPEC defines each pattern with a description and a mitigation recommendation. CAPEC's motivation is that developing "attack patterns help[s] categorize attacks in a meaningful way in an effort to provide a coherent way of teaching designers and developers how their systems may be attacked and how they can effectively defend them [21]".

2 Confidentiality Techniques

Confidentiality is the ability to hide information from those people unauthorized to view it. Bishop [8] defines confidentiality as the "concealment of information or resources"; and Whitman and Mattord [31] define confidentiality as "the quality or state of information that prevents disclosure or exposure to unauthorized individuals or systems". The goal of confidentiality, according to Daswani et al. [12] is to "keep the contents of a transient communication or data on temporary or persistent storage secret". Confidentiality gives rise to questions about privacy and the usage of collected information. Privacy, as defined by Whitman and Mattord [31], is the "state of being free from unauthorized observation" and reflects the users ability to exercise control over how their collected personal information is used by third-parties. Bellovin [4,5] and Androulaki et al. [2] are well known for their research in computing privacy.

Current confidentiality research encompasses applications' leaking information unknowingly to the user. This research domain studies applications that both deliberately leak data as well as naively potentially leak data. If an application is malicious, it will purposefully leak private data from the mobile device without receiving adequate approval from the user. If an application does not require permissions to access its exported activities, exported content providers and exported services, it may unknowingly support a malicious application's request to collect and/or leak private data. In addition, Android allows multiple applications, if they are signed by the same certificate, to share the same Linux user ID on an Android device allowing the applications to share information and resources. In such a situation, a combination of applications can be used to collect and/or leak private data unknowingly from the user.

Many static analysis research techniques exist for confidentiality malware detection concerns. This category of attacks falls under the CAPEC category 118, "Gather Information". We have created tables to house these intermediate findings which are posted online. Some of the earlier tools were replaced with newer research using either new techniques or more precise techniques for malware analysis. A table housed online at www.technologyinthepark.com/FTC2019/ presents the novelty of each tool and indicates if they are still publicly available informing community as to the main contributions claimed by each paper.

We also examined research for confidentiality weakness detection. Weakness detection can be used at compile time to identify vulnerabilities before the application is released; or, it can be used by researchers to discover vulnerabilities in released applications. We found six research papers in this category falling mainly in the CWE category 200 "Information Exposure". We analyze the papers by date of publication starting with the most recent publication first. Our intermediate findings are listed online.

3   Integrity Techniques

Integrity is the ability to ensure that processed data is an accurate and unchanged representation of the original secure information. Bishop [8] defines *integrity* as the "trustworthiness of data or resources, and it is usually phrased in terms of preventing improper or unauthorized change"; and Whitman and Mattord [31] define integrity as "the quality or state of being whole, complete, and uncorrupted". The goal of integrity, according to Daswani et al. [12] is to "not [let] a third party ... be able to modify the contents". In general, static analysis in this domain are either for malware abuse detection or application weakness detection.

Static analysis with respect to malware detection contains a category of attacks which fall under the CAPEC category 233 "Privilege Escalation". In privilege escalation, adversaries exploit a weakness enabling them to elevate their privilege and perform an action that they are not supposed to be authorized to perform. The following research paper was specifically geared for this domain; however, some of the techniques used in the earlier confidentiality category extend to integrity. A table housed online at www.technologyinthepark. com/FTC2019/ presents the novelty of each tool and indicates if they are still publicly available informing community as to the main contributions claimed by each paper.

Static analysis research techniques exist for integrity concerns that can be done by a developer during application development (i.e., pre-deployment) or after development to discover if an application is susceptible to attack (e.g., security analysis). These techniques, not aimed at finding malware, are usually geared at "hardening" applications to lower the susceptibility of their being compromised by an adversary. This category of attacks falls under the CWE category 441, "Unintended Proxy, Intermediary or Confused Deputy". The following fifteen research papers each cover a different aspect of this domain. Starting with the most recent publication first, we analyze the papers by date of publication. Again, we present the findings online.

4   Availability Techniques

Availability is the assurance that information systems are readily accessible to the authorized viewer at all times. Bishop [8] defines *availability* as the "ability to use the information or resource desired"; and Whitman and Mattord [31] define availability as, "the quality or state of information characterized by being accessible and correctly formatted for use without interference or obstruction". The goal of integrity, according to Daswani et al. [12] is to "respond to users' requests in a reasonable timeframe".

We were unable to find any complete papers researching Android application availability in entirety from the perspective of security. A few papers mentioned either a particular case where their technique found a particular availability concern or stated they have done an analysis closely related to availability but without a discussion of security. Availability concerns have not not been the focus of any one paper to date.

5   Generalizable Techniques

Generalizable tools examine an Android applications for more than one weakness or malware patterns. Tools that typically fall into this category include sandboxes and well-developed weakness analyses. Current general analysis research includes proprietary static-analysis tools such as Coverity [7], Fortify [18] and Klocwork [25]. Coverity, Klocwork and Fortify all include CWE references in reports generated from code analysis. In all cases, the tools have also found faults that are not, yet, labeled by the CWE. Research in this domain is typically either geared for malware detection or weakness detection. Static analysis research techniques which are used to find malicious generalized security concerns. This category of attacks falls under the CAPEC category 115, "Deceptive Interactions" and CAPEC category 118, "Gather Information". A table housed online at www.technologyinthepark.com/FTC2019/ presents the novelty of each tool and indicates if they are still publicly available informing community as to the main contributions claimed by each paper.

Static analysis for weakness detection cover different aspects of security: malware frequently uses the same APIs, instrumenting applications can aid in the security analysis process to help log data before it is encrypted, machine learning can be used by different algorithms to identify malware on large-scale analysis, among others. We analyze the papers by date of publication and posted our analysis online.

6   Other Polyhedral Techniques

According to Daswani et al. [12] there are other security goals. Authentication, authorization and accountability, among others, fall into this category. After examining the research, typically research falls into two categories abuse detection and abuse weakness detection. For abuse detection, the category of attacks falls under other CAPEC categories (e.g., CAPEC-225 "Exploitation of Authentication", CAPEC-232, "Exploitation of Authorization", CAPEC-262, "Manipulate Resources", CAPEC-527 "Manipulate System Users", etc.). We found only one tool geared specifically for this category: *Short* as presented in our tables online.

Many static analysis research techniques exist for detecting weaknesses in other polyhedral security concerns. We found four research papers in this domain: *CredMiner, PlayDrone, Static Analysis for Extracting Permission Checks of a Large-Scale Framework* and *PScout*. This category of attacks falls under other CWE categories (e.g., CWE-522 "Insufficiently Protected Credentials", CWE-272 "Least Privilege Violation", etc.). We found three tools geared specifically for this category (*CredMiner*, *PlayDrone* and *PScout*) and one paper. A table housed online at www.technologyinthepark.com/FTC2019/ presents the novelty of each tool and indicates if they are still publicly available informing community as to the main contributions claimed by each paper. In one case, we found that the open source tool is great for a static analysis reference; however, the crawler no longer works with Google Play since Google has changed their API.

7   Research Findings

There are books that discuss attacks and defense for Android (e.g., Elenkov [14] and Dubey and Misra [13]). Our research has found that most static analysis research broadly considers security mechanisms. This paper is the first, to our knowledge, to specifically examine techniques used in recent static analysis publications in top research venues for Android applications and categorize the security issues they address to understand at a broader level research strengths, specifically where in the code are researchers looking at security?.

7.1   Static Analysis Techniques Findings

The findings of the static-analysis techniques used in Android mobile security analysis are diverse. Our research indicates that there are many attack surfaces which could benefit from static analysis. In addition, research over time shows that analysis can be modified to capture a more precise and/or faster analysis.

Static Analysis Strengths. Certain categories of faults have been highly researched, such as detecting applications that are over privileged, leaking data and exposing data through the intercomponent call mechanism. We found a plethora of research on these topics in the published research literature. Other research for domains such as exposing user tokens, not correctly verifying SSL certificates and re-writing malicious programs are less researched. These other domains are extremely important in the Android security community as they completely break the security trust chain (e.g., accepting any certificate to establish an SSL connection, leaving in user tokens so that anyone can log into AWS and/or OAuth portals, etc.).

Static Analysis Limitations. We found four main static analysis limitations. First, we found that the published research is overlooking attack surfaces. Second, we found that static analysis is relying on many internal representations of code which may propagate limitations. Third, we found that many cloud service provider regularly change their APIs making analysis methodologies quickly outdated. Fourth, we found that static-analysis tools which rely on other tools propagate the limitations of their reliant tools to their own analysis. We discuss these four issues in depth in the following paragraphs.

First, we found that much current published research examines the same malware attacks, threats and vulnerabilities, leaving many unexamined open doors unexamined in security from a static analysis perspective. For example, virtually no papers are thoroughly considering availability concerns, manipulating system users, database concerns, general counter measure hardening mechanisms, dynamic analysis preparations, standard application threats, native code invocation and user-defined library concerns, among other threats and attacks listed by Mitre. The significance of ignoring these issues leaves other legitimate attack surfaces unanalyzed. Using a metaphor of trying to secure a house, leaving attack surfaces unanalyzed and/or unsecured would be like using the best possible security camera to monitor activity on the attack surface of the front door but leaving the side door, another completely legitimate attack surface, completely unlocked. This directed focus to certain attack surfaces and attack vectors leaving some attack surfaces and attack vectors without extensive research may be due to the fact that there is very little overarching research characterizing and analyzing the overall state of the Android mobile security field. This article's contribution is to identify this meta-analysis gap in Android mobile security analysis by compiling the current research and mapping it with known security attack vectors and weaknesses. The next subsection, Sect. 7.2, works to correlate current published research with MITRE's current online at www.technologyinthepark.com/ FTC2019/ weakness taxonomy, the CWE, and attack taxonomy, the CAPEC, to discover what mobile security issues remain largely unresearched.

Second, many of the existing techniques rely on Java dataflow libraries (e.g., *Soot* and *WALA*). These libraries are exceptionally good resources as they have been extensively researched to capture Java control flow. They do, however, have analysis limitations (e.g. keeping current, not having exact Android *DEX* code representations, precision, etc.) that propagate to those using them in the Android domain, adding additional limitations to static analysis research. In some cases, the researchers noted that analysis was skipped due to the fact the Android application could not be transformed into the *Soot* or *WALA* internal representations. Noting these library limitations, some researchers (e.g. *AnaDroid, SmartDroid*, etc.) have created their data flow analysis based on the intermediate *smali* code representation which is an assembly-level representation for Android *Dalvik* byte code. This methodology now faces limitations as Google has changed *Dalvik* bytecode into *ART* (Android *Run Time*) bytecode. Although there is not drastic changes between the *Dalvik* bytecode and the *ART* bytecode (as *ART* must currently maintain backward compatibility), there *are* differences. Lastly, in addition, static analysis usually, due to path complexity, can only consider a certain finite values, which we will express with the value k. Any hacker, knowing the particular kth value, can bury malware at a deeper level to thwart analysis techniques.

Third, Google Android and many cloud service providers regularly change their APIs. When the providers change their structure of their coding, it is difficult for static analysis freeware techniques to keep up with ever changing dynamic environment. In fact, most research, once developed, is not maintained.

In many cases, the lack of maintenance represents the flux of the research community. In fact, similarly we found that some researchers are not maintaining their evaluations benchmarks (e.g., Android Malware Gnome Project, etc.). It is currently noted on the Malware Gnome Project that the flux of researchers caused the maintenance to stop on the benchmark in 2015.

Fourth, many of the static-analysis tools we analyzed relied on other tools for part of their analysis. Similar to the internal library propagation issues, the analysis tools which rely on other tools propagate the weaknesses of their reliant tools. This is an almost unavoidable issue due to the flux in the research community at large.

7.2   Graphs of Domain Coverage Findings

Understanding the security domains covered by existing static-analysis techniques and tools helps identify limitations and openings in attack vector research. Consequently, Schmeelk [26] categorized each security paper using the analysis in the research as determined by the definitions of Mitre's online at www.technologyinthepark.com/FTC2019/ vulnerability taxonomy and attack taxonomy. Mitre's taxonomies were created to create a common dialog and number scheme for all known security issues—both weaknesses and actual attacks.

Mitre's CAPEC. The CAPEC view on "Mechanisms of Attack" [20] lists sixteen categories shown in Table 1. These categories are malware attack surfaces. Examining and expanding on the categories of attacks given by CAPEC help developers, researchers and security analysts secure all attack vectors associated with Android applications. Securing a some attack surfaces leaving other attack surfaces completely unanalyzed and potentially unsecured is completely unacceptable from a security perspective.

Each of these high-level categories has additional sub-categories. Security research in the static analysis domain would benefit with mapping their research to specific malware discoveries as shown on the CAPEC. Without a specific mapping of events knowledge of the usefulness of a technique is limited to both ease of use for security analysts and importance. In addition, the lack of both universally maintained benchmarks and open source projects reduces the analysis comparison metrics.

Schmeelk [26] found that the majority of malware research fit within two of the above categories with outlying research fitting into eight other categories as can be seen in Fig. 1 from [26]. Interestingly, most of the malware research is examining CAPEC-118 and CAPEC-156. A few research studies have considered the following: CAPEC-225, CAPEC-210, CAPEC-262, CAPEC-232, CAPEC-119, and CAPEC-526. Examining the research analysis techniques shows that there is limited research of the following attack surfaces: CAPEC-172, CAPEC-223, CAPEC-255, CAPEC-281, CAPEC-436, and CAPEC-527. These statistics help inform the security research community to shape the research into attack surface breadth and depth. They indicate one form of strengths and weaknesses

Table 1. CAPEC "Mechanisms of Attack"

Mechanisms of Attack
Gather information (118)
Deplete resources (119)
Injection (152)
Deceptive interactions (156)
Manipulate timing and state (172)
Abuse of functionality (210)
Probabilistic techniques (223)
Exploitation of authentication (225)
Exploitation of authorization (232)
Manipulate data structures (255)
Manipulate resources (262)
Analyze target (281)
Gain physical access (436)
Malicious code execution (525)
Alter system components (526)
Manipulate system users (527)

in the malware static analysis methodologies, directly answering the research motivation questions. The community should carefully consider the Mitre attack surface taxonomy when designing analysis tools and techniques to ensure that Android applications have coverage over more attack surfaces.

Mitre's CWE. There are over 1,000 CWE identifiers in existence, according to a Mitre representative. Schmeelk [26] reported that the majority of vulnerability research of the publications we analyzed, fell within eleven of the 1,000+ weakness categories. The eleven weaknesses we identified can be seen in Fig. 2 from [26]. These statistics help inform the security research community to shape the research into weakness analysis both by breadth and depth. The community should carefully consider the Mitre weakness taxonomy when designing analysis tools and techniques to ensure that Android applications have coverage over more weaknesses. These statistics indicate one form of strengths and weaknesses in the malware prevention static analysis methodologies, directly answering the research motivation questions.

NIST's Bugs Framework. The research shows that in the Bugs Framework, six current classes capture the static analysis research, as seen in Fig. 3. The static analysis also raises some questions about five findings that may be of the BF four main elements of a bug: causes, attribute, consequences and sites of bugs. The six current classes captured in the research are: IEX (matching CWE-200:

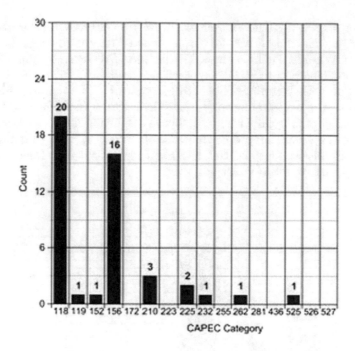

Fig. 1. The fraction of CAPEC categories researched in Android analysis across the publications [26].

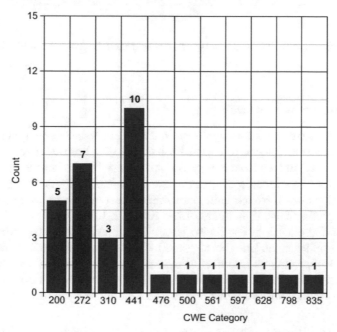

Fig. 2. The fraction of CWE categories researched in Android analysis across the publications.

Information Exposure), CRY (matching CWE-310 CATEGORY: Cryptographic Issues), AUT (matching CWE-441: Unintended Proxy or Intermediary), PTR (matching CWE-476: NULL Pointer Dereference), WOP (matching CWE-597: Use of Wrong Operator in String Comparison), and ARG (matching CWE-628: Function Call with Incorrectly Specified Argument). The non-matching CWE comparisons may not directly map into NIST BF as individual classes; they may be another of four main elements. These are: CWE-798: Use of Hard-coded Credentials, CWE-835: Loop with Unreachable Exit Condition ('Infinite Loop'), CWE-500: Public Static Field Not Marked Final, CWE-561: Dead Code, and CWE-272: Least Privilege Violation.

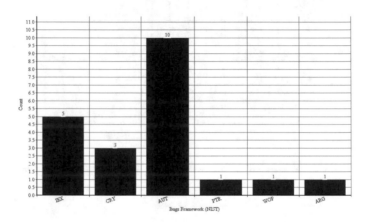

Fig. 3. The fraction of NIST BF categories researched in Android analysis across the publications.

7.3   Openings in Research

We found that there are many openings in static analysis research, answering the research motivation, for Android security based on the limitations of current tools as well as overarching Android concerns not discussed and overarching static analysis goals. Openings include countermeasure hardening, dynamic updating, efficiency, nonstandard development, precision, and dynamic analysis preparation. It is an interesting fact, that many of the research tools used to help locate vulnerabilities in applications could also be used by hackers to locate vulnerable and exploitable applications. Google has provided developers with a *Security Tips* web page to advise developers into overall best practices [17] which they should consider when developing Android applications. It covers many, but not all the attack mechanisms/vulnerabilities which Mitre's taxonomies characterize.

Technique Evaluation. We found that each tool is using a different evaluation strategy. To address this concern, we systematically studied all the research papers evaluation techniques. A review of the appendix shows that it is next to impossible, with the current state of research, to have a mathematical comparison between techniques. In many cases, the research is not open source or has changes with the services offered by Google. Since there is a combination of different evaluation techniques in the community, lack of a maintained benchmarks (e.g. Malware Gnome Project is no longer maintained as of December 21, 2015 [19,32]) and lack of open source research it is difficult to broadly mathematically evaluate methodologies.

Overall from a security analyst's viewpoint, some techniques require more direct interaction and resources from an analyst. For example, instrumenting an operating system may be more difficult for an analyst as it requires building a customized Android operating system. For another example, creating a dual analysis setting may not be feasible for some security analysts where first an application is statically analyzed and then the application is dynamically analyzed. Interestingly, some of the techniques used instrumentation and dynamic analysis only as a method to confirm their static analysis findings. Instrumentation is somewhat difficult to implement and is an excellent accuracy confirmation. Some of the techniques result in instant malware prevention/detection and other techniques require deeper inspection for malware prevention/detection. In addition, some techniques can be automated and other techniques require some reliance on user intervention.

Evaluation overall used directly by the researchers in their publications had a high standard deviation. Some research papers used instrumentation and/or dynamic analysis for evaluation purposes. Other research papers used benchmarks and applications from application stores for evaluation purposes. At least one research paper used simulations for evaluation purposes. Finally, a few research papers where entirely theoretical. Our findings show the broad standard deviation of evaluation methodologies being used across the field.

7.4   Research Limitations

Many of the security papers which report on their applications on static analysis do not provide open-source tools or have quickly become outdated with the fast pace of new Android operating system versions. As such, it is difficult to pin-point exactly how tools differ specifically or even benchmark their soundness/completeness. These inherent domain limitations are reflected in this meta-research study.

7.5   Future Work

Future work in this domain is to further develop mobile software security assurance classifications while focusing on Android. In addition, future work involves analyzing proprietary static analysis tools. Finally, clearer classification must be accomplished to further guide penetration testers, risk analysts and software developers.

8   Conclusions

In conclusion, an mobile applications have many attack surfaces which can be exploited; and, current mobile malware prevention, detection and mitigation static analyses are only considering a subset of attack mechanisms and vulnerabilities leaving surfaces wide-open for attack. The CAPEC, CWE and BF categorization inform the developer community, research community and security analyst community which attack mechanisms and vulnerabilities are not being considered. It is extremely important for the community to consider the NIST and Mitre databases when formalizing their research to inform the developer community, researcher community, penetration testing community, forensics community and security analyst community as to which attack surfaces and attack mechanisms are both being addressed and important.

References

1. Aho, A.V., Lam, M.S., Sethi, R., Ullman, J.D.: Compilers: Principles, Techniques, and Tools with Gradiance, 2nd edn. Addison-Wesley Publishing Company, USA (2007)
2. Androulaki, E., Bellovin, S.M.: A secure and privacy-preserving targeted ad-system. In: Proceedings of the 14th International Conference on Financial Cryptograpy and Data Security, FC 2010, pp. 123–135. Springer-Verlag, Heidelberg (2010)
3. Arzt, S., Rasthofer, S., Fritz, C., Bodden, E., Bartel, A., Klein, J., Le Traon, Y., Octeau, D., McDaniel, P.: Flowdroid: precise context, flow, field, object-sensitive and lifecycle-aware taint analysis for android apps. In: SIGPLAN Notices, vol. 49, issue 6 (2014)
4. Bellovin, S.M.: The puzzle of privacy. IEEE Secur. Priv. 6(5), 88–88 (2008)
5. Bellovin, S.M.: Identity and security. IEEE Secur. Priv. 8(2), 88–88 (2010)
6. Bellovin, S.M.: Thinking Security: Stopping Next Year's Hackers. Addison-Wesley, Boston (2016)
7. Bessey, A., Block, K., Chelf, B., Chou, A., Fulton, B., Hallem, S., Henri-Gros, C., Kamsky, A., McPeak, S., Engler, D.: A few billion lines of code later: using static analysis to find bugs in the real world. Commun. ACM 53(2), 66–75 (2010)
8. Bishop, M.: Introduction to Computer Security. Addison-Wesley Professional, Bishop (2004)
9. Bojanova, I., Black, P.E., Yesha, Y.: Cryptography classes in bugs framework (Bf): encryption bugs (ENC), verification bugs (VRF), and key management bugs (KMN). In: 2017 IEEE 28th Annual Software Technology Conference (STC), pp. 1–8, September 2017
10. Bojanova, I., Black, P.E., Yesha, Y., Wu, Y.: The bugs framework (BF): a structured approach to express bugs. In: 2016 IEEE International Conference on Software Quality, Reliability and Security (QRS), pp. 175–182, August 2016
11. Darpai. Darpa's automated program analysis for cybersecurity (APAC). http://www.darpa.mil/program/automated-program-analysis-for-cybersecurity. Accessed 7 September 2015
12. Daswani, N., Kern, C., Kesavan, A.: Foundations of Security: What Every Programmer Needs to Know. Apress, Berkely (2007)

13. Dubey, A., Misra, A.: Android Security: Attacks and Defenses. Auerbach Publications, Boca Raton (2013)
14. Elenkov, N.: Android Security Internals: An In-Depth Guide to Android's Security Architecture. No Starch Press, San Francisco (2015)
15. Erturk, E.: A case study in open source software security and privacy: Android adware. In: 2012 World Congress on Internet Security (WorldCIS), June 2012
16. Felt, A.P., Finifter, M., Chin, E., Hanna, S., Wagner, D.: A survey of mobile malware in the wild. In: Proceedings of the 1st ACM Workshop on Security and Privacy in Smartphones and Mobile Devices, SPSM 2011, pp 3–14, New York, NY, USA. ACM (2011)
17. Google. Android security tips. https://developer.android.com/training/articles/security-tips.html. Accessed 3 July 2016
18. L. Hewlett-Packard Development Company. HP fortify. http://www8.hp.com/us/en/software-solutions/application-security/index.html. Accessed 21 September 2015
19. Jiang, X., Zhou, Y.: Android Malware. Springer Publishing Company, Incorporated (2013)
20. MITRE. Capec view: Mechanisms of attack. https://capec.mitre.org/data/definitions/1000.html. Accessed 10 October 2015
21. Mitre. Common attack pattern enumeration and classification (capec^{TM}). http://capec.mitre.org/. Accessed 7 September 2015
22. Mitre. Common weakness enumeration (CWE). https://cwe.mitre.org/. Accessed 7 September 2015
23. NIST. National vulnerability database. https://nvd.nist.gov. Accessed 3 June 2016
24. Reisinger, D: Android shipments in 2014 exceed 1 billion for first time. CNET, January 2015
25. I. Rogue Wave Software. Klocworkr. http://www.klocwork.com/. Accessed 21 September 2015
26. Schmeelk, S.: Where are we looking? Understanding android static analysis techniques. In: 2019 IEEE International Conference on Services Computing, July 2019
27. Schmeelk, S., Yang, J., Aho, A.: Android malware static analysis techniques. In: Proceedings of the 10th Annual Cyber and Information Security Research Conference, CISR 2015, pp. 5:1–5:8, New York, NY, USA. ACM (2015)
28. Sufatrio, D., Tan, J.J., Chua, T.-W., Thing, V.L.L.: Securing android: a survey, taxonomy, and challenges. ACM Comput. Surv. **47**(4), 581–5845 (2015)
29. Verizon. Verizon mobile security index 2019 report. https://enterprise.verizon.com/resources/reports/msi-2019-report.pdf. Accessed 24 June 019
30. Verizon. Verizon mobile security index 2019 report - executive summary. https://enterprise.verizon.com/content/dam/resources/reports/msi-2019-exec-summary.pdf. Accessed 24 June 2019
31. Whitman, M.E., Mattord, H.J.: Roadmap to Information Security: For IT and Infosec Managers, 1st edn. Delmar Learning, Clifton Park (2011)
32. Zhou, Y., Jiang, X.: Dissecting android malware: characterization and evolution. In: Proceedings of the 2012 IEEE Symposium on Security and Privacy, SP 2012, pp. 95–109, Washington, DC, USA. IEEE Computer Society (2012)

Decentralized Autonomous Video Copyright Protection

Qifeng Chen, Shiyu Zhang[(⊠)], and Wilson Wei

Lino Network, Cupertino, USA
shiyu@lino.network

Abstract. We design a decentralized autonomous protocol for video copyright protection. Our protocol enables copyright protection for videos on the blockchain where users can submit proposals against videos that potentially infringe copyright, and a copyright committee can vote on the proposal for consensus. Penalty and reward are carefully designed in the protocol to ensure long-term stability. Furthermore, we propose a deep learning based approach for fast and accurate similar video detection, which will help fight against video plagiarism. Our quantitative evaluation shows that our approach can achieve high detection rate for similar videos, and run in real time.

Keywords: Blockchain · Copyright · Video

1 Introduction

Blockchains have shown potential to revolutionize multiple areas including payment systems and content sharing platforms. Recently, blockchains for video sharing and live streaming have emerged in large numbers, such as Lino Network[1] and DLive[2]. Blockchains provide a powerful platform for content creators to monetize their videos. However, a challenge for blockchains for video monetization has arisen naturally: how can we have copyright protection for videos in a decentralized fashion? Copyright protection is an important feature for content blockchains to be sustainable in the long run. Furthermore, another technical challenge in video copyright protection is partial copy detection which aims to find segments of videos that are duplicate or similar in a large video database. Here, we are interested in the frame-level similar video detection where we find pairs of similar video frames. Showing matched pairs videos frames makes it easy for users to detect copyright infringement.

Enforcing copyright protection on the blockchains is not trivial due to the decentralization nature. Traditionally, copyright is protected through central authorities which decide whether certain content creators have published some contents that violate copyright. However, there should not exist a central judge to trust on the decentralized blockchain.

---

[1] https://lino.network.
[2] https://dlive.tv.

© Springer Nature Switzerland AG 2020
K. Arai et al. (Eds.): FTC 2019, AISC 1070, pp. 484–492, 2020.
https://doi.org/10.1007/978-3-030-32523-7_33

A blockchain is public ledger that can be not manipulated, and thus it is ideal to record the permanent content (or hash of the content) on the blockchain. Most existing blockchains on copyright protection only focus on recording content and claiming ownership of the content. However, in case of copyright infringement, however these blockchains usually do not enforce should be also enforced on the blockchain. Further, there is no efficient AI algorithms to automatically detect content infringement, especially partial video copy.

The problem of decentralized video copyright protection on the blockchain remains an open question. There exist partial copy detection methods and datasets in a centralized environment. VCDB [1] is a dataset 9,236 pairs of partial copies for partial copy detection in over 100,000 videos. Partial copy detection for videos aims to determine if segments of a query video have similar copies in a video database.

In this paper, we present an approach for to decentralized autonomous video copyright protection. The contribution of our approach is two-fold. First, we present a deep learning based approach for partial copy detection in a large video database. We extract high-level features for each video from the VGG-19 network pretrained on the ImageNet dataset [2,3]. The extracted features enabled efficient nearest neighbor search in the high-dimensional feature space for similar video frame detection. Second, we propose an autonomous copyright adjudication on video pairs that may involve copyright protection. A selected committee is responsible for the judgment. The committee analyzes all sources of evidence and reach a consensus on whether a certain video violates the copyright of another video. If that is the case, the video will be given a tag indicating copyright infringement.

2   Related Work

Blockchain for copyright protection. Most existing blockchains for IP protection only focus on claiming the ownership of digital content by recording the content or its hash on the blockchain with an unfungible time stamp. However, this is not the whole story content creators need. It is useful to content creators to claim that they are the authors of the content, but it is also important to penalize users that violate the copyright and detect similar videos. Most of the blockchains do not take into consideration similar video detection and the governance for copyright. In this paper, we present an approach to solve these problems.

Video Copyright protection. There are prior datasets and approaches for near duplicate or partial copy. Wu et al. [4] created a dataset of a 12,790 videos retrieved based on 24 popular queries on Google, Yahoo, and YouTube for near-duplicate video retrieval. Near-duplicate video retrieval is simpler than partial copy detection because two video are considered near-duplicate if the two videos are essentially the same except minor difference in resolution or color distribution.

3   Overview

There are two major problems we are solving: decentralized copyright protection and video partial copy detection. The decentralized copyright protection is used to decide if certain content violates copyright and video partial copy detection is used by users to detect videos that may infringe copyright.

How can we determine if certain content violates the copyright of another content? We design a decentralized protocol to resolve this problem. In the protocol for decentralized content sharing, we have multiple types of participants for our decentralized copyright protection: content creators, app developers, end users, and copyright committees. The content creators can publish their content on the blockchain but at the same are required to agree with the copyright license. The app developers provide the applications for users to interact with the blockchain, and also provide user-friendly interface for similar content detection. The role of the end users is simply to consume videos and check that potentially infringe the copyright of other contents. If some video is believed to violate copyright, the end users can submit a proposal to the copyright committee for further review. Finally, the copyright committees need to review the proposals raised by the users and reach a consensus on the proposals by voting in a week. If a content is determined to infringe copyright, its content providers will be penalized in terms of reward and reputation. In summary, the overall flow of our protocol is:

- users consume videos with detected similar videos;
- users submit proposals on copyright infringement with a 7-day deposit;
- high-priority proposals are resolved by the copyright committee; and
- penalty is applied to users who publish videos that infringe copyright.

The second question is how we can detect similar video contents efficiently in a large video database. We develop an AI algorithm that leverages deep features pretrained on the ImageNet dataset [3] which contains more than one million images in one thousand categories. Then we use the deep features for each frame for similar video detection. We can further reduce the feature dimensions by principal component analysis (PCA) [5]. Some infrastructure providers need to provide an API for app developers to detect similar videos given a query video by running an AI program.

4   Decentralized Copyright Protection

4.1   Content Creators

Content creators can use blockchain as powerful tool to claim ownership of their digital content with a unique time stamp. Once the content (hash) is recorded on the blockchain. The record on the blockchain would be a non-fungible evidence in case of copyright infringement disputes.

To prevent someone who shares copyright-protected content from earning valuable coins, all the donation and (minted) reward should be paid only after

7 days instead of instant payment. If someone finds that the content may violate copyright and submitted a proposed, then the expected earning will be delayed until the copyright committed voted on the proposal in 7 days. If the content indeed violates copyright, then all the earning will be revoked: the donation will be sent back the donors and reward will be put back to the reward pool. In addition, the reputation of users who share copyright-protected content will suffer and low reputation will affect future earning: less reward or longer time to receive payment.

4.2 App Developers

App developers play an important role in protecting copyright of all the videos. They are responsible to provide interface to playing videos for users and at the same time display similar video content on the internet if users need. Infrastructure providers provide an easy-to-use API for app developers to show similar videos. Similar video content detection is done at the frame level so that users do not need fast forward or rewind to watch the similar video frames. The detected similar video frames are compressed as a short video so that users can see similar videos without sacrificing much download bandwidth.

4.3 End Users

End users can verify if some video violates other copyright-protected videos. While they are watching the videos of their interests, they can optionally watch similar videos at the matched frame level. If they believe some video has infringed the copyright, they can submit a proposal to the copyright committee for further review. To avoid malicious users spamming the proposal pool, users need to make a deposits locked for seven days when submitting proposal. The deposit of a proposal indicates the priority of the proposal and proposals of high priorities will be reviewed first by the copyright committee.

4.4 Copyright Committee

A copyright committee is composed of multiple members selected by users based on the proof-of-stake principles. Each vote is valid for up to a year. The role of a committee member is to determine if certain content has violated the copyright given a proposal submitted by a user. The final decision on the proposal is based on the majority vote.

A member will be removed if he or she does not perform the duty: the member does not vote on the proposals he or she is supposed to vote twice in a month. If a member consistently make a irresponsible voting, users can also submit a proposal to impeach the member.

To incentivize the copyright committee to perform the duty, a rewarding pool (from the inflation, about 0.1%) can be distributed to the copyright committee members. To keep the number of proposals the copyright committee can work on reasonable, only the top the $K = 500$ proposals will be selected for the committee to vote for each week.

Sometimes there is misunderstanding or unclear information in the proposal, the committee members can actively discuss with the corresponding content creators, users to clarify the details before making the final decision.

5   Similar Video Detection

The similar video detection is powerful tool for copyright protection. The task is technically equivalent partial copy detection in a large video dataset. A video that violates copyright is one where a segment of the video is a (modified) copy from another copyright-protected content. In a decentralized autonomous organization for video sharing, it is essential to provide a tool that given a query video, so that similar videos can be retrieved efficiently and users can help protect copyright. In this section, we introduce a method that uses deep features extracted from videos and perform fast approximate nearest neighbor search [6] in a constructed feature space to retrieve similar videos. Our approach is capable of retrieve similar videos at the level of matched frames.

5.1   Deep Feature Extraction

Deep features extracted from a pretrained convolutional neural network (CNN) can be used as a perceptual measurement [7–9]. We use the set of deep features extracted from the frames of a video as the signature of the video. Specifically, we use the last fully connected layer of the VGG-19 network [2] for feature extraction. The feature layer we use has a fixed dimension of 4096. To save computational time, we sample a frame every second and compute a feature vector for the sampled frames. For a video of an hour, the cost to store all features is $3600 \times 4096 = 14.7M$. For a video V, we denote $f(V)$ as the set of deep features extracted for the sampled frames.

5.2   Dimensionality Reduction

We can further apply principal component analysis (PCA) to reduce any dimension we want. Given a list of feature vectors F, we compute its principal component coefficients C and scores S:

$$S \times C^T = F. \tag{1}$$

Here C is an orthogonal matrix and the each column contains coefficients for one principal component. For a feature vector x, we can convert it to a coordinate x' the PCA space by computing

$$x' = x \times C. \tag{2}$$

To reduce the feature dimension to k, we can simply take the first k elements in x' as the feature.

Fig. 1. Detected similar video frames given a query video frame (on the top left). Our approach can effectively retrieve similar videos at the frame level and place them in a 5×5 grid. In this example, roughly half of the video frames are relevant to the query video frame, and the other half are shown as they are top nearest neighbors in the deep feature space.

We compute C and S on the VCDB dataset [1], and then the dimensionality reduction can be applied to any video frames, including those unseen in the VCDB dataset.

5.3   Nearest Neighbor Search

Given a query video Q, we search for similar videos in a large video database $\mathcal{D} = \{V_i\}$. To enable efficient search, we first build a kd tree T on the features for all the videos, $F = \{f(V_i)\}$. We use the popular library VLFeat [10] to build the kd tree. The kd tree can be built once a day as more and more videos are added to database. Then for each frame feature $x \in f(Q)$, we find the approximate k nearest neighbors in F efficiently with the precomputed kd tree T. Each feature corresponds a frame in a video and the k matched features for x provide the similar video frames in the video database $\mathcal{D}$. We also use the approximate nearest neighbor search algorithm provided by VLFeat where we restrict the maximum number of comparisons to be $2k$ for fast searching.

Table 1. The performance of the presented algorithm. In different feature dimensions in the PCA space or the original VGG-19 space, we show the recall in partial copy detection, the time for constructing the kd tree, and query speed in terms of frames per second (fps). The statistics are collected under our setting of showing videos in a 5×5 grid to the users.

Feature space	PCA					Original
Dimension	16	32	64	128	256	4096
Recall	70.5%	72.8%	73.9%	74.1%	74.2%	79.2%
Kd tree (second)	0.22	0.43	0.53	0.77	1.46	26.21
Query speed (fps)	9298	6072	2680	1458	778	57

6   Experiments

We conduct our experiments on the public available VCDB dataset for partial copy detection in a large video dataset. There are nearly ten thousands pairs of found similar segments in 528 core videos. The core videos have total video length of about 27 h.

For each video, we sample one frame per second and compute the last 4096-dimensional fully connected layer from the VGG-19 network as the feature for each sampled frame. It takes 0.05 s to compute the VGG-19 features on a Titan Xp GPU. We also reduce the feature dimensions and analyze the performance with different feature lengths. For each frame, we extract up to 24 similar frames from other videos and align them with the query frame in a 5×5 grid. Figure 1 is an example where we show the query frame and similar retrieved frames from other videos.

6.1   Evaluation

We analyze the performance of our propose method in terms of recall and speed. Recall here means the the percentage of similar video copies found in our model. We conducted the experiment in a setting where a query video and similar videos are shown in a 5×5 grid. A similar partial copy is detected if we show a pair of corresponding video frames with a precision within 3 s.

Our result is summarized in Table 1. Our approach can achieve recall of 78.2% in our decentralized video copyright protection framework. Due to lack of publicly available code by Jiang et al. [1], we do not have a direct comparison with their approach. As a reference, the best recall they get is less than 60% in their paper.

In terms of speed, it takes 0.22 s to 26.21 s to construct the kd tree for 528 videos, depending on the feature dimensions. The kd tree is only constructed once a day so it is not needed to be real time. The query speed ranges from 57 fps to 9298 fps, depending on the number of feature dimensions. We find that performance starts to saturate when the dimension is reduced to 64%.

We also conduct a stress test on the performance. We add one million distraction features which are constructed by adding random noise to the mean feature vector in the PCA space. The random noise follows the distribution between $[-\sigma, \sigma]$ where σ is the standard deviation of all the PCA values. In the stress test with 16-d PCA space, the running time for constructing the kd tree is 4.21 s, the query speed is 565 fps, and the recall is slightly reduced to 69.7% from 70.5%.

Now we estimate the cost for a video database with 1 million hours. Our approach can approximately process 50-h videos with one hour GPU time. So totally it costs about $20,000 GPU hours.

7   Discussion

So far we have presented a unified framework for decentralized autonomous video copyright protection solution. We first propose a decentralized autonomous protocol to determine if certain video violates copyright of other videos given a proposal raised by some users. Secondly, we present a real-time approach for similar video detection based on deep features. Evaluation shows that our approach can effectively detect similar videos with high recall.

One potential extension for handling tons of proposals is having copyright committee subgroups. We can divide the copyright committees into N subgroups, and each proposal is randomly sent to one of of the subgroups. Suppose each subgroup can handle K a week proposals, we have totally resolve $N \times K$ proposals in total each week.

References

1. Jiang, Y., Jiang, Y., Wang, J.,: VCDB: a large-scale database for partial copy detection in videos. In: 13th European Conference on Computer Vision - ECCV 2014, Zurich, Switzerland, 6–12 September 2014, Proceedings. Part IV, pp. 357–371 (2014)
2. Simonyan, K., Zisserman, A.: Very deep convolutional networks for large-scale image recognition. In: International Conference on Learning Representations (ICLR) (2015)
3. Russakovsky, O., Deng, J., Su, H., Krause, J., Satheesh, S., Ma, S., Huang, Z., Karpathy, A., Khosla, A., Bernstein, M.S., Berg, A.C., Li, F.: Imagenet large scale visual recognition challenge. Int. J. Comput. Vis. **115**(3), 211–252 (2015)
4. Wu, X., Hauptmann, A.G., Ngo, C.: Practical elimination of near-duplicates from web video search. In: Proceedings of the 15th International Conference on Multimedia 2007, Augsburg, Germany, 24–29 September 2007, pp. 218–227 (2007)
5. Wold, S., Esbensen, K., Geladi, P.: Principal component analysis. Chemometr. Intell. Lab. Syst. **2**(1), 37–52 (1987). Proceedings of the Multivariate Statistical Workshop for Geologists and Geochemists. http://www.sciencedirect.com/science/article/pii/0169743987800849
6. Arya, S., Mount, D.M., Netanyahu, N.S., Silverman, R., Wu, A.Y.: An optimal algorithm for approximate nearest neighbor searching fixed dimensions. J. ACM **45**(6), 891–923 (1998). https://doi.org/10.1145/293347.293348

7. Johnson, J., Alahi, A., Fei-Fei, L.: Perceptual losses for real-time style transfer and super-resolution. In: 4th European Conference on Computer Vision - ECCV 2016, Amsterdam, The Netherlands, 11–14 October 2016, Proceedings, Part II, pp. 694–711 (2016)

8. Chen, Q., Koltun, V.: Photographic image synthesis with cascaded refinement networks. In: IEEE International Conference on Computer Vision, ICCV 2017, Venice, Italy, 22–29 October, 2017, pp. 1520–1529 (2017)

9. Zhang, R., Isola, P., Efros, A.A., Shechtman, E., Wang, O.: The unreasonable effectiveness of deep features as a perceptual metric. *CoRR*, vol. abs/1801.03924 (2018)

10. Vedaldi, A., Fulkerson, B.: Vlfeat: an open and portable library of computer vision algorithms. In: Proceedings of the 18th International Conference on Multimedia 2010, Firenze, Italy, 25–29 October 2010, pp. 1469–1472 (2010)

A ZigBee Based Architecture
for Public Safety Communication
in Hurricane Scenario

Imtiaz Parvez, Yemeserach Mekonnen, and Arif I. Sarwat$^{(\boxtimes)}$

Florida International University, Miami, FL 33174, USA
{iparv001,ymeko001,asarwat}@fiu.edu

Abstract. Communication failure due to a power outage and infrastructure breakdown is a common phenomenon during the hurricane scenario, which hampers the post-disaster rescue operation. For this, an alternative and effective communication infrastructure are needed which can be deployed instantly with limited resources. As such, this paper proposes a ZigBee based ad-hock communication infrastructure for hurricane and post-hurricane scenarios. In this architecture, ZigBee based hop-to-hop communication will be used to communicate among the rescue team and victims, which can be powered by energy sources such as car batteries. Since each message will contain a geo-tagging, the rescue team can track the location of victims along with the normal communication with them. In addition, the performance of the communication infrastructure is further evaluated during and after the hurricane scenarios as well. Since the ZigBee is a low power device and can use power from energy sources such as a car battery, the proposed communication architecture can be a potential solution for a hurricane scenario.

Keywords: Ad-hoc network · Public safety communication · Channel modeling · ZigBee · Geo-tagging · Reliable communication

1 Introduction

One of the most direct and major impacts of a natural disaster like a hurricane is the sudden and wide-scale interruption of communication infrastructure. Communication network failure has the capacity to wipe out access to standard cellular or landlines communications, including access to Internet and satellite based emergency communication devices. This has an enormous disadvantage in deploying emergency response to the affected areas that are at the crossroad of life and death. This further complicates rescue efforts by making it impossible to locate and coordinate emergency efforts. The main cause of communication failure during a hurricane is the power outage [1,2]. Besides that, the communication networks might fail either as a result of physical damage to the infrastructure or due to network congestion. Hurricane winds and floodwater can damage network

The work is an outcome of the research supported by the U.S. National Science Foundation under the grant RIPS-1441223 and CAREER-1553494.

© Springer Nature Switzerland AG 2020
K. Arai et al. (Eds.): FTC 2019, AISC 1070, pp. 493–510, 2020.
https://doi.org/10.1007/978-3-030-32523-7_34

components, and cables creating vulnerability to the communication infrastructure. Wireless links similarly are prone to transmission disruption as a result of wavelength signal cut off due to heavy rain and wind. Network congestion is an additional concern during natural disaster situations as the levels of data traffic are higher. Aggregation hubs are often failure points for congested networks. This happens when data from various sources go into a central processing point and create bottlenecks. At this point, communication becomes severely limited and can even cut off completely.

A consequence of communication failure can prevent the emergency response from providing aid, causing major loss of life, infrastructure, and property damages. In addition, it makes it hard to expedite and coordinate the disaster relief process without knowing any information. Unfortunately, natural disasters such as hurricanes will be a constant occurrence in parts of the United States, therefore creating a resilient or alternative communication infrastructure is a necessity. Every decade, 18 different hurricanes on average are estimated to strike the United States [3,4].

Preventative measures to mitigate communication failures have been studied and researched. One method presented is diversifying network paths which establish and rely on two or more network connectivity utilizing various types of technology following different paths. This diminishes the chances of losing information from all connections simultaneously. A perfect example would be the use of one or more wireless links maintaining the existing fiber-optic cable connection. It goes without saying that the pathways must be secured and data processors must be checked to ensure no interruption.

The other effective alternative is the use of ad-hoc networks which is quicker to deploy and launch after the disaster. These networks can be launched using a wireless transport device on the top of a mobile platform such as Cellular On Wheels (COW) or Cellular On Light Trucks (COLT), to allow communications. There are various ad-hoc communication paradigms currently used in the disaster recovery network such as Mobile Ad-hoc Networks (MANETs), Wireless Mesh Networks (WMNs), Vehicular Ad-hoc Networks (VANETs), opportunistic or Delay Tolerant Networks (DTNs) and Wireless Sensor Networks (WSN) [5,6]. Only some of them have been built and evaluated in terms of efficiency as some require an infrastructure. MANETs and DTNs are the most well-studied models to allow communications in disaster applications as they use ad-hoc connections among mobile devices [7–9].

In this paper, a ZigBee based ad-hoc network architecture is proposed for disaster and post-disaster scenarios during the hurricane. In this architecture, the users will communicate to rescue teams or government authorities using an emergency ZigBee module. Since the Zigbee module can be installed in a car powered by the car battery, the range of communication can be extended with the help of hop-to-hop communication using nearby vehicles. The rescue team will roam the disaster areas in a facilitated vehicle. If the vehicle is within a specific range, a message sent from the emergency module will be received. Each message will also convey geo-tagging so that the rescue team can track the message sender.

Furthermore, within a range, the module can be used for voice communication [10]. This paper evaluates the performance of the proposed architecture during the hurricane and post-hurricane scenarios. The proposed communication infrastructure will be helpful for rescue operations during and after the hurricane mitigating communication interruption. The overall contributions of this work are as follows:

(1) It proposes a ZigBee based ad-hoc network for hurricane scenarios powered by the car battery for the first time.
(2) Since the architecture uses power sources, such as a car battery, it is easily deployable without regular electrical power sources.
(3) Since each house will have a communication module installed in its car, hop-to-hop communication will enable the long range communication.
(4) Finally, the performance of this system is evaluated and presented by using channel conditions during and after the hurricane scenario.
(5) The proposed communication infrastructure will be helpful for life saving rescue operation during and after the hurricane scenario.

The rest of this paper is organized as follows: Sect. 2 provide the literature review. In Sect. 3, the proposed communication architecture has been presented. Section 4 elicits the system model in detail. The simulation results for the communication system during and after the hurricane is illustrated in Sect. 5. Finally, a brief conclusion is included in Sect. 6.

2   Literature Review

The attention in vehicular network research has been growing exponentially over the last few years. This paper focuses on a ZigBee based ad-hoc vehicle to vehicle communication i.e. VANETs. VANETs with high mobility provides real-time communication functionality during and post-disaster recovery. VANETs can use either a broadcast type of communication or a multi-hop communication enabled with routing protocol for data transmission [11]. Broadcast communication for an emergency scenario in vehicular networks was studied in [12]. There are two main wireless communication types in VANETs: Vehicle to Vehicle communication (V2V), and Vehicle to Infrastructure (V2I) [13]. The most applicable type during a disaster scenario is V2V since in the V2I scenario the infrastructure will be destroyed during the disaster. V2V communication allows short and medium range communication supporting short message delivery with fast transmission [14]. This communication scheme allows vehicles to talk to disseminate safety and critical information among each other. V2V uses either GPS or short range communication technologies like Bluetooth, ZigBee, infrared and Ultra-Wideband (UWB) to transmit messages [15]. In general, VANET has attractive characteristics such as no power constraints, being able to support different network densities, and high computational ability. However, it also comes with its own challenges in signal fading, security issues and bandwidth limitations [13, 14, 16–20]. Hence, most current research works revolve around these challenges and further areas in routing protocol design and connectivity as mentioned in Table 1. A location-based VANETs, "RescueMe", for disaster recovery is proposed in [21] to support post-disaster rescue planning. This methodology securely stores user location information using the existing V2V and V2I communication to facilitate the rescue mission.

Previous research works on VANETs for disaster scenario applications focus on two major areas. The first one focuses on novel broadcasting and routing schemes that are feasible for the scenarios with a consequential effect in enhancements of emergency message dissemination. For instance, in [22] a positive orthogonal code is proposed to disseminate a transmission pattern for broadcast messages and the performance was evaluated as well. A model for deriving the packet delivery delay between disconnected vehicles was presented in [23], where it is longer in multi-hop disconnected communications. The second research area focuses on the usage of single or hybrid emerging wireless network technologies such as ZigBee, LTE, 5G and their performance evaluation [24–28]. Although not for disaster application, a ZigBee-WiFi VANET paradigm is explored in [29] to broadcast unsafe road conditions messages in rural areas. Different short range communication protocols were compared in [30] for V2V communication that can be employed in traffic congestion, road safety, and accident response cases. In [31], ZigBee was employed in V2I communication to locate positional coordinates to develop an effective emergency response system in a post-accident scenario.

WiFi and Bluetooth technologies are widely utilized in the current mobile device and vehicle interfaces. Although it offers a decent combination of throughput and energy efficiency, it takes a large amount of bandwidth and energy in times of contention and avoiding a collision which can be evidenced in a densely populated small area. ZigBee is an emerging wireless communication technology that supports low-cost, low-power and short-range wireless communication. As technology becomes more advanced, ZigBee similar to Bluetooth can be embedded in mobile devices and vehicles [32].

Most of the applications in the VANETs are focused on two segments traffic management, road safety, accident mitigation, and post-accident management [2,33–36]. Insignificant work has been done in the use of ZigBee based V2V communication platforms for the application of natural disaster response. Therefore understanding the research gap, this paper proposes a novel architecture during a hurricane scenario and post-hurricane condition. The simulation performance of data transmission for a line of sight (LOS) and non-line of sight (NLOS) communication are presented. The throughput, latency, and reliability of data packets are evaluated under different hurricane wind speeds. This system model assumes that the Zigbee module is integrated with a car and is powered by a car battery.

Table 1. Summary of Ad-hoc communication schemes.

Ad-hoc technologies	Communica tion scheme	Features	Disaster application	Existing work	Challenges & future work
MANETs [37–41]	Broadcast & multi-hop communication via routing protocol	– Broadcast will be suitable for warning messages and alarm – Multi-hop will be suitable in disseminating information to other different nodes from a central unit data from another node to the central unit	– Broadcast will be suitable for warning messages and alarm – Multi-hop will be suitable in disseminating information to other different nodes from a central unit	– Focused on the design of different routing protocols utilizing various communication patterns – Optimizat- ion of broadcast scheme based on dissimilarity coefficient	Few works on actual implementation and real experimentation of all proposed routing scheme
VANETs [42–46]	Broadcast communication and multi-hop communication via routing protocol	Similar to MANET but communication design requires fast medium access for quick transmission	Vehicle to Vehicle (V2V) communication are the main topology for disaster scenarios	– Novel broadcasting and routing procedures for emergency message. distribution between vehicles – Designing a schema when in disaster situation in the VANET infrastructure	Not enough work on the comparison of all proposed VANETs
DTNs [8, 47–51]	Only one way of communication scheme among nodes	Data are sent in bundles. Data are first split in different bundles and nodes wait until encountering another node to deliver the data	Great for low density, high mobility networks where MANETs wouldn't withhold	– Evaluating existing routing protocol like Epidemic, PRoPHET, MaxProp , TTR etc. – Proposing new routing protocols with energy efficiency like PropTTR	– Forwarding schemes while being energy efficient is to be achieved yet – A more automated system for deployment
WSNs [7, 52–56]	Centralized networks with cluster nodes	Network usually have a central node that collects data from all other nodes	Can be used in sensing, detection, warning and alerting roles	– Improving disaster prediction and alert systems – Improving node deployment for post disaster recovery using multi-agent systems	– Challenges in the deployment of WSNs in disaster areas – Use of crowd sensing and smartphone applications along sensor technologies
WMNs [9, 57–60]	Similar to MANETs communication	Routing protocol design different from MANET	Used as a support network to afford internet or connectivity	Proposing different mesh network topologies for connectivity	Covering more ground for connectivity in isolated zones

3   System Architecture

In this section, a communication network architecture is proposed considering the hurricane scenario. In this case, a Manhattan grid residential area is proposed where the distance between two houses is 30 m. Each house has at least one car. It assumes that the power outage occurs during the hurricane and post-hurricane scenario.

In this architecture, it is assumed that each car has an integrated ZigBee module that can be used for voice or data communication. Since the range of ZigBee is maximum 90 m, neighboring cars will relay data/voice to the destination. As a result, if power outage happens after the hurricane, ZigBee module can get power from the car battery. An architecture of the proposed communication structure is illustrated in Fig. 1. The users of ZigBee module may get in contact with a rescue team via message or voice. Each voice contains a geo-location tagging. Since Zig-Bee has a range of 90 m, the user's message or voice call will be received by a rescue vehicle once it reaches within 90 m range. By interpreting a message or after conversation with affected people, the rescue team can get the people to a safe zone. ZigBee is used as the communication standard for this architecture.

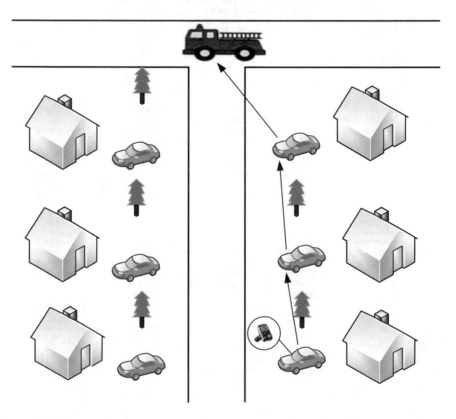

Fig. 1. Architecture of the ZigBee based Adhoc communication infrastructure for hurricane scenario

3.1   ZigBee

ZigBee is a low power, short range, low throughput, and energy efficient communication protocol that is suitable for wireless ad-hoc networks. The operational range of ZigBee varies from 30 m–90 m and it can support up to 64,000 nodes depending on the module. In addition, it can send data at a long distance using a mesh network, where intermediate nodes relay the data packet.

ZigBee modules comply with the IEEE 802.15.4 standard which defines the physical layers (PHY and MAC layers) and the link layers, whereas the upper layers- Network & Security, and Application layers are characterized by ZigBee alliance. In addition, the modules add node discovery and duplicated packet detection functionality. The application layer consists of manufacturing specific application profiles and public application profiles. In the network layer, the ad-hoc on-demand distant vector routing protocol is utilized. If security is enabled, 128 bit AES is used between its nodes. The MAC frame is composed of a header, payload, and footer (MFR). MAC frames can be categorized into four types: (1) Data, (2) Beacon, (3) acknowledgment, and (4) MAC command. These MAC frames are used to establish connections within the Personal Area Network (PAN) exchanging system information. The lowest level of ZigBee layers is the PHY layer. It is responsible for activation and deactivation of radio transceivers, energy detection within the channel, link quality indicator, selection of channel, and data transmission and reception. ZigBee can work on three different bands. In the 868 MHz band, there is only one channel and it can support a data rate of 20 Kbps. In the 915 MHz band, there are 10 channels and a maximum achievable data rate is 40 Kbps. In both bands, BPSK is used as the modulation technique. The 2.4 GHz band consists of 16 channels and it can support data rates up to 250 Kbps. The modulation technique used in this band is offset quadrature phase shift keying (OQPSK). The frequency nomenclature of ZigBee is tabulated in Table 2.

Table 2. IEEE 802.15.4 frequency nomenclature.

Frequency band	Number of channels	Center frequency	Channel spacing
868 MHz	1	868.3 MHz $k = 0$	0
915 MHz	10	$906 + 2(k - 1)$ MHz $k = 1, 2, 3...10$	2 MHz
2.5 GHz	16	$2405 + 5(k - 1)$ MHz $k = 11, 12...26$	5 MHz

ZigBee network defines three different device types: coordinator, router, and end devices. The coordinator assigns a PAN ID and channel for the network and allows routers and end devices to join the network. End devices join a router or coordinator with the intent to transmit data. ZigBee modules can adopt a star, tree and peer-to-peer topology to create a network. In star topology, a central node that acts as the coordinator is linked to all the other nodes using either a

MAC or network address. The central node also known as the coordinator collects all the data coming from the network nodes which are the end devices. A tree network has a top node with a branch structure where a message travels up the coordinator to the router and to the end device. Peer-to-peer, also referred to as mesh topology, has a similar connectivity scheme. In this study, mesh topology is proposed for the communication network.

In a mesh topology, packets pass through multiple hops to reach their destination, creating a multihop network. Since ZigBee supports mesh routing, data can be relayed through the intermediate node to the destination. The transmission power at the 868 MHz band is 25 mW. On the other hand, the maximum transmission power at 900 MHz and 2.4 GHz is 1000 mW. ZigBee utilizes a collision sense multiple access/collision avoidance (CSMA/CA) mechanism to access the network. When a packet comes to a node, the MAC layer initiates two variables- the number of back-off tries and the exponential back-off with a minimum value of 3. The MAC layer generates a random value within the range $[0, 2^{BE} - 1]$ and sets the delay accordingly. When $BE = 0$, the MAC senses the channel. If the channel is found idle, the packet is transmitted, otherwise, the value of BE and NB is increased by 1. If the value of NB is less than the maximum number of back-off (N_{Bmax}), the whole process is repeated for the transmission of the packet. Otherwise, the transmission attempt is discarded.

4   Communication Topology

Let us consider, a ZigBee module sends data packets to the destination using a mesh topology. In this architecture, let's assume that there is a M number of ZigBee nodes. Zigbee nodes use ad hoc On-demand Distance Vector Routing (ADOV) protocol for routing data packets.

Let us consider, a ZigBee module A communicates with rescue car D as illustrated in Fig. 2. The module A uses hops B and C for connecting with D. This path will be only activated when A, B, C, and D are within the range (i.e. 30 m–90 m) among themselves.

The path loss of data communication between any two hops such as $A - B$ is given by the below formula:

$$PL(dB) = 20 \log_{10}(d) + 20 \log_{10}(f) + 32.45 \tag{1}$$

The probability of sensing carrier for the first time Φ, probability of finding channel busy during the clear channel assessment 1 (CCA1) Ψ, and the probability of finding channel busy during the clear channel assessment 2 (CCA2) Ω of any ZigBee module such as A are given [61] by:

$$\Phi = \left(\frac{1 - y^{m+1}}{1 - y} \right) \left(\frac{1 - z^{n+1}}{1 - z} \right) b_{0,0,0} \tag{2}$$

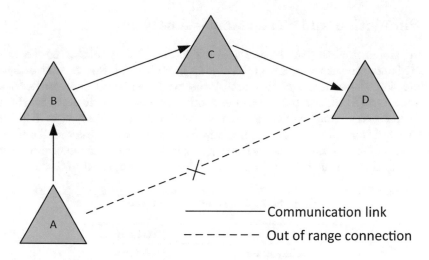

Fig. 2. Packet routing in a mesh topology network

$$\Psi = \&\left(l + \frac{M(1-\rho_0)\Phi(1-(1-\rho)\Phi)^{M-1}}{1-(1-(1-\rho_0)\Phi)^M}l_{\text{ACK}}\right)$$
$$\&\left(1-(1-(1-\rho_0)\Phi)^M\right) \quad (3)$$
$$\&\left(1-\Psi\right)\left(1-\omega\right)$$

$$\Omega = \&\frac{1-(1-(1-\rho_0)\Phi)^{M-1}+M(1-\rho_0)}{2-(1-(1-\rho_0)\Phi)^M+M(1-\rho_0)}\cdots$$
$$\&\frac{\Phi(1-(1-\rho_0)\Phi)^{M-1}}{\Phi(1-(1-\rho_0)\Phi)^{M-1}} \quad (4)$$

where,

$y = \Phi + (1-\Phi)\Omega$

$z = P_{\text{fail}}(1-y^{m+1})$

$m = $ Maximum back off number

$n = $ Maximum number of frame retries

$l = $ Length of data frame in slots

$l_{\text{ACK}} = $ Length of data frame in slots

$b_{0,0,0} = $ The states of Markov chain with normalized condition

The probability of collision of data sent by module A is given by

$$P_c = 1 - (1-\Phi)^{M-1} \quad (5)$$

For packet loss due to radio and channel over a region is P_{tx}, the probability of successful bit revived by ZigBee module B is:

$$P = (1-P_c)(1-P_{\text{tx}}) \quad (6)$$

5   Simulation and Performance Analysis

For the simulation, the Manhattan grid residential area is considered where the distance between the two houses is 30 m [62]. Each house has a car with a ZigBee module. For the simulation, two conditions are considered: (1) post-hurricane scenario, and (2) during the hurricane condition. For a hurricane scenario, the ultra-wideband channel model parameter is utilized that is derived in [63]. For a normal condition, a log-normal channel model is used for the residential area [64]. LOS and NLOS communication is further considered for performance evaluation. The simulation parameter for the ZigBee network is tabulated in Table 3.

Table 3. ZigBee simulation parameter

References	Techniques
Frequency band	2.4 GHz
Number of frames	51
Window length	8
Maximum frame retries	3
Maximum CSMA back-offs	4
Unit back off period	320 µs
Frame size	800 bits
Overhead size	48 bits
Maximum ACK wait duration	1920 µs
Sensing time	128 µs
Preamble rate	40 bits
Frame length	808 bits
Maximum distance	100 m
Minimum distance	30 m

5.1   Performance for LOS Communication

The throughput performance of LOS communication under a hurricane and normal condition is illustrated in Fig. 3. As the transmitted bits are incremented, the throughput decreases. However, after a certain amount of load (bits), the throughput becomes almost constant. Similarly, it is noted that throughput drops significantly for a hurricane condition compared to normal conditions. With the increase of wind speed, the throughput continues to drop as well. This is due to the degraded channel condition as a result of the high wind speeds.

Figure 4 illustrates the latency during the hurricane and normal conditions. With the increase of loads (i.e. bits), the latency increases. Furthermore, hurricane conditions demonstrate higher latency or wait time compared to normal conditions. As wind speed increases, latency is increased. This is due to

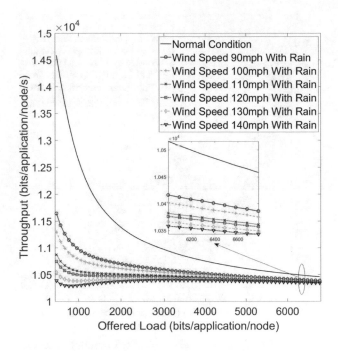

Fig. 3. Throughput of packets for LOS communication

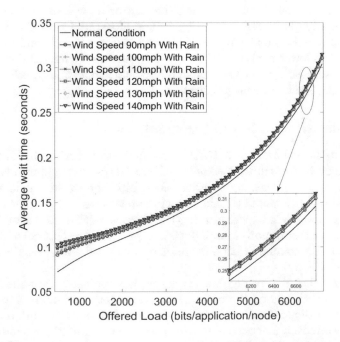

Fig. 4. Average wait time of packets for LOS communication

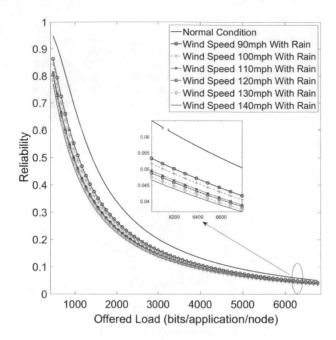

Fig. 5. Reliability of packets for LOS communication

the degraded channel condition during the high wind speeds. The reliability of data transmission is illustrated in Fig. 5. It is illustrated that the reliability of packets decreases for hurricane scenario compared to normal conditions. Higher wind speed has lower packet reliability due to increased channel degradation at higher wind speeds. For both normal and hurricane conditions, packet reliability decreases with the increment of packet load.

5.2   Performance for NLOS Communication

The throughput performance of NLOS communication during the hurricane and the normal condition is illustrated in Fig. 6. For NLoS, throughput increases with the packet load increment, however after 5000 bits/application/node, it becomes almost constant. It is noted that higher wind speeds result in higher throughput compared to lower wind speeds during the hurricane. This is because of higher wind speeds facilitate more robust data transmission for NLOS condition. The latency of NLOS communication during the hurricane and normal condition is illustrated in Fig. 7.

It is noted that the average wait time/latency increases with the increase of packet load. Furthermore, the normal condition has lower latency compared to hurricane conditions. This is due to the degraded channel condition during the hurricane. Moreover, hurricanes have higher latency for lower wind speeds in comparison to higher wind speeds. This is because higher wind speed facilitates more robust data transmission for NLOS conditions. The packet transmission

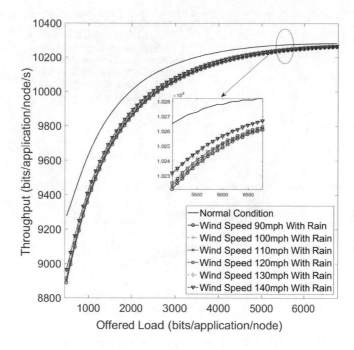

Fig. 6. Throughput of packets for NLOS communication

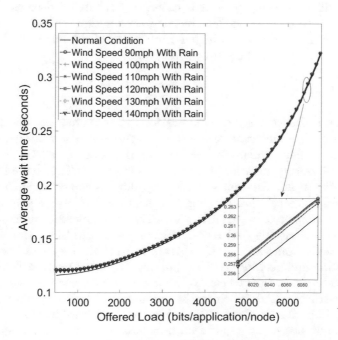

Fig. 7. Average wait time of packets for NLOS communication

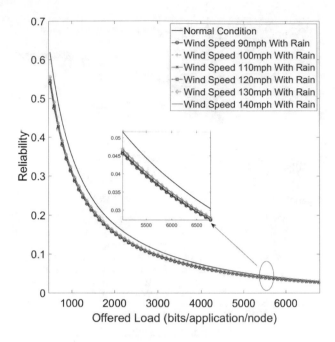

Fig. 8. Reliability of packets for NLOS communication

reliability is illustrated in Fig. 8. It is noted that reliability decreases with load size increment. Hurricane condition has degraded packet reliability compared to normal conditions. Moreover, a higher wind speed of the hurricane has slightly better reliability compared to lower wind speed.

6   Conclusion

In this paper, a ZigBee based communication architecture is proposed for rescue operations during a hurricane scenario. Considering a power outage and in the absence of a regular cellular network, the ZigBee module can be used to communicate with rescue operation teams. In this case, each ZigBee module can use the power of a car and can send data through hop-to-hop communication. Since ZigBee messages carry a geo-tagging, it can be used to rescue trapped victims during the hurricane. The performance of the communication infrastructure is also assessed for ongoing and post-hurricane scenarios. Since the ZigBee is a low power device and can use power from sources such as a car battery, this communication architecture can be a potential solution for hurricane scenario. The distance between the two cars is the bottleneck for the proposed communication infrastructure. Additionally, the communication infrastructure has been validated only by the simulation. Therefore, in the future, hardware implementation of the proposed communication infrastructure on a hurricane simulation facility (i.e. Wall of Wind facility of Florida) would be the final step that would further validate the communication infrastructure.

References

1. Sarwat, A.I., Sundararajan, A., Parvez, I., Moghaddami, M., Moghadasi, A.: Toward a smart city of interdependent critical infrastructure networks. In: Sustainable Interdependent Networks, pp. 21–45. Springer (2018)
2. Sundararajan, A., Olowu, T.O., Wei, L., Rahman, S., Sarwat, A.I.: A case study on the effects of partial solar eclipse on distributed photovoltaic systems and management areas. arXiv preprint arxiv:1905.11883 (2019)
3. Khawaja, W., Guvenc, I., Chowdhury, A.: Ultra-wideband channel modeling for hurricanes. In: IEEE 86th Vehicular Technology Conference (VTC-Fall), 1–6 September (2017)
4. National hurricane center. http://www.nhc.noaa.gov/pastdec.shtml
5. Rosas, E., Hidalgo, N., Gil-Costa, V., Bonacic, C., Marin, M., Senger, H., Arantes, L., Marcondes, C., Marin, O.: Survey on simulation for mobile ad-hoc communication for disaster scenarios. J. Comput. Sci. Technol. **31**(2), 326–349 (2016). https://doi.org/10.1007/s11390-016-1630-x
6. Gielen, M.: Ad hoc networking using wi-fi during natural disasters: overview and improvements. In: 17th Twente Student Conference on IT, vol. 17 (2012)
7. Yick, J., Mukherjee, B., Ghosal, D.: Wireless sensor network survey. Comput. Netw. **52**(12), 2292–2330 (2008)
8. Raffelsberger, C., Hellwagner, H.: Overview of hybrid MANET-DTN networking and its potential for emergency response operations. Electron. Commun. EASST **56**, 505–510 (2013)
9. Bruno, R., Conti, M., Gregori, E.: Mesh networks: commodity multihop ad hoc networks. IEEE Commun. Mag. **43**(3), 123–131 (2005)
10. Wang, C., Sohraby, K., Jana, R., Ji, L., Daneshmand, M.: Voice communications over ZigBee networks. IEEE Commun. Mag. **46**(1), 121–127 (2008)
11. Chuang, M.-C., Chen, M.C.: Deep: density-aware emergency message extension protocol for vanets. IEEE Trans. Wirel. Commun. **12**(10), 4983–4993 (2013)
12. Farnoud, F., Valaee, S.: Reliable broadcast of safety messages in vehicular ad hoc networks. In: IEEE INFOCOM, pp. 226–234 (2009)
13. Tanuja, K., Sushma, K., Bharathi, M., Arun, K.: A survey on vanet technologies. Int. J. Comput. Appl. **121**(18), 1–9 (2015)
14. Al-Sultan, S., Al-Doori, M.M., Al-Bayatti, A.H., Zedan, H.: A comprehensive survey on vehicular ad hoc network. J. Netw. Comput. Appl. **37**, 380–392 (2014)
15. Jafari, H., Mahmoudi, M., Rastegar, H., Rabiee, A., Naderi, M.H., Kazemi, F.: Using wide-area signals to improve the inter-area mode damping performance of static VAR compensators. In: 2018 IEEE Texas Power and Energy Conference (TPEC), pp. 1–6, February 2018
16. Kumar, V., Mishra, S., Chand, N.: Applications of vanets: present and future. Commun. Netw. **5**(01), 12 (2013)
17. Wei, L., Sundararajan, A., Sarwat, A., Biswas, S., Ibrahim, E.: A distributed intelligent framework for electricity theft detection using benford's law and stackelberg game. In: Resilience Week, pp. 5–11, September 2017
18. Parvez, I., Islam, A., Kaleem, F.: A key management-based two-level encryption method for AMI. In: 2014 IEEE PES General Meeting—Conference Exposition, pp. 1–5. July 2014
19. Parvez, I., Sarwat, A.I., Wei, L., Sundararajan, A.: Securing metering infrastructure of smart grid: a machine learning and localization based key management approach. Energies **9**(9), 691 (2016). https://www.mdpi.com/1996-1073/9/9/691

20. Parvez, I., Abdul, F., Sarwat, A.I.: A location based key management system for advanced metering infrastructure of smart grid. In: 2016 IEEE Green Technologies Conference (GreenTech), pp. 62–67, April 2016

21. Sun, J., Zhu, X., Zhang, C., Fang, Y.: Rescueme: location-based secure and dependable vanets for disaster rescue. IEEE J. Sel. Areas Commun. **29**(3), 659–669 (2011)

22. Wang, Y.: ZigBee-assisted ad-hoc networking of multi-interface mobile devices (2012)

23. Wisitpongphan, N., Bai, F., Mudalige, P., Sadekar, V., Tonguz, O.: Routing in sparse vehicular ad hoc wireless networks. IEEE J. Sel. Areas Commun. **25**(8), 1538–1556 (2007)

24. Mekonnen, Y., Haque, M., Parvez, I., Moghadasi, A., Sarwat, A.: LTE and WiFi coexistence in unlicensed spectrum with application to smart grid: a review. In: IEEE/PES Transmission and Distribution Conference and Exposition (T&D), pp. 1–5. IEEE (2018)

25. Parvez, I., Sarwat, A.I.: A spectrum sharing based metering infrastructure for smart grid utilizing LTE and WiFi. Adv. Sci. Technol. Eng. Syst. J. **4**(2), 70–77 (2019)

26. Parvez, I., Jamei, M., Sundararajan, A., Sarwat, A.I.: RSS based loop-free compass routing protocol for data communication in advanced metering infrastructure (AMI) of smart grid. In: 2014 IEEE Symposium on Computational Intelligence Applications in Smart Grid (CIASG), pp. 1–6, December 2014

27. Sarwat, A.I., Sundararajan, A., Parvez, I.: Trends and future directions of research for smart grid iot sensor networks. In: Proceedings of International Symposium on Sensor Networks, Systems and Security, May 2018

28. Parvez, I., Sriyananda, M., Güvenç, I., Bennis, M., Sarwat, A.: Cbrs spectrum sharing between LTE-U and WiFi: a multiarmed bandit approach. Mob. Inf. Syst. **2016** (2016)

29. Reina, D.G., Coca, J.M.L., Askalani, M., Toral, S.L., Barrero, F., Asimakopoulou, E., Sotiriadis, S., Bessis, N.: A survey on ad hoc networks for disaster scenarios. In: 2014 International Conference on Intelligent Networking and Collaborative Systems, pp. 433–438, September 2014

30. Rawat, D.B., Bista, B.B., Yan, G., Olariu, S.: Vehicle-to-vehicle connectivity and communication framework for vehicular ad-hoc networks. In: 2014 Eighth International Conference on Complex, Intelligent and Software Intensive Systems, pp. 44–49. IEEE (2014

31. Bhargav, K.K., Singhal, R.: ZigBee based vanets for accident rescue missions in 3G wcdma networks. In: 2013 IEEE Global Humanitarian Technology Conference: South Asia Satellite (GHTC-SAS), pp. 310–313. IEEE (2013)

32. Shree, K.L., Penubaku, L., Nandihal, G.: A novel approach of using security enabled ZigBee in vehicular communication. In: 2016 IEEE International Conference on Computational Intelligence and Computing Research (ICCIC), pp. 1–5. IEEE (2016)

33. Olowu, T.O., Jafari, M., Sarwat, A.I.: A multi-objective optimization technique for volt-var control with high PV penetration using genetic algorithm. In: 2018 North American Power Symposium (NAPS), pp. 1–6, September 2018

34. Jafari, M., Olowu, T.O., Sarwat, A.I.: Optimal smart inverters volt-var curve selection with a multi-objective volt-var optimization using evolutionary algorithm approach. In: 2018 North American Power Symposium (NAPS), pp. 1–6, September 2018

35. Onibonoje, M.O., Olowu, T.O.: Real-time remote monitoring and automated control of granary environmental factors using wireless sensor network. In: 2017 IEEE International Conference on Power, Control, Signals and Instrumentation Engineering (ICPCSI), pp. 113–118. IEEE (2017)
36. Rahman, S., Moghaddami, M., Sarwat, A.I., Olowu, T., Jafaritalarposhti, M.: Flicker estimation associated with PV integrated distribution network. In: SoutheastCon 2018, pp. 1–6, April 2018
37. Chlamtac, I., Conti, M., Liu, J.J.-N.: Mobile ad hoc networking: imperatives and challenges. Ad hoc Netw. **1**(1), 13–64 (2003)
38. Johansson, P., Larsson, T., Hedman, N., Mielczarek, B., Degermark, M.: Scenario-based performance analysis of routing protocols for mobile ad-hoc networks. In: Proceedings of the 5th annual ACM/IEEE International Conference on Mobile Computing and Networking, pp. 195–206. ACM (1999)
39. Reina, D., Toral, S.L., Barrero, F., Bessis, N., Asimakopoulou, E.: Evaluation of ad hoc networks in disaster scenarios. In: 2011 third International Conference on Intelligent Networking and Collaborative Systems, pp. 759–764. IEEE (2011)
40. Raffelsberger, C., Hellwagner, H.: Evaluation of MANET routing protocols in a realistic emergency response scenario. In: Proceedings of the 10th International Workshop on Intelligent Solutions in Embedded Systems, pp. 88–92. IEEE (2012)
41. Macone, D., Oddi, G., Pietrabissa, A.: MQ-routing: Mobility-, GPS-and energy-aware routing protocol in manets for disaster relief scenarios. Ad Hoc Netw. **11**(3), 861–878 (2013)
42. Lin, Y.-W., Chen, Y.-S., Lee, S.-L.: Routing protocols in vehicular ad hoc networks: a survey and future perspectives. J. Inf. Sci. Eng. **26**(3), 913–932 (2010)
43. Fasolo, E., Zanella, A., Zorzi, M.: An effective broadcast scheme for alert message propagation in vehicular ad hoc networks. In: 2006 IEEE International Conference on Communications, vol. 9, pp. 3960–3965. IEEE (2006)
44. Peng, J., Cheng, L.: A distributed mac scheme for emergency message dissemination in vehicular ad hoc networks. IEEE Trans. Veh. Technol. **56**(6), 3300–3308 (2007)
45. Lee, D., Bai, S., Kwak, D., Jung, J.: Enhanced selective forwarding scheme for alert message propagation in vehicular ad hoc networks. Int. J. Automot. Technol. **12**(2), 251 (2011)
46. Lee, J.-F., Wang, C.-S., Chuang, M.-C.: Fast and reliable emergency message dissemination mechanism in vehicular ad hoc networks. In: 2010 IEEE Wireless Communication and Networking Conference, pp. 1–6. IEEE (2010)
47. Fall, K.: A delay-tolerant network architecture for challenged internets. In: Proceedings of the 2003 Conference on Applications, Technologies, Architectures, and Protocols for Computer Communications, pp. 27–34. ACM (2003)
48. Aschenbruck, N., Gerhards-Padilla, E., Martini, P.: Modeling mobility in disaster area scenarios. Perform. Eval. **66**(12), 773–790 (2009)
49. Saha, S., Sheldekar, A., Mukherjee, A., Nandi, S., et al.: Post disaster management using delay tolerant network. In: Recent Trends in Wireless and Mobile Networks, pp. 170–184. Springer (2011)
50. Keränen, A., Ott, J., Kärkkäinen, T.: The one simulator for dtn protocol evaluation. In: Proceedings of the 2nd International Conference on Simulation Tools and Techniques, p. 55. ICST (Institute for Computer Sciences, Social-Informatics and Telecommunications Engineering) (2009)
51. MartíN-Campillo, A., Crowcroft, J., Yoneki, E., Martí, R.: Evaluating opportunistic networks in disaster scenarios. J. Netw. Comput. Appl. **36**(2), 870–880 (2013)

52. Mekonnen, Y., Burton, L., Sarwat, A., Bhansali, S.: Iot sensor network approach for smart farming: an application in food, energy and water system. In: 2018 IEEE Global Humanitarian Technology Conference (GHTC), pp. 1–5. IEEE (2018)
53. Bahrepour, M., Meratnia, N., Poel, M., Taghikhaki, Z., Havinga, P.J.: Distributed event detection in wireless sensor networks for disaster management. In: 2010 International Conference on Intelligent Networking and Collaborative Systems, pp. 507–512. IEEE (2010)
54. Cayirci, E., Coplu, T.: Sendrom: sensor networks for disaster relief operations management. Wirel. Netw. **13**(3), 409–423 (2007)
55. Miyazaki, T., Kawano, R., Endo, Y., Shitara, D.: A sensor network for surveillance of disaster-hit region. In: 2009 4th International Symposium on Wireless Pervasive Computing, pp. 1–6. IEEE (2009)
56. Saha, S., Matsumoto, M.: A framework for data collection and wireless sensor network protocol for disaster management. In: 2007 2nd International Conference on Communication Systems Software and Middleware, pp. 1–6. IEEE (2007)
57. Suzuki, H., Kaneko, Y., Mase, K., Yamazaki, S., Makino, H.: An ad hoc network in the sky, skymesh, for large-scale disaster recovery. In: IEEE Vehicular Technology Conference, pp. 1–5. IEEE (2006)
58. Shibata, Y., Sato, Y., Ogasawara, N., Chiba, G., Takahata, K.: A new ballooned wireless mesh network system for disaster use. In: 2009 International Conference on Advanced Information Networking and Applications, pp. 816–821. IEEE (2009)
59. Dilmaghani, R.B., Rao, R.R.: Hybrid wireless mesh network with application to emergency scenarios. J. Softw. **3**(2), 52–60 (2008)
60. Suzuki, T., Shibata, Y.: Autonomous power supplied wireless mesh network for disaster information system. In: 2010 International Conference on Broadband, Wireless Computing, Communication and Applications, pp. 88–93. IEEE (2010)
61. Park, P., Marco, P.D., Soldati, P., Fischione, C., Johansson, K.H.: A generalized markov chain model for effective analysis of slotted IEEE 802.15.4. In: 2009 IEEE 6th International Conference on Mobile Adhoc and Sensor Systems, pp. 130–139, October 2009
62. Parvez, I., Islam, N., Rupasinghe, N., Sarwat, A.I., Güvenç, İ.: LAA-based LTE and ZigBee coexistence for unlicensed-band smart grid communications. In: SoutheastCon 2016, pp. 1–6. IEEE (2016)
63. Khawaja, W., Guvenc, I., Chowdhury, A.: Ultra-wideband channel modeling for hurricanes. In: 2017 IEEE 86th Vehicular Technology Conference (VTC-Fall), pp. 1–6. IEEE (2017)
64. Molisch, A.F.: Ultrawideband propagation channels-theory, measurement, and modeling. IEEE Trans. Veh. Technol. **54**(5), 1528–1545 (2005)

An Efficient Singularity Detector Network for Fingerprint Images

Geetika Arora[2(✉)], C. Jinshong Hwang[1], Kamlesh Tiwari[2],
and Phalguni Gupta[3]

[1] Department of Computer Science, Texas State University,
San Marcos, TX 78666-4616, USA
cjhwang@txstate.edu
[2] Department of CSIS, Birla Institute of Technology and Science Pilani,
Pilani 333031, Rajasthan, India
{p2016406,kamlesh.tiwari}@pilani.bits-pilani.ac.in
[3] Department of Computer Science, Indian Institute of Technology Kanpur,
Kanpur 208016, UP, India
pg@iitk.ac.in

Abstract. Singular point related to a fingerprint is a special location that has been used for long to classify the fingerprint in specific pre-defined categories. It is also useful for the alignment of a fingerprint for better matching and comparison. This paper proposes a novel end-to-end deep learning model that incorporate pre-processing, enhancement, segmentation and localization for accurate and efficient retrieval of a singular point. The proposed model has been tested on two databases *viz.* FVC2002 DB2_A and FPL05K and achieved a Correct Detection Rate of 96.12% and 91.1% respectively which is better than any other state-of-the-art technique.

Keywords: Fingerprint · Singular points · Core point · Encoder · Decoder

1 Introduction

Biometrics have been very successful in replacing and complementing traditional authentication mechanisms such as PIN or password. This is because of their ability to provide higher level of security. They are always available with the user and a user cannot deny his participation in authentication once done. Various physiological and behavioral traits have been explored such as face, fingerprint, ear, iris, knuckle, palm, signature, voice *etc.* However, fingerprint has been found to most acceptable in the society. This is because of the convenience it offers during the authentication process. Fingerprint is an impression that gets developed on a surface touched by the upper part of a human finger [4]. It is widely believed that the fingerprints are unique for each finger. An automatic authentication system based on fingerprint needs a matching module in its core which would extract reliable features from the two fingerprints presented for the comparison and generates a similarity/dissimilarity score. Feature extraction is one of the crucial step. It requires fingerprint alignment that can be done based on some pivot points. Core or singular points are such key points.

© Springer Nature Switzerland AG 2020
K. Arai et al. (Eds.): FTC 2019, AISC 1070, pp. 511–518, 2020.
https://doi.org/10.1007/978-3-030-32523-7_35

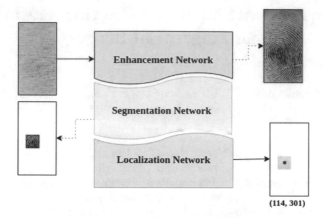

Fig. 1. Block diagram of the proposed network model (output of logical steps are shown using dotted arrows).

There are two ways one can interact with the fingerprint recognition system. First, when the user provides his identity and fingerprint for the verification. In this setting, a saved template corresponding to the identity of the user is retrieved from the database and is one-to-one match with the newly provided one. Second, when the user only provides his fingerprint and the system has to scan the complete database to determine whose fingerprint it is. In this setting, number of comparisons needed is equal to the size of the database [7]. With the increase in the size of the database, that happens by increasing the adaptability of the system, the response of the identification system worsens. To sustain the system performance, it is needed to somehow reduce the comparison over the complete database by employing a technique called as indexing. Indexing involves registration and the singular points are helpful for this purpose. Singular points have high curvature radius as shown in Fig. 5.

An approach employing Poincare index and multi-scale detection algorithm has been proposed in [2]. The Poincare index finds out the probable blocks for the presence of singular point. This approach has an advantage of speed as the Poincare index is not calculated at every pixel and only for the probable areas. However, this approach may result in false detection. This limitation has been addressed in [5] which has proposed a modified Poincaré index and tested it on FVC2000 DB1 database. Gupta *et al.* has proposed a hybrid technique employing directional filtering, orientation field and Poincare index for singular point detection. It has been tested on Biostar, FVC2004 DB1 and FVC2004 DB2 database. The advantage with this approach is that it was robust to situations where the singular point is not visible due to occlusion.

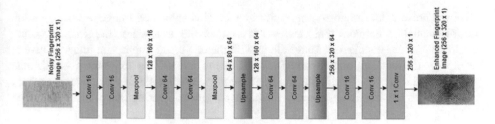

Fig. 2. Enhancement network architecture

Another approach using orientation field has been proposed in [12]. It essentially generates a walking directional field (WDF) from the orientation field. A walking algorithm has also been proposed that uses the generated WDF to move to the singular point location instead of scanning the whole fingerprint image. The approach has been tested on FVC2000 DB2, FVC2002 DB2 and FVC2004 DB2 database. This approach has been enhanced in [3] wherein after walking to the singular point, the detected location has been further refined using a proposed enhanced method applying mean-shift concept.

Meandering energy potential has been used for the detection of core points in [8]. This approach does not require any knowledge of the structure of the fingerprint beforehand. It has been evaluated on FVC2002 database. A modified k-curvature method has been utilized to find the high-curvature area of fingerprint ridges in [10]. The ridge's bending energy has been used to detect the location of the reference point. The approach has been evaluated on FVC2002 and FVC2004 datasets. Min *et al.* employed a local search method to detect the core point. The approach uses a fixed-size sliding window that scans the regions of the fingerprint locally for the presence of the core point. A deep-learning based approach has been proposed in [6]. The approach uses Faster-RCNN network to generate region proposals that could contain the singular point. Among all the proposed regions, top-100 regions were considered for final search. This approach has been tested on FVC2002 DB1A. In this paper, the problem of singularity detection for fingerprint images is addressed by using deep learning architecture. The approach provides end-to-end solution that integrates essential parts of enhancement, segmentation and localization into one and outperforms by adjusting the parameters in a collaborative way. The paper is organized as follows; next section provides details of the proposed model and highlights three important sub components; enhancement, segmentation and localization network. Subsequent section explains about the experimental setting and the results obtained on the two databases. Conclusions are presented at the end.

2 Proposed Approach

The aim of this work is to find the location of the singular point in the given fingerprint image. This is achieved by using three components in the system namely, enhancement network, segmentation network and the localization network. The first network removes noise from the given fingerprint image. This enhanced image is passed on to the segmentation network that creates a mask over the area where the singular point could be found. This masked image along with the enhanced fingerprint image is given as input to the localization network. The localization network regresses the masked area to reach to the location of singular point. These networks are stacked one over the other and performs localization in one go. The block diagram depicting this process is shown in Fig. 1

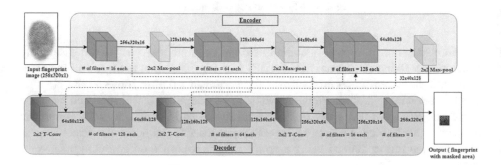

Fig. 3. Architecture of the segmentation network

Enhancement Network. The enhancement network removes noise and improves visual appearance of the fingerprint images. This is an encoder-decoder network which is trained by minimizing the loss between original and its artificially tampered image. The model consists of convolution blocks having filter size of 3×3 and ReLU activation function followed by a max-pool layer. After this, up-sampling is done by using a 2×2 up-sampling operation. The architecture of the enhancement network is shown in Fig. 2.

The Segmentation Network. The segmentation network follows an encoder-decoder type of architecture. It takes the enhanced image as input and outputs a block of size 41×41 marking the probable region for the presence of the singular point. It is again an encoder-decoder network containing a series of convolution layers followed by a max pool layer of 2×2 size. The encoder part performs down-sampling of features. This feature is then up-sampled using transposed-convolution layer of 2×2 size. The output of the up-sampling layer is merged with the feature map produced by the corresponding max-pool layer using the merge connections. Merge connections eliminates the problem of vanishing gradient during learning. The architecture of the network is shown in Fig. 3. In this figure, the orange blocks and dotted arrows represent 3×3 convolution and merge connections, respectively.

The Localization Network. The Localization network is a basic convolutional neural network model. It consists of four convolutional blocks, with each block containing two convolution layers followed by a max-pool layer. The blocks are followed by a flattening layer and two fully connected layers. The enhanced image and the segmented image becomes the input to this network. This network outputs the predicted (x, y) coordinates of the singular point. It essentially performs regression on the proposed region of interest and outputs the location of the singular point in the fingerprint image as shown in Fig. 4.

The proposed network has been trained using the binary cross entropy and mean squared error loss. The cross entropy loss is back-propagated by the segmentation network to learn the location of the probable region where the singular point could be present. The localization of the singular point is learned by back-propagating the mean squared error to the whole network.

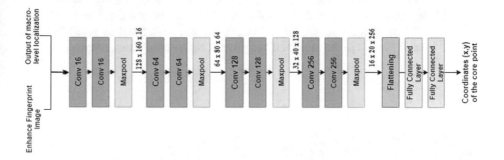

Fig. 4. Architecture of the localization network

3 Results

This section explains the database used for testing the proposed approach followed by the evaluation parameters. The observations are listed afterwards.

3.1 Database

We have used two databases for testing our proposed approach. First one is the standard database that has been widely used for singular point localization. The other is collected in-house. We have generated the ground truth for both the datasets by manually marking the core point in every image and storing the (x, y) coordinates. Description of both the databases has been given below.

FVC2002 Db2_A. This database [1] is publicly available and contains 800 fingerprint images. These image shave been collected from 100 subjects. The fingerprints have huge intra-class variation among different fingerprint impressions of the same subject. This is due to difference in placement, pressure *etc*.

FPL05K. This database has been collected from 855 subjects using three different sensors. Each subject has provided six fingerprint impressions of the same finger. We have generated the ground truth for 5,000 images from this dataset; some of which are shown in Fig. 5.

3.2 Evaluation Parameters

Correct Detection Rate (CDR) has been used to measure the accuracy of the proposed method in detecting the location of the singular point. The output of the proposed model are the (x, y) coordinates of the singular point. The location of the singular point is considered to be correctly detected if the L2 distance between the predicted and the actual coordinate is less than 20 pixels. Mathematically it can be defined as given in Eq. 1, where P_c and A_c refers to the predicted and ground truth coordinates, respectively.

$$\sqrt{(P_c.x - A_c.x)^2 + (P_c.y - A_c.y)^2} \leq 20 \; pixels \qquad (1)$$

Fig. 5. Singular point in a fingerprint image. First four and next four are from FVC2002_2A and FPL05K database respectively.

3.3 Observations

The proposed model achieves correct detection rate of 96.12% and 91.1% on FVC2002 DB2_A and in-house database FPL05K respectively. The obtained results have been compared with the other state-of-the-art techniques and is given in Table 1. We can observe from this table that the proposed approach performs better than [6] which is better than other previous proposed techniques.

Table 1. Results. Comparison with other State-Of-The-Art approaches on FVC-2002-DB2A.

Database: FVC-2002-DB2A

Technique	Description	TDR
Zhou *et.al*. [11]	Orientation values along a Circle (DORIC)	95.95%
Xie *et.al*. [9]	Inconsistency feature	90.00%
Tiwari *et.al*. [8]	Meandering energy potential (MEP)	95.75%
Liu *et.al*. [6]	Faster R-CNN	96.03%
Proposed	Deep Learning	96.12%

Database: FPL05K

Technique	Description	TDR
Proposed	Deep Learning	91.1%

4   Conclusion

This paper addresses the problem of singularity detection in fingerprint images. These points are also called as core points and have been found useful in indexing and classification of fingerprints. The paper proposes a deep learning based architecture that helps to localize singular point in one-go. Proposes model is efficient and more accurate as compared to any other model proposed in the literature. Experiments have been conducted on two databases, FVC2002 DB2_a and FPL05K that contain 800 and 5,000 fingerprint images. It has been observed that the proposed model has correct detection rate of 96.12% and 91.1% on the two databases respectively. It has been observed that the end-to-end learning using a deep neural network not only outperforms state-of-the-art detection techniques but has also been found to be efficient.

References

1. The FVC2002 database. http://bias.csr.unibo.it/fvc2002/
2. Bo, J., Ping, T.H., Lan, X.M.: Fingerprint singular point detection algorithm by poincaré index. WSEAS Trans. Syst. **7**(12), 1453–1462 (2008)
3. Guo, X., Zhu, E., Yin, J.: A fast and accurate method for detecting fingerprint reference point. Neural Comput. Appl. **29**(1), 21–31 (2018)
4. Gupta, P., Tiwari, K., Arora, G.: Fingerprint indexing schemes - a survey. Neurocomputing **335**, 352–365 (2018)
5. Iwasokun, G.B., Akinyokun, O.C.: Fingerprint singular point detection based on modified poincare index method. Int. J. Sig. Proces. Image Process. Pattern Recognit. **7**(5), 259–272 (2014)
6. Liu, Y., Zhou, B., Han, C., Guo, T., Qin, J.: A method for singular points detection based on faster-RCNN. Appl. Sci. **8**(10), 1853 (2018)
7. Tiwari, K., Gupta, P.: Indexing fingerprint database with minutiae based coaxial Gaussian track code and quantized lookup table. In: IEEE International Conference on Image Processing (ICIP), pp. 4773–4777. IEEE (2015)

8. Tiwari, K., Gupta, P.: Meandering energy potential to locate singular point of fingerprint. In: International Conference on Biometrics (ICB), pp. 1–6. IEEE (2016)

9. Xie, S.J., Yoo, H.M., Park, D.S., Yoon, S.: Fingerprint reference point determination based on a novel ridgeline feature. In: 17th IEEE International Conference on Image Processing (ICIP), pp. 3073–3076. IEEE (2010)

10. Zacharias, G.C., Nair, M.S., Lal, P.S.: Fingerprint reference point identification based on chain encoded discrete curvature and bending energy. Pattern Anal. Appl. 20(1), 253–267 (2017)

11. Zhou, J., Chen, F., Gu, J.: A novel algorithm for detecting singular points from fingerprint images. IEEE Trans. Pattern Anal. Mach. Intell. 31(7), 1239–1250 (2009)

12. Zhu, E., Guo, X., Yin, J.: Walking to singular points of fingerprints. Pattern Recognit. 56, 116–128 (2016)

An Implementation of Smart Contracts by Integrating BIM and Blockchain

Alireza Shojaei[1]($\boxtimes$), Ian Flood[2], Hashem Izadi Moud[2],
Mohsen Hatami[2], and Xun Zhang[2]

[1] Building Construction Science, College of Architecture, Art and Design,
Mississippi State University, Starkville, MS 39762, USA
`shojaei@caad.msstate.edu`
[2] M. E. Rinker, Sr. School of Construction Management, University of Florida,
P.O. Box 115703, Gainesville, FL 32611, USA
`{flood,izadimoud,mohsen.hatami,xzz0032}@ufl.edu`

Abstract. This paper presents an implementation of smart contracts by integrating BIM and blockchain. After briefly reviewing the literature regarding blockchain technology, smart contracts, and integration of the BIM and blockchain, a blockchain network using Hyperledger fabric is proposed and tested to govern a sample construction project. The proposed framework does not utilize the cryptocurrency aspect of the blockchain as the payment form. Instead, it discusses the integration of the current banking system and the use of fiat currencies in transactions. The results show that blockchain is a viable system for governing construction project contracts by automating the consequences of each transaction and maintaining a tamper proof record of project progress, which would be valuable in any kind of dispute resolution. The blockchain network developed in this study implements the smart contract as its network operation logic. As a result, the notion of having to translate all the traditional contract clauses to the computer program is shown to be unnecessary and to some extent not suitable for construction, due to the complexity, fluidity, and high uncertainties involved in each project.

Keywords: Building Information Modeling · Blockchain · Smart contract ·
BIM · Cyber-physical space

1 Introduction

The construction industry is filled with legal disputes. Disputes can result from a wide range of issues such as ambiguity in the terms of the contract, late payments, late or underperforming delivery of work, to name but a few. One of the proposed solutions to improve the current working environment is smart contracts. A smart contract can be defined as a computer program with if/then structure which administers the contract clauses. Once a party executes its tasks, it asks for inspection, and upon verification, a payment would automatically be issued. The advantage of such a system is its clarity and enforceability which may result in smoother contract execution and a significant reduction in disputes. Building Information Modeling (BIM) due to its data-intensive

© Springer Nature Switzerland AG 2020
K. Arai et al. (Eds.): FTC 2019, AISC 1070, pp. 519–527, 2020.
https://doi.org/10.1007/978-3-030-32523-7_36

nature and the level of details presented in an appropriate model is an excellent way to tie different sections of the work to a smart contract. However, a secure and proper link between the BIM model, the physical work, and the smart contract is needed to make such a system viable. Blockchain has been successfully implemented in supply chain management in different industries to connect stakeholders and facilitate storage and sharing of information. This study aims to test the feasibility of blockchain technology as the link between the BIM model and the physical world with the implementation of smart contracts as the business logic of the blockchain network.

The rest of this paper is organized as follows. First, an overview of the blockchain technology, smart contracts, and the current state of knowledge regarding the integration of BIM and blockchain is presented. Then, a blockchain network for connecting the BIM model to the physical world based on the Hyperledger Fabric platform is discussed with a presentation of a sample implementation of smart contracts that integrates BIM and blockchain. At last, the paper concludes with discussing the results, research limitations, and future direction for research.

2 Blockchain Technology

Enforcement of trust throughout the construction industry is always challenging. One way to ensure trust among all parties is by using blockchain technology in construction contracts. Blockchain is a database technology that provides smart and coded schemes to verify and store transactions throughout chains of communications. Blockchain is a Distributed Ledger Technology (DLT), which is simply a database of transactions. Blockchain transactions are stored on different nodes of the network, which makes it a fully decentralized system. Transactions are regularly synchronized, keeping the system up to date at all times. A blockchain network is secure as it uses cryptography to store and transfer all the transactions throughout its decentralized network and the data is tamper proof due to the consensus system and chain-like data sequence [1, 2]. In order for a transaction to be executed, all the nodes in the network need to process that transaction. The fact that DLTs are not owned by specific parties, are fully decentralized and coded, makes them a suitable platform for contracts. Trust in contracts is instilled in people, organizations, and authorities who are prone to error, corruption, and misuse. However, trust in DLTs are based on coded transactions that are not owned by anyone and executed through a chain of distributed nodes [1, 2]. Blockchain technology is widely used in cryptocurrencies, most notably bitcoin. While the first version of blockchain started with cryptocurrencies, blockchain technology has profoundly evolved throughout the years. Blockchain 1.0 was the basis of the Bitcoin, blockchain 2.0 introduced the idea of smart contracts through blockchain with Ethereum (a blockchain-based distributed computing platform). Blockchain 3.0 goes beyond smart contracts and tries to apply blockchain to areas such as healthcare, culture, and government.

Enforcement of trust, distribution of power among different parties (nodes) and its cryptographic nature have made blockchain technology a promising platform for smart contracts within the Architecture, Engineering, and Construction (AEC) domain. Also, the full digitalization of transactions provides a door to the future of transactions which the AEC industry needs to move towards it [1, 3, 4]. The characteristics of blockchain along with its evolution through introducing different versions that can easily accommodate different needs helped spread the application of blockchain across multiple disciplines. Currently, blockchain is being used for record keeping, supply chain transparency, smart grid distribution, money tracking, voting, and insurance industries [5]. While numerous advantages have been mentioned for blockchain technology, there are some downsides that might slow down its adoption in the AEC industry. For example, depending on the consensus algorithm in use, confirming a transaction might take a long time. Also, applying blockchain technology in the AEC industry, which is fairly technology-resistant, might inhibit the adoption of this new approach.

3 Smart Contracts

Smart contracts are computer protocols that verify, simplify, and enforce the performance or negotiation of a contract or eliminate the unforeseen clauses in the contract [6]. Smart contracts comprise several transactions taking place between verified parties; they usually vary widely in scale and complexity and are executed by computer codes [1]. According to Sklaroff [7], a transition from human-language contracts to technology-based system contracts creates new disorganizations. These issues arise from the following features of smart contracts: (1) automation, which requires all agreements be formed by fully-defined terms; (2) decentralization, which requires the verification of job performance by third parties; and (3) anonymity, which reduces the dependency of the contract on the commercial context it is being used. As a result, a semi-automated system is the likely outcome in the short and medium term [8]. In order to encode the parameters of a contract by a programmer, smart contracts commonly omit the necessity of administrative staff and expenses. By automating the execution of the contract, it can be interpreted that smart contracts are legal self-help agreements outside the obligation of the law. Subsequently, computer codes are used to write them, which are not legal languages of contract law.

The ultimate mission of smart contracts is to replace traditional contracts with a new method of technological contract. Through this process, smart contracts enable companies to operate with significantly fewer disputes and need for law enforcement or court intervention. They can even negotiate with other parties autonomously (or other parties' smart contracts) [7]. The AEC industry needs to adapt to the digital age effectively in order to improve its productivity. One of the most beneficial aspects of using smart contracts is that understanding the contract and its implications would be much easier than reading the common contractual language. Furthermore, it is not necessary to know how the smart contract works in order to use and trust it. The users of smart contracts are not supposed to know the coding structure and complicated algorithms needed for it to work. An excellent example of this advancement in daily life is a car which can be used without knowing the details of how its internal parts work [9].

According to Lynch [10], the construction industry has several issues regarding its reputation, such as a lack of trust, being unfair to subcontractors, and a slow rate of change in comparison with other industries. Due to the lack of flexibility in smart contracts, a significant challenge will remain in the technology's scalability [7]. Also, because of the nature of construction projects regarding existing numerous uncertainties in each project, smart contracts in their current state are not perceived as suitable for complex contracts [11]. Smart contracts for construction purposes have been considered concerning various components including the automation of contract, the execution, programming, payment, and certification [12, 13]. According to Fox [12], by using smart contracts, the scope, type, and size of a contracts' dispute will be significantly reduced. By using smart contracts, the dependency on intermediaries such as lawyers and cost estimators would be decreased, which will result in both time and cost savings.

4   BIM and Blockchain

As a digital model that integrates many sources, Building Information Modeling (BIM) provides an exchange platform of information that allows different disciplines to collaborate and solve design and construction issues. To protect the digital intellectual properties of different stakeholders, legal information management is critical in case of disputes and litigation among the numerous and fine-grained contributors involved in the collaboration process [14]. It is widely accepted that the legal uncertainties and model ownership management are the main barriers to BIM adoption [15]. The implementation of BIM needs more trust among different stakeholders in the design and construction industry [16].

According to Barnett [17], the main areas of research on the use of blockchain in construction include: designing a common data environment, automated dispute resolution, and automated regulation and compliance. The strength of blockchain is to solve the issues of trust, which has the potential to bridge the barriers in BIM adoption [2]. Ramage [18] points out that the centralized ledger of standard forms of a contract results in an expensive and time-consuming process. Mason [9] proposes to build intelligent contracts between different participants and to develop a trust-based model to enhance the legal reliability of BIM. The author believes that as a distributed ledger technology, blockchain can be used as the basic infrastructure that helps smart contracts be executed without the common payment systems such as banks [9].

Previous studies cite several pieces of evidence in both academia and industry that support blockchain integration with BIM [3, 19]. Mathews [16] believes that since blockchain is an efficient way to solve the problem of trust, its combination with BIM allows innovation to evolve and challenge the boundaries, and hierarchies of the industry. Turk and Klinc [2] point out that blockchain has the potential to address some of the issues that discourage the design and construction industry from using BIM, including confidentiality, provenance tracking, disintermediation, non-repudiation, multiparty aggregation, traceable inter-organizational recordkeeping, change tracing, data ownership, etc. Blockchain integration with BIM and other rapidly advancing technologies will push the industry towards leaner procurement systems with improved collaboration. As a result, there will be more control and transparency regarding the cost and duration of the projects by eliminating intermediate parties, which also reduces the overall cost of projects. Mathews et al. [20] hold the view that BIM and now blockchain offer real solutions in pushing forward the oncoming digital transformation. The authors also express that the implementation of blockchain will impact the building and construction industries just as it did the finance, insurance, health, and education sectors [20].

5 Smart Contract Implementation Model

Public blockchain networks that require every node to execute every transaction and maintain the ledger while controlling the consensus have their own limitations. For instance, they cannot scale properly, and they cannot support private and confidential transactions and contracts. Due to the nature of projects in construction and the importance of transaction confidentiality, it is more reasonable to look for private rather than public permission based blockchain solution in the AEC industry. In this study, a private, permission based blockchain, using Hyperledger fabric [21] is deployed, which is modular, scalable, and secure. In contrast to the common blockchain networks such as Bitcoin, where the network members are secret, and transactions are public, in the framework proposed and tested in this research the network members are known, but the transactions are secret (unless you have permission to view it). In other words, the network is private, and the membership is controlled. As a result, the business partners know who they are dealing with, but the detail of each transaction is known only to the parties involved in that transaction.

The literature reviewed in the smart contract section shows that the perception of smart contracts is a hard coded computer program that executes the contract clauses. In the implementation used in this study, the smart contract is basically the logic governing the network that is encoded within itself; it also relies on peers for verification. In other words, the method used here is to some extent can be considered as a semi-automated contract where the execution and verification of obligations are done by peers inside the network, but the blockchain network controls the consequence of their actions upon verification. The advantage of such a system is that in addition to being flexible and versatile enough to scale to any construction project, it has consequence-based autonomy and tamper proof record keeping which are a significant improvement to the current state of practice in the AEC industry. It is also worth mentioning that in adopting smart contract and blockchain in industrial applications, there is no need to use the cryptocurrency aspect of the blockchain, and monetary compensations can be executed through traditional channels such as electronic deposits. Figure 1 shows the network structure proposed and tested in this study where seven participant types are defined, namely, client, architect/engineer, General Contractor (GC), regulators, inspectors, suppliers, and sub-contractors. More than one peer is assigned the role of sub-contractor, regulator, supplier, and inspector. The types of transaction requests and endorsement ability (verification of other transactions) that each peer has is different based on their role in the network. For instance, the GC can request a transaction for the work executed, but only the inspectors can endorse and approve that transaction.

Figure 2 shows the sample BIM model used in this study and highlights a column on the third floor of the building under consideration. By assigning a unique ID to each

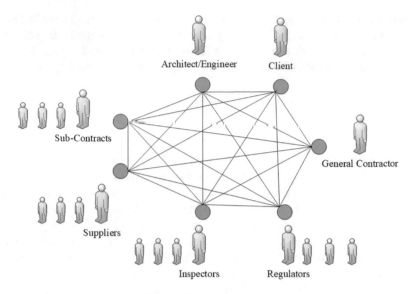

Fig. 1. Network structure.

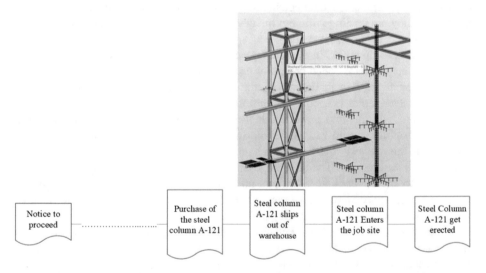

Fig. 2. BIM structure and the corresponding blockchain of the column A-121

element as their identifiers, the BIM model is tied to the physical world by the blockchain. The highlighted column is IDed as A-121. The linked list of blocks are also presented in Fig. 2, where each block describes a transaction that is already endorsed by the appropriate parties and now is part of the blockchain. The workflow process starts with the client requesting the notice to proceed transaction for the project and upon endorsement by the GC, and appropriate regulatory peers, the first block is created. The blockchain continues to grow as the project progresses. To better understand the process, the blocks related to the column A-121 will be discussed in more detail. The process starts with the Subcontractor/GC requesting a transaction for purchasing the column. Upon endorsement by the supplier, that block is created. The next transaction would be requested by the supplier when the column is shipped and endorsed by its peers. Then, upon delivery of the material to the job site, the supplier requests another transaction that relates to the delivery of the column and request for the payment. Upon the endorsement of this transaction by the GC, the fund would be transferred to the supplier, and also the appropriate funds would be transferred from the client to the GC for the material on site. The GC requests the last transaction in this case upon the erection of the column, which then needs to be endorsed by the inspector.

The use of smart contracts in the construction industry and especially implementing it by blockchain is a novel approach. In a recent study, Lou et al. [22] discussed automating construction payments through a smart contract based blockchain framework. While both this study and Lou et al. [22] uses permission-based blockchain to facilitate smart contract enforcement and automate the payment process, this study departs in integrating BIM model into the contract and create a cyber-physical space for administrating the project through the blockchain network.

6 Conclusion

The smart contract implementation model discussed in this paper is generic and versatile in a manner that can be applied to any construction project with minimal adjustments, such as updating the payment amounts. The framework can be easily matched to the structure of any project by adjusting the parties involved, type of transactions and endorsements, and the value of each transaction. However, the structure of the blockchain network designed in this research reflects a design-bid-build delivery system at the construction stage. Further research is needed to adopt blockchain technology to other procurement methods and also possibly to extend its application to the design and operation stages of the building lifecycle. Also, the smart contract implementation and consensus system used in this study are by no means optimal, and it is only adopted as a starting point to show the feasibility of the approach. Further work is needed to identify the optimal network structure and configuration that matches the characteristics of the AEC industry. This paper proposed a conceptual framework with a limited empirical test to test its viability. More extensive empirical tests and refinement is needed to polish the framework and make it operational. The structure of the network in this study is designed from the perspective that the project is an organization and all the parties involved play their part as a node in that organization. The Hyperledger fabric allows multiple organizations to operate in a network or multiple networks to interact with each other. As a result, one direction for future research would be modeling the construction process where each party is defined as an organization with its own network and the project workflow is modeled through the interaction between the different blockchain networks. Overall, it is shown how blockchain can be used in a semi-automated to implement smart contracts in a construction process. In a way, it can be used to link the physical and digital world (BIM

Model) to maintain and control the cyber-physical space. Also, more automation can be achieved by integrating IoT, RFID, and other sensory technologies where departure, arrival, and inspection of the material, and workmanship can be streamlined.

References

1. Lamb, K.: Blockchain and Smart Contracts: What the AEC sector needs to know (2018)
2. Turk, Ž., Klinc, R.: Potentials of blockchain technology for construction management. Procedia Eng. **196**(June), 638–645 (2017)
3. Shojaei, A.: Exploring applications of blockchain technology in the construction industry. In: Interdependence Between Structural Engineering and Construction Management (2019)
4. Lia, J., Greenwood, D., Kassem, M.: Blockchain in the built environment: analysing current applications and developing an emergent framework. In: Proceedings of the Creative Construction Conference, pp. 59–66 (2018)
5. Johansson, J., Nilsson, C.: How the blockchain technology can enhance sustainability for contractors within the construction industry (2018)
6. Meitinger, T.H.: Smart contracts. Informatik-Spektrum **40**(4), 371–375 (2017)
7. Sklaroff, J.M.: Smart contracts and the cost of inflexibility. U. Pa. L. Rev. **166**, 263 (2017)
8. Mason, J.: Smart contracts in construction: views and perceptions of stakeholders. In: Proceedings of FIG Conference, May 2018
9. Mason, J.: Intelligent contracts and the construction industry. J. Leg. Aff. Disput. Resolut. Eng. Constr. **9**(3), 04517012 (2017)
10. Lynch, P.: HGCRA: re-addressing the balance of power between main contractors and subcontractors. Msc, National Academy for Dispute Resolution (2011)
11. Gabert, H., Grönlund, H.: Blockchain and smart contracts in the Swedish construction industry (2018)
12. Fox, S.: Why construction needs smart contracts | NBS. https://www.thenbs.com/knowledge/why-construction-needs-smart-contracts. Accessed 05 Jul 2019
13. Ahmadisheykhsarmast, S., Sonmez, R.: Smart contracts in construction industry. In: 5th International Project & Construction Management Conference, pp. 767–774 (2018)
14. Thomas, L.W.: Legal issues surrounding the use of digital intellectual property on design and construction projects, no. 58 (2013)
15. Redmond, A., Hore, A., Alshawi, M., West, R.: Exploring how information exchanges can be enhanced through Cloud BIM. Autom. Constr. **24**, 175–183 (2012)
16. Mathews, M.: Building information modeling technology and blockchain – IEBC. https://iebc.co/bim-and-blockchain/. Accessed 05 Jul 2019
17. Barnett, J.: Blockchain for BIM | Smart Contracts Lawyer. http://www.jeremybarnett.co.uk/blochain-for-bim-smart-contracts-lawyer. Accessed 05 Jul 2019
18. Ramage, M.: BIM and blockchain. in construction. https://constructible.trimble.com/construction-industry/from-bim-to-blockchain-in-construction-what-you-need-to-know. Accessed 05 Jul 2019
19. Wang, J., Wu, P., Wang, X., Shou, W.: The outlook of blockchain technology for construction engineering management. Front. Eng. Manag. **4**(1), 67 (2017)

20. Mathews, M., Robles, D., Bowe, B.: BIM+Blockchain: a solution to the trust problem in collaboration? In: BIM+Blockchain: A Solution to the Trust Problem in Collaboration?, p. 11 (2017)
21. Androulaki, E., et al.: Hyperledger fabric: a distributed operating system for permissioned blockchains. In: Proceedings of the Thirteenth EuroSys Conference, p. 30 (2018)
22. Luo, H., Das, M., Wang, J., Cheng, J.C.P.: Construction Payment Automation through Smart Contract-based Blockchain Framework (2019)

Are Esports the Next Big Varsity Draw?
An Exploratory Study of 522 University
e-Athletes

Thierry Karsenti[(✉)]

University of Montreal, Montreal, Canada
thierry.karsenti@umontreal.ca

Abstract. The spectacular rise of esports has persuaded the Paris 2024 bid team to consider adding them to the lineup. Esports are winning social and professional recognition and gaining a foothold at universities. This article reviews the fledging research on esports and presents the conclusions of a study on the esports practices of 522 university e-athletes. The four-fold aim was to: (1) determine their training methods and conditions, (2) describe their physical training routines, (3) describe their e-sport practices and experiences, and (4) better assess the potential for esports to achieve recognition as legitimate sports. Relationships between esports and academia were also examined. The originality of this study lies in the investigation of a forward-thinking research field that is poised to soar.

Keywords: Electronic sports · Esports · Competitive sports · Professional gaming · Professional gamers · e-Athlete

1 Introduction

On October 19, 1972, almost half a century ago, and weeks before the first Pong console was released to the public, the world's first esports event was held at the Stanford Artificial Intelligence Lab in Los Altos, California. As immortalized in the magazine *Rolling Stone*, the winner was named the Intergalactic Spacewar Champion for that year. Since then the video gaming culture has spread far and wide and has even crashed the gates of schools and universities. Researchers have long agreed that children develop and learn best when they explore, discover, and play [7, 8, 17, 34]. In a similar vein, video games for classroom teaching and extracurricular activities have shown positive effects on learning [5, 21, 31].

As we near the close of the second decade of the third millennium, schools are experimenting with esports as an innovative approach to collaborative learning. Video gamers are going for the gold at organized varsity tournaments. Rival teams battle on popular games like Fortnite, League of Legends, Counter-Strike, Call of Duty, Overwatch, and Madden NFL. And just like other athletes, these e-athletes draw live audiences in the thousands and maintain an international following of millions of fans. This esports boom is due in large part to the availability of streaming platforms such as Twitch (http://twitch.tv).

© Springer Nature Switzerland AG 2020
K. Arai et al. (Eds.): FTC 2019, AISC 1070, pp. 528–541, 2020.
https://doi.org/10.1007/978-3-030-32523-7_37

In South Korea, which enjoys a sophisticated fiber-optic infrastructure, professional videogaming is a national pastime [32]. In a culture where Koreans watch 24-h esports on TV, gaming is regarded as a valuable social activity. In North America, Robert Morris University blazed a trail in 2014 when it introduced esports as a scholarship-funded sport. Currently, over 50 North American universities and colleges offer competitive esports programs. Like other athletes, esports teams receive financial as well as nutritional and psychological support. And now Paris is "deep in talks" about including esports as a demonstration sport at the 2024 Olympics (BBC SPORT, April 2018).[1]

Still, there is something about this new craze that seems unnatural, and questions spring to mind. One worry for educators is that esports are carving up more and more space out of university life. Another concern is that esports might not actually be a true sport, so e-athletes might not be true athletes. According to Parry [33], esports are not "*an institutionalized, rule-governed contest of human physical skill,*" even though some countries recognize them as such [31]. Regardless, as varsity e-sports teams appear more frequently on the world stage, the impact of esports on university students should be systematically investigated and clarified.

We begin by presenting the results of a study of 522 international e-athletes attending university. Our four main study objectives were to: (1) determine their training methods and conditions (e.g., hours of coaching), (2) describe their physical training routines, (3) describe their e-sports practices (e.g., gaming conditions, collaboration, and competition), and (4) better assess the potential for esports to be recognized as a legitimate sport. The originality of this study lies in the investigation of a forward-thinking research field that is poised to soar.

2 Review of the Literature on Esports

The impassioned debate on esports has waxed furious since the possibility of demonstration sport status at the 2024 Paris Olympics. But first, what exactly are esports, and who plays them? We reviewed the literature to gain an understanding of what esports are and how e-athletes practice and experience them.

2.1 Are e-Sports like Regular Sports, but in a Different Form?

We're talking about a very popular pastime that has entered the mainstream [40]. The numbers don't lie: 1.8 billion people worldwide play video games, and over 100 million of them compete in esports tournaments (Pinault 2018). No doubt, this is what inspired Paris, host of the 2024 Summer Olympic Games, to discuss the possibility of introducing esports as a demonstration sport [20]. Note also that "*competitive video gaming will be a medal event at the 2022 Asian Games*" (BBC SPORT, April 2018). (see Footnote 1) Happily for esports, they have the wind at their back. Professional esports leagues are springing up all over [25], (see Footnote 1) and dedicated TV channels

---

[1] BBC Sport: https://www.bbc.com/sport/olympics/43893891.

broadcast tournaments 24/7. This enthusiastic reception presents a puzzling contrast with the real difficulties that some other sports have had earning legitimacy, such as extreme sports. Even more mind-boggling is the fact that high-profile companies like Orange (France)[2] and Adidas are underwriting esports, leaving more traditional sports like soccer in the shade. But the thorniest question of all remains: are esports "true" sports? Or are they just amusing virtual diversions that have nothing in common with physical sports? Even worse, could they simply be vehicles for product placement or cunning marketing ploys [35]? Xue, Pu, Hawzen, and Newman [44] respond to these questions with the following definition of esports: *"any organized multi-player video game competition, where individuals and teams assemble in stadia and arenas to compete in sanctioned, real-time, broadly streamed, financially incentivized, and widely-attended tournament events."* The term "esports" has found its way into *Dictionary.com* as: *"Noun 1. (usually used with a plural verb) competitive tournaments of video games, especially among professional gamers. Adjective 2. Of or relating to esports."* Nevertheless, Jenny *et al.* [20] argue that, according to historical definitions of sport, *"esports include play and competition, are organized by rules, require skill, and have a broad following. However, esports currently lacks great physicality and institutionalization,"* such that, *"a refinement of the definition of sport [...] will need to occur before esports are totally accepted."* Jenny et al.'s [20] study is one of the rare few to compare esports with traditional sports. Some North American universities consider competitive video games like League of Legends and Halo: Reach as legitimate team sports because, like soccer and hockey players, e-athletes have to react instantaneously to strategic situations. For instance, in Quebec (Canada), the Université de Montréal added an esports team to its roster in 2017, on an equal footing with its hockey, football, swimming, and other sports teams. According to Doran [10], e-athletes have many of the same qualities that traditional athletes possess, such as *"tenacity, critical thinking, teamwork, communication skills, and a constant desire to get better at what you are doing,"* (although similar characteristics may be found in artists, musicians, writers, and other creative types). Jenny *et al.* [20] beg to differ, arguing that, *"the common connection between the terms video games and esports continues to equate esports with gaming—a word which can be perceived to be on a lower level than sport."* Ironically, basketball, hockey, and other such sports are now available as simulated video games, for example FIFA Games and NHL19 [14]. So, even as increasing numbers of universities are setting up esports programs, the question refuses to go away: are esports truly sports? We turn now to another compelling question: what do e-athletes gain from playing esports?

2.2   Learning Through Esports

Many authors, including Martončik and his colleagues [28, 29], feel that esports go far beyond mere entertainment. They should be viewed and conceived as practices that, besides being enjoyable, are above all beneficial for the players, both physically and psychologically. For example, studies have shown that players of popular online video

---

[2] *THE ESPORTS OBSERVER*: https://esportsobserver.com/team-vitality-orange-partnership/.

games experience less social anxiety and solitude in virtual compared to real-world environments. And as in traditional sports, being on a team or participating as a member of a gaming group or community can be psychologically positive. It satisfies the need to belong, and for team leaders, it bestows a feeling of power [29]. Other authors have found associations between, for example, increased cortical thickness in the brain and professional online gaming [18]. Still, gamers are said to play mainly for the social, fun, and performance aspects [12, 15, 19, 30]. In parallel, studies have demonstrated numerous benefits of video gaming, including better communicative, cooperative, and spatial representation skills as well as improved academic competencies in math and other subjects [1, 11]. It turns out that, like other sports, gaming is not only fun, but as a bonus it can help you learn and perform. Therefore, as esports continue to close in on respectability, and with Olympic stature in sight, this emerging form of sport merits careful attention. The benefits (and drawbacks) for e-athletes should be examined, especially when gaming is being embraced by educators. For instance, colleges and universities are championing video gaming teams that compete in intercollegiate tournaments. Studies are beginning to explore the associated benefits (e.g., new types of scholarships, expanded athletic departments, revenue generation, greater student diversity), along with the drawbacks, including health problems (eye fatigue, overuse injuries, lack of fitness, excessive stimulant use) and academic issues (failure to attend athletic practices leading to scholarship loss) [9, 23, 39]. Some schools are trying to address these issues and raise the profile of e-athletes. At Matane College, in Canada, esports teams are required to do physical training and keep up with their schoolwork [37]. Again in Canada, Arvida School a pilot project that combines the usual school subjects with digital literacy, team esports, and physical activity [38]. Designed to help combat social isolation in elementary and high school gamers, the program has become the envy of Canadian schools.

3   Research Question

Although the research on esports is thriving (Fig. 1), the results are generally dispersed among the different game genres that attract distinct subcultures [14]. Some of the categories are "multiplayer online battle arenas," "first-person shooters," "real time strategy," "collectible card games," and "sports games" [13, 14]. These subcultures generally have their own hierarchical leagues, ladders, and tournaments as well as sponsors, just like other types of sports. All this makes it possible to examine the issues from different perspectives, including the tools, players, subcultures, and organizations.

In this article, we add to the research by focusing on the structured practices of e-athletes. Play has long been thought to promote learning [34, 43], and so has digital gaming [1–3]. Various authors have sought to determine similarities and differences between play and gaming [6, 42]. In the present study, we wanted to explore how esports can or could help players learn. The value of esports fluctuates according to the stakeholder: although many people dread the negative impacts of incessant gaming, young people are crazy about gaming, some countries regard it as a respectable pastime, some cultures are obsessed with it, businesses appreciate the profitability, and educators are cautiously hopeful about the learning potential. At a time when certain

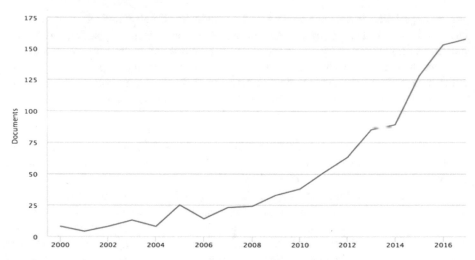

Fig. 1. Graphic representation of publications that address esports (or e-sport, e-sports) in the 2000–2017 Scopus database.

physical sports are also struggling to achieve legitimacy and the influence of video gaming is a contentious topic, we wanted to settle some questions about this new breed of sports. How to account for this seemingly anomalous practice at school? What are the features that set esports apart from regular sports? Do e-athletes train? And if so, how? In teams? Alone? Do they work on mental and physical preparation? And importantly, do they think of esports as a "real" sport, or a legitimate varsity sport? The main objective was to shed scientific light on the status of esports: is it truly a sport in the sense of an organized activity that includes competition, strategies, codes, values, and specific practices?

4 Method

To respond to these questions and to get to the bottom of what esports are and how e-athletes behave, particularly at competitions, we gathered data during the 2018 academic term.

4.1 Participants

A total of 522 e-athletes participated in the study. They resided in Canada (68.44%), France (3.35%), the United States (8.68%), Hungary (10.5%), the United Kingdom (1.58%), and other countries (8.68%). Of the participants who responded to the question on sex, 11% were women and 89% were men. Average age was 20.4 years.

4.2   Data Collection

We adopted an innovative and exhaustive data collection strategy. We used three main data collection methods: (1) an online survey questionnaire (522 respondents), (2) individual semi-directed online interviews with members of esports teams who agreed to be interviewed ($n = 22$), and (3) three online (and face-face) interviews with an esports team ($n = 8$). The team was chosen for its frequent appearances at official competitions and for its high-profile success and recognition in the esports community. In addition, we conducted "think aloud" interviews [22] with the team and its manager during competitions and training sessions. As a complementary method, we videotaped (350 min) their competitions and training sessions to examine the immersion experience.

4.3   Data Treatment and Analysis

We applied a mixed analysis to the questionnaire data, which included responses to Likert-rated items and open questions. The quantitative analysis comprised descriptive statistics using SPSS (version 23) and the online survey tool SurveyMonkey (http://surveymonkey.com). We extended and enriched these initial results with a qualitative analysis of the open responses using QDA Miner [36], a widely used mixed-methods qualitative text analysis software [22]. Thus, we applied a content analysis [24] with semi-open coding in relation to the main research objectives. The individual and group interviews and the videotaped observations were subjected to content analysis as well, according to L'Écuyer [24] and Miles and Huberman [16] and using QDA Miner [36].

4.4   Methodological Strengths and Limitations

The methods used in this study constitute a main strength of this study. The open and closed responses and the individual interviews in the online survey were combined with face-to-face group interviews and observations of an e-athlete team in action. This enhanced the results and enabled triangulation. Nevertheless, this approach has some limitations. Not least was the fact that the results were based on the participants' perceptions. We attempted to compensate by building a large sample ($n = 522$) and by applying a variety of data collection methods. To reduce the potential methodological bias, we systematically compared the responses among respondent types to identify discrepancies as necessary. A further limitation was that the sampling was not random: the objective not being to achieve a representative subsample, we chose to use a convenience sample. For all practical purposes, it would be impossible to capture a predictable sample with an online survey. Another strength of this study lies in the collaboration of several esports stakeholders, including members of university and provincial esports associations and organizers of esports events (e.g., Lan ETS 2018, which drew over 2,000 participants).

5 Main Results

The results provided valuable insights into the behaviors of varsity e-athletes during training and official competitions.

5.1 Experienced Players for Stable and Recognized Practices

The results in Fig. 2 show that the most often played game was Overwatch (35% of players), closely followed by League of Legends (24%).

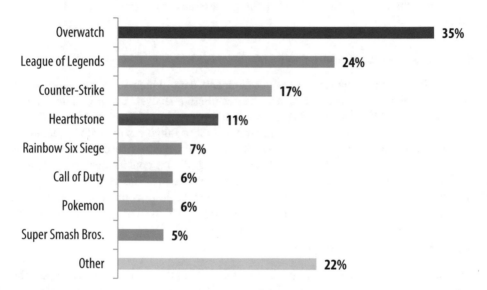

Fig. 2. Distribution of the most played games

Over 70% of the participants said that they had participated in an esports competition, and some had competed for many years. A few (12.04%) had over 10 years of experience, and a non-negligible proportion (23.81%) had from 2 to 4 years of experience. These were not occasional players, which concur with Hamari and Sjöblom (2017), who describe the esports world as a coordinated system of competitive gaming that is structured as leagues, ladders, and tournaments run by sports organizations and often sponsored by businesses. The main attraction for playing esports was the competition (25.63%). Second comes the emotional satisfaction of playing with their friends (20.49%), followed by having fun by themselves or with friends (18.35%). Making a pile of money through competitive gaming was only a minor consideration (8.39%).

Apparently, competitive gaming was also time-consuming (Fig. 3). Between training and competing, the players spent many hours daily facing a screen. Almost 65% spent from 12 to 35 h a week training on their own. In addition, the majority

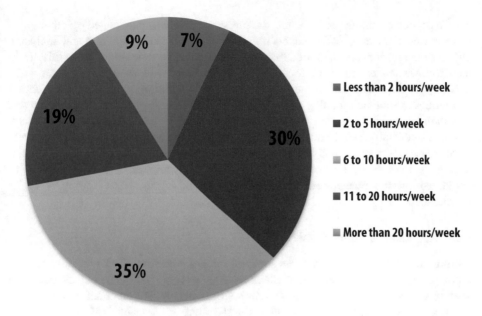

Fig. 3. Team training per week

(83%) spent from 2 to 30 h a week training with their team. Overall, more time was spent on team than individual training, highlighting the collective nature of esports.

These results were confirmed by the individual interviews. One team member said that in order to prepare for competitions, it was necessary to train at least five hours per night, five days in a row. Training was of primary importance for these e-athletes. One player recounted attending a training camp with 5-day, 10-h per day sessions.

Figure 4 shows the types of training done in teams, revealing two major differences with the individual training. The first is in terms of strategy, which was strongly emphasized in the team sessions. Communication was also stressed, whereas dexterity (repeated movements without adversity) received less attention.

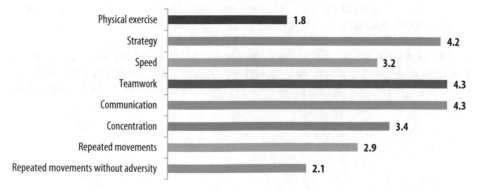

Fig. 4. Content of team training sessions

To optimize gaming performance, the physical grind was punctuated by other types of exercises. For example, the players honed their mental skills (thinking and analysis), although the respondents admitted that not everybody was scrupulous about this. They also did written exercises to focus on objectives, as well as discussion exercises designed to take them out of their comfort zone. In addition, they used visualization techniques to spot their gaming strengths and weaknesses. A problem for many players was their lack of ability to fully focus on key aspects of the task. To offset this problem, e-athletes generally look to other team members or staff for support, and many players have trainers, therapists, analysts, coaches, or even managers (e.g., [32]). The manager acts as a middleman between the coach, staff, players, and the organization, as explained by the manager of the esports team at an interview during a major tournament.

5.2   Are Esports Educational?

During the individual training sessions, a large proportion (33%) of the players reported that what seemed most effective was working on technical aspects and spending lots of time practicing. In other words, the most important thing was to practice the actual gaming moves. Interestingly, although over half (56%) the players did without external aids (e.g., check lists, stopwatches, timers) during their games, only about a third (29%) never consulted some kind of external adviser. Notably, the majority (73%) of players used mental and technical preparation strategies before competitions, such as a match game plan or player position assignments. In this sense, esports matches resemble other sports matches. For example, soccer players prepare in similar ways. The above-mentioned manager weighed in on this:

It's really important for the players to work on their mindset. They have to be bullet-proof. A player who thinks he's at the top of his game stops learning. The key is to learn constantly. Even the world's best players train all the time. They want to learn and find ways to improve [...]. The players do regular physical training. They exercise, because in order to do intense intellectual gaming, they've got to be in shape.[3]

According to this manager, the players prepared physically for competitive events. However, the players were not so sure. They stressed the social side: the teams ate together and spent lots of time training together, talking things over, and waiting for the tournament venue to fill up. In short, the team was a highly social community of e-athletes who happened to be competing in an educational setting.

As esports continue to gain ground and recognition, it becomes more important to examine their impact on the players. And now that universities are latching on to esports, it is essential to assess their educational value. In this sense, when the e-athletes were asked what they took away from esports (Fig. 5), almost half (42%) said that they

---

[3] Translated from the French transcript.

learned about teamwork. Some improved in terms of perseverance (29%), personal discipline (12%), and cognitive and intellectual skills (11%). A few also mentioned other benefits: better physical skills (7%), better tactical and strategic skills (6%), ethical awareness (4%), enjoyment (3%), autonomy (1%), and self-knowledge (1%).

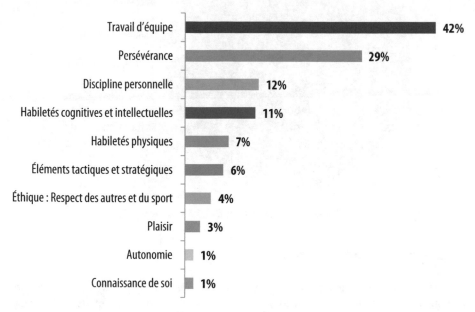

Fig. 5. Benefits of esports

As to the question of whether or not esports are a "true" sport, not to mention a potential Olympic demonstration sport, the manager of the team that we interviewed offered this comment: if the Olympics can feature games that aren't very physically demanding (curling is a case in point), he didn't see why they couldn't include esports too, because after all, what is a sport, really, other than a rule-based game where the players use strategies? (see Footnote 3) An e-athlete added that, even if one is reluctant to think of esports as a true sport, because there's no actual "action," they're highly strategic and extremely complicated. (see Footnote 3) Another team manager pointed out that although the players don't need all that much strength or physical agility, they need finely honed reflexes, hand-eye coordination, and a lot of other talent to perform well. (see Footnote 3) Notwithstanding, recognition of esports seems to be an uphill battle. The players themselves did not see eye to eye on this point: 48% of our respondents were not convinced that esports were a true sport (Fig. 6), and almost 20% felt that it probably wasn't a true sport.

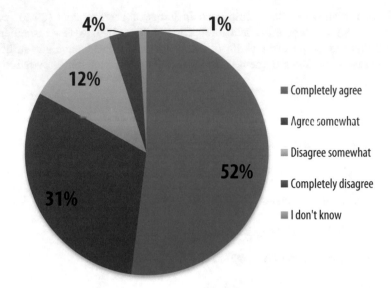

Fig. 6. Percentages of e-athletes who considered esports to be a true sport.

6   Discussion and Conclusion

Esports are wildly popular. As official competitions multiply, esports are gaining acceptance as a "true" sport, with possible demonstration status at the 2024 Paris Olympics [26, 27]. At the same time, universities are recruiting a new breed of students to compete in varsity esports tournaments. This rapid upswing has been facilitated by the tremendous advances made in mobile technology [41]. Gaming has clambered up the social rungs and is well within reach of status as a legitimate communal pastime [4]. Meanwhile, the production of educational videos, also referred to as serious play or serious games, has emerged as a major cultural industry. Are esports poised to be the next big varsity draw? In an attempt to answer this question, we asked 522 university e-athletes and some esports team coaches about their training practices, and we made comparisons with traditional, more physically demanding sports.

Our aim was not to demonstrate that esports are, or should be, officially recognized as legitimate varsity sports. Instead, we wanted to get a better sense of esports as a discipline, including training routines and the tournament experience. The main take-away is that e-athletes are dedicated players who train just as hard as any other athletes do. They strive to excel at an activity that is, like other sports, governed by standards and organizations, has coaches and managers, and emphasizes teamwork. Of the survey respondents, only 20% did not consider esports to be a "true" sport. In future studies, it would be informative to measure the physiological impact of esports practices on e-athletes, for example, in terms of heart rate and caloric consumption. This would enable useful comparisons with more traditional sports.

The effects on emotional and cognitive engagement could also be assessed, to allow estimating the educational effect and the value of esports as a university activity. Electroencephalography (EEG) studies using mobile headsets (e.g., Emotiv)[4] could also be conducted to compare competitive gaming with traditional, standardized sports activities.

For esports to gain wider acceptance and legitimacy, however, the players themselves need to be on board. The finding that not all our surveyed players were convinced that esports were true sports merits further investigation.

References

1. Amato, E.A.: Les utilités du jeu vidéo sérieux: finalités, discours et mises en corrélation. La revue canadienne de l'apprentissage et de la technologie **37**(2) (2011). https://doi.org/10.21432/T25C70

2. Berry, V.: Loisirs numériques et communautés virtuelles: des espaces d'apprentissage? In: Brougère, G., Ulmann, A.-L. (dir.) Apprendre de la vie quotidienne, Chapitre 11, pp. 143–153. Presses Universitaires de France, Paris (2009). https://doi.org/10.3917/puf.broug.2009.01.0143

3. Berry, V.: Jouer pour apprendre: est-ce bien sérieux? Réflexions théoriques sur les relations entre jeu (vidéo) et apprentissage. La revue canadienne de l'apprentissage et de la technologie, **37**(2) (2011). https://doi.org/10.21432/T2959X

4. Bouchard, L.: Jeux vidéo aux Musées de la civilisation: témoins du changement social. THEMA. La revue des Musées de la civilisation (2), 128–134 (2015). http://thema.mcq.org

5. Bugmann, J.: Apprendre en jouant: du jeu sérieux au socle commun de connaissances et de compétences. Thèse de doctorat, Université de Cergy-Pontoise, France (2016). l'archive TeL. http://tel.archives-ouvertes.fr

6. Buzy-Christmann, D., Filippo, L.D., Goria, S., Thévenot, P.: Correspondances et contrastes entre jeux traditionnels et jeux numériques. Sciences du jeu (5) (2016). https://doi.org/10.4000/sdj.547

7. Caillois, R.: Les Jeux et les hommes: le masque et le vertige. Gallimard, Paris (1958)

8. Dewey, J.: Experience and Education. Collier, New York (1963)

9. DiFrancisco-Donoghue, J., Balentine, J.R.: Collegiate eSport: where do we fit in? Curr. Sports Med. Reports **17**(4), 117–118 (2018). https://doi.org/10.1249/JSR.0000000000000477

10. Doran, L.: How "eSports" is changing the college sports scene, 27 février 2017. InsideSources. http://insidesources.com

11. Ferguson, C.J., Garza, A., Jerabeck, J., Ramos, R., Galindo, M.: Not worth the fuss after all? Cross-sectional and prospective data on violent video game influences on aggression, visuospatial cognition and mathematics ability in a sample of youth. J. Youth Adolesc. **42**(1), 109–122 (2013). https://doi.org/10.1007/s10964-012-9803-6

12. Frostling-Henningsson, M.: First-person shooter games as a way of connecting to people: "Brothers in Blood". CyberPsychol. Behav. **12**(5), 557–562 (2009). https://doi.org/10.1089/cpb.2008.0345

---

[4] http://emotiv.com.

13. Funk, D.C., Pizzo, A.D., Baker, B.J.: eSport management: embracing eSport education and research opportunities. Sport Manage. Rev. **21**(1), 7–13 (2018). https://doi.org/10.1016/j.smr.2017.07.008

14. Hamari, J., Sjöblom, M.: What is eSports and why do people watch it? (SSRN Scholarly Paper No. ID 2686182) (2017). Social Science Research Network. http://ssrn.com

15. Hobler, M.: Shoot first, ask questions later: motivations of a women's gaming clan. In: Presentation at the Annual Meeting of the International Communication Association, San Francisco, CA, May 2007. http://www.allacademic.com

16. Miles, M.B., Huberman, A.M.: Analyse des données qualitatives. De Boeck Supérieur, Bruxelles (2003)

17. Huizinga, J.: Homo Ludens: A Study of the Play-Element in Culture. Beacon, Boston (1971)

18. Hyun, G.J., Shin, Y.W., Kim, B.-N., Cheong, J.H., Jin, S.N., Han, D.H.: Increased cortical thickness in professional on-line gamers. Psychiatry Inv. **10**(4), 388–392 (2013). https://doi.org/10.4306/pi.2013.10.4.388

19. Jansz, J., Martens, L.: Gaming at a LAN event: the social context of playing video games. New Media Soc. **7**(3), 333–355 (2005). https://doi.org/10.1177/1461444805052280

20. Jenny, S.E., Manning, R.D., Keiper, M.C., Olrich, T.W.: Virtual(ly) athletes: where eSports fit within the definition of "sport"? Quest **69**(1), 1–18 (2017). https://doi.org/10.1080/00336297.2016.1144517

21. Joly-Lavoie, A., Yelle, F.: Le jeu vidéo pour enseigner l'histoire: synthèse d'une approche théorique et pratique. TRACES **3**(54), 19–24 (2016). LeDidacticien. http://ledidacticien.com

22. Karsenti, T., Savoie-Zajc, L.: La recherche en éducation: étapes et approches, 4th edn. Presses de l'Université de Montréal, Montréal (2018)

23. Keiper, M.C., Manning, R.D., Jenny, S., Olrich, T., Croft, C.: No reason to LoL at LoL: the addition of esports to intercollegiate athletic departments. J. Study Sports Athletes Educ. **11**(2), 143–160 (2017). https://doi.org/10.1080/19357397.2017.1316001

24. L'Écuyer, R.: Méthodologie de l'analyse développementale de contenu. Méthode GPS et concept de soi. Presses de l'Université du Québec, Sainte-Foy (1990)

25. Le Monde: Paris se dit ouvert à l'idée d'inclure l'e-sport dans les Jeux olympiques de 2024, 10 août 2017a. http://lemonde.fr

26. Le Monde: L'e-sport va-t-il devenir une discipline olympique?, 7 novembre 2017b. http://lemonde.fr

27. Le Monde: L'e-sport à grandes foulées, 30 janvier 2018. http://lemonde.fr

28. Martončik, M.: e-Sports: playing just for fun or playing to satisfy life goals? Comput. Hum. Behav. **48**, 208–211 (2015). https://doi.org/10.1016/j.chb.2015.01.056

29. Martončik, M., Lokša, J.: Do World of Warcraft (MMORPG) players experience less loneliness and social anxiety in online world (virtual environment) than in real world (offline)? Comput. Hum. Behav. **56**(C), 127–134 (2016). https://doi.org/10.1016/j.chb.2015.11.035

30. Müller-Lietzkow, J.: Sport im Jahr 2050: E-sport! Oder: Ist e-sport sport? Merz Wissenschaft 50 Jg 6(102–112) (2006)

31. Natkin, S.: Du ludo-éducatif aux jeux vidéo éducatifs. Les dossiers de l'ingénierie éducative (65), 12–15 (2009). http://www2.cndp.fr/DossiersIE

32. Paberz, C.: Le jeu vidéo comme sport en Corée du Sud? Hermès, La Revue **1**(62), 48–51 (2012). http://www.cairn.info/revue-hermes-la-revue.htm

33. Parry, J.: E-sports are not sports. Sport Ethics Philos. (2018). https://doi.org/10.1080/17511321.2018.1489419

34. Piaget, J.: La formation du symbole chez l'enfant – imitation, jeu et rêve – image et représentation, 2nd edn. Delachaux et Niestlé, Neuchâtel (1959)

35. Pinault, A.: Éducation peu physique. Médium **2**(55), 57–73 (2018). https://doi.org/10.3917/mediu.055.0057
36. QDA Miner (version 5). Provalis Research. http://provalisresearch.com
37. Radio-Canada: Le sport électronique, là pour rester au Cégep de Matane, 30 avril 2017a. http://ici.radio-canada.ca
38. Radio-Canada: Le sport électronique comme programme scolaire à Arvida, 8 mai 2017b. http://ici.radio-canada.ca
39. Schaeperkoetter, C.C., Mays, J., Hyland, S.T., Wilkerson, Z., Oja, B., Krueger, K., Christian, R., Bass, J.R.: The "new" student-athlete: an exploratory examination of scholarship eSports players. J. Intercollegiate Sport **10**(1), 1–21 (2017). https://doi.org/10.1123/jis.2016-0011
40. St-Pierre, R.: Des jeux vidéo pour l'apprentissage? Facteurs de motivation et de jouabilité issus du game design. DistanceS **1**(12), 4–26 (2010). http://distances.teluq.ca
41. Ter Minassian, H., Boutet, M.: Les jeux vidéo dans les routines quotidiennes. Espace populations sociétés **2015**(1–2) (2015). https://doi.org/10.4000/eps.5989
42. Virole, B.: Du bon usage des jeux vidéo. Enfances Psy. **1**(26), 67–72 (2005). https://doi.org/10.3917/ep.026.0067
43. Winnicott, D.W.: Jeu et réalité: l'espace potentiel (Monod, C., Pontalis, J.B. (trad.); préf. de J.-B. Pontalis). Gallimard, Paris, France (1975)
44. Xue, H., Pu, H., Hawzen, M., Newman, J.: E-sports management? Institutional logics, professional sports, emerging E-sports field. In: 20-Minute Oral Presentation at the 31st North American Society for Sport Management Conference, Orlando, FL, USA (2016)

Using Maximum Weighted Cliques in the Detection of Sub-communities' Behaviors in OSNs

Izzat Alsmadi[1(✉)] and Chuck Easttom[2]

[1] Texas A&M University, San Antonio, TX 78224, USA
ialsmadi@tamusa.edu
[2] Capitol Technology University,
11301 Springfield Road, Laurel, MD 20708, USA
chuck@chuckeasttom.com

Abstract. The extraction of clique relations in social networks can show information related to groups and how those groups are formed or interact with each other. In one pending application, we are showing how such "detection of abnormal cliques' behaviors" can be possibly used to detect sub-communities' behaviors based on information from Online Social Networks (OSNs). In social networks, a clique represents a group of people where every member in the clique is a friend with every other member in the clique. Those cliques can have containment relation with each other where large cliques can contain small size cliques. This is why most algorithms in this scope focus on finding the maximum clique. In our approach, we evaluated adding the weight factor to clique algorithm to show more insights about the level of involvement of users in the clique. In this regard clique activities are not like those in group discussions where an activity is posted by one user and is visible by all others. Our algorithm calculates the overall weight of the clique based on individual edges. Users post frequent activities. Their clique members, just like other friends, may or may not interact with all those activities.

Keywords: Security in online social networks · OSN groups detection · Weighted cliques · Groups collaboration

1 Introduction

There are numerous applications for tracking and measuring relationships in Online Social Networks (OSN's), [26, 28]. One application is to determine relationships that affect marketing, [5]. Another application is to study the influence of emotionally charged information such as social or political memes [23]. However, social network relationships are also important for investigative applications, [14]. Whether the investigating agency is a law enforcement agency or an intelligence gathering agency the needs for an investigative application are identical. Counter-terrorism is one area of investigation in which analysis of social media interactions can be very important and produce significant, actionable intelligence [10, 21, 22, 25].

© Springer Nature Switzerland AG 2020
K. Arai et al. (Eds.): FTC 2019, AISC 1070, pp. 542–551, 2020.
https://doi.org/10.1007/978-3-030-32523-7_38

There are diverse approaches to analyzing social media interactions, regardless of the investigative intent [3, 31]. Some depend on large scale data scaping and mining in order to parse data. Other methodologies involve statistical analysis of the interactions. A common means of analyzing social media is the application of graph theory [27, 33].

The approach we used was to focus on OSN cliques. In the social sciences a clique is some group that has interaction, and some common interest [31]. That interest can be social, political, economic, in fact any interest that brings together individuals such that they have some periodic, but regular interaction, can form a clique. The extraction of clique relations in social networks can show information related to groups and how those groups are formed or interact with each other. In one pending application, we are showing how such "detection of abnormal cliques' behaviors" can be possibly used to detect terrorists' attacks based on information from OSNs.

In social networks, a clique represents a group of people where every member in the clique is a friend with every other member in the clique. Several recent contributions tried to extract knowledge related to cliques in OSNs (e.g. [1, 2, 6, 11, 12, 17, 18, 20]).

1.1 Maximum Cliques, Weighted Cliques and Maximum Weighted Cliques

A clique, C, (in an undirected graph G of (V, E): Vertices and Edges), is a subset of the vertices where between each two vertices, there is an edge (i.e. a relation). In the scope of groups collaborations, $\omega(G)$, is the size of a largest clique or maximal clique of G. For an arbitrary graph:

$$\omega(G) \geq \sum_{i=1}^{n} \frac{1}{n - d_i}$$

where d_i is the degree of a graph vertex i

Several papers indicated that maximum cliques' problem is an NP-complete problem (e.g. [7, 8, 15, 16, 28, 29] etc.).

A weighted clique is the total weight of weighted maximum clique [24]. Based on this definition, calculating weighted clique takes two consecutive steps:

For a particular graph, calculate maximum clique in that graph, and then:

1. For that particular clique, calculate the overall weight of the edges in the clique.

In this formula, while for a particular graph, the maximum clique is dynamic and may change from one time to another, however, the weighted clique will always be tied to the maximum clique.

We introduced (Maximum weighted clique) to be always dynamic and not tied to the maximum clique. This means that for a particular graph, the maximum clique and the maximum weighted clique may refer to different subsets in the same graph. In order to show the difference, assume Fig. 1 below indicates a hypothetical graph subset with

3 cliques (1,2,4,5,7,11), (1,5,6,8), and (3,9,10). In this graph and edges-weights shown in the right:

- Maximum clique is (1,2,4,5,7,11).
- Weighted clique is 2.84.
- Maximum weighted clique is 5.4 (for a clique that is not a maximum clique).

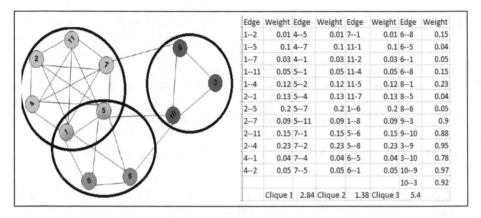

Edge	Weight	Edge	Weight	Edge	Weight	Edge	Weight
1--2	0.01	4--5	0.01	7--1	0.01	6--8	0.15
1--5	0.1	4--7	0.1	11-1	0.1	6--5	0.04
1--7	0.03	4--1	0.03	11-2	0.03	6--1	0.05
1--11	0.05	5--1	0.05	11-4	0.05	6--8	0.15
1-4	0.12	5--2	0.12	11-5	0.12	8--1	0.23
2--1	0.13	5--4	0.13	11-7	0.13	8--5	0.04
2--5	0.2	5--7	0.2	1--6	0.2	8--6	0.05
2--7	0.09	5--11	0.09	1--8	0.09	9--3	0.9
2--11	0.15	7--1	0.15	5--6	0.15	9--10	0.88
2--4	0.23	7--2	0.23	5--8	0.23	3--9	0.95
4--1	0.04	7--4	0.04	6--5	0.04	3--10	0.78
4--2	0.05	7--5	0.05	6--1	0.05	10--9	0.97
						10--3	0.92
Clique 1	2.84	Clique 2	1.38	Clique 3	5.4		

Fig. 1. Three hypothetical cliques and their maximum weighted cliques

A couple of observations on our definition of: maximum weighted clique:

- This is a direct graph where edges are directional and for example value of edge-weight 1–5 is 0.1 and is different from the value of edge 5–1, (0.05). This is particularly true for relations in Online Social Networks (OSNs) where relations and not symmetric. A typical example is between a fanatic and their celebrity, edge will be very high only from one direction (which indicates high interactions from the fanatic side with celebrity activities).
- In our hypothetic example, that may look odd but true, a small size clique may have an overall stronger weighted clique than a larger clique.
- If we consider relations dynamics, maximum cliques can vary dynamically as so their maximum weighted cliques. We hypothesize that a significant increase of maximum weighted clique for a particular clique beyond their historical norms can be used to predict "abnormal behaviors".
- Using this method detection of some "detection avoidance methods" is possible. For example, a clique may suddenly delete an edge to avoid detection. However, as we evaluated directed-edges, a single edge removal will not change the structure of the clique and will only cause that edge value to be zero.

2 Methods and Approaches

The focus of this study is on how to analyze information gathered from online social networks. The process of acquiring that information is not part of this study. There are a range of tools and techniques available for acquiring such data. One such tool that has been applied in many areas of social medial data gathering, is BuzzSumo, [13]. This tool has been used to analyze fake news and its relation to medical data, [32]. ViralWoot is another tool that can be used to gather data from social media, [30]. The specific tool or technique for gathering the data is not relevant to this current study. Our focus is on analyzing the data once it has been collected. We evaluated in previous contributions models to quantify interactions between friends in OSNs [4, 9]. The goal is to encourage quality content sharing and distribution.

For investigative purposes, particularly in intelligence gathering, the primary focus is on any deviation from established norms. For example, terrorists' networks will often have an increase in interactions prior to an operation. Therefore, we focused on an approach that identified such anomalous social interactions. We propose a method to detect "abnormal" clique behaviors through OSNs. We will collect cliques' networks from major OSNs such as: Facebook, Twitter and LinkedIn. We will use a weighted clique metric that we developed to be as a baseline for estimating average interactions between any OSN clique members. This weighted clique metric will be periodically (e.g. once a week), assessed and updated. A separate monitoring system will frequently (e.g. once every 8 h) read the current weighted clique metric. The early alert will be triggered if this instance metrics is significantly (e.g. >25%) higher than average weighted clique.

For example, say, for a Facebook clique of 6 individuals, weighted clique reading (as of the last month) is 13. While reading most recent weighted clique, if it shows a weighted clique of 20 (which is more than 25% increase from 13), an alert will be triggered to investigate the increase.

The usage of Cliques is popular in graph-based analysis areas (e.g. social networks, bioinformatics, etc.) in order to understand connections and trust relations between graph node members.

As an algorithm, Clique calculates the maximum number of nodes in the graph in which every node in the Clique is a friend to all other nodes in the Clique.

Here are the major tasks toward this goal:

- We introduced a new "weighted clique" algorithm. We showed also how this algorithm can be used to extract different types or aspects of knowledge in Bioinformatics or OSNs. This was the first task to accomplish in this goal.
- As we acknowledged that different OSNs have different "models" of interactions between the different individuals, we plan to create several concrete models, based on the availability of time and resources for one or more of the following OSNs: Facebook, Twitter and LinkedIn.
- A major task, in terms of time and resources, in this goal is data collection and analysis for cliques' interactions. We started collecting data, based on our model from Facebook and Twitter. We need to extend the collection and analysis process to collect a "significantly" large enough dataset.

In this research we introduced the following three intuitive concepts:

1. Weighted edges (i.e. strength of bonds between nodes that interact with each other). Typically, such relation is considered only from a binary perspective (i.e. exist:1 or does not exist: zero). We showed that the mere existence of the relation can be misleading in some cases and a significant amount of information is needed to know the weight of level of such relation, if exists.
2. Weighted Friendships: Based on each node or user (e.g. in OSNs) edges or connections, weighted friendship can calculate a friendship value based on the overall values of weighted edges. We showed in our paper, some applications for such value.
3. Weighted Cliques: As described earlier the mere existence of relation between either one or a group of nodes, may not be enough to understand social interactions. Based on weighted edges, we introduced the concept of weighted cliques to show the strength of bond between clique members. Figure 2 shows the different between normal cliques count and our proposed weighted or normalized or weighted clique.

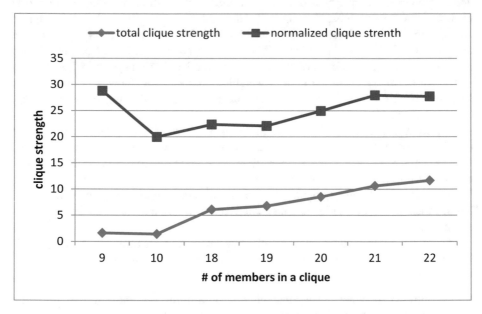

Fig. 2. The comparison between total and normalized/weighted clique strengths

The extraction of clique relations in social networks can show information related to groups and how those groups are formed or interact with each other. In social networks, a clique represents a group of people where every member in the clique is a friend with every other member in the clique. Those cliques can have containment relation with each other where large cliques can contain small size cliques. This is why most algorithms in this scope focus on finding the maximum clique. In our approach, we evaluated adding the weight factor to clique algorithm to show more insights about the level of involvement of users in the clique. In this regard clique activities are not like those in group discussions where an activity is posted by one user and is visible by all others. Our algorithm calculates the overall weight of the clique based on individual edges. Users post frequent activities. Their clique members just like other friends, may or may not interact with all those activities.

We showed in Table 1, the largest cliques in the dataset. The first column includes IDs of node members in the clique. The second column includes the number of members in each clique. The third column includes the clique strength as the summation of all weighted edges in the clique. The last column represents the weighted clique strength which divides the total clique by the number of edges in the clique. Normal clique strength value is constantly increasing with the number of nodes or edges in the clique. However, the weighted clique value eliminates the dependency on the number of nodes or edges. As such, we can see for example that the smallest clique in the table has the highest weighted clique.

Table 1. Total and weighted cliques

Clique members	No	Total clique	Weighted clique %
221, 215, 190, 187, 177, 176, 171, 151	8	1.61	28.77
221, 215, 190, 187, 177, 175, 171, 170, 151	9	1.43	19.96
221, 215, 190, 181, 177, 176, 172, 171, 169, 165, 164, 162, 157, 156, 153, 152, 151	17	6.08	22.36
221, 215, 190, 181, 177, 175, 172, 171, 170, 169, 165, 164, 162, 157, 156, 153, 152, 151	18	6.75	22.08
221, 215, 181, 177, 175, 172, 171, 170, 169, 168, 166, 165, 164, 162, 157, 156, 153, 152, 151	19	8.52	24.94
216, 209, 206, 193, 192, 191, 180, 179, 175, 169, 167, 160, 158, 151, 98, 20, 18, 9, 4, 2	20	10.62	27.96
216, 209, 193, 192, 191, 180, 179, 175, 170, 169, 167, 163, 160, 158, 151, 98, 20, 18, 9, 4, 2	21	11.64	27.73

This is not a trend however as the next highest weighted clique is the one with 20 nodes. This indicates that trust or interaction between members in the first clique is the highest. When we study social networks at a larger context, highest and lowest weighted cliques can be of special interests. We can also look at variations in clique strength over a certain period of time. For example, a clique sudden increase of the weighted clique in a certain month over a period of time can trigger further investigations of what could cause such significant sudden increase (e.g. a clique collaborative activity or event).

2.1   Enhancing the Methodology

It should be readily apparent that the current methodology can be easily enhanced or modified if needed. For example, the current weighting threshold is set at a specific value. It would be relatively easy to adjust the method such that the weighting is dependent on specific statistical values that can vary over time. Such statistical values could include the variance in weighting over a period in time, expressed as with the following simple formula:

$$\Sigma_{t=1}^{n} s2/\Delta T$$

This measures the change in variance in relation to a change in time (delta T). This is one example of how statistical analysis can be applied to the clique analysis to provide either more information, or more granular information.

The current methodology can also be combined with graph theory to model a given clique. The clique would be represented as vertices in the graph, with the connections being edges (or arcs). Each edge would be weighted based on the strength of the connection. That strength could be measured via frequency of connectivity, number of messages per unit of time, or any similar measurement that is pertinent to the investigation in question. Then the variation in not only activity for the clique as a whole, but also for the individual edges could be monitored. This would allow the detection in variations of sub groups within the clique. These sub groups would be represented as sub graphs of the cliques' larger graph.

3   Model Evaluation

In order to evaluate our weighted clique model in OSNs and its our ability to detect "abnormal groups dynamics", we built a historical dataset based on the initial dataset described in [19]. The volume of interactions between the friends in the network is calculated as totals for the given period of the collection process.

Following are our general steps to construct the model dataset.

- Original Facebook dataset described in [19] is used is the baseline dataset.
- The number of nodes or Facebook users as well as the number of edges between the members will be frozen through our one-week historical analysis. This is an assumption to fix the overall structure of the model where no changes on the nodes or relations will occur throughout the assessment. This assumption is to simplify the model demonstration, but the model and algorithms do not exclude any frequent changes in the network structure.
- In our model, we will give to the different Facebook activities the same weight in the relation (e.g. a Like, Comment, Post, etc.). This is another assumption in the model to simplify its demonstration that can be adjusted based on the different OSNs and the nature of the different activities.
- We will randomly vary volumes of activities between network users between (25% of their baseline up to 300% of their baseline). The goal is to simulate users' actual behaviors in OSN and also demo cliques' dynamics.

Figure 3 shows a small sample of our model output on some selected cliques. We can observe the followings based on Fig. 3:

N-edges	N-nodes	list-edges....	Clique Weight											
3	3	1	2	66.88889										
6	4	1	2	4	74.72222									
10	5	1	2	1	4	61.33333								
15	6	1	2	3	2	4	44.44444							
21	7	1	2	1	6	2	7	38.63492						
28	8	1	2	2	1	4	3	7	34.30952					
38	9	1	2	1	5	2	8	3	7	34.40741				
46	10	1	2	10	1	9	2	6	4	9	33.16296			
56	11	1	2	10	4	1	5	3	11	5	9	32.39394		
67	12	1	2	10	1	8	2	10	3	11	6	9	8	10

N-edges	N-nodes	list-edges....	Clique Weight											
3	3	1	2	136.3333										
6	4	1	2	4	142.8333									
10	5	1	2	1	4	106.5								
15	6	1	2	3	2	4	74.8							
21	7	1	2	1	6	2	7	67.09524						
28	8	1	2	2	1	4	3	7	60.42857					
38	9	1	2	1	5	2	8	3	7	61.58333				
46	10	1	2	10	1	9	2	6	4	9	59.31111			
56	11	1	2	10	4	1	5	3	11	5	9	57.45455		
67	12	1	2	10	1	8	2	10	3	11	6	9	8	10

N-edges	N-nodes	list-edges....	Clique Weight											
3	3	1	2	114.3333										
6	4	1	2	4	102.5									
10	5	1	2	1	4	86.3								
15	6	1	2	3	2	4	63.8							
21	7	1	2	1	6	2	7							
28	8	1	2	2	1	4	3	50.53571						
38	9	1	2	1	5	2	8	7	52.33333					
46	10	1	2	10	1	9	2	4	9	50.82222				
56	11	1	2	10	4	1	5	11	5	9	46.01818			
67	12	1	2	10	1	8	2	3	11	6	9	8	10	

Fig. 3. A sample output of our model

- As our model uses weighted clique values rather than clique or maximum clique values, the overall strength of the clique shows at the end of each record is independent from the clique size.
- Our model shows undirect edges/relations. This means that for two-friends, we are considering only one edge-value that can reflect one-way strength of the friendship. This was just an assumption in the model to simplify its demonstration and the model can be extended to consider or assume directed-edges or the two-way strength of the relation. The total number of the edges in the clique will then be doubles (N*N − 1), rather than now (N*N − 1)/2. As at the end, we are considering the overall weighted strength of the clique, this will not make a significant change.
- The 3rd figure part indicates three consecutive days in the selected network. It can be used to monitor gradual or sudden increase in the overall weighted strength of communication of the clique. This can be used to trigger further deeper focused analysis of that particular clique.
- Our model is built purely on statistics; it does not require looking at the content of communication activities between clique members. This can alleviate issues related to privacy, legal concerns and performance.

4 Conclusion

In this paper, we proposed a statistical approach to detect or alert for cliques' abnormal behaviors' using Online Social Networks (OSNs). This can be part of national security alert system that does not violate users' privacies as it does not need to look into users' contents (i.e. posted activities, friends, private messages, etc.). Due to its focus only on statistical assessments, the system also balances between security issues with performance and the impact that such systems may cause to OSNs.

As was noted in this paper, clique analysis can be integrated with other methodologies. For example, the weighting can be represented with graph theory. The individuals are the nodes in the graph, the connections are the edges or arcs, and the weighting can be represented in a weighted graph. It is also possible to integrate additional statistical analysis into the clique analysis, as was discussed in this paper. Both of these enhancements to the currently described clique analysis technique are avenues for further research.

References

1. Acemoglu, D., Bimpikis, K., Ozdaglar, A.: Dynamics of information exchange in endogenous social networks. Theor. Econ. **9**(1), 41–97 (2014)
2. Agarwal, V., Bharadwaj, K.K.: A collaborative filtering framework for friends recommendation in social networks based on interaction intensity and adaptive user similarity. Soc. Netw. Anal. Min. **3**(3), 359–379 (2013)
3. Alhajj, R., Rokne, J.: Encyclopedia of Social Network Analysis and Mining. Springer Publishing Company, New York (2014)
4. Alsmadi, I., Xu, D., Cho, J.-H.: Interaction-Based Reputation Model in Online Social Networks (2016)
5. Ang, L.: Community relationship management and social media. J. Database Mark. Cust. Strateg. Manage. **18**(1), 31–38 (2011)
6. Basuchowdhuri, P., Anand, S., Srivastava, D.R., Mishra, K., Saha, S.K.: Detection of communities in social networks using spanning tree. In: Advanced Computing, Networking and Informatics, vol. 2, pp. 589–597. Springer, Cham (2014)
7. Butman, A., Hermelin, D., Lewenstein, M., Rawitz, D.: Optimization problems in multiple-interval graphs. In: Proceedings of the Eighteenth Annual ACM-SIAM Symposium on Discrete Algorithms, SODA 2007, pp. 268–277 (2007)
8. Jansen, B.M.P.: Constrained bipartite vertex cover: the easy kernel is essentially tight. In: Proceedings of 33rd STACS, LIPIcs, vol. 47, pp. 45:1–45:13 (2016). https://doi.org/10.4230/LIPIcs.STACS.2016.45
9. Cho, J.-H., Alsmadi, I., Xu, D.: Privacy and social capital in online social networks. In: 2016 IEEE Global Communications Conference (GLOBECOM). IEEE (2016)
10. Choudhary, P., Singh, U.: A survey on social network analysis for counter-terrorism. Int. J. Comput. Appl. **112**(9), 24–29 (2015)
11. Comandur, S., Gupta, R., Roughgarden, T.: Counting small cliques in social networks via triangle-preserving decompositions (No. SAND2014-1516C). Sandia National Laboratories (SNL-CA), Livermore, CA (United States) (2014)
12. Cotterell, J.: Social Networks in Youth and Adolescence. Routledge, New York (2013)

13. Dicerto, S.: A new model for source text analysis in translation. In: Multimodal Pragmatics and Translation, pp. 1–14. Palgrave Macmillan, Cham (2018)
14. Duijn, P.A., Klerks, P.P.: Social network analysis applied to criminal networks: recent developments in Dutch law enforcement. In: Networks and Network Analysis for Defence and Security, pp. 121–159. Springer, Cham (2014)
15. Francis, M.C., Gonçalves, D., Ochem, P.: The maximum clique problem in multiple interval graphs. Algorithmica **71**(4), 812–836 (2015)
16. Guo, L., Zhang, C., Fang, Y.: A trust-based privacy-preserving friend recommendation scheme for online social networks. IEEE Trans. Dependable Secure Comput. **12**(4), 413–427 (2015)
17. Hao, F., Yau, S.S., Min, G., Yang, L.T.: Detecting k-balanced trusted cliques in signed social networks. IEEE Internet Comput. **18**(2), 24–31 (2014)
18. Hao, F., Park, D.S., Min, G., Jeong, Y.S., Park, J.H.: k-cliques mining in dynamic social networks based on triadic formal concept analysis. Neurocomputing **209**, 57–66 (2016)
19. Himel, D., Ali, M., Hashem, T.: User interaction based community detection in online social networks. In; The 19th International Conference on Database Systems for Advanced Applications (DASFAA), Bali, Indonesia, p. 580 (2014)
20. Hunter, R.F., McAneney, H., Davis, M., Tully, M.A., Valente, T.W., Kee, F.: "Hidden" social networks in behavior change interventions. J. Inf. **105**(3), 513–516 (2015)
21. Ishengoma, F.R.: Online social networks and terrorism 2.0 in developing countries (2014). arXiv preprint arXiv:1410.0531
22. Kirby, A.: The London bombers as "self-starters": a case study in indigenous radicalization and the emergence of autonomous cliques. Stud. Confl. Terror. **30**(5), 415–428 (2007)
23. Kramer, A.D., Guillory, J.E., Hancock, J.T.: Experimental evidence of massive-scale emotional contagion through social networks. In: Proceedings of the National Academy of Sciences, 201320040 (2014)
24. Kumlander, D.: A new exact algorithm for the maximum-weight clique problem based on a heuristic vertex-coloring and a backtrack search. In: Proceedings of The Forth International Conference on Engineering Computational Technology. Civil-Comp Press, pp. 202–208 (2004)
25. Leistedt, S.J.: Behavioural aspects of terrorism. Forensic Sci. Int. **228**(1–3), 21–27 (2013)
26. Malhotra, A., Totti, L., Meira Jr., W., Kumaraguru, P., Almeida, V.: Studying user footprints in different online social networks. In: Proceedings of the 2012 International Conference on Advances in Social Networks Analysis and Mining (ASONAM 2012), pp. 1065–1070. IEEE Computer Society (2012)
27. Mitrou, L., Kandias, M., Stavrou, V., Gritzalis, D.: Social media profiling: a panopticon or omnipticon tool? In: Proceedings of the 6th Conference of the Surveillance Studies Network, April 2014
28. Papadopoulos, S., Kompatsiaris, Y., Vakali, A., Spyridonos, P.: Community detection in social media. Data Min. Knowl. Discov. **24**(3), 515–554 (2012)
29. Paulusma, D., Picouleau, C., Ries, B.: Reducing the clique and chromatic number via edge contractions and vertex deletions, In: Proceeding of ISCO 2016, LNCS, vol. 9849, pp. 38–49 (2016)
30. Simmhan, Y.: L1: Introduction (Doctoral dissertation, Department of Computational and Data Sciences© Department of Computational and Data Science, IISc) (2017)
31. Scott, J.: Social network Analysis. SAGE, London (2017)
32. Waszak, P.M., Kasprzycka-Waszak, W., Kubanek, A.: The spread of medical fake news in social media–The pilot quantitative study. Health Policy and Technology (2018)
33. Zafarani, R., Abbasi, M.A., Liu, H.: Social media mining: an introduction. Cambridge University Press, Cambridge (2014)

Gamification in Enterprise Systems: A Literature Review

Changiz Hosseini and Moutaz Haddara[✉]

Kristiania University College, 0186 Oslo, Norway
moutaz.haddara@kristiania.no

Abstract. Currently, gamification is gaining a considerable popularity within the information systems field. Gamification concepts have various potentials in enhancing and utilizing the implementation and use of information systems within organizations. Thus, this literature review aims to summarize gamification utilization and application in enterprise systems' (ES) context. ES are known to be large and complex applications to work with, and with the help of gamification concepts it may aid users to be more engaged and work efficiently to utilize these systems to their fullest extent. This article also highlights potential research avenues for academics and ideas for practitioners.

Keywords: Enterprise systems · Gamification · Literature review

1 Introduction

In 2017, the gaming industry had a revenue of 108.4 billion dollars [1]. Over 65% American households play games, with the average age being 35 years old, and almost evenly split on both female and male players [2]. How can we translate this dedication to gaming in a non-gaming context? In our case; the different enterprise systems in organizations. Enter gamification, which is a tool to increase engagement from employees at work, by utilizing design principles from gaming [2–5]. Gamification has had a surge of popularity in the information systems (IS) field, and there has been a push in using these concepts in enterprise systems [2]. Reason for the rising popularity is theorized to be due to cheaper technology and the success and prevalence of the game medium [5]. Seaborn and Fels [5] conducted a comprehensive survey of Gamification theory and came to conclude that there is no standard for the Gamification concept, but the general idea is to apply game elements and mechanics in a non-gaming context. Their result from the survey pointed to that Gamification is a distinctive concept on its own merit but is still under active development. Other important takeaway is that Gamification has two main ingredients; it is used for non-gaming domain and is inspired from games, especially the game elements. It is important to note that the gamification concept must never cross to become a fully-fledged game [4, 5]. Below is a table of the various game elements that can be used as part of a Gamification concept, taken from the seminal book written by Kapp [4].

© Springer Nature Switzerland AG 2020
K. Arai et al. (Eds.): FTC 2019, AISC 1070, pp. 552–562, 2020.
https://doi.org/10.1007/978-3-030-32523-7_39

Enterprise System (ES) implementations are one of the largest projects an organization can undertake, both regarding budget and time [6, 7]. It shakes up the routines of employees and requires them to adapt to a new work routine. Numerous implementations fail due to lack of employee engagement. It is also no secret that these enterprise systems are sometimes viewed as "bureaucratic" to work with, and many employees don't feel engaged in their work [8, 9]. This coupled with the complexity and difficult use of Enterprise Systems leads to decreased productivity and employees not utilizing these systems to their fullest extent [8, 10]. No matter how perfect the technical systems implementation is, it is in the end up to how good the users are at using that system. Gamification can have great potential in Enterprise Systems' context, as these software are known to being efficient and optimize business processes alike [8]. With the introduction of gamification in these systems, it can help elevate this facet. This by regulating people's behavior in a non-game context for strategic purposes [2, 4, 8]. Enterprise Systems in the end are simply used by people, therefore it can be beneficial to examine how Gamification is applied today, and what potential future usages it can have.

The objective of this paper is to conduct a systematic review on gamification applications in the main enterprise systems. The articles reviewed are categorized based on which enterprise system the gamification concepts are applied to; i.e. enterprise resource planning (ERP), customer relationship management (CRM), knowledge management systems (KMS) and supply chain management (SCM) systems. Thus, three research questions guide this paper:

RQ1: What Gamification concepts are applied to which Enterprise Systems?
RQ2: Effectiveness of Gamification concepts in Enterprise Systems?
RQ3: Other potential uses of Gamification in Enterprise Systems?

The rest of the paper is structured as follows: Sect. 2 demonstrates the research methodology utilized for the literature review. Section 3 summarizes the overview of the articles reviewed and presents the main findings. Section 4 provides a discussion on the main use of gamification in ES and potential future research avenues for both practitioners and academics. And finally, a conclusion is provided in Sect. 5.

2 Research Methodology

Literature reviews are important to create a backbone for further advancement of knowledge. Doing a survey of existing literature on any topic, might uncover areas for further research and also states which areas of research are mature or saturated [11]. In this article, a systematic literature review was done to accumulate and gain an overview of existing literature on the domain of gamification in enterprise systems context. This research followed the process and guidelines recommended by Brereton et al. [12] as illustrated in Fig. 1.

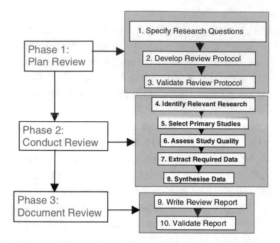

Fig. 1. Systematic literature review process [12]

This review process first started with the construction of research questions and then followed the steps provided in the figure above.

1. First, a search was conducted on gamification literature, specifically what gamification is as a concept and its established theory. For this objective, the authors purposefully selected seminal papers and books (containing comprehensive information about gamification) which were the top cited.
2. Second search was focused on finding papers that are relevant to gamification and enterprise systems, this was done through Google Scholar first. With a set of different keywords based on which enterprise. Namely, supply chain management, SCM, enterprise systems, ES, enterprise resource planning, ERP, knowledge management systems, KMS, customer relationship management, and CRM.
3. Another secondary search on ACM Digital Library was conducted, as the number of papers specifically addressing gamification within the different ES systems was very low.
4. Additional search was done through scanning the reference list of collected articles to potentially find more relevant articles.
5. As soon as the articles were accumulated, all abstracts were read by authors to ensure relevancy, and discard those which didn't fit the review's objectives.
6. We then extracted from the papers the main themes that each paper addresses. Additionally, we mapped the game elements presented in Table 1 with the reviewed papers.

No boundaries were set for the publishing period (at the time of writing this article in January 2019), as the term Gamification is quite recent, especially in the context of ES. Results from the searches resulted in few relevant articles to the research questions

Table 1. List of game elements [4]

Nr	Game elements	Description
1	Abstractions of concepts and reality	Abstracted reality gives inexperienced people to engage with concepts such as business strategy, processes etc.
2	Goals	Adds purpose, focus and measurable outcomes
3	Rules	Split into 4 types: Operational (How is the game played), Constitutive/Fundamental Rules (Rules only needed to be known by the designer), Implicit/Behavioral Rules (social rules in the game) and Instructional Rules (rules the designer wants the players to know after playing the game)
4	Conflict, competition, cooperation	Conflict: Challenge provided by an opponent (real person or from the game) Competition: Compete against each other towards a common goal Cooperation: Work together towards a common goal
5	Time	Motivating factor for player urgency, such as completing a customer support task. Similar use for consequences, not meeting a deadline etc.
6	Reward structures	Badges, Points and other similar rewards structures to show progress and provide reward for players
7	Feedback	Needs to be frequent and instant, and to promote correct behavior in the game
8	Levels	Players progress from one level to another, mission-oriented structure. Helps make the game not feel "directionless"
9	Storytelling	Give context to the tasks done in the game
10	The hero's journey	Story made up of characters, plot, tension and resolution
11	Curve of interest	Flow and sequences of events that keep users engaged
12	Aesthetics	Visual presentation to provide both feedback and pleasure
13	Replay or do over	Permission to "fail", giving opportunity to explore and experiment in the game

of this research. Thus, no strict parameters were put on the article selection criteria, except that the reviewed articles must be peer-reviewed. In addition, articles which did not specify the application of gamification in specific enterprise system and were discussing enterprise systems in general were discarded from this review. When it comes to the review structure, this paper followed the recommendations of Webster and Watson [11] and followed a concept-centric matrix. This was to provide a better ability to synthesize the literature. All of the articles were organized by which enterprise system was in focus, grouping all gamification and ERP articles in one category, gamification and CRM in another etc. Further, several sub topics were identified based on the purpose of the gamification concept (e.g. user training, increasing motivation for knowledge sharing etc.), and finally which game elements were used in each case.

3 Findings and Articles Overview

This section presents an overview of the articles, categorized by type of enterprise system they address. In total 24 articles were reviewed. The majority of the reviewed articles were targeting ERP and KMS systems. Below Table 2 provides an overview on all the articles and game elements used within these articles. The intention of this section is not to go in depth in each of the articles, but rather to present an overview of how gamification has been applied in different ES contexts.

Table 2. Article categorization

Enterprise system	Usage	Articles	Game elements discussed/Used (*from* Table 1)
ERP	User training and teaching	[13–20]	1–4, 6–8, 11
	Other	[21, 22]	1–4, 6–8, 12
CRM	Measuring potential use	[23, 24]	2–3, 6–7
	Other	[25]	1–4, 6–8
KMS	Motivation for knowledge sharing	[3, 26–33]	2–4, 6–8
SCM	Teaching concepts	[34, 35]	1–5, 6–8, 12–13

3.1 ERP

ERP are complex systems encompassing and integrating an organization's business functions (finance, HR, procurement, production, etc.). Serving as a central focal point for information and providing support for cross-functional business processes across the company [36]. For many, these systems can be intimidating to use and navigate, especially for beginners. This may also be an issue for veteran users if they are using cloud ERP, because there will be software updates that may rearrange or change the navigation in the ERP [37].

User Training and Teaching. Most ERP implementation failures are almost exclusively an organization problem rather than a technical one [6, 7]. The ERP implementation may be successful as far the technical part, but it won't matter if the users are not using the system. As Alcivar and Abed [13] put it "the benefits attributed to an ERP implementation can be lost without an effective user training". Therefore, several studies [10, 13, 16, 18] mention that user training is one of the paramount success factors for implementation. Since it can reduce user resistance, smooth out the starting phase and that uneducated users can be detrimental, since they will start to work around the system. Gamification has been used heavily to enhance this success factor for ERP implementation, as most of the studies related to gamification and ERP are focused on user training [13–20]. Most common ERP training methods don't provide meaningful learning to the employees, and the satisfaction rates ranges from neutral to low [13]. Many studies use this as point to try to extract more valuable ROI for user training, by implementing Gamification elements such as feedback, reward structures and goals.

Many studies measured the effectiveness of Gamification by conducting user studies [13, 14, 17–20]. They had two focus groups, one went through the traditional ERP user training, while the other group went through a gamified ERP user training method. Afterwards had the participants conduct questionnaire or interview afterwards. Results unanimously showing that using gamification had increased satisfaction, productivity, morality and engagement compared to the traditional way. Though one study does argue that with the introduction of game elements, that users may be more focused on "winning the game" rather than being focused on learning [15]. Other studies [14, 19] show that introducing competition and cooperation game elements have the potential to further enhance user training even more, since the can compensate for lack skills.

Other Uses. Not many articles were found that utilized gamification outside of the user training domain. Few articles investigated the potential of Gamification in ERP context outside of ERP training. El-Telbany and Elragal [21] undertook a design science research to develop a gamified process for the ERP life cycle implementation, to further improve and mitigate problems that ERP life cycles may have (social and technical). As they put it: "ERP implementations are classified as one of the most expensive business information technologies in the corporate world" [21]. Their findings were that most phases in the ERP life cycle phase has the potential to be affected by Gamification. Herzig et al. [22] conducted a user experimental study, where they tested user acceptance of a prototype gamified ERP system, measuring employees' reaction and enjoyment in using a gamified ERP compared to a traditional one. Their findings showing improvements in areas such as software enjoyment, perceived ease of use and flow experience. At first glance, the results are quite positive but mention the limitations in their study, and suggest further research with an improved prototype in much larger work setting.

3.2   CRM

CRM manages interactions with potential and existing customers an organization has. With the aim of making customers happy and satisfied.

The few articles found regarding Gamification and CRM, is about measuring the interest for a gamified CRM system and how to introduce a gamified CRM system in an organization [23, 24]. Initially, there seems to be no "obvious" way to utilize Gamification in a CRM system as there is for ERP (training) and KMS (increasing user participation), however, still some studies investigated the application of gamification techniques in several CRM related scenarios.

Measuring Potential Use. One study conducted interviews with Sales Representatives about their interest for a gamified CRM system for day to day use. Resulting most showing enthusiasm towards it, but their biggest concern is that making sure the gamification element aligns with their business objective, this concern falls in line with established Gamification theory, stressing the importance of aligning Gamification concepts with business objectives, this sentiment supported by established Gamification theory [3–5]. Author in [24] introduced a conceptual paper to introduce Gamification concepts in CRM systems, specifically the customer service domain. They claim there is an untapped potential here with the introduction of Gamification because as they point out in their article, customer service and support agents have one of the least pleasing jobs in modern economy, as they chiefly deal with unhappy and angry customers, and are under a continuous time pressure to quickly resolve issues [24]. Additionally, they also argue that Gamification is the perfect antidote for this dilemma [24].

Other. Jantke et al. [25] developed an AR game for a castle company in Germany. The idea is that you can explore the castle in an Augmented Reality game, which in return would incentivize these players to visit the castles in real life. Resulting in attracting customers and increased engagement for the castle company.

3.3   KMS

Knowledge Management Systems aim to capture, document, assimilate and share organization knowledge. These systems are important for any organization since knowledge is one of the main assets for any organization [26, 27].

Managing knowledge through implementing organization processes such as: construction, transfer, implementation and evaluation of knowledge [27]. Many employees show reluctance to share their knowledge because the system is perceived as "boring" to use. It is seen as one of key challenges faced in knowledge management in general [2, 26, 30]. Systems exist as platforms for users to do the processes such as: construction, transfer, implement and evaluate knowledge [26, 27], but user lack the motivation and engagement to do so [31, 38].

Motivation for Knowledge Sharing. Effective knowledge sharing heavily relies on employee motivation [38], Gamification is being used to combat this challenge, evident by all of the studies found for this review [3, 26–33]. Ďuriník's [39] findings show that gamification for KMS is very effective and game elements such as badges, reward structures and feedback are amongst the most used important elements to be used. He also makes point that Gamification is not just simply adding points and badges, but require precise and well thought out game design that support the system as a whole.

Supply Chain Management is the activities required for to manage the flow of goods throughout an organization. Involving procurement, production and distribution. Articles found related to SCM and Gamification exclusively focused on teaching the concepts of SCM with help of Gamification elements such as; abstractions of concepts, rules, feedback and goals [34, 35].

Teaching Concepts. Both papers found in this area focused on creating a gamified concept to teach the basic and complex strategies of SCM to students and professionals alike [34, 35]. As both studies concluded with using gamified concepts they managed to teach both basic and fairly complex strategies and concepts quite effectively. This started a chain reaction, the more they learned through teaching, the better they managed to leverage this knowledge. This is corresponds to another study that argues that learning more about SCM strategies in real-world context, can aid ERPsim players to make better decisions [19].

4 Discussion and Future Research Avenues

In total, 24 articles were reviewed related to Gamification and the different Enterprise Systems. With the earliest article stemming from 2012. Our findings show that, the research output is considerably low, especially in the scope of examining it from many different ES such as ERP, SCM, KMS and CRM.

In some studies, it was also challenging to determine explicitly which type of game elements they used for their gamification strategy. It is evident that gamification in ES context, is still a young topic, and needs further research to be carried out to advance our knowledge.

ERP. Majority of the articles regarding ERP gamification elements solely focused on user training. To get increased return of investment for the money spent on training, but not a lot of has been focused on having a truly gamified ERP for day to day use, instead of "just" training, with the exception of one paper by Herzig et al. [2].

Since all ERP user training gamification were highly effective and successful, the next iteration of this step is to start experimenting with a fully-fledged gamified ERP system. Important to note that this may be challenging, since the scope is much larger, and important to align gamification objectives with business objectives. This has potential to create a whole new type of engagement amongst employees and users as many in general feel disengaged when working with such big and massive systems.

CRM. As mentioned before, support and service are seen as monotonous tasks done in an organization, and have great potential to increase employee satisfaction/engagement if enhanced. The idea to use badges, rewards and levels, can have the same effect as the argued in another article [30], which states that a gamified KMS can create self-worth within an organization. In a CRM context, the more you establish yourself as a high-level case solver, the more bragging rights and leverage you will have in the company culture. There is also potential to involve a competition game element, such as who manages to close cases the most within a time period. There is also potential here for more experimental use of gamification, such as how Jantke et al. [25] conducted it with an AR game. This type of experimentation with Gamification can be strategies used to attract new customers and increase engagement, and excitement for existing customers.

SCM. The few studies reviewed solely focused on teaching the concepts of SCM. It has been measured to be effective, but not much more than that. Not much Gamification strategy is discussed in this domain, but there might be a potential here to use GPS based gamification concepts. AR GPS based games are increasing with popularity, with apps like Pokemon GO, having a staggering amount of users.

KMS. All studies conducted in this domain, exclusively focused on increasing employee motivation. KMS seems to have developed further in implementing gamification to incentivize here. Schpakova et al. [32] argues that by reducing Gamification strategy for KMS to only enhance employee motivation is only staying on the top surface of it's true potential. their users to part take in knowledge sharing in return for social badges, levels progressions which can be used for social bragging and "respect" amongst the company culture. But that's the only thing Gamification has been used for.

General. Most gamification usages are used for training and teaching of ES concepts, not many studies focused on having a gamified ERP system for day to day use. This would be the true next evolution for Gamification in Enterprise Systems. In essence, making people forget that they are "working" and instead "having fun" by playing a game.

5 Conclusions

Gamification is not really leveraged to its full capacity in enterprise systems yet. Many studies focused on using Gamification for mostly training and teaching, but there are opportunities to apply Gamification strategies in different ways. It is highlighted that research in this area is still fresh and potential here is waiting to be unlocked. Starting to introduce a fully-fledged gamified enterprise system may change the enterprise landscape in a big way.

This paper aimed at providing insights for academics and industry practitioners alike. It highlights the gamification applications area in enterprise systems for the practitioners, and see where they can take the next step, and for academics, it sheds light that there is limited amount of research done about gamification in enterprise system context.

References

1. Batchelor, J.: Games industry generated $108.4Bn in revenues in 2017 (2018). https://www.gamesindustry.biz/articles/2018-01-31-games-industry-generated-usd108-4bn-in-revenues-in-2017
2. Herzig, P., Ameling, M., Schill, A.: A generic platform for enterprise gamification. In: 2012 Joint Working IEEE/IFIP Conference on Software Architecture and European Conference on Software Architecture, pp. 219–223. IEEE (2012)
3. Deterding, S., Dixon, D., Khaled, R., Nacke, L.: From game design elements to gamefulness: defining "Gamification." In: Proceedings of the 15th International Academic MindTrek Conference: Envisioning Future Media Environments, pp. 9–15. ACM, New York (2011). https://doi.org/10.1145/2181037.2181040
4. Kapp, K.M.: The Gamification of Learning and Instruction: Game-Based Methods and Strategies for Training and Education. Wiley, Hoboken (2012)
5. Seaborn, K., Fels, D.I.: Gamification in theory and action: a survey. Int. J. Hum. Comput. Stud. **74**, 14–31 (2015)
6. Davenport, T.H.: Putting the enterprise into the enterprise system. Harvard Bus. Rev. **76** (1998)
7. Haddara, M.: ERP selection: a case of a multinational enterprise. Inf. Resour. Manage. J. (2016)
8. Augustin, K., Thiebes, S., Lins, S., Linden, R., Basten, D.: Are we playing yet? A review of gamified enterprise systems. In: PACIS, p. 2 (2016)
9. Haddara, M., Hetlevik, T.: Investigating the effectiveness of traditional support structures & self-organizing entities within the ERP shakedown phase. Procedia Comput. Sci. **100**, 507–516 (2016)

10. Haddara, M., Moen, H.: User resistance in ERP implementations: a literature review. Procedia Comput. Sci. **121**, 859–865 (2017)
11. Webster, J., Watson, R.T.: Analyzing the past to prepare for the future: writing a literature review. MIS Q. **26**(2), xiii–xxiii (2002)
12. Brereton, P., Kitchenham, B.A., Budgen, D., Turner, M., Khalil, M.: Lessons from applying the systematic literature review process within the software engineering domain. J. Syst. Softw. **80**, 571–583 (2007)
13. Alcivar, I., Abad, A.G.: Design and evaluation of a gamified system for ERP training. Comput. Hum. Behav. **58**, 109–118 (2016)
14. Heričko, M., Rajšp, A., Horng-Jyh, P.W., Beranič, T.: Using a simulation game approach to introduce ERP concepts–a case study. In: International Conference on Knowledge Management in Organizations, pp. 119–132. Springer (2017)
15. Horng-Jyh, P.W.: Learning Enterprise Resource Planning (ERP) through business simulation game. In: Proceedings of the The 11th International Knowledge Management in Organizations Conference on The changing face of Knowledge Management Impacting Society, p. 5. ACM (2016)
16. Hwang, M., Cruthirds, K.: Impact of an ERP simulation game on online learning. Int. J. Manage. Educ. **15**, 60–66 (2017)
17. Kuem, J., Wu, J., Kwak, D.-H., Deng, S., Srite, M.: Socio cognitive and affective processing in the context of team-based gamified ERP training: reflective and impulsive model. In: Proceedings of the Midwest Association for Information Systems Annual Conference (2016)
18. Rajšp, A., Horng-Jyh, P.W., Beranič, T., Heričko, M.: Impact of an introductory ERP simulation game on the students' perception of SAP usability. In: International Workshop on Learning Technology for Education in Cloud, pp. 48–58. Springer (2018)
19. Sepehr, S., Head, M.: Competition as an element of gamification for learning: an exploratory longitudinal investigation. In: Proceedings of the First International Conference on Gameful Design, Research, and Applications, pp. 2–9. ACM (2013)
20. Wu, H.-J.P., Beng, S.T., Heričko, M., Beranič, T.: Knowledge creation activity system for learning ERP concepts through simulation game. In: International Conference on Knowledge Management in Organizations, pp. 133–143. Springer (2017)
21. El-Telbany, O., Elragal, A.: Gamification of enterprise systems: a lifecycle approach. Procedia Comput. Sci. **121**, 106–114 (2017)
22. Herzig, P., Strahringer, S., Ameling, M.: Gamification of ERP systems-Exploring gamification effects on user acceptance constructs. In: Multikonferenz Wirtschaftsinformatik, pp. 793–804. GITO Braunschweig (2012)
23. Carignan, J., Kennedy, S.L.: Case study: identifying gamification opportunities in sales applications. In: International Conference of Design, User Experience, and Usability, pp. 501–507. Springer (2013)
24. Makanawala, P., Godara, J., Goldwasser, E., Le, H.: Applying gamification in customer service application to improve agents' efficiency and satisfaction. In: International Conference of Design, User Experience, and Usability, pp. 548–557. Springer (2013)
25. Jantke, K.P., Krebs, J., Santoso, M.: Game amusement & CRM: castle Scharfenstein AR case study. In: 2014 IEEE 3rd Global Conference on Consumer Electronics (GCCE), pp. 488–491. IEEE (2014)
26. Elm, D., Kappen, D.L., Tondello, G.F., Nacke, L.E.: CLEVER: gamification and enterprise knowledge learning. In: Proceedings of the 2016 Annual Symposium on Computer-Human Interaction in Play Companion Extended Abstracts, pp. 141–148. ACM (2016)
27. Jurado, J.L., Fernandez, A., Collazos, C.A.: Applying gamification in the context of knowledge management. In: Proceedings of the 15th International Conference on Knowledge Technologies and Data-Driven Business, p. 43. ACM (2015)

28. Lee, C.-S., Foo, J.-J., Chan, P.-Y., Hor, W.-K., Chan, E.-K., et al.: A knowledge management-extended gamified customer relationship management system. In: 2017 International Conference on Soft Computing, Intelligent System and Information Technology (ICSIIT), pp. 341–346. IEEE (2017)
29. Meder, M., Plumbaum, T., De Luca, E.W., Albayrak, S., DAI-Labor, T.: Gamification: a semantic approach for user driven knowledge conservation. In: LWA, pp. 265–268 (2011)
30. Paul, P.V.: Knowledge management using gamification. Int. J. Adv. Sci. Res. Dev. **3**, 35–39 (2016)
31. Rinc, S., et al.: Integrating gamification with knowledge management. In: Management, Knowledge and Learning, International Conference, pp. 997–1003 (2014)
32. Shpakova, A., Dörfler, V., MacBryde, J.: Changing the game: a case for gamifying knowledge management. World J. Sci. Technol. Sustain. Dev. **14**, 143–154 (2017)
33. Swacha, J.: Gamification in knowledge management: motivating for knowledge sharing. Pol. J. Manag. Stud. **12**, 150–160 (2015)
34. Lau, A.K.: Teaching supply chain management using a modified beer game: an action learning approach. Int. J. Logist. Res. Appl. **18**, 62–81 (2015)
35. Uhlmann, T.S., Battaiola, A.L.: Applications of a roleplaying game for qualitative simulation and cooperative situations related to supply chain management. In: International Conference on HCI in Business, pp. 429–439. Springer (2014)
36. Demi, S., Haddara, M.: Do cloud ERP systems retire? An ERP lifecycle perspective. Procedia Comput. Sci. **138**, 587–594 (2018)
37. Bjelland, E., Haddara, M.: Evolution of ERP systems in the cloud: a study on system updates. Systems **6**, 22 (2018)
38. Swacha, J.: Gamification in enterprise information systems: what, why and how. In: 2016 Federated Conference on Computer Science and Information Systems (FedCSIS), pp. 1229–1233. IEEE (2016)
39. Ďuriník, M.: Gamification in knowledge management systems. Cent. Eur. J. Manage. **1** (2015)

An Experiment with Link Prediction in Social Network: Two New Link Prediction Methods

Ahmad Rawashdeh$^{(\boxtimes)}$

Department of Computer Science and Math,
University of Central Missouri, Lee's Summit, MO, USA
arawashdeh@ucmo.edu

Abstract. This paper investigates link prediction methods in social networks (Facebook) and discusses two new link prediction methods. The two methods are structural based methods, which implies they are not based on any content of user profiles, but instead they are based on connections of the users in the social network. These two methods are the Common Neighbors of Neighbors and Node Connectivity prediction methods. The first introduced method can be considered as an extension to the Common Neighbors link prediction method. The second method is based on average connections of neighbors. Both methods are discussed in this paper and have been used in experiment. Additionally, Formulas, explanation, Pseudocode and an example is included about some preexisting methods of link prediction and the introduced methods considered in this paper. This paper also includes detailed applications of link predictions methods, and experimental results that compare the new methods with well-known methods in link prediction. Results show better performance for the proposed link prediction method when applied to a friendship Facebook dataset, which is characterized in this paper. The experiment is described in details and the results (precision and number of positives) shows superiority of proposed methods in terms of performance over well-known link prediction methods.

Keywords: Link prediction · Social networks · Common Neighbors · Common Neighbors of Neighbors · Facebook · Recommender system

1 Introduction

Link prediction is one of the main features of networking web sites, particularly social networks. Users of social networks tend to rely on the website to suggest friends.

Social networks are represented as a graph of nodes and edges. Where nodes represent individuals and links represents friendship. With the huge number of social network users, users tend to be more confused as to whom they should connect to. This is due to the nature of social network profile, which is sometimes limited in visibility (controlled by privacy), and the difficulty to know about people by only looking at the public view of their profiles. Therefore, as a solution, friends' recommendation emerged as a form for helping people to connect with their friends or for making new friends.

© Springer Nature Switzerland AG 2020
K. Arai et al. (Eds.): FTC 2019, AISC 1070, pp. 563–581, 2020.
https://doi.org/10.1007/978-3-030-32523-7_40

As of nowadays, Facebook[1] recommend friends to users. Co-authorship websites display recommended authors to users [1]. The recommendation of Friends in Facebook is based on the common or mutual friends (structure of the graph of users) and this is the most common reason for suggesting friends [2]. This is can be observed from the Mutual friends' information (list of links to users' profiles) displayed under each recommended user profiles. A semantic similarity of profiles in social network was investigated in [3]. This paper focus is only the structural similarity, which is the similarity of users based only on the connection of the profiles as opposed to profile content similarity. This structural similarity is included in the graph of social network.

As it has been mentioned, Social networks are represented as a graph of nodes (vertices) and edges (connections). Where nodes represent the users and their profiles, and edges represents their friends. By analyzing the connections of two pair of users, one of end users can be recommended as a new connection to the other user. This is a link prediction, in the sense that friends' recommendation will turn into a link (connection) once accepted. It is predicted by the website based on the structure of the social network graph.

Several link prediction algorithms were proposed and studied in the literature [4]. Additionally, several link prediction methods were compared with LinkGyb method for link prediction in [5]. In [6] and [7], authors investigated a machine learning classifier to predict links. The earlier, constructed a feature vector from topological information and node attributes and was evaluated on a co-authorship dataset. The later evaluated the algorithm on 10 different datasets. Also, the research in [8] evaluated two new proposed methods, local (CNFN) and global (KatzGF) on co-authorship dataset (DBLP). The focus of this paper is different as it is using a structural non-machine learning (classifiers) methods to predict links. Moreover, this paper introduces new link prediction methods that can recommend friends which can't be recommended by preexisting methods of link predictions.

In the remaining sections, the link prediction problem is defined followed by applications of link prediction, which is discussed in detail. Next is classification of link prediction methods. After that the formulas, example, pseudo code and explanation of each link prediction method is considered in the paper. Then experiment including how to sample and evaluate link prediction. Next, is a description of the Facebook Dataset, which was used in the experiment. Following that are the results and discussion including results produced when completely random and semi random sampling are used. Finally, is conclusion and future work.

2   Definition of Link Prediction

If a graph (network) is defined as $G = (V, E)$, where V is the set of vertices and E is the set of edges. Given a snapshot of the graph, link prediction can be defined as to predict the probability of the link between two nodes (there is a link or will be a link). These two nodes belong to the set V [8, 9].

---

[1] https://www.facebook.com/.

Link prediction can be classified into two problems: infer missing links and predict links that will be added to an observed network in the future. This paper falls under the first type of link prediction [9].

Figures 1 and 2 are used to illustrate the link prediction definition. Each is a graph of four nodes. At time T = 0, as it is shown in the Fig. 1, user 1 is connected to user 2 and user 3, and user 4 is connected to user 2 and 3. Then at T = 1, in Fig. 2, we can predict a link between user 1 and user 4 (common neighbors predict this link).

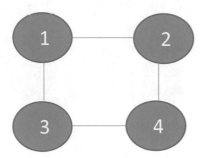

Fig. 1. Graph example at time T = 0. The graph contains four nodes. Each node represent a user in social network. Node 1 is connected to node 2 and node 3. Node 4 is connected to node 2 and 3.

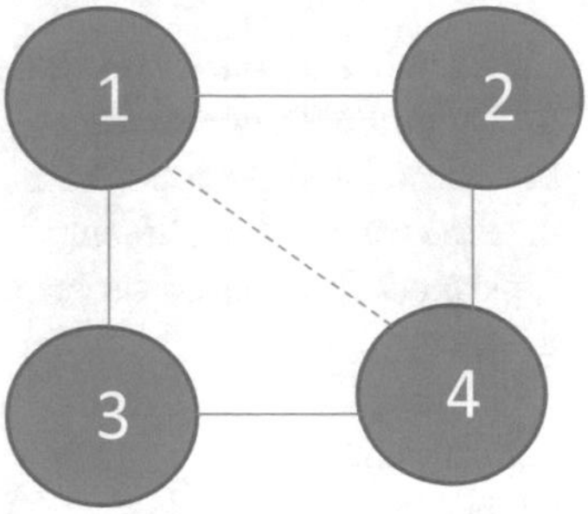

Fig. 2. Graph example at time T = 1. A link between user 1 and user 4 is predicted to be created

3   Application of Link Predictions

Link prediction has many applications ranging from Recommender system in social networks to find possible friends, and in biology to find possible interactions between pairs of protein in a protein-protein interaction network, to scientific co-authorship network (DBLP[2]). Additionally, one interesting application is Entity resolution in Database systems in what is known as record linkage or deduplication. Applications in social network analysis is to analyze the structure of criminal or terrorist networks; also, application in transportation networks and telecommunication networks [10].

Another application in social network is sign prediction such as inferring attitude of a specific user towards another user using the information extracted from positive and negative relations that are in the vicinity [11]. Application also includes predict future conflict and individual preferences [12]. Link prediction can also be used in aircraft route planning [13]. Moreover In security domain link prediction can assist in identifying hidden groups of terrorists or criminal or even compromise social networks user's privacy [14–16]. Link prediction has many applications in domains outside social networks. Example is in sensor networks [16].

---

[2] https://dblp.uni-trier.de/.

In addition to Biology (protein interaction, metabolic network), social science, routing, and security, it has application in medicine in disease-gene networks [9, 17].

4 Classification of Link Prediction Methods

Link prediction methods can be classified into different category reflecting the nature of the method.

Link prediction can be grouped into [4]:

(a) Link prediction based on node neighborhood:
    a. Common Neighbors
    b. Jaccard and Adamic/Adar
    c. Preferential Attachment
(b) Link prediction based on ensembles of all path:
    a. Katz
    b. Hitting time
    c. Simrank
(c) Higher level approaches:
    a. Low-rank approximation
    b. Unseen bigrams
    c. Clustering

The study by Martinez, V., et al. [10] is the most comprehensive study I encountered about the link prediction methods that have been proposed in terms of the number of methods and network employed. The authors grouped link prediction into the following:

a. Local Approaches:
Common neighbors, adamic-adar index (AA), the resource allocation index (RA), resource allocation based on common neighbors interaction (RA_CNI). The preferential attachment index (PA), the jaccard index (JA), the salton index (SA), the sornesen index (SO), the Hub Depressed Index (HDI), the local Leicht-Holme-Newman Index (LLHN), the individual attraction index (IA), mutual information, functional similarity weibht (FSW)

b. Global Approaches:
Negated Shortest Path (NSP), The Katz Index (KI), the global Leicht-Holme-Newman Index (GLHN), Random Walks (RW), Random Walks with Restart (RWR), Flow Propagation (FP), SimRank (SR),
Pseudoinverse of Laplacian Matrix (PLM), Random Forest Kernel Index (RFK), the Blondel Index (BI)

c. Quasi-Local Approaches:
The Local Path Index (LPI), Local Random Walk (LRW), Superposed Random Walk (SRW), Third-Order Resource Allocation based on common neighbors interaction (ORA-CNI), FriendLink (FL), ProFlow Predictor (PFP)

d. Algorithmic Methods:
   Classifier-based method, metaheuristic-based methods, factorization-based methods
e. Preprocessing Methods:
   Low-rank approximation, unseen bigrams, filtering.

5 Link Prediction Methods

In this section of the paper, four different type of information are given. First, the formula of each link prediction, discussed and considered in the paper, is included. Followed by an example of calculating the score using each links prediction method. Then a brief explanation of the algorithm followed by the step-by-step pseudocode of the algorithm. Note: in the following formulas, the symbol $\daleth(x)$ denotes the direct neighbors (friends) of node (user x). From the seven link prediction methods below, 6 and 7 are the proposed methods.

5.1 Formulas of Link Prediction Methods

The formula for links predictions considered in this paper are as follows:

1. Common neighbors [4]

$$\text{Score}(x \text{ and } y) = |\daleth(x) \cap \daleth(y)| \tag{1}$$

2. Jaccard's coefficient [4]

$$\text{Score}(x \text{ and } y) = \frac{|\daleth(x) \cap \daleth(y)|}{|\daleth(x) \cup \daleth(y)|} \tag{2}$$

3. Adamic/adar [4]

$$\text{Score}(x \text{ and } y) = \sum_{z \in |\daleth(x)| \cap |\daleth(y)|} \frac{1|}{\log |\daleth(z)|} \tag{3}$$

4. Preferential attachment [4]

$$\text{Score}(x \text{ and } y) = |\daleth(x)| \cdot |\daleth(y)| \tag{4}$$

5. Random

$$\text{Score}(\text{user } x \text{ and user } y) = \text{true or false depending on the binary random value computed using Math.Random.} \tag{5}$$

6. Neighbors Connectivity, Hybrid (proposed)
   (a) If Common Neighbors in (1) gives a Score(x and y) >= 1, use that score
   (b) Else use the following formula:

$$\text{Score}(x, \text{ and } y) = \text{Average degree of neighbors of neighbors of user } x$$
$$+ \text{Average degree of neighbors of neighbors of user } y$$
$$= \frac{\sum_{z \in \rceil(x)} |\rceil(z)|}{|\rceil(x)|} + \frac{\sum_{z \in \rceil(y)} |\rceil(z)|}{|\rceil(y)|}$$

$$(6)$$

7. Common Neighbors of Neighbors is calculated as follows (proposed)

$$\text{Score}(x \text{ and } y) = |(\text{neighbors}(\text{neighbors}(x)) \cap (\text{neighbors}(\text{neighbors}(y))|$$
$$= |\rceil(\rceil(x)) \cap (\rceil(\rceil y))|$$

$$(7)$$

5.2 Link Prediction Example

The graph in Fig. 3 is an example included to show how each link prediction method calculates the score between the two nodes, which will be used to predict the link. The link considered is between node 1 and node 11. Different link prediction methods consider different factors in calculating the score. All of these methods are structural based method as mentioned earlier, which means only the topology of the graph is considered, no content or profile information is used.

We are going to discuss each link prediction above in more detail.

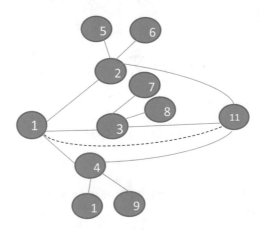

Fig. 3. An example graph

Common Neighbors. The score of Common Neighbors when calculated using Fig. 3 on node 1 and 11 is 3. There are three common nodes between node 1 and node 11, which are node 2, 3 and 4. See Eq. (1) in Sect. 5.1.

Jaccard. The score of Jaccard when calculated using Fig. 3 on node 1 and 11 is $3/3 = 1$. Because there are three common neighbors and no other neighbors for node 1 and 11. So 3 is common that neighbors over the number of all neighbors of node 1 and 11 which is also 3. See Eq. (2) in Sect. 5.1.

Preferential Attachment. The score of Preferential Attachment when calculated using Fig. 3 on node 1 and 11 is 9. Because the number of neighbors of node 1 is 3. In addition, the number of neighbors of node 11 is 3. So, $3 * 3 = 9$. See Eq. (3) in Sect. 5.1.

Adamic Adar. The score of the Adamic Adar when calculated using Fig. 3 is 0.903. The common neighbors between node 1 and 11 are nodes 2, 3, and 4. The adamic adar uses the neighbors of the common neighbors of node 1 and 11. See Eq. (4) in Sect. 5.1.
Given that $|\sqcap(x)| \cap |\sqcap(y)| = |\sqcap(1)| \cap |\sqcap(11)| = \{2, 3, 4\}$. Then,

$$\sum_{z \in |\sqcap(x)| \cap |\sqcap(y)|} \frac{1|}{\log |\sqcap(z)|} = \sum_{z \in \{2,3,4\}} \frac{1|}{\log |\sqcap(z)|}$$

$$= \frac{1|}{\log |\sqcap(2)|} + \frac{1|}{\log |\sqcap(3)|} + \frac{1|}{\log |\sqcap(4)|}$$

$$= \frac{1}{\log 2} + \frac{1}{\log 2} + \frac{1}{\log 2} = 0.903$$

Random. The score of random when calculated using Fig. 3 on node 1 and 11 is simply true or false based on the random value computed within the range of 0 (inclusive) and 2 (exclusive). Therefore, every time random is calculated on two nodes, it will either predict a link or not depending on the binary random value computed using the Math.Random and Random.Next class and method respectively. See Eq. (5) in Sect. 5.1.

Neighbors Connectivity (Avg Degree). The score of Neighbors connectivity when calculated using Fig. 3 on node 1 and 11 is 4. For node 1, it has 3 neighbors, which are 2, 3 and 4. Each has 2 neighbors. So, in total 6 neighbors. When dividing that by 3, which is the number of neighbors of node 1, we get 2. The average degree of those neighbors without including node 1 and 11 is Average degree of neighbors of neighbors of user x which is 2. The same calculated for node 11 which will results in 2. So, add the two terms $2 + 2 = 4$. See Eq. (6) in Sect. 5.1.

Explanation: Based on popular neighbors. The more the direct neighbors are connected (high degree), the more is the chance to predict link between node 1 and node 11. Based on connectivity of neighbors of node x and node y. This not only consider shared neighbors between x and y, but also their popularity/connectivity. If connected to social neighbors, then should predict link. It is not a match between x and y but connect the social to the node under consideration.

Common Neighbors of Neighbors. The score of Common Neighbors of Neighbors is 6. Because the neighbors of node 1 are node 2, 3, and 4, which are the same neighbors of node 11. The neighbors of node 2 are 5 and 6. The neighbors of node 3 are 7 and 8. The neighbors of node 4 are 9 and 10. Therefore, in total, the number of common neighbors of the neighbors of node 1 and 11 are 6. See Eq. (7) in Sect. 5.1.

Explanation: One thing that might arise when reading about Common neighbors is that why not consider common neighbors of friends, or neighbors of neighbors. Let's assume there exist two individuals, x and y. both has a friend who knows person A. It may not take that much of a time for both of x and y to come closer in the graph. The friend of X, who knows A, and the friend of Y, who also knows A, may get connected. So as a way of speeding this process. If the link is predicted between person x and y, that might be a good recommendation.

5.3   Link Prediction Algorithm Pseudocode

Below is the pseudocode corresponding to each link prediction method discussed before and considered in the experiment. In the pseudocodes that follows, *Algorithm_predicted* is dictionary data structure. It stores the score calculated by the *Algorithm* on each pair of nodes. CCNeighbors_predicted is of type Dictionary<int, SortedList<int,int[]>>. The remaining data structures, including Jaccard_predicted are of type Dictionary<int, SortedList<int,double>>. Random Algorithm uses Random_predictedLinks, which is of type Dictionary<int,SortedList<int,bool>>. "g" is an object of class Graph. neighbors_of_node is a member method from class Graph that returns the neighbors of a node as a List<int> object.

Algorithm 1: common neighbor
1. For i = 1 to NumberOfNodes
2.   Nneighbors1 = g.neighbors_of_node(i)
3.   For j = 1 to NumberOfNodes
4.   If i == j continue
5.   Nneighbors2 = g.neighbors_of_node(j)
6.   Score = | nneighbors1 ∩ nneighbors2 |
7.   if score != 0
8.   Totalpredicted++
9.   Print "score " + i + " and " + j + score
10.   Set   the   value   of CCNeighbors_predictedLinks[i] to j and score

Algorithm 2: Jaccard
1.   For i = 1 to NumberOfNodes
2.   Nneighbors1 = g.neighbors_of_node(i)
3.   For j = 1 to NumberOfNodes
4.   If i == j continue
5.   Nneighbors2 = g.neighbors_of_node(j)
6.   num = | nneighbors1 ∩ nneighbors2 |
7.   Den = | nneighbors1 ∪ nneighbors2 |
8.   Score = num / den
9.   if score > 0
10.   Totalpredicted++
11.   Print "score " + i + " and " + j + "is " + score
12.   Set the value of Jaccard_predictedLinks[i] to j and score

Algorithm 3: Preferential Attachment
1.   For i = 1 to NumberOfNodes
2.   Nneighbors1 = g.neighbors_of_node(i)
3.   For j = 1 to NumberOfNodes
4.   If i == j continue
5.   Nneighbors2 = g.neighbors_of_node(j)
6.   ne1 = | nneighbors1 |
7.   ne2 = | nneighbors2 |
8.   Score = ne1 * ne2
9.   if score > 0
10.   Totalpredicted++
11.   Print "score " + i + " and " + j + "is " + score
12.   Set the value of Preferetnial_predictedLinkns[i] to j and score

Algorithm 4: Adamic Adar
1.   For i = 1 to NumberOfNodes
2.   Nneighbors1 = g.neighbors_of_node(i)
3.   For j = 1 to NumberOfNodes
4.   If i == j continue
5.   Nneighbors2 = g.neighbors_of_node(j)
6.   CommonNeighbors = nneighbors1 ∩ nneighbors2
7.   Foreach cn in CommonNeighbors
8.   Score += (1 / Math.Log(g.NodeNeighbors(cn).Count));
9.   if score > 0
10.   Totalpredicted++
11.   Print "score of " + i + " and " + j + "is " + score
12.   Set the value of Adamic_predictedLinks[i] to j and score

---

Algorithm 5: Random

1. For i = 1 to NumberOfNodes
2.   Nneighbors1 = g.neighbors_of_node(i)
3.   For j = 1 to NumberOfNodes
4.   If i == j continue
5.   Nneighbors2 = g.neighbors_of_node(j)
6.   Random r = new Random()
7.   Int value = r.next(0,2)
8.   Bool linkExist = false;
9.   If value == 1
10.   Totalpredicted++
11.   Print "score of " + i + " and " + j + "is " + score
12.   Set the value of Random_predictedLinks[i] to j and score

---

Algorithm 6: Neighbors Connectivity

1. For i = 1 to NumberOfNodes
2.   Nneighbors1 = g.neighbors_of_node(i)
3.   For j = 1 to NumberOfNodes
4.   If i == j continue
5.   End if
6.   Nneighbors2 = g.neighbors_of_node(j)
7.   CCScore = |nneighbors1 ∩ nneighbors2|
8.   If CCScore > 0
9.   Score = CCScore
10.   Else
11.   For ind = 0 to Nneighbors1.count
12.   Neighbors_Neighbors = g.neighbors_of_node(Nneighbors1[ind])
13.   For m =0 to Neighbors_Neighbors.count
14.   If Neighbors_Neighbors[m] == i || Neighbors_Neighbors[m] == j
15.   Continue
16.   End if
17.   If Neighbors_Neighbors.count – 2 > 0
18.   Sum1 += Neighbors_Neighbors.Count – 2
19.   avgNeighborsConn1 += Neighbors_Neigbors.Count – 2
20.   Else
21.   Sum1 += 0
22.   avgNeighborsConn1 += 0
23.   End if
24.   End for
25.   End for
26.   For ind = 0 to Nneighbors2.count
27.   Neighbors_Neighbors = g.neighbors_of_node(Nneighbors2[ind])
28.   For m =0 to Neighbors_Neighbors.count
29.   If Neighbors_Neighbors[m] == i || Neighbors_Neighbors[m] == j
30.   Continue
31.   End if
32.   If Neighbors_Neighbors.count – 2 > 0
33.   Sum2 += Neighbors_Neighbors.Count – 2

34.	avgNeighborsConn2 += Neighbors_Neigbors.Count – 2
35.	Else
36.	Sum2 += 0
37.	avgNeighborsConn2 += 0
38.	End if
39.	End for
40.	End for
41.	End if
42.	If Nnieghbors1.count != 0
43.	If Nnieghbors2.count != 0
44.	normalizedSocre = (sum1 / Nneighbors1.count()) + (sum2/Nneighbors2.count())
45.	Else
46.	normalizedScore = (sum1/ Nneighbors1.count())
47.	End if
48.	Else
49.	If Nneighbors2.count != 0
50.	normalizedScore = (sum2 / Nneighbors2.count())
51.	Else
52.	normalizedScore = 0
53.	End if
54.	End if
55.	If *noramlziedScore* > 0
56.	Totalpredicted ++
57.	End if
58.	Print "score of " + i " and " + j + " is " + normalizedScore
59.	Set the value of of NeighborsConn_predictedLinks[i] to j + *normalizedScore*

Algorithm 7: Common Neighbors of Neighbors

1.	For i = 1 to NumberOfNodes
2.	Nneighbors1 = g.neighbors_of_node(i)
3.	For j = 1 to NumberOfNodes
4.	If i == j continue
5.	Nneighbors2 = g.neighbors_of_node(j)
6.	AllNeighborsOfNode1 = { }
7.	AllNcighborsOfNode2 = { }
8.	Foreach n in Nneighbors1
9.	AllNeighborsOfNode1 = ((from n2 in g.NodeNeighbors(n) select n2).Concat(AllNeighborsOfNode1)).ToArray()
10.	AllNeighborsOfNode2 = ((from n2 in g.NodeNeighbors(n) select n2).Concat(AllNieghborsOfNode2)).ToArray();
11.	score = (((from n in AllNeighborsOfNode1 select n).Intersect((from n in AllNieghborsOfNode2 select n))).Distinct()).Count();
12.	if score > 0
13.	Totalpredicted++
14.	End if
15.	Set the value of CCNeighborsOfNeighbors_predictedLinks[i] to j and score

6 Experiment

C# .net was used to implement all required data structures and algorithms and to produce results. Excel was used to convert numbers into charts.

6.1 How to Sample and Evaluate Link Prediction

Due to computational complexity, test sampling is common in link prediction. Some of the ways sampling is performed is as follows [9]:

1. Select at random (completely random)
2. Any number of potential methods can select edge that present a sufficient amount of information a long a particular domain. For example, selecting only edge where each member vertex has a degree of at least 2 (semi random sampling)

In this paper, results of both methods were used. Note: testing set size is not the actual size of the entire testing test, but the number of nodes in the testing data. The experiment was conducted as follows:

(1) First the graph data is read from the dataset file.
(2) The program runs and predict links (compute the score for all pair of nodes) according to the corresponding criteria of each predicting algorithms (Common Nieghbors, Jaccard, Preferential Attachment, Adamic Adar, Random, Neighbors Connectivity, and Common Neighbors of Neighbors). There are 2699 total nodes as indicated in Table 1. Each algorithm computes a score, given two pair of nodes, and this score determines whether a link is predicted between the pair of nodes or not. For Common Neighbors and Common Neighbors of Neighbors, if the score is >= 1, then link is predicted. For other except random, if score > 0 then link is predicted. For random, a binary value (0, and 1) is generated at random, if the value generated is 1, then a link is predicted, otherwise no link is predicted.
(3) Then the precision of each algorithm is computed as follows:
    A number of pair of nodes is selected at random (candidate links) from the entire set of nodes. Total of TestingSize X (TestingSize-1) is selected. Where TestingSize is input. Each selected pair of nodes is checked with the predicted links by each algorithm, if the algorithm predicts a link between this pair, then a counter variable, *positive*, is incremented for that algorithms. After all pair of nodes are considered, the precision is calculated as $(\frac{positive}{testingSize \times (testingSize-1)}) \times 100$.
(4) The experiment is repeated with different TestingSize values. The following values were used:
    Testingsize = {540 (20%), 699(26%), 1079(40%), 1349(50%), 1699(63%)}.
(5) Results are saved into files. Then they are plotted into chart to show the performance (precision and number of positives) of each predicting algorithm.

7   Facebook Dataset

The Friendship Facebook dataset was used to compare between the link prediction methods. I chose this dataset because of the undeniable influence of social network, specifically Facebook, on social life nowadays, and because link prediction is considered a core features of those networks. Table 1 shows some information about the dataset.

The Facebook Friendship dataset is a two-column data, where the first column is the source nodes id, and the second column is the target node id. For instance, the first row in the dataset is as follows:

12
13

That means Facebook user whose id is 1 is a friend of Facebook user whose id is 2 and 3. For more information about the dataset, one can refer to the dataset website listed in the table.

Figure 4 shows the Facebook Friendship dataset as a graph of nodes and edges. Where nodes represent users and edges represent friendship links. The nodes are the black circles, and the links are the red lines between the nodes. The figure was generated using Gephi[3] Software for graph analysis. The Facebook dataset was converted into .cvs format then opened using Gephi software. These steps were necessary because Gephi accepts files with certain extensions including .cvs.

Table 1. Facebook dataset characteristic

Dataset url	http://konect.cc/networks/ego-facebook/
Info about dataset from Gephi Software	
original	nodes: 2888 edges: 2981 directed graph the dataset is not completed
Actual	it is supposed to contains data about 2981 but it contains only data about 2699 (nodes after node 2699 have no connections)
Average path length	Diameter: 5 Radius: 0 Average Path length: 2.3744337546917467
Average clustering coefficient	Average Clustering Coefficient: 0.014 The Average Clustering Coefficient is the mean value of individual coefficients
Page rank	Epsilon = 0.001 Probability = 0.85
Average degree	Average Degree: 1.032
Connected component	Number of Weakly Connected Components: 1 Number of Strongly Connected Components: 2888

---

[3] Gephi: https://gephi.org/.

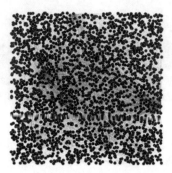

Fig. 4. The Facebook dataset

More experiments on additional dataset were conducted in [9] and [14], which show results for four (Facebook, Enron, DBLP, and Condmat) and 10 different datasets, respectively.

8   Results and Discussion

All the figures included in the following two sections show the results from the experiments. As it has been mentioned in the experiment section (Sect. 6), the experiment was produced for different testing sizes. The x-axis in Figs. 5, 6, 7, 8, 9 and 10, shows the testing sizes: 20%, 26%, 40%, 50%, and 63%.

8.1   Completely Random Sampling

The first set of figures represent the results when completely random sampling is used. Figure 5 shows the precision of Common Neighbors of Neighbors vs. precision of Common Neighbors. As it can be seen from the figure, The Common Neighbors of Neighbors has a higher precision than Common Neighbors. Figure 6 shows the number of positive predictions by Common Neighbors and Common Neighbors of Neighbors. Again, the number of positives by Common Neighbors of Neighbors is higher than the Common Neighbors. for example, when testing size is 1699 (63%) Common Neighbors of Neighbors predict 62 positive links, while Common Neighbors predicts 47 positives.

Figure 7 shows the Precision of four link prediction algorithms, Common Neighbors, Common Neighbors of Neighbors, Jaccard, and Adamic/Adar. As it can be shown, the Common Neighbors of Neighbors has a higher precision than the remaining three. Figure 8 also shows a higher number of positive predictions by Common Neighbors of Neighbors when compared with Common Neighbors, Jaccard, and Adamic/Adar.

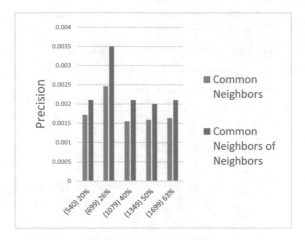

Fig. 5. Precision by Common Neighbors and Common Neighbors of Neighbors (Completely random sampling)

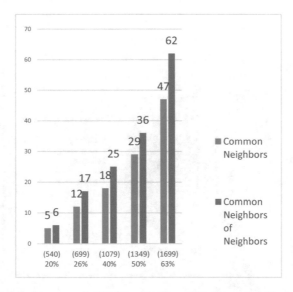

Fig. 6. Number of positives predictions by Common Neighbors and Common Neighbors of Neighbors (Completely random sampling)

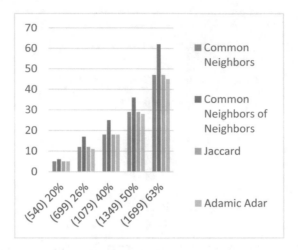

Fig. 7. Precision of four algorithms: Common Neighbors of Neighbors, Common Neighbors, Jaccard, Adamic/Adar

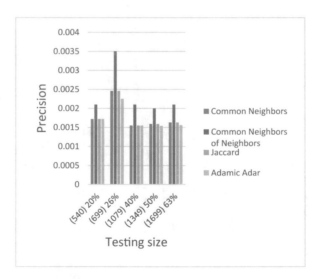

Fig. 8. Number of positive predictions by four algorithms: Common Neighbors of Neighbors, Common Neighbors, Jaccard, Adamic/Adar

8.2 Selective Semi Random Sampling

The second set of figures represent the results when semi random sampling is used with criteria of selecting edges whose any of its end nodes has degree > 2. Figures 9 and 10 shows the better performance of Common Neighbors of Neighbors when compared with performance of Common Neighbors even with semi random sampling.

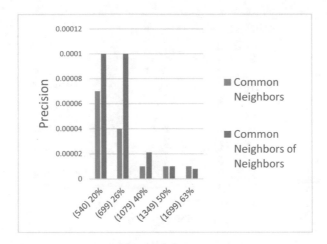

Fig. 9. Precision by common neighbors and Common Neighbors of Neighbors (Semi random sampling)

Figure 11 shows a comparison of the precision of Neighbors connectivity and Random. Neighbors Connectivity has not less precision than random. Also, the number of positives by Random and Neighbors Connectivity is almost equal as indicated by Fig. 12. Both Figures were produced when semi random sampling was used. It can be concluded that Neighbors Connectivity has a higher precision as much high as Random prediction. Also, the number of positive predictions is higher by Neighbors Connectivity.

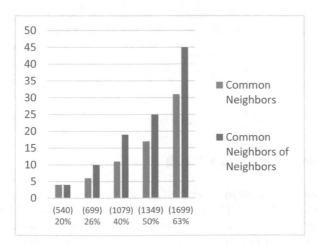

Fig. 10. Number of positive predictions by Common Neighbors and Common Neighbors of Neighbors (semi random sampling)

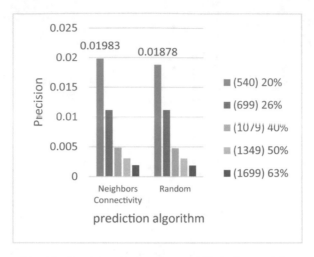

Fig. 11. Precisions by Random and Node Connectivity

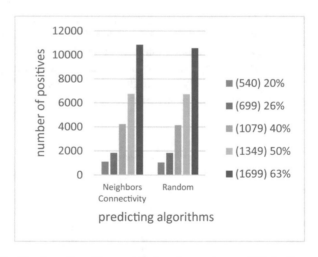

Fig. 12. Number of positives predictions by random and Node Connectivity

9   Conclusion and Future Work

This paper is an experiment with link prediction of several known methods as well as two new methods of link prediction applied on Facebook Dataset. The first proposed method, Common Neighbors of Neighbors, produces better results than Jaccard, Common Neighbors, and Adamic/Adar. The next proposed method, Node Connectivity produces results better than random predictor does. Future work can be done in continuing to extend the research by applying the link prediction methods on more datasets. In addition, Node profile can be used to assist predicting the links even though few papers already investigated that idea.

References

1. Gimenes, G.P., Gualdron, H., Raddo, T.R., Fernando Rodrigues Jr., J.: Supervised-learning link recommendation in the dblp co-authoring network. In: 2014 IEEE International Conference on Pervasive Computing and Communication Workshops, PERCOM WORKSHOPS, pp. 563–568 (2014)
2. Facebook. Help Center - Finding Friends and People You May Know. https://www.facebook.com/help/www/336320879782850. Accessed 20 Jan 2019
3. Rawashdeh, A., Rawashdeh, M., Díaz, I., Ralescu, A.: Measures of semantic similarity of nodes in a social network. In: Information Processing and Management of Uncertainty in Knowledge-Based Systems, Montpellier, France, pp. 76–85. Springer, Cham (2014)
4. Liben-Nowell, D., Kleinberg, J.: The link prediction problem for social network. J. Am. Soc. Inf. Sci. Technol. **58**(7), 1019–1031 (2007)
5. Nandi, G., Das, A.: An efficient link prediction technique in social networks based on node neighborhoods. (IJACSA) Int. J. Adv. Comput. Sci. Appl. **9**(6), 257–266 (2018)
6. Liang, Y., Huang, L., Wang, Z.: Link prediction in social network based on local information and attributes of nodes. J. Phys. Conf. Ser. **887**(1), 012043 (2017)
7. Fire, M., Tenenboim-Chekina, L., Puzis, R., Lesser, O., Rokach, L., Elovici, Y.: Computationally efficient link prediction in a variety of social networks. ACM Trans. Intell. Syst. Technol. **5**(1), 1–25 (2013)
8. Dong, L., Li, Y., Yin, H., Le, H., Rui, M.: The algorithm of link prediction on social network. Math. Probl. Eng. **2013**, 7 (2013)
9. Yang, Y., Lichtenwalter, R., Chawla, N.V.: Evaluating link prediction methods. Knowl. Inf. Syst. **45**(3), 751–782 (2015)
10. Martinez, V., Berzal, F., Cubero, J.-C.: A survey of link prediction in complex networks. ACM Comput. Surv. (CSUR) **49**, 1–33 (2017)
11. Jalili, M., Orouskhani, Y., Asgari, M., Alipourfard, N., Perc, M.: Link prediction in multiplex online social networks. R. Soc. Open Sci. **4**(2), 160863 (2017)
12. Yang, J., Zhan, X.: Revealing how network structure affects accuracy of link prediction. Eur. Phys. J. B **90**, 157 (2017)
13. Wang, T., He, X., Zhou, M., Fu, Z.: Link prediction in evolving networks based on popularity of nodes. Sci. Rep. **7**(1), 7147 (2017)
14. Fire, M., Tenenboim-Chekina, L., Puzis, R., Lesser, O., Rokach, L., Elovici, Y.: Computationally efficient link prediction in a variety of social networks. ACM Trans. Intell. Syst. Technol. (TIST) **5**(1), 10 (2013)
15. Hasan, M., Chaoji, V., Salem, S., Zaki, M.: Link prediction using supervised learning. In: Proceedings of the Workshop on Link Discovery: Issues, Approaches (2005)
16. Ibrahim, N.M.A., Chen, L.: Link prediction in dynamic social networks by integrating different types of information. Appl. Intell. **42**(4), 738–750 (2015)
17. Dai, C., Chen, L., Li, B.: Link prediction based on sampling in complex networks. Appl. Intell. **47**(1), 1–12 (2017)

Dynamic User Modeling in the Context of Customer Churn Prediction in Loyalty Program Marketing

Ishani Chakraborty[(⊠)]

University of California, Santa Cruz, USA
ischakra@ucsc.edu

Abstract. Daily human activities are dictated by real-life motivators such as schedules, locations, tastes and expectations. The similarity in the motivators can give rise to similarity in behavior. Thus, similarity in motivators can create a latent group among restaurant goers. In this paper such latent groups are identified and profiles for every latent group of restaurant goers are discovered/generated. User is modeled dynamically and HMM is used for user modeling based of the individual user's weekly restaurant visit history. The weekly restaurant visit data used in this paper are real industrial data, collected from a collaborating restaurant marketing company which promotes restaurants through offering redeemable gifts. Group profiling of the users can help predicting the possibility of churn of a particular user group or segment of users for user retention purposes.

Keywords: HMM · User modeling · Churn modeling · Loyalty Programs

1 Introduction

This paper proposes to apply Hidden Markov Model for dynamically modeling user behavior in the context of Loyalty Program domain. So to understand the context of the work, it is important to understand what is Loyalty Program. Lets assume, an user flew with airlines X and had dinner at restaurant Y. Now, the question we want to address the question is, how will the airlines X and the restaurant Y will make the user fly with airlines X again and dine in restaurant Y again; in short, we want to know how a business would retain their customers. One answer is that the business will retain customers by providing good service. But there is competition in the market and everyone gives good service. A better answer is that customer can be retained by building relationship; by showing the business-users that they are valued customers. This is the philosophy of Loyalty Programs. One of the ways to build relationships is giving rewards to valued customers. Loyalty program retains existing customers by giving rewards in many forms: discounts, free items, points, etc.

© Springer Nature Switzerland AG 2020
K. Arai et al. (Eds.): FTC 2019, AISC 1070, pp. 582–588, 2020.
https://doi.org/10.1007/978-3-030-32523-7_41

Targeting Individual users is expensive. On the other hand, targeting all of them by a common item is ineffective. A middle ground is to target similar users by launching a campaign and to do that a group or segment of users must be formed. The simplest way is to create a segment of users based on gross demographic feature selection criteria. But a more effective approach is to find the Latent groups that already exist in the users. To find these groups we cluster users by their behavior footprints. Rather than grouping users based on some aggregate statistics crudely measuring user behavior, such as user's buying history or reward-response history (Fig. 1), it is more effective to treat the user behavior sequences (User visit data or rewards response data or buying behaviour data) as generated by some distribution (generative model) and cluster the distributions. Clustering with aggregate feature is not sufficient because, first, aggregate feature such as mean will heavily depend on the length of the data. Secondly, for any Sparse sequence mean will be very close to zero. So discriminating/distinguishing one sequence from another will be very difficult. So, we need a generative model and need to infer a distribution which is generating such sequences as the data. The following are business scenarios where the proposed algorithm can be applied.

2   Business Scenario I

One important question for an advertising company running a loyalty program is when to launch a retention campaign for about-to-be-lapsed user? These campaigns are run to send rewards to the customers who are going to lapse. These rewards can be personalised [4] to make them more attractive to the user. The question that remains however is selecting an user segment which is about to be lapsed at a point of time. So along with a recommendation of the rewards/coupons, the recommender system of a B2B marketing company should generate a recommendation about the segment of the targeted customers. For example, when the user has become less frequent in the system and needs some encouragement in the form of coupon or gift; this is the time when a reward can be really effective for retaining existing customers. But generating rewards is done by a campaign process and to run a campaign when every individual user is about to drop off is resource-intensive on the marketing company. So it is important to strike a middle ground by selecting a group/segment of user instead of an individual user. This can be done by investigating the existence of latent groups among the users based on their behavior features and create a group profile for each group which will be representative of user behavior of that group of users. The recommender system should decide when this "representative user" is about to drop off of the loyalty program and recommend the corresponding user group as target group for the campaign at that point in time. So the contribution of this paper is user group profiling for recommending a target user segment at a point of time based on user behavior history.

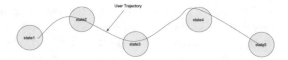

Fig. 1. User trajectory in loyalty systems

3   Business Scenario II

Another business goal of a marketing company is to increase footfall or traffic of customers in a store/restaurant for which they are running the loyalty program. For increasing the footfall promotional campaigns are run. During Campaigns rewards/gifts/coupons are sent to the users to encourage them to visit certain stores. In this business scenario it is important to know the weekly store visit pattern of the individual user so that the user could be encouraged (by sending rewards) to increase her visit frequency. As it is resource-intensive to run campaigns for individual user, we need to discover a latent group among the users based on their behavior pattern and run the tailor-made campaign for that user segment. The goal of this paper is user group profiling for such latent groups among the users based on their store visit pattern so that they can be recommended as a target user segment to the business.

4   Literature Review

4.1   Discovering Groups Among Users

The existing literature related to user group profiling does not necessarily look for latent groups that naturally existed among the users, but somehow manages to put some users together as a group for eliciting a preference of the group of users for certain product or item given the data about individual user's preference for that product or item [10]. For example in a rating based system, the objective of the existing literature is to generate a group rating for an item from individual ratings for that item. In the existing literature, the problem has been approached in two different ways. One way is aggregating the ratings of the users' group and generating a single rating for the group [5]. Another way is to create a representative user for the user group and predicting the rating by that representative user for the given item [7].

The group of users thus selected by the existing literature lack certain characteristics. First, these people may not really be connected in any intrinsic manner, they are apparently just random people thrown together. For example, MUSICFX [6] selects music channels for the music to be played in a fitness center. Based on the preferences that have been previously specified by the members who are currently working out, the system chooses one of 91 possible music channels, including some randomness in the process.

INTRIGUE [2] recommends tourist attractions for heterogeneous groups of tourists that include relatively homogeneous subgroups (e.g. "children").

However, the people working out on a particular day and time has no connection between their tastes in music! Similarly, random tourists making a tour plan have no inherent connection. Hence, they share no commonality that a group should share. In this paper, a group is a latent group, sharing similar behavior pattern. Latent groups are discovered by hierarchical Bayesian modeling of the user-visit data. Hence, they are more connected as a group and thus a more fit candidate for group profiling.

Another difference between the existing literature and this paper is that the objective of the existing literature is to predict a group rating for some item; the user is the recommender. However, in this paper user group is being recommended based on their group profile as a target group.

In case of a social network based user base, the existence of social groups is investigated [9] for estimating the probability of churning out of closely connected users from a loyalty program. The group among the users has been determined based on their social connectedness. This process does use an intrinsic connection to define a group, but it is doubtful whether the churn behavior is at all impacted by social connection. In this paper, based on user behavior data we aim to find latent groups and to create a group user profile, which is a novel idea compared to these papers.

4.2   Sequence Clustering

Sequence is an ordered list of things or events. In particular to this work we are concerned about sequence of events which are ordered by time, i.e., time series sequences. Even if we are discussing time series sequences, for reasons discussed in the introduction, the entries of the sequence are assumed to be independent of each other. So the LDA algorithm is ideologically close to this algorithm as it treats every entry of a sequence to be independent of another. The documents can be treated as a sequence and the topics as the clusters. LDA associates each word of a document to one topic and models each document as a mixture of topics. In this work, not only a sequence of user is a mixture, but each entry of the sequence is generated by a mixture of distributions. Since unlike a sequence of words of a document, the user-visit sequence is very sparse, i.e., mostly the value of each entry is zero, modeling each entry as generated by a single cluster is not as effective as it is in case of a document. Also in case of sparse sequence, Poisson distribution is a much better choice than a multinomial distribution which is used in LDA.

In general sequence classification or clustering techniques can be divided into two categories. The first technique is where the sequence itself is treated as a vector or its features are extracted and those features constitute a vector. Then some similarity measure is used to cluster those vectors. Sequence Clustering by similarity measures has been mostly applied to RNA or DNA sequences, the similarity measure being edit distance, hamming distance, longest common sub-sequence. Another way of representing and clustering sequences is done by suffix tree [1,11]. A variant of this feature extraction technique is modeling the sequence as a collection of independent entries, generated by a Poisson distribution or multinomial distribution and then computing a dissimilarity matrix. A RNA sequence clustering model was proposed using Poisson model [DM Witten]. Berninger et al. (2008) propose a method for computing a dissimilarity matrix using sequencing data. They assume that each observation is drawn from a multinomial distribution, and they test whether or not the multinomial parameters for each pair of observations are equal. Anders and Huber (2010) propose a variance-stabilizing transformation based on the negative binomial model, and suggest performing standard clustering procedures on the transformed data–for instance, one could perform hierarchical clustering after computing the squared Euclidean distances between the transformed observations.

This above technique of treating sequences as vectors is not applicable to our work. Firstly, the length of the sequences are unequal. Secondly, the entries does not represent any feature as the entries in a vector does.

The second technique of sequence modeling is using the structural information of a sequence and model it as a time-series sequence such that the generation of a later entry is dependent on former entries. These are probabilistic models, most common among them is built using HMM, of which one of the best papers is [7]. Microsoft paper uses Markov chains to model and cluster click user navigation pattern [3]. The following link discusses Microsoft's model for sequence clustering [8]. We also used HMM to model user behaviour in this paper.

5   Data Description

The data is collected from a marketing company which promotes loyalty programs for its client company. Each record contains user id, the total number of restaurant visits by the user, number of visits every week for 113 weeks and the interval between visits. In this paper, 'number of visits every week' is the user behavior that has been modeled. Almost 20000 in 25000 total number of users visited at most 7 times in 113 weeks. However, there are at least 100 users who visited almost every week. From the business perspective these 100 people are very important and cannot be treated as outliers. This skewness is the reason why a clustering technique which tries to get equal size clusters, such as K-means is ineffective because all the cluster centers merge to one. We have taken a less sparse subsection of the data for the experiment, but still our algorithm also is impacted by the skewness. Thus it is important to effectively initialize the algorithm and also it is important to prune it at certain step when all the cluster centers are not too close.

From the viewpoint of sequence modeling every User's data is a sequence made of zeros and positive numbers, each entry representing the number of user visits every week. The data is very sparse and skewed because very few users visited very frequently and most of the users visited below seven times in the whole observation period. There are huge number of zeroes in the user visit Sequences since there is no restaurant visits most of the weeks. For the experiment we have selected users who meet at least a visit threshold, empirically determined. Still for all users, zero is the value for maximum entries.

Fig. 2. HMM clustering of sequences

6   Model: Hidden Markov Model

Consider a data set D consisting of N sequences, $D = \{Y_1, Y_2, \ldots, Y_N\}$ and $Y_i = \{y_{i1}, y_{i2}, \ldots, y_{iL}\}$ is a sparse sequence of length L_i of small positive numbers. The problem addressed here is to find N latent groups in the data. The data used is described in detail in the business scenario section.N sequences are sequences of restaurant visits of N users. Thus y_{iw} is the number of times the ith user visited the restaurant in the wth week. Since the number of user-visits every week is either zero or a small positive number, Zero being the most common entry, we assume that every user-visit sequence i is generated from a Poisson distribution with mean-visits λ_{il}. We further assume that each entry of the user-visit sequences are generated from L latent states associated to k clusters. So the hidden layer of the HMM can be in one of the K distinct states indicating K clusters. *theta* is the transition probability from one cluster to another which is a parameter to infer (see Fig. 2 and Table 1).

Table 1. Example of user affiliation to five clusters

λ	θ
0.31	9.6e-4
0.05	0.3
0.07	0.56
0.3	1.03
0.09	4.0

7   Conclusion

We have modeled the user behavior using HMM and can see that there are five distinct groups among the users representing number of user visits. Every user sequence is partially affiliated to these clusters and this affiliation level is given by θ.

References

1. Agarwal, C.: Data Classification Algorithms and Applications. Chapman and Hall, London (2014)

2. Ardissono, L., Goy, A., Petrone, G., Segnan, M., Torasso, P.: Intrigue: personalized recommendation of tourist attractions for desktop and hand held devices. Appl. Artif. Intell. **17**(8–9), 687–714 (2003)

3. Cadez, I., Heckerman, D., Meek, C., Smyth, P.: Visualization of navigation patterns on a website using model based clustering. In: KDD 2000 Proceedings of the Sixth ACM SIGKDD International Conference on Knowledge Discovery and Data Mining, pp. 280–284 (2001)

4. Chakraborty, I.: Hierarchical Bayesian modeling for clustering sparse sequences in the context of user profiling in customer loyalty program (2018)

5. Masthoff, J., Luckin, R.: Future TV: adaptive instruction in your living room. In: Proceedings of the Workshop Associated with the Intelligent Tutoring Systems Conference, ITS 2002 (2002)

6. McCarthy, J., Anagnost, T.: MUSICFX: an arbiter of group preferences for computer supported collaborative workouts. In: Proceedings of the ACM 1998 Conference on Computer Supported Cooperative Work, CSCW 1998, pp. 363–372 (1998)

7. McCarthy, K., Salamó, M., McGinty, L., Smyth, B.: CATS: a synchronous approach to collaborative group recommendation. In: Proceedings of the Nineteenth International Florida Artificial Intelligence Research Society Conference, pp. 86–91 (2006)

8. Microsoft: BWorld Robot Control Software (2017). https://docs.microsoft.com/en-us/sql/analysis-services/data-mining/microsoft-sequence-clustering-algorithm-technical-reference?view=sql-server-2017/

9. Richter, Y., Yom-Tov, E., Slonim, N.: Predicting customer churn in mobile networks through analysis of social groups. In: Proceedings of the 2010 SIAM International Conference on Data Mining (SDM 2010), pp. 732–741 (2010)

10. Senot, C., Kostadinov, D., Bouzid, M., Picault, J., Aghasaryan, A., Bernier, C.: Analysis of strategies for building group profiles. In: De Bra, P., Kobsa, A., Chin, D. (eds.) User Modeling, Adaptation, and Personalization. LNCS, vol. 6075, pp. 40–51. Springer, Heidelberg (2010)

11. Xing, Z., Pei, J., Keogh, E.: A brief survey on sequence classification. ACM SIGKDD Explor. Newsl. **12**(1), 40–48 (2010)

Buddha Bot: The Exploration of Embodied Spiritual Machine in Chatbot

Pat Pataranutaporn[1,2,3(✉)], Bank Ngamarunchot[2,3], Korakot Chaovavanich[1,3],
Sornchai Chatwiriyachai[2,3], Potiwat Ngamkajornwiwat[2,3],
Nutchanon Ninyawcc[2,3], and Werasak Surareungchai[2,3]

[1] Massachusetts Institute of Technology (MIT), Cambridge, MA 02139, USA
patpat@media.mit.edu
[2] Futuristic Research Cluster (FREAK Lab), Bangkok 10150, Thailand
[3] True Corporation Public Co., Ltd., Bangkok 10150, Thailand

Abstract. We presented the concept of "Embodied Spiritual Machine",
a type of technology that embody the known religious/spiritual figure by
learning the figure's remaining artifacts. We implemented Buddha Bot as
an example of the embodied spiritual machine using NLP and semantic
computing. Through online participatory study, we presented our preliminary findings on the user dialogues and feedback of the Buddha Bot.
The results gave us the insights to the perceptions and experiences of
the users on the embodied spiritual machine.

Keywords: Embodied spiritual machine · Chatbot · Buddhism ·
Religious · Conversational agent

1 Introduction

The Pew Foundation's Internet and American Life Project has reported that
64% of the 128 million US internet users have used technology for religious purposes [8,13]. For a decade, Intel researchers, Genevieve Bell, have urged the
Human-Computer Interaction (HCI) research community to explore the interaction between humanity, spirituality, and technology [1,3]. Still, the intersections
of the areas are still underexplored. We identified two existing circumstances in
which technology play a critical role in the practice of faith.

First, the "techno-spirituality": the uses of technology to support spiritual
and religious activities [2]. Researchers have studied Techno-spiritual practices
including of online mediation video [6], and mobile phones for enhancing the
bonds between members in the religious community [17], spiritually oriented
mobile application [7], and the discussion of religious content on digital platforms
[2,5]. This technology served as the tools for empowering religious and spiritual
institutions by better engaging and reaching out to their followers.

Second, the "computational theocracy": Ian Bogost argued that the complexity level of modern technology had reached the point where human put faith into

© Springer Nature Switzerland AG 2020
K. Arai et al. (Eds.): FTC 2019, AISC 1070, pp. 589–595, 2020.
https://doi.org/10.1007/978-3-030-32523-7_42

the system in navigating everyday lifestyle, thus created the notion of "Computational theocracy", where technology became the new religious and spiritual
institution [4,10]. Furthermore, there has been attempts to build "Divine AI",
the artificial intelligence that acts as a new spiritual entity [12]. This kind of
technology serves the same functionality as the normal religious institutions
by providing social cohesion, religious-based morals, and meaning to existential
questions.

This paper presents the idea of "Embodied Spiritual Machine", the kind
of technology that exists between the first and second categories. Similar
to computational theocracy, the embodied spiritual machine itself is a reli-
gious/spiritual entity; however, it relies on the existing context and institution
through the embodiment of knowledge, personality, and behavior of the known
religious/spiritual figure by learning the remaining artifacts of the figure. Examples of the embodied spiritual machine included robots that portrayed the religious figures such as God-Jesus Robot – a desktop robot for divination and life
guidance [18], Minda – the humanoid robot that was modeled after Kannon
Bodhisattva, the Buddhist Goddess of Mercy for teaching scriptures [14], and
more [19].

2 Methodology

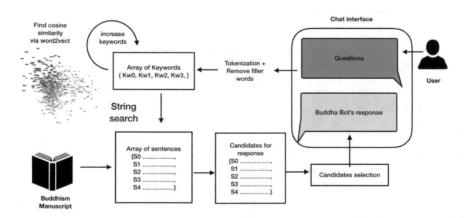

Fig. 1. The architecture of Buddha Bot system.

We explored the concept of the embodied spiritual machine by developing
Buddha Bot, a chatbot that learn and personify the manuscripts of Buddhism
to answer questions from the users. The architecture of Buddha Bot system is
shown in Fig. 1. The rationales behind the chatbot interface are:

(1) The conversational agent is one of the most promising user interfaces for
    various digital transactions with a growing interests from HCI researchers
    and industries [11].

(2) The conversational agent can create the sense of an "embodied conversational character" [9,20], a notion that the agent can personify human-human interaction, allowing the user to have trust in the agent.
(3) The prior exploratory study on natural language processing (NLP) for giving answers from religious texts demonstrates a promising opportunity for using the conversational agent in the spiritual and religious context [16].

In our experiment, we used the online participatory research approach by inviting the social media followers of our research group's social media page to interact with the Buddha Bot. The invitation went viral, and we evaluated over 2,000 conversations between the bot and the participants online to understand the human's experience and perception toward the embodied spiritual machine. The chat interface of Buddha Bot is shown in Fig. 2.

Fig. 2. The chat interface of Buddha Bot.

2.1 Designing the Buddha Bot

Buddha Bot platform is designed to be an online space where the user can communicate with the digital embodiment of the Buddha. The visual narrative and the avatar image of the Buddha Bot is a combination of symbols from Buddhism and glitch aesthetics to communicate the notion of computation and spirituality. The opening question when the user enters the platform (the initial prompt) is "What is enlightenment?" to guide the user to type other relevant questions.

2.2 Creating the Responses

When the user types a question to the platform, the Buddha Bot algorithm tokenizes the input sentence, and then removes any filler words. The remaining set of keywords is then used as a search query to find candidate responses in the corpus of Buddhist manuscripts: "The Eightfold Path". In the case that the keywords do not match any sentences in the corpus, the algorithm would expand the search query by assigning spatial word vector to each word based on the pre-trained 5,000 dimensional vectors of Word2Vec [15] and then perform spatial similarity calculation to find semantically similar words to add to the query. This process of creating a set of candidate responses from the corpus via keyword search and expansion would loop until the number of cycle reaches threshold. The candidate responses were then randomly selected to diversify the answers from the Buddha Bot.

2.3   Conversation Analysis and Feedback

We used unsupervised clustering approach to explore the conversational topics in the user's dialogues. We tokenized the conversational texts, removed stop words, and then assigned word vectors from word2vec pre-trained vectors from Google News corpus. The word vectors were clustered into groups using K-Means clustering algorithm and then evaluated on the number of classes by the researchers. The classes created by this unsupervised process were then used to perform keyword matching to cluster the conversations into themes. We launched a preliminary survey questions on the user's experience inside the platform after the user have exchanged more than 6 dialogues with the Buddha Bot. To mitigate any confusion, we created another conversation agent with a different avatar and visual appearance that would ask the survey questions instead of the Buddha Bot. We perform axial coding on the conversational texts to group the user's feedback.

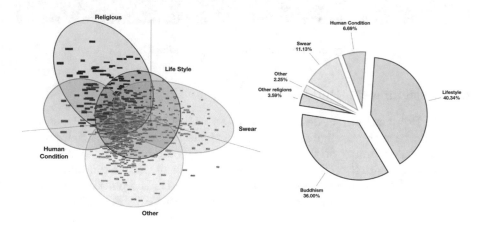

Fig. 3. Results from the conversation clustering using K-Means unsupervised learning.

3   Results and Discussion

Based on 2,716 conversations from 805 participants, the average conversational session is between 3–4 dialogues (SD = 2.81), and each conversation has an average word count of 3.46 (SD = 2.79). The result from K-Means unsupervised clustering on the keywords from the dialogues shows 4 themes in the conversations:

religious (Buddhism: Enlightenment, Buddha, Nirvana, Buddhism, Dhamma, ... and other religions: Allah, God, Jesus, ...), lifestyles (crush, boyfriend, photography, food, supermarket, ...), human condition (love, karma, sin, sad, happiness, ...), swear (Wow, Wtf, ohhhh, ahh, hummm). By matching keywords from each theme to classify the conversation at the sentence level, we identify that the most common theme in the participant's dialogues in order are lifestyle (40.34%), Buddhism (36%), swear (11.13%), human condition (6.69%), other religions (3.59%), and other (2.25%). Results from the conversation clustering using K-Means unsupervised learning are shown in Fig. 3.

For the most common themes, the participants tended to ask about two things: general lifestyle and their personal problems. For the lifestyle related conversations, example questions include "How to forgive and focus", "What should I do with my girlfriend and enemy". For Buddhism-related questions, there are questions about the specific ideas in Buddhism such as "What are the five perceptions?". Most of the questions are in the form of "what is X", where "X" are: reincarnation, hell, karma, reality, nirvana, transcendence, etc.

Based on the open-ended survey questions, we found that 37.42% of the users react positively to the Buddha Bot, 28.22% respond negatively, and the other one-third (33.31%) gave either neutral or suggestive reactions. For the positive and neutral responses, the participants usually expressed their feelings and satisfactory without providing any reasons. In contrast, the responses from the negative or suggestive types are more informative. The majority of them relates to the accuracy of the bot. In their opinions, some of the bot's responses were not relevant enough to their questions.

4   Conclusion and Future Works

The Buddha Bot experiment was preliminarily conducted to explore the idea of the embodied spiritual machine, which demonstrates how technology could interact with human by embodying a spiritual and religious figure. We purposed some future directions that can be taken to further explore the concept.

4.1   Efficiency Improvement

We analyzed the feedback from participants and found that some of the answers were irrelevant to the questions. This challenge is due to the fact that: First, the corpus text was written in a different context from today. Therefore, the most relevant answer from the corpus might still be distant from the expectation of the user. This presents an opportunity to apply a generative language synthesis model to create new responses for the questions. And second, when the user asked about the topics not covered in the corpus text, the used of word vectors could be ambiguous. Consider that the bot is embodying the figure of a religious or spiritual entity, this scenario poses an interesting design decision whether the bot should simply not respond, respond with an excuse for not knowing the answer, or answer with the most relevant answer.

4.2   Multiple Embodiment

It is possible to have more than one figure for the bot to embody at once. This presents an interesting opportunity for the HCI community to engage and study how technology could be used as the agent for mediating interfaith and inter-religious activities.

4.3   Mixed Methods Studies

Mixed methods studies are crucial to understand the deeper dimensions of the Buddha Bot as the embodied spiritual machine. In this paper, we presented the use of social media as a platform for launching the study as well as collecting large preliminary responses. However, the mixed methods studies on both online and in person experiment could be the next step to further investigate the spiritual relationship between human and technology.

References

1. Bell, G.: The age of auspicious computing? Interactions **11**(5), 76–77 (2004)
2. Bell, G.: No more SMS from Jesus: Ubicomp, religion and techno-spiritual practices. In: International Conference on Ubiquitous Computing, pp. 141–158. Springer (2006)
3. Bell, G.: Messy futures: culture, technology and research. In: Proceedings of the SIGCHI Conference on Human Factors in Computing Systems. ACM (2010)
4. Bogost, I.: The cathedral of computation. The Atlantic **15**, (2015)
5. Brasher, B.E.: Give Me That Online Religion. Wiley, London (2001)
6. Buie, E., Blythe, M.: Meditations on youtube. In: Proceedings of the 6th International Conference on Designing Pleasurable Products and Interfaces, pp. 41–50. ACM (2013)
7. Buie, E., Blythe, M.: Spirituality: there's an app for that! (but not a lot of research). In: Extended Abstracts on Human Factors in Computing Systems, CHI 2013, pp. 2315–2324. ACM (2013)
8. Campbell, H.: Making space for religion in internet studies. Inf. Soc. **21**(4), 309–315 (2005)
9. Cassell, J., Bickmore, T., Billinghurst, M., Campbell, L., Chang, K., Vilhjalmsson, H., Yan, H.: Embodiment in conversational interfaces. In: Human Factors in Computing Systems (CHI 1999), pp. 520–527 (1999)
10. Finn, E.: What Algorithms Want: Imagination in the Age of Computing. MIT Press, Cambridge (2017)
11. Følstad, A., Brandtzæg, P.B.: Chatbots and the new world of HCI. Interactions **24**(4), 38–42 (2017)
12. Harris, M.: Inside the first church of artificial intelligence — backchannel, February 2018. https://www.wired.com/story/anthony-levandowski-artificial-intelligence-religion/
13. Hoover, S., Clark, L.S., Rainie, L.: Faith online: 64% of wired Americans have used the internet for spiritual or religious information. Pew Internet and American Life Project (2004)

14. McGleenon, B.: Robot 'GOD': AI version of Buddhist deity to preach in Japanese temple (2019). https://www.express.co.uk/news/world/1091915/Japan-robot-God-robot-buddhist-Kyoto-temple-Kannon-Bodhisattva

15. Mikolov, T., Chen, K., Corrado, G., Dean, J.: Efficient estimation of word representations in vector space. arXiv preprint arXiv:1301.3781 (2013)

16. Shawar, A., Atwell, E.: An Arabic chatbot giving answers from the Qur'an. In: Proceedings of TALN04: XI Conference sur le Traitement Automatique des Langues Naturelles, vol. 2, pp. 197–202. ATALA (2004)

17. Sterling, R., Zimmerman, J.: Shared moments: opportunities for mobile phones in religious participation. In: Proceedings of the 2007 Conference on Designing Pleasurable Products and Interfaces, pp. 490–494. ACM (2007)

18. Wallace, M.: Museum of curious toys and games: Bandai's "god-jesus" robot (1984) (2018). https://medium.com/@margaretw/museum-of-curious-toys-games-bandais-god-jesus-robot-1984-162722d38a9

19. Williams, R.J.: The Buddha in the Machine: Art, Technology, and the Meeting of East and West. Yale University Press, London (2014)

20. Zamora, J.: I'm sorry, Dave, I'm afraid I can't do that: Chatbot perception and expectations. In: Proceedings of the 5th International Conference on Human Agent Interaction, pp. 253–260. ACM (2017)

My Personal Brand and My Web Presence: Mining Digital Footprints and Analyzing Personas in the World of IOT and Digital Citizenry

Fawzi BenMessaoud[✉], Taryn Elizabeth Husted,
Dwight William Hall, Holly Nichole Handlon,
and Niranjan Valmik Kshirsagar

IU School of Informatics and Computing,
Indiana University Purdue University Indianapolis, Indianapolis, USA
{fawzbenm, thusted, dwwhall, hhandlon, nvkshirs}@iu.edu

Abstract. Personal Branding is a way to elevate oneself in a new marketing concept. For many, building and managing a Personal Brand is about value, mission, image, and vision. However, with the shift in the marketplace and the rise of the Internet of Things (IoT), Personal Branding has become a data function and summation of one's Digital Footprint, which can impact how people are seen by Databots and how others observe them. Subsequently, there has been an upsurge in the utilization of deep machine learning and predictive analytics to profile as well as select talents using their Personal Brand as a data function of their Web Presence/Digital Citizenry. To find a solution, a survey was conducted to examine the number of students who are not fully aware of their Web Presence/Digital Citizenry and how their social and professional networks can and will be used to profile them. After identifying this problem, the idea of building and managing a Personal Brand was explored. To solve this problem, a Personal Branding tool that allows students to develop and promote their Personal Brand using a Digital Passport to express, share, and manage a multi-dimensional presentation of all their academic and non-academic achievements.

Keywords: Digital Citizenry · Web Presence · Digital Footprints ·
Personal Branding · Internet of Things · Predictive analytics ·
Machine learning · Social Media manager · Web 2.0

1 Introduction

Traditionally, the term, brands, has been associated with businesses, products, services, or organizations. As technology continues to grow, however, brands can be extended to also include humans. The rise of using Professional Social Networks, such as LinkedIn, is an example of brands extending to humans. As Hood, Robles, and Hopkins [1] mentioned, a surfeit of hiring managers looked at potential job candidates' Social Media profiles, such as Facebook, throughout the process of recruitment and hiring.

© Springer Nature Switzerland AG 2020
K. Arai et al. (Eds.): FTC 2019, AISC 1070, pp. 596–604, 2020.
https://doi.org/10.1007/978-3-030-32523-7_43

Personal Branding has been an integral part of marketing since a plethora of companies research their potential employees through their digital presence [1]. Likewise, the value of students' degrees and how they are applicable to employers has become an integral discussion among businesses [5]. Because of this, managing one's Digital Footprint has become imperative. A Digital Footprint can show digital traces that have been left by people as they use various services on the internet [2]. Similarly, a person's Digital Footprint can be analyzed to gain insights on individuals [2]. As the job market continuously changes and further shifts toward technology driven methods, it becomes crucial that people realize the importance of their Personal Brand and Digital Footprint as well as the impact that it can have on their career.

The purpose of this research is based on the hypothesis that people with more professional accounts have an increased awareness of their Personal Brand and Digital Footprint. Alternatively, the other part of the hypothesis is that the people who have fewer professional accounts have a decreased awareness of their Personal Brand and Digital Footprint. To test this hypothesis, a survey was conducted. The respondents were scrutinized to gauge their awareness of their Personal Brand, the number of Social Media accounts that they use, the number of Professional Social Network accounts that they utilize, and their willingness to use a technology that would allow them to manage their Personal Brand and Digital Footprint. Furthermore, the survey gathered respondents' inclination to use an application that would allow them to build a Digital Passport to hold their academic and non-academic achievements, experiences, skills, interests, aptitudes, and career highlights to share with potential employers.

In addition to the survey, a literature review was also conducted. From this research, it became apparent that professionalism among students is paramount. As Chretien, Goldman, Beckman, and Kind [3] noted in their article, *It's Your Own Risk: Medical Perspectives on Online Professionalism,* differentiating between professional and unprofessional behavior is difficult. While some areas in which to delineate what could be considered as unprofessional are easier to understand, trying to find a balance can be difficult [3].

Additionally, linked to one's personal brand is their Web Presence, particularly in the context of Social Media [4]. The significance of one's Personal Brand is indubitably important since it can have such a high impact on an individual's career [4]. In order to allow students to realize this, an application was conceptualized for them to be able to check their Web Presence as well as Personal Brand and Digital Footprint. Moreover, the students are also able to upload information about themselves such as their resume, portfolio that denotes what they learned, and a video. Another solution for the problem that students face as their Personal Brand and Digital Citizenry becomes scrutinized by potential employers is to also create a space where students have a digital space, or a "vault", for their information and be able to have a Digital Passport. Similar technology, such as Google Docs or Microsoft Dropbox, also use a similar method to save and share files, but the "vault" feature of this application would allow students to have a specific space for their information, which would have extra security features for more sensitive and restricted document files and personal information.

To build a professional Personal Brand and Digital Footprint, adhering to the concept of value, mission, image, and vision is paramount. When posting information about oneself, it is important to weigh the value of the information as well as the mission of this data. Similarly, building an image and having a vision is equally as significant since they put an individual's goals into the forefront. The formula based off this is:

$$\int (Personal\ Brand) = \sum_{i=1}^{\infty} DFP(Image + Value + Mission + Vision)^t$$

Fig. 1. Personal Brand formula

Each of these values is fundamental to a person's Personal Brand. The meaning of this formula is that Personal Brand is equal to the sum of an individual's Digital Footprint, which is a conglomeration of image plus value plus mission plus vision.

With the rise of the usage of Databots and deep machine learning in the World Wide Web 2.0, predictive learning has become popular since a plethora of information can be about users based off the online data that Databots recover [6]. However, depending on the information found, it could negatively impact a person since Databots only see facts and not the possible reason for the result. For example, if a person were to post a picture of them going to a restaurant that also had an image of them standing next to a bar, the Databots could see that and make the conclusion that the individual frequents bars, which would harm someone's credibility with potential employers. Due to this, the need to find a solution that allows students to be able to build and manage their Personal Brand and Digital Footprint has become necessary.

2   Methods

In choosing a method of study, a survey was chosen in order to reach the group of respondents that would be using this application. Likewise, the survey approach was chosen because it met the criteria of this project since it needed input from the targeted user group. The study was conducted through Google Forms and dispersing the links through classes at Indiana University Purdue University, Indianapolis where students would take the survey and give feedback. By giving the survey to students, the research was able to include the opinion of the primary users of the application. The survey itself included questions that asked students to opine on their awareness level of their Personal Brand, the number of Social Media accounts, as well as the amount of Professional Social Network, accounts that they had, and their willingness to use an application like the Personal Branding and Digital Citizenry system.

The survey was left open for several weeks. The hope was that, by leaving it open for a longer period of time, more people would be able to participate in the research. The analytics provided by the Google Forms were used to analyze the information that the participants contributed to make the conclusions. The survey itself was split into different sections; each section being applicable to three topics that were being tested. By doing this, the team was able to form a working knowledge of the level of familiarity that the participants had for the topic. For example, the questions, "How much are you aware of the Personal Brand?", was a question that was asked to gauge the level of awareness that participants had of this topic. Additionally, this was asked in order to gain insight into what the general public may or may not know about the subject.

3   Findings

The data that was collected from the surveys held a number of multifarious patterns that were found in the process of analyzation. The survey had 55 volunteer participants, which enabled the team to gather a surfeit of information that was applicable to the topic.

The primary interest was in determining if people considered themselves a Personal Brand. From the results, which can be seen in Fig. 1, the majority of the respondents answered positively, "Yes". 62% considered themselves a Personal Brand while 38% did not believe that they were a Personal Brand.

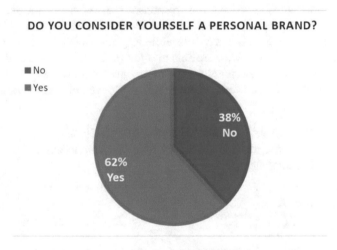

Fig. 2. Do you consider yourself a Personal Brand

Another question that was asked was "How much are you aware of the Personal Brand?" This was asked in order to gauge the respondents' level of awareness regarding Personal Brand. The results indicated that 51% of the respondents answered that they were "Somehow Aware". A small percentile (23%) responded that they were "Very Aware". However, 13% of the individuals who took the survey noted that they were either "Not Sure What My Personal Brand is" or they "Have No Idea". The breakdown of the results can be seen in Fig. 2 as it shows a histogram representation of the respondents' answers.

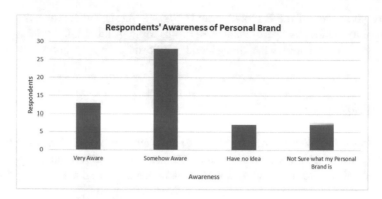

Fig. 3. Respondents' awareness of Personal Brand

Interestingly, when asked regarding the number of Social Media accounts, such as Facebook or Instagram, that the respondents have used in the past or are still using, the majority (53%) answered that they had fewer than five Social Media accounts. Furthermore, the area with the second most responses was between five, but less than 10 (33%). However, the preponderance of individuals (84%) stated that they had less than five Professional Social Network accounts while very few stated that they had more than five (7%). From these results, we were able to infer, based on Fig. 3, that the number of students who had a greater awareness of their Personal Brand and Digital Footprint had more Professional Social Network accounts than those who are not as aware.

Respondents Awareness of Personal Brand & Digital Footprint Compared to Number of Professional and Social Media Accounts

		Number of Professional Media Accounts				
Respondents Awareness of Personal Brand & Digital Footprint	Number of Social Media Accounts	I have several but not sure how many	Less than 5	More than 5 less than 10	More than 10	None
Have no idea	Less than 5		9.09%			1.82%
Not sure what my Personal Brand is	Less than 5		7.27%			
	None		1.82%			
Somehow aware	I have several but not sur..		1.82%			
	Less than 5		21.82%			
	More than 5 less than 10	1.82%	12.73%	5.45%		1.82%
	More than 10			1.82%		
	None		3.64%			
Very aware	Less than 5		12.73%			
	More than 5 less than 10		10.91%			
	More than 10				3.64%	
	None		1.82%			

% of Total Respondents broken down by Number of Professional Media Accounts vs. Respondents Awareness of Personal Brand & Digital Footprint and Number of Social Media Accounts . Percents are based on the whole table.

Fig. 4. Respondents' awareness of Personal Brand & Digital Footprint compared to number of Professional and Social Media accounts chart

In addition to looking at Social Media accounts, respondents' willingness to use an application that would allow them to build and/or manage their Personal Brand, Web Presence, and Digital Footprint was also evaluated. Based on these findings, the respondents had an inclination to utilize an application that would allow them to build and manage their Personal Brand as well as other related activities. Of the results, 71% were positive. From this and comparing the respondents' awareness of their Personal Brand and Digital Footprint, it can be suggested that students who have more incli-nation toward using our application also have more awareness in these areas as it can be seen in Fig. 4.

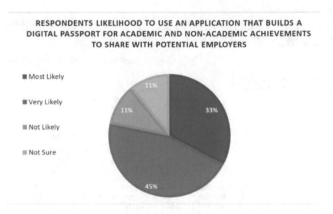

Fig. 5. Likelihood to use application that builds a Digital Passport for academic & non-academic achievements to share with potential employers

The results shown in Fig. 5 are closely correlated with the number of people who are aware of their Personal Brand and Digital Footprints. Based on the results shown in the pie chart above, it was noticeable that as a person's awareness level of their Personal Brand and Digital Footprint rose, so did their likelihood of using an appli-cation that would allow them to make a Digital Passport Application.

As an effort to study how likely users would be to use an application that would allow them to manage their Personal Brand and Digital Footprint, a question was included about this to gauge their level of interest. Over 70% of the respondents answered positively that they would be likely to use an application like the Digital Passport as it can be seen in Fig. 6. It can also be seen how the respondents are aware of their Personal Brand and Digital Footprint in conjunction to their willingness to use an application like this. Figure 7 shows the likelihood of using digital passport application compared to awareness of personal brand and digital footprint.

Likelihood to use Application Compared to Awareness of Personal Brand and Digital
Footprint

Fig. 6. Likelihood to use application compared to awareness of Personal Brand & Digital Footprint.

Likelihood of Using Digital Passport Application Compared to Awareness of Personal Brand & Digital Footprint

Likelihood of Respondent Using an Application That Builds a Digital Passport That Holds Academic and Non-Academic Achievements	Respondents Awareness of Personal Brand & Digital Footprint			
	Have no idea	Not sure what my Personal Brand is	Somehow aware	Very aware
Very likely	1.82%	5.45%	27.27%	10.91%
Most likely	7.27%		12.73%	12.73%
Not likely	1.82%	3.64%	3.64%	1.82%
Not sure			7.27%	3.64%

% of Total Respondents broken down by Respondents Awareness of Personal Brand & Digital Footprint vs. Likelihood of Respondent Using an Application That Builds a Digital Passport That Holds Academic and Non-Academic Achievements. Percents are based on the whole table.

Fig. 7. Likelihood of using Digital Passport Application compared to awareness of Personal Brand & Digital Footprint

Another question gathered students' opinions regarding whether or not they believed that an application that allows them to store, manage, and share their projects like a Capstone is needed. Based on the answers, 82% believed that there should be an application like this. To view a visual representation of this data, see Fig. 8.

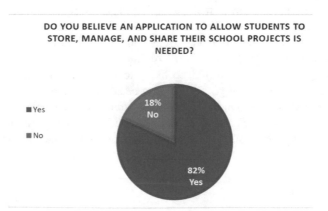

Fig. 8. Do you believe an application to allow students to store, manage, and share their school projects is needed.

4   Solution

Based on the research gathered, there appears to be an acute need for an application that allows students to be able to build and manage their Personal Brand, Web Presence, and Digital Footprint. Because of this, the development of the Personal Branding and Digital Citizenry (PBDC) system is important for students and universities. The system provides both schools and professional organization with a Capstone and Career Fair management application that allows the school to schedule and manage secure access to view and rate students' capstone or final course projects. Similarly, the PBDC is also a powerful Personal Branding tool that allows students to develop and promote their own Personal Brand. Students can also measure and manage their career using a Digital Passport to express, share, as well as manage a multi-dimensional presentation of all their academic and non-academic achievements, experiences, skills, interests, aptitudes, and career highlights.

Similarly, another part of the solution is to create a Digital Citizenry tool in addendum to the PBDC system that allows students to become fully aware of their Web Presence and manage, facilitate, and monitor their Social and Professional Networks with a built-in Digital Footprint Analyzer and their own Social Media manager to supervise their Web Presence.

MyLibrary, a personal and smart digital object repository, is another part of the PBDC system that interfaces with the campus' Learning Management Systems. The MyLibrary application enables students to build, deposit, and maintain their own virtual library shelves of their favorited course materials that they wish to save for future use. Inside the MyLibrary application, a Digital Vault is part of the system that would be hidden in students' virtual library shelves which would allow them to safeguard and protect their digital assets, private information, and documents. For added security, Blockchain would be added to the Digital Vault.

Moreover, the PBDC solution also offers perpetual efficacy due to its unique nature. The application is built with a combined 360° approach for students and aspiring professionals to be accountable for their career. Through this, recruiters, admissions, and employers can in turn, accountably attract, manage, retain, and grow talent more efficiently. This can be done by providing best-in-class Personal Branding & Digital Citizenry management support for students and aspiring career professional, and best-in-class management support for academic admins, career services, recruiters/admission, and employers; both campus and professional organizations will change the way students manage their Personal Brand and careers.

Other notable features of the PBDC is that it would enable students to create a more inclusive portfolio is the built-in Digital Rights Management (DRM), Copyright Tracking with Audit Trail and Version Control, and the ability to hold and manage all types of files. Likewise, students can look at metadata for searching, browsing, and discovery as well as configurable workflows with user roles and permissions. They can also use access-controlled collections with customizable classification and web service interfaces for various types of Learning Management Tools such as Canvas or Blackboard.

5 Conclusion

As technology continues to grow, the importance of managing one's Personal Brand and Digital Footprint becomes paramount. Students' actions online matter as well as their Web Presence. Previously, brands have been limited to various products or business, but not, brands can encompass humans. The current era is one of technology, which increases the importance of Professional Social Networks and peoples' presence on those networks. Because of this, the survey that was conducted found out that the majority of the respondents had some awareness of their Personal Brand. Additionally, the individuals who had more awareness also appeared to have more Professional Social Network accounts than those who had less awareness of their Personal Brand. Due to these facts, our team concluded that the initial hypothesis that stated that people with more professional accounts have an increased awareness of their Personal Brand and Digital Footprint appeared to be correct based off of the correlations between the data. The need for an application like the PBDC system is becoming more important as the influence of the World Wide Web 2.0 becomes more prevalent in society.

References

1. Hood, K.M., Robles, M., Hopkins, C.D.: Personal branding and social media for students in today's competitive job market. J. Res. Bus. Educ. (2014). https://www.questia.com/library/journal/1P3-3948508701/personal-branding-and-social-media-for-students-in. Accessed 4 Apr 2019
2. Hinds, J., Joinson, A.N.: What demographic attributes do our digital footprints reveal? A systematic review. PLoS ONE (2018). https://doi.org/10.1371/journal.pone.0207112. Accessed 4 Apr 2019
3. Chretien, K.C., Goldman, E.F., Beckman, L., Kind, T.: It's your own risk: medical students' perspectives on online professionalism. Acad. Med. (2010). https://www.ncbi.nlm.nih.gov/pubmed/20881708. Accessed 4 Apr 2019
4. Kawano, Y., Obu, Y., Kishimoto, Y., Yamaguchi, T., Nunohiro, E., Yonekura, T.: A personal branding for university students by practical use of social media. In: 15th International Conference on Network-Based Information Systems, September 2012. https://ieeexplore.ieee.org/document/6354941. Accessed 4 Apr 2019
5. Staton, M.: The degree is doomed. Harvard Bus. Rev. (2014). https://hbr.org/2014/01/the-degree-is-doomed. Accessed 4 Apr 2019
6. O'Dell, C.: AI, machine learning, and the basics of predictive analytics for process management. In: Predictive Analytics World, August 2018. https://www.predictiveanalyticsworld.com/patimes/ai-machine-learning-and-the-basics-of-predictive-analytics-for-process-management/9632/. Accessed 4 Apr 2019

Formalizing Graph Database and Graph Warehouse for On-Line Analytical Processing in Social Networks

Frank S. C. Tseng[1($\boxtimes$)] and Annie Y. H. Chou[2]

[1] Department of Information Management,
National Kaohsiung University of Science and Technology,
Kaohsiung, Taiwan, ROC
imfrank@nkust.edu.tw
[2] Department of Computer & Information Science,
ROC Military Academy, Kaohsiung, Taiwan, ROC
imyhchou@gmail.com

Abstract. Graph databases have been widely employed for representing connected pieces of information for different kind of domains. The data model embraces relationships as a core aspect to connect objects, and organizes everything into a network for efficient query processing of versatile applications, e.g., on-line social networking, metropolitan traffic modeling, marketing channels simulations or even counterterrorism analysis. As an emerging technology for encoding network structures, graph databases are also widely used as an infrastructure for social network analytics, which help us understand some phenomenon or hidden knowledge in buzz marketing, technology trends or public issues regarding social behaviors. Although many graph database management systems have been developed, there are still no formal definitions for theoretical graph database modeling. In this paper, we will present a formal definition for graph database model, extend the concept of data warehouse into graph warehouse, and define the basic elements of a graph warehouse for the development and derivation of graph-based multi-dimensional business intelligence through on-line analytical processing (OLAP) on graph databases.

Keywords: Graph database · Graph warehousing · Social-related business intelligence · Social networking

1 Introduction

Social media proliferate drastically and create a global channel as discussed in [6], intermixed with a bunch of information, knowledge, and concepts regarding products information, social behaviors, or even recreational activity recommendations, which cultivate a lot of influential internet celebrities in the cyber-community. To understand the cyber-community, on one hand, for resolving social problems, like counter-terrorism, cyber war or cyberbully; or even, on the other hand, for developing agile management skills for human welfare, we need an internal structure for encoding social networks.

© Springer Nature Switzerland AG 2020
K. Arai et al. (Eds.): FTC 2019, AISC 1070, pp. 605–618, 2020.
https://doi.org/10.1007/978-3-030-32523-7_44

As indicated by Vicknair *et al.* [9] and Ho, Wu and Liu [3], netizen and their relationships or activities in social networks can be seamlessly expressed in most of the matured graph database management systems. When cyber-communities are transformed into graph database structure, the opinion leaders of various areas or cyber warriors can be identified through an inquiry process by comparing their in-between behaviors and relationships. Besides, as relational data warehouse technologies has been well developed, it is also fruitful to combine these technologies to conduct on-line analytical processing (OLAP) for exploring the business intelligence hidden in social networks, namely *social business intelligence*, by collaborating graph database and relational data warehousing technologies in a seamless manner.

However, many of the well-developed relational technologies have not been formally extended or adjusted for their counterparts in graph databases. In this paper, we will present a formal definition for graph database model, extend the concept of data warehouse into graph warehouse, and present the formal definitions of the basic elements of graph warehouse. Graph warehouses, unlike graph database management systems, include extensive semantics regarding relationships between objects, provide efficient feature extraction and indexing functionalities and offer flexible summarization from different perspectives for grouping or clustering subgraphs to provide a more subtle and elegant access to socially-related business intelligence.

2   Formal Graph Data Modeling

2.1   Mathematical Graph Definition

The theoretical base of graph data model stems from the graph theory. Bondy and Murty [1] defined a graph as $G = (V, E, \psi)$, where

1. $V = \{v_1, v_2, \ldots, v_i, \ldots, v_n\}$ denotes a non-empty set of vertices (or nodes),
2. $E = \{e_1, e_2, \ldots, e_i, \ldots, e_m\}$ contains a set of edges, disjointing from V.
3. ψ is an incidence function that associates with each edge of G, an unordered pair of vertices of G. If e is an edge and u and v are vertices such $\psi(e) = (u, v)$, then e is said to join u and v; the vertices u and v are called the ends of e.

Edges can be regarded as directed or undirected. For directed edges $(u, v) \neq (v, u)$; otherwise $(u, v) = (v, u)$. For an undirected edge (u, v), a line is used to connect two nodes when drawing the graph, i.e., $u - v$. For a directed edge (u, v), an arrow will be depicted to connect two nodes, $u \to v$, to express the direction, when drawing the graph.

2.2   A Formal Definition of Graph Database Model

By respectively regarding nodes and edges as objects and relationships, we group objects of the same category into *object types* and relationships of the same type into *relationship types*, and extend the mathematical *graph* definition by adding schemas (consisting a set of attributes) to these types for our graph database modeling. Besides, as there are possibly multiple relationships between two objects in a graph database, the model should be a labeled multi-digraph, such that each label represents a relationship with associated attribute values; each object has its own attribute values; labels of the same type can be classified into the same relationship types; and objects of the same type can be grouped into an object type, which can be defined as follows.

Definition 1: A *graph database* is a multi-digraph with labeled vertices and arcs. Formally it is a 7-tuple $G = (\Sigma_O, \Sigma_R, O, \mathcal{R}, f_s, f_t, \Psi)$, where

1. Σ_o is a set of universal unique identifiers (UUIDs) of all the object instances in G.
2. Σ_R is a set of UUID pairs (s, t) of all the relationship instances, such that $s, t \in \Sigma_O$.
3. $O = \{O_1(\mathbb{A}_1), O_2(\mathbb{A}_2), \ldots, O_i(\mathbb{A}_i), \ldots, O_n(\mathbb{A}_n)\}$ represents a set of *object types* O_i with schema $\mathbb{A}_i = (A_1, A_2, \ldots, A_{d(i)})$ of *degree* $d(i)$, such that each $O_i(\mathbb{A}_i) = \{o_1, o_2, \ldots, o_k\}$ contains a set of objects of type O_i, and $o_j = (a_1, a_2, \ldots, a_{d(i)})$ represents an object instance with the universal unique identifier (UUID) j.
4. $\mathcal{R} = \{R_1(\mathbb{B}_{R_1}), R_2(\mathbb{B}_{R_2}), \ldots, R_i(\mathbb{B}_{R_i}), \ldots, R_m(\mathbb{B}_{R_m})\}$ represents a set of *relationship types* R_i with schema $\mathbb{B}_{R_i} = (B_1, B_2, \ldots, B_{e(i)})$ of degree $e(i)$, such that each $R_i(\mathbb{B}_{R_i}) = \cup_{p,q \in \Sigma_O} \{r_{(p,q)}\}$ denotes a set of relationships of type R_i, and $r_{(s, t)} = (s, t, b_1, b_2, \ldots, b_{e(i)})$, $b_i \in B_i$, is a relationship instance for a pair of UUIDs (s, t), s and $t \in \Sigma_O$. $\mathbb{B}_{R_i}$ can be null, which means that R_i has no attributes and can also be denoted as $R_i(\varnothing)$.
5. $f_s: \Sigma_R \rightarrow \Sigma_O$ and $f_t: \Sigma_R \rightarrow \Sigma_O$ are two maps indicating the *source* and *target* object UUIDs of a relationship UUID pair.
6. $\Psi = \{\psi_{R1}, \psi_{R2}, \ldots, \psi_{Ri}, \ldots, \psi_{Rm}\}$ represents a set of maps, such that $\psi_{R_i}: \Sigma_R \rightarrow \mathbb{B}_{Ri}$ is a map returning the tuple of attribute values $(b_1, b_2, \ldots, b_{e(i)})$ of a relationship (s, t) in Σ_R. For example, a relationship type *Colleague* with schema (*affiliation*) contains a set of relationships between each pair of objects in *Person*. Then, $\psi_{Colleague}(o_1, o_2) = \psi_{Colleague}(1, 2) = (\text{TSMC})$ represents that o_1 and o_2 have a *Colleague* relationship with attribute *affiliation* = 'TSMC'.   □

A *graph database*, in this definition, contains a collection of different types of objects (e.g., a person or an affiliation) and their multilateral relationships of different relationship types (e.g., friend or spouse relationships). Both objects and relationships may be associated with different number of attributes. To support network analytics, contemporary full-fledged graph database management systems (GDBMS) are equipped with specific query language constructs for deriving these attributes, relationships, or even any kind of transitive closures in networks. Users can pose their query statements by expressing pattern matching or multi-hop navigation in social networks very easily [2, 3].

2.3   An Illustrative Example

In this section, we use the famous Krackhardt kite graph [4] as depicted in Fig. 1 to illustrate the formal definition of our graph database model. The database contains 10 persons with 18 relationships of 4 different types, namely *Colleague*, *Roommate*, *Friend*, and *Classmate*.

Based on Definition 1, the graph in Fig. 1, $G = (\Sigma_O, \Sigma_R, O, R, f_s, f_t, \Psi)$, where

1. $\Sigma_O = \{1, 2, 3, 4, 5, 6, 7, 8, 9, 10\}$. There are 10 UUIDs for ten persons.
2. $\Sigma_R = \{ (1,2), (2,3), (3,4), (3,5), (4,5), (4,6), (4,7), (4,9), (5,7), (5,8), (5,10),$
   $(6,7), (7,8), (6,9), (7,9), (7,10), (8,10), (9,10)\}$. There are 18 UUID pairs for the
   relationships between two persons.
3. $O = \{O_1(\mathbb{A}_1)\} = \{Person(Name, Gender, City)\}$, where $Person(Name, Gender,$
   $City) = \{(1, Mint, M, Taipei), (2, Luna, F, New Taipei), (3, Lora, F, Taipei),$
   $(4, Mary, F, Taichung), (5, Tom, M, Changhua), (6, May, F, Tainan),$
   $(7, Ling, F, Tainan), (8, Ren, M, Kaohsiung), (9, John, M, Tainan), (10, Wen,$
   $F, Kaohsiung)\}$ is an object type of degree 3, containing 10 objects.
4. $R = \{R_1(\mathbb{B}_{R_1}), R_2(\mathbb{B}_{R_2}), R_3(\mathbb{B}_3), R_4(\mathbb{B}_4)\} = \{Colleague(Unit),$ $Roommate(\varnothing),$
   $Friend(\varnothing), Classmate(\varnothing)\}$,   where $Colleague(\text{Unit}) = \{(1,$ 2,   TSMC)$\}$,
   $Roommate(\varnothing) = \{(2,$ 3$)$, $(4,$ 6$)$, $(4,$ 7$)$, $(4,$ 9$)$, $(6,$ 7$)$, $(6,$ 9$)$, $(7,$ 9$)\}$,
   $Friend(\varnothing) = \{(3, 4), (3, 5), (4, 5), (9, 10)\}$, and $Classmate(\varnothing) = \{(5, 7), (5, 8),$
   $(5, 10), (7, 8), (7, 10), (8, 10)\}$.
5. $f_s: \Sigma_R \to \Sigma_O$ and $f_t: \Sigma_R \to \Sigma_O$ are two maps indicating the *source* and *target* objects
   of a relationship. Therefore, we obtain $f_s((1,2)) = 1$ and $f_t((1,2)) = 2$.
6. $\Psi = \{\Psi_{Colleague}, \Psi_{Roommate}, \Psi_{Friend}, \Psi_{Classmate}\}$ represents a set of maps, such that
   $\Psi_{Colleague}((1,2)) = (\text{TSMC})$, is a map returning the tuple of attribute values
   $(b_1, b_2, \ldots, b_{e(i)})$ of a relationship (s, t) in Σ_R.

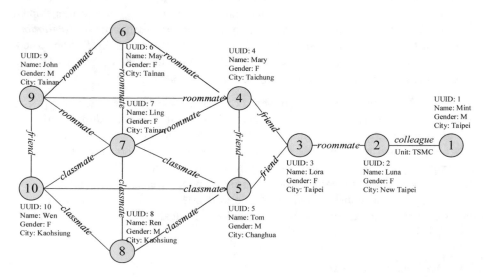

Fig. 1. An example Krackhardt kite graph database G.

3 Multi-dimensional Graph Warehouse Modeling

To extend the relational data warehouse technology for graph database, we modify our previous work [7, 8] to define the related definitions about dimensions, entity tuples, relationship tuples and graph cubes for graph warehousing.

Definition 2: A *dimension D* is a tree structure with h levels, $h \geq 1$, used for representing the hierarchical relationships among a set of attribute values. A vertex in a dimension D is called a *member*, and each internal node contains a special child, called *summary member*, and denoted '*', which is used for denoting the summary of the children of an internal node.

The dimensions in a graph database can be distinguished into the following types:

1. *Resource dimension*. A dimension created based on the buzzwords, keywords or tags from the shared resources in a social network.
2. *Metadata dimension*. A dimension contains metadata of the shared resources in a social network, like language, domain name, or Uniform Resource Identifier (URI), date-of-publish.
3. *Profile dimension*. A dimension created based on netizens' public profile, like e-mail, affiliation, gender, or location.
4. *Relationship dimension*. A dimension contains keywords correspond to the hierarchical relationships of netizen, based on the relationship type or relationship attributes.

To simplify our discussion, we mainly concerns resource dimensions in the following examples.

Definition 3: For a *dimension D*, the *ith-level member set*, denoted $D(i)$, is defined as $D(i) = \{a \mid a$ is a member in the ith level of D, but a is not a summary member$\}$. We use $D(0)$ to denote the union of all non-summary members in D, which is the union of all i-th level member sets in D. That is, $D(0) = \cup_{1 \leq i \leq h} D(i)$, where h is the height of D. Each $D(i)$ has a specific name, generally called the ith-level name.

Practically, a dimension can be constructed from data organized in tabular format, such that a level corresponds to an attribute and the attribute names relate to the corresponding level names. We illustrate an example as follows.

Table 1. A relation **Region** for constructing dimension R

Area	City
North	Taipei
North	New Taipei
Middle	Taichung
Middle	Changhwa
South	Tainan
South	Kaohsiung

Example 1: Suppose we have a table named **Region** storing the cities, and their corresponding areas, in Taiwan as show in Table 1. Then, we may construct a dimension R as depicted in Fig. 2, where the top level corresponds to the dimension itself (commonly denoted '(*All Region*)'), and Levels 2 and 3 are derived from the attributes *Area* and *City*, respectively. The starred nodes are summary members.

That is, the summary member in Level 1 has the same meaning as *all regions in Taiwan*, which represents {*North, Middle, South*}. Also, the summary members under *North, Middle* and *South* mean the same as *North, Middle* and *South* which denote {*Taipei, New Taipei*}, {*Taichung, Changhwa*} and {*Tainan, Kaohsiung*}, respectively. By omitting all the summary members, Fig. 2 can be illustrated concisely in Fig. 3, such that $R(1) = \{(All\ Region)\}$ (or $R(1) = \{Taiwan\}$), $R(2) = \{North, Middle, South\}$, $R(3) = \{Taipei,\ New\ Taipei,\ Taichung,\ Changhwa,\ Tainan,\ Kaohsiung\}$ and $R(0) = \{(All\ Region),\ North,\ Middle,\ South,\ Taipei,\ New\ Taipei,\ Taichung,\ Changhwa,\ Tainan,\ Kaohsiung\}$.

In Fig. 4, we depict another dimension C for representing a categorization of computer, communication, and consumer electronic products.

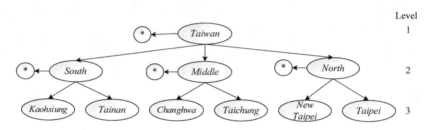

Fig. 2. The dimension of R about *Taiwan*.

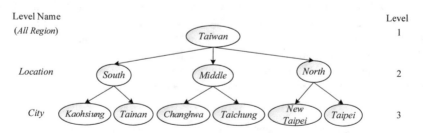

Fig. 3. A concise illustration of dimension R.

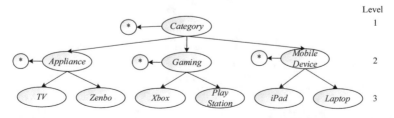

Fig. 4. The dimension C about a categorization of general electronic products.

For a dimension D, two basic operations called *drill-down* and *roll-up* can be defined for easy browsing as follows. By drilling down and rolling up in dimensions, users can browse a graph cube by different granularities to obtain fine-grained or coarse-grained insights of the netizen relationships between different levels.

Definition 4: For a *dimension D*, expanding an internal vertex to obtain all of its children is called *drill-down*, and shrinking a set of children to obtain their common parent is called *roll-up*.

To construct a graph data warehouse, a graph (i.e., graph database) and its subgraphs should be assigned with a unique identifier for multi-dimensional indexing. We define the concept of graph index as follows.

Definition 5: For a graph $G = (\Sigma_O, \Sigma_R, O, \mathcal{R}, f_s, f_t, \Psi)$ with a unique identifier id_G, the *graph index* of G defined on n dimensions $(D_1, D_2, ..., D_n)$ is denoted $x = (id_G, K_G)$, where $K_G = (K_1, K_2, ..., K_i, ..., K_n)$ is an n-tuple of attribute value sets, such that each K_i contains a set of attribute values, and for all keywords $k_{ij} \in K_i$, $k_{ij} \in D_i(0)$, $1 \leq i \leq n$.

For simplicity, the first and second components of a graph index $x = (id_G, K_G)$ will be denoted x^1 and x^2 (i.e., $x^1 = id_G$ and $x^2 = K_G$), respectively, where id_G can be regarded as a handle pointing to a graph query statement (like Cypher in Neo4j, or Transact-SQL in SQL Server 2017) returning the graph G. The graph query statement is actually a function of $K_G = (K_1, K_2, ..., K_i, ..., K_n)$, which can be implemented or defined based on the needed vertices and relationships information selected by $K_1, K_2, ..., K_i, ..., K_n$ for a specific application.

When all $|K_i| = 1$, the graph index is also called a *base graph index*, and each K_i can be denoted by the unique element for concise representation. In such case, a $K_G = (\{k_1\}, \{k_2\}, ..., \{k_i\}, ..., \{k_n\})$ can be abbreviated as $K_G = (k_1, k_2, ..., k_i, ..., k_n)$. If there is at least one K_i, such that $|K_i| > 1$, and if $|K_j| = 1$, for all $i \neq j$, then the graph index is also called a *composite graph index*. If there are some K_i, such that $|K_i| = 0$, then the graph index is also called a *degenerate graph index*. In our work, a degenerate graph index with some $|K_i| = 0$ will be generalized by using the top level member set of the corresponding dimension, i.e., $D_i(1)$ or '*', to replace the missing keyword set K_i.

Example 2: Following Example 1, an example graph index defined on the dimensions (R, C) may be $x = (G001, (Kaohsiung, XBOX))$, where G001 is the unique identifier of the graph G consisting of customers located in *Kaohsiung* who brought *XBOX*, and their multilateral relationships (e.g., friend relationship), as depicted in Fig. 5. We deliberately separate male and female customers in the graph, as *gender* may be used as another dimension. That is, for another graph index $y = (G011, (Kaohsiung, XBOX, Male))$, the corresponding graph contains the left dotted part only.

The basic component of a graph cube is a *cell*, which is defined as follows.

Definition 6: A *cell* defined on n dimensions $(D_1, D_2, ..., D_n)$ for a graph G is denoted $c = (t_c, x)$, where $t_c = (c_1, c_2, ..., c_i, ..., c_n), c_i \in D_i(0) \cup \{ \text{'*'} \}$, $1 \leq i \leq n$, and x is a graph index of the form $x = (id_G, (K_1, K_2, ..., K_n))$, where id_G is the unique identifier graph G, and $K_i \cap D_i(0) \neq \varnothing$, $1 \leq i \leq n$. That is, a cell c contains a subgraph of G, generated by a graph query statement based on $(K_1, K_2, ..., K_n)$, and the subgraph is pointed by id_G.

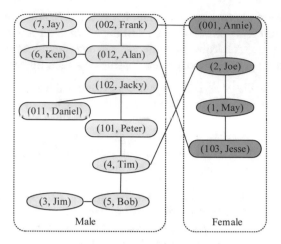

Fig. 5. An example graph indexed by (G001, ({*Kaohsiung*}, {*XBOX*})).

Definition 7: A *cell* $c = (t_c, x)$, where $t_c = (c_1, c_2, ..., c_i, ..., c_n)$, defined on n dimensions $(D_1, D_2, ..., D_n)$ is called an *m-d cell*, $0 \leq m \leq n$, if and only if there are exactly m non-summary members c_i (i.e., $c_i \neq$ '*'). When $m = n$ and $c_i \in D_i(h_i)$, where h_i denotes the height of D_i, for all $1 \leq i \leq n$, then c is called a *base cell*; otherwise, c is called a *non-base cell*.

Definition 8: An n-dimensional i–d cell $a = ((a_1, a_2, ..., a_n), x_a)$ is a *parent* of another n-dimensional j-d cell $b = ((b_1, b_2, ..., b_n), x_b)$, if and only if the following conditions hold:

1. $i = j - 1$,
2. There is exactly one k, such that a_k is the parent of b_k in D_k and $a_l = b_l$ for all $l \neq k$, $1 \leq l \leq n$.
3. The graph indexed by x_b is a subgraph of that indexed by x_a.

Definition 9: A *graph cube* $GC = (G, (D_1, D_2, ..., D_n))$ for $G = (\Sigma_O, \Sigma_R, O, \mathcal{R}, f_s, f_t, \Psi)$ defined on n dimensions $(D_1, D_2, ..., D_n)$, is a cube composed of all cells $c_i = (t_{c_i}, x_i)$ with $t_{c_i} \in \times_{1 \leq i \leq n} D_j(0)$ and all of the graph indexed by x_i is a subgraph of G.

A sample illustration of a graph cube $GC = (G, (R, C, T))$ is shown in Fig. 6, where T is a *Time* dimension, and R and C are the dimensions respectively depicted in Figs. 3 and 4.

Each cell in Fig. 6 points to a subgraph consisting of objects and their relationships defined by the intersected dimension members of all engaged dimensions. For example, the cells a, c, and e respectively link to three sub-graphs regarding the social network netizen in *Taipei* (in *North* Area), *Taichung* (in *Middle* Area), and *Tainan* (in *South* Area), who unilaterally posted comments regarding *TV* on the same day (e.g., '2017/08/08' in the *Time* dimension), with their *friend* relationships. The system can generate these subgraphs for users when the tuple (TV, {Taipei, Taichung, Tainan}, 2017/08/08) has been selected as a filter on that graph cube. In contrast, in a traditional relational cube structure, such cells just store three numbers regarding the amounts of TVs bought by Taipei, Taichung and Tainan customers on 2017/08/08.

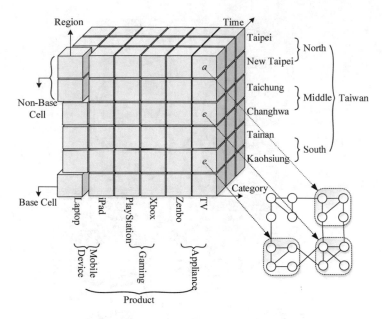

Fig. 6. A multi-dimensional graph cube structure example *GC*.

4 Graph Cube Visualization and Summarization

After creating a graph cube, all dimensions or attributes of vertices and relationships can now be used to calculate summarization for social network analytics. For example, based on the two views from **Type**, which contains {*Employee, Customer*}, and Gender (i.e., {Male, Female}) as shown in Fig. 7, it is easily to derive the summarization of comments posted by *Taipei* and *Kaohsiung* netizen, and another summarization based on gender, respectively (shown in Fig. 8).

Furthermore, if users want to view two dimensional summarization, suppose the chosen dimensions are Gender × Type, (i.e., {Male, Female} × {*Employee, Customer*} = {(Male, *Employee*), (Female, *Employee*), (Male, *Customer*), (Female, *Customer*)}), then the result can be derived as Fig. 9 illustrates. The detailed relationships in Fig. 7 can be stored internally in the system, and employed for the implementation of an operation corresponding to traditional DRILL-THROUGH option in multi-dimensional query language like MDX [5] or MD^{2}X [7]. Such that users can trace from the summary result of Fig. 8 to reach the details illustrated in Fig. 7, respectively.

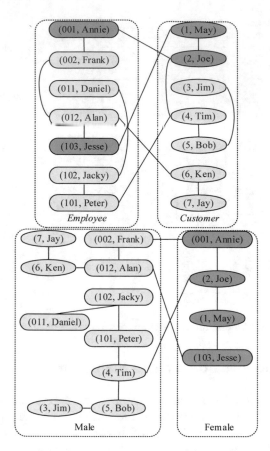

Fig. 7. Two views of the *Friend* relationships from the type and gender of person.

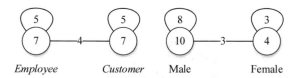

Fig. 8. Two summarizations of *friend* relationships of Fig. 7.

Another unfriendly view regarding the located cities of person Type (i.e., {*Employee, Customer*}) can be found in Fig. 10, which can help us calculate the summarization for the Type-*City* relationships (Fig. 11). The view concerning the located city of different gender has been presented in Fig. 12, which can help us compute the summarization for the *Gender-City* relationships (Fig. 13).

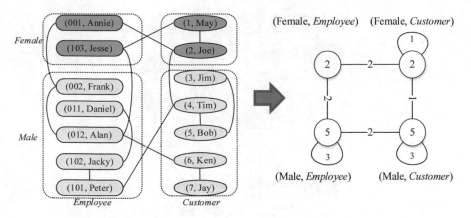

Fig. 9. The summarizations of Gender × Region.

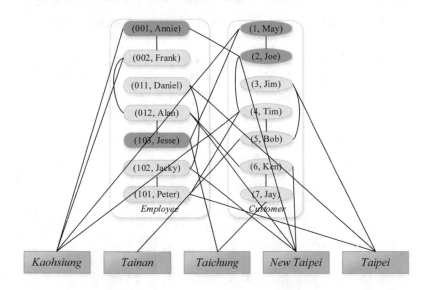

Fig. 10. A view of the type-*City* relationships.

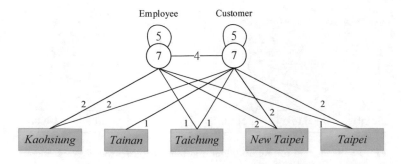

Fig. 11. The summarization of type-*City* relationships.

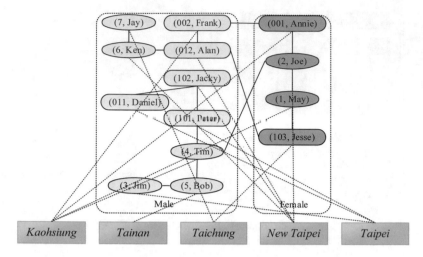

Fig. 12. A View of the located *City* relationships of different gender.

The drill-down and roll-up operations defined in Definitions 3 and 4 can also be processed easily in graph cubes. For example, Fig. 14 shows two sub-graphs of rolling up to one level (along the dimension *R*) of Figs. 11 and 13, respectively.

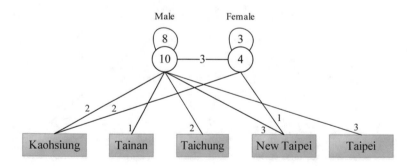

Fig. 13. The summarizations of Gender × City.

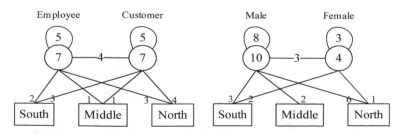

Fig. 14. One level rolling up (along the Dimension *R*) for Figs. 11 and 13.

Nowadays, so many fan pages has been created in social networks, like Facebook or Instagram, to collect stakeholders' comments, spread promotion information, or conduct sentiment analysis of valuable customers, together with their friends or followers. These posted comments or resources can be processed, organized and interweaved into graph cube structures for social network analytics using the concept of graph data warehousing [8] to create values for short-term analysis or long-term planning. These features are rich enough to help users derive social business intelligence on graph databases substantially. The created values can be systematically utilized for internal knowledge management, and can be broadcasted to related netizen with value-added feedbacks, which makes the whole process a virtuous cycle.

5 Conclusion and Future Directions

5.1 Conclusion

While graph databases emerge for storing networked objects and their interrelationships, we still need more studies to be conducted to explore and extend the scope of graph-based business intelligence. In this paper, we advocate the importance of constructing graph warehouses to support social business intelligence, and formally define the concept of a graph database, and its multi-dimensional structures for graph warehousing. When social network netizen, resources, together with their relationships, are properly warehoused, users can use such structure to perform *ad hoc* on-line analytical processing (OLAP) over the network in a graph warehouse, which is just as the way users can perform OLAP over summarized data in traditional data warehouse.

The applications of graph warehousing are versatile and diverse. For instance, it may help administrators organize the relationships between stakeholders, like customers, company staffs, popular products, and enterprise partners, such that by using related profile data or resources as the dimensions, their inter-relationships can be extracted and browsed instantly. As graph databases can be used to model on-line social network streaming, which is becoming an influential media for the analysis of counter-terrorism, agile management, marketing and service design thinking for any kind of business activities, our work help to pave a way for pivoting people-centric or topic-centric applications for social resource integration and business intelligence in various domains.

When social relationships and resources are warehoused with timing considerations, then users can trace the relationships and their evolution based on some criteria along the time dimension directly. Besides, graph clustering can be achieved through visualizations and users can develop value-added graph summarization tools to decompose a social network into subgraphs based on predefined dimensions, or gather up a cluster of related graphs for different perspectives. Therefore, graph warehousing compensates the deficiencies of data warehousing or even document warehousing for social network applications, such that the knowledge management of people's network, machine's network or even object's network can be properly resolved.

5.2   Future Directions

In our future work, we will propose an ETL (Extraction, Transformation and Loading) architecture for graph warehousing. The preliminary components may differ from that in traditional data warehousing as the input sources are on-line social media sites instead of enterprise databases.

Besides, as we have pointed out that, for a graph index $x = (id_G, K_G)$, id_G is actually a handle pointing to a graph query statement returning the query result, which is a subgraph of GC. The graph query statement is a function of $K_G = (K_1, K_2, ..., K_i, ..., K_n)$, to be defined or systematically generated to retrieve the needed vertices and relationships information selected by $K_1, K_2, ..., K_i, ..., K_n$. We will explore the rules or model regarding how to systematically define or generate the query statements in the future. Finally, since the construction and visualization of a graph warehouse need to scan a large amount of nodes and relationships, which is a task prone to be time-consuming, the parallel or high performance architecture for such process will be further investigated.

Acknowledgment. This work is partially supported by the Ministry of Science and Technology, TAIWAN, ROC, under contract No.: MOST 107-2410-H-992-016-MY2.

References

1. Bondy, J.A., Murty, U.S.R.: Graph Theory with Applications. The Macmillan Press Ltd., London (1976)
2. Easley, D., Kleinberg, J.: Networks, Crowds, and Markets: Reasoning about a Highly Connected World. Cambridge University Press, Cambridge (2010)
3. Ho, L.Y., Wu, J.J., Liu, P.: Distributed graph database for large-scale social computing. In: Proceedings of the 5th International Conference on Cloud Computing (2012)
4. Krackhardt, D.: Assessing the political landscape: structure, cognition, and power in organizations. Admin. Sci. Q. **35**(2), 342–369 (1990)
5. Spofford, G., Harinath, S., Webb, C., Huang, D.H., Civardi, F.: MDX Solutions—With Microsoft SQL Server Analysis Services 2005 and Hyperion Essbase, 2nd edn., Wiley (2006)
6. Tan, W., Blake, M.B., Saleh, I., Dustdar, S.: Social-network-sourced big data analytics. IEEE Internet Comput. **17**(5), 62–69 (2013)
7. Tseng, F.S.C.: Design of a multi-dimensional query expression for document warehouses. Inf. Sci. **174**(1–2), 55–79 (2005)
8. Tseng, F.S.C., Chou, A.Y.H.: The concept of document warehousing for multi-dimensional modeling of textual-based business intelligence. Decis. Support Syst. **42**(2), 727–744 (2006)
9. Vicknair, C., Macias, M., Zhao, Z., Nan, X., Chen, Y., Wilkins, D.: A comparison of a graph database and a relational database: a data provenance perspective. In: Proceedings of the 48th ACM SE Annual Southeast Regional Conference, pp. 42–48 (2010)

Blending Six Sigma and Software Development

Hassan Pournaghshband$^{(\boxtimes)}$

Kennesaw State University, Kennesaw, GA, USA
hpournag@kennesaw.edu

Abstract. While Six Sigma and other quality methodologies have been used extensively and quite effectively in the manufacturing sector for decades, the adoption rate in other industries such as the software development industry has been a bit slower. However, in recent years, research literature clearly suggests that the trend line is definitely on the upswing for software practitioners who see the value of applying Six Sigma and other proven quality methodologies in better managing software development projects, which have notoriously been prone to delays in migrating to production. We contemplate that the melding of these two methodologies will result in a sum that is greater than its two parts. In this paper, we elaborate on current issues/challenges of Six Sigma utilization in software development and management and explain problems that the software project managers encounter when dealing with them. We also investigate a broad collection of toolkits that are typically used in the Six Sigma framework and propose what tools, in each Six Sigma DMAIC (*define, measure, analyze, improve, and control*) phase are better candidates to consider and recuperate for software development and management.

Keywords: Process improvement · Six Sigma · Software development

1 Six Sigma: Overview

Six Sigma is a highly structured methodology for improving processes and solving problems. Six Sigma utilization and other quality methodologies have proved to be very effective in the manufacturing sector for the past three decades. A research study about the most frequently occurring causes of software defects, which in fact is the main reason why an industry is interested in using Six Sigma in a process for defects prevention, is conducted by authors in [1]. Table 1 shows the causes of defects for the software development process that was sampled in their study.

The intentions of the Six Sigma methodology are mainly to reduce process variation, improve the quality of production processes, making them more consistent in terms of the end results achieved in order to at least satisfy and preferably exceed customers' expectations, decrease the number of defects, and improve the capability of processes that are used to produce products or services. It is a fact-based, data-driven philosophy of improvement that clearly values defect prevention over defect detention. In other words, in the software development, the idea of prevention is a powerful concept as it certainly trumps defect detention and subsequent follow-up resolution, especially if the software defect is identified by the customer as opposed to the software company. In terms of driving quality in software, this is a critical notion that we feel

© Springer Nature Switzerland AG 2020
K. Arai et al. (Eds.): FTC 2019, AISC 1070, pp. 619–624, 2020.
https://doi.org/10.1007/978-3-030-32523-7_45

Table 1. Causes of defects for software development process

Cumulative percentage cutoff			80%
#	Causes	Defects	Cumulative %
1	Wrong requirements	50	46.3%
2	Miscommunication	30	74.1%
3	Last minute changes	15	88.0%
4	Unrealistic timeframe	5	92.6%
5	Untrained developers	3	95.4%
6	Human factors	1	96.3%
7	Poor design & logic	1	97.2%
8	Machine error	1	98.1%
9	Lack of testing	1	99.1%
10	Delivery	1	100.0%

Source: http://papers.ssrn.com/sol3/papers.cfm?abstr-act_id=2624188

makes Six Sigma an ideal quality methodology that should be more consistently incorporated into every software development project's overall methodology. Six Sigma is based on the DMAIC (Define, Measure, Analyze, Improve, and Control) cycle for process improvement. In graphical form, the DMAIC approach is represented well with the diagram shown in Fig. 1, which explains the progressive steps within each of the five phases of this particular quality methodology [2].

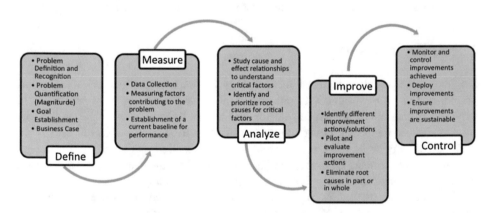

Fig. 1. Graphical representation of the DMAIC

Currently, software development methodologies and project planning techniques are not producing reliable software anywhere near the Six Sigma desired state of 3.4 defects per million opportunities [2]. To alleviate this shortcoming we need to get to the root causes that are directly leading to a larger number of defects than what we would ideally want when looking through a Six Sigma lens. The root cause analysis tools that are often associated with Six Sigma can and often are tremendously beneficial in uncovering the root causes that are creating the software defects or the variations in the software development process, regardless of what type of software development technique is being used. The two quality tools that come to mind when investigating and analyzing root cause are Ishikawa Cause & Effect or "Fishbone" Diagrams, and Five Why's Analysis.

Some samples of these two root cause analysis tools are shown in Fig. 2.

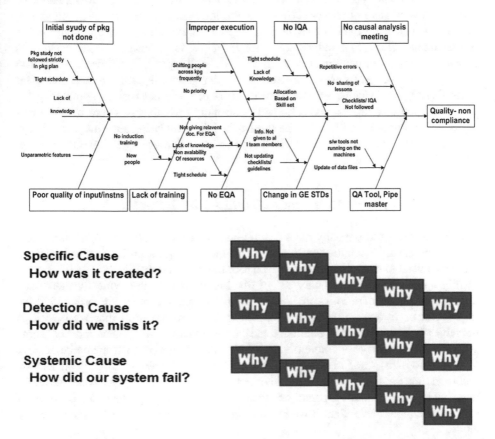

Fig. 2. "Fishbone" Diagram and Five Why's Analysis. (Source: http://ssrn.com/abstract=2624-188, http://www.qualitydigest.com/inside/quality-insider-article/leveraging-five-whys.html#)

As it is discussed in [5], similar issues can be prevented from recurring on subsequent software development projects, by actively using root cause analysis as part of the software project planning methodology which is helpful from a software project planning perspective by utilizing and actually incorporating lessons learned from previous software development projects. There is no doubt that the Six Sigma quality methodology supports learning from previous mistakes in order to improve the process going forward, and conducting thorough root cause analysis when a software defect has been incurred can help target the true root cause for that issue.

2   Six Sigma and Software Project Planning

A decent amount of research on whether Six Sigma should be applied to the development of software products has been done by the author in [3, 4]. Meanwhile, he sensibly covers major phases of software development such as integration, deployment, and maintenance. He found these three principles to be of note when applying the Six Sigma methodology to building software: Principle #1 - measure customer-related metrics only (i.e. use combinatory metrics to cover all topics), Principle #2 - adjust to moving targets (i.e. your goals may need to be changed: accept change and manage it accordingly), and Principle #3 - enforce measurement (do not rush in completing the job to enforce meeting targets). The author strongly feels that software project managers will struggle mightily with the third principle, mainly because they are always held accountable to meeting targets, i.e. is the software delivered with quality, on time, and under or at budget.

3   Six Sigma: Common List of Toolkits by DMAIC Phase

Six Sigma initiatives typically have a different and broader selection of various tools that can be used. Depending on the nature of the Six Sigma project, not all of these tools may need to be employed for a given job. In reviewing this list below, it becomes readily apparent that this laundry list of Six Sigma tools is not typically mentioned during the course of a software development project. However, in order to drive improvements in any software development project, we contend that we owe it to ourselves to blend these two methodologies, i.e. software development and Six Sigma – in order to try to drive out more of the variation that we find in software defect results. We want to satisfy our customers with any software development project, and by incorporating Six Sigma and blend this quality technique into the software development and project planning paradigm, then we are certainly putting our chances of success at a higher level than if we just used a software project planning methodology alone.

We contemplate that the melding of these two methodologies will result in a sum that is greater than its two parts. Table 2 shows a broad collection of toolkits that are typically used in the Six Sigma framework and even more importantly, in which Six Sigma DMAIC phase are each of these tools best utilized to drive defects down and preferably out of the equation. By examining these tools and their characteristics, and investigating what specific type of characteristic a tool should possess to be used effectively for each phase of DMAIC, we are convinced that most of them maybe used (though not effectively) for any field involved with one or more of the *defining, measuring, analyzing, improving,* and *controlling.* However, considering the nature of the software development and management we have come up with Table 3 below, listing those tools that would be better candidates for this task.

Table 2. Toolkits that are typically used in the Six Sigma framework

DMAIC phase	Tools
Define	• Project Charter
	• Project Evaluation
	• Critical to Quality (CTQ) Characteristics
	• Software Failure Mode and Effects Analysis (SWFMEA)
	• Software Quality Function Deployment (SWQFD)
	• Kano Analysis
	• Process Map
	• Affinity Diagram
Measure	• Gage Repeatability & Reproducibility (R&R)
	• Process Capability Analysis
	• Process Yield Analysis
Analyze	• Pareto Analysis
	• Cause and Effect Diagram
	• Correlation Analysis
	• Regression Analysis
	• Two Sample T-test
	• ANOVA
	• Earned Value Analysis
	• Scatter Plots
	• Line Chart
Improve	• Cause and Effect Diagram
	• Software Failure Mode and Effects Analysis (SWFMEA)
	• Cost Benefit Analysis
Control	• SWFMEA (Control)
	• Statistical Process Control Charts
	• Project Assessment
	• Project Summary

In fact, some of these tools such as Earned Value Analysis and Project Evaluation are overlapping with software project management while others have the typical characteristics of a tool satisfying the corresponding DMAIC phase. Needless to say, for each of the tools listed in Table 3, further comprehensive research is needed for figuring out what specific activities can be handled by the tool along with its benefits and shortcoming. Two points that we can make here, first, how can we expect to achieve a much higher level of success in developing software if we continue to use the exact same methodology over and over again. And second, isn't this the definition of insanity – using the same methodology but expecting different results? Perhaps, it is time for us to try an enhanced software development and project planning methodology – one that at least incorporates many of the aspects of Six Sigma so that we can attempt to significantly reduce defects and lower the variation in the process results so that the delivered software is more consistent, ceteris paribus.

Table 3. Six Sigma potential tools for software development and management

DMAIC phase	Tools
Define	• Project Evaluation
	• Software Failure Mode and Effects Analysis (SWFMEA)
	• Software Quality Function Deployment (SWQFD)
Measure	• Process Capability Analysis
	• Process Yield Analysis
Analyze	• Cause and Effect Diagram
	• Earned Value Analysis
Improve	• Cause and Effect Diagram
	• Software Failure Mode and Effects Analysis (SWFMEA)
	• Cost Benefit Analysis
Control	• SWFMEA (Control)
	• Project Assessment

4   Conclusions and Future Work

In this paper, we examined the current issues/challenges of Six Sigma utilization in software development and management and discussed problems that the software project managers encounter when dealing with them. In addition, we showed the common list of toolkits by DMAIS phase and proposed a subset of those tools that would contain the characteristics of the ones that would make good candidates for software development and management.

Mapping the characteristics of each tool with specific activities of software development will be a significant step toward this study. This is what authors are presently studying, along with the benefits and shortcoming of each tool when used for software development and management.

References

1. Chauhan, Y., Belokar, R.M.: Six Sigma in project management for software companies, 28 June 2015. SSRN. http://ssrn.com/abstract=2624188 or http://dx.doi.org/10.2139/ssrn. 2624188
2. http://agilealliance.org/files/4814/0509/9273/ExperienceReport.2014.AmrKhalifa.pdf
3. Fehlmann, T.: Six Sigma for software. Http://citeseerx.ist.psu.edu/viewdoc/download?doi= 10.1.1.91.6736&rep=rep1&type=pdf. N.p., n.d. Web
4. Fehlmann, T.: Statistical process control for software development – Six Sigma for software revisited. Http://www.e-p-o.com/downloads/sixsigmarevisited.pdf. N.p., n.d. Web
5. Pournaghshband, H., Watson, J.: Should Six Sigma be incorporated into software development & project management? In: Proceedings of the International Conference on Computational Science and Computational Intelligence, December 2017

Toys That Mobilize: Past, Present and Future of Phygital Playful Technology

Katriina Heljakka[1] and Pirita Ihamäki[2](✉)

[1] University of Turku, Pori, Finland
katriina.heljakka@utu.fi
[2] Prizztech Ltd., Pori, Finland
pirita.ihamaki@prizz.fi

Abstract. This exploratory paper focuses on the technological development of the toy medium with an interest in toys' capacity to mobilize *homo ludens*, the playing human. By conducting an extensive literary review on the history and the present of mobile play objects, the study demonstrates how toys have developed from early moving and mechanical automata to playthings that move by themselves through in-built computerized components. The interest is two-fold: By analyzing the historical trajectory of mobile toys, the authors highlight the role of both toys and their players as participants in technologically mediated, phygital play. This hybrid form of playing combines the physicality of playthings with both mechanical and digital features. The results of the review show how toys—character toys in particular—have transformed from entertaining, self-moving spectacles to educational machines that mobilize the player both physically and geographically. Based on the results of the literary review, the authors suggest a continuum that visualizes the development of the types of toys that afford mobility in play of the past, present, and future. The paper concludes with the observation that phygital, playful technologies, such as character toys, have the capacity to influence human well-being in its various dimensions. By making them mobile, toys as a medium invite engagement with phygital playful technologies to enhance physical, cognitive, and social well-being.

Keywords: Toys · Toy mobility · Phygitals · Smart toys · Well-being

1   Introduction: From Mechanical to Phygital Toy Mobility

It has been stated that "the access, the mobility, and the ability to effect change are what will make the future so different from the present" [1]. In this article, a mobile continuum for toys focusing on the past, present, and future is presented, and the ways toys have developed from self-moving automata to become a *mobilizing media* of the 21st century using phygital playful technology have been investigated. Phygital playful technology refers to hybridity in play objects that combines physical playthings with both mechanical and digital affordances. During the past decennia, phygitality has come to be known as a *mobilizing* technology implemented in the design of many new toys.

So far, phygital technologies that combine physical objects and environments with digital and mobile devices and systems have, for example, been studied from the perspectives of games [2, 3], heritage communication [4], and education [5]. In this

© Springer Nature Switzerland AG 2020
K. Arai et al. (Eds.): FTC 2019, AISC 1070, pp. 625–640, 2020.
https://doi.org/10.1007/978-3-030-32523-7_46

paper, the focus is on spontaneous and unstructured play with phygital technologies. The decrease in unstructured playtime can present serious issues for the cognitive, emotional, physical, and social development of children [6]. Again, play with phygital toys can prompt creative play patterns among players. Therefore, the way children play is to some extent influenced by the industry creating new types of play-objects and experiences, such as the toys focused on in this paper, which integrate physical and digital elements known as phygital play-objects [7].

The evolution of playthings enhanced by technologies is inevitable with the invention of new and interesting toys and games being created and released every year. Hybrid, or as referred to in this paper, *phygital* toys and play, challenge a strict dichotomy between physical and digital, non-connected and connected toys by resting on the interaction of material and digital. In other words, traditional, physical toys are merging with digital, "smart" toys [8]. Yelland (2015) has stated that "making new technologies available alongside traditional materials (e.g. blocks) enables and extends playful explorations" [9]. The concept of hybridity, or rather, phygitality, is of essential importance when considering the latest developments around play objects that move by themselves, and even more so that increasingly encourage physical (even geographical) movement of their players. One observation that has guided the interest towards the mobilizing tendency of toys is the insight of the importance of computerized technologies as drivers of human well-being in its various dimensions. Phygital toys represent a growth industry, and they can make a positive contribution to well-being facilitated by toy mobility, discussed in this paper.

There are many types of movement in relation to the physical plaything: mobility in toys may be addressed by considering toys that move by themselves—the historical category of *automata*—or by considering toys as artifacts that are, in some way, made mobile by their players. This exploratory paper sheds light on the development of both of these areas—the toys that are *given* mobility by their designers and manufacturers, and the toys that have the *capacity to mobilize* their users as players. Next, a brief overview of the history of mobile toys is provided.

Automata and Mechanical Toys
Toys as playthings that move by themselves have always intrigued inventors. According to the curators of the *Of Toys and Men* exhibition that was showcased in the early 2010s in European cities, "moving toys [were] developed in the 1800s side-by-side with advances in clockwork construction and the technology of metal plate punching. Lots of cheap mechanical toys were manufactured in Europe in the late 19th century, representing various aspects of life, such as professions, the circus, animals, music and transport. The earliest manufacturers were French and German" [10].

Jaffé stated that "The addition of the simplest mechanism can dramatically transform a static toy into one that moves, introducing new elements of surprise and unpredictability that change and expand the scope for the play activity. This use of mechanics and movements in toy making is not new, and some techniques are almost as old as the first toys themselves" [11, p. 172]. Jaffé further explained that the early historical toys, such as toy animals with wheels or wings from Egypt, Mexico, Russia, India, and Japan, did not only "represent a history of technology and the human fascination with movement, but they also reflect attitudes to the role of God and the meaning of life" [11, p. 173].

In the early days of commercial toys, the first technologically advanced toys came to be known as automata. An *automaton* refers to a self-moving machine, a popular toy-type that predicted the birth of the category of smart toys of today. Technologically-mediated interaction, another affordance of many toys of the present, was already a design feature of playthings of the early 1800s. For example, in 1823, Johann Maelzel, the inventor of the metronome, patented the first talking doll in Paris, which said "mama" and "papa". Again, during this time in Britain, John Garsad took out a patent for "Automaton Figures and Appliances for Operating the same" for a moving horse and rider [11, p. 179].

Phygital Playful Technology
Toys as an area of playthings have always represented a field of artifacts that have invited artistic and design experiments. For example, it is known that major influencers of their time, da Vinci and Picasso, invented and designed toys. As demonstrated, from the 1800s onwards, commercial toys were produced using wood and metal and sometimes in combination with fabrics, such as dolls and animal figures, and were given movement by mechanical features. In a digitalizing world, it is then of no surprise that the creatives of the industrial toy world have been open to the idea to merge the physicality of toys with the digitality of computerized systems, and later, the connectivity of online environments. Among the earliest digital toys are Mattel Auto Race and the Little Professor, both released in 1976.

The concept of using technology in a way that aligns with the physical world, providing unique hybrid and interactive experiences for the user, has also been referred to as "phygital". As Pesce already noted in 2000, "the insubstantial world of cyberspace and the world of real objects are beginning to intersect and influence one another" [12, p. 232]. In 2013, *Play it! The Global Toy Magazine* published by the world's largest toy fair, Nürnberg Spielwarenmesse, stated that "classic and digital play ideas merge to become the toys of tomorrow." Technological features in playthings, as well as digitalized toys, are believed to be here to stay, and the current mindset seems to support that "iToys do not pose a threat, but should instead be seen as a complement to classical toys" [13]. After this idea was communicated in 2013, hundreds of new phygital toys, combining the physicality of the three-dimensional toy with digitality of computerized systems, have been introduced to the toy market. This paper aims to highlight this development by focusing on one aspect of phygital toys: their mobility.

The research questions are as follows:

- How does playful phygital technology, such as physical toy characters, mobilize players in the 21st century?
- How do connected toys, such as the Internet of Toys (IoToys, like many current coding toys), encourage players to become physically mobile?

This paper is structured as follows. In Sect. 2, the methods of the study are explained and the research material is introduced. Section 3 defines the concept of mobile toys and toys that mobilize their players. In Sect. 4, the results of the study are presented by demonstrating how it is possible to understand the past and present landscape of toy mobility. Moreover, in this section, a classification of mobilizing toy technologies is proposed by introducing the *Toy Mobility Continuum*. The following section summarizes and discusses the findings of the study. The final section, Sect. 6 concludes the paper and suggests ideas for future research in the area of mobilizing toys.

2 Method

2.1 Research Design

To investigate the past, present, and future of the toys that mobilize players, the authors conducted an extensive literature review on the historical trajectory of toys with mechanically mobilizing affordances as well as playthings with phygitally emerging mobilizing capacities.

In this study, the triangulation approach was used, and mixed methods were applied. Previous empirical study materials [14–19] were used in combination with a literature review to develop a synthesis of mobile toys and toys that mobilize to create a continuum for mobile toys focusing on the past, present, and future.

This section is based on a review of existing literature from articles that are published in reputed journals and conference papers and focus on the mobility toys and toys that mobilize. A data set of references available for or related to the domain of mobile toys and toys that mobilize was obtained through an online search of select databases, such as Emerald, Google Scholar, Research Gate, Academia Edu, and ACM. The keywords used to perform the search included "mobile toys," "robot toys," "smart toys," "Internet of Toys," and "toys that mobilize," which were used to obtain a list of references pertaining to literature available on the subject.

It is important to note that the analysis excluded research papers that refer to patents of technologically-enhanced toy designs. Moreover, the analysis did not include descriptions or specifications of toys' technological affordances that allow movement. Therefore, the aspect of toy mobility has mostly been considered with the player perspective in mind. For current mobile toys to be relevant to the inquiry, references were found in relation to how they have been envisioned to function in play and thus afford movement in relation to the player as either a spectator of the plaything or a co-player that is "set in motion" by the plaything.

A sample of the literature used for the review is provided in Fig. 1.

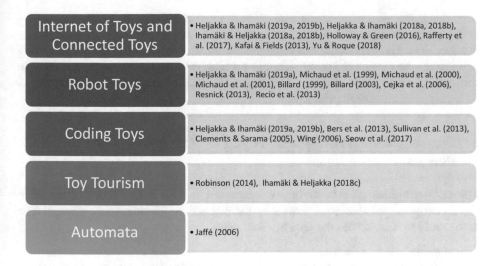

Internet of Toys and Connected Toys	• Heljakka & Ihamäki (2019a, 2019b), Heljakka & Ihamäki (2018a, 2018b), Ihamäki & Heljakka (2018a, 2018b), Holloway & Green (2016), Rafferty et al. (2017), Kafai & Fields (2013), Yu & Roque (2018)
Robot Toys	• Heljakka & Ihamäki (2019a), Michaud et al. (1999), Michaud et al. (2000), Michaud et al. (2001), Billard (1999), Billard (2003), Cejka et al. (2006), Resnick (2013), Recio et al. (2013)
Coding Toys	• Heljakka & Ihamäki (2019a, 2019b), Bers et al. (2013), Sullivan et al. (2013), Clements & Sarama (2005), Wing (2006), Seow et al. (2017)
Toy Tourism	• Robinson (2014), Ihamäki & Heljakka (2018c)
Automata	• Jaffé (2006)

Fig. 1. Literature review sample of toy mobility and toys that mobilize players.

2.2 Convenience Sampling Method: Earlier Empirical Studies

The convenience sampling method was used because accessible literature that was reviewed in combination with a series of empirical studies was selected. These multiple, empirical studies included voluntary teachers and preschool-aged children at a Scandinavian kindergarten who participated in research exploring play patterns related to state-of-the-art phygital toys, such as IoToys and coding toys. This paper presents a synthesis of all the authors' empirical materials from earlier studies, and the studies that were supported by new research material related to scholarly contributions to the field were extended. Moreover, the newest toy fair-related marketing materials acquired from the toy fairs in Nuremberg and New York from 2019 were employed to understand the potential future of mobile toys and toys that are capable of mobilizing players.

In the case studies conducted by the authors in earlier stages of research, the toy marketers, preschool teachers, and parents of preschool children who participated in playtests with a focus on various toys' affordances, were analyzed [15, 18]. For example, three types of research materials were employed. First, toy marketer's descriptions of three IoToys through websites and digital and printed marketing materials were collected. Second, a semi-structured survey for preschool teachers was administered, and third, a semi-structured survey for the parents of the preschool children was conducted. The toys under investigation fulfill the criteria of IoToys: they are "smart," and their connectivity usually occurs through mobile devices. In some cases, smart toys contain their own computers (e.g., the CogniToys Dino and Fisher-Price's Smart Toy Bear.) The toys were chosen based on their age appropriateness, gender-neutrality as character types of toys, and their availability on Amazon US (in August 2017). Moreover, one reason for selecting these IoToys was the awards they have received. For example, the CogniToys Dino has received the Silver Honor 2016 Parent's Choice Award. To analyze the research materials, a content analysis was performed. The goal of a content analysis is to provide knowledge and understanding of the phenomenon under study [20]. It provides the researcher with the possibility to conduct a close reading of the data through a systematic classification process of coding and identifying themes and patterns. In this process, researchers immerse themselves in the data to allow new insights to emerge [21]. The method is also described as inductive category development [22].

3   Previous Research: Mobile Toys/Toys That Mobilize Players

This section outlines some traces of development of the toys that have been designed in the recent past and that either move by themselves or mobilize their players. The categorization begins with the area of robot toys.

Toy Mobility – Robot Toys
In recent years, new learning standards and best practices for integrating technology into early childhood education have gained increasing interest [23]. To teach technology and engineering to young children in a developmentally appropriate way, robotics and computer programming initiatives have grown in popularity among early education researchers and educators [24]. Since 1998, the Department of Electrical Engineering and Computer Engineering of the Université de Sherbrooke have been designing a variety of mobile robotic toys with the goal of using them as pedagogical tools for children suffering from autism or other developmental disorders [25–27]. Previous research has shown that the field of robotics holds special potential for early education by facilitating both cognitive and social development [28]. For example, recent research suggests that children as young as four years old can successfully build and program simple robotics projects while learning a range of engineering and robotics concepts in the process [29, 30]. In a study by Billard, which focuses on the novel application of Robota as a laboratory platform in an introductory class to robotics taught at the undergraduate level, they used a set of 10 Robota robots. Robota is a doll-shaped robot toy, which was developed with the goal of investigating social skills, and it is a human-shaped robot that children can have social interactions with [31]. In this study, the class of students learned to program the robot's micro-controller and the robot-PC interface using both vision and speech synthesizing. The lab session's aim was to develop games suitable for normal and disabled children [32].

Using coding toys and computer programming in early education has been claimed to support *cognitive* and *social* development. For example, studies with the text-based language *Logo* have shown that computer programming can help young children with a variety of cognitive skills, including number sense, language skills, and visual memory [33]. It is notable that many robotics activities do not involve sitting alone in front of a computer; rather, they encourage physically *mobile play*: robotic manipulatives allow children to develop fine motor skills and hand-eye coordination while also engaging in collaboration and teamwork [33]—a finding that the current case study with the Dash toy robot also supports.

Moreover, robots have been used as a part of playful learning trough facilitation of *movement*. For example, Recio et al. used the Nao robot to assist a physiotherapist in modeling the movements performed during physiotherapist interventions. The study showed that they programmed the Nao robot so that it could perform nine exercises that were previously specified by the physiotherapists of the assisted living facility. The workshop consisted of one-to-one sessions with 13 participants. The results showed that Nao's movements seemed to be a highly relevant aspect to take into consideration when observing playful interactions with it [34].

Toy Mobility – Internet of Toys
Young children's growing fascination with interactive play is itself one of the clearest signs that the digital era is well underway for toys. As Negroponte claims, "it is almost genetic in its nature, in that each generation will become more digital than the preceding one" [1]. Children's media spaces and practices with the current Internet of Toys (IoToys) are digitally integrated to enable possibilities to learn through the mobility and flow of both content and data. The IoToys can interact in interesting ways, creating novel, appealing, and meaningful communication experiences and play situations using speech, sounds, visual cues, and movements; however, there is a subcategory of the IoToys, which seemingly affords more movement of the players than others—coding toys. For example, when children learn a programming language with the IoToys, they are not "just learning code, they are coding to learn" [35, 36]. As a consequence of IoToys being fitted with sensors and connecting to networks, these toys gain "new skills (affordances) that are present in new forms of communication" [37]. For example, with coding features, toys like the Dash Robot make children move physically [14]. The IoToys can react to various physical inputs (sounds, images, touch, and movements), which means that they track children's behavior and measure the surrounding environments and can interact with children on a personalized basis (e.g., by replying to their queries). In this way, the IoToys have become "media that has not been mediated before" [Ibid.]. *Mobile playful learning*, then, can be understood by analyzing contemporary smart toys such as toys with coding features, and those that allow players to code the toys' actions. Coding toys are usually toys that players mobilize through programming. In this way, mobility includes considerable potential to be put into play. For instance, Dash encourages children to move physically, such as running after the toy.

There are three general properties of IoToys. IoToys are: (1) *pervasive*—a smart, connected toy may follow a child through every activity (mobility toys or toys mobilized for players); they are (2) *social*—social aspects, such as multiplayer aspects, with one-to-one (for example, Wonder Workshop's Dash), one-to many (Wonder Workshop's Dash has a "Play by Wonder Workshop" website for more activities, challenges, and pro-tips for playing with Dash), and *promote many-to-many relations* (the Blockly app for Dash provides the opportunity to share one's own coding tips with the Dash robot with the community); and they are (3) *connected*—the IoToys may connect and communicate with other toys, devices, and services through networks. "Many adults might believe these connected toys turn children into virtual drives down the information highway and anticipate a future where popular culture will be accommodated to their own deepest desires" [38–40].

Toys That Mobilize the Player – Toy Tourism

Currently, 21st century advances have provided connected play for both adults and children of all ages [41]. Kafai and Fields propose that "connected playthings are in the digital playground of the twenty-first century" [42, p. 3]. Children are now using online and offline and local and global connections in dynamic ways and a child may even play synchronously in a virtual world with others in ways that distinguish the nexus between private and public worlds [43]. One example is *toy tourism*—the mobilizing of physical toys. Toy tourism can be categorized as free-form and open-ended and thus as a creative play practice and as a form of play structured to be goal-oriented and therefore game-like. Toy tourism is defined as follows: toy tourism occurs when toys travel either as companions of their owners or "single-handedly" as organized by toy travel agencies [44].

In addition, physical toys are also made mobile through games, such as Geocaching. Connected play can involve using the Internet while playing with the physical toy and moving the toy and can also occur in the context of toy tourism all over the world, such as with the Travel Bugs in the Geocaching game. Geocaching is a high-tech, worldwide treasure hunt game, where a player hides a cache for others to find. Geocaching is now available in more than 185 countries. Travel Bugs are metal dog tags typically attached to another object, such as a toy. Each Travel Bug has its own unique tracking number stapled on it and has its own "diary" (a website) that follows its movements. The idea is to move a toy through the geocaches, where people can take the toy and another player can take the toy and change it to another cache. The tracking number of Travel Bugs is used as proof by the user that they found the item, and it also doubles as a way for the user to locate the personal website of the Travel Bug. Players move the toys, and their movement is recorded on a website. The idea is that by picking up and dropping off Travel Bug Trackables and reporting their movements on the Travel Bugs individual website, one is mirroring the Travel Bug's real-world adventures [45].

4 Results

4.1 Mobile Toys

In this exploratory study, the goal was to understand toy mobility with its related experiences of players that move beyond the existing genres of computer-based experiences, which already exist with digital games and virtual worlds. By focusing on toy mobility and toys that mobilize players, the authors demonstrated key findings of previous research.

The results of the study showed that toy mobility contributes to a broader culture of play and edutainment, in which robot toys and IoToys, including coding toys develop new ways of thinking about toy computing and new ways of thinking about mobile playful learning. These new smart toys, robot toys, coding toys, and IoToys increase the understanding of the mobility of toys and are capable of encouraging players to both learn and become physically mobile and in this way, potentially enhance well-being.

Our study highlights how a player can interact with a robot toy, and when an activity such as coding is unlimited, it can continue as long as the player is interested. Some of the key values of robot toys are tactile exploration, expressivity, acting to change the state of the robot, the expressive potentialities of the toy robot through the use of colors and visual patterns (for example, the CogniToy Dino has a light that represents the mood of the dinosaur-shaped toy robot), shape and movement (such as the coding of the Dash robot for moving along a certain trail and letting other children try to catch it), the sense of touch in combination with proximity, and the use of the remote control to modify a robot's behavior. These findings are similar to those related to the Roball toy [46], which is a spherical robot capable of navigating in all types of environments without becoming stuck in some places or falling to the side. The interaction with Roball occurs through vocal messages and movements patterns, such as spinning, shaking, or pushing. According to a case study of Michaud et al., the majority of children were trying to catch Roball, to grab it, or to touch the robot [46].

Furthermore, the results showed that coding toys [47] can be programmed to respond differently to situations. For example, the IoToys can also learn over time through users' actions and can change the ways they respond to the world, generating more sophisticated interactions and unpredictable situations that can help capture and retain the player's interest. Therefore, IoToys encourage players to move. For example, in our case study, one preschooler coded Dash, and others became physically mobile by running and making a tunnel with their legs and allowing Dash to move through the tunnel [14]. This shows how children become creative in finding their own way to play with IoToys, which also mobilizes players. Similar results were observed by Yu and Roque [48], and the researchers mention that the toy robot Dot (Dash's companion) can cause the robot Dash to move, light up, or play a sound. They found that children program instructions based on different conditions or events through three main approaches: (1) sensors are usually embedded in physical robots to sense different environmental conditions so that the robots can be programmed to react accordingly; (2) for physical robots without sensors, maps are usually provided for children to decide how the robots will react to different conditions on a map; and (3) for virtual platforms, a series of commands or special effects, such as animations, sounds, and light effects, could be triggered when sprites interact without other objects on the screen. For example, the Curlybots can be synchronized to run together and in ScratchJr, several sprites can be programmed to move, play, make sounds, and change sizes at the same time [48].

Players can control the motion of several coding toys simultaneously or can program the motions, lights, and sound effects at the same time. For example, with the Dash and Dot toy robots employed in our own research, players can program the toys to interact with each other.

As the results have shown, toy computing is a recently developing concept in which traditional toys enter a new area of computer research using service computing technologies [49]. Therefore, in this context, a toy is a physical embodiment artifact that acts as a child's user interface for toy computing services in a cloud. The IoToys as an example of contemporary mobile toys can also capture a player's physical activity state (e.g., walking, running, etc.) and store personalized information (e.g., location, activity pattern, etc.) through a camera, microphone, Global Positioning System (GPS), and various sensors, such as facial recognition or sound detection [50].

4.2 Toys That Mobilize Players

According to the findings on our study on toys that mobilize players, following the development of historical automata, or self-moving toys, the commercial toy industry has developed a multitude of phygital play objects that have been given affordances related to sound, light, and movement. Many of these toys represent character toys, or toys with a face, such as dolls, action figures, or soft toys, with an emphasis on the idea of potential anthropomorphization. "The long-established fascination with automata and the need to animate anthropomorphic figures are still current," Deborah Jaffé, author of *The History of Toys*, wrote in 2006 [11].

As shown, a player does not always keep his/her toys in the intimacy of the play-room in one's living spaces. The results of this study showed that non-digital (character) toys, such as dolls, action figures or plushies can also mobilize players by making them geographically mobile by traveling with their toys, or setting them out to travel. *Toy tourism* may be divided into the two categories of either traveling with toys, or travel agencies moving others' toys all over the world and playing the Geocaching game, where players move toys (Travel Bugs) from certain caches to other caches. In Geo-caching, the main idea is to become mobile with the toy all over the world. This means that the toy is moving with different players, in this case while playing the geocaching game. For example, the Dr. Geocacher Travel Bug has moved 4041 km [51].

Outside of Geocaching, toys mobilize players by motivating them to more creative play than game-like competitions. One important facet of this activity is (visual) sto-rytelling focusing on the toys travels. According to Robinson (2014), a traveling toy may have its own website or blog, a Twitter following, or a Facebook page. Travel agents for toys have appeared in the past decade as well as hosts who take the toy traveling and posts images – and the toy – back to the owner. Robinson's study indicated that the Flickr photo service has 139 specific groups identified by Traveling Toys parameters, and the highest number of images uploaded and shared was 31,000 Trav-eling Toys (based on her survey in 2012) [52]. In this way, toys urge their players to travel and bring the toys along on the journeys. Similar play patterns can be detected among adult toy players [53, p. 285]. According to the findings, in order for toys to become mobile/mobilize, the traveling toys must be made *portable*: mobility in terms of toy play behavior relates to the affordances of toys. For example, toy company Kenner made its Star Wars figures pocket-sized so they could be carried around and played with wherever the player might choose [54, p. 213]. Since the beginnings of these massively popular action figures, many character toys are designed with mobility in mind. In other words, the players should be able to travel with their character toys, and in this way, expand the play environment beyond the ordinary living space of their players.

4.3 Classification of Mobilizing Technologies: The Toy Mobility Continuum

The focus of this section is the proposition of a structured classification of the identified mechanical, phygital, and digital realities of mobilizing technologies in relation to toys. This classification summarizes the main differences between these realities according to the two dimensions and is applicable to the design of mobility in toys. It also allows for comparisons between reality and toy technologies. There are three areas (mechanical, phygital and digital toy mobility and analogue/digital toy tourism), which are referred to as "dimensions," that relate either to a correlation of interactive toys that mobilize the player or a player traveling with toys.

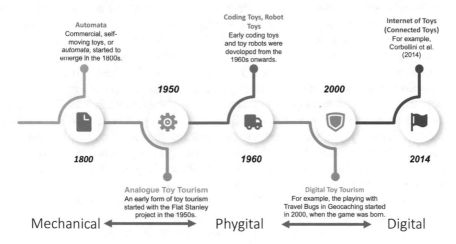

Fig. 2. Toy Mobility Continuum—a classification of toy-based mobilizing technologies.

Based on the findings, the Toy Mobility Continuum is proposed (see Fig. 2), which suggests a timeline of the development of phygital playful technology with mobilizing tendencies: toys that move by themselves or that mobilize their players. On one end of this timeline are the mechanical automata, which demonstrate toy inventors', designers' and artists' desire to create self-moving toys. On the other end of continuum are the contemporary, connected toys, which represent both physical play objects as well as digitally connected playthings. Both the self-moving toys and the connected toys of the present have a relation to phygitality, one particularly due to their mechanics and the other due to their digitality.

5   Discussion

This exploratory paper has focused on the technological development of the toy medium with an interest in toys' capacity to mobilize *homo ludens*, the playing human. By conducting an extensive literary review on the history and the present of mobile play objects, the way toys have developed from early moving and mechanical automata to playthings that move by themselves through in-built computerized components have been demonstrated. The interest of the paper was two-fold: By analyzing the historical trajectory of mobile toys, the authors highlighted the role of both toys and their players as participants in technologically mediated, phygital play. This hybrid form of playing combines the physicality of playthings with both mechanical and digital features.

Second, through a mixed methods approach, the findings of the literary review have been synthesized with findings of the authors' earlier empirical research conducted with preschool-aged children and their teachers with different types of contemporary interactive toys; toy robots and IoToys.

The results of the study have illustrated how toys, character toys in particular, have transformed from entertaining, self-moving spectacles to educational machines that mobilize the player both physically and geographically. Based on the results of the study, the literary review, and previous empirical research, the *Toy Mobility Continuum* that visualizes the development of phygital playful technology and the types of toys that afford mobility in play of the past, present, and future has been proposed.

The limitation of the study is its conceptual approach—the authors based the findings of the paper on earlier studies and their own previous research as other studies in the field are limited. For example, there is no quantitative analysis of similar studies. Moreover, this study synthesized several case studies conducted earlier, but they are limited to Finnish participants only. However, although the present study only represents a firsthand approach in understanding the concept of toy mobility, it also provides an important contribution to understanding the ways that toys are a medium that mobilize in many ways. On the one hand, toys are given mobile affordances by their designers and makers. On the other hand, they are made mobile through the activities of the players, or make their players mobile, even by urging them to travel with their toys. Following these leads, our aim is that the tentative study presented in the paper at hand, will inspire ongoing and future studies of toy mobility and even generate new ideas regarding, for example, how toy mobility can be utilized as a starting point for innovating new products and services not only within the industries of play but also within the larger context of the experience economy interested in various forms of physical, cognitive, and social well-being, such as in education, and the health and tourism industries.

6   Conclusions and Future Studies

The main contribution of this study is that it highlights the mobility aspect of toys in many ways. Based on the findings, it was concluded that phygital, playful technologies, such as non-digital character toys, robot toys, or connected IoToys toys, have the capacity to influence human well-being in various dimensions. By making players mobile, toys as a medium that invite engagement with phygital playful technologies have the potential to enhance physical, cognitive, and social well-being. Therefore, other researchers in the field of toys, both traditional and technologically-enhanced, are able to develop their understanding of toys' capacities in relation to mobility, mostly their mobilizing tendencies, which can be employed much more in the design of future toy-to-human interactions. More than before, toys can be used to increase human well-being by designing specific affordances into toys that result in well-being effects, such as physical (through exercise and geographical movement), cognitive (through educational benefits), and social (through social object play by playing together) well-being.

The novelty value of the paper is in presenting the Toy Mobility Continuum that emerged from the mixed method approach combining an extensive literature review with empirical research on contemporary toy play cultures with children, mature players, and transgenerational toy audiences. These observations would not have been possible to make through theoretical research only but required a creative approach: a synthesis of earlier work, both theoretical and empirical. Although extensive and unique in its research design, the limitation of this study is that it focuses on *character toys*, meaning dolls, action figures, and soft toys that in some cases have been turned into social robots. Therefore, a multitude of mobile toys and toys with mobilizing potential, such as toy vehicles were excluded in the analysis. Nevertheless, the study presented in this paper functions as an example of a firsthand approach and start for a scholarly dialogue on the mobilizing tendency of toys. To conclude, the main contribution of the paper is once more accentuated. By conducting this study, the way the pure physicality of the mechanically mobile toy has transformed into phygitality with a mobilizing capacity has been highlighted. It is assumed that the role of digital, social, and mobile technologies in particular will impact future toy design in unforeseeable ways. The future of toy play will involve an upswing in the integration of advanced, visual, and augmented technologies. For instance, the ubiquity of technology will mean more physical but screenless playthings with "hidden technologies" that through their phygitality continue to mobilize *homo ludens*, the playing human, in many more ways. Thus, to conclude, it is predicted that future toys will continue to influence well-being in more ways than before. One area of this development focuses on the physical mobility aspect of interactions with toys.

References

1. Negroponte, N.: Being Digital. Hodder and Stoughton, London (1995)
2. Lupetti, M., Piumatti, G., Rossetto, F.: Phygital play HRI in a new gaming scenario. In: 7th International Conference on Intelligent Technologies for Interactive Entertainment (INTETAIN). IEEE (2015)
3. Tsekleves, E., Gradinar, A., Darby, A., Smith, M.: Active parks: "Phygital" urban games for sedentary and older people. In: Games for Health 2014, pp. 140–143. Springer, Wiesbaden (2014)
4. Nofal, E., Reffat, R.M., Vande Moere, A.: Phygital heritage: an approach for heritage communication. In: Proceedings of the 3rd Immersive Learning Research Network Conference (iLRN 2017), Coimbra, Portugal, pp. 26–29 (2017)
5. Schreibman, S., Papadopoulos, C., Hughes, B., Rooney, N., Brennan, C., Mac Caba, F., Healy, H.: Phygital augmentations for enhancing history teaching and learning at school. In: DH (2017)
6. McManus, C.: Exploring the phygital: an assessment of modern play objects. Master of Design at Massey University, Wellington, New Zealand. https://mro.massey.ac.nz/bitstream/handle/10179/12449/02_whole.pdf?sequence=2&isAllowed=y. Accessed 19 Mar 2019
7. Trautman, T.: The Trouble with Toys. https://www.newyorker.com/business/currency/the-trouble-with-toys. Accessed 19 Mar 2019
8. Tyni, H., Kultima, A., Mäyrä, F.: Dimensions of hybrid in playful products. In: Proceedings of International Conference on Making Sense of Converging Media, p. 237. ACM (2013)

9. Yelland, N., Moyles, J.: The Excellence of Play, 4th edn. Open University Press, Berkshire (2015)
10. Of Toys and Men, Helsinki Art Museum Tennis Palace 24.2.–20.5.2012 exhibition press materials (2012)
11. Jaffé, D.: The History of Toys. From Spinning Tops to Robots. Sutton Publishing Limited, England (2006)
12. Pesce, M.: How Technology is Transforming Our Imagination. The Ballantine Publishing Group, New York (2000)
13. "Evolution, not revolution" in Play it! The Global Toy Magazine, Spielwarenmesse International Toy Fair Nürnberg, 30 January–4 February 2013, and "International Toy Fair 2013 shows the future of playing: classic and digital too," Spielwarenmesse International Toy Fair publication 30 January–4 February 2013. (press information)
14. Heljakka, K., Ihamäki, P.: Ready, Steady, Move! coding toys, preschoolers and mobile playful learning. In: HCI 2019, 21st International Conference on Human-Computer Interaction, Orlando, Florida, USA, 26–31 July 2019
15. Ihamäki, P., Heljakka, K.: Smart, skilled and connected in the 21st century: educational promises of Internet of Toys (IoToys). In: Hawaii University International Conferences, ART, Humanities, Social, Sciences & Education, 3–6 January, Price Waikiki Hotel, Honolulu, Hawaii (2018)
16. Ihamäki, P., Heljakka, K.: The Internet of Toys, connectedness and character-based play in early education. In: Arai, K., Bhatia, R., Kapoor, S. (eds.) Proceedings of the Future Technologies Conference, FTC 2018. Advances in Intelligent Systems and Computing, vol. 880, pp. 1079–1096. Springer, Cham (2018)
17. Heljakka, K., Ihamäki, P.: Verkottunut esineleikki osana esiopetusta: Lelujen Internet leikillisen oppimisen välineenä. Lähikuva 31(2), 29–49 (2018)
18. Heljakka, K., Ihamäki, P.: Preschoolers learning with the internet of toys: form toy-based edutainment to transmedia literacy. Seminar.net Int. J. Media Technol. Lifelong Learn. 14 (1), 86–102 (2018)
19. Heljakka, K., Ihamäki, P.: Persuasive toy friend and preschoolers: playtesting IoToys. In: Mascheroni, G., Holloway, D. (eds.) Internet of Toys–Practices, Affordances and the Political Economy of Children's Smart Play, pp. 159–178. Palgrave Macmillan (2019)
20. Downe-Wambolt, B.: Content analysis: method, applications and issues. Health Care Women Int. 13, 313–321 (1992)
21. Kondraki, N.L., Wellman, N.S.: Content analysis: review of methods and their application in nutrition education. J. Nutr. Educ. Behav. 34, 224–230 (2002)
22. Mayring, P.: Qualitative content analysis. Forum Qualit. Soc. Res. 1(2). http://www.qualitative-research.net/fqs-texte/2-00/02-00mayring-e.htm. Accessed 19 Mar 2019
23. NAEYC & Fred Rogers Center for Early Learning and Children's Media: Technology and interactive media as tools in early childhood pragmas serving children from birth through age 8. Joint position statement. NAEYC: Latrobe, PA: Fred Rogers Center for Early Learning at Saint Vincent College, Washington, DC. http://www.naeyc.org/files/naeyc/file/positions/PS_technology_WEB2.pdf. Accessed 8 Feb 2019
24. Bers, M.: Blocks to Robots: Learning with Technology in the Early Childhood Classroom. Teachers College Press, New York (2008)
25. Michaud, F., Lucas, M., Lachiver, G., Clavet, A., Dirand, J.M., Boutin, N., Mabilleau, P., Descôteaux, J.: Using ROBUS in Electrical and Computer Engineering education. In: Proceedings American Society for Engineering Education, Chalotte (1999)
26. Michaud, F., Clavet, A., Lachiever, G., Lucas, M.: Designing toy robots to help autistic children – an open design project for Electrical and Computer Engineering education. In: Proceedings American Society for Engineering Education, St. Louis (2000)

27. Michaud, F., Clavet, A.: Organization of the RoboToy contest. In: Proceedings American Society for Engineering Education, Albuquerque (2001)
28. Cejka, E., Rogers, C., Portsmore, M.: Kindergarten robotics: using robotics to motivate math, science, and engineering literacy in elementary school. Int. J. Eng. Educ. **22**(4), 711–722 (2006)
29. Sullivan, A., Kazakoff, E.R., Bers, M.U.: The wheels on the bot go round and round: robotics curriculum in pre-kindergarten. J. Inf. Technol. Educ. Innov. Pract. **12**, 203–219 (2013)
30. Clements, D.H., Sarama, J.: Young children and technology: what's Appropriate? In: Masalski, W.J. (ed.) Technology-Supported Mathematics Learning Environments (Sixty-Seventh Yearbook), pp. 51–73. NCTM, Reston (2005)
31. Billard, A.: DRAMA, a connectionist architecture for online learning and control of autonomous robots: experiments on learning of a synthetic proto-language with a doll robot. Ind. Robot. **26**(1), 56–66 (1999). https://doi.org/10.1108/01439919910250232
32. Billard, A.: Robota: clever toy and educational tool. Robot. Auton. Syst. **42**, 259–269 (2003). https://doi.org/10.1016/s0921-8890(02)00380-9
33. Wing, J.: Computational thinking. Commun. ACM **49**(3), 33–35 (2006)
34. Recio, D.L., Segura, E.M., Segura, L.M., Waern, A.: The NAO models for the elderly. In: Proceedings of the 8th ACM/IEEE International Conference on Human-Robot Interaction, pp. 187–188 (2013). http://mobilelifecentre.org/sites/default/files/NAO_HRI%2BE%2BE.pdf. Accessed 19 Mar 2019
35. Resnick, M.: Learn to code, code to learn. How programming prepares kids for more than math. EdSurge **8** (2013). https://www.edsurge.com/news/2013-05-08-learn-to-code-code-to-learn. Accessed 25 Feb 2019
36. Holloway, D., Green, L.: The Internet of toys. Commun. Res. Pract. **2**(4), 506–519 (2016). https://doi.org/10.1080/22041451.2016.1266124. Accessed 3 Mar 2019
37. Bunz, M., Meikle, G.: The Internet of Things. Polity Press, Cambridge (2018)
38. Kline, S., Dyer-Witheford, N., de Peuter, G.: Digital Play, The Interaction of Technology, Culture, and Marketing, p. 14. McGill-Queen's University Press, Montreal & Kinston, London Ithaca (2003)
39. That, E.I.: Context data model privacy. In: PRIME Standardization Workshop, IBM Zurich, p. 6 (2006)
40. Rafferty, L., Hung, P.C.K., Fantinato, M., Peres, S.M., Iqbal, F., Kuo, S.-Y., Huang, S.-C.: Towards a privacy rule conceptual model for smart toys. In: Proceedings of the 50th Hawaii International Conference on Systems Sciences (HICCS), Hawaii, pp. 1226–1235 (2017). https://scholarspace.manoa.hawaii.edu/bitstream/10125/41299/paper0150.pdf. Accessed 19 Mar 2019
41. Rossow, A.: Playing in the digital age: 2019 CES opener. In: "Techup" Talks 21st Century Toy Market, CES 2019, Emerging Tech, News (2019). https://gritdaily.com/techup-talks-21st-century-toy-market/. Accessed 28 Feb 2019
42. Kafai, Y., Fields, D.: Connected Play: Tweens in a Virtual World. Mitt Press, Cambridge (2013)
43. Marsh, J.: Online and offline play. In: Burn, A., Richards, C. (eds.) Children's Games in the New Media Age, pp. 109–312. Ashgate, Cambridge (2014)
44. Ihamäki, P., Heljakka, K.: "Travel Bugs": toys traveling socially through geocaching. In: DIGRA 2018: The 11th Digital Games Research Association Conference, The Game is the Message, Turin, Italy, 25–28 July 2018
45. Ihamäki, P., Heljakka, K.: The sigrid-secrets geocaching trail: influencing well-being through a gamified art experience. In: GamiFIN Conference 2017, Pori, Finland, 9–10 May 2017

46. Michaud, F., Caron, S.: Roball–an autonomous toy-rolling robot. In: Proceedings of the Workshop on Interactive Robotics and Entertainment (2000)
47. Seow, P., Looi, C.-K., Wadhwa, B., Wu, L., Liu, L.: Computational thinking and coding initiatives in Singapore. In: Kong, S.C., Sheldon, J., Li, K.Y. (eds.) Proceedings of International Conference on Computational Thinking Education 2017, Hong Kong, China, pp. 164–167. The Education University of Hong Kong (2017)
48. Yu, J., Roque, R.: A survey of computational kits for young children. In: IDC 2018, Trondheim, Norway, 19–22 June 2018, pp. 289–299 (2018), https://doi.org/10.1145/3202185.3202738
49. Hung, P.C.K.: Mobile services for toy computing. Springer International Series on Application and Trends in Computer Science. Springer, Cham (2015)
50. Valadao, C.T., Alves, S.F.R., Goulart, C.M., Bastos-Filho, T.F.: Robot toys for children with disabilities. In: Tang, J., Hung, P. (eds.) Computing in Smart Toys. International Series on Computer Entertainment and Media Technology, pp. 55–84. Springer, Cham (2017). https://doi.org/10.1007/978-3-319-62072-5_5
51. Dr. Geocacher, Travelbug in Geocaching. https://www.geocaching.com/track/details.aspx?tracker=TB6RNPN. Accessed 3 Mar 2019
52. Robison, S.: Toys on the move: vicarious travel, imagination and the case of traveling toy mascots. In: Lean, G., Staif, R., Waterton, E. (eds.) Travel and Imagination, 1st edn., p. 153. Routledge, London (2014)
53. Heljakka, K.: Principles of adult play(fulness)–from wow to flow to glow. A Ph.D. Dissertation for the School of Arts, Design and Architecture. Aalto University, Helsinki (2013)
54. Geraghty, L.: Aging toys and players: fan identity and cultural capital. In: Kapell, M.V., Shelton, J.L. (eds.) Finding the Force of the Star Wars Franchise. Fans, Merchandise and Critiques, p. 213. Peter Lang Publishing, New York (1996)

Unified and Stable Project: "Ushering in the Future"

Mohamed E. Fayad[1]([⊠]), Gaurav Kuppa[1], and David Hamu[2]

[1] San Jose State University, San Jose, CA, USA
{m.fayad,gaurav.kuppa}@sjsu.edu
[2] Liberty Consulting, St. Petersburg, AZ, USA
dave.hamu@gmail.com

Abstract. Why do we need projects? What is the need to carry out a project? What is the ultimate-goal of any project? These are a few thoughts that crop up in one's mind while trying to develop and implement any project. The motivations of this paper are to establish the unified functional and non-functional requirements of AnyProject for the first time. Our goal is to enable unification for AnyProject pattern leading to the creation of a stable pattern language and to enable usability for applications across numerous domains and apply for an unlimited number of scenarios. The idea is to compare the existing or traditional model (TM) of a project and software stability model (SSM) of any project using non-functional requirements as criteria for comparison, evaluation, and measurements. The significant findings are: (1) The TM can only be applied to one scenario, unlike the AnyProject Pattern, which can be effectively applied to unlimited scenarios; (2) The TM requires high maintenance costs and limited scalability. On the other hand, the SSM definitely cuts down on costs because it is cognitive knowledge; and (3) In the measurability, the TM fails to model the essential properties of AnyProject. However, the stable model delineates essential properties. As a result, the SSM enables stability and unlimited applicability.

Keywords: Project definitions · TM · SSM · UML · Abstraction · Applicability · Impact · AnyProject stable design pattern

1 Introduction

A project is defined as an individual or collaborative enterprise that is carefully planned and designed to achieve a goal [1]. Projects have great importance in all domains because innovation and discovery are initiated by a large undertaking in the form of a project. The rapid integration of multidisciplinary fields with technology brings about a new era of collaboration and fusion of ideas that have never been experienced before. Due to the breadth of possible projects, the conventional definitions of "project" are incomplete. Not to worry, herein, we provide a complete and widely applicable definition for any project.

Existing or traditional models (TM) of a project are representations that are vastly different for each deployment in an application context. Consequently, TMs have

© Springer Nature Switzerland AG 2020
K. Arai et al. (Eds.): FTC 2019, AISC 1070, pp. 641–651, 2020.
https://doi.org/10.1007/978-3-030-32523-7_47

proven to be ineffective in achieving any reuse even across a very similar application context, and much of the work will need to be reinvented for subsequent projects. Unnecessary development and maintenance time are spent developing new models. In comparison, a software stability model (SSM) [2–5] defines and uses high-level relationships to illustrate relationships between components that can be used to describe the core aspects of any project. Using the concepts of Enduring Business Themes (EBTs) and Business Objects (BOs), we can construct the SSM. We compare the TM of a project and the SSM of any project pattern using non-functional requirements, the critical attributes and criteria of a system application, as criteria for comparison, evaluation, and measurements.

This paper is organized by 5 major sections: Sect. 2 provides scenarios of any project in use; Sect. 3 illustrates corresponding TM; Sects. 4 to 8 portrays a weighted comparison between the stable, unified model and traditional models based on non-functional requirements and qualitative and quantitative measurements; Sect. 9 presents further discussion about AnyProject, and Sect. 10 presents a conclusion.

2   The Problem

In a world driven by productivity, innovation is heavily project-based. Unfortunately, there is no clarity as to what a project is. Every definition of a project portrays a different aspect of a project. This leaves no cohesive, unified definition for any project. More specifically, all definitions of a project do not cohesively piece together what a project is, nor do the conventional definitions identify the functional requirements, and how to use a project, or convey the true nonfunctional requirements.

In contemporary business and science domains, a project is an individual or collaborative enterprise, possibly involving research or design, that is carefully planned to achieve an aim and goal [6]. On the other hand, a project is also defined as a set of interrelated tasks to be executed over a fixed period and within certain cost constraints and other limitations [7].

The subtle but glaring differences between definitions exist in many forms. Beyond the semantic differences, the two definitions are different in what they convey. The first definition identifies a project to be an "undertaking with a goal" while the second definition determines a project to be "a set of interrelated tasks with limitations." We have not established a universal definition for all projects. The problem extends to all existing definitions of a project. More broadly, there is no unified and applicable definition for AnyProject and this paper aims to illustrate that.

3   Context

Scenario #1: The Manhattan Project
The Manhattan Project is a famous example. It was an effort during World War II in the United States to develop the first nuclear weapon. This top-secret project lasted from 1939 until 1945 and utilized various resources under the direction of General Leslie R. Groves, Deputy Chief of Construction of the U.S. Army Corps of Engineers.

It resulted successfully the production of nuclear bombs used in WW II The Manhattan Project was an effort to make the nuclear bomb done by the US Government and the US Army in the middle of the 20th century. After spending considerable time and money and by utilization of advanced laboratories and knowledgeable personnel, the project was completed successfully in 1945 [9].

Scenario #2: Boeing B-29 Super-fortress

"The Boeing B-29 Super-fortress is a four-engine propeller-driven heavy bomber designed by Boeing, which was flown primarily by the United States during World War II and the Korean War. It was one of the largest aircraft operational during World War II and featured state-of-the-art technology. Including design and production – at over \$3 billion – it was the most expensive weapons project in the war, exceeding the \$1.9 billion cost of the Manhattan Project—using the value of dollars in 1945. Innovations introduced included a pressurized cabin, dual-wheeled, tricycle landing gear, and an analog computer-controlled fire-control system directing four remote machine gun turrets that could be operated by a single gunner and a fire-control officer. A manned tail gun installation was semi-remote" [8].

4   Traditional Model

The following traditional models which are business as usual in Figs. 1 and 2 will illustrate the two specific projects one scenario per model.

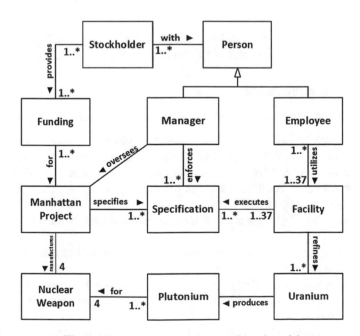

Fig. 1. The manhattan project traditional model

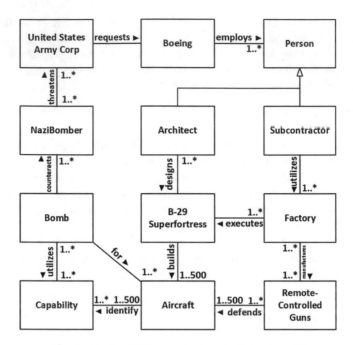

Fig. 2. Boeing B-29 superfortress traditional model

5 Stable Model

The stable project pattern is unified and stable to build on top of its unlimited applications, whereas TM is only suitable for a single application.

5.1 Functional Requirements

Figure 3 illustrates the functional requirements of any project:

1. Need is the goal of AnyProject, i.e., it satisfies one more many (AnyProject).
2. One or more project (AnyProject) contains one or more tasks (AnyTask).
3. One or more task (AnyTask) reaches one or more milestone (AnyMilestone).
4. One or more party (AnyParty) such as human, organization, etc., defines or manages Need for a project.
5. One or more party (AnyParty) defines one or more criteria (AnyCriteria) for the project.
6. One or more party (AnyParty) examines one or more milestones (AnyMilestone) defined for a project.
7. One or more project (AnyProject) requires one or more resources (AnyResource).
8. One or more party (AnyParty) provides/controls one or more resources (AnyResource).
9. One or more criteria (AnyCriteria) influences one or more (AnyResource).

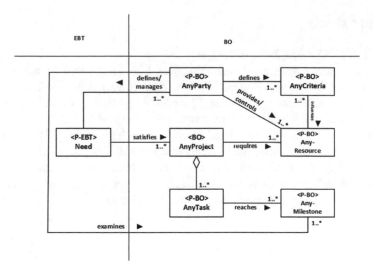

Fig. 3. AnyProject stable design pattern

5.2   Non-functional Requirements

In the following section, we present the non-functional requirements of AnyProject:

1. **Practical:** AnyProject should be practical in completing AnyTask as a part of satisfying the ultimate Need of AnyProject. A practical project is a very important quality factor because it highlights the usefulness of a project in the world and the innovation that the project embodies. Also, a project's practicality is correlated with its critical value to a larger movement. In total, a project must be practical, useful, and provide some critical value to a larger movement in the world.

2. **Doable:** AnyProject is doable if it has AnyTask and AnyMilestone laid out with AnyParty who are responsible for management and execution. In other words, a project must have a systematic approach to achieving a specific milestone. This results in a project that can be done through the given resources and limitations that inherently exist.

3. **Manageable:** AnyProject must have AnyParty who manages the project and enforces AnyCriteria to ensure quality output. A project is deemed manageable if its criteria are measurably satisfied. The constant evaluation and re-calibration of the project's goals, criteria, and outlook result in a manageable project.

4. **Growth:** AnyProject must lay out several AnyTasks that will progressively pile up and satisfy AnyMilestone. In other words, the advancement of a project must lie in the productive nature of completing tasks such that the larger milestone is met. In the same way, these completions of milestones will satisfy the need.

6   Applicability

Application 1: The Manhattan Project

The United States (AnyParty), United Kingdom (AnyParty), and Canada (AnyParty) worked on The Manhattan Project (Fig. 4) (AnyProject) from 1942 to 1946. This project was a $2 billion (AnyResource) undertaking during the World War II in attempts to create the world's first nuclear weapon (AnyMilestone) before Germany's nuclear weapon project was completed (AnyCriteria). The Military Major (AnyParty) managed this top-secret project and employed up to 130,000 people (AnyResource) at 37 facilities (AnyResource) across the nation. These employees conducted research on enriching and separating uranium (AnyTask), producing plutonium from uranium (AnyTask), and gathering intelligence on the German nuclear weapon project (AnyTask).

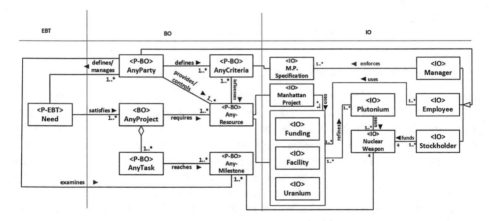

Fig. 4. The manhattan project application

Application 2: Boeing B-29 Superfortress Project

Boeing (AnyParty) began to design the B-29 Superfortress (AnyProject) (Fig. 5) in 1938 in response to a request from the United States Army Air Corps (AnyParty). This request detailed that the aircraft should be capable of delivering 20,000 lb of bombs (AnyCriteria) from a distance of 2667 miles (AnyCriteria), and is capable of flying at a speed of 400 mph (AnyCriteria). Henry H. Arnold (AnyParty) led the production of a new bomber flight to counter the production of Nazi nuclear bombers (AnyMilestone). After getting the proposal approved, Boeing received a production order for 500 aircraft (AnyMilestone). The manufacturing process involved operating plants at four different locations and employing thousands of subcontractors. The B-29 Superfortress had many new features such as remote-controlled guns (AnyCriteria) and pressurized and connected crew areas (AnyCriteria).

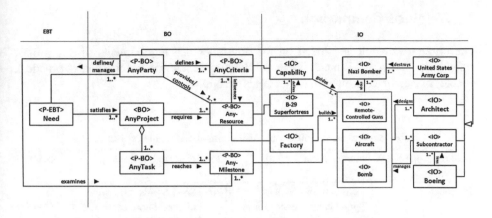

Fig. 5. Boeing B-29 superfortress project application

7 Scenarios in Table

See Table 1

Table 1. Five scenario table

EBT	BOs	App-1 – Directing a Movie	App-2 – Presidential Campaign	App-3 – Southern Border Wall	App-4 – Building Automatic Cars	App-5 – Hacking a Bank
Need	AnyParty	Director Pixar Studio	United States Voters Congress	President Senate Congress	Tesla United States	Wells Fargo Hacker
	AnyCriteria	Profit	Voter Popularity Political Experience	Protection Support	Air Bags	Stealth
	AnyProject	Movie	Campaign	Wall	Automatic Cars	Hack
	AnyResource	Performers Money Movie Set	Money Political Support Endorsements	Money Voter Support	Technology Car	Password
	AnyTask	Direct Movie	Voter Rally Town Hall	Construction	Testing	Virus
	AnyMilestone	Filming Production	Win Election	Approval	Release of Car	Infiltration

8   Weighted Comparison

The criteria were selected based on the non-functional requirements of any project which are shown in Table 2. Each of the non-functional requirements, practical, doable, manageable, growth, have an equal weighting.

Table 2. Weighted comparison of traditional and stable project model

Criteria/Property	W	Description for TM	TM W	Description for SSM	SSMW
Practical	25	TM has elements of practicality because it is constructed from tangible objects, which leads to the easy implementation of a scenario	20	SSM builds the foundation of core knowledge, which applies to all scenarios of AnyProject; therefore, the SSM is more practical	23
Doable	25	TM is doable because its construction is based on tangibility. Tangible objects can be constructed and changed	18	SSM builds on top of the defined core knowledge which extends to the IOs, which are tangible objects that can be easily implemented	25
Manageable	25	The TM is based on tangible objects, and the design of the model is incomplete because it is not based on the functional and nonfunctional requirements	2	As a result of the core knowledge, the SSM has a great conceptual understanding of AnyProject. Therefore, the SSM can control, execute and manage a project more effectively	24
Growth	25	The TM does not show the outlook and growth of the project over time. The TM cannot grow because it can only be applied to one scenario. Therefore, the high maintenance costs and the single-use patterns stunt the growth of the TM	2	SSM is designed to deal with dynamic parts because the BOs can seamlessly be implemented without disrupting the entire structure of the pattern language. Therefore, there are lower maintenance costs, which enable the system to grow more effectively	25
Total	100	–	42	–	97

8.1   Measurability Study

The total number of methods in any system can be calculated using the formula (Table 3):

Table 3. Summary of comparison of traditional and stable model

Feature	TM	SSM
Number of tangible classes	11–12	7
Number of inheritances	1	0
Number of attributes per class	1–7	3–6
Number of operations per class	0–6	1–5
Number of applications	1	Unlimited

T = C * M; where,
T = total number of operations
C = total number of classes
M = number of methods per class

Traditional Project Model 1
C = 11
M = 4
T = 11 * 4 = 44
Traditional Project Model 2
C = 12
M = 3
T = 12 * 3 = 36
Stable Project Model
C = 7
M = 3
T = 7 * 2 = 14

The Stable Project Design Pattern is more applicable compared to the TM because it is less complex, as evidenced by the fewer operations in the Stable Project Design Pattern. The stable model based on three level architecture achieves a more detailed understanding of the problem requirements. Moreover, it achieves this deep representation with far less complexity than TM. The Traditional Project Model includes tangible classes which make it vulnerable in the event of any change.

8.2   Qualitative Measurements

8.2.1   Growth

The following analysis will use growth as the criteria to determine the better model between TM and SSM. Since the SSM is completely applicable, it enables greater growth because it can adapt to situations, new technologies, and innovation much easier than a TM. Additionally, there will be greater growth because of reduced maintenance costs because the SSM core knowledge does not change. On the other hand, TM will not nurture growth because it is only useful for a single scenario. Just like a single-use plastic water bottle, a TM is built to serve its purpose only once. Therefore, the long-term costs of using the TM are higher than the Stable Project Design Pattern. It follows that the Stable Project Design Pattern has a higher ceiling and more potential than the TM.

9   Discussions and Analysis

9.1   Abstraction

The abstraction process emphasizes pattern reusability and usability. Reusability refers to utilizing the same model in various scenarios, reuse in different contexts, thus it proves to be cost-effective. As the qualitative and quantitative measurements clearly show, SSM is much more effective at creating a stable and reusable design pattern language for AnyProject. The reusability factor and the count of reused classes speak to how the SSM abstracts the patterns more efficiently than the TM. This study appeals to what we know about classes in the TM since all the classes are defined for a scenario. Hence, for each different scenario, TM must be developed anew. However, with software stability model, we have the freedom to plug in any number of IOs to the Bos to make the model fit into different scenarios.

Additionally, the nonfunctional requirements abstracted within each class in the SSM creates a functioning pattern language. The comparative study demonstrates the superiority of the SSM over the TM, as evidenced by the nonfunctional requirements, which serve as the evaluation and measurement metrics. The effective abstraction of the nonfunctional requirements and pattern language shows how the SSM is more stable and reusable.

9.2   Applicability

TM does not have a wide range of applicability. Since a TM is constructed in the business as usual fashion, it is created for a specific scenario, only. There is little to no consideration for wide applicability beyond the imminent scenario. In contrast, the SSM is constructed with applicability in mind. The process of constructing an SSM is rigorous in its attempt to encompass all scenarios of AnyProject. The core knowledge of the AnyProject SSM is extended to include the specific IOs of each scenario, as shown in the applicability. This core knowledge is extensible to any scenario of AnyProject. Applicability exists when you have less complexity in classes such that you can represent the entire system without having a large number of operations that are susceptible to error TM 1 and 2 have 33 and 36 classes, respectively, whereas SSM only has 14 methods. This analysis shows that the Stable Project Design Pattern is more applicable than the TM as it can be utilized in many more scenarios as a result of its applicability. This applicability leads to immense prosperity for the AnyProject implementation.

9.3 Impacts

Traditional Model has zero impact in terms of usability and costs. Since its applicability is very limited, and the idea of doing a project is not defined, to obtain more and exact information, considerable rework is required, which involves money as well as cost. On the other hand, the stability model proves to be a clear winner in terms of usability and costs. Its impact is much higher because costs are significantly reduced due to the design patterns involved and it supports reusability too. Thus, using the software stability model as a design model results in a higher success rate, which is targeted towards achieving goals of completing a project.

10 Conclusion

In conclusion, we could say that software stability model helps obtain the enduring concept dynamically, i.e., it has wide applicability for different scenario where it is applied. Thus, the enduring concepts help build a solid foundation for solutions and implement a better way of software stability model. Hence, the stability model is preferred as a design model for any application.

Also, we recognize that contemporary approaches to project execution tend to shun the creation of analytical artifacts and abundant documentation. This is referred to as to this as NBUA (No Big Upfront Anything) [10]. With the implicit reusability of AnyProject Stability Model, the approach will reduce the creation of redundant project artifacts while employing the requisite analysis to complete the project successfully.

References

1. "Project," Oxford English Dictionary, Oxford, United Kingdom, Oxford University Press, Oxford (2012)
2. Fayad, M.E., Altman, A.: An introduction to software stability. Commun. ACM **44**(9), 95–98 (2001)
3. Fayad, M.E.: Accomplishing software stability. Commun. ACM **45**(1), 95–98 (2002)
4. Fayad, M.E.: How to deal with software stability. Commun. ACM **45**(4), 109–112 (2002)
5. Fayad, M.E., Sanchez, H., Hegde, S., Basia, A., Vakil, A.: Software Patterns, Knowledge Maps, and Domain Analysis. Auerbach Publications, Boca Raton, 422 p., December 2015
6. Compare: An individual or collaborative enterprise that is carefully planned to achieve a particular aim. Oxford English Dictionary, Oxford, United Kingdom, Oxford University Press, Oxford (2012)
7. What is a project? definition and meaning, 19 April 2016. BusinessDictionary.com
8. Boeing B-29 Superfortress, 28 March 2019. https://en.wikipedia.org/wiki/Boeing_B-29_Superfortress. Accessed 08 April 2019
9. Kelly, C.C.: The Manhattan Project: The Birth of the Atomic Bomb in the Words of Its Creators, Eyewitnesses, and Historians. Black Dog & Leventhal Publishers, Incorporated - History - 495 p. (2009)
10. Big Design Up Front, 28 March 2019. https://en.wikipedia.org/wiki/Big_Design_Up_Front. Accessed 08 April 2019

Global Logistics in the Era of Industry 4.0

Ayodeji Dennis Adeitan[1](✉), Clinton Aigbavboa[2],
and Emmanuel Emem-Obong Agbenyeku[3]

[1] Postgraduate School of Operations Management, University of Johannesburg,
Johannesburg, South Africa
adeitandennis@gmail.com
[2] Department of Construction Management and Quantity Surveying,
University of Johannesburg, Johannesburg, South Africa
caigbavboa@uj.ac.za
[3] Department of Chemical and Civil Engineering, University of Johannesburg,
Johannesburg, South Africa
kobitha2003@yahoo.com

Abstract. This study aims at determining the impact and level of implementation of technology drivers such as automatic identification technology, communication technology, and information technology to Nigerian logistics firm. A structured questionnaire was used to collect necessary data from respondents. The research hypothesis was tested using the Pearson product moment correlation coefficient to justify results of the survey. A total of 90 logistics companies were selected in Lagos Nigeria to represent the 50% of the population which is urbanized. The result from the Pearson correlation coefficient shows that there was a strong negative correlation between implementation of technologies and Nigeria logistics chain. The result also revealed that implementation levels of Barcode, Electronic Data Interchange (EDI), Geographical Positioning System (GPS), Enterprise Resource Planning (ERP), and Distribution Requirement Planning (DRP) are very high whereas Radio Frequency Identification (RFID), Voice Recognition Technology (VRT), Very Small Aperture Terminal (VSAT), Geographical Information System (GIS), Automated Guided Vehicle System (AGVS), and Automated Inventory Tracking System (AITS) are very low in Nigerian logistics firm. The main barriers to full implementation of the technology tools to Nigerian logistics firm are lack of awareness on IT tools, poor logistics standard, cost of implementation of IT infrastructure and government policies. From the findings of the survey, logistics firm in Nigeria should fully implement technology tools in their logistics operations.

Keywords: Pearson correlation · Communication technology · Information technology · Logistics · Identification technology

1 Introduction

The application of technology to global logistics operations namely, acquisition, communication, warehousing, retailing, and transportation are vital to creating new competitive advantages for organizations [1]. According to Mohan [2], the importance

© Springer Nature Switzerland AG 2020
K. Arai et al. (Eds.): FTC 2019, AISC 1070, pp. 652–660, 2020.
https://doi.org/10.1007/978-3-030-32523-7_48

of technology is reflected in logistics through reduction in inventory, less in delivery schedule and a significant reduction in damages. This means organizations main concern is in the selection of efficient and effective technology for his various logistic activities such as data gathering, processing & analyzing with a high degree of performance, reliability, and accuracy. Filho [3] reports that the modern business comprises of growing supply chain, global logistics, global competition and a boost in customers expected value. Wong et al. [4] defines information technology as a meaningful factor to heighten the supply chain performance index, and with its power to provide timely, efficient, accurate and credible information, between firms. Bhandari [5] describes "Technology" as a vehicle to strengthen supply chain management competitiveness and exploit by choosing the right technology for various logistics activities. In order for current firms or organizations to be competitive, it must learn to become flexible, highly operational and more determined to integrate new technologies in delivering value to consumers and partners.

According to Price Water House Coopers [6], wholesale and retail are an essential sector which accounted for 18.8% of GDP in Nigeria economy. The population size, rapid urbanization (about 50%) and economic development are the reasons why investors are targeting Nigeria. Most of the logistics firms are under constant pressure from their suppliers and customers to remodel both their primitive styles of operation and organization to replace them with modern network that help boost the speed of information flows [7]. Over the years, Nigeria has been rated poorly in logistics performance index due to the following reasons; high cost of procuring IT infrastructure, time delays, bottlenecks for global freight, poor tracking & tracing proficiencies, poor logistics standard, clumsy customs policies, and transportation risk. Based on these discoveries, this paper aims at the impact of technology on global logistics in the Nigeria context. Furthermore, the level of the impact is determined by examining the implementation level of these technology tools to logistics in Nigeria.

1.1 Research Hypothesis

H_o: There is no significant relationship between the implementation of technologies and Nigeria logistics supply chain ($r = 0$).
H_i: There is a significant relationship between the implementation of technologies and Nigeria logistics supply chain ($r \neq 0$).

2 Literature Review

The literature review has identified some of the drivers of technology used in global logistics. The technology tools in global logistics are classified in form of automatic identification technology, radio frequency identification (RFID), and information and communication technology (ICT).

Krstev et al. [8] gave a comprehensive reappraisal of new market trends that have meaningful advantage on logistics, among the market trends discussed by the author are innovation in the technology field, RFID technology, automated financial processes, and EDI system communications. The results suggest that the insistent to finance in new technology are high and will continue to grow. Susanna et al. [1] have investigated the role of ICT in boosting competitive advantages of a firm. Results generated a model for ICT use in logistics by applying four determining factors: individual, organizational, technological or innovation and environmental. Ayantoyinbo [9] have assessed the impact of ICT on performance of consignment distribution. The literature highlighted the importance in the application of computers, internet and communication technology in transportation, freight distribution, warehousing, and materials procurement. Gurung [10] conducted a survey of works of literature on academic logistics journals and practitioner journals. The aim of the survey is to detect the impact of information technologies (IT) on logistics. The result revealed that academic researchers are aware of new technologies in practice such as RFID. It was also revealed that development in technologies offer a proper business model and sources of competitive advantage for companies. Finally, the adoption of new technologies tools proves a challenge and opportunities for the companies. IBM Global Business Services [11] described the activity involves in modifying worldwide logistics for key advantage in emerging and competitive markets. The result suggested that the complication of logistics management in emerging business areas ultimately add to the overall landed price of the corresponding goods. Therefore, the redesigning method of supply chain operations to logistics management efficiency in trending markets is a basic dimension of a long-term business strategy. Lastly, the business strategy must include flexibility, responsiveness, and resiliency. Irefin et al. [12] highlighted the cause preventing the full acceptance of ICT in SMEs in Nigeria. The result of the research shows that cost is a decisive factor for Small and Medium enterprises (SMEs) in adopting ICT. Information and communication technology infrastructure; government support; management backing and business size are other critical determinants. Wilson et al. [13] investigated the determining result of information technology on a performance of logistics firms in Nairobi. Quantitative data were collected from 10 firms in the logistics industry in Nairobi. A model was developed based on the data retrieved. The result showed that four decisive components have a considerable function to play in the performance of logistic industry in Nairobi. Nixon [14] conducted a study on the operation efficiency and awareness of Information and communication technology in E-logistics firms. The result showed that organizations should advance their operation efficiency, through constant implementation of ICT to raise their service capacity in an E-business context. Oyebamiji and Florence [15] investigated information technology and Its Effect on Performance of Logistics Firms in Nigeria. The study used purposive sampling technique in selecting ten logistics organizations in Lagos, Nigeria. Random sampling was used in selecting eight respondents from each logistics firm in Lagos, making it eighty respondents as the sample size for the study. Results revealed that the tracking and security system, usage of IT for the customer service delivery system and information integration have a positive and beneficial influence on the logistics firm's performance.

2.1 Impact of Technology in Nigerian Global Logistic

Efforts towards diversifying Nigeria's economy cannot be complete without reviving the courier, logistics and transport industry through practical policies and adequate managerial supervisions. Business day online [16] revealed that Nigeria logistics performance in global trade is central to economic growth and market competitiveness. It implies that Countries logistics sector is now recognized as one of the driving pillars of economic growth. The logistics growth is accounted for 18.8% of GDP in Nigeria economy and this is due to population size, rapid urbanization, and economic expansion. Lagos state is the economic center of business in Nigeria, it is a good model of Nigerian scenario. According to logistics Performance Index [17], Nigeria's logistics firm has witnessed a drastic fall in the past two years due to low rate in tracking & tracing capacity, clearance process, a high cost of procuring good IT infrastructure, and time delays. The 200 billion strong industry, which is ranked 70 in 2014 is now being ranked 90 out of 160 countries in 2017 report and remain 4% due to the underperformance of the sector.

Furthermore, most logistics firms in Nigeria still adopts the primitive styles of operations for information gathering due to the high cost of operating IT facilities or replacing their low technology to modern day facilities. The growth of consumer industries in Lagos is able to be done through urbanization (about 50%) and it is expected to be the 12th most populous city in the world by 2025 [18]. This implies that technology integration in logistics and manufacturing firms coupled with high rates of urbanization will ensure that Nigeria grows into an industrial and services-based economy. Technologies improve efficiency and quality in information gathering of logistics industry have been highlighted by some authors such as Gurung [10], Wilson et al. [13], Susanna et al. [1], and Florence and Oyebamiji [15].

Some of the identified technologies tools that can impact logistics performance in Nigeria include; Barcoding, Radio Frequency Identification (RFID), Voice Recognition Technology (VRT), Electronic Data Interchange (EDI), Geographical Positioning System (GPS), Geographical Information System (GIS), Web-Based Tracking (WBT), Automated Guided Vehicle System (AGVS) and Automated Inventory Tracking System (AITS). The next section of the study will ascertain the current levels of implementation of technology to Nigeria logistics firms.

3   Research Methodology

A questionnaire was designed and structured to collect data on the impact of technology tools on selected Nigeria logistics companies. A total of 90 logistics companies were selected in Lagos state. The state was chosen as the location of the survey because 50% of the population is urbanized and it is where most Nigeria logistics firm's headquarters or branches are situated. A total of 90 questionnaires were administered, 62 useful questionnaires retrieved while 28 were not retrieved. This was considered a fair response rate. The respondent Logistics firms are into retailing, warehousing, transportation and courier services. The questions and the terms used in the questionnaire were properly explained. Based on this, Questionnaire with unclear information was considered invalid for the survey and if doubts exist, telephone confirmations were made.

3.1   Data Analysis

The method for data analysis for this survey will be the statistical process, which include statistical tables, and percentages. The research hypothesis was tested using the Pearson product moment correlation coefficient to justify results of the survey. The results from the survey are shown in Table 1.

Table 1. The result from the survey indicating types of technology and percentages of implementation

S/N	Types of Technology	Variation		Percentage (%)	
		YES	NO	YES	NO
1	Radio Frequency Identification	4	58	6	94
2	Very Small Aperture Terminal	10	52	16	84
3	Electronic Data Interchange	52	10	84	16
4	Voice Recognition Technology	8	54	13	87
5	Barcode	57	5	92	8
6	Automated Inventory Tracking System	9	53	15	85
7	Distribution Requirement Planning	51	11	82	18
8	Enterprise Resource Planning	48	14	77	23
9	Automated Guided Vehicle System	6	56	10	90
10	Web Based Tracking	41	21	66	34
11	Geographical Information System	8	54	13	87
12	Geographical Positioning System	55	7	89	11

3.2   Results and Discussions

According to Bhandari [5], technologies being used in international and domestic logistics can be classified into three categories namely: automatic identification technology, communication technology, and information technology. The type of technology listed in Table 1 is composed of each category. The questionnaire was designed to determine the level of implementation using these technologies.

3.2.1   Automatic Identification Technology

The technologies in this category are mainly implemented by the people at the retailing, warehousing, courier, manufacturing and shipping logistics firms. It assists at reducing human error, paper work, reducing production lead time and increasing logistics system productivity through speed, accuracy, and reliability. The survey in Table 1, shows that the implementation levels of Bar Code are high in this category with a percentage score of 92% compared with RFI and VRT both with percentage scores of 6% and 13% respectively. Even though RFI and VRT are viewed as the next step in the evolution of barcoding and identification market, the level of implementation in Nigeria logistics system is still low. The study agrees with the findings of Ayantoyinbo [9], which reiterates that most of the technologies in use by most of the logistics firms are from the low technology, few of them use high technology for information gathering.

3.2.2 Communication Technology

The technologies listed in this category are mainly used for flow of information, data collection, and trace & tracking consignments from one place to another. It helps at reduction in transaction cost due to paperless operations. As shown in Table 1, GPS (89%), EDI (84%), WBT (66%) are the most practiced technology in this category whereas GIS (13%), AVGS (10%) and VSAT (16%) implementation are very low within Nigeria logistics firm. The low result attributed to the level of awareness of these technologies importance to logistics. Another factor might be the cost of procuring and maintenance of these technologies. The GPS, EDI and WBT technology high implementation in Nigeria logistics is due to its ability in tracing and tracking of consignment and low cost in planning the dispatch schedule. Therefore, this agrees with the findings of Wilson [13], which confirms that the tracking and security system helps in cost reduction, boost customer confidence and improves the company profitability. Finally, the full integration of GPS, EDI, WBT, GIS, AVGS, and VSAT in Nigeria logistics community will have a positive and significant influence on its global competitiveness.

3.2.3 Information Technology

The technology tool in this category is mainly implemented by the people at the customer service, inventory management, shipment planning, warehousing and retail stores. It aids in reducing freight cost, reduction in inventory, and encompasses all the business operations in logistics firms. As shown in Table 1, the implementation level of DRP and ERP are high in this category with a percentage score of 82% and 77% compared with AITS with a low percentage score of 15%. AITS implementation is low because the technology is still new to Nigerian firms, some selected few retail giants used this tool in controlling their inventory. DRP and ERP technologies implementation rate is high in Nigeria logistics organizations because it improves organizational performance, helps in the integration of all business operations and reduction in inventory level. This is reiterated in the findings of Mzoughi et al. [19], which confirms that integration of ICT tools in a logistics supply chain has the effective factor on organizational performance and global competitive advantage.

3.2.4 Scatter Plot of the Data

In order to test for the strength of Pearson's coefficient of linear correlation, a scatter plot of the data variables will be studied to determine if there is a trend in the relationship. As shown in Fig. 1, the scatter plot implies that as the variable X increases, variable Y decreases and as variable X decreases, variable Y increases. It can also be seen from the scatter plot that the points are along a negative straight line, so there is a strong negative linear relationship between the two variables.

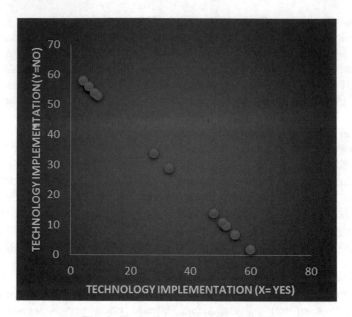

Fig. 1. Scatter plot of the data (X&Y)

3.2.5 Hypothesis Testing

The hypothesis for this study will be tested using the Pearson product – moment correlation coefficient (r). The level of significance is 5%. The responses from the Questionnaire in Table 1, was used for the testing. Also, the statistical analysis of the questions used for correlation coefficient is shown in Table 2.

The formula is represented by:

$$r = \frac{N\sum xy - \sum x \sum y}{\sqrt{N\sum x^2 - (\sum x)^2}\sqrt{N\sum y^2 - (\sum y)^2}} \tag{1}$$

Where; n is the sample size, x represents YES, y represent NO

$$N = 12, \sum x = 349, \sum y = 395, \sum xy = 5713$$
$$\sum x^2 = 15925, \sum y^2 = 18777$$
$$\left(\sum x\right)^2 = 121801, \left(\sum y\right)^2 = 156025$$
$$r = \frac{12\,(5713) - (349)(395)}{\sqrt{12(15925) - 121801}\,\sqrt{12(18777) - 156025}}$$
$$r = -1$$

The degrees of freedom = N−2 = 12−2 = 10

Table 2. Statistical analysis of the questions used for correlation coefficient

Tools	X	Y	XY
1	4	58	232
2	10	52	520
3	52	10	520
4	8	54	432
5	57	5	285
6	9	53	477
7	51	11	561
8	48	14	672
9	6	56	336
10	41	21	861
11	8	54	432
12	55	7	385

Pearson's correlation coefficient has a value between -1 (perfect negative correlation) and 1 (perfect positive correlation). From the Correlations table, it can be seen that the correlation coefficient (r) equals -1, $P < 0.005$ indicating a strong negative relationship.

Testing the hypothesis indicated that the obtained correlation coefficient (r) must not be equivalent to zero to be significant. Therefore, if the correlation coefficient (r) equals 0, we reject the null hypothesis (H_o) and accept the alternative hypothesis (H_i). We therefore accept the alternative hypothesis because there was a strong negative correlation between implementation of technologies and Nigeria logistics chain.

4   Conclusion

The impact of full implementation of technology on global logistics through exchange of goods internationally and domestically has enabled developed nations to be a major leader in supply chains. Nigerian logistics firms' barriers in competing globally include lack of awareness on IT tools, poor logistics standard, cost of implementation of IT infrastructure and government policies. Therefore, the study investigates the implementation of technology tools within Nigerian logistics firms. Out of the 12 tools investigated, the study discovered that the level of implementation of some technologies tools such as Barcode, EDI, GPS, ERP, DRP are very high whereas RFI, VRT, VSAT, GIS, AGVS, and AITS have very low level of implementation. Finally, the study shows that full implementation of right technology tools to logistics chain will be bring about competitive advantages for organizations.

References

1. Susana, G.A., Joao, F., Joao, L.: The role of logistics' information and communication technologies in promoting competitive advantages of the firm. Munich Personal RePEc Archive, MPRA Paper No. 1359 (2018)
2. Mohan, C.J.: The impact of logistic management on global competitiveness. Int. J. Bus. Manag. Invention 2(3), 39–42 (2013). 2319–8028
3. Filho IBMF. E-Business: The challenges of a new business strategic model for the brazilian companies. In: Proceedings of the 21th Conference of the Production and Operations Management Society, POM-2001, March 30–April 2, Orlando Fl
4. Wong, W.P., Keng, L.S., Mark, G.: Innovation and productivity: insights from Malaysia's logistics industry. Int. J. Logistics Res. Appl. 19(4), 318–331 (2016)
5. Bhandari, R.: Impact of technology on logistics and supply chain management. J. Bus. Manag. (IOSR-JBM), 19–24 (2017)
6. Price Water House Coopers. Africa Gearing Up: Nigeria's vision 20:2020 aims for it to one of the world's top 20 economies by 2020 (2012). www.pwc.com/
7. Somuyiwa, A.O., Adebayo, T.: Analysis of information and communication technologies (ICT) usage on logistics activities of manufacturing companies in Southwestern Nigeria. J. Emerg. Trends Econ. Manag. Sci. (JETEMS) 2(1), 66–72 (2011)
8. Krstev, A., Donev, A., Krstev, D.: Information technology in logistics: advantages, challenges and opportunity for efficiency from problem decision in different activities. In: SGEM, Albena, Bulgaria (2011)
9. Ayantoyinbo, B.B.: Assessing the impact of information and communication technology (ICT) on the performance of freight distribution. Eur. J. Logistics Purchasing Supply Chain Manag. 3(4), 18–29 (2015)
10. Gurung, A.: A survey of information technologies in logistics management. In: Proceedings of the Decision Sciences Institute (2006)
11. IBM Global Business Services. Transforming global logistics for strategic advantage in emerging markets (2006). https://www.ibm.com. Accessed 4 Nov 2018
12. Irefin, I.A., Abdul-Azeez, I.A., Tijani, A.A.: An investigative study of the factors affecting the adoption of information and communication technology in small and medium scale enterprises in Nigeria. Aust. J. Bus. Manag. Res. 2(02), 01–09 (2012)
13. Wilson, M.N., Iravo, M.A., Tirimba, O.I., Ombui, K.: Effects on information technology on performance of logistics firms in Nairobi County. Int. J. Sci. Res. Publ. 5(4), 1–26 (2015)
14. Nixon, M.: Innovations in logistics technology: generating top-line value and bottom-line ROI. World Trade 14, 62–64 (2001)
15. Oyebamiji, F.F.: Information technology and its effect on performance of logistics firms in Nigeria. Asian Res. J. Arts Soc. Sci. 6(1), 1–11 (2018)
16. Business day online. Nigeria logistics performance in global trade is central to economic growth and market competitiveness (2018). https://www.businessdayonline.com. Accessed 3 Dec 2018
17. Logistics Performance Index. The International Logistics Performance Score Card (2018). www.Ipiworldbank.org
18. Obi, O.: Leveraging Technology: Nigerian companies can make profit in logistics industry (2017). https://www.vanguardngr.com. Accessed 3 Dec 2018
19. Mzoughi, N., Bahri, N., Ghachem, M.: Impact of supply chain management and ERP on organizational performance and competitive advantage: case of Tunisian companies. J. Global Inf. Technol. Manag. 2(17), 24–46 (2008)

Enabling Empirical Research: A Corpus of Large-Scale Python Systems

Safwan Omari[✉] and Gina Martinez

Department of Computer and Mathematical Sciences,
Lewis University, Romeoville, IL, USA
{omarisa,martingi}@lewisu.edu

Abstract. The Python programming language has been picking up traction in Industry for the past few years in virtually all application domains. Python is known for its high calibre and passionate community of developers. Empirical research on Python systems has potential to promote a healthy environment, where claims and beliefs held by the community are supported by data. To facilitate such research, a corpus of 132 open source python projects have been identified, basic information, quality as well as complexity metrics has been collected and organized into CSV files. Collectively, the list consists of 36, 635 python modules, 59, 532 classes, 253, 954 methods and 84, 892 functions. Projects in the selected list span various application domains including Web/APIs, Scientific Computing, Security and more.

Keywords: Python · Corpus · Empirical research · Quality metrics · Complexity metrics

1 Introduction

The Python programming language has been picking up popularity over the past few years according to several indices that track and measure programming languages popularity [7,13]. Python provides a rich set of evolving primitives and open-source frameworks that support various application domains such as Web/API, Data science, security and much more. The rise of Python has caught the attention of the research community and triggered a slew of research efforts addressing various aspects of the language and its evolution [3].

Among other research efforts, empirical research in software engineering has great potential in bridging the gap between Academia and Industry. Rooted in extensive measurements and use of sound statistical models, empirical research has the potential in supporting or refuting many of the unsupported claims often made and beliefs held sacred by various communities of developers. This promotes a more data-driven approach to best practices and well-informed decision making in Industry, which invites more of constructive discussions and less of ideological debates.

© Springer Nature Switzerland AG 2020
K. Arai et al. (Eds.): FTC 2019, AISC 1070, pp. 661–669, 2020.
https://doi.org/10.1007/978-3-030-32523-7_49

Statically typed languages such as Java and C have enjoyed the bulk of empirical research in software engineering [6,10,11,14], we believe that Python exhibits unique internal characteristics [5]. This warrants a fresh wave of empirical research to address different aspects of the language and process related practices in the community [1,2,6,8,11]. Robustness of Statistical conclusions and quality of models presented in empirical research is largely impacted by use of consistent and representative software artifacts [12,15]. In this paper, we build upon previous efforts [12,15] and present a large-scale Python project data-set to further and facilitate empirical research in software engineering. Selection criteria is carefully defined to yield popular projects with long development history and many contributors.

Corpus consists of 132 Python projects, necessary meta-data including repository URLs, description, release tags, and several others are collected and organized in csv files. Furthermore, several quality and complexity metrics have been collected and included. Collectively, corpus consists of 36,635 modules and more than 5.7 million Python lines of code. Average system is 4 years old with at least 26 releases. 50% of corpus systems exhibit a Pylint quality score of 8 out of 10 or higher, and 50% of projects showed an average McCabe's cyclomatic complexity score of 7.

The rest of the paper is organized as follows, Sect. 2 presents related work; corpus design goals, selection criteria and contents are discussed in Sect. 3; Sect. 4 presents several quality as well as complexity metrics and provides summary statistics; and finally paper is concluded and future work is discussed in Sect. 5.

2   Related Work

Python's rise as one of the mainstream languages in Industry is attracting increasing interest from the research community, which studies all aspects of the language features and evolution [1,2,4,6,8,9,11,13]. While more traditional languages such as Java and C/C++ has enjoyed the bulk of research community attention [6,10,11], empirical research on Python is still in its infancy. Similar to previous research efforts in [12], a corpus of Python systems is proposed to help advance and facilitate future empirical research. Proposed corpus builds on similar corpus proposed in [12], by constructing a larger set of Python systems, which are carefully selected to include large-scale systems. Furthermore, an extensive set of meta-data and metrics are collected and organized in csv files for easy access and replication. Finally, a thorough description of various system characteristics is presented, the goal is to provide as much context as possible around statistical results drawn based off the corpus.

3   Python Corpus

3.1   Design Goals

Software engineering empirical research targets a wide range of statistical hypothesis. The type of hypothesis under study plays a primary role in data-set

```
ajenti ansible astropy autobahn-python aws-cli beets biopython blaze bokeh boto3 buildbot bup
calibre celery certbot ckan cobbler curator compose conda CouchPotatoServer cryptography cython
django django-allauth django-blog-zinnia django-cms django-extensions django-haystack django-
oscar django-rest-framework django-shop django-tastypie docker-py edx-platform elastalert
electrum erpnext eve fail2ban faker falcon flask fonttools gensim gevent glances google-cloud-
python headphones home-assistant hue ipython jinja jupyterhub kafka-python kivy livestreamer
luigi matplotlib mongoengine mezzanine mitmproxy mongo-python-driver mopidy mrjob munki mypy
networkx paramiko nova numba nupic OctoPrint orange3 pandas peewee pelican picard pika Pillow
pip platform_development play1 psutil pwntools pyethereum pyinstaller pymc3 pyramid pyzmq
qutebrowser ranger raven-python readthedocs.org redash you-get robotframework s3cmd saleor salt
scikit-image scikit-learn scipy scrapy securedrop security_monkey sentry Sick-Beard SiCKRAGE
spaCy spinnaker sqlalchemy sqlmap st2 supervisor swift sympy synapse Theano thumbor tornado
tribler troposphere twisted w3af wagtail WeasyPrint web2py aiohttp werkzeug youtube-dl zipline
```

Fig. 1. Python project names: total of 132 projects.

design criteria. For example, testing hypothesis about early stages of software development life-cycle, requires a data-set of artifacts in their early states of evolution, whereas, doing same for latter stages of software life-cycles, should be performed on a corpus of projects with several years of development history and multiple release cycles. We seek to facilitate and support empirical research on large-scale and mature Python Software Projects. Therefore, selection criteria in this work are crafted to include popular projects with long development history and multiple release cycles. To ensure easy access to the exact source code, all necessary project meta-data and information are collected and included in corpus. Furthermore, several complexity as well as quality metrics were generated and included. Having good understanding of corpus characteristics is essential to verify necessary assumptions and appropriateness of corpus for target study.

To meet stated goals, the authors chose to include open-source projects hosted in Github and collected needed meta-data to enable future researchers to obtain and reconstruct exact copy of software artifacts (i.e., URL and release tags). For quality and complexity metrics, Pylint and radon are used. The latter generates raw and object-oriented complexity metrics as well as McCabe's complexity, whereas, the former generates metrics about code style convention violations as well as code smells.

3.2 Selection Criteria

The total number of Python projects in Github is upward of 608K projects. In this work, a two-phase approach is used to construct a corpus of projects that satisfy stated goals. In *Phase-I*, Github REST searching API is used to filter projects based on primary programming language, date of creation, total stars received and date of last commit to the project repository. Phase-I filtering is meant to define bare-minimum thresholds in terms of age and popularity. Parameters are set as follows: created before $= 2015 - 01 - 01$, stars $>= 600$ and date of last commit $>= 2018 - 07 - 01$. This resulted in a total of 928 Python projects. A manual investigation of the initial list revealed several repositories that were not in fact Python systems, these systems were eliminated. The following are some examples:

- Public lists (json): list of Members of the US Congress since 1789.
- Personal web-sites.
- Lists of frameworks and libraries: awesome-python, awesone-machine-learning.
- Documentation: PEPs, Raspberry Pi, Python packaging user guide, Ansible playbooks for Wordpress, scipy-lecture-notes, etc.
- Configuration lists (json).

To elaborate on the goal of selecting the most mature projects with a rich development history, in *Phase-II*, the following criteria are developed and projects that fall in top 50th percentile are selected:

1. *Number of stars* received and *watchers*: Project must be popular and of interest to a large population. 50th percentile thresholds for stars received and number of watchers are found to be 1457 and 99, respectively.
2. *Number of commits*: Project must have a long development and evolution history. 50th percentile threshold is found to be 1067 commits.
3. *Number of contributors*: Project must be developed by a large community of developers. This is essential for corpus population to capture and exhibit typical team dynamics characteristics. 50th percentile threshold is found to be 22 contributors.
4. *Number of releases*: Project must be mature and has gone through multiple release cycles and actively in use. 50th percentile is found to be 26 releases.

3.3   Corpus Contents

Fig. 2. Word cloud based on topic keywords

The total number of projects that satisfied initial Phase-I criteria and ranked in top 50th percentile came out to be 132 Python projects. For each project, the following meta-data is collected: *name, description, topics, age, repository URL, number of commits, number of watchers, stars received, number of releases, release tag. Using repository URL and specifying release tag*, future researchers should be able to an clone exact copy of source code for project in corpus (`git clone -b release-tag URL`). Figure 1 shows the final list of repositories. Table 1 shows basic summary statistics of corpus projects.

Table 1. Summary corpus meta-data

Project Age (years, months)			
Min	Max	Median	Mean
4y, 2m (elastalert)	10y, 3m (plat-form_development)	7y, 3m	7y, 5m
Number of commits			
Min	Max	Median	Mean
1261 (django-tastypie)	101,678 (salt)	10,255	6,784
Number of stars received on Github			
Min	Max	Median	Mean
1,463 (fonttools)	45,187 (youtube-dl)	6,863	4,016
Number of watchers on Github			
Min	Max	Median	Mean
103 (s3cmd)	2,238 (flask)	394	252
Number of contributors on Github			
Min	Max	Median	Mean
49 (WeasyPrint)	443 (matplotlib)	207	192
Number of releases			
Min	Max	Median	Mean
26 (django-tastypie, flask, zipline)	1114 (edx-platform)	132	74

As depicted in the word cloud in Fig. 2, projects span a wide array of application domains. Word cloud is based on topic keywords stated for each one of the projects. It is found that 44 projects were lacking any topic keywords on Github, for those, topic keywords were compensated with keywords from project description. Keywords were obtained by applying stemming and removing of stop words from full description text. Word cloud is generated using Pro Word Cloud add-ins plugin in MS Word. Web, data science as well as security applications and frameworks are dominant themes. Top frequency words are: django: 14 times, web: 9 times, frameworks: 7, library: 7 times, data science: 5 times, http: 5 times

4   Internal Characteristics

In addition to basic project information, several internal source code metrics have been collected. Metrics are grouped into two categories: quality and complexity metrics. Such metrics shed light on internal code quality of data-set and help future researchers assess the appropriateness of corpus for the problem at hand. Python module (file) is used as a unit for collecting quality and complexity metric. Modules that are not part of the core system are excluded from quality and complexity calculations, including unit tests, scripts and example code, if any. We believe that code in these modules exhibit different quality and complexity characteristics from the core system. Module-level metrics are aggregated in a corresponding project-level metric. With the exception of McCabe's cyclomatic complexity and Pylint score, project-level metric is simply calculated as the sum of all corresponding module-level metrics. For example, the total number of convention violations at the project-level equals to the sum of module-level convention violations for all modules in the project. Details on McCabe's complexity and Pylint score are presented later in the section.

4.1   Quality Metrics

Table 2. Summary corpus quality metrics

Number of refactor violations			
Min	Max	Median	Mean
3 (munki)	10,149 (ansible)	386	872
Number of style violations			
Min	Max	Median	Mean
6 (home-assistant)	375,317 (calibre)	1,951	10,235
Pylint score			
Min	Max	Median	Mean
−10.3 (fonttools)	9.9 (home-assistant)	7.9	6.9

We use Pylint to generate several quality metrics of selected projects. Pylint is a static analyzer that identifies five classes of quality concerns: **R**efactor, **C**ode Style, **W**arnings, **E**rrors and **F**atal. In this study, we focus our attention on Refactor and Code Style as primary quality metrics. Refactors are concerned with locating code smells and measuring how often it occurs in Python code, whereas, code style reports on project's adherence to a well-accepted set of code conventions and standards. Pylint recognizes and reports on 34 unique code style violations and 24 code smells. Example convention violations include bad variable names, missing docstring and using `type()` rather than idiomatic `isinstance()` for type checking. Example refactor violations include too many ancestors and too few public methods.

Besides raw number of **R**efactor and **C**onvention violations in a Python module, Pylint calculates module-level score as $10 - (C + R)/S * 10$, where C represents number of convention violations, R represents number of refactors and S represents number of logical Python lines of code in the module. Project-level score applies same equation using total C, R and S across all modules in the project. Scripts has been developed to run and collect Pylint metrics per module and calculate project-level score. Table 2 provides summary quality statistics of corpus.

4.2   Complexity Metrics

Radon is used to obtain raw and complexity metrics including source lines of codes, number of classes, number of methods, number of functions and McCabe's cyclomatic complexity per Python module. With the exception of McCabe's complexity, project-level metric is calculated as sum of corresponding module-level one. For example, number of classes in a project is calculated as sum of classes in all modules of the project. Radon reports McCabe's complexity score per methods and functions. We define module-level McCabe's complexity as the maximum complexity among all methods and functions defined in the module. We believe the module's complexity is dictated by its most complex function or method. Furthermore, Project-level McCabe's complexity is calculated as average McCabe's score among all modules in the project. Table 3 reports summary statistics of complexity metrics for projects in corpus.

Using pylint score and McCabe's complexity score as proxies for project quality and complexity, respectively, Fig. 3 depicts an empirical cumulative distributive function for Pylint score in Fig. 3(a), and McCabe's cyclomatic complexity in Fig. 3(b). Almost 50% of projects in corpus have a Pylint score of 8 or below. 80% of projects have a complexity score of less than 11.

Table 3. Summary corpus complexity metrics

Source lines of code			
Min	Max	Median	Mean
934 (raven-python)	723,480 (ansible)	17,465	45,400
Number of classes			
Min	Max	Median	Mean
14 (securedrop)	5,893 (hue)	241	476
Number of methods			
Min	Max	Median	Mean
37 (securedrop)	26,842 (hue)	910	2,031
Number of functions			
Min	Max	Median	Mean
11 (pika)	15,337 (salt)	268	679
McCabe's cyclomatic complexity			
Min	Max	Median	Mean
1.3 (django-allauth)	29.0 (WeasyPrint)	6.8	7.9

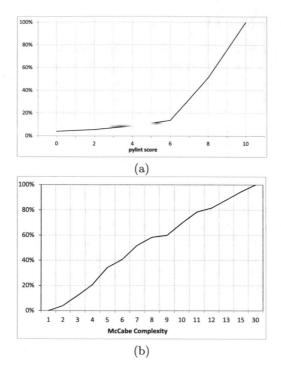

Fig. 3. Cumulative distribution function, (a) Pylint score, (b) McCabe's cyclomatic complexity.

5   Summary and Future Work

A corpus of 132 Python projects are constructed and organized into csv files. Collectively, corpus consists of 36,635 python modules, 59,532 classes, 253,954 methods and 84,892 functions. To provide future researchers with insights into corpus internal characteristics, several quality as well as complexity metrics has been aggregated and included in the corpus. For future work, the authors are planning to pursue construction of specialized corpora to facilitate empirical research on niche Python systems populations, in addition to collection and inclusion of more metrics such as Halstead metric, Maintainability Index, etc.

References

1. An empirical study of dynamic types for python projects. In: 8th International Conference (SATE)), November 2018
2. Akerblom, B., Wrigstad, T.: Measuring polymorphism in Python programs. In: Proceedings of the 11th Symposium on Dynamic Languages, DLS 2015. ACM (2015)

3. Alexandru, C.V., Merchante, J.J., Panichella, S., Proksch, S., Gall, H.C., Robles, G.: On the usage of Pythonic idioms. In: Proceedings of the 2018 ACM SIGPLAN International Symposium on New Ideas, New Paradigms, and Reflections on Programming and Software, Onward! 2018. ACM (2018)

4. Chen, Z., Ma, W., Lin, W., Chen, L., Xu, B.: Tracking down dynamic feature code changes against Python software evolution. In: 2016 Third International Conference on Trustworthy Systems and their Applications (TSA), September 2016

5. Destefanis, G., Ortu, M., Porru, S., Swift, S., Marchesi, M.: A statistical comparison of Java and Python software metric properties. In: 2016 IEEE/ACM 7th International Workshop on Emerging Trends in Software Metrics (WETSoM), May 2016

6. Destefanis, G., Counsell, S., Concas, G., Tonelli, R.: Software metrics in Agile Software: an empirical study. In: Agile Processes in Software Engineering and Extreme Programming, pp. 157–170. Springer, Heidelberg (2014)

7. Guo, P.: Python is now the most popular introductory teaching language at top U.S. universities (2014). https://cacm.acm.org/blogs/blog-cacm/176450-python-is-now-the-most-popular-introductory-teaching-language-at-top-u-s-universities/fulltext

8. Lin, W., Chen, Z., Ma, W., Chen, L., Xu, L., Xu, B.: An empirical study on the characteristics of Python fine-grained source code change types. In: 2016 IEEE International Conference on Software Maintenance and Evolution (ICSME), October 2016

9. Malloy, B.A., Power, J.F.: Quantifying the transition from Python 2 to 3: an empirical study of Python applications. In: 2017 ACM/IEEE International Symposium on Empirical Software Engineering and Measurement (ESEM), November 2017

10. Nagappan, N., Ball, T., Zeller, A.: Mining metrics to predict component failures. In: Proceedings of the 28th International Conference on Software Engineering, ICSE 2006. ACM (2006)

11. Nanz, S., Furia, C.A.: A comparative study of programming languages in Rosetta code. In: 2015 IEEE/ACM 37th IEEE International Conference on Software Engineering, vol. 1, May 2015

12. Orrú, M., Tempero, E.D., Marchesi, M., Tonelli, R., Destefanis, G.: A curated benchmark collection of Python systems for empirical studies on software engineering. In: Proceedings of the 11th International Conference on Predictive Models and Data Analytics in Software Engineering, PROMISE 2015. ACM (2015)

13. The software quality company. Python is TOIBE's programming language of the year 2018 (2019)

14. In, H., Lee, T., Lee, J.B.: A study of different coding styles affecting code readability. Int. J. Softw. Eng. Appl. $\mathbf{7}$(5), 413–422 (2013)

15. Tempero, E., Anslow, C., Dietrich, J., Han, T., Li, J., Lumpe, M., Melton, H., Noble, J.: The Qualitas Corpus: a curated collection of Java code for empirical studies. In: 2010 Asia Pacific Software Engineering Conference (2010)

M-Health Android Application Using Firebase, Google APIs and Supervised Machine Learning

Charu Nigam[✉], Mahima Narang, Nisha Chaurasia, and Aparajita Nanda

Department of CS and IT, Jaypee Institute of Information Technology, Noida, Sector 62, Uttar Pradesh, India
charu8812@yahoo.com, mahimanarang7@gmail.com, {nisha.chaurasia, aparajita.nanda}@jiit.ac.in

Abstract. With the increasing use of mobile technology across the globe, android applications are of significant use, including the health sector. For efficient quick results while looking up for medical help in case of accidents or medicines and blood requirement emergencies, one can use such mobile-health application. The proposed paper focuses on making advantageous use of Android and Machine Learning to provide an integrated platform for delivering health-related information on major healthcare units such as hospitals, pharmacies and blood banks. These facilities are provided through the developed app in which the data is customized according to the factors viz. user's location, their distance to nearby health units obtained through Google Maps API, availability of required resources such as number of beds in hospitals and the ratings which are predicted by applying Multinomial Naive Bayes technique on the customers' reviews. The list of the health units, their reviews and count of beds in a hospital, medicines in pharmacies or blood types in blood banks are all stored in the Firebase.

Keywords: Android · Firebase · Google maps APIs · Multinomial Naive Bayes

1 Introduction

With exponentially increasing mobile utilization and innovation empowering us greatly, versatile medicinal services applications offer astounding chances to enhance one's wellbeing, security and in some sense readiness to common illnesses. Because of their moderateness, accessibility and the conveyance of smart devices, wellbeing applications have prominent chances. The unique characteristics of m-health apps are simplicity, customizability and immediacy for instant results. The ever increasing rat race of ease-of-access of resources which people are facing in such a highly competitive world, one can easily fall victim to health issues and accidents. Also, it is to be noted that only 3.7% of total GDP is spent on health.

There are a lot of cases of accidental situations like fires, road accidents or natural calamities such as earthquakes witnessed. The incidents like the breaking of fire or occurrence of any natural calamity also lead to health issues. The above factors have

© Springer Nature Switzerland AG 2020
K. Arai et al. (Eds.): FTC 2019, AISC 1070, pp. 670–681, 2020.
https://doi.org/10.1007/978-3-030-32523-7_50

motivated researchers to develop apps using Android to deal with health related issues and assist people in case of emergencies. The pre-built apps although have features which shows nearby health units, but they lack other characteristic features such as showing the most optimised result based on availability of resources and ratings. This plays crucial role; if suppose you are on a trip and fall sick, or meet with an accident at an unknown place, or a pregnant lady has to undergo immediate delivery, here in case of emergency, one can't waste their critical time hoping around different health units. Hence, the proposed app intends to provide solution to the above problem; keeping in mind the two main factors of the project, viz.

1. To help users to search for most appropriate (i.e. optimised) health units during emergency based on availability of resources.
2. To allow people to update information about the availability of resources of various health units and share/get guidance from peers.

The above point brings forward the concept of the proposed community based app.

It assists the users to look for nearby hospitals, at the same time getting the best ones according to the ratings obtained through their reviews. In addition, it fetches the nearby pharmacy and blood banks. The users can fetch and update information about the health units such as availability of beds in hospitals, medicines in pharmacies and blood types in blood banks. The token IDs for app users are generated and along with this their respective locations are stored in the Firebase database. Firebase is a cloud based server used to store data in JSON (JavaScript Object Notation) format. Using Google Maps and Places APIs, the list of nearby health care units (hospitals or pharmacies or blood banks) is displayed based on their ratings and availability of beds or medicines and blood types which is all stored in the NoSQL database.

The ratings are predicted by training and testing the reviews database using Multinomial Naïve Bayes algorithm and connected with Android using Flask. The users can update the beds or medicines or blood bags availability and also add review for the health care unit based on their experience (in restricted privileged environment).

Discussion about related relevant papers has been done in Sect. 2. Section 3 details the proposed m-health in terms of system functionality with technical details. The experimental details of proposed m-health app are described in Sect. 4 followed by conclusion in Sect. 5.

2   Related Work

There exist some applications that have been developed using similar technologies or researches based on the utility of mobile-health applications in the community. They contain different functionalities and solutions for different problems in the field of medical health care that are executed through smartphones.

One such phone application is proposed in [1] where the details of patients are stored on Firebase which helps the concerned in generating prescription and required doctors. The author in [2] presents the development of a mobile application for an emergency response system which helps rescue service provider in determining the shortest route to incident location and nearby hospitals to it.

Another such application is in [3] whose search engine can search hospitals, available doctors and medicines via category, name, and location, blood donation camps. Their evaluation of the comparison between the situation with and without using Health Care Android Application has favored the utilization of m-health apps. However, the above apps focus on just one health unit. Moreover, then don't consider integrated factors like distance, resources and ratings in providing optimized result.

The author in [7] focused on the development of an app that shall improve the health care system in Bangladesh through smartphones by involving features such as hospitals suggestion, cabin booking and appointment scheduling. Their survey showed that such medical apps are found useful by the people as they are convenient and time-saving.

The authors' motivation is to implement some of the significant features in the emergency situations or general health care, such as listing of the nearby hospitals or pharmacies or blood banks. The listing of these health units is based on the distance and factors such as availability of the beds in hospitals which is sorted in the order of the ratings that are produced on the basis of reviews given by the users. In [8], the authors have proposed the module for the prediction of star ratings based on the reviews. The reviews are tokenized and the generated feature vectors are passed to Multinomial Naïve Bayes theorem for rating prediction. The experiments in [9] showed that that the TFIDF (Term Frequency-Inversion Document Frequency) conversion to the data greatly improves the results for MNB than the simple Naïve Bayes theorem.

Here, in this proposed application, the authors have applied stemming on the words (reviews) for vector generation that undergo supervised learning module to predict ratings for the hospitals, pharmacies and blood banks. Based on the factors of distance, availability of resources (beds/medicines/blood bags) and the ratings, the health units are listed upon the request of the user for the optimized result.

3   Proposed Model

The overview of the proposed model is described in Fig. 1, which facilitates different functions that are provided through the developed app. The data is further customized according to the factors viz. user's location, their distance to nearby health units obtained through Google Maps API, availability of required resources such as number of beds in hospitals and the ratings. Based on the factors of distance, availability of resources (beds/medicines/blood bags) and the ratings, the health units are listed upon the request of the user for the optimized result. The details of each functions are described in subsequent sub-modules.

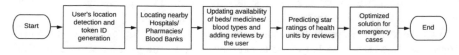

Fig. 1. Flowchart of the proposed model

3.1 Token ID Generation and User's Location Detection

Once the user install's the app, token ID of that particular device gets created which gets stored in the Firebase [4, 5] as illustrated in Fig. 3.

After the generation of token IDs, the nearby health units (hospitals/bloodbanks/pharmacies) can be fetched on users' requests with respect to their devices' token ID using Google Places API as shown in Fig. 4. The distance up to which the nearby health units are to be located is passed as parameter to the Geofire to retrieve and store the closely located health units. The algorithm applied to generate effective solution is as follows:

i. Sorting of the health units according to the distance which is automatically done using Google APIs (it generates list of health units sorted according to nearest distance from the user's device).

ii. Comparison on the basis of availability of resources. The unit having higher availability of resources is given priority else if two nearest unit have same availability of resources then the unit having higher rating among the two is given priority. Figure 2 below shows the step wise details of the algorithm used to fetch the most optimized result.

```
┌─────────────┐     ┌─────────────┐     ┌─────────────┐     ┌─────────────┐
│ Sorting By  │ ──▶ │  Filter by  │ ──▶ │ Sorting By  │ ──▶ │  Relevant   │
│  Distance   │     │  Available  │     │   Ratings   │     │   Results   │
└─────────────┘     └─────────────┘     └─────────────┘     └─────────────┘
```

Fig. 2. Flowchart for optimized result for hospitals in emergency case

The data is stored in Firebase along with the number of beds in case of hospitals, medicines in case of pharmacies and type of blood in blood banks, which is shown in Figs. 4 and 5.

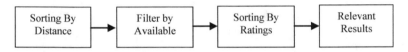

- cYB5hzKP74o:APA91bEd4Oyohz-KW2ssMxWpo91dCx-7wOKOiX5w-RaOgKJeF0KrIjwCbeXQH0G-W(
- d2saHo8qqsw:APA91bEXg6K3gdMSNZb2Ofq8cJ0zoOMZGmV_70bD6airvrr6OYxO58Kf12WH3mcns
- eG11Huo8qVg:APA91bE6wJ9qAVR1WLR0AABzixElmCz0zqP_g_Bj04djDGevZ_92P5-4hDw7QPp8p5(
- da7OwMt1gDl:APA91bENZDqKTnLe2MjlVNG1o6llFHHOUFps7aXQYu5caCPlBELRfRlgqx8QN76R9QF

Fig. 3. Tokens generated for different users

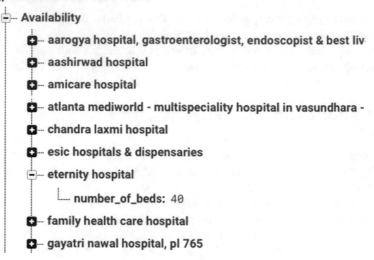

Fig. 4. Nearby hospitals stored in Firebase

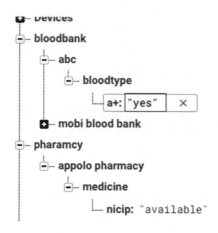

Fig. 5. Nearby blood banks and pharmacies stored in Firebase

3.2 Updating Availability of Beds/Medicines/Blood-Type

Since the app is community based, therefore it gives flexibility to its users to update the availability of beds in hospitals, medicines in pharmacies and type of blood in blood banks on visiting any of the above health units for helping out other app users to know updates of the availability of these resources in real time [6] as demonstrated in Fig. 4. Besides this, the user can write review of the health unit they visit which helps in predicting the star rating, shown in Fig. 6.

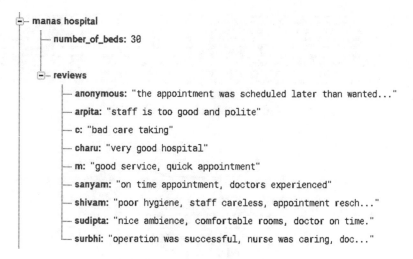

Fig. 6. Database of users' provided reviews.

3.3   Star Rating Prediction

For predicting the ratings from the reviews given by users, TF-IDF vectorization is used followed by classification of data using Multinomial Naïve Bayes theorem. TF-IDF stands for term frequency-inverse document frequency.

$$\text{Tf-Idf} = \text{Tf} \times \text{Idf}, \text{ where}$$

$$\text{Idf} = \text{Log}_2(\text{D/N}) \text{ and Tf } (w_i)$$
$$= \text{Total no. of word } w_i \text{ present in the entire document.}$$

IDF is the logarithmic ratio of the total number of documents (D) to the number of documents (N) in which that word (w_i) is present [13]. For predicting the rating of health-care units, each document represents the reviews of a hospital, blood banks or pharmacies. Since the reviews are composed by users, there may be errors or certain words and punctuation marks that won't be required to classify the data. Thus, tokenisation of the documents will be required [7]. Thus, the steps followed are:

   i. Removal of Punctuations
  ii. Removing Stopwords
 iii. Tokenizing
 iv. Stemming

Figure 7 below shows the output of the reviews generated after implementation of text cleansing (steps of which are mentioned above).

```
Out[7]: ["noida sector 62 hospital bad experience ent dr. anurag jain cheater staff since last friday
         follow dr. urgent treatment dr. continuously refer expensive lab test even n't consider day b
         ack lab report refer another doctor clearance spent ... view approx 11 000 within two visit m
         other condition problem increase even n't suggest proper medicine rhey refer andriscopy bypsy
         test tps team told approx bill 90 000 already spent 10 000 means spent money lab test even ti
         ll date give medicine immediately relief leave hospital immediately visit another hospital im
         megency treatment never suggest anyone visit hospital make money nothing else even n't take p
         ain happen patient condition bad entire team staff cheater want money punish badly view less"
         .
```

Fig. 7. Output generated as a result of text cleansing

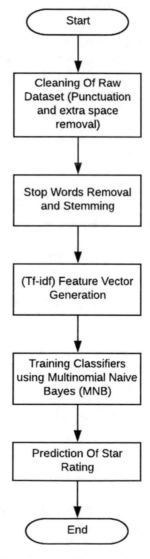

Fig. 8. Flow diagram for prediction of ratings from reviews

Wordnet Lemmatizer can be used for stemming purpose. The then generated root words are converted into tf-idf vectors. As stated in [10] Multinomial Naïve Bayes can be improved substantially by applying a TFIDF (term frequency-inverse document frequency) transformation to the word features and normalizing the resulting feature vectors to the average vector length observed in the data. Figure 8 below shows the flowchart depicting the steps followed to predict the star ratings starting from text cleansing, followed by (Tf-idf) Feature Vector Generation and finally applying Multinomial Naïve Bayes theorem.

Training and Rating Prediction
(Multinomial Naive Bayes Approach)

Multinomial Naive Bayes is a supervised learning technique, in which a new document is classified by allocating one or more class labels from a fixed set of pre-defined classes [8]. It is easy to implement and is efficient in computations.

For the objective of predicting the ratings of the hospitals, pharmacies or blood banks, Multinomial Naive Bayes algorithm (MNB) [5] has been used to train the datasets that are further used to generate the ratings based on the review text.

MNB considers that the distributions of words are done by applying a specific parametric model, and the parameters can be estimated through the training data. The following equation represents the MNB model [14]:

$$P(c|d) = \frac{P(c)\pi_{i=1}^{n}P((w_i|c)^{f_i}}{P(d)}$$

Where, P(c) is the probability that a document with class c may happen in the document collections, n is the number of unique words in document d, P(wi|c) is the conditional probability that a word wi may occur in a document d given the class value c, fi shows the count of occurrences of a word wi in document d.

The star ratings generated for respective health unit is displayed along with their names as shown in Fig. 12.

```
In [12]:  from sklearn.naive_bayes import MultinomialNB

          clf = MultinomialNB()
          clf.fit(train_transformed1, y_train)
          clf.score(test_transformed1, y_test)

Out[12]:  0.8220773171867722
```

Fig. 9. MNB accuracy (82.2%)

Flask

For the ratings to be generated using Machine Learning, one needs to connect it with Android which can be done using Flask. Flask is a very famous web framework which provides simplicity and flexibility by implementing a bare-minimum web server [11], it creates a local host that sends generated result from Python to Android in JSON form, from where data is extracted and displayed on the app's UI (User Interface).

The python files of flask and code to train and test data can be deployed on platforms such as python anywhere or run on Jupyter notebook. The reviews are sent for tokenization and vectorization. The MNB model trains the dataset of around 10,000 hospital reviews obtained through a website by web crawling [12].

Figure 10 shows the data under training which includes the reviews by the users and the respective ratings given.

```
In [2]:    1  df=pd.read_csv('workbook.csv',encoding='ISO-8859-1')
           2  df
```

10724	I had an appointment in 2015 to get my jaw ali...	1
10725	My husband had a stroke in March. From the mom...	5
10726	I saw an ENT consultant on 18th November 2015....	1
10727	I would like to say a huge thank you to everyo...	5
10728	After four years of appointments and getting n...	1
10729	My husband was admitted to Lincoln Hospital in...	3
10730	I gave birth to my baby boy at lincoln hospita...	5
10731	I have been seeing consultants from the ENT de...	5
10732	I am always uneasy having procedures done at L...	2
10733	Appalling - from when I gave birth to my first...	1
10734	You will quite likely have a very long wait on...	2

10735 rows × 2 columns

Fig. 10. Training set for generating ratings

The testing dataset i.e. the users' reviews are tested and corresponding ratings are predicted for each review of a hospital using Multinomial Naïve Bayes algorithm and the average rating is the new star rating predicted for hospital. Similar is applicable for all other hospitals and pharmacies and blood banks.

4   Results and Outcomes

Shown below are the examples of outputs of the application. Figure 11 displays the user's current location in the map. Figure 12 displays the list of hospitals on user's request along with the ratings predicted by the reviews and the 'update' option to add a review and update availability of beds shown in the Fig. 14. Similarly the list of pharmacies and blood banks can be displayed. Figure 13 displays those health units in the list, marked on the map from where the user can direct to Google Maps for directions to the nearby health unit.

To display the higher rated health units, Multinomial Naïve Bayes algorithm applied after TF-IDF vectorization of the reviews, shows an accuracy of 82.2%. For time-saving, the optimized result of the respective health unit required by the user is triggered in the form of notification as shown in Fig. 15, and for his convenience clicking on which the user is directed to Google Maps that shows the route from user's current location to the resultant health unit.

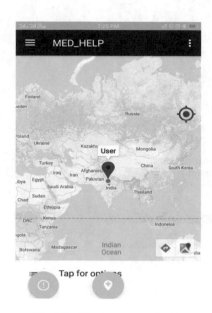

Fig. 11. User's current location

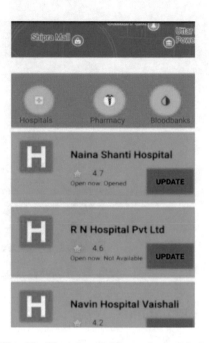

Fig. 12. List of available nearby hospitals

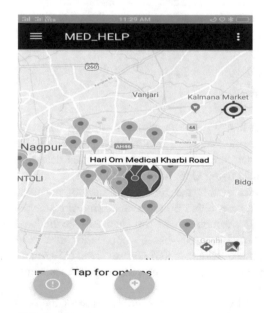

Fig. 13. Nearby hospitals with respect to the user as visible on Google maps

Fig. 14. User input for the availability of beds and hospitals

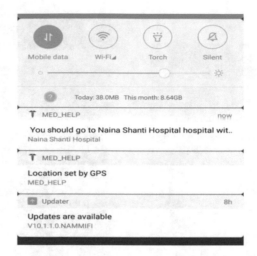

Fig. 15. Notification showing the optimized result

5 Limitations

There are a few limitations however. One of them being the requirement of internet. The application requires internet connectivity to run and provide real time results. Another drawback is on the authentication part on the availability of beds or medicines at pharmacies. The availability of these resources need to be authorized by the respective health unit administration for better/accurate results. For the prediction of ratings, the request goes to the server, so the results are reliable on the functioning of the server as well.

6 Conclusion and Future Scope

The uniqueness of the app lies in the fact that it gives an effective solution based on three factors that are distance, availability and ratings of health units. The procedure followed is: (i) Generating the token for each user who installs the app, then detecting the location of user and the nearby hospitals using Google Places API. To optimize it further by showing the nearby health units, within the specified radius of the user, "GEOFIRE" is used. (ii) The filtration of data is based on the availability of resources for which data is stored in "FIREBASE" through which the data can be updated dynamically. Following the concept of community-based app, the user can himself/herself update the availability of these resources and add reviews. (iii) Again, the filtration mechanism is applied to sort the result according to ratings of these units generated using Multinomial Naive Bayes theorem.

The authors desire to extend this work by expanding the information of the hospitals with the knowledge of the doctors in their respective specialized service areas and then predicting the ratings for the same thus showing the optimized results after the combination of distance and reviews. Also, adding the facility for the user to predict the diseases by entering their symptoms. The list of doctors with the specialization in the treatment of those diseases can be retrieved from the database along with their reviews by the patients and current locations and doctor's availability status. If available, the user can be notified of the subscribed doctor's status.

References

1. Rahmi, A., Piarsa, N.I., Buana, P.: FinDoctor–interactive android clinic geographical information system using firebase and Google maps API. Int. J. New Technol. Res. **3**(7), 8–12 (2017)
2. Akram, A., Anjum, M., Rehman, M., Bukhary, H., Amir, H., Qaisar, R.: Android life savior: an integrated emergency response system. In: 8th International Conference on Information Technology (ICIT). IEEE, Amman, Jordan (2017)
3. Vaswani, N., Patel, V., Saheta, A., Shah, S., Shah, S.: Advanced and structured system to ease medical assistance. Int. J. Innov. Res. Sci. Eng Technol. **6**(10), 19826–19831 (2017)
4. Imteaj, A., Hossain, M.: A smartphone based application to improve the health care system of Bangladesh. In: International Conference on Medical Engineering, Health Informatics and Technology (MediTec). IEEE, Dhaka, Bangladesh (2016)
5. Reddy, C., Kumar, K., Keshav, J., Prasad, B., Agarwal, S.: Prediction of star ratings from online reviews. In: Proceedings of the 2017 IEEE Region 10 Conference (TENCON), pp. 1857–1861. IEEE, Penang, Malaysia (2017)
6. Kibriya, A., Frank, E., Pfahringer, B., Holmes, G.: Multinomial naive Bayes for text categorization revisited. In: Webb, G.I., Yu, X., (eds.) AI 2004: Advances in Artificial Intelligence. AI 2004. Lecture Notes in Computer Science, vol. 3339, pp. 48–499. Springer, Heidelberg (2004)
7. Firebase Realtime Database. Firebase. https://firebase.google.com/docs/database/. Accessed 12 November 2018
8. Leavitt, N.: Will NoSQL databases live up to their promise? Computer **2**(43), 12–14 (2010)
9. Alsalemi, A., Alhomsi, Y., Disi, M., Ahmed, I., Bensaali, F., Amira, A.: Real-time communication network using firebase cloud IoT platform for ECMO simulation. In: IEEE International Conference on Internet of Things (iThings) and IEEE Green Computing and Communications (GreenCom) and IEEE Cyber, Physical and Social Computing (CPSCom) and IEEE Smart Data(SmartData). IEEE, Exeter, UK (2017)
10. Lu, M., Wen, S., Xiao, Y., Tian, P., Wang, F.: The design and implementation of configurable news collection system based on web crawler. In: 3rd IEEE International Conference on Computer and Communications (ICCC), pp. 2812–2816. IEEE, Chengdu, China (2018)
11. Vogel, P., Klooster, T., Andrikopoulos, V., Lungu, M.: A low-effort analytics platform for visualizing evolving flask-based python web services. In: IEEE Working Conference on Software Visualization (VISSOFT). IEEE, Shanghai, China (2017)
12. Dahiwale, P., Raghuwanshi, M., Malik, L.: Design of improved focused web crawler by analyzing semantic nature of URL and anchor text. In: 9th International Conference on Industrial and Information Systems (ICIIS), pp. 1–6. IEEE, Gwalior, India (2014)
13. Using TF-IDF to Determine Word Relevance in Document Queries. https://www.cs.rutgers.edu/~mlittman/courses/ml03/iCML03/papers/ramos.pdf
14. Su, J., Shirabad, J.S., Matwin, S.: Large scale text classification using semi-supervised multinomial naive Bayes. In: Proceedings of the 28th International Conference on Machine Learning, Bellevue, WA, USA (2011)

ICTs Adoption in SMEs Located in Rural Regions: Case Study of Northern of Portugal and Castela and Leão (Spain)

João Paulo Pereira^(⊠), Valeriia Ostritsova, and João Pereira

Instituto Politécnico de Bragança,
Campus de Santa Apolónia, 5300-253 Bragança, Portugal
jprp@ipb.pt, valeriyaostricova@mail.ru,
jpbp2012@hotmail.com

Abstract. ICTs are one of the most important values for the economic growth. Technology, information, and knowledge have an important impact in peripheral regions when they want to participate in the global market. Exist a lot of discussion about the digital divide between countries but little in relation to the regions and especially Less Favoured Regions (LFR). There is clear evidence that these internal disparities can lead to the continued marginalization of people and regions "disconnected" from global information networks, which support the modern economy and causing strong inequalities. By other side, data available about the ICT use are normally from central regions and not Rural and Less Favoured Regions. With this study, we made a study to understand the ICT use by small enterprises and entrepreneurs from peripheral regions from Portugal and Spain: 5 Spanish regions (León, Zamora, Salamanca, Valladolid, Ávila) and 2 Portuguese (Alto Tâmega e Terras de Trás-os-Montes). For that, 433 surveys were carried out (170 in the Portuguese regions and 263 in Spanish areas), included in project COMPETIC (COMPETIC (0381_COMPETIC_2_E) - Co-financed by the European Fund for Regional Development (FEDER) by the Program Interreg V-A Espanha-Portugal (POCTEP) 2014-2020).

Keywords: ICT · Peripheral regions · Micro-enterprises · Entrepreneurs

1 Introduction

Information and Communication Technologies (ICT) are considered fundamental to the development process of any country, region or city. The delay in the incorporation of ICT in the development of any region can be a great loss for people, communities and companies. In this sense, the variable "time" can be decisive, since the delay can mean "delay" in the development and the quality of life of the people, especially to the localities of the peripheral regions and the less favored regions (LFRs) [1–7].

ICTs, because of the development opportunities they offer, are an essential factor in the creation of rural development policy. Despite its additional confirmation at the regional level, through some projects that are implemented in rural areas, it is necessary to insist on the advantages and benefits that can be obtained from its use, for the integration of rural spaces into a new type of society, which has been called the information or knowledge society [8].

© Springer Nature Switzerland AG 2020
K. Arai et al. (Eds.): FTC 2019, AISC 1070, pp. 682–691, 2020.
https://doi.org/10.1007/978-3-030-32523-7_51

Local development is a complex and multidimensional process in which ICT is one aspect that has greater influence and determination on others. Thus, if the process of local ICT development differs in different locations, they will not be able to achieve similar or proportional results in the long term [7].

The literature review shows the importance of ICT for SMEs in rural regions but exists a lack of studies that compare different regions. The main objective of the work are to conduct a comparative description of the selected areas, to determine the level of the current state of ICT, to bring proposals for improving this level. For this study, it was decided to study 5 Spanish regions (Lean, Zamora, Salamanca, Valladolid, Avila) and 2 Portuguese (Alto Tâmega and Terras de Trás-os-Montes).

In the case of Spain, there are 109,449 companies (in 2018), with 98.3% having 10 or less employees. In this sense, the existence of companies with more than 10 employees is less than on a regional or national scale. However, the small size of companies is a characteristic that defines the Spanish business structure. With regard to the rural provincial municipalities under study, Valladolid and Lean agglutinate two out of three companies (64.7%). On the other hand, if we exclude Avila, then in the industrial sector will be a higher percentage of presence, especially in León and Zamora (11.8% and 11.7% respectively). The construction also has an important weight, mainly in Ávila, with 21.7%. Services have a markedly lower presence, while commerce and hospitality have weights similar to the weights of the reference provinces.

In Portugal, micro-enterprises accounted for almost 96% of SMEs. Micro, small and medium-sized enterprises (SMEs) in the non-financial sector accounted for 99.9% of the total business sector, and in 2010, their number exceeded 59 167 the one verified at the beginning of the period (1 083 901). Microenterprises, which accounted for nearly 96% of the number of SMEs throughout the series, were highlighted. The largest companies by activity were 1,082, and in 2010 only 55 more than in 2004 [9].

Thus, this study intends to evaluate the level of use of ICT by self-employed entrepreneurs and micro-enterprises in the region of Castile and León, Alto Tâmega and Terras de Trás-os-Montes.

2   Definition of the Study

2.1   Methodology

The study was carried out through a quantitative field work in the Spanish regions of Ávila, León, Salamanca, Valladolid and Zamora, and in the Portuguese regions of Alto Tâmega and Terras de Trás os Montes. In Spain, 263 telephone surveys were conducted of companies with 10 or fewer employees, located in municipalities with less than 5,000 inhabitants in the provinces of Castile and León selected: Leão, Zamora, Salamanca, Valladolid, and Ávila. In Portugal, 170 surveys were conducted, 57 in Alto Tâmega and 114 in Terra de Trás-os-Montes. Field work was conducted from 24 September to 16 November 2018. Sample design: random sample by area, type of company and sector.

In Portugal, the sample size was 170 surveys. Sampling error: ± 7.52% for global data, taking into account the infinite universe, 95.5% for a confidence level and estimates of equally likely categories (p = q = 50%). In Spain, the sample size was 263 surveys. Sampling error: ± 6.04% for global data, taking into account the infinite universe, 95.5% for a confidence level and estimates of equally likely categories (p = q = 50%).

2.2   Sample Distribution

The sample was defined in order to obtain representativeness by sectors and types of companies (Table 1).

Table 1. Sample distribution (%)

	Leão	Zamora	Salamanca	Valladolid	Ávila	Alto Tâmega	Terras Trás os Montes
Agriculture sector	10,9	20,0	13,5	14,3	10,0	7,0	9,7
Industry	23,7	24,0	23,1	26,8	28,0	28,1	26,5
Construction	7,3	6,0	5,8	5,4	6,0	5,3	3,5
Trade	9,1	16,0	24,9	8,9	14,0	8,8	9,7
Tourism	18,2	10,0	9,6	14,3	20,0	12,3	8,0
Services	12,7	10,0	7,7	10,7	6,0	14,0	14,2
Social services and education	3,6	4	7,7	10,7	4	12,3	12,4
Other services	14,5	1	7,7	8,9	12	12,3	15,9
Base	55	50	52	56	50	57	113

3   Results

Next, we present the results and analysis of data obtained in successful 433 surveys.

In the two Portuguese regions, there is a higher perception of the level of ICT use: 24.3% indicate that it is high or very high and only 19.44% in all of the Spanish provinces (Fig. 1).

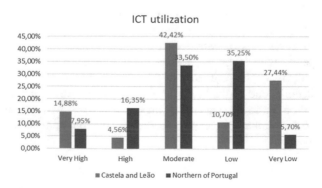

Fig. 1. ICT use in the explored regions.

However, when the analysis falls to the provincial and regional levels, the data show a high heterogeneity: in Avila the percentage is 22.0%, in Salamanca 21.1% and in León only 14.5%, while in the Terras de Trás-os-Montes 31.0% have a positive outlook.

Table 2 shows the percentage of use by province and region, denoting that the large % is moderate.

Table 2. Level of ICT use in enterprises by region

	Very high	Quite high	Moderate	Low	Very low
Leão	12,7%	1,0%	52,8%	9,1%	23,6%
Zamora	16,0%	4,0%	38,0%	8,0%	34,0%
Salamanca	19,0%	1,9%	42,4%	17,3%	19,2%
Valladolid	12,5%	7,1%	42,9%	7,1%	30,4%
Ávila	14,0%	8,0%	36,0%	12,0%	30,0%
Alto Tâmega	8,8%	8,8%	35,1%	40,4%	7,0%
Terras de Trás-os-Montes	7,1%	23,9%	31,9%	30,1%	4,4%

With regard to existing ICTs, the following are noteworthy in province of Castela and Leão:

(1) Laptops (70.2% versus 56.7% in Northern Portugal)
(2) Automatic presence control systems (17.7% versus 5.9% in Northern Portugal)
(3) Backup systems (56.8% versus 42.1% in Northern Portugal)
(4) Web (55.0% versus 46.2%)

On the contrary, in the companies of Alto Tâmega and Terras de Trás-os-Montes, the following ICTs are more present (Table 3):

(1) Document scanning system: 55.6% compared to 43.4%
(2) Cloud Computing: 32.2% compared to 9.3%
(3) Social networks: 63.2% compared to 53.2%
(4) Open source software: 59.7% versus 11.6%

Table 3. ICT in companies

	Provinces of CeL	Total Northern Portugal
Hardware		
Desktop	77,9%	71,4%
Portable	70,2%	56,7%
Mobile devices	90,2%	87,7%
Data storage solutions	49,4%	43,9%
Automatic presence control system	17,7%	5,9%

(*continued*)

Table 3. (*continued*)

	Provinces of CeL	Total Northern Portugal
Video projectors	11,9%	8,2%
Backup system	56,8%	42,1%
Scanning		
Document scanning system	43,4%	55,6%
Access and use of the Internet		
Internet access	93,3%	94,7%
Intranet		
Intranet	10,5%	13,5%
Cloud		
Cloud computing	9,3%	32,2%
Social networks and others		
Social networks, positioning in search engines or other tools to advertise	53,2%	63,2%
Web site		
Web site	55,0%	46,2%
Open source software		
Open source software	11,6%	59,7%

With regard to security, a higher level of protection is observed among companies in the Spanish provinces. With regard to availability of information, it is available only to authorized persons (see Table 4).

Table 4. Security policies by region

	Total selected provinces of CeL	Total areas of Northern Portugal
The information is only available to authorized people	72,3%	93,5%
The operating system is up to date	78,9%	62,4%
Always use original software	64,5%	61,2%
Perform regular backups	67,7%	55,3%
Anti-virus and anti-malware are reliable	80,1%	50,6%
Antivirus updates are performed daily	64,5%	50,6%
Personal and critical data are adequately protected	79,5%	40,6%
Firewall is enabled	64,7%	37,1%
There are up-to-date security, contingency, and recovery plans	49,0%	14,7%

In view of the difficulty of implement ICT, there are differences between the individual provinces of Castile and León and northern Portugal. While in the latter case the maintenance costs (69.4%) and the lack of human capital (55.3%) are allocated mainly in Spanish, the first is the lack of infrastructure (see Table 5).

Table 5. Problems with the implementation of ICT

	Total selected provinces of CeL	Total areas of Northern Portugal
Lack of ICT infrastructures (for example: Internet access)	31,5%	2,9%
Lack of human capital with knowledge of ICT	25,9%	55,3%
Lack of government support	22,7%	7,6%
Maintenance costs	22,3%	69,4%
Difficulties financing	19,2%	41,2%
The solutions are little adapted to the needs of the company	14,9%	2,4%
Safety concerns	14,2%	22,4%
Suppliers and customers do not use ICT	11,4%	20,6%
Lack of awareness about ICT benefits	9,9%	10,0%
Service providers offer little confidence.	8,7%	2,4%
Difficulty of integration between systems	7,5%	4,7%
I do not know	5,7%	0,0%

In the northern region of Portugal, the estimated use of ICT between Alto Tâmega and Terra de Trás-os-Montes is low or very low (38.8%). These numbers are lower in Terras de Trás-os-Montes (34.5%) than in Alto Tâmega (47.4%).

However, more than half of the explored companies (53.5%) indicate that they plan to invest in ICT in the next two years; 50.9% in the Alto Tâmega region and 54.9% in the Trás-os-Montes lands. The most implemented ICT are Internet access (95.3%) and mobile devices (88.2%). However, the least common in companies are automatic presence control systems (5.9%) and video projectors (8.2%). The majority of explored companies (73.5%) didn't participate in subcontracting activities in the field of ICT. Among companies that have access to the Internet, 44.4% of companies indicate that they have reduced the speed above 100 Mbps, although 39.5% believe that they do not know how fast they are. Businesses use the Internet to:

(1) Email: 96.9%
(2) Information demand: 82.1%
(3) Banking services: 71.0%
(4) Send/receive information from public organizations: 62.3%

As for web pages, only 46.5% indicate their presence, below is functionality:

(1) Catalog or price lists: 74.7%.
(2) Online orders: 36.7%.
(3) Site security certificate: 30.4%.

Among companies who don't have a web page, 46.0% say that this is due to maintenance costs, and 35.6% say that this is because of the suppliers and customers don't have it.

The main difficulties identified by these companies are mostly internal: the cost of maintenance (69.4%) and the lack of human capital with ICT knowledge (55.3%). Barriers that can be viewed as external were mentioned by a very small percentage.

Most companies consider ICT positive and have important benefits. The most valuable are:

(1) Improved performance: 68.2%
(2) Improved customer service: 62.9%
(3) Sales increase: 54.1%
(4) Contact networks and new business opportunities: 42.9%
(5) Competitive advantages: 42.4%

The Spanish region of Castilla and León is slightly below the national average for the use of new technologies: for example, only 2.98% have employees specializing in ICT.

Among companies based in municipalities with less than 5,000 inhabitants, there is no positive opinion about the introduction of ICT. Thus, in rural municipalities, the percentage of companies with low or very low use of ICT (36.9%) is higher compared to those that are defined as high or very low high (18.8%).

The best prospect is observed among entrepreneurs, where 28.1% consider it high or very high. By sector, services (28.5%) and agriculture (22.7%) stand out, as there is a broader perception of the use of ICT, while in the trade half of the explored companies believe that it is low or very low (50.8%).

On the other hand, it is noted that among those companies that believe that they have greater use of ICT, there are great prospects for investment in the coming years.

The study shows that mobile devices (90.2%) and Internet access (93.3%) are present in most companies. And only 9.3% got access to cloud computing, and 10.5% got access to the intranet. However, there are differences between the types of enterprises: entrepreneurs and micro enterprises use different technologies to a greater extent, while among self-employed people their presence is smaller.

There are large differences in equipment by sector: in the services sector there are more various hardware elements, with the exception of office computers, which are more common in industry and in the agricultural sector, mobile devices are common in agricultural activities and industry, video projectors in tourism. This latest sector also highlights the presence of social networks and the network.

The vast majority of companies located in rural municipalities have access to the Internet (93.7%). As for 6.3% who do not use, the main reason is the cost of either connecting (18.6%) or maintenance (26.6%).

Among Internet companies, DSL technology is the most commonly used access (55.9%). Fiber is used only by 12.8%. In this sense, it has already been observed that in rural areas, and in Spain as a whole, this type of technology is the least popular.

A high level of ignorance among executives of companies about the speed of connection under the contract was noted: 46.2% indicated that they did not know what the speed should be. 38.0% indicate that it is less than 10 Mbps.

With regard to the use of the Internet, most companies use it in the usual way: more than 70% manage e-mail, information search, banking services, as well as sending or receiving information about products. On the other hand, only 15.3% use the Internet for video conferencing.

Regarding the use of the Internet for relations with the authorities (65.5% of companies with access to the Internet), most of them used to receive forms (62.4%) involved in social security processes (57.1%) and obtain information from web pages (49.9%).

55.0% of the companies have web page, and the functionalities that have implemented are as follows:

(1) Availability of catalogs or price lists: 58.1%
(2) Requests (or reservations) online: 41.2%
(3) Web site security certification: 32.5%
(4) Online order tracking: 27.7%

The 45.0% that does not have it, the main reasons are:

(1) Providers and customers do not have a Web page: 17.7%
(2) Online transactions are not common in the sector: 15.7%
(3) online marketing is not common in the sector: 13.2%

11.6% of companies located in rural municipalities in the analyzed provinces use open source software, and the most common is browsing the Internet. This type of software reaches a higher level of use among entrepreneurs (20.6%) and in the service sector (24.1%), the province of Valladolid (18.5%) and among companies that indicate a very high level of ICT use (13.9%).

The highest level of ignorance is observed in: self-employed workers (37.1%), tourism (51.9%), Salamanca (42.6%). In this sense, 67.5% don't have any management tool (ERP, CRM, BSC, SCM, BI, etc.). The most frequently used tools: ERP (17.3%) and workflow (13.2%).

It is confirmed that 60.2% of companies indicated some difficulties with the introduction of ICT. The main barriers are both internal and external factors in relation to the company. Thus, they emphasize: lack of ICT infrastructure, lack of human capital, lack of government support and maintenance costs. In general, it is entrepreneurs who report the fewest problems, and most of all indicate those who are not self-employed. Companies are interested in implementing security measures: 89.2% took some measures.

The most frequent are: antivirus (80.1%), personal data protection (79.5%), an updated operating system (78.9%), access to information accessible with authorization (72.3%) and backup (67, 7%). Security is implemented to a greater extent among entrepreneurs (an average of 6.3) and among microenterprises (6.75); However, the rate among self-employed is noticeably lower: 5.30.

An important aspect is that the size of the company determines the implementation of security policies. Thus, in companies where there are no workers, only 84.3% have such a percentage, which increases to 91.7% among those who have workers.

In particular, among the explored companies, 45.3% indicated that they had sub-contracted activities in the field of ICT.

The highest level of subcontracting is observed among microenterprises (52.4%), in the service sector (53.8%), in the province of Leon (50.0%).

Most companies believe that scanning is a good thing, and that it has important benefits:

(1) Improved operation: 61.5%
(2) Improved customer service: 61.5%
(3) Contact networks and new business opportunities: 52.2%.
(4) ICT helps penetrate new markets: 48.5%
(5) Increase market share: 47.0%
(6) Increase sales: 43.9%

4 Conclusions

Further analysis of the results leads to some general conclusions, emphasizing the following.

Most companies and entrepreneurs based in rural areas of the explored areas consider that their level of ICT use is low or very low, and only one in five qualifies as high or low. However, this perception is more optimistic in Portuguese areas, especially in Terras de Trás-os-Montes.

In the analysis on specific technologies, the integration of more common ICTs (access to the Internet, mobile devices and computers, both office and portable) is at a high level. Regarding security, a higher level of protection is observed among companies in the Spanish provinces, except that information is only available to duly authorized, which is more present in Portuguese areas.

In relation to the difficulties to implement ICT, there are differences between the individual provinces of Castile and Leon and the provinces of the north of Portugal.

A high percentage of companies see positive aspects in ICT, both in Spain and Portugal. In both countries, the improvement of the company's operations and customer services (two-thirds in both cases) is particularly important.

The lack of skilled human resources to undertake digitization means that many companies, especially the smaller ones, choose to outsource ICT maintenance provision.

Acknowledgments. COMPETIC Project - Support to entrepreneurs, self-employed and micro-enterprises in rural areas to create and develop their businesses taking advantage of ICT opportunities (operation 0381_COMPETIC_2_E).

UNIAG, R & D unit funded by the FCT - English Foundation for the Development of Science and Technology, Ministry of Science, Technology and Higher Education. Project n. UID/GES/4752/2019.

References

1. Pereira, J.P.: ICTs and the development of LFRs. In: 5th Conference of the Portuguese Association of Information Systems, Lisbon, pp. 1–12 (2004)
2. Pereira, J.P.: Telecommunications infrastructures in the peripheral regions: the role of public funds. In: 6th Conference of the Portuguese Association of Information Systems, Bragança, Portugal, pp. 20–33 (2005)
3. Pereira, J.P., Teixeira, F.: Information Society in the Peripheral Regions: The Case of the Alto Trás-os-Montes Region (Portugal) First Educational Innovation Days (2006)
4. Pereira, J.P., Teixeira, F.: Information Society in Peripheral Regions: The case of "Trás-os-Monte"- Portugal CONTECSI, São Paulo, Brazil (2006)
5. Pereira, J.P.R.: Broadband access and digital divide. In: Rocha, Á., Correia, A., Adeli, H., Reis, L., Mendonça Teixeira M. (eds.) New Advances in Information Systems and Technologies, vol. 2, pp. 363–368. Springer International Publishing, Cham (2016)
6. Pereira, J.P.: Decision support system for evaluation of organizational management structures: methods and models. In: IEEE International Conference on Intelligent Systems 2018, Funchal, Portugal (2018)
7. Lang, T., Görmar, F.: Regional and Local Development, Times of Polarization, Palgrave Macmillan, Singapore (2019)
8. Cornford, R., Richardson, M.S., Marques, P., Gillespie, A.: Transformation of Regional Societies through ICTs, State (s) of the Art (s) (2006)
9. National Institute of Statistics. www.ine.pt

Kaizen 4.0 Towards an Integrated Framework for the Lean-Industry 4.0 Transformation

Peter Burggräf[1], Carolin Lorber[2], Andreas Pyka[2], Johannes Wagner[1], and Tim Weißer[1(✉)]

[1] University of Siegen, Paul-Bonatz Str. 9-11, 57076 Siegen, Germany
tim.weisser@uni-siegen.de
[2] University of Hohenheim, Wollgrasweg 23, 70599 Hohenheim, Germany
Carolin.Lorber@uni-hohenheim.de

Abstract. Due to increasing competitive pressure, growing customer demands and rising desire for individuality in the product design, companies face rising demands on the efficiency of their production systems. This challenges the alignment of established philosophies such as Lean and Industry 4.0 and requires respective strategy derivations to be translated in concrete concepts, methods and technology projects on an operative level. Building on the work of Spath *et al.* 2017 and Schuh *et al.* 2017, a model based approach is presented which contributes to a reduction of complexity in challenging decision-making at transformations from Lean to Digitization and Industry 4.0. Perspectives given by the development of a maturity model allow considerations of the overall system and target configuration for both local and temporal decisions. This leads to an integrated view of Lean and Industry 4.0 as complementary paradigms ranging from a strategic to an operational level. As the complexity of the developed framework rises with the number of considered concepts, methods and technologies as well as the target configuration, an examination of further complexity-reducing methods needs to be performed. Our approach sheds light on how interrelations between Lean Production and Industry 4.0 occur when considering an evolutionary transformation. Furthermore, it shows how companies can evaluate concepts, methods and technologies on an operative level to develop their production systems in the context of Lean and Industry 4.0. Moreover, our model offers proceedings to develop their Lean Production systems in the context of Industry 4.0.

Keywords: Digital transformation · Digitization · Lean production · Industry 4.0 · Capability maturity model · Technology assessment

1 Introduction

The value creation system of companies, especially within the production system of automotive companies, needs to be highly flexible and versatile, due to increasing competitive pressure, growing customer demands and rising desire for individuality in the field of product design. As a result of increasing requirements in flexibility, speed, profitability and efficacy, the overall targets regarding costs, time, quality and sustainability represent a key motivation not only for Lean Management approaches but

© Springer Nature Switzerland AG 2020
K. Arai et al. (Eds.): FTC 2019, AISC 1070, pp. 692–709, 2020.
https://doi.org/10.1007/978-3-030-32523-7_52

also for Industry 4.0 approaches [1]. Companies face the challenge of aligning their Lean and Industry 4.0 strategies and transferring respective strategy derivations into an operative level in terms of concrete concepts, methods and technology projects. However, appropriate guidelines and models uniting Lean and Industry 4.0 on an operative decision level are still missing. This paper aims to close this gap by providing a range of contributions for the establishment of such standards. The first section conduces to a critical appreciation of related research as well as the definition of corresponding objectives. Maturity model creation including evaluation criteria is being described in section two and concluded in section four. Section three provides a procedural perspective to the field of observation. Concluding, a brief discussion and summary will be given in section five at the end of this paper.

1.1 Lean and Industry 4.0

Womack and Jones define five core principles of lean: First, the identification of the customer defined value; second, the continuous improvement of the value stream; third, the elimination of waste in order to convert the value flow smoothly; fourth, the synchronization of information flow and customer demand to activate a demand driven pull system; fifth, the perfection of all processes, products and services [2].

Concentrating on these core lean principles enables manufacturing processes as well as whole organizations to achieve increasingly high levels of efficiency and to compete at a low cost level while maintaining high speed of delivery and gaining optimal quality [3]. In order to accomplish a Lean Production system, lean processes need to be established. Those can be defined as "an integrated socio-technical system whose main objective is to eliminate waste by concurrently reducing or minimizing supplier, customer, and internal variability" [4]. The invention of lean production having taken place more than 50 years ago, however, it seems that the concept/management technique has reached its limit. In the future, production decoupled from market requirements will become more and more relevant, as deviations in market requirements are already in conflict with leveled capacity utilization [5, 6].

Furthermore, due to the increase of international supply chains, the environment became more and more complex, entailing the establishment of intelligent products and production systems [7]. The necessity of lean as an evolving concept has already been described by Hines *et al.* [8]. The goal of companies in the 21th century is to verify, examine and perfect proven technologies continuously and to intelligently combine them with new methods (e.g. Industry 4.0). Industry 4.0 is a term for the increased integration of information and communication technologies with the aim to optimize value chains by implementing an autonomously controlled, dynamic production [9, 10]. It is commonly considered as representing the fourth industrial revolution, a new level of organization and control of the entire value chain beyond the product life cycle. Kolberg and Zühlke suggest that the successful merge of Lean and Industry 4.0 can be achieved on four different operational levels [9]:

Smart Operator: Reduction of time from failure occurrence to failure notification by real time error detection and direct employee localization through CPS (Cyber Physical Systems).

Smart Product: Collection of process data for the analysis during and after the production tailored to each product through the utilization of sensors and CPS.

Smart Machine: Collection of process data by every machine with connectable sensors and CPS to achieve fast and flexible predictive maintenance.

Smart Planer: Decentralization of planning processes using working stations integrated in CPS to negotiate cycle times and thus find the optimum between highest possible capacity utilization per working station and continuous flow of goods.

1.2   Research Dedicated to the Field of Observation

Shortly after the technology driven Industry 4.0 paradigm emerged, researchers tended to examine potential ties to the traditional best practice driven paradigm of lean production. Whereas a consensus concerning the complementarity of both paradigms has been developed so far, concrete cooperative specifications are still subject to current research. Therefore, the subsequently used classification of scientific approaches serves to allocate the field of research chronologically.

Compatibility of the Lean and Industry 4.0 Paradigms. Examinations originate at the executive principles of the respective paradigm. Via reciprocal comparison of those principles (geared to operational functions), there to some extent different goals become apparent, as e.g. Roy *et al.* [11] and Quasdorff and Bracht [12] demonstrate. Therefore, an exhaustive conformity of both paradigms cannot be assumed. Ganschar *et al.* discuss the issue with the discrepancy between FIFO/Flow (Lean) and a decentral, case-based control (I4.0) as part of a difference between the goals of standardization and complete information of technological pervasion [13]. Despite the determination of certain characteristic differences and partial incoherency, most researchers see a benefit in the complimentary application of the two paradigms (especially [9, 11, 14]; Cf. a detailed record on this in [1]). Leveraging Industry 4.0, higher Lean maturity levels are expected, as well as with the aid of Lean, a reasonable integration of Industry 4.0 is possible. Accordance prevails with the assumption that process orientation and especially lean production represent a key prerequisite for Industry 4.0.

Paradigm Transformation. Besides the consensus of general complementarity, a plethora of research focuses on the transformation process onto Industry 4.0. In this regard, there are different approaches trying to enhance contemporary lean methods unidirectionally using technologies of Industry 4.0 (cf. [9, 15]). Sanders and his colleagues respond in a very detailed manner to ten measurable lean manufacturing dimensions developed by Shah and Ward [4] and describe the respective capabilities of Industry 4.0. Beyond that, Metternich *et al.* provide a first proposal onto capability maturity levels with the attempt of arranging the principles and methods of the paradigms in a hierarchical stage model from Lean across Digitization to Industry 4.0 [16]. Concerning specified transformation options, approaches merely concentrate on a production technological view so far, which is only preliminary to the value added/production systematical view (cf. [17–19], with low consideration of lean, respectively). Also concerning maturity based assessment, dedicated Lean maturity models are unintegratedly confronted with dedicated Industry 4.0 maturity models: e.g. [20, 21], finishing lean optimized and [22], with the SIMMI 4.0 maturity model, starting at the digitization level.

Semi-integrated Approaches. Consecutive to (II), semi-integrated approaches exist that are oriented on specific methods. Meudt *et al.* expand the value stream mapping to a 4.0 level by the integration of a data view [23]. Rauch *et al.* focus on operational functions, in detail on the transformation from lean product development to smart product development [24]. Likewise, semi-integrated approaches represent a sub-step in the direction of a holistic paradigm integration and a detailed elaboration of an artifact thereof. Schuh *et al.* develop an Industry 4.0 maturity index, also classified as being semi integrated, because their approach includes both an applicable procedure and accurately detailed maturity model, both giving direction to the research at hand [25]. Though dignifying lean in terms of organizational and cultural contributions, complementarities to established principles and methods of lean manufacturing are however not examined.

Fully Integrated Approaches. Burggräf *et al.* propose a fully integrated procedure model emanating from a company's value stream analysis [26]. Through the identification of mission critical lean principles, a digitized value stream is designed by serving the three perspectives Human-Technology, Human-Organization and Technology-Organization (to the H-M-T approach cf. also [19]) with applications and related technologies from Industry 4.0. Burggräf *et al.* conclude their work by questioning evaluation and comparison methods for digital solutions. Regarding this, Spath *et al.* provide with 'Productionassessment 4.0' a first integrated approach for a maturity level based representation and consequently measurability of the transformation from lean to Industry 4.0 [27]. Their model allows an assessment of the production systems' maturity within 33 criteria with up to eight sub-criteria each. Additional to the traditional development fields of lean production (e.g. standardization, continuous improvement), scopes of Industry 4.0 are reasonably integrated (e.g. M2M, digital image of the production system). A further allocation is achieved by aligning strategy, processes and value stream, organization, methods and tools, as well as labor. However, the model of Spath *et al.* does not exceed the allocation of general development fields and a proposal for potential maturity levels.

The integration and further development of Burggräf *et al.*'s procedure model in terms of bidirectionality, as well as Spath *et al.*'s assessment model in terms of operationalization are thus subject to the subsequently presented work under consideration of [26, 27]:

- Operationalization and value oriented model usability
- Decision base for transformation interfaces
- Transfer and individual application of artefacts
- Bidirectional paradigm view within a reference procedure model

Subsumable as framework consisting of procedure and evaluation model for the lean Industry 4.0 transformation is missing.

2 Methodology and Aim of Research

The explanations above yield three main questions that the presented works attempts to answer:

1. Which *interrelations* occur between lean production and Industry 4.0 considering an evolutionary *transformation*?
2. How can companies *evaluate* concepts, methods and technologies on an operative level to develop the production system in the context of lean and Industry 4.0?
3. How can companies *proceed* to develop their lean production systems in the context of Industry 4.0?

Interrelations result in our field of observation from the diversity of concepts, methods and technologies, serving similar requirements in different areas of a company, while being individually adapted and embedded into the local system environment. With regard to this circumstance, a multi-dimensional model intends to contribute in particular to a complexity reduction in the given context. Contemplating possible interdependencies, the continuity of transformation processes has to be likewise considered, requiring a temporal perspective of the evaluation of concepts, methods and tools. To answer questions (1) and (2), we suggest the development of a multi-dimensional capability maturity model, which depicts the spatial factor of interdependencies as well as the temporal factor of transformation processes. The model development process performed in our research follows the procedure of Becker *et al.*, proposing a reference procedure for maturity model development focusing on information technology [28]. Commencing with a problem definition, a review of hitherto relevant approaches to the subject matter is now complemented by a development strategy. Thus, the further procedure consists of an iterative maturity model development. In addition to a plausibility evaluation, a quantitative evaluation regarding specific assessment criteria for deployment is performed in a different paper (cf. explanations on related research). The iteration of model development is carried out across each of the following levels: Design level selection, approach selection, design of model section, and testing. As maturity models according to [28] only allow a systems' posterior appraisal, our maturity model is complemented by a procedure model. Through this, the consideration of adequate concepts, methods and technologies (C/M/T) referring to the system configuration shall be enabled a priori. The procedure model development is based on the idea of hierarchic views as an Enterprise Architecture principle (cf. [29–31]). Passing through the procedure model, the maturity model in turn allows an evaluation on each view/level, which is why both models have to be seen interconnected.

3 Results: A Multi-dimensional Capability Maturity Model

3.1 Maturity Model Development

Field of Observation. As shown above, the synthesis of lean production and Industry 4.0 is not comprehensively discussed. It can however be asserted that the transformation to Industry 4.0 is based on a lean production and develops in an evolutionary process of single elements in a systems' configuration. Referring to a further development of the production assessment model of [27], modelling is particularly performed under consideration of an operative assessment and decision based scope and rests upon the foundations of [32] and [28] on capability maturity model development.

Model Development. Starting at an ontology, describing diverse development fields derived from a company's vision or strategy, Lean Production and Industry 4.0 are company-specifically integrated as paradigms using associated principles (cf. also enterprise architecture frameworks). The latter is then differentiated in characteristic sections (partitions) in the field of observation (objects to be designed), as e.g. "real-time mapping of production processes" or "labor organization".

These sections, representing an operational function and an associated paradigm, can be detailed in artifacts at various aggregation levels, depending on the scope and complexity of the particular section. The artifact with the highest level of detail contains concepts, methods and technologies, which contribute as elements to the

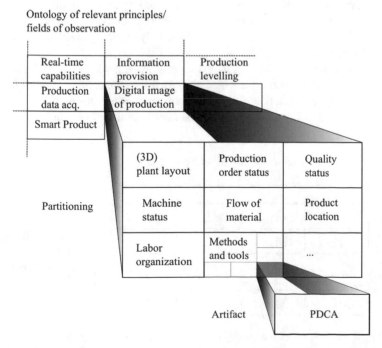

Fig. 1. Exemplary itemization for hierarchical domains within the capability maturity model, extending the model of Spath et al. 2017.

particular maturity level of the major artifact. As an example, KANBAN is a method to realize the pull principle as a concept and is supported by (electronic) KANBAN-cards as the technology. Because concepts, methods and technologies are highly integrated, meaning they interact or rely on adjacent artefacts, interdependencies occur between the artefacts.

Maturity Level Development. Parallel to the model development, maturity levels (ML) can be derived from each abstraction layer, representing successive steps for the achievement of objectives (see Fig. 2). The ML's specificity thereby depends on the design criteria of the particular field of observation respective artifact (based on meta-maturity levels). The initial highest degree of aggregation for ML's in the production is provided by the four following global levels and the definition of their production paradigm (focus on capabilities):

Lean := Subsumable as perfection in the disciplines clustered by the Lean philosophy as for example exhaustive exploitation of naturally available (from planning imperfection) potentials in the spatial-temporal control of products and operational resources [33].

Digitization := Represented by the computerization of repetitive or long lasting manufacturing tasks enabling high quality at high efficiency, as well as connectivity of operational technology combined to a digital representation and execution of operational functionalities [25].

Industry 4.0 := Visibility as up-to-date digital model enriched by sensor data achieving a company's digital shadow (e.g. with realtime KPIs and Dashboards). Transparency: Contextualized semantic linking and aggregation of data "to support complex and rapid decision-making". Predictive capacity enabling automated and adaptive decision-making in terms of data based real time optimization in contrast to determined rule/process based decision making [25].

Target configuration := Operational optimum system-configuration of subsequent maturity levels aspired by the company or a partition, concerning economic viability and the interest of all stakeholders. Referring to individual business venture success, this does not require overreaching accomplishments in the aforegoing maturity levels.

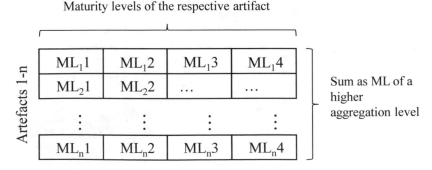

Maturity levels of the respective artifact

Fig. 2. Schematic of the capability maturity levels across aggregation niveaus.

The definition and quantity of capability maturity levels is yet very abstract on the lower levels of aggregation (e.g. shop floor transparency for automatic, personalized and timely information of employees about relevant events in the labor organization). This requires a separate set of evaluation criteria for all artifacts to be included in the artifact configuration during a transformation process. The elaboration of the criteria will be carried out in the following sections. For a further specification of the model, in particular with regard to the problem definition, we identify four mission critical building blocks, including specific subcriteria to be applied to the C/M/T: synergies and interdependence, trajectory, sustainability and leanness. Each C/M/T is represented by one specific/individual artefact that is in the focus of the maturity assessment.

Synergies and Interdependence Between the Artefacts' C/M/T. Artefacts as artificial constructs originate from a complexity reduction in modelling. However, their fit into the holistic information system landscape has to be considered thoroughly (for the artifact as result of research and development, cf. [34] and especially [35]). Synergies and interdependencies occurring between artifacts thus need to be harmonized. Although synergies do not contribute to any buildup of competencies, they derive from existing internal competence and therefore enable improvement at low risk [36].

For example, an artifact configuration emanating from a successful pilot project into contiguous departments/zones causes interdependencies between different artifacts to arise from mutual influences and dependencies amongst elements from concept, method and technology. Thus it is important to consider the impact on contiguous artifacts, as well as the overall configuration while manipulating an artifact element. Interdependencies can likewise lead to positive effects increasing the maturity level of surrounding artifacts but also result in reciprocal restrictions or even exclusion (see Fig. 3). For instance the adoption of a new technology to record operational data, which has no interface or is incompatible to other departments/zones inside the value stream. Approaches for a methodical assessment of interdependency and synergy within the C/M/T portfolio only exist rudimentarily. Although generic methods exist in the related field of technology assessment and product development (e.g. Multiple Domain Matrix or Design Structure Matrix; see [37, 38]) they have not been transferred to the field of observation at hand, so far.

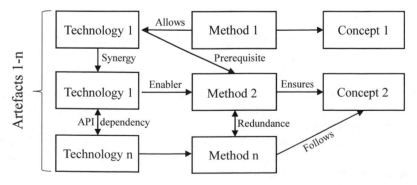

Fig. 3. Exemplary artefact interdependencies across concepts, methods and technologies.

C/M/T – Trajectory Development. Within an artifact, it makes sense to describe the development process as a vector to the target configuration, whose trajectory ideally points to a dedicated target point. The vector itself consists of C/M/Ts and is therefore directionally influenced when a new C/M/T is applied. In terms of a transformation process to Industry 4.0, process and vector describe the evolutionary trajectory.

With this kind of perspective, it becomes possible to evaluate potential improvements not only quantitatively (e.g. monetary) but also qualitatively with regard to possible influences on the trajectory and the transformation process to a target configuration. If a value stream analysis under consideration of information flow is set as a target, an organizational optimized, manual method, for instance, would lead to a deviation of the target vector. However, any methodological update via a data-based information density would shift the target vector to the target point as a building block within the target configuration.

Assuming that a Lean maturity model represents the (area specific) optimal configuration, an intersection to Industry 4.0 exists initially with the digitization of the optimized processes of that configuration. With the application of Industry 4.0, the target convergence needs to be realigned and evaluated, but this must be done starting from the Lean C/M/Ts. The validity of the one-piece flow as opposed to a decentralized, flexible production system serves as an example.

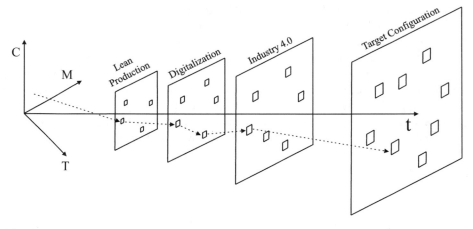

Fig. 4. Trajectory of the target vector in dependency to the aspired configuration, based on [25] and [19].

C/M/T's Future Viability as a Function of Achievable Maturity Levels. An assessment of the maximum achievable maturity level by applying a C/M/T involves projecting the specific C/M/T to its ultimate stage. For a vector-based viewpoint, this describes the vector length and, at the same time, the achievable proximity to a target point. For example a new visualization tool based on MS Excel would therefore only achieve a very small step, compared to web applications, as they might be developed much further, e.g. by continuous integration of web services.

The explanation above leads to a concept of strong analogies to the established S-Curve model for technology lifecycles and specifies it with concepts and methods, especially on at the operational level. The latter implies that decisions must be made in the constant awareness of the known innovator dilemma [39].

C/M/T's Continuous Improvement Phase. Since no C/M/T is implemented at the highest maturity level, a continuous improvement process must be performed. This process represents the distance between the target points (see Fig. 4). There are four implicit phases within the continuous improvement process - sensitization, start, implementation and stabilization [40]:

- Sensitization: Not yet implemented, but already communicated
- Start: Ready for implementation and responsibilities defined
- Implementation: C/M/T is productive
- Stabilization: C/M/T is productive and optimized

For instance a new visualization tool that was implemented six months ago (implementation phase, almost stabilization phase) can hardly be compared to a tool that was implemented two days ago (starting phase).

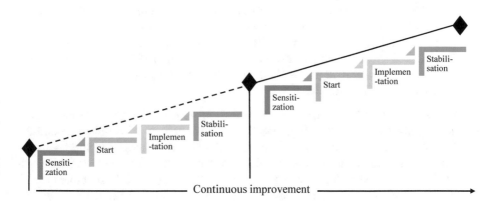

Fig. 5. Continuous improvement steps, based on [40].

C/M/T's Leanness. Lean is already described in the maturity level as an abstract level. In addition to transparency, standardization, and optimization/continuous improvement, the principles tact, flow, pull, zero defects and zero waste need to be taken into consideration, when evaluating the artefact leanness of C/M/Ts. The C/M/T was evaluated from two perspectives of leanness. First, if the C/M/T itself is lean and, secondly, to what extent can the C/M/T contribute to increasing the leanness of the system. To evaluate whether a C/M/T is lean, the performance capacity, the effort of method maintenance and the technology acceptance need to be taken into consideration.

Performance Capacity (Time, Cost and Quality): The time, cost and quality triangle derives from the project management research field and is used in this article to determine the artifact capacity (capacity management). Time, cost, and quality are interlinked, so concentrating on one point of the triangle affects the other two points [41]. Therefore, we propose the evaluation of the three categories time, cost and quality, including their interrelations.

> *Time* is defined as the temporal expenditure that is necessary for the use of the C/M/T.
>
> *Costs* are divided into short-term (development), mid-term (introduction) and long-term costs (maintenance).
>
> *Quality* is defined as the responsiveness to customer demands and encompasses every element that has an impact on customer satisfaction such as dimensional accuracy, product performance, specifications or basic functionalities [42]. When the production quality of the product rises, production time as well as the production costs increase initially. Short-termed, time, cost and quality as success critical factors of a production system, may outweigh the subsequently explained criteria, however long-termed they scale moderately.

Effort of Usage and Maintenance: Regarding the cost-benefit ratio of a method or technology, two key factors need to be taken into consideration. At first the people, in particular their effort to maintain the method or technology and second, the software itself, particularly "waste" within software development and handling. Especially when decisions about potential digital replacements are pending, the effort of usage and maintenance has to be scrutinized.

> *People* are the center of resources, information, process design, decision making and organizational energy. Organization structures have to be centered around the flow of value, not on functional expertise. For the assessment and the degree of depletion it is essential that the digital information is presented and handled dedicated to the people who are adding value to the system [43]. It is therefore crucial, that new methods or technologies do not cause more effort for the people than established ones.
>
> *Waste* in software development consists of seven different categories: extra features and requirements, extra iterations, information gathering, unexposed defects during tests, waiting, handoffs [43]. The usage of multiple software systems additionally leads to interface- and architecture complexity. Thus it has to be ensured that these seven categories of waste are fundamentally eliminated.

Technology Acceptance: With regard to the target configuration, the technology acceptance model of Davis is used to evaluate, whether the usage of the C/M/T is evaluated positive or negative [44]. The use of the C/M/T is dependent on the intention of the user [44]. The intention or attitude towards the usage results from the perceived usefulness as well as the perceived ease of use. These two categories can be influenced by different factors such as job relevance, output quality, subjective norms or personal experience [45].

For example a visualization tool based on MS Excel may seem quickly to be introduced on a low cost level, but it lacks a lot of the already introduced criteria (e.g. efficiency or future viability), compared to a web application. The Excel tool has furthermore restrictions concerning usability and compatibility, which decreases the method maintenance by creating work around (waste) that are not necessary in the web application. The technology acceptance for prevalent MS Excel may be high in the beginning, but regarding its limited usability the acceptance will decrease with rising application complexity. To evaluate to which degree the C/M/T can help to increase the leanness of the system, it has to be taken into consideration how a new level of tact, flow, pull, zero defects or zero waste is achieved. For instance production activities of one specific area are mapped on a physical board with printed sheets containing basic information about the machines that are used in the system (e.g. overall equipment effectiveness), people that are working in the area (e.g. shift schedule and accidents at work) and the key performance indicators (e.g. output per hour), contain high transparency but also a high effort of usage and maintenance. Therefore the introduction of a manually handled Excel tool that unites the information about the machines, the people and the key performance indicators on one Excel sheet that is digitally accessible is a first step in increasing the leanness of the whole system. A web application that unites all information automatically without needing a person to summarize the information that is digitally accessible will be a next step.

Model Consolidation. The successive conceptualizations of the components of the maturity models are now recapitulated and presented as an overall model in Fig. 6. The nature of concepts, methods and technologies influences the direction of a target vector, which in an ideal case, is aligned to a point of the systems' target configuration. The vectors' ascend thereby serves a better visualization, however the orientation has to be understood as solely dependent of the spatial and temporal location of the target point. As the latter is not achievable by implication, a continuous process of improvement (CiP) is necessary, which is represented by the distance to the target point and can be evaluated by the maturity levels. Kaizen 4.0 then describes an aggregated view to improve artifacts that integrates both Lean and Industry 4.0 to obtain the target configuration (see Fig. 1 for example aggregation levels). As a C/M/T (depending on the aggregation level) enters an artifact, it passes through the phases depicted in Fig. 5. There are several potential reasons for a C/M/T shift and the question of whether and when is also the core topic of this work as well as what makes Δt therefore the focal point of the model. As already discussed, a shift may either arise from saturated process optimizations pulling new approaches, or from C/M/T pushed towards apparent inefficiencies. For the assessment of single C/M/T-elements concerning their integration in the overall system configuration and to determine their potential for an achievement of objectives, leanness plays an important role. Deviations within the trajectory may occur, caused by temporary target configurations, either requiring interim solutions or economic optima. Synergies and interdependence across multiple artefacts are shown in between the maturity levels, as their influence directly relates to adapting, facilitating, or interlocking of adjoining capabilities.

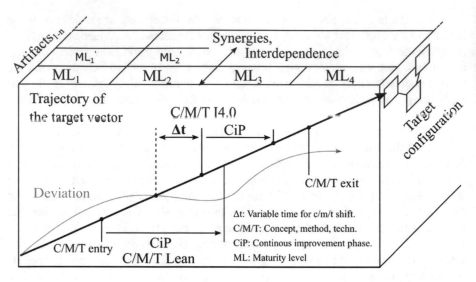

Fig. 6. Schematic diagram of the functionality of the paradigm integrated capability maturity model

3.2   Procedure Model Development

Responding to the maturity model for an identification of the need for action, beyond the scope of Lean production and the evaluation of potential for improvement, a procedure model has to be developed, which enables the 'how' within a sustainable transformation towards of a new paradigm. Since the operationalization of both Paradigms is of highest perceptibility and manageability, presupposed principles and concepts are often neglected. This may also be a reason of the matter's maturity itself and a previously missing long-term strategic approach. At the operational level, however, the visibility of methods and technologies is just as high as the demand for improvements that lead to problematic (limited) approaches:

Problem of the Unidirectional Approach. Either lean principles or Industry 4.0 principles manifesting in a range of lean development technologies, or a boost of qualified Industry 4.0 technologies to lean methods and tools. Even if the initial paradigm is necessarily considered (see above), this can ultimately lead to contradictions with the opposite paradigm. An example would be the big data-driven establishment of redundant data lakes distributed across a company. Although each may later be kept as lean as possible for the use cases it serves, a missing comprehensive architecture (also caused by the low Industry 4.0 principles' maturity), will violate the mature lean principle of waste avoidance.

Problem of the Bottom-Up, Least Common Denominator Approach. Paradigm synthesis initiated on technology/process level based on "best match" decisions. Even though a technology induced reengineering of lean is performed, disregarding the paradigms' methods and principles, may lead to problems on the long term oriented, strategic level. The prime example and source of this paper, are fast digitization gains with technologies with little potential for development and thus outdated in a long-term Industry 4.0 strategy.

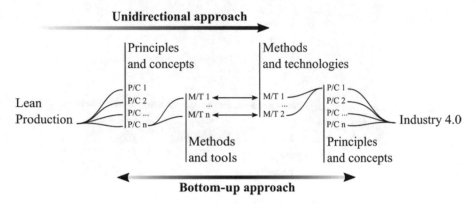

Fig. 7. Unidirectional and bottom-up approach as a result of operational visibility of methods and technologies

The problems arising from (I) and (II) are shown in Fig. 7 and demonstrate the need of a holistic view of both paradigms in the development of a procedure model. The conceptualization therefore refers to the holistic perspective of established Enterprise Architecture Frameworks (EAF) and incorporates the accomplishments of Henderson's Strategic Alignment Model [46]. The latter requires the strategic adaption of business and IT strategy to processes and infrastructure as well as the functional integration of IT and business. The corresponding top-down approach starts with a synchronization at the principle level and enables further synchronization on the successive levels of methods and technologies (see EAF business-, application- and technology view). At the principle level (resp. concept level), supra-principles such as the elimination of waste can remain unchallenged, while sub-principles compete. The resulting consensus allows synchronization at the method level without methods that use outdated principles. Conducted methods that exist on both sides, however, need to be re-validated, possibly combined or discarded if outdated. Technology or tools must then essentially serve the consolidated method base and must be evaluated under the considerations given with the maturity model (e.g. trajectory). Figure 8 outlines the explanations on the holistic procedure model. The completion of each process step within the procedure model demands an evaluation under the perspectives given by the maturity model. Both models are therefore seen as strongly interwoven.

3.3   Limitation

As the complexity of the developed framework increases with the number of considered C/M/T's as well as the target configuration, a further investigation of complexity reducing methods has to be carried out. Although the framework can be fully applied

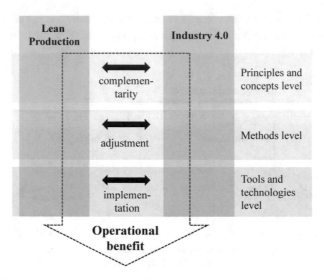

Fig. 8. Procedure Model for the lean - industry 4.0 transformation

less complex subsystems, such as production sites, departments or zones, an integration into the overall system is from an economic perspective obligatory, especially with regard to cross-artifact interdependencies. An essential approach is already found in the principles of architectural frameworks and includes the standardization, multiple/re(-) use of artefacts from a repository. A second approach is the application of decision support methods dedicated to highly cross-linked systems: Design Structure- and Multiple Domain Matrix (continuative to this, see [37, 38]).

4   Conclusion and Prospects

Building on the work of [27] and [25] a model based approach is presented, which contributes to a complexity reduction for challenging decision-making at transformation interfaces from Lean to digitization and Industry 4.0.

Perspectives given by the construction of the maturity model allow considerations of the overall system- and target configuration at local/temporal operational decisions. Based on the developed evaluation criteria, the decision-making process can be deployed user-friendly. The model recognizes Lean, in particular the standardization of processes, as the basis for Industry 4.0. With regard to open questions concerning the compatibility of Lean and Industry 4.0 concerning e.g. decision processes, people centricity versus decentralised responsibility by the use of artificial intelligence, further empirical validations of the model must be carried out. A methodologically integrated application and the validation of the model, in particular the operational evaluation criteria, will be subject to further examination and will be published in following papers. Although the development of the framework was dedicated to the integration of the Lean and Industry 4.0 paradigms, its generic structure also allows for its application on transformation processes without a concrete paradigm reference. Project, technology, and change management are mentioned here exemplary at different levels. A parameterization or a further mathematical description of the model would allow use for simulations. For this, the vector character of the model has to be elaborated. Further investigations should also focus on the organizational and management perspective of the subject matter in terms of employee centricity, as a success factor of the lean philosophy and a major bias against digitization and Industry 4.0. In addition, the ambidexterity of Lean and Industry 4.0 needs to be explored. With Industry 4.0 requiring standardized processes as a basis for sensible crosslinking and automation, on the one hand, as well as Industry 4.0 supporting leanness by gathering process data to enhance the transparency within the production system, on the other hand, the chicken-and-egg problem becomes obvious. Thus, a further aspect that needs to be explored within this transformational process is if Lean remains as a basis of a value added production systems and potentially only the tools are changing from e.g. Ishikawa Diagrams to advanced analytics or from Kanban to intelligent carriers.

5 Future Work

The research carried out in this article is based on the analysis of operational digitization measures in the automotive industry. It became clear that the digitization of similar (lean production) processes using different methods and technologies was carried out in discrete departments. This leads to a high variance of the performance and integrability of these methods and technologies. This fact is included in the descriptive aspects (e.g. dependencies) of our maturity model. For subsequent research, the performance and integrability of these already implemented artifacts will be assessed using a set of operational a priori evaluation criteria. Subsequent validation can underpin and extend the process model proposed in this paper with an operational decision support component. As the advancement of digitization in industry is a major problem and, not least, it often affects traditional lean production structures, we also encourage other researchers to explore this field of observation in future research.

References

1. Dombrowski, U., Richter, T., Krenkel, P.: Interdependencies of Industrie 4.0 & lean production systems. A use cases analysis. Procedia Manuf. **11**, 1061–1068 (2017)
2. Womack, J.P., Jones, D.T.: Lean Thinking. Ballast abwerfen, Unternehmensgewinn steigern, 1st edn. Campus Verlag (2004)
3. Rosário Cabrita, M.d., Duarte, S., Carvalho, H., Cruz-Machado, V.: Integration of lean, agile, resilient and green paradigms in a business model perspective: theoretical foundations. IFAC-PapersOnLine **49**, 1306–1311 (2016)
4. Shah, R., Ward, P.T.: Defining and developing measures of lean production. J. Oper. Manag. **25**, 785–805 (2007)
5. Erlach, K.: Value Stream Design. Springer, Heidelberg (2013)
6. Dickmann, P.: Schlanker Materialfluss. Mit Lean Production, Kanban und Innovationen. VDI-Buch (2007)

7. Vyatkin, V., Salcic, Z., Roop, P., Fitzgerald, J.: Now that's smart! EEE Ind. Electron. Mag. **1**, 17–29 (2007)
8. Hines, P., Holweg, M., Rich, N.: Learning to evolve: a review of contemporary lean thinking. Int J. Oper. Prod. Manag. **24**, 994–1011 (2004)
9. Kolberg, D., Zühlke, D.: Lean automation enabled by Industry 4.0 technologies. IFAC-PapersOnLine **48**, 1870–1875 (2015)
10. Bitkom e.V., VDMA e.V., ZVEI e.V.: Implementation strategy Industrie 4.0. Report on the results of the Industrie 4.0 platform (2016)
11. Roy, D., Mittag, P., Baumeister, M.: Industrie 4.0 - Einfluss der Digitalisierung auf die fünf Lean-Prinzipien. Prod. Manag. (2), 27–30 (2015)
12. Quasdorff, O., Bracht, U.: Die Lean Factory. Basis für den Erfolg von Digitaler Fabrik und Industrie 4.0. ZWF **111**(12), 843–846 (2016)
13. Ganschar, O., Gerlach, S., Hämmerle, M., Krause, T., Schlund, S.: Produktionsarbeit der Zukunft - Industrie 4.0. Fraunhofer Verlag (2013)
14. Wagner, T., Herrmann, C., Thiede, S.: Industry 4.0 impacts on lean production systems. Procedia CIRP **63**, 125–131 (2017)
15. Sanders, A., Elangeswaran, C., Wulfsberg, J.: Industry 4.0 implies lean manufacturing. Research activities in industry 4.0 function as enablers for lean manufacturing. JIEM (2016)
16. Metternich, J., Müller, M., Meudt, T., Schaede, C.: Lean 4.0 – zwischen Widerspruch und Vision. ZWF **112**, 612–615 (2017)
17. Baena, F., Guarin, A., Mora, J., Sauza, J., Retat, S.: Learning factory. The path to Industry 4.0. Procedia Manuf. **9**, 73–80 (2017)
18. Erol, S., Schuhmacher, A., Sihn, W.: Strategic guidance towards Industry 4.0. A three-stage process model. In: International Conference on Competitive Manufacturing COMA, Stellenbosch University, South Africa, 27–29 January 2016, pp. 495–500 (2016)
19. Morlock, F., Wienbruch, T., Leineweber, S., Kreimeier, D., Kuhlenkötter, B.: Industrie 4.0-Transformation für produzierende Unternehmen. ZWF (2016)
20. Nightingale, D., Mize, J.H.: Development of a lean enterprise transformation maturity model. Inf. Knowl. Syst. Manag. **3**, 15–30 (2002)
21. Maasouman, M.A., Demirli, K.: Assessment of lean maturity level in manufacturing cells. IFAC-PapersOnLine **28**, 1876–1881 (2015)
22. Leyh, C., Schäffer, T., Bley, K., Forstenhäusler, S.: SIMMI 4.0 – a maturity model for classifying the enterprise-wide IT and software landscape focusing on Industry 4.0. In: Federated Conference on Computer Science and Information Systems, 11–14 September 2016, pp. 1297–1302. IEEE (2016)
23. Meudt, T., Metternich, J., Abele, E.: Value stream mapping 4.0. Holistic examination of value stream and information logistics in production. CIRP Ann. **66**, 413–416 (2017)
24. Rauch, E., Dallasega, P., Matt, D.T.: The way from lean product development (LPD) to smart product development (SPD). Procedia CIRP **50**, 26–31 (2016)
25. Schuh, G., Anderl, R., Gausemeier, J., ten Hompel, M., Wahlster, W.: Industrie 4.0 Maturity Index. Managing the Digital Transformation of Companies (acatech STUDY). Herbert Utz Verlag, Munic (2017)
26. Burggräf, P., Dannapfel, M., Voet, H., Bök, P.-B., Uelpenich, J., Hoppe, J.: Digital transformation of lean production: systematic approach for the determination of digitally pervasive value chains (2017)
27. Spath, D., Schlund, S., Pokorni, B., Berthold, M.: Produktionsassessment 4.0. Integrierte Bewertung variantenreicher Einzel- und Kleinserienfertigung in den Bereichen Lean Management und Industrie 4.0. In: Koether, R. (ed.) Lean Production fur die variantenreiche Einzelfertigung. Flexibilitat wird zum neuen Standard, 1st edn., pp. 45–68. Springer Gabler (2017)

28. Becker, J., Knackstedt, R., Pöppelbuß, J.: Developing maturity models for IT management. Bus. Inf. Syst, Eng (2009)
29. Josey, A.: The TOGAF® Standard, Version 9.2. A pocket guide. TOGAF series. The Open Group, Berkshire (2018)
30. Sowa, J.F., Zachman, J.A.: Extending and formalizing the framework for information systems architecture. IBM Syst. J. **31**, 590–616 (1992)
31. Zachman, J.A.: A framework for information systems architecture. IBM Syst. J. **26**, 276 (1987)
32. de Bruin, T., Freeze, R., Kaulkarni, U., Rosemann, M.: Understanding the main phases of developing a maturity assessment model. In: Campbell, B. (ed.) Proceedings of the 16th Australasian Conference on Information Systems (ACIS 2005), 30 November–2 December 2005, Sydney. Australasian Chapter of the Association for Information Systems, Sydney (2005)
33. Karlsson, C., Åhlström, P.: Assessing changes towards lean production. Int. J. Oper. Prod. Manag. **16**, 24–41 (1996)
34. Hevner, A.R., March, S.T., Park, J., Ram, S.: Design science research in information systems research. MIS Q. **28**(1), 75–105 (2004)
35. Simon, H.A.: The Sciences of the Artificial. MIT Press, Cambridge (1996)
36. Eversheim, W. (ed.): Innovation Management for Technical Products. Systematic and Integrated Product Development and Production Planning. RWTHedition. Springer, Berlin (2009)
37. Eppinger, S.D.: Design Structure Matrix Methods and Applications (Engineering Systems). MIT Press, Cambridge (2016)
38. Bartolomei, J.E., Hastings, D.E., de Neufville, R., Rhodes, D.H.: Engineering systems multiple-domain matrix. An organizing framework for modeling large-scale complex systems. Syst. Eng. **15**, 41–61 (2012)
39. Christensen, C.M.: The Innovator's Dilemma. The Management of Innovation and Change Series. Harvard Business School Press, Boston (1999)
40. Kostka, C., Kostka, S.: Der kontinuierliche Verbesserungsprozess. Methoden des KVP, 6th edn. Pocket-Power (2013)
41. Atkinson, R.: Project management: cost, time and quality, two best guesses and a phenomenon, its time to accept other success criteria. Int. J. Project Manag. **17**, 337–342 (1999)
42. Pil, F.K., Fujimoto, T.: Lean and reflective production: the dynamic nature of production models. Int. J. Prod. Res. **45**, 3741–3761 (2007)
43. Poppendieck, M., Poppendieck, T.: Lean Software Development. An Agile Toolkit. The Agile Software Development Series (2003)
44. Davis, F.D.: Perceived usefulness, perceived ease of use, and user acceptance of information technology. MIS Q. **13**, 319–340 (1989)
45. Venkatesh, V., Davis, F.D.: A theoretical extension of the technology acceptance model: four longitudinal field studies. Manag. Sci. **46**, 186–204 (2000)
46. Henderson, J.C., Venkatraman, H.: Strategic alignment. Leveraging information technology for transforming organizations. IBM Syst. J. **32**, 472–484 (1993)

Energy Efficient Power Management Modes for Smartphone Battery Life Conservation

Evelyn Sowells-Boone[✉], Rushit Dave, Brinta Chowdhury, and DeWayne Brown

North Carolina A&T State University, Greensboro, NC 27411, USA
sowells@ncat.edu

Abstract. Smartphone usage has dramatically increased over the past 10 years. This is due in part to the increased functionality of the portable device. Smartphone users depend on the instant access to limitless information made available by using this portable device. The downside to this growing trend is the smartphone's limitation of battery power. To address this growing concern, designers implement smartphone power management techniques which can be hardware-based or software-based optimizations. We have chosen software level optimizations because these techniques are more robust. This paper introduces a novel customizable power management scheme that prioritizes smartphone application access based on the users' usage profile to conserve battery life. The results are more than promising because the user's power savings increase as the amount of time the smartphone operates in the customized power management mode increases.

Keywords: Smartphone · Power management · Energy efficiency · Low power design · Applications

1 Introduction

Energy conservation has become an increasingly important issue among modern digital circuit designers. As the evolution of digital technology takes us into the 21st century coupled with ground breaking system performance, the power consumed by these circuits are at record highs. In fact, power dissipation or energy loss in the form of heat is reaching levels comparable to nuclear reactors. The negative affect associated with the power dissipation compromises or in many cases, impair chip reliability and life expectancy [1].

A portable electronic device is characterized as a battery-powered device weighing less than two pounds. Tables, e-readers and headphones are considered portable electronic devices but the smartphone is by far the most popular. The advancements of smartphone functionalities have allowed it to become an indispensable part of our daily life. In fact, smartphones are replacing personal computers. From 2007–2017, the number of smartphones sold worldwide was 1.54 billion and it is projected that by 2020 this number will be 1.7 billion [2]. By 2021, 40% of the world population is predicted to own a smartphone [3]. If you closely examine the features incorporated in the smartphones, you will notice that they use RAMs comparable to computers, the

K. Arai et al. (Eds.): FTC 2019, AISC 1070, pp. 710–716, 2020.
https://doi.org/10.1007/978-3-030-32523-7_53

ROMs is in the scale of Gigabytes and they use a variety of sensors. Additionally, by the end of the first quarter of 2018, smartphone users have the options of choose among 3.8 million applications in the marketplace like Google Play and Apple application store respectively [4]. The downside to this growing trend of increased usage is the smartphone's limitation of battery power. Currently, smartphone users are faced with limited battery lifetime or limited battery interaction which is an inverse process. The more we use the smartphones, the less battery life our smartphones have.

In this paper, we present a novel approach to address this growing concern of conserving battery life. There has been several hardware-based and software-based optimization techniques introduced over the years to combat this growing concern. Hardware level modifications have been highly successful. However, research has proven that techniques that are effective for one smartphone architecture may be ineffective for different architecture [5]. Software-based optimization techniques are more robust which removes platform restrictions. Smartphone applications, a key area for software-based optimizations, are the focus of this paper.

Researcher and design engineers have developed various methods for optimizing power usage in smartphones. To check the battery consumption in smartphones, authors in [6] developed eDoctor and they found that 47.9% issues causes battery drain are related to applications. To address this problem researchers of [7] developed android mechanism named TAMER to control the background task that causes wakeup calls, wake-locks, broad cast receivers and service invocations. Researchers also proposed ways to mitigate the trade-off between display density and power consumption. In [8], they proposed Dynamic Resolution Scaling which adjust user-interfaced resolution on the screen based on the viewing distance while keeping user experience unaffected. They developed a hardware-based prototype, but they may have an issue while integrating with smartphones with different architecture. Our research has discovered a technique that uses artificial intelligence to conserve battery power.

The demand for energy efficient applications are increasing daily. Presently, more ground-breaking smartphone applications with power hungry advances like GPS, 5G and Wi-Fi are being developed and used globally. These applications require more power than standard applications like messaging and email services. Millennials frequently utilize video streaming applications like FaceTime and YouTube which quickly drain the battery's power. However, if the user realizes how their smartphone's battery is used, they can implement a power saving technique to extend battery life.

There are also a significant number of applications that are embedded in Android smartphones which further increase battery power drainage. Many of these embedded applications are installed for two reasons: Standard functionality such as messaging, email, calendar and maps or as it very well may be found in consent of the operation system. Most of the end clients experience issues figuring out which applications are more power efficient. Thus, application designers have a motivation to create efficient smartphone programming. Their primary obstacle is deciding the effect of programming plan choices on battery power efficiency; however that challenge can be overcome.

The limit of smartphone battery is constrained due to the size and weight of the battery. Therefore, the energy optimization of the smartphone is basic for their usefulness. Currently, smartphones have numerous functionalities in addition to making telephone calls. They act as cameras, video recorders, and GPS's just to name a few. This further lends itself to battery power efficiency for smartphone vitality proficiency. Customarily, cellphones that do not utilize any intelligent applications can operate for a few days without charging. However, when utilized effectively using applications, a smartphone may operate about 24 h without charging the battery. Smartphones require more energy than traditional cellphones regardless of whether intelligent applications are seldom utilized. Although cellphone battery technology has advanced, the life span of the smartphone battery is shorter in contrast to a standard cellphone.

To combat this growing concern of power utilization optimizations for smartphones, we propose customizable power management modes to prioritize smartphone application access based on the users' needs to conserve battery life. Although Android and Apple have introduced generic low power modes and Doze settings, they do not consider an individual's personal smartphone usage profile and customize application management dynamically to best fit the user's needs. Frequent Interactive Training (FIT) allows the user to define the priority level of applications; thus, introducing their unique prioritized low power modes.

2 Project Description

The goal of FIT is to proactively extend smartphone battery life. While most commercial power conservation modes are activated when the battery has reached less than 20% of its power, FIT is designed to allow the user to predefine how power is always managed and can be updated at any given time. Traditional settings' features allow users to access applications installed on a smartphone to set notifications rules, location services access and background refresh level of usage. These features have introduced some level of customization for power management. However, the complexity involved with making these changes for an extensive amount of installed applications can be daunting. This can be particularly challenging for some demographics that are not as technologically savvy.

FIT is simple. When you download an application to your smartphone, the application is assigned to select power modes based on the user's response to control questions and daily calendar schedule. The series of questions allow the user to create three levels of power conservation and have these different levels automatically implemented during different times of the day and/or according to the user's calendar schedule. The power management modes are:

- Power management mode 1- Ultra-low power, Do Not Disturb (DND): Only priority level 1 applications can use system resources and select contacts may communicate via text or call. All other applications are suspended.
- Power management mode 2- Low power, Limited Access (LA): Priority levels 1 and 2 applications can use system resources and contacts may communicate via text or call. No Bluetooth, locations services or streaming.
- Power management mode 3, All Access (AA): Priority level 1, 2, and 3 applications can use system resources and all calls, text, notifications may communicate. No application limitation.

These power management modes coupled with user defined power mode scheduling creates a novel approach to smartphone battery conservation. Figure 1 illustrates a daily activity scheduler for customization of smartphone application system resource access. To date, there is no literature or applications which can autonomously adjust power management modes according to users' daily routine.

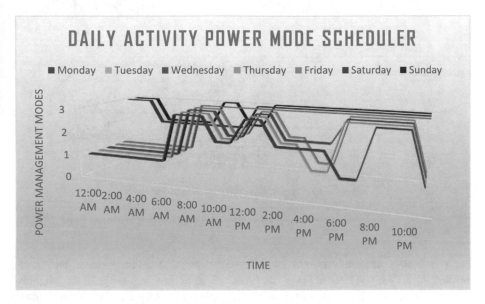

Fig. 1. Daily activity power mode scheduler

2.1 Features

Our application is synchronized with users' email account and calendars. Then chooses power mode based upon the user's calendar schedule.

The selection of power saving modes are determined by the users' daily schedule. For example, during the time a student is in a class or an important meeting, services like Wi-Fi, LTE mobile data, Bluetooth, location service, or any other power consuming running application can be disabled to save battery power. The FIT application can work on these afore mentioned services whenever the user's calendar has any class, meetings or any other occasion that the user is not actively using the smartphone (see Fig. 2).

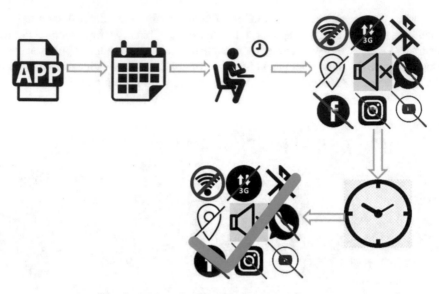

Fig. 2. FIT application automated decision steps

3 Results

For our testing and analysis, we used TracFone android smart phones. The model number is STALA502DCP. Some features which it has are as follows Memory-2 GB/16 ROM and SD support up to 32 GB, quad-core A53 1.1 GHz with Android 8.1 Oreo, Display- size-5.43 inch, and Battery- standby(3G): 547 h.

Table 1. Battery power saving while using FIT

Day of the week	Power usage	Power savings
Monday	63%	37%
Tuesday	66%	34%
Wednesday	69%	31%
Thursday	66%	34%
Friday	80%	20%
Saturday	88%	12%
Sunday	90%	10%
Average weekly total	75%	**25%**

When using the activity schedule shown for one week or 168 h in Fig. 1, we have found substantial power savings. Note that during 80 of those hours, the smartphone is operating in the low power management mode 2 (Limited Access) or ultra-low power management mode 1 (Do Not Disturb) yielding daily battery power savings of up to 37% shown in Table 1. The weekly savings are impressive as well with an average weekly battery power saving of 25%. Since this customizable approach is unique for each user, the power saving results are unique as well. The results are more than promising because the user's power savings increase as the amount of time the smartphone operates in the customized low or ultra-low power management mode increases.

4 Conclusion

The main goal of this project is to increase battery life efficiency by adjusting the smartphone's power usage automatically using FIT power management settings. Our research is still works in progress but demonstrates proof of concept that we can create a smartphone that can independently operate and save power. Our preliminary results shown above are more than promising. Modern research is fueled by the need to make smart devices smarter and make life easier with smarter cars, smarter homes, smarter grids and so forth. Our research is one step in the evolution of smart technologies and the IoT. On the other hand, the slow advancements in battery efficiency technology is one of the biggest concerns. Our research addresses the two problems together: one is saving power, and another is giving intelligence to the portable devices.

This project advances knowledge in energy efficient electronic system design and implementation and will be introduced in the low power design chapter of an electronic technology digital system design undergraduate course. The project's outcome contributes to the state of the art in digital system design through understanding how efficient power system design strategies improve portable device performance and reliability, thereby leading to a national model tailored toward power aware electronic system designs. The focus of this research is the development of power-aware algorithms and techniques for electronic system optimization. Other research areas benefiting from this sort of work include high performance computing and computational science.

References

1. Sowells, E., Seay, C., Brown, D.: A novel technique for low power electronic system design. In: The Proceedings of the American Society of Engineering Education 123rd Annual Conference, New Orleans, LA, 26–29 June 2016. Paper #14820. https://peer.asee.org/26385
2. Number of smartphones sold to end users worldwide from 2007 to 2017 (in million units)' (2018), Statista. https://www.statista.com/statistics/263437/global-smartphone-sales-toend-users-since-2007/
3. Smartphones industry: Statistics & Facts' (2018). Statista. https://www.statista.com/topics/840/smartphones/
4. Number of apps available in leading app stores as of 3rd quarter 2018' (2018). Statista. https://www.statista.com/statistics/276623/number-of-apps-availablein-leading-app-stores/
5. Zaman, N., Almusalli, F.A., Review: smartphones power consumption and energy saving techniques. In: Proceedings of 2017 International Conference on Innovations in Electrical Engineering and Computational Technologies, pp 1–7. IEEE (2017)

6. Ma, X., Huang, P., Jin, X., Wang, P., Park, S., Shen, D., Zhou, Y., Saul, L.K., Voelker, G.M.: Edoctor: automatically diagnosing abnormal battery drain issues on smartphones. In: NSDI 2013, vol. 13, pp. 57–70, 2 April 2013
7. Martins, M., Cappos, J., Fonseca, R.: Selectively taming background android apps to improve battery lifetime. In: USENIX Annual Technical Conference, pp. 563–575, 8 July 2015
8. He, S., Liu, Y., Zhou, H.: Optimizing smartphone power consumption through dynamic resolution scaling. In: Proceedings of the 21st Annual International Conference on Mobile Computing and Networking, pp. 27–39, 7 September 2015. ACM (2015)

Providing Some Minimum Guarantee for Real-Time Secondary Users in Cognitive Radio Sensor Networks

Changa Andrew[1], Tonny Bulega[1], and Michael Okopa[2(✉)]

[1] Makerere University, Kampala, Uganda
achanga85@gmail.com, tbulega@cit.mak.ac.ug
[2] Faculty of Science and Technology, Cavendish University Uganda,
Kampala, Uganda
mokopa@cavendish.ac.ug

Abstract. Previous studies of real-time traffic in Cognitive Radio Sensor Network (CRSN) were done using reservation-based method and absolute priority method. These studies were done under the assumption that primary user's traffic arrive at a low rate. However, under high arrival rate of primary user traffic, real-time secondary user traffic is starved. In this paper, we developed an analytical model of source-to-sink delay in a Cognitive radio sensor network (CRSN) that differentiates the service of users into primary and secondary and further partitions secondary user's traffic into real-time and best effort. During service, primary traffic is given priority over real-time secondary user traffic which in turn has higher priority over best effort secondary traffic. Furthermore, a threshold is introduced on the number of primary user packets served to reduce the starvation of real-time secondary users under high arrival rate of primary user traffic. The numerical results obtained from the derived models show that real-time secondary users experience a reduction in source-to-sink delay as a result of introducing a threshold on the number of primary user packets served without over degrading the service of primary user traffic.

Keywords: Absolute priority · Best effort · Real-time · Source-to-sink delay

1 Introduction

Wireless sensor networks (WSNs) are special networks with large number of nodes consisting of embedded processors, sensors and radios. A typical sensor network consists of a large number of sensor nodes deployed either inside the phenomenon of interest or close to it. According to Romer *et al.* [1] a wireless sensor network is an ad hoc multi-hop network which consists of large number of tiny homogeneous sensor nodes that are resource-constrained, mostly immobile and randomly deployed in the area of interest.

© Springer Nature Switzerland AG 2020
K. Arai et al. (Eds.): FTC 2019, AISC 1070, pp. 717–730, 2020.
https://doi.org/10.1007/978-3-030-32523-7_54

The primary purpose of sensor networks is to provide users access to information gathered by the spatially distributed sensors, rather than enabling end-to-end communication between pairs of nodes as in other large-scale networks such as the Internet or wireless mesh networks.

Sensor nodes can collaborate on real-time monitoring, sensing, collecting network distribution of the various environments within the region or monitoring object information [?] Due to limited transmission range of sensor nodes, the sensory data are delivered to a processing center, called sink node, via multi-hop communication meaning that each sensor node plays the dual role of being a data originator and a data router. This information is then processed to obtain useful data, which is then sent to the user [3].

To reduce delay in data delivery that happens through multi-hop communication, sensor nodes form clusters and a cluster head is selected for each cluster [4]. Instead of the data being relayed by every node along a particular path, it is passed from the source node to the cluster head, hence reducing the number of nodes through which data is relayed.

WSNs work in the license-free band, hence vulnerable to suffer from heavy interference caused by other networks sharing the same spectrum [2,3]. To overcome this challenge, cognitive radio network has been suggested as a promising approach to reduce interference [5]. Cognitive radio (CR) networks provide efficient utilization of the radio spectrum and highly reliable communication to users whenever and wherever needed. In the event that the licensed owner also referred to as primary user (PU) is not utilizing the channel, it can be used by the unlicensed user also called secondary user (SU) without jeopardizing the service quality of the primary user [6].

Combining WSN and cognitive radio can overcome the challenge of providing users access to information as well as reducing interference caused by other networks sharing the same spectrum.

Providing minimum guarantee in cognitive radio sensor networks (CRSNs) is of great importance for example in health care, a packet indicating an abnormal event of a patient should reach the doctor as soon as possible. In environmental monitoring, a wireless smoke sensor should provide real-time recognition of the smoke.

Two methods have been proposed for prioritizing the real-time traffic over the best effort (BE) traffic in the Cognitive Radio Sensor Network (CRSN), that is reservation-based method [7] and an absolute priority method [8,9]. In the reservation-based method [8], each type of traffic can only be served when a channel is available during the pre-allocated time intervals. On the other hand, in the absolute priority method [9], the real-time packets can be served whenever a channel is available, and the BE packets can only be served when there is no buffered real-time packets. However, under high arrival rate of primary user packets, real-time secondary user packets are starved.

In this paper, we propose an analytical model that differentiates the service of users while at the same time delaying the primary user packets so as to serve real-time secondary user in a CRSN based on absolute priority.

The model provides minimum guarantee to real-time secondary user packets under high arrival rate of primary user packets. The study further evaluates how the proposed model improves the performance of real-time secondary user packets without appreciably degrading the performance of primary user packets.

The rest of the paper is organized as follows: in the next section, we present the system models. We evaluate the proposed models in Sect. 3, and finally conclude the paper in Sect. 4.

2   System Models

We consider a cognitive radio sensor network where the available frequency channels are assumed to be error-free. Co-channel interference between different clusters is avoided in two ways. First, a different set of candidate channels is assigned to neighboring clusters if the number of candidate channels is sufficiently large. Secondly, if neighboring clusters have to share the same set of candidate channels, their cluster heads (CHs) may sense the channels in different orders so that they find different available channels with a high probability. If neighboring clusters have to share the same frequency channel, simultaneous transmissions can be avoided by carefully coordinating the timelines of the clusters using similar models as in [10,11].

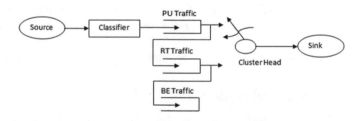

Fig. 1. Single-cluster queue model

Figures 1 and 2 shows the single-cluster queue model and multi-cluster queue model respectively. For the case of single-cluster, traffic traverses only one cluster head before reaching the sink, while for the case of multicluster, traffic traverses two or more cluster heads before reaching the sink.

In multi-cluster transmissions, both real-time and best effort data collected by the CH from the sensors are forwarded to the next hop CH and further to the sink. The CRSN opportunistically accesses vacant channels in a spectrum. Each cluster requires only one available frequency channel at any time due to the fact that the CH has only one radio for data communications. The CRSN is assumed to have primary user which has one type of traffic, whereas the secondary user has both real-time and best effort traffic. In order to keep short the source-to-sink delay, real-time traffic collected by the CH will be given priority over best effort traffic. We assume the prioritization of real-time traffic is only done at the

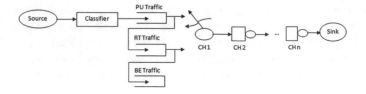

Fig. 2. Multi-cluster queue model.

first cluster-head and the rest of the cluster-heads perform only the forwarding function along a particular path to the sink.

In addition to collecting data from the sensors, the CH is also responsible for sensing available channels from a number of candidate channels, allocating radio resources, and sending control signals to the sensors. The sensor nodes have to switch between different channels depending on channel availability.

The arrival of all packets to each node is assumed to follow a Poisson process with mean arrival rate, λ per node due to the random order of arrival of packets. The service time at each node is assumed to be exponentially distributed. Buffer capacity of each node is assumed to be finite. Both real-time traffic and best effort (BE) data traffic can be served, but the real-time traffic is given a higher priority. In order to achieve small transmission delay, the real-time traffic is served with contention-free transmissions using the IEEE 802.15.4 MAC protocol, which is commonly used for WSNs [7]. On the other hand, BE traffic is given some minimum guarantee to be transmitted only under high arrival rate of secondary user real-time traffic.

In the next section, we derive the expression for source-to-sink delay.

2.1 Source-to-Sink Delay Models

Source-to-sink delay consists of the following delays; transmission delay, propagation delay, and queuing delay. Of these components, queuing delay is the most difficult to model. We use Jackson's theorem [2] to model the delay experienced by the Cluster-Heads (CHs) in a multi clustered network. Jackson's theorem states that in a network of queues, each node is an independent queuing system, with a Poisson input determined by the principles of partitioning, merging, and tandem queuing. The theorem is based on three assumptions:

(i) The queuing network consists of m nodes, each of which provides an independent exponential service.
(ii) Items arriving from outside the system to any one of the nodes arrive with a Poisson rate.
(iii) Once served at a node, a packet goes to one of the other nodes with a fixed probability, or it goes out of the system.

We model each node using the $M/M/1/k$ queue system, and the mean delay at each node is then added to derive the total node delays. Due to space limitation the study only considered the delay for a single cluster.

2.2 Expression for Source-to-Sink Delay for Primary User's Packets Without Threshold

The Source-to-sink delay of primary user packets consists of the following delays: transmission delay, propagation delay, and queuing delay due to primary users.

(i) Transmission delay is given by the formula, L/R, where L is the packet length and R is the transmission rate.
(ii) The propagation delay is given by d/s, where d is the total distance between the nodes and s is the propagation speed of the link [12].
(iii) The queuing delay is made up of two components; the delay due to the service of a particular primary packet, and the waiting time in queue of a primary user packet.

The queuing delay experienced by a primary user packet in a single cluster can be modified from [13] to become:

$$\frac{1}{E[\tau_P]} + \frac{\lambda E[\tau_P^2]}{2(1 - \lambda_P/E[\tau_P])}$$

Therefore, the source-to-sink delay for primary user's packets without threshold can be deduced as:

$$E[SSD] = \frac{L}{R} + \frac{d}{s} + \left(\frac{1}{E[\tau_P]} + \frac{\lambda E[\tau_P^2]}{2(1 - \lambda_P/E[\tau_P])} \right) \tag{1}$$

2.3 Expression for Source-to-Sink Delay for Secondary User's Packets Without Threshold

The Source-to-sink delay of secondary user packets consists of the following delays: transmission delay, propagation delay, queuing delay due to primary user's packets and queuing delay due to secondary user's packets.

(i) Transmission delay is given by the formula, L/R, where L is the packet length and R is the transmission rate.
(ii) The propagation delay is given by d/s, where d is the total distance between the nodes and s is the propagation speed of the link [12].
(iii) The queueing delay due to the service of the primary user packets found in the queue.
(iv) The queueing delay due to the service of a particular secondary user packet, and the waiting time in queue of a secondary user packet.

The queuing delay experienced by the secondary user packet in a single cluster is given as:

$$\frac{1}{E[\tau_R]} + \frac{\lambda E[\tau_R^2]}{2(1 - \lambda_R/E[\tau_R])}$$

Therefore, the source-to-sink delay for real-time secondary user packet without threshold can be deduced as:

$$E[SSD] = \frac{L}{R} + \frac{d}{s} +$$

$$\sum_{i=1}^{k} \left(\frac{\lambda E[\tau_{P_i}^2]}{2(1 - \lambda_{P_i}/E[\tau_{P_i}])} \right) + \frac{1}{E[\tau_R]} + \frac{\lambda E[\tau_R^2]}{2(1 - \lambda_R/E[\tau_R])} \tag{2}$$

where k is the number of primary user packets served before serving the secondary user packets.

2.4   Source-to-Sink Delay with Threshold

Under high arrival rate of primary user packets, real-time secondary user packets can be starved. Therefore, we employ a threshold on the number of primary user packets served before serving real-time secondary user packets. In doing this, real-time secondary user packets receive minimum guarantee, that is, they can still receive service even when there are primary user packets in the queue. Next, we derive expressions for source-to-sink delay when thresholds are applied to reduce the starvation of real-time secondary user packets as a result of service of primary user packets.

2.4.1   Source-to-Sink Delay for Primary User's Packets with Threshold

The Source-to-sink delay of a primary user packet consists of the following delays: Transmission delay, propagation delay, the waiting time in queue of a primary user packet, and the waiting time due to service of real-time secondary user packets which do not have to wait for the service of other primary user packets. We consider the worst case scenario where the primary user packet finds a real-time secondary user packet in service.

Therefore, the queuing delay experienced by the primary user packet in one cluster is given as:

$$E[T_{P_Q}] = \left(\frac{\lambda E[\tau_{P_i}^2]}{2(1 - \lambda_{P_i}/E[\tau_{P_i}])} + \sum_{i=1}^{k} \frac{\lambda E[\tau_{R_i}^2]}{2(1 - \lambda_{R_i}/E[\tau_{R_i}])} \right) \tag{3}$$

where k is the number of real-time secondary user packets that have to be served before serving primary user packets.

The source-to-sink delay, E[SSD] of a primary user packet with threshold is therefore:

$$E[SSD] = \frac{L}{R} + \frac{d}{s} +$$

$$\frac{1}{E[\tau_P]} + \frac{\lambda E[\tau_P^2]}{2(1 - \lambda_P/E[\tau_P])} + \sum_{i=1}^{k} \frac{\lambda E[\tau_{R_i}^2]}{2(1 - \lambda_{R_i}/E[\tau_{R_i}])} \tag{4}$$

2.4.2 Source-to-Sink Delay for Secondary User Packets with Threshold.

In this case, the source-to-sink delay of real-time secondary user packets consist of the following delays: Transmission delay, propagation delay, the queuing delay due to some primary user packets found in queue and queuing delay due to real-time secondary user packets. The total queuing delay experienced by packets in one cluster is therefore given as:

$$E[T_{R_Q}] = \left(\sum_{i=0}^{k-(i+1)} \frac{\lambda E[\tau_{P_i}^2]}{2(1 - \lambda_{P_i}/E[\tau_{P_i}])} \right) + \frac{1}{E[\tau_R]} + \frac{\lambda E[\tau_R^2]}{2(1 - \lambda_R/E[\tau_R])} \quad (5)$$

The source-to-sink delay, E[SSD] of real-time secondary user packets with threshold is, therefore:

$$E[SSD] = \frac{L}{R} + \frac{d}{s} +$$

$$\left(\sum_{i=0}^{k-(i+1)} \frac{\lambda E[\tau_{P_i}^2]}{2(1 - \lambda_{P_i}/E[\tau_{P_i}])} \right) + \left(\frac{1}{E[\tau_R]} + \frac{\lambda E[\tau_R^2]}{2(1 - \lambda_R/E[\tau_R])} \right) \quad (6)$$

The performance of BE secondary traffic is the same, whether the threshold is applied or not. This is because BE secondary user traffic packets have to wait for the queue of primary user and real-time secondary user to be empty before receiving service.

In the next section, we present numerical results showing the performance of the source-to-sink delay models.

3 Performance Evaluation

In this section, we use the derived models to evaluate its performance. In particular, the variation of source-to-sink delay with arrival rate of packets are analyzed. In each case, we consider primary user, real-time secondary user traffic, and secondary user best effort traffic.

3.1 Model Parameters

Table 1 shows the basic mathematical symbols used in the analysis.

Table 2 shows the hypothetical parameters used in the analysis which is consistent with parameters used in literature [14].

Table 1. Basic mathematical symbols used in the analysis

Parameter	Meaning
ρ_R	Load due to Real-time secondary user packets
ρ_{BE}	Load due to best effort secondary user packets
λ_R	Arrival rate of real-time secondary user packets
λ_{BE}	Arrival rate of best effort secondary user packets
ρ	Load

Table 2. Evaluation parameters

Parameter	Value
Number of cluster heads	1
Packet inter arrival time	260 ms
Average service rate	10 packets/second
Average packet length	7.5 Mbits
Transmission rate	1.5 Mbps
Distance between two nodes	2 km
Propagation speed	$3 * 10^8$ ms^{-1}

3.1.1   Variation with Load

In this case the increase in delay of primary users and decrease in delay of real-time secondary users with increase in load is investigated. In doing this, Eqs. 1, 4 and 2 and 6 are used to plot graphs of source-to-sink delay as a function load due to primary user's packets.

Figure 3 shows a graph of source-to-sink delay for primary user as a function of primary user load when the load due to real-time secondary user and BE secondary user is fixed. It is observed that source-to-sink delay increases with increase in load due to primary user's packets. It is also observed that the higher

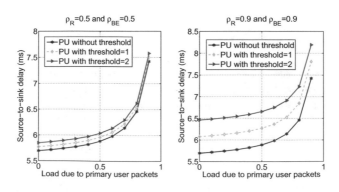

Fig. 3. Source-to-sink delay for primary user as a function of primary user load

the threshold, the higher the source-to-sink delay. For example for low load of real-time and BE secondary users, that is $\rho = 0.5$, when the load due to primary user is 0.5 the source-to-sink delay increases by 0.02 ms when the threshold is one, while when the threshold is two the source-to-sink delay increases by 0.03 ms. On the other hand, when the load of real-time and BE secondary users is put at $\rho = 0.9$ and the load due to primary user is 0.5 the source-to-sink delay increases by 0.77 ms when the threshold is one, while when the threshold is two the source-to-sink delay increases by 0.96 ms. The source-to-sink delay of primary user packets also increase with increase in load due to real-time and BE secondary user's packets.

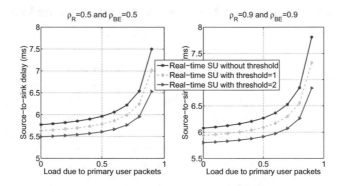

Fig. 4. Source-to-sink delay for real-time secondary user as a function of primary user load

Figure 4 shows a graph of source-to-sink delay for real-time secondary user as a function of primary user load when the load due to real-time secondary user and BE secondary user is fixed. It is observed that source-to-sink delay increases with increase in load due to primary user's packets. It is also observed that the higher the threshold, the higher the source-to-sink delay reduces. For example for low load of real-time and BE secondary users, that is $\rho = 0.5$, when the load due to primary user is 0.5 the source-to-sink delay reduces by 0.02 ms when the threshold is one, while when the threshold is two the source-to-sink delay reduces by 0.08 ms. On the other hand, when the load due to real-time and BE secondary users is put at $\rho = 0.9$ and the load due to primary user is 0.5 the source-to-sink delay reduces by 0.15 ms when the threshold is one, while when the threshold is two the source-to-sink delay reduces by 0.66 ms. The source-to-sink delay of real-time secondary user packets reduce further with increase in load due to real-time and BE secondary user's packets.

3.1.2   Variation with Arrival Rate

In this section, the increase in delay of primary user's packets and decrease in delay of real-time secondary user's packets with increase in arrival rate of primary user's packets is investigated. In doing this, Eqs. 1, 4 and 2 and 6 were used to plot graphs of source-to-sink delay as a function of arrival rate of primary user's packets.

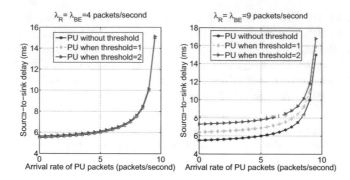

Fig. 5. Source-to-sink delay for primary user as a function of arrival rate of primary user traffic

It is observed from Fig. 5 that source-to-sink delay for primary user generally increases with increase in arrival rate of primary user traffic. It is also observed that at low arrival rate (4 packets/second) of real-time and BE secondary user's packets, the performance of primary user packets are almost the same when the threshold is applied and when the threshold is not applied. This implies that introducing a threshold when the arrival rate of primary users is low has no significant effect on the performance of the system. However, when the arrival rate of real-time and BE secondary user's packets are increased to 9 packets/second, the source-to-sink delay of primary user traffic increases by 1 ms when the threshold is one and by 2 ms when the threshold is two at the primary user arrival rate of 5 packets/second.

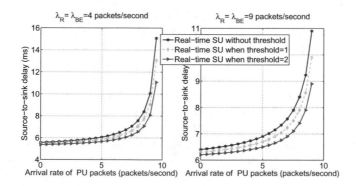

Fig. 6. Source-to-sink delay for real-time secondary user as a function of arrival rate of primary user's packets

It is observed from Fig. 6 that source-to-sink delay for real-time secondary user generally increases with increase in arrival rate of primary user packets. It is also observed that at low arrival rate (4 packets/second) of real-time and BE secondary user's packets, the reduction in source-to-sink delay is low. However, when the arrival rate of real-time and BE secondary user's packets is increased to 9 packets/second, the source-to-sink delay of real-time secondary user packets increase by 0.2 ms when the threshold is one and by 0.4 ms when the threshold is two. The reduction in source-to-sink delay is more pronounced at high arrival rate of real-time and BE secondary user packets.

3.2   Tradeoff

In this section, the tradeoff between the increase in source-to-sink delay experienced by primary user traffic and the reduction in source-to-sink delay experienced by real-time secondary users as a result of introducing a threshold is investigated. In doing this, Eqs. 4 and 6 are used to plot graphs 7 to 10.

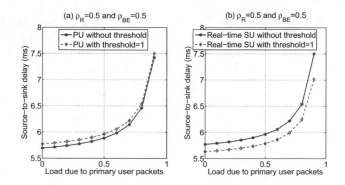

Fig. 7. Source-to-sink delay as a function of primary user load, at low secondary user load

Figure 7(a) shows a graph of source-to-sink delay for primary user as a function of primary user load, when the load due to real-time and non BE secondary user traffic are fixed at low load, $\rho = 0.5$, while Fig. 7(b) shows a graph of source-to-sink delay for real-time secondary user as a function of primary user load, when the load due to real-time and BE secondary user traffic are fixed at low load, $\rho = 0.5$. It is observed that source-to-sink delay generally increases with increase in primary user load. It is also observed that the increase in source-to-sink delay experienced by primary user is 0.1 ms at primary user load of 0.5 while the decrease in source-to-sink delay experienced by real-time secondary user is 0.3 ms at the same primary user load. Therefore, the reduction in source-to-sink delay experienced by real-time secondary user packets is more than the increase in source-to-sink delay experienced by primary user packets at low load.

Figure 8(a) shows a graph of source-to-sink delay for primary user as a function of primary user load, when the load due to real-time and BE secondary user traffic are fixed at high load, $\rho = 0.9$, while Fig. 8(b) shows a graph of source-to-sink delay for real-time secondary user as a function of primary user

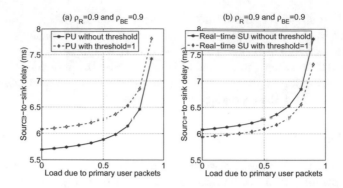

Fig. 8. Source-to-sink delay as a function of primary user load, at high secondary user load

load, when the load due to real-time and BE secondary user traffic are fixed at high load, $\rho = 0.9$. It is observed that the source-to-sink delay generally increases with increase in primary user load. It is also observed that the increase in source-to-sink delay experienced by primary user is 0.6 ms at a primary user load of 0.5 while the decrease in source-to-sink delay experienced by real-time secondary user is 0.2 ms at the same primary user load. Therefore, the reduction in source-to-sink delay experienced by real-time secondary user packets is lower than the increase in source-to-sink delay experienced by primary user packets at high load.

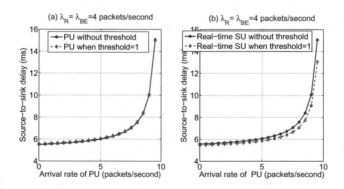

Fig. 9. Source-to-sink delay as a function of arrival rate of primary user packets

It is observed from Fig. 9(a) that for low arrival rate of real-time and BE secondary users traffic, the increase in source-to-sink delay experienced by primary users is low compared to the reduction in source-to-sink delay experienced by real-time secondary users observed in Fig. 9(b). For example, at primary user arrival rate of 5 packets/second, the source-to-sink delay experienced by primary user is the same with or without the threshold, while for real-time secondary user, there is a reduction in source-to-sink delay of 0.1 ms at the same arrival rate.

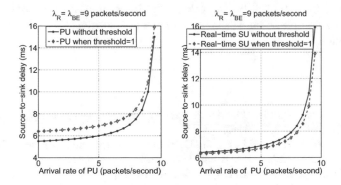

Fig. 10. Source-to-sink delay as a function of arrival rate of primary user packets

It is observed from Fig. 10 that for high arrival rate of real-time and BE secondary user's packets, the increase in source-to-sink delay experienced by primary users packets is about 0.8 ms compared to the reduction in source-to-sink delay experienced by real-time secondary user which is approximately 0.4 ms. Therefore, it can be concluded that the reduction in source-to-sink delay experienced by real-time secondary users is achieved at no cost at low arrival rate of real-time and BE secondary user traffic. However, the reduction in source-to-sink delay experienced by real-time traffic at high arrival rate of real-time and BE secondary user packets is less than the increase in source-to-sink delay experienced by primary user packets.

4 Conclusion

An analytical model of source-to-sink delay is developed for a CRSN that differentiates the service of users into primary and secondary and further partitions secondary user into real-time and BE. During service, the first priority is given to primary user, second priority to real-time secondary user and third priority is given to BE secondary user.

It is observed that real-time secondary user packets experience a reduction in source-to-sink delay as a result of introducing a threshold on the number of primary user packets served. On the other hand, the primary user packets experience a higher source-to-sink delay as a result of the introduction of the threshold. The reduction in source-to-sink delay experienced by real-time secondary user is lower than the increase in source-to-sink delay experienced by primary user traffic at high arrival rate. However, at low arrival rate of packets into the system, the reduction in source-to-sink delay experienced by real-time secondary user is higher than the increase in source-to-sink delay experienced by primary user traffic.

Therefore it can be concluded that service differentiation and inclusion of threshold can improve the performance of real-time secondary user packets at an appreciably low degradation to the primary user.

However, the potential bottleneck to the implementation of service differentiation is the computational overhead necessary to identify which type of user to give service.

References

1. Romer, K., Mattern, F.: The design space of wireless sensor networks. Proc. IEEE Wirel. Commun. **11**(6), 54–61 (2004)
2. Angrisani, L., Bertocco, M., Fortin, D., Sona, A.: Experimental study of coexistence issues between IEEE 802.11b and IEEE 802.15.4 wireless networks. Proc. IEEE Trans. Instrum. Measur. **57**(8), 1514–1523 (2008)
3. Pollin, S., Tan, I., Hodge, B., Chun, C., Bahai, A.: Harmful co-existence between 802.15.4 and 802.11: a measurement-based study. In: 3rd International Conference on Cognitive Radio Oriented Wireless Networks and Communications (Crown-Com), pp. 1–6 (2008)
4. Tang, S.: An analytical traffic flow model for cluster-based wireless sensor networks. In: Proceedings of IEEE (2006)
5. FCC: Report of the spectrum efficiency working group. FCC Spectrum Policy Task Force, Technical report, November 2002
6. Garhwal, A., Bhattacharya, P.P.: Dynamic Spectrum Access in Cognitive Radio: a brief review, pp. 149–153 (2011)
7. Liang, Z., Feng, S., Zhao, D., Shen, X.: Delay performance analysis for supporting real-time traffic in a cognitive radio sensor network. Proc. IEEE Trans. Wirel. Commun. **10**(1), 325–335 (2011)
8. Liang, Z., Zhao, D.: Quality of service performance of a cognitive radio sensor network. In: Proceedings of IEEE International Conference on Communications (ICC), Cape Town, South Africa, May 2010
9. Feng, S., Zhao, D.: Design and performance analysis of a cognitive radio sensor network. In: Proceedings of IEEE Wireless Communications and Networking Conference (WCNC), Sydney, Australia, April 2010
10. Zhang, F., Todd, T.D., Zhao, D., Kezys, V.: Power saving access points for IEEE 802.11 wireless network infrastructure. Proc. IEEE Trans. Mob. Comput. **5**(2), 144–156 (2006)
11. Zhao, D., Zou, J., Todd, T.D.: Admission control with load balancing in IEEE 802.11-based ESS mesh networks. Proc. Wirel. Netw. **13**, 351–359 (2007)
12. Kurose, J.F., Ross, K.W.: Computer Networking: A Top-Down Approach Featuring the Internet (2000)
13. Kleinrock, L.: Queueing Systems, Volume I&II. Computer Applications. John Wiley & Sons (1976)
14. Kaur, P., Khosla, A., Uddin, M.: Markovian queuing model for dynamic spectrum allocation in centralized architecture for cognitive radios. Proc. Int. J. Eng. Technol. IJET **3**(1), 96–101 (2011)

Sparse Signal Reconstruction by Batch Orthogonal Matching Pursuit

Lichun Li[✉] and Feng Wei

Information Engineering University, Zhenzhou 450000, Henan, China
llichun23@outlook.com

Abstract. A new reconstruction algorithm called BaOMP (Batch Orthogonal Matching Pursuit) for *analog-to-information conversion* (AIC) system which is used to sample the sparse bandlimited signals is proposed in this paper. The main idea of the new algorithm is to find a batch of possible non-zero coefficients at each iteration step. This idea may sound like StOMP, but the way to find the non-zero coefficients is quite different from both StOMP and OMP. The residuals and the squared numerical characteristics got from sampled signal are used to find the most possible non-zero coefficients at each iteration. The threshold for the judgement is derived in this paper. Compared with StOMP which derives the threshold based on the supposition that the sensing matrix is 'typical'/'random', BaOMP runs well no matter the sensing matrix is typical or not. This is important for the reconstruction from the compressed measurements sampled by AIC system which means the sensing matrix may not be typical as StOMP has supposed. After the locating of the non-zero coefficients, the value of the non-zero coefficients can be solved. The analysis and the computational experiments show that this approach can not only recover the signal with high probability but also be significantly faster (in terms of computation time) than competitive methods.

Keywords: Compressed sensing · Convex optimization · Sparse reconstruction · Chi-squared distribution · Adaptive threshold

1 Introduction

For compressed sensing, the efficient and accurate reconstruction algorithm is always a hot topic for researchers. The process of taking compressive measurements in a noisy environment can be modeled as

$$\mathbf{y} = \mathbf{A}\mathbf{x} + \mathbf{n} \tag{1}$$

where $\mathbf{A}$ is called "sensing matrix" which is $m \times N$ matrix, $m < N$ and it is normalized that $\sum_{j=1}^{N} A_{i,j}^2 = 1$. $\mathbf{x}$ is a K-sparse N-vector which means it has at most K non-zero entries. $\mathbf{n} \in \mathbb{R}^m$ is a Gaussian white noise $\mathbf{n} \sim \aleph(0, \sigma_n^2 I)$.

© Springer Nature Switzerland AG 2020
K. Arai et al. (Eds.): FTC 2019, AISC 1070, pp. 731–744, 2020.
https://doi.org/10.1007/978-3-030-32523-7_55

The reconstruction is to find the approximation of the unknown $\mathbf{x}$ given $\mathbf{A}$ and $\mathbf{y}$. Since $m < N$, the problem seems to be ill-posed. However, if the signal is sparse, the exact reconstruction is possible under certain condition. The recovery method is to solve the convex problem.

$$\min_{\mathbf{x}} \|\mathbf{x}\|_1 \ \ subject\ to\ \|\mathbf{y} - \mathbf{Ax}\|_2^2 \leqslant \lambda \tag{2}$$

The initial work in CS solved (2) using convex relaxation methods. The most famous is basis pursuit denoising (BPDN) [1]. However, these convex programs are extremely computationally demanding when dealing with large scale signals. Therefore, lower cost iterative algorithms were developed. Up to now, there are several classes of reconstruction algorithms: (1) Convex relaxation method: such as Gradient Projection for sparse reconstruction (GPSK) [2], Basis Pursuit (BP) [1], Least Angle Regression (LARS) [3], etc. This kind of algorithm introduces relaxation factor to balance between the reconstruction error and the signal sparsity. Though the reconstruction error of the convex relaxation method is small, its complexity is too high to deal with large scale data. (2) Iterative greedy algorithm: such as the Orthogonal Matching Pursuit (OMP) [4], the Regularized OMP (ROMP) [5]and the Stagewise OMP(StOMP) [6], compressive sampling matching pursuit [7], approximate message passing [8,9], and iterative soft-thresholding [10,11], etc. This kind of algorithm reconstructs the signal by iterative greedy method and tries to find balance between the complexity and the reconstruction error. (3) Combination algorithm: This kind of algorithm including Gradient Pursuit (GP) [12], Conjugate GP (CGP) [13] and the expander graphs [14], etc. combines the above two methods.

For large scale problem, the iterative greedy algorithms are more attractive for their balance between the complexity and the accuracy. For example, StOMP has fixed number of stages and outperforms OMP in complexity. However, the threshold should be carefully chosen otherwise the algorithm will fail. Besides, for those sensing matrix which may not be typical random, the algorithm will fail for the setting of the threshold based on the proposal of 'typical'/'random' sensing matrix. The BaOMP proposed in this paper develops a new filter to choose the indices which are related to the non-zero coefficients and derives the threshold with less constraints to the sensing matrix. Though BaOMP does not has fixed number of stages, its total iteration number is less than one fifth of OMP according to the numerical experiments. The reason is that at each iteration stage a batch of indices is chosen while OMP chooses only one index at each stage. The robustness of BaOMP is nearly the same as OMP which means it can be used more widely than StOMP.

The main contributions of this paper are as follows:
(1) The squared numeral characteristics are proposed to filter the most possible non-zero coefficients indices. These numeral characteristics contain the non-negative linear combination of the squared non-zero coefficients which makes the discrimination of the active indices from those related to zero easy.
(2) A set of adaptive thresholds for each stage are derived. By setting adaptive thresholds, the algorithm can find a batch of most possible indices at each stage

which will accelerate the convergence of the searching procedure. Notations used in this paper are as follows. Boldface letters are reserved for vectors and matrices. $\mathbb{R}$ and $\mathbb{C}$ denote the sets of real and complex numbers, respectively. $|\cdot|$ denotes the amplitude of a scalar or cardinality of a set. $\|\cdot\|_{\ell_0}, \|\cdot\|_{\ell_1}$ and $\|\cdot\|_{\ell_2}$ denote the ℓ_0, ℓ_1 and ℓ_2 norms respectively. $\mathbf{A}^T$ and $\mathbf{A}^H$ are the matrix transpose and conjugate transpose of $\mathbf{A}$ respectively. a^* is the conjugate of a. x_k is the kth entry of a vector $\mathbf{x}$. $E[\cdot]$ denotes expectation and $\tilde{f}$ is an estimation of f.

The rest of the paper is organized as follows. In Sect. 2, the background of the AIC system is introduced. The mathematical model and the procedure of the proposed approach are formulated in Sect. 3. In Sect. 4, the performance of the algorithm is analyzed. Simulation results are shown in Sect. 5. And Sect. 6 concludes the paper.

2   Background of the AIC System

The recovery algorithm BaOMP can be used in such AIC system as proposed in [14] or [15] which samples the bandlimited sparse signals. The block diagram for the system is displayed in Fig. 1. Suppose that the input signal is $s(t)$ and the demodulation signal is $p(t)$ which is a high-rate pseudo-noise sequence. The compressive measurement y_i is sampled every $1/m$ s. After m samples are taken, the accumulator is reset. In summary

$$y_i = \int_0^{\frac{i}{m}} s(t)p(t)dt, \quad i = 0, 1, \ldots m \tag{3}$$

Let $N/2$ be the positive integer that exceeds the highest frequency present in the continuous-time signal $s(t)$. Then the bandlimited signal $s(t)$ can be expressed in the form of:

$$s(t) = \sum_{k=1}^{K} \alpha_k e^{j2\pi f_k t} \quad \begin{matrix} t \in [0, 1) \\ f_k \subset \{0, \pm 1, \ldots \pm (N/2 - 1), N/2\} \end{matrix} \tag{4}$$

Here, α_k is a set of complex-valued amplitudes. And $K \ll N$ is the number of the active tones.

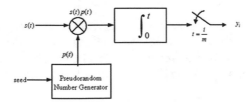

Fig. 1. Block diagram for the AIC system

The demodulation signal $p(t)$ switches between the levels ± 1 randomly and its rate is at or above the Nyquist frequency rate of the signal $s(t)$, i.e.

$$p(t) = \varepsilon_n \ t \in \left[\frac{n}{N}, \frac{n+1}{N}\right), n = 0, 1 \ldots N - 1 \tag{5}$$

Since the output signal of the integrator $y(t) = \int_0^t s(t)p(t)dt$ then

$$
\begin{aligned}
y_i &= \int_0^{\frac{i}{m}} \sum_{k=1}^{K} \alpha_k e^{j2\pi f_k t} p(t)dt \\
&= \sum_{k=1}^{K} \sum_{n=0}^{iN/m} \varepsilon_n \int_{(n-1)/N}^{n/N} \alpha_k e^{j2\pi f_k t} dt \\
&= \sum_{n=0}^{iN/m} \varepsilon_n \sum_{k=1}^{K} \frac{\alpha_k(e^{-j2\pi f_k/N} - 1)}{(2\pi j f_k)} e^{-j2\pi f_k n/N}
\end{aligned}
\tag{6}
$$

Assuming that N can be divided by m and $r = N/m$, let $x_i = \frac{\alpha_i(e^{-2\pi j f_i/N}-1)}{2\pi j f_i}$, $i = 0, 1, \ldots N - 1$, $\mathbf{x} = [x_0, x_1, x_{N-1}]^T$ and $F = \left[e^{-2\pi j f_i n/N}\right]_{n,f_i}$, then

$$\mathbf{Y} = \mathbf{HFx} \tag{7}$$

where

$$
\mathbf{H} = \begin{bmatrix}
\varepsilon_0 \cdots \varepsilon_{q-1} & & \\
\varepsilon_0 & \cdots & \varepsilon_q \cdots \varepsilon_{2q-1} \\
\vdots & & \vdots \\
\varepsilon_0 & & \cdots \\
& \cdots & \cdots \varepsilon_{mq-1}
\end{bmatrix}
\tag{8}
$$

If we let $\mathbf{A} = \mathbf{HF}$, then $\mathbf{Y}$ can be expressed as the same form as (1). It is proved in [8] that if $m \geq bK \log(N/K)$ (b is a small non-negative constant) and $\mathbf{A}$ satisfies RIP (restricted isometry property) constraint, (1) can be solved with high probability. The recovery method is to solve the convex problem.

$$\min_{\mathbf{x}} \|\mathbf{x}\|_1 \ subject \ to \ \|\mathbf{y} - \mathbf{Ax}\|_2^2 \leqslant \lambda \tag{9}$$

3   Proposed Approach

As mentioned above, the main idea of the proposed algorithm is to find the most possible locations of the non-zero coefficients by filtering out the most possible indexes related to the non-zero coefficients. Here, the squared numeral characteristics which are used as the basis for filtering are got from the diagonal elements of $\mathbf{A}^H \mathbf{YY}^H \mathbf{A}$.

From (7), we have

$$
\begin{aligned}
\mathbf{R}_{A\_Y} &= \mathbf{A}^H \mathbf{YY}^H \mathbf{A} \\
&= \mathbf{A}^H \mathbf{A} \mathbf{xx}^H \mathbf{A}^H \mathbf{A} \\
&= \mathbf{R}_A \mathbf{R}_x \mathbf{R}_A^H
\end{aligned}
\tag{10}
$$

where $\mathbf{R_A} = \mathbf{A}^H\mathbf{A}$ and $\mathbf{R}_x = \mathbf{x}\mathbf{x}^H$. As $\mathbf{A}$ satisfies RIP(restricted isometry property) constraint, all subsets of K columns taken from A are in fact nearly orthogonal and then the $\mathbf{R_A}$ is diagonally-dominant matrix and can be written as:

$$\mathbf{R_A} = \begin{bmatrix} 1 & \rho_{12} & \cdots & \rho_{1N} \\ \rho_{21}^* & 1 & \cdots & \rho_{2N} \\ \vdots & \vdots & & \vdots \\ \rho_{N1}^* & \cdots & \cdots & 1 \end{bmatrix}. \tag{11}$$

where, $|\rho_{ij}| = |A_i^H A_j|$.

For $\mathbf{R}_x$, since all K active tones are independent, it is a diagonal matrix with only K non-zeros among the N diagonal elements. Suppose that the set of the K indices for the non-zero coefficients is J_0, then the non-zero diagonal elements of $\mathbf{R}_x$ can be written as $x_k^2, k \in J_0$. Let $\mathbf{r} = diag(\mathbf{R_{A\_Y}})$. r_i, $i = 1, 2, \cdots N$ denotes the ith element of $\mathbf{r}$. Then from (10), r_i is

$$r_i = \sum_{k \in J_0} x_k^2 \rho_{ik}^2 \tag{12}$$

When $i = j, j \in J_0$,the element is

$$r_j = x_j^2 + \sum_{\substack{k \in J_0 \\ k \neq j}} x_k^2 \rho_{jk}^2 \tag{13}$$

From (12) and (13), it can be found that all diagonal elements of $\mathbf{R_{A\_Y}}$ are non-negative and when an index is related to a non-zero coefficient, the squared value of the non-zero coefficient is polluted by a non-negative value. If a suitable threshold is set at each step, then a batch of non-zero coefficients which is bigger than the threshold will be found.

Without loss of generality, the threshold for the first iteration stage will be derived first. Let $|x_{\max}| = \max_{k \in J_0} |x_k|$ and suppose that when $x_k = x_{\max}, k = l, l \in J_0$. Then for those $i \notin J_0$

$$\sum_{\substack{k \in J_0 \\ i \notin J_0}} x_k^2 \rho_{ik}^2 \leq x_{\max}^2 \sum_{\substack{k \in J_0 \\ i \notin J_0}} \rho_{ik}^2 \tag{14}$$

When $k \neq i$, the distribution of the correlation coefficients $\rho_{ik} \in (-1,1), 1 \leq i$, $k \leq N$ can be considered as independent normal distribution $N(0, \sigma_c^2)$ within $(-1, 1)$ approximately. Then the characteristics of the sum $\sum_{\substack{k \in J_0 \\ i \notin J_0}} \rho_{ik}^2$ in (14) can be analyzed based on Chi-Squared distribution with K degrees of freedom. The α-quantile of the sum can be got by searching the quantile table. When $K$is big enough $(K \geq 45)$,

$$\chi_\alpha^2 \approx \frac{\sigma_c^2}{2}(u_0 + \sqrt{2K-1})^2 \tag{15}$$

where, u_0 is the α-quantile of the standard normal distribution. For the derivation of the threshold, we use (15) as an approximation and set u_0 according to the 3σ rule that $\alpha = 1 - 0.9973$.

Then (14) can be rewritten as

$$\sum_{\substack{k \in J_0 \\ i \notin J_0}} x_k^2 \rho_{ik}^2 \leq x_{\max}^2 \sum_{\substack{k \in J_0 \\ i \notin J_0}} \rho_{ik}^2 \tag{16}$$

$$\leq x_{\max}^2 \chi_\alpha^2 \approx x_{\max}^2 \sigma_c^2 (3 - \sqrt{2K-1})^2 / 2$$

with high probability (about 0.9973). Suppose $r_{\max} = \max_{1 \leq i \leq N} |r_i|$, then

$$r_{\max} \geq x_{\max}^2 + \sum_{\substack{k \in J_0 \\ k \neq l}} x_k^2 \rho_{lk}^2 \tag{17}$$

If we set threshold to be

$$\tau = r_{\max} \cdot \frac{\sigma_c^2 (3 - \sqrt{2K-1})^2}{2} \tag{18}$$

For $i \notin J_0$, we have

$$\tau = r_{\max} \cdot \frac{\sigma_c^2 (3 - \sqrt{2K-1})^2}{2}$$
$$> x_{\max}^2 \frac{\sigma_c^2 (3 - \sqrt{2K-1})^2}{2} \tag{19}$$

From (16), it is of high probability that the right part of (19) satisfies

$$x_{\max}^2 \frac{\sigma_c^2 (3 - \sqrt{2K-1})^2}{2} \geq x_{\max}^2 \sum_{\substack{k \in J_0 \\ i \notin J_0}} \rho_{ik}^2 \tag{20}$$

where $i \in \{1, 2, \ldots N\}, i \notin J_0$. Considering both (13), (14) and (20), it can be concluded that only for $i \in J_0$, the r_i has the possibility to be larger than the threshold τ. So in the first stage, those indices which are bigger than τ are chosen as the non-zero locations. After remove the influence of these big coefficients from Y, the new threshold from the residual for the next stage can be derived in the similar way. For the qth stage, $r_{\max}^{(q)}$ represents the maximum r_i in the q stage. The threshold for the qth stage is

$$\tau^{(q)} = r_{\max}^{(q)} \frac{\sigma_c^2 (3 - \sqrt{2K-1})^2}{2} \geq x_{\max}^2 \sum_{k \in J_0} \rho_{ik}^2 \tag{21}$$

Though chi-squared distribution is used to derive the thresholds here, it does not mean that the distribution of the correlation coefficients $\rho_{ik} \in (-1, 1), 1 \leq i, k \leq N$ should obey chi-squared distribution strictly. A loose compliance with the chi-squared distribution is enough for (21) to be satisfied with high probability. The procedure of the algorithm is as follows:

(1) Initialization. Let $L^{(0)} = 0$, $Rf^{(0)} = Y$ and the initial index set $J^{(0)} = \Phi$. Set the iteration stage number $q = 1$.

(2) Get $\mathbf{R_{A\_Y}}$ according to (9) by using $\mathbf{Rf}^{(q)}$ instead of $\mathbf{Y}$.

(3) Find $r_{\max}^{(q)}$ and set the threshold

$$\tau^{(q)} = r_{\max}^{(q)} \frac{\sigma_c^2(3 - \sqrt{2K-1})^2}{2}$$

(4) Find the indices which are bigger than $\tau^{(q)}$ and form the new subset $J_s^{(q)}$ that denotes the new batch of active indexes found in stage q. Then update the active index set as $J^{(q)} = J^{(q-1)} \cup J_s^{(q)}$.

(5) Let $\mathbf{B}^{(q)}$ be the submatrix of $\mathbf{A}$ that is composed by the columns whose indexes are in the set of $J^{(q)}$ and $\mathbf{a}_j$ is the jth column of matrix $\mathbf{A}$, then $\mathbf{B}^{(q)} = \{\mathbf{a}_j, j \in J^{(q)}\}$.

(6) Remove the interference of the biggest coefficients that are found from the former stages. The residual is

$$\mathbf{Rf}^{(q)} = \mathbf{Y} - \mathbf{B}^{(q)}(\mathbf{B}^{(q)})^+\mathbf{Y} \tag{22}$$

(7) If the ending condition is satisfied, stop. Otherwise go back to (2).

Here the ending condition is whether the residual is small enough or the iteration steps exceeds the presetting number.

4   Performance Analysis

The number of the active indices that can be found in each stage is crucial for the speed of the algorithm. In this section, it will be discussed mainly. Without loss of generality, the number of the active indices which are found in the first stage is discussed. Suppose that $|x_k|^2, k \in J_0$ are uniformly distributed between $(0, |x_{\max}|^2]$. For the jth active index, only when $r_j = x_j^2 + \sum_{\substack{k \in J_0 \\ k \neq j}} x_k^2 \rho_{jk}^2$ is bigger than the threshold $\tau = r_{\max}. \frac{\sigma_c^2(3-\sqrt{2K-1})^2}{2}$, the index j can be chosen. That is

$$x_j^2 + \sum_{\substack{k \in J_0 \\ k \neq j}} x_k^2 \rho_{jk}^2 > r_{\max}. \frac{\sigma_c^2(3-\sqrt{2K-1})^2}{2}$$

$$\Rightarrow x_j^2 > r_{\max}. \frac{\sigma_c^2(3-\sqrt{2K-1})^2}{2} - \sum_{\substack{k \in J_0 \\ k \neq j}} x_k^2 \rho_{jk}^2 \tag{23}$$

Since $\sum_{\substack{k \in J_0 \\ k \neq j}} x_k^2 \rho_{jk}^2 \geq 0$, the least probability for its to be found will be

$$P_{d\_least} = p(x_j^2 > r_{\max}. \frac{\sigma_c^2(3 - \sqrt{2K-1})^2}{2})$$

$$= \frac{|x_{\max}|^2 - r_{\max}.\sigma_c^2(3 - \sqrt{2K-1})^2/2}{|x_{\max}|^2} \tag{24}$$

To make (24) easy to be analyzed, some approximations are made to r_{max}. Let

$$r_{max} \approx |x_{max}|^2 + \sum_{\substack{k \in J_0 \\ k \neq l}} x_k^2 \rho_{lk}^2$$

$$< |x_{max}|^2 + |x_{max}|^2 \frac{\sigma_c^2(3 - \sqrt{2(K-1)-1})^2}{2}$$

(25)

(25) is a reasonable approximation according to (13), (14) and (16). Then

$$p_{d\_least} > 1 - \frac{\sigma_c^2(3 - \sqrt{2K-1})^2}{2} \cdot \left[1 + \frac{\sigma_c^2(3 - \sqrt{2(K-1)-1})^2}{2}\right]$$

(26)

The number of the batch of active indices which are found in the first stage will approximately be

$$K_f^{(1)} > K p_{d\_least}$$

$$\approx K \left\{1 - \frac{\sigma_c^2(3 - \sqrt{2K-1})^2}{2}\left[1 + \frac{\sigma_c^2(3 - \sqrt{2(K-1)-1})^2}{2}\right]\right\}$$

(27)

In the similar way, suppose that there are $K_f^{(q-1)}$ active indices found in the former $(q-1)$ stages, then the number of the indices found in the qth stage will approximately be

$$(K - K_f^{(q-1)})p_{d\_least}$$

(28)

When the compressed samples are polluted by noise, according to (1), (9) can be written as

$$\mathbf{R_{A\_Y}} = \mathbf{A}^H \mathbf{Y} \mathbf{Y}^H \mathbf{A} + \mathbf{A}^H \mathbf{n} \mathbf{n}^H \mathbf{A}$$

$$= \mathbf{A}^H \mathbf{A} x x^H \mathbf{A}^H \mathbf{A} + \mathbf{A}^H \mathbf{R}_n \mathbf{A}$$

(29)

$$= \mathbf{R_A} \mathbf{R_x} \mathbf{R}_A^H + \mathbf{A}^H \mathbf{R}_n \mathbf{A}$$

where $\mathbf{R}_n = \mathbf{n}\mathbf{n}^H$. The diagonal numeric characteristic is

$$r_i = \sum_{k \in J_0} x_k^2 \rho_{ik}^2 + \sum_{l=1}^{m} n_l^2 a_{li}^2$$

(30)

where n_l is the lth element of $\mathbf{n}$.
When $i = j, j \in J_0$

$$r_j = x_j^2 + \sum_{\substack{k \in J_0 \\ k \neq j}} x_k^2 \rho_{jk}^2 + \sum_{l=1}^{m} n_l^2 a_{lj}^2$$

(31)

Since $\sum\limits_{l=1}^{m} n_l^2 a_{lj}^2 \geq 0$ and $E\left[\sum\limits_{l=1}^{m} n_l^2 a_{lj}^2\right] = \sum\limits_{l=1}^{m} E\left[n_l^2\right] a_{lj}^2 = \sigma_n^2$, (25) is changed to

$$r_{\max} \approx |x_{\max}|^2 + \sum_{\substack{k \in J_0 \\ k \neq l}} x_k^2 \rho_{lk}^2 + \sigma_n^2$$

$$< |x_{\max}|^2 + |x_{\max}|^2 \frac{\sigma_c^2(3 - \sqrt{2(K-1)-1})^2}{2} + \sigma_n^2 \tag{32}$$

Define $SNR_{\max} \triangleq |x_{\max}^2|/\sigma_n^2$,then

$$p_{d\_least} > 1 - \sigma_c^2(3 - \sqrt{2K-1})^2/2 \bullet$$
$$\left[1 + \sigma_c^2(3 - \sqrt{2(K-1)-1})^2/2 + \frac{1}{SNR_{\max}}\right] \tag{33}$$

As the iteration goes on, $SNR_{\max}$ will descend, which means the probability of founding the active indices will be smaller in the following stage, which may slow down the speed of the algorithm. However, for a signal to be detected, the SNR will not be too small. So the influence will be limited. Noticing that the algorithm may fail when $\frac{\sigma_c^2(3-\sqrt{2K-1})^2}{2} > 1$ and this situation may happen when the compressed ratio m/N is too small for the sparsity K. Simulation results for the stability test disclose that the probability of recovery is high when $m \geq K \log(N/K)$, which is in accordance with the conclusion in [8]. In addition, for OMP, when the compressed data are polluted by noise, the detection of the small indices under the noise level may become difficult for the interference of the noise may weaken the value that is used for the detection. The reason is that the interference of the noise has the same probability to be positive or negative. Whereas, according to (30), the interference of the noise is always positive, it will not weaken the squared diagonal characteristics and the probability to find the small indices will be higher than that in OMP. Simulations results also support the above conclusions.

As in each stage, a batch of active indices are found in contrast to the only one found in OMP, the iteration number will be smaller than the OMP, which will help to accelerate the algorithm. Besides the analysis of the number of the indices found in each step, we will also analyze the complexity in each stage. The main purpose in step (2) is to get the diagonal numeric characteristic $r_i, i = 1, 2 \ldots N$. So there is no need to calculate all the elements in $\mathbf{R}_{A\_Y}$. Instead, $r_i = \left[(A^H Y)_i\right]^2, i = 1, 2 \ldots N$. The complexity to work out the diagonal elements is $O(mN)$. The complexity of the searching procedure to find the active indices in step (4) is $O(N)$. In step (6), the complexity to produce the pseudo inverse is $O(m^3)$. Thus, in each stage the total complexity is $O(mN + N + m^3)$.

5 Numeral Simulation Results

In this section, the proposed algorithm is evaluated compared with OMP, BPDN and Lasso in terms of the stability, the squared reconstruction error and the efficiency. The impact of the noise to the performance of recovery is also studied.

The continuous-time bandlimited sparse signal used here is generated by randomly choosing K active tones in the band and the amplitudes of the non-zero coefficients are also randomly chosen.

5.1   Stability

In this simulation, the stability of the algorithm with different compressed ratio at different sparse degree K is studied. The active K tones frequencies are randomly chosen from the band of $[0, 128]MHz$. Whereas, the related amplitudes of the tones are uniformly distributed between $[0, 2]$. The Nyquist sampling rate is $256\,MHz$. m compressed samples within 6ms are used for the recovery. Thus, $N = 1536$ and the sensing matrix is of size $m \times 1536$. The squared recovery error is defined as $err = \frac{\|s(n) - \tilde{s}(n)\|_2^2}{N}$, where $\tilde{s}(n)$ is the recovered signal. For BP algorithm, the maximum iterations, the parameter λ and the error tolerance are set to 600, $\lambda = 0$ and 10^{-10} separately. For OMP, the termination condition is $lambdaStop = 10^{(-4)}$ and the maximum number of iterations to perform is 100. The matrix $\mathbf{H}$ is from USE(uniform spherical ensemble). All the experiments were run on Matlab code and the code of the competitive algorithms were from Sparse Lab with URL http://sparselab.stanford.edu. The competitive algorithm StOMP was also tested, but unfortunately, it failed for all the sparse degrees and compressed ratios we set. The reason may be that though the matrix $\mathbf{H}$ is from USE, the matrix $\mathbf{A} = \mathbf{HF}$ is not so 'random' and thus the residual MAI may not have approximate Gaussianity. From Fig. 2, it can be found that BaOMP has the similar stability with OMP and the performance degrades abruptly when the compressed ratio is too low for the sparsity. The reason is that when the compressed ratio is too small versus the sparsity K, the value $\frac{\sigma_c^2(3 - \sqrt{2K-1})^2}{2}$ may be larger than 1.   (2) Comparison of running time.

Tables 1, 2, 3 and 4 illustrate the running time and the iterative times that different algorithms took to recover the signal during the above experiments. The computer used is a standard PC with Intel core i7 and 4G memory running Window 7 system.

It is shown that the running time of BaOMP is much shorter than other competitive algorithms when $m \geq K \log(N/K)$ is satisfied. The iterative times of BaOMP are less than 1/8 than that of OMP at most of the time, which improves the running efficiency of the algorithm.

5.2   Performance Comparison with Noisy Data

To study the influence of the noise, the performance of the algorithm in terms of different standard deviation σ_n of the noise was tested. The signal was generated in the same way with the above experiments and $K = 40$. The measurements taken within 6ms were used. The parameter m was set as $m = 384$.

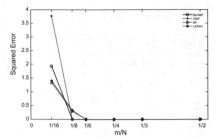

(a) The squared recovery errors vs the compressed ratio when $K = 20$.

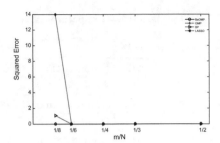

(b) The squared recovery errors vs the compressed ratio when $K = 45$.

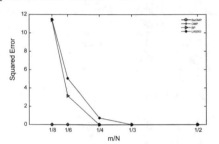

(c) The squared recovery errors vs the compressed ratio when $K = 60$.

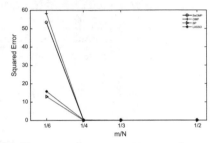

(d) The squared recovery errors vs the compressed ratio when $K = 80$.

Fig. 2. Stability Comparison

Table 1. Comparison of running time and iterations for K=20

m/N		BaOMP	OMP	BP	LASSO
1/8	Running time	0.0073	0.0227	0.3093	0.0613
	Iterative times	3	20	600	24
1/6	Running time	0.0063	0.0272	0.3648	0.0486
	Iterative times	3	20	600	19
1/4	Running time	0.0093	0.0346	0.5144	0.0511
	Iterative times	3	20	600	19
1/3	Running time	0.0066	0.032	0.688	0.0527
	Iterative times	2	20	600	19
1/2	Running time	0.0071	0.0381	0.946	0.0622
	Iterative times	2	20	600	19

As one of the comparative algorithms, BPDN was used instead of BP. The maximum iterations, the parameter λ and the error tolerance for BPDN are set to 600, $\lambda = 0.5$ and 10^{-10} separately. The parameters for the other comparative algorithms were the same with the above experiments. 100 times runnings to get the mean squared reconstruction error. Figure 3 shows the results. It can be found

Table 2. Comparison of running time and iterations for K=45

m/N		BaOMP	OMP	BP	LASSO
1/8	Running time	0.0073	0.0227	0.3093	0.0613
	Iterative times	3	20	600	24
1/6	Running time	0.0063	0.0272	0.3648	0.0486
	Iterative times	3	20	600	19
1/4	Running time	0.0093	0.0346	0.5144	0.0511
	Iterative times	3	20	600	19
1/3	Running time	0.0066	0.032	0.688	0.0527
	Iterative times	2	20	600	19
1/2	Running time	0.0071	0.0381	0.946	0.0622
	Iterative times	2	20	600	19

Table 3. Comparison of running time and iterations for K=60

m/N		BaOMP	OMP	BP	LASSO
1/8	Running time	0.0146	0.0261	0.499	0.2264
	Iterative times	11	48	600	310
1/6	Running time	0.0147	0.0283	0.4224	0.0867
	Iterative times	8	46	600	95
1/4	Running time	0.0096	0.0325	0.5205	0.0657
	Iterative times	5	46	600	53
1/3	Running time	0.0155	0.0516	0.7857	0.0763
	Iterative times	4	46	600	46
1/2	Running time	0.016	0.0603	1.2031	0.0909
	Iterative times	4	46	600	46

Table 4. Comparison of running time and iterations for K=80

m/N		BaOMP	OMP	BP	LASSO
1/8	Running time	0.051	0.0424	0.4277	0.2626
	Iterative times	98	100	600	334
1/6	Running time	0.0279	0.0413	0.5842	0.4571
	Iterative times	25	100	600	442
1/4	Running time	0.02	0.0513	0.6735	1.4539
	Iterative times	14	83	600	763
1/3	Running time	0.0196	0.0642	0.8014	0.1666
	Iterative times	8	81	600	1629
1/2	Running time	0.0226	0.096	1.287	0.1364
	Iterative times	7	81	600	89

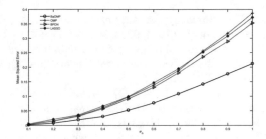

Fig. 3. The mean squared reconstruction errors vs the standard deviation of the noise when $K = 40$.

from Fig. 3 that the performance of BaOMP are better than the comparative algorithms even including BPDN. There are two reasons for it. One reason has been mentioned in Sect. 6 that for BaOMP algorithm, the noise interference is always positive according to (30) and it will not weaken the squared diagonal characteristics. The other reason is that as more than one index are chosen at each iteration, all the interference of these indices will be removed from the residual, which will make the detection for the small indices in the next step more accurate.

6   Conclusion

In this paper, BaOMP is proposed for the reconstruction of sparse bandlimited signal. The squared diagonal characteristics got from the compressed data is used to filter a batch of active indices in each iteration. And a set of adaptive thresholds in each iteration are derived based on the analysis of the covariance matrix $\mathbf{R_A}$. Both the theoretical analysis and the numeral simulations show that the proposed algorithm runs much faster than comparative algorithms and meanwhile it is not fussy about sensing matrix. As long as the sensing matrix satisfies RIP roughly, the proposed approach works well. Simulations also show that the stability of BaOMP is comparable with that of the standard greedy algorithm OMP.

   BaOMP can be extended to noisy data naturally. It outperforms both ℓ_1 minimization and standard greedy algorithm OMP when the compressed data are polluted by noise.

References

1. Chen, S., Donoho, D., Saunders, M.: Atomic decomposition by basis pursuit. SIAM Rev. **43**(1), 129–159 (2001). https://doi.org/10.1016/0022-2836(81)90087-5
2. Figueiredo, M.A.T., Nowak, R.D., Wright, S.J.: Gradient projection for sparse reconstruction: application to compressed sensing and other inverse problems. J. Sel. Top. Signal Process. **1**(4), 586–598 (2007). https://doi.org/10.1137/S003614450037906X. Special Issue on Convex Optimization Methods for Signal Processing

3. Efron, B., Hastie, T., Johnstone, I., et al.: Least angle regression. Ann. Stat. **32**(2), 1–48 (2004). https://doi.org/10.1214/009053604000000067

4. Tropp, J.A., Gilbert, A.C.: Signal recovery from random measurements via orthogonal matchinh pursuit. IEEE Trans. Inform. Theory **53**(12), 4655–4666 (2007). https://doi.org/10.1109/TIT.2007.909108

5. Dohono, D.L., Tsaig, Y., Drori, I., Starck, J.-L.Z.: Sparse solution of underdetermined linear equations by stagewise orthogonal matching pursuit. IEEE Trans. Inf. Theory. **58**(2), 1094–1121 (2012). https://doi.org/10.1109/TIT.2011.2173241

6. Mallat, S.G., Zhang, Z.: Matching pursuits with time-frequency dictionaries. IEEE Trans. Signal Process. **41**(12), 3397–3415 (1993). https://doi.org/10.1109/78.258082

7. Donoho, D.L., Maleki, A., Montanari, A.: Message-passing algorithms for compressed sensing. Proc. Nat. Acad. Sci. **106**(45), 18914–18919 (2009). https://doi.org/10.1109/ITWKSPS.2010.5503193

8. Parker, J.T., Schniter, P., Cevher, V.: Bilinear generalized approximate message passing. IEEE Trans. Signal Process. **22**(22), 5854–5867 (2014). https://doi.org/10.1109/TSP.2014.2357773

9. Blumensath, T., Davies, M.E.: Iterative hard thresholding for compressed sensing. Appl. Comput. Harmon. Anal. **27**(3), 265–274 (2009). https://doi.org/10.1016/j.acha.2009.04.002

10. Maleki, A., Donoho, D.L.: Optimally tuned iterative reconstruction algorithms for compressed sensing. IEEE J. Sel. Top. Signal Process. **4**(2), 330–341 (2010). https://doi.org/10.1109/JSTSP.2009.2039176

11. Blumensath, T., Davies, M.E.: Gradient pursuit. IEEE Trans. Signal Process. **56**(6), 2370–2382 (2008). https://doi.org/10.1109/TSP.2007.916124

12. Fioretti, L., Mazzucchelli, P., Bienati, N.: Stagewise conjugate gradient pursuit for seismic trace interpolation. In: 77th EAGE Conference and Exhibition, June 2015. https://doi.org/10.3997/2214-4609.201412980

13. Xu, W., Hassibi, B.: Efficient compressive sensing with deterministic guarantees using expander graphs. In: Proceedings of IEEE Information Theory Workshop, Lake Tahoe, September 2007. https://doi.org/10.1109/ITW.2007.4313110

14. Tropp, J.A., Laska, J.N., Duarte, M.F., Romberg, J.K., Baraniuk, R.G.: Beyond nyquist: efficient sampling of sparse bandlimited signals. IEEE Trans. Inf. Theory **56**(1), 520–544 (2010). https://doi.org/10.1109/TIT.2009.2034811

15. Kirolos, S., Laska, J., Wakin, M., Duarte, M., Baron, D., Ragheb, T., Massoud, Y., Baraniuk, R.: Analog-to-information conversion via random demodulation. In: Proceedings of 2006 IEEE Dallasa/CAS Workshop on Design Application, Integration and Software, October 2006. https://doi.org/10.1109/DCAS.2006.321036

Performance of Cooperative System Based on LDPC Codes in Wireless Optical Communication

Ibrahima Gueye, Ibra Dioum, K. Wane Keita, Idy Diop$^{(\boxtimes)}$,
Papis Ndiaye, Moussa Diallo, and Sidi Mohamed Farssi

Department of Computer Science, Polytechnic Institute (ESP),
Universite Cheikh Anta Diop, Dakar, Senegal
idy.diop@esp.sn

Abstract. This paper proposes the use of the technique of cooperative systems in wireless optical communication. Several obstacles can degrade the performance of information transmission in wireless optical links. These include atmospheric turbulence in communication links greater than 1 km, the rocking of buildings, wireless optical channel fading and flickering. Several studies have shown that the use of cooperative systems in this communication technique has better performance compared to conventional systems. To improve the performance of the transmission in the optical links we propose to place a DF (Decode and Forward) relay between the transmitter and the receiver, using the LDPC (Low-density Parity-check) codes at the relay level. Another innovation of this article is the introduction of the series relays to connect two very distant sites by also using LDPC at the level of each relay. Our method achieves a remarkable efficiency in terms of BER (Bit Error Rate) compared to some previous work.

Keywords: Decode and Forward · LDPC code · Wireless optical communication

1 Introduction

Wireless optical communication consists of transmitting information via a light emitting diode or a laser diode to transfer the free space [1]. WOC (Wireless Optical Communication) systems, several obstacles can prevent the transmission of information among which we can cite: ambient light (incandescent lamp, fluorescent lamp, daylight) their limited ranges, the rate of performance errors in the presence of atmospheric turbulence in communication links greater than 1 km [1]. To overcome these drawbacks we use relays with these performances through the use of spatial diversity technique. This notion of relay has been stated by Van der Meulen [2]. Various technique of cooperation has been proposed in the literature. The most used techniques are Amplify and Forward (AF) and Decode and Forward (DF) [3]. Under AF, the relay retains the signal information received from the source by performing only one amplification operation. However, the noise of the channel is amplified. Under DF, the relay first decodes the source signal and re-codes it to produce

© Springer Nature Switzerland AG 2020
K. Arai et al. (Eds.): FTC 2019, AISC 1070, pp. 745–757, 2020.
https://doi.org/10.1007/978-3-030-32523-7_56

additional parity bits at the destination. Since the DF mode can eliminate noise, it provides a substantial improvement in performance over AF.

In wireless optics, the use of relays have been studied [1] in order to optimize the location of parallel and series relays in WOCs by using a relational optimization algorithm. Cooperative optical communication has also been used in the field of VLC (visible light communication). In [4] a new cooperative transmission technique (a multi-point cooperative transmission technique) and a reception scheme in VLCs has been proposed. This cooperation technique provides for improvements and transmission reliability of information in indoor environments, such as corridors, laboratories, shops, or conference rooms. A 5G backhaul relay-supported cooperative communication system with a focus on reliability and 5G backhaul network resources was also proposed in [5]. They also introduced the architecture of parallel Free Space Optical (FSO) systems to increase throughput [6]. They considered a multi-jump cooperative FSO system with DF parallel relays [7].

LDPC codes are block correcting codes linear. Proposed in 1962 by Gallager [8], these codes did not immediately attract much interest from the information theory community. It was not until the 1990s that the Gallager codes were rediscovered in particular by MacKay [9] who brought to light the interest of the LDPC (Low-density Parity-check) codes using Tanner's representations and recursive approach [10]. The principle of the LDPC codes is based on the parity check matrix, one of the simplest control systems called the VRC (Vertical Redundancy Check). In this simple scheme of coding, an extra bit (the parity bit) is added to the K data bits to form the size code word $N = K + 1$. Each LDPC code is defined by a unique parity matrix H, of dimension $m \times N$. The particularity of the LDPC codes is that the parity matrix is hollow, that is to say that it contains only a very small number of non-zero elements, while it has a lot of '0'. This makes it possible to have a simplified decoding scheme and thus a fast decoding compared to the other types of codes in usual blocks.

A classical representation of the matrix H is done by a Tanner graph, also called a bipartite graph, like that presented in Fig. 1.

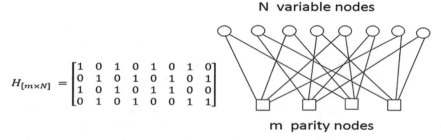

$$H_{[m \times N]} = \begin{bmatrix} 1 & 0 & 1 & 0 & 1 & 0 & 1 & 0 \\ 0 & 1 & 0 & 1 & 0 & 1 & 0 & 1 \\ 1 & 0 & 1 & 0 & 1 & 1 & 0 & 0 \\ 0 & 1 & 0 & 1 & 0 & 0 & 1 & 1 \end{bmatrix}$$

N variable nodes

m parity nodes

Fig. 1. Equivalent representation of the parity matrix by the Tanner graph

The contribution of this paper is to propose a cooperative DF system in wireless optical communication based on LDPC codes at each relay.

This document is organized as follows: in Sect. 2, Preliminaries and the System Model are introduced. Belief Propagation (BP) decoding is discussed in Sect. 3. Analysis of simulation results is provided in Sect. 4. Finally, conclusions are drawn in Sect. 5.

2 Model System

2.1 Ratings

In what follows, we use letters, such as H, to designate a single parity matrix. We use G to design the matrix generating the code, H' the sub-matrix, I represents the identity matrix. We use the notation for x_1^N to designate a vector of dimension N ($x_1, x_2 \ldots x_N$).

2.2 The LDPC Codes

2.2.1 General Principle

For the generation of LDPC codes, the principle is to build from the parity matrix H the generator matrix of the code G. By the property of the linear codes, the generating matrix of the code admits a systematic representation. The procedure is as follows to determine it:

- Transformation of the matrix H in the systematic form $H' = [I.A]$, where I represents the identity matrix
- Calculation of $G = [A^T.I]$

This transformation in the systematic form is done using eliminations along the Gaussian pivot. At the end of the transformation, the original matrix H must be subjected to the rearrangement of the columns according to the operations carried out during the elimination of Gauss. Two possible cases may occur at the end of the transformation:

- H' has no dependent lines: in this case the transformation results in an identity matrix of size $m \times m$. Therefore A is of sizer $(m \times N - m)$ and G of size $(N - m \times N)$,
- H' has at least 2 dependent lines: in this case, the transformation makes all lines appear to 0. These lines are deleted and the transformed matrix has fewer rows. Let L be the number of dependent lines of H', the transformation results in I of size $(m - L \times m - L)$. A is then of size $(m - L \times N - (m - L))$, and G of size $(N - (m - L) \times N)$.

In all cases, consider that G is of dimension $K \times N$, by putting $K = N - (m - L)$. This means that for a word to be coded U of K bits, one obtains a coded word $C_{[1 \times N]} = U_{[1 \times K]}.G_{[K \times N]}$ of size N. The size of the packets of bits to be sent will therefore depend on the matrix H used and the dependence of its lines.

Once $H(m \times N)$ and $G(K \times N)$ are constructed, we can verify that by construction:

$$G.H^T = 0 \tag{1}$$

Equation (1) is the equation of parity in matrix form. It is equivalent to the set of K parity equations that characterize the LDPC code (N; K). In this case, if a coded word $C = U.G$ is received without error, then $C.H^T = 0$, since $C.H^T = (U.G).H^T = U.(G.H^T)$ and $G.H^T = 0$.

Thus, the matrix H is used to check the parity $v.H^T = 0$ where v is the received word. If the result is non-zero, an error is detected and must be corrected. Since decoding is based on the structure of the parity matrix, the construction of the parity matrix is a very important step in the generation of a code. The length of the code is $N = 2^n, n > 0$

$$Encoding: \qquad x_1^N = u_1^N G_N \tag{2}$$

u_1^N is the message to be sent (uncoded bits), x_1^N is the code word.

A criterion having a significant influence on the convergence of the decoding is the presence of short cycles in the parity matrix. A cycle of order S is characterized by the fact that S elements at '1' in the matrix are on the same $S/2$ rows and columns so that the exchanges between the control and corresponding variable nodes do not evolve and do not converge. Indeed, if errors occur on the bits participating in the cycle, they can "compensate" and not be corrected. This phenomenon is illustrated in the decoding section of this chapter. The size of the shortest cycle in the parity matrix (also called "girth") is thus an important characteristic since the performances will be all the better as this minimal cycle is great.

2.3   Study Model

We consider a typical three-node relay channel (Fig. 2). It consists of a source node, a relay node, and a destination node, which are designated S, R, and D, respectively. The transmission on the channel is divided into two time slots. In the first time slot, the source encodes the information in LDPC code and sends it by broadcast to the relay and to the destination. In the second time slot, the relay processes the received signal during the first time slot and transmits the signal to the destination while the source node remains silent. In addition, the S-D channel, the R-D channel and the S-R channel are all assumed to be lognormal probability density channels. For simplicity, we assume that each node has a single antenna and the modulation scheme is PPM (Pulse Position Modulation). At the source node, the modulated signal is noted x_1^N, where N is the length of the coded signal. At the relay node, the signal received from the source and the re-transmitted signal at the destination are denoted $y_{1,SR}^N$ and w_1^N respectively. $y_{1,SR}^N$ is given by:

$$y_{1,SR}^N = \sqrt{P_s} h_{1,SR}^N x_1^N + n_{1,SR}^N \tag{3}$$

Where P_s is the transmission power of the source and $n_{1,SR}^N$ is the noise of the S-R and $h_{1,SR}^N$ represents the channel parameters between the source and the relays.

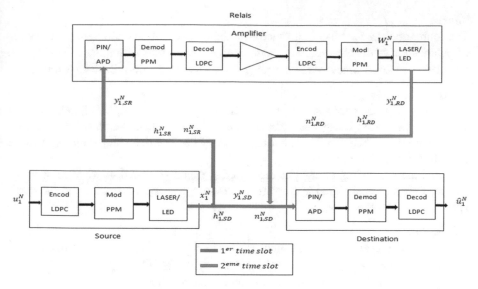

Fig. 2. Our cooperative scheme1

At the destination node, the signals received from the source and the relay are designated $y_{1,SD}^N$ and $y_{1,RD}^N$ respectively and are given by:

$$y_{1,SD}^N = \sqrt{P_S} h_{1,SD}^N x_1^N + n_{1,SD}^N \tag{4}$$

$$y_{1,RD}^N = \sqrt{P_R} h_{1,RD}^N w_1^N + n_{1,RD}^N \tag{5}$$

Where w_1^N is the signal after processing at relay R.

P_R is the transmission power of the relay, $n_{1,SD}^N$ and $n_{1,RD}^N$ are the noise of the S-D et R-D channels respectively, $h_{1,SD}^N$ and $h_{1,RD}^N$ designates the channel parameters between S-D and R-D. Throughout this part we will assume that $P_R = P_S = 1$.

2.4 Our Decode and Forward Scheme Based on LDPC Codes

The Decode and Forward or DF is today one of the best cooperation techniques implemented at relay level in half-duplex mode when the state of the Source-Relay channel is better than that of the Source-Destination channel ($SNR_{S-R} > SNR_{S-D}$).

In our model, we denote by M the message to send of length $N = 2^n (n \geq 0)$ composed of K bits of information and $(N - K)$ frozen bits. Soit ˆ be the estimate of M. The LDPC coding of the information bits at the source is carried out.

Source

The source encodes the bits of information and the frozen bits according to the principle of polarization of the channel in x_1^N containing k bits of information and $N - K$ frozen bits. This x_1^N is then broadcast to the relay and to the destination for the 1st time slot.

Relay

At the relay after the conversion of the optical signal into an electrical signal, the latter will be demodulated using the PPM demodulation and then decoded by an LDPC decoder using the belief propagation algorithm (BP). The decoding principle will be described in the next section.

After these decoding operations the signal is then re-encoded, modulated as at the source. This electrical signal is converted into an optical signal and sent to the destination at the 2nd time slot.

Destination

The destination receives the two signals $y_{1,RD}^N$ and $y_{1,SD}^N$ coming from different channels (order of diversity equal to 2) combines them by the MRC (Maximum Ratio Combining) method before the final decoding and demodulation.

With the parameters used in schema1, we proposed a second schema (Fig. 3) that cooperates in wireless optical communication to increase the range. We used allied relays by a direct path from the source to the destination. The source encodes the information bits using the LDPC codes, which is modulated using PPM modulation and sends it to the first relay. At the relay, after the conversion of the optical signal into an electrical signal, the latter will be demodulated using the PPM demodulation and then decoded by an LDPC decoder using the belief propagation algorithm (BP). The decoding principle will be described in Sect. 3. After these decoding operations the signal is then re-encoded as at the source. This electrical signal is converted into an

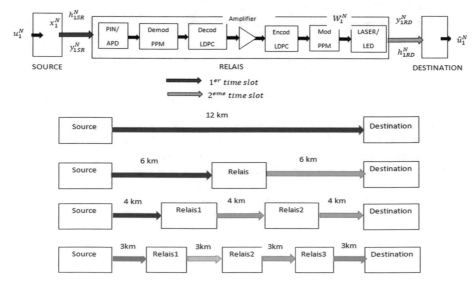

Fig. 3. Our cooperative scheme2

optical signal and sent to the next relay. This process continues until the information finally reaches the destination.

3   Algorithm Belief Propagation

In general, the decoding of the LDPC codes is done using a soft decoding algorithm (BP) or rather derivatives of the BP method.

The flexible iterative decoding algorithm initially presented by Gallager [11], then reviewed by Mackay [12] in the framework of graph theory, is known as belief propagation algorithm (Belief Propagation (BP)). The principle of belief propagation is the direct application of the Bayes rule on each bit of a parity equation. The parity check calculates an estimate of each bit. These estimates, forming messages propagating on the branches of the graph, are then iteratively exchanged in order to compute a posteriori information on each bit. In the case of a propagation of belief on a graph without cycle, the exchanged messages are independent, which leads to the simple and exact calculation of the posterior probabilities: the algorithm is in this case optimal. If the factor graph presents cycles, the assumption of independent messages is no longer valid. However, the more the graph is hollow (that is, the less the parity check matrix is dense), the more the approximation of a graph without a cycle becomes valid. It is therefore under this hypothesis that the decoding algorithm is described. So, BP is an information exchange algorithm between data nodes and control nodes associated across the branches. It can be broken down into several stages:

- A first phase consists of calculating the messages propagating from a data node to a control node (Fig. 4).
- A second step calculates messages generated at the control nodes.
- Once all messages are updated, they are propagated from the control nodes to the data nodes (Fig. 5).
- Finally, after a certain number of iterations, the posterior information associated with each data node is updated before the decision is made.

For the rest, we will note m_{vc} messages propagating from a data node to a control node, m_{cv} is used to designate the messages coming from a control node and transmitted to a data node. The update of the m_{cv} messages from the data node v to the iteration i is calculated as follows (Fig. 4):

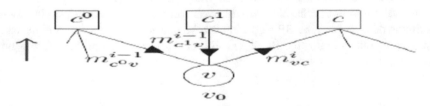

Fig. 4. Illustration of the update of messages propagating from a data node to a control node m_{vc}

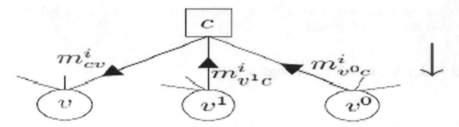

Fig. 5. Illustration of the update of messages propagating from a control node to a m_{cv} data node

$$m^i_{cv} = v_0 + \sum c' \in c_v / cm^{i-1}_{c'v} \tag{6}$$

Where v_0 represents the log-likelihood ratio resulting from the observation y_v at the output of the channel:

$$v_0 = \ln \frac{pr(y_v|v = 0)}{pr(y_v|v = 1)} \tag{7}$$

And where c_v is the set of control nodes connected to the data node v. At the first iteration, the messages from the control nodes are null.

The second step of the belief propagation algorithm is to update the output messages of a control node. The m_{cv} messages are calculated at iteration i in the following way (Fig. 5):

$$m^i_{cv} = 2 \tan h^{-1} \left(\Pi_{(v' \in c_c|v)} \tan h \left(\frac{m^{i-1}_{v'c}}{2} \right) \right) \tag{8}$$

Where c_c represents the set of data nodes connected to the control node c. After each iteration, a decision can be made on the posterior information A^i_v associated with the data node v:

$$A^i_{cv} = v_0 + \sum c' \in c_v / cm^i_{c'v} \tag{9}$$

The iterative process is stopped after a maximum number of iterations. But we can also stop the iterative process by using the null syndrome condition.

The BP algorithm is the one we will use in this work. Nevertheless, in practice, algorithms derived from the BP algorithm can be considered. These algorithms can be seen as a simplification of the BP algorithm, and therefore suboptimal (always under the assumption of a graph without cycle).

4 Simulations and Results

- The Please Simulation parameters;
- Length of the code word LDPC is N = 1008;
- Information bits K = 504;
- The source message is a binary sequence of length (N − K), with bits 0 and 1 distributed evenly;
- The decoding results are compared to the source sequence and calculate the BER (Bit Error Rate) which is the number of error bits divided by the total number of bits transmitted;
- PPM codes n bits of PCM in a pulse that occupies one of the time slots;
- The Log-normal model is used in the case of the low turbulence regime with its variance $\sigma_x^2 < 0$.

To evaluate the performance of our DF schema based on LDPC codes at the relay level we will use comparison systems. The previously described systems were simulated with Matlab. We present in Figs. 6 and 7, the bit error rates (BER) as a function of the signal-to-noise ratio in the case of a scheme with or without cooperation (Fig. 6) and in the case of a scheme with or without cooperation, with or without LDPC codes (Fig. 7).

In Fig. 6 we compared a cooperative scheme and a scheme without cooperation.

It can be seen that, if the SRN is set at 12 dB, the non-cooperative system has a larger bit error rate (BER) of about $10^{-0.75}$ and the cooperative scheme has the lowest bit error rate (BER) $10^{-1.4}$. In Fig. 7, we compared a non-cooperative scheme and a

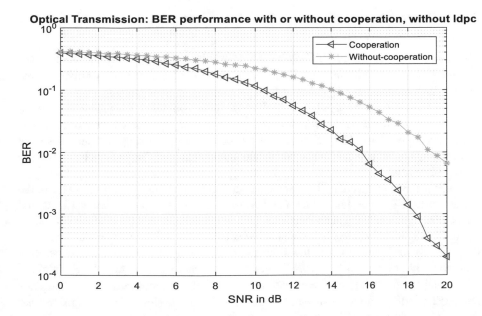

Fig. 6. Comparing our schema against a non-cooperative schema without LDPC codes

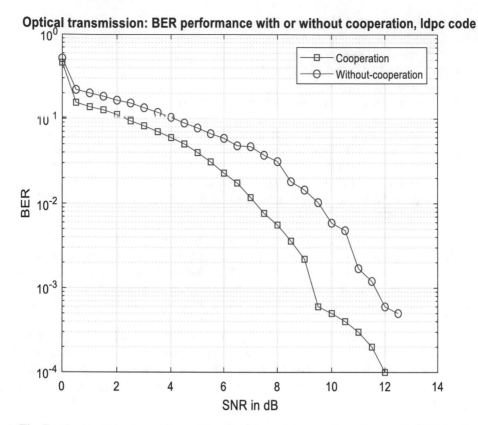

Fig. 7. Comparison of our schema (Fig. 2) with a non-cooperative scheme with LDPC code.

cooperative scheme using LDPC codes. If the SNR is set at 12 dB, the non-cooperative scheme has a larger bit error rate (BER) of about $10^{-3.3}$ and our cooperative scheme has the lowest bit error rate (BER) 10^{-4}.

This means that cooperative optical communication with or without the use of LDPC codes further enhances performance over conventional communications systems (without cooperation). This is because at the destination, the signals from the source and relay are combined to increase resistance against channel fluctuations by exploiting spatial diversity.

In Fig. 8, we present the bit error rates as a function of the signal-to-noise ratio, the impact of LDCP codes for two cooperative schemes (cooperation without LDPC and cooperation with LDPC (our scheme)).

In Fig. 8, we compared two cooperative scheme using the LDPC codes on one of the schemes (our scheme). It is found that, if the SRN is set at 12 dB, the cooperation scheme without the use of the LDPC codes has a larger bit error rate (BER) of about $10^{-1.3}$ and the cooperative scheme with the use LDPC codes have the lowest bit error rate (BER) = 10^{-4}. So we realize the use of LDPC codes is significant. This means that these codes make it possible to achieve a better performance in BER.

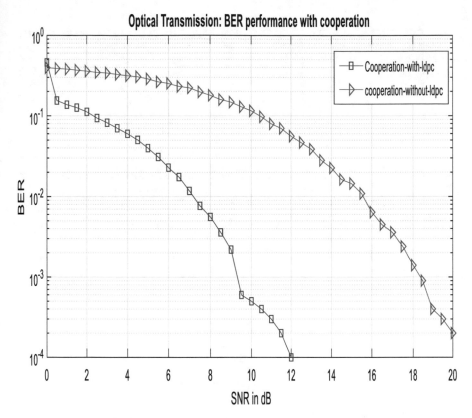

Fig. 8. The impact of LDPC codes with cooperation.

The following simulation aims to study the impact of using serial relays to increase the range. Figure 9 shows the bit error rates (BER) as a function of the signal-to-noise ratio in the case of a schema with several serial relays using the LDPC codes at the relays.

Figure 9 allows us to appreciate in comparison with a transmission without relay, the impact of the use of one or more relays on wireless optical transmission using the LDPC codes at each relay. For an SNR of 8 dB we have a BER of 0.1 for 4 relays, 0.28 for 3 relays, 0.45 for 2 relays and 0.5 without relays. So to increase the distance, we realize that the use of LDPC codes at each relay increases the performance of the network. So the more we increase the distance, the more we increase the relays, the higher the performance of the network.

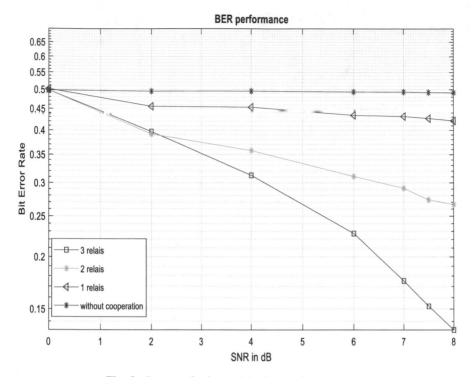

Fig. 9. Impact of using serial relays to increase range

5   Conclusion

The In this paper, we focused on the use of cooperative systems combined with error correcting codes especially LDPC codes in wireless optical communication to mitigate turbulence-induced fading. Two cooperation schemes have been proposed on the basis of the LDPC codes. We compared a cooperative scheme and a scheme without cooperation and the results show that for a SNR of 12 dB we have a gain of $10^{-0.65}$. We have also compared two cooperative schemes using the LDPC codes on one of the schemes (our diagram), the results show that for a SNR of 12 dB we have a gain of $10^{-2.7}$. In the end we also looked at the impact of the use of the relays in series combined with the LDPC codes to increase the range. The results of our simulations show the use of cooperative systems combined error correcting codes can mitigate turbulence-induced fading. In future work, we would like to investigate the introduction of wavelength division multiplexing techniques to increase the capacity of cooperative optical communication but also MIMO cooperative. Polar codes associated with spatio-temporal transmission techniques already proven in the case of mobile networks can be adapted to communication cooperative optics to increase diversity gain and reach.

References

1. Jabeena, A., Jayabarathi, T., Gupta, P.R., Hazarika, G., Nibedan: Cooperative wireless optical communication system using IWO based optimal relay placement. In: International Conference on Advances in Electrical Engineering (ICAEE) (2017)
2. van der Meulen, E.C.: Three-terminal communication channels. Adv. Appl. Probab. **3**, 120–154 (1971)
3. Laneman, J.N., Tse, D.N.C., Wornell, G.W.: Cooperative diversity in wireless networks: efficient protocol and outage behavior. IEEE Trans. Inf. Theory **50**(12), 3062–3080 (2004)
4. Guzman, B.G., Serrano, A.L., Jimenez, V.P.G.: Cooperative optical wireless transmission for improving performance in indoor scenarios for visible light communications. IEEE Trans. Consum. Electron. **61**(4), 393–401 (2015)
5. Song, S., Liu, Y., Song, Q., Guo, L.: Relay selection and link scheduling in cooperative free-space optical backhauling of 5G small cells. In: IEEE/CIC International Conference on Communications in China (ICCC) (2017)
6. Kashani, M.A., Uysal, M.: Outage performance and diversity gain analysis of free-space optical multi-hop parallel relaying. J. Opt. Commun. Networking **5**(8), 901–909 (2013)
7. Uyasl, M., Safari, M.: Relay-assisted free space optical communication. IEEE Trans. Wireless Commun. **7**(12), 5441–5449 (2008)
8. Gallager, R.G.: Low Density Parity Check Codes. Monograph. MIT Press, Cambridge (1963)
9. MacKay, D.J.C.: Good error-correcting codes based on very sparse matrices. IEEE Trans. Inf. Theory **45**, 399 (1999)
10. Tanner, R.: A recursive approach to low complexity codes. IEEE Trans. Inf. Theory **27**(5), 533–547 (1981)
11. Gallager, R.G.: Low-density parity-check codes. Ph.D. dissertation (1963)
12. MacKay, D.J.C., Neal, R.M.: Near Shannon limit performance of low density parity-check codes. Electron. Lett. (1996). 191 192 bibliographie

Studying the Impacts of the Renewable Energy Integration in Telecommunication Systems: A Case Study in Lome

Koffi A. Dotche[1]([✉]), Adekunlé A. Salami[1], Koffi M. Kodjo[1],
François Sekyere[2], and Koffi-Sa Bedja[1]

[1] Department of Electrical Engineering, High National College of Engineering
(ENSI), University of Lome, Lome, Togo
kdotche2004@gmail.com, akim_salami@yahoo.fr,
rig_kodjo@yahoo.fr, eugenebedja@yahoo.fr
[2] Department of Electrical Technology, Faculty of Technical Education,
University of Education Winneba, Kumasi, Ghana
fsekyere@gmail.com

Abstract. The green energy is foreseen to be a prominent resource to abate the rate of the dioxide of carbon. The technique of the switching off transmission nodes in the absence of requested service has addressed (projected) the research to reduce their energy consumption and carbon dioxide footprint. A transmission node has two kinds of circuits: transmission and non-transmission. The non-transmission circuit takes into account the treatment, cooling, energy conversion and room lighting systems. The transmission circuit supports the amplification and power supply system of the transceivers. A clean direct continues energy supply source could be used to power these telecommunications' equipment in order to reduce the unnecessary conversion, and reduce power losses that are resulted. The consumption of the static power of the microcells is normally low. However, for a fixed telecommunication system, this energy consumption is present at the access network level. This article evaluates the impact of the renewable energy integration (REI) in telecommunication systems. It intends further to promote the introduction of the green energy and the reduction of power consumption by the switchover into the sleep mode of the transmission nodes based on the threshold of the active number of users in a service perimeter. Radio parameters of a LTE system are considered with some heterogeneous cells. A multitask optimization problem is formulated by assuming Gaussian random transmission of mobile users. A resolution with Pareto optimum front methods is used. The partial results indicate a shrink when the idle users are more than the active ones. However, the right hand plane shows the feasible solutions with a positive energy efficiency value.

Keywords: Green energy · Green communication · Multi-objective problem · Pareto optimum method

© Springer Nature Switzerland AG 2020
K. Arai et al. (Eds.): FTC 2019, AISC 1070, pp. 758–780, 2020.
https://doi.org/10.1007/978-3-030-32523-7_57

1 Introduction

The telecommunication systems give a support to the local economy and at the nation level, a real backbone for the development. They operate in a limited spectrum [1, 2] and have a mandate to satisfy the increasing growth of customers. This translates a high power consumption subsequently yielding a lot of the dioxide of carbon, CO_2, emission. The survey paper in [3] has pointed out that the Information Communication and Technology (ICT) industries have had a power consumption about 3% of the global energy demand. These systems were responsible for about 2% of the CO2 emissions in the world. The study went on arguing that there was a need to optimize the power consumption of such systems in order to abate the global climatic warming. It was projected that the number of base stations would be approximating the 12 million all over the world, in 2020 [4]. The Global System for Mobile communication Association (GSMA) has indicated in 2009 the necessity for including the renewable energy for a green communication. The goal was to provide some communication facilities to the remote areas and reduce the dioxide of carbon emission. Several studies were conducted within many projects in the scope of the energy reduction in future wireless systems with the objective of achieving a green energy communication reposing on energy harvesting. These projects were, namely, 'Optimizing Power Efficiency in mobile RAdio Networks (OPERANet), Towards Real Energy-efficient Network Design (TREND), Energy Aware Radio and neTwork tecHnologies (EARTH), Cognitive Radio and Cooperative strategies for Power (C2POWER), GreenTouch' a type of a project for power saving in multi-standard wireless devices, [5, 6], etc.

The energy consumption minimization goes with the management of the spectrum [7]. Many approaches have been adopted for the spectral efficiency such the heterogeneous networks, wider radio channels, femtocells, Wi-Fi offload, solution. The area spectral efficiency could be optimized on one hand by the advanced physical layer techniques [8]. These techniques also aim at combatting the radio channel impairments [9] and in improving the spectral efficiency of the baseband signal at expense of the energy efficiency. Improving the energy efficiency may depend on the Medium Access Control (MAC) protocols [10]. MAC protocols coordinate the channel access of each individual device and optimize their power consumption [11]. The minimization of global cost of the energy was evaluated for mobile users within a group of BTSs by including the available cost of energy in random allocation [12]. The minimization of the energy of the macrocell was considered with the dilemma of power transmission and the discontinuity protocol in frame transmission period. The BSs after serving all users in its service area switch to sleep mode. The problem of optimization has been formulated and its resolution by the help of Karush-Kuhn-Tucker (KKT) initial conditions. The results have shown an improvement of an energy saving about 40% was achieved by the proposed algorithm as compared to that of discontinuity (DTx) protocol used in Long Term Evolution LTE [13]. In LTE, the power consumption minimization is achieved by the use of a hybrid of three modulation schemes [14] such as 64 Quadrature Amplitude Modulation (QAM), 16-QAM and Quadrature Phase Shift Keying (QPSK) modulation. In [15], the paper examined the power saving of a mobile terminal by exploiting the relay stations' deployment in cellular network. The authors made the usage of a genetic algorithm to find the optimized mobile transmit power and the bandwidth allocation. A simulation method was adopted. The results indicated that by increasing the relay-station coverage would ultimately increase the operational cost. It was argued that the optimal way of using the relays would be to set their radius about 50–70% to that of the macro-cell radius in hierarchical network.

The need to integrate the renewable energy into telecommunication system has been a concern for many experts and agencies alike the GSMA and the International telecommunication union (ITU). There is a huge potential of the renewable energy resources in Africa, particularly in the West African countries, and the mobile cellular networks are well established but there is a tangible discrepancy between the rate of telecommunication systems' facilities penetration and that of the electricity. A policy of encouraging the mobile networks operators and the ICTs infrastructure sites being equipped with a micro grid and the possibility of including the local community energy demand will definitively boost the rate of the electricity and data universal access. In sensitive areas such as an airport, the use of many supply sources will enforce the power security, with the integration of the renewable energy sources.

This paper proposes the integration of the renewable energy into telecommunication systems in order to cater for the universal access in electricity and data.

The contributions of the paper are as follows:

- It proposes a new heterogeneous architecture of a dense cellular network. This architecture is applicable to any data network.
- It formulates the energy efficiency as function of the injected green energy in order to schedule the users depending on the available green energy; and solves the optimization problem using Pareto methods.
- It introduces a hybrid micro-grid power supply for an heterogeneous network, to serve as an electrification backbone for achieving a universal access policy for electricity.

The rest of the paper is organized as follows: the Sect. 2 formulates the mathematical derivation of the energy efficiency. The Sect. 3 proposes the problem resolution. The Sect. 4 presents the methodology for the problem simulation. The Sect. 5 focuses on the presentation of the results, and their analysis. The last Sect. 6, closes the paper with a succinct conclusion and further research directions.

2 Model

The system consists of one macro-cell, pico-cells, three relays, and femto-cells as shown in Fig. 1.

2.1 Problem Description

A multi-transmitters' system of a single macrocell-radius with a flat fading channel is considered as shown in Fig. 1. The femto-, pico-, and relay-cells are deployed to support the central base station (CBS) to form a typical dense tier heterogeneous network. The type-I relays are selected. They are non-transparent relays. The transmitters could be viewed as remote antenna unit (RAI). Users are served by the nearest transmitter (RAI, cell) and the channel state information (CSI) is known by the BSs. The relays, femto, pico and macro-cell antenna are independent transmitters. The virtual MIMO coordination of the nodes is assumed to support the weak receiver.

All the nodes' links are in the direct line of sight, and that of the mobile users in a non-line of sight. The transmitters are empowered by a micro-grid and the renewable energy sources are used to reduce the emission of the dioxide of carbon.

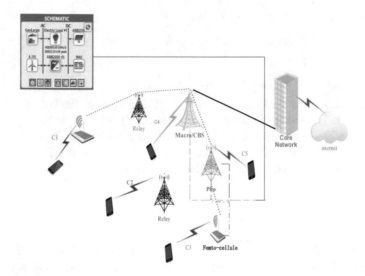

Fig. 1. The system model of the study

2.2   Problem Formulation

The propagation model, $p_l\left(\vec{i}, \vec{k}\right)$ between a mobile user, $\vec{i}$, and a node $\vec{k}$ is given as:

$$p_{l,i,k} = r_{i,k}^{-\alpha} \times \chi$$
$$P_{L,i,k}(\text{dB}) = 10 \log_{10} r^{\alpha} + 10 \log_{10} \chi \tag{1}$$

$r_{i,k}$ is the separating distance, and r_k radius of the serving node k (macro-, pico-, femto-cellule and relay station);

χ is the medium attenuation, $\chi = 10^{x/10}$, where $\chi \sim N(0, \sigma^2)$, is the standard deviation of the shadowing, having a variance in the range of 6–12 dB.

α is the path loss exponent, $\alpha > 2$, for a medium that impairs highly the signal, for a non-direct line of sight signal, typical case of a mobile user signal to the serving node; and $\alpha \leq 2$, for line of sight signal, applicable to communication among for nodes. The maximum radius, r_{Max}, is that of the macro-cell such that:

$$r_\lambda \leq \sum_{k=1}^{K} r_\lambda \leq r_{Max} \tag{2}$$

for the path macrocell-relay (r_{ra})-picocell (r_p):

$$2r_{ra} + 2r_p \leq r_{Max} \tag{3}$$

with $r_{ra} < r_{Max}$, and $d\left(\vec{k_b}, \vec{k_{ra}}\right) < d\left(\vec{k_b}, \vec{k_f}\right) < d\left(\vec{k_b}, \vec{m}\right)$.

Mobile users are locked on the various cells based upon the signal quality measurement level. The mode of access to the various cell is based on the signal to noise interference ratio, $SINR_i$ level. Four floors are defined such as $SINR_{ra}$, $SINR_m$, $SINR_p$ and $SINR_f$ respectively for the relay, macro-, pico- and femto-cell. The probability for an active cell f_{ac} of the active node k, [16], is given as:

$$f_{ac} = 1 - \left(1 + \frac{1}{\mu.\rho}\right)^{-\mu} \tag{4}$$

with the BS density, λ_b-user density, λ_u the BS-User density ratio $\rho \triangleq \frac{\lambda_b}{\lambda_u}$ and μ is a constant related to the cell-size distribution obtained through data fitting. A value of 3.5 was provided in [17]. It should be noted that, the f_{ac} can be regarded as the probability of the active base station to the total number of base stations. The probability of a non-active cell, f_{id}, is given as:

$$f_{id} = \left(1 + \frac{1}{\mu.\rho}\right)^{-\mu} \tag{5}$$

using the Bayes's formula [18], the conditional probability of the transitional cell's status, f_{tr}, is given as:

$$f_{tr} = p(f_a|f_{id}) = \frac{P(f_a|f_{id}).P(f_a)}{P(f_a|f_{id}).P(f_{id}) + P(f_{id}|f_a^c).P(f_{id}^c)} \tag{6}$$

Several power consumption modelling are discussed in [19, 20]. The linear model for the power consumption at an active node, $P_{i,k}$, is as follows:

$$P_{i,k} = \frac{1}{\eta} P_{t,i} + \alpha P_c + P_0 \tag{7}$$

η is the efficiency of power amplifier, p_t, the transmit power of the node, α the number of antennas at the node, the circuit power P_c, and P_0 is the non-transmission power consumption including P_{DP} baseband signal processing, P_{Conv}, AC-DC converter; P_{Cool}, cooling; P_{LB}, the lighting system; E_{bb}, battery backup (uncharged); etc. However, this model does not cater for the traffic data, but the EARTH's has made a provision for the traffic. It formulates the power consumption model, $P_{i,k}$, at a node as the traffic dependency, ΔP the slope of the load dependent; N_{Trx}, the number of transceiver; P_0, the non-transmission power; P_{sleep}, the minimum power consumption in the idle mode, and P_{out} the output power transmission. This model is given as:

$$P_{i,k} = \begin{cases} N_{Trx} * (P_0 + \Delta P . P_{out}) \\ \\ N_{Trx} * P_{sleep} \end{cases} \tag{8}$$

In this study, the power consumption, $P_{i,k}$, of an active node, k, for the downlink service to a user, i, is formulated as:

$$P_{i,k} = \begin{cases} P_{Ti,k} + P_o + P_c \\ \\ N_{Trx} * P_{sleep} \end{cases} \tag{9}$$

where, $P_{Ti,k}$ is the downlink transmission dependent, P_o is the uplink transmission dependent that includes the data processing, non-transmission, cooling, lighting, etc. and P_c, the circuit power, which is function of the radio units. The non-transmission power, P_0, is given by

$$P_0 = P_{DP} + P_{Conv} + \sum_{i}^{m} P_{Cool_i} + \sum_{j}^{n} P_{LB_j} + E_{bb} + P_{link} \tag{10}$$

the transmit power for the downlink service, $P_{Ti,k}$ to serve a user attached to the cell, k, can be calculated by:

$$P_{Ti,k} = \lambda_b . f_{ac} . \Delta P . \frac{P_{out}}{\eta} \tag{11}$$

The signal to noise plus interference, SINR, $\gamma_{i,k_{DI}}$, for a detectable communication in a network, which comprises a macro-, pico-, and femto-cell is given in [21] as:

$$\gamma_{i,k_{DI}} = \frac{P_{t,k} . P_{L,i,k}^{-1}}{I_{m,i,k} + N_0} \tag{12}$$

which can be written by

$$\gamma_{i,k_{Dl}} = \frac{\lambda_b . f_{ac} . \Delta P . \frac{P_{out}}{\eta}}{I_{m,i,k} + N_0} * r^{-\alpha} \times \chi \geq \gamma_0 \tag{13}$$

I_m the interference due to multiple access and other low nodes on the user, i. γ_0 is the minimum threshold for the a detectable communication.

Let N users be distributed homogeneously accordingly to the distance such that, the noise plus interference ratio $I_{m,i,k}$, at MS_i located in the cell, k, is given as:

$$I_{m,i,k} = \sum_{\substack{j \neq i}}^{N} I_0 + \sum_{\substack{j \neq k}}^{N} \sum_{Lp} \sum_{j \neq i} P_{Tj,k} + N_0 \tag{14}$$

with $I_0 = g_{j,k} . p_{1j,k}^{-1}$ and $P_{Tj,k} = g_{j,k} * \chi * r^{-\alpha}$; $p_{t,k.}$ is the transmission power spectral density, of the node or the cell, k; N_0 is the power spectral of the additive white Gaussian noise, Lp is the number of existing low power nodes, including the relay. Its value is 3.

It comes that the Shannon's data rate [13] of each user MS_i being served in the cell, k, $\eta_{SE_k}(i)$, is given as:

$$\eta_{SE_k}(i) = \log_2\left(1 + \gamma_{i,k_{Dl}}\right) \tag{15}$$

A. For *one cell*,

The spectral efficiency, SE, η_{A_k} can be written as:

$$\eta_{A_k} = \sum_{i}^{A} x_{i,k} . \eta_{SE_k}(i) + \sum_{j}^{B} \left(1 - x_{j,k}\right) . \eta_{SE_k}(j) \tag{16}$$

where $x_{i,k}$ is a binary parameter, it equals 1 for a connected user and otherwise 0.

$$\eta_{A_k} = \sum \sum \eta_{SE}(i,j) = \eta_{A_{active}} + \eta_{A_{idle}} \tag{17}$$

The energy efficiency, EE, $\gamma_{EE}(A)$ in a given area is defined as the data throughput, total spectral efficiency, (η_A) of users within the cell area A over the total transmit power and accounting for the circuit power and non-transmission power consumption of inactive cells (idle transmitters in sleep mode). For the downlink, EE, $\gamma_{EE_{Dl}}(A_k)$, it is as:

$$\gamma_{EE_{Dl}}(A_k) = \frac{\eta_{A_k}}{P_t(A_k) + p_c(A_k)} = \frac{\eta_{A_{active}}}{P_{To}} + \frac{\eta_{A_{idle}}}{P_{To}} \tag{18}$$

the quantities in (18) are as follows:

$P_t(A_k) = P_{i,k}$ the power consumption of the cell,

$p_c(A_k) = \lambda_b(P_0 + f_{id}.P_c)$, is related to the total non-transmission power consumption from inactive nodes and, P_c the circuit power. The total power consumption of the system, P_{To}, is given as:

$$P_{To} = P_{i,k} + \lambda_b(P_0 + f_{id}.P_c) \tag{19}$$

The energy efficiency, EE, $\gamma_{EE}(A)$ is the ratio of the system total throughput (η_A) over the total energy consumption, which is expressed as:

$$\gamma_{EE_{Dl}}(A_k) = \frac{\eta_{A_k}}{P_{T\,i,k} + \lambda_b(P_0 + f_{id}.P_c)} \tag{20}$$

it comes

$$\gamma_{EE_{Dl}}(A_k) = \frac{\sum_i^A x_{i,k}.\eta_{SE_k}(i)}{\left(\lambda_b.f_{ac}.\Delta P.\frac{P_{out}}{\eta}\right) + \lambda_b(P_0 + f_{id}.P_c)} \tag{21}$$

The SE and EE maximization of the system Λ, is formulated by

$$\Lambda \begin{cases} Max\,\eta_A \\ \\ Max\,\gamma_{EE}(A) \end{cases} \tag{22}$$

B. In a *multi-transmitter system*,

The power consumption, P_{To}, accounting for the backhaul loss, P_{Bach}, is given as:

$$P_{To} = a_{i,k} \sum_{k=1}^{K_a} n_k.P_k.f_{ac} + P_{Bach} + (1 - a_{i,k}) \sum_{s=1}^{K_S} n_s.P_{k_{sl}}.f_{id} \tag{23}$$

where $K = K_a + K_s$ is the number of base station types used in the network (the subscript a, and s means respectively active and sleep), n_k is the total number of BSs of a specific type k-th, P_k is the power consumption of a BS of type k, and the f_k the fractional probability of usage in active mode. P_{Bach} is the backhaul loss, $P_{k_{sl}}$ the power consumption in sleep mode.

$a_{i,k} \in \{0, 1\}$ be a binary variables, which takes the value of 1 if BS, k, is on, and the value 0, if it is in standby mode, S the number of nodes in sleep mode. The energy dissipated in the system, E_{diss}, is given as:

$$E_{diss} = \int P_{To}dt \tag{24}$$

Some accurate results could be obtained with (24), but the power demand profile of each of the nodes consumption is difficult to be retrieved unless the daily cell history is available. It is recommendable to use some seasonal data in such situations.

The available generated power of the grid is given as:

$$P_{Ro,k} = k_o.C_{Cst} + (1 - s_0)(1 - k_0)P_g \tag{25a}$$

k_0 is the proportion of the injected green power, s_0 is a binary number that takes a value of 0 if the constant supply, C_{Cst}, is off, otherwise 1. This C_{Cst} can be a diesel generator, any other power generation from a non clean energy. The factor $(1 - s_0)$, implies that the green generator cannot be switched on while the main power supply, C_{Cst}, is on. P_g, the green injected power, into the system, that aims at reducing the dioxide of carbon emission.

$$P_{Ro,k} = P_{T0} + \delta P \tag{25b}$$

δP the other low power consumptions nodes

The rate of the dioxide of carbon T_{GHG}, which is associated to a power generation in [22] is as:

$$T_{GHG} = \sum_i^K F_{i,k}P_{Ro,k} \tag{26a}$$

$F_{i,k}$ is the factor of emission of the power technology used; $P_{Ro,k}$, the power generation supply at a node.

The computation of the green metric is given as:

$$\rho_g = \frac{\sum_i^k \eta_{A_{i,k}}}{\sum_i^K F_{i,k}P_{R0,k}} \tag{26b}$$

Following the energy conservation law, it follows:

$$s_0k_o.C_{Cst} + (1 - s_0)(1 - k_0)P_{gr} - \left(P_{T0} + \delta P - \frac{E_{bb}}{T}\right) = 0 \tag{27}$$

k_0 is the proportion of the injected green power, s_0 is a binary number that takes a value of 0 if the constant supply, C_{Cst}, is off, otherwise 1. C_{Cst} can be any other form of power generations than a clean energy.

The stored energy, E_{bb}, can be derived as:

$$E_{bb} = \int (s_0k_o.C_{Cst} + (1 - s_0)(1 - k_0)P_g).dt \tag{28}$$

The total injected green power, P_{gr}, comes as function of the power generation:

$$P_g = \frac{T.(P_{To}) - E_{bb} - T * [s_0 k_o . C_{Cst}]}{T.(1 - s_0)(1 - k_0)} \tag{29}$$

The total spectral efficiency (SE) of the system, η_A, is expressed by:

$$\eta_A = \sum_{k=1}^{K} \sum_{i=1}^{A} x_{i,k}.\eta_{SE_{i,k}}(i) + \sum_{k=1}^{K} \sum_{j=1}^{B} (1 - x_{j,k}).\eta_{SE_{j,k}}(j) \tag{30}$$

and the system downlink energy efficiency (EE), $\gamma_{EE_{Dl}}(A)$, is calculated:

$$\gamma_{EE_{Dl}}(A) = \frac{\eta_{A_{active}}}{P_{To}} + \frac{\eta_{A_{idle}}}{P_{To}} \tag{31a}$$

With respect to the active users, it comes as

$$\gamma_{EE_{Dl}}(A) = \frac{\sum_{k=1}^{K} \sum_{i=1}^{A} x_{i,k}.\eta_{SE_{i,k}}(i)}{a_{i,k} \sum_{k=1}^{K_a} n_k.P_k.f_{ac} + P_{Bach}} \tag{31b}$$

and the inactive users

$$\gamma_{EE_{Dl}}(A) = \frac{\sum_{k=1}^{K} \sum_{j=1}^{B} (1 - x_{j,k}).\eta_{SE_{j,k}}(j)}{a_{i,k} \sum_{k=1}^{K_a} n_k.P_k.f_{ac} + P_{Bach}} \tag{31c}$$

The maximization of the EE, $\gamma_{EE_{Dl}}$, and the SE, η_A, for the system is given as:

$$\Lambda \begin{cases} Max \ \eta_A \\ \\ Max \ \gamma_{EE}(A) \end{cases} \tag{32}$$

subject to

$$\sum_{i=1}^{M} P_{i,k} \leq P_{R0}^{max} \tag{33}$$

$$A + B = M \tag{34}$$

$$\eta_A \leq T_0 \tag{35}$$

$$P_g \leq P_{T0}(1 + \delta P) \leq P_{Ro,k} \tag{36}$$

$$\sum_{k}^{K} \sum_{i=1}^{M} \eta_{SE_{i,k}} \leq T_0^{max} \tag{37}$$

$$\sum_{i=1}^{M} C_{i,k} \leq T^{max} \tag{38}$$

It would ideally be more convenient to consider the idle users as unscheduled and awaited data. T_0 is the total throughput of the network and T is the total time to scheduled users.

3   Problem Resolution

In this section the analytical, and Pareto methods are addressed.

3.1   Analytical Resolution for Finding the Rate of the Injected Green Power

The analytical resolution can be adopted to find the accurate rate of the injected green power of the (31a, 31b, 31c) into the system.

The total injected green power, P_g, comes as:

$$P_g = \frac{1}{T} \left[\frac{T.(P_{To})}{(1 - s_0)(1 - k_0)} - \frac{T.s_0 k_o.C_{Cst}}{(1 - s_0)(1 - k_0)} - \frac{E_{bb}}{(1 - s_0)(1 - k_0)} \right] \tag{39}$$

Firstly, assuming $s_0 = 1$, this implies that the main supply is on. This is meaning that, the green power will tend to an infinite value. Thus setting the optimal value of k_0, will not be important.

Secondly, when the $s_0 = 0$; this implies that the main supply is off,

$$P_g = \frac{T.(P_{To}) - E_{bb}}{T * (1 - s_0)(1 - k_0)} \tag{40}$$

setting the optimal value for k_0, means k_0 must be close to 1 so that the injected green energy in the system may also be approaching the infinity value, and yielding a load dependency.

Nevertheless, if a decision is to run on the maximum green energy together with other generating sources, then the total injected green power, P_{gr}, comes as:

$$P_g = \frac{T.(P_{To}) - E_{bb} - T.s_0 k_o.C_{Cst}}{T * (1 - k_0)} \tag{41}$$

For $s_0 = 1$, the other generation power source is on, for k_0, means k_0 must be very close to 1 and the green energy injected into the system also approaches the infinity value. To find the optimal value of the green energy in (41), the Hôpital's method can be used.

$$\frac{d(P_g)}{dk_0} = \frac{d}{dk_o}\left(\frac{T.(P_{To}) - E_{bb} - T.s_0 k_o.C_{Cst}}{T*(1-k_0)}\right) \tag{42}$$

This comes

$$\frac{d(P_g)}{dk_0} = \left(\frac{T.(P_{To}) - E_{bb} - T.s_0.C_{Cst}}{-T}\right) \tag{43}$$

This result is independent of k_0, which implies that the injected rate of the renewable energy into a telecommunication system that does not matter rather the time of the service utilisation, which becomes delicate.

3.2   The Pareto Method for Finding the Optimal Energy Efficiency as Function of the Injected Green Power

In other to find the Pareto set front (PSF) as solution of the multi-objective problem, two set of variables are considered the active and idle users.

Let A denote the active users a class of population, and B that of the idle users a class of population.

$$A \cup B = C$$

where C is the total population of users under a servicing area.

Let $P_\epsilon(Y)$ be the set of all $\in$-approximate Pareto fronts of Y.

Definition 1 ($\in$-dominance). Let, $b \in Y$. a is said to $\in$-dominant b for some $\epsilon > 0$, denoted as $a >_\epsilon b$ if

$$\in * a_i \geq b_i \qquad \forall i \in \{1, \ldots, k\}$$

Definition 2 ($\in$-approximate Pareto Front). Let $Y \subseteq \mathbb{R}^{+k}$ be a set of vectors and $\in \geq 1$. Then a set Y, is called an $\in$-approximate Pareto front of Y, if any vector $b \in Y$ is $\in$-denominated by at least one vector $b \in Y_\epsilon$

$$\forall b \in Y, \ \exists a \in Y_\epsilon \qquad a >_\epsilon b$$

Definition 3 ($\in$-Pareto fronts). Let $Y \subseteq \mathbb{R}^{+m}$ be a set of vectors and $\epsilon > 0$. Then a set
$Y_\epsilon^* \subseteq Y$ is called an ϵ-Pareto of Y;
If Y_ϵ^* is an ϵ-approximate Pareto set of Y, i.e. $Y_\epsilon^* \in P_\epsilon(Y)$, and
Y_ϵ^* contains Pareto points of Y only, i.e. $Y_\epsilon^* \subseteq Y^*$.
The set of all $\in$-Pareto fronts of Y is denoted as $P_\epsilon^*(Y)$.

The problem (32) can be solved by the use of the Karush-Kuhn Tucker initial conditions [23]. However, assuming q^* be an optimal solution of the (32), with the Dinkelbach's theorem [24], it comes as:

$$q^* = \frac{\eta\left(P_g^*\right)}{\gamma_{EE}\left(P_g^*\right)} = \max \frac{\eta\left(P_g^*\right)}{\gamma_{EE}\left(P_g^*\right)} \tag{44}$$

with P_g^* the optimum point of the generated green energy.

The point P_g^* is a Pareto optimum, if there is a feasible solution for all the objectives.

The Eq. (44) is the ratio of a concave function and an affine function, thus is a quasi-concave function. The point $(\eta\left(P_g^*\right), \gamma_{EE}\left(P_g^*\right))$ is a Pareto optimal solution to (32), if there is a feasible solution for at least one objective function $\{\eta\left(P_g\right), \gamma_{EE}\left(P_g\right)\}$, to (32) such that $\eta\left(P_g\right) > \eta\left(P_g^*\right)$ and $\gamma_{EE}\left(P_g\right) < \gamma_{EE}\left(P_g^*\right)$.

From (44), it comes that $\eta(q^*) = \max\left[\eta\left(P_g^*\right) - q^* \cdot \gamma_{EE}\left(P_g^*\right)\right]$ and the optimal value $q^* = \frac{\eta\left(P_g^*\right)}{\gamma_{EE}\left(P_g^*\right)} \Leftrightarrow \eta(q^*) = \max\left[\eta\left(P_g^*\right) - q^* \cdot \gamma_{EE}\left(P_g^*\right)\right] = 0$.

The sufficient condition for q^* to be a Pareto optimal solution is that:

$$q^* = \frac{\eta\left(P_g^*\right)}{\gamma_{EE}\left(P_g^*\right)} \geq \frac{\eta\left(P_g\right)}{\gamma_{EE}\left(P_g\right)} \qquad \forall P_g \in \Lambda$$

denoted,

$$\eta\left(P_g\right) - q^* \gamma_{EE}\left(P_g^*\right) = 0$$

and $\eta\left(P_g\right) - q^* \gamma_{EE}\left(P_g^*\right) \leq 0$ when the quantity $\eta\left(P_g\right) - q^* \gamma_{EE}\left(P_g^*\right)$ reaches the maximum value, when $P_g = P_g^*$ and $\max \eta\left(P_g\right) - q^* \gamma_{EE}\left(P_g^*\right) = 0$.

The corresponding optimum value of the injected green energy P_g^* is the optimal solution of the problem (32), if only $\eta\left(P_g\right) - q^* \gamma_{EE}\left(P_g^*\right) = 0 \ \forall P_g \in \Lambda$, this yields $\eta\left(P_g\right) - q^* \gamma_{EE}\left(P_g\right) \leq \eta\left(P_g^*\right) - q^* \gamma_{EE}\left(P_g^*\right) \leq 0$; thus $\forall P_g \in \Lambda$, $q^* = \frac{\eta\left(P_g^*\right)}{\gamma_{EE}\left(P_g^*\right)}$ and $q^* \geq \frac{\eta\left(P_g^*\right)}{\gamma_{EE}\left(P_g^*\right)}$.

It can therefore be concluded that the equivalent problem in (32) has the same optimal solution with the original problem in (31a, 31b, 31c).

Algorithm : Power Consumption Minimization of the System
1. set the initial the number of Users; and active Nodes $K = K_a$
2. Initialization of the maximum iteration indices T_{max}, the t=1 and the stopping test tolerance ξ
Compute the energy efficiency
3. While $\left|EE^{(t)} - EE^{(t-1)}\right| < \xi$ or $t \leq T_{max}$ do
4. 1. Update the power consumption of the base station $P_{i,k}{}^{(t+1)}$
    2. set $t = t + 1$ and compute $EE^{(t+1)}$
    3. Compute the green metric

$$\rho_g = \frac{\sum_i^k \eta_{A_{i,k}}}{\sum_i^K F_{i,k} \, P_{R0,k}}$$

7. End

4 Methodology and Materials of the Problem Simulation

The validation method has consisted in performing some simulation based upon a linear programing approach. The Matrix Laboratory (MatLab) software was used for the simulation platform. A random power transmission of the users was adopted. The nodes' power transmission was closely chosen to be conformed to the LTE radio parameters. These parameters are given in the Table 1. The cell power consumption [5, 6] was chosen relative to the various cell sizes.

Table 1. The radio parameters

Description		Symbol	LTE
Number of users		N	150
Constant related to the cell-size distribution		a	0.6
User density		$\mu = u$	0.95
BS density		$\lambda_b = b$	0.33
BS-User density		d = b/u	
Variance of AWGN channel		D	25
Environment fading		h	1
Operating frequency		f	2700 MHz
Base station transmit-power (CBS)	Power SISO amplifier	P_{Amp}	300 W
		η	6.67%
		P_{Tx}	20 W
			43 dBm
The circuit power of transceivers		P_c	100
The non-transmission power consumption		P_o	
Digital signal processor		P_{dsp}	100

(continued)

Table 1. (*continued*)

Description	Symbol	LTE
Signal generator	P_{gen}	384
AC-DC converter	P_{conv}	100
Backhaul link equipment	P_{link}	80
Air-Conditioner	P_{cool}	690

The wind data in Lome is shown in Fig. 2.

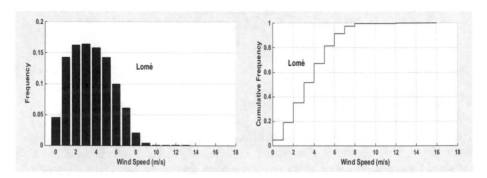

Fig. 2. The wind speed data of Lome

The wind data indicate that the average wind speed is around 4 m/s. The monthly average solar irradiation shows that it is about 4.4 kWh/m^2/day in Lome (see Fig. 3).

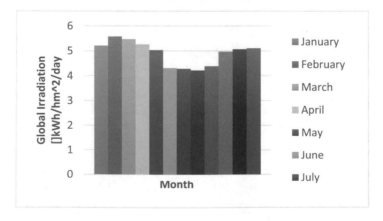

Fig. 3. Monthly solar irradiation of Lome

The climatic data were obtained from the national meteorological center in Lome. The wind data indicate that the average wind speed is around 4 m/s. The monthly average solar irradiation shows that it is about 4.4 kWh/m²/day in Lome.

5 Results and Discussion

This section presents the simulation results. A family of subplots is illustrated in Figs. 4, 5 and 6. The first and the fourth subplots describe the SNR against the EE. It is shown that the energy efficiency reaches a maximum point before it decreases. Similar observations are made in Fig. 4 (3shots are made), the graphs are typically not the same but display the same trend due to the random transmission of the users. The subplot 3 illustrates the ASE against the EE.

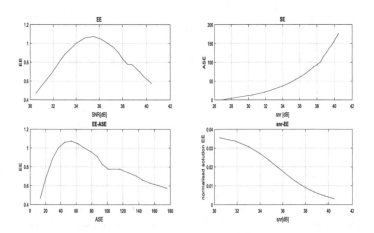

Fig. 4. The area spectral efficiency against the Energy Efficiency 1st shot

However, when implementing the Pareto optimal solution algorithm, the obtained data are shown in Fig. 7. From the resolution method with Pareto fronts, the partial results may indicate a shrink when the idle users are more than the active. The right plane in Fig. 7 may indicate the feasible solutions, since solution cannot be negative. It should be pointed out that (18) is feasible since the nodes are dynamically solicited. The solutions may indicate that the function $g(q)$ is negative when q approaches the infinite, and $g(q)$ is positive when q tends to the negative infinite. $f(q)$ is a convex function with respect to q.

In Fig. 8, the user spectral efficiency and the CBS spectral efficiency are computed against the system energy efficiency, a result of 42% is achieved for the system energy efficiency and close to 38% for the user equipment with the proposed algorithm.

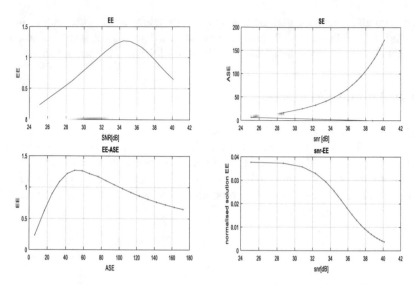

Fig. 5. The area spectral efficiency against the energy efficiency (2nd shot)

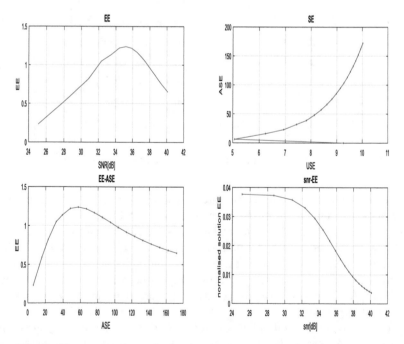

Fig. 6. The area spectral efficiency against the energy efficiency (3rd shot)

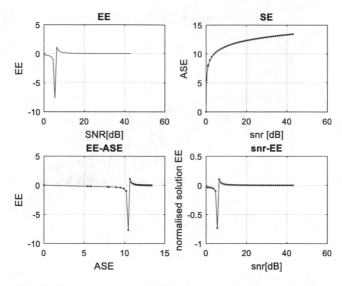

Fig. 7. The area spectral efficiency against the energy efficiency

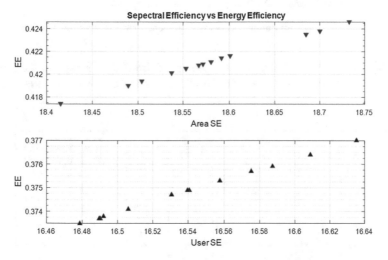

Fig. 8. The optimal points EE vs SE

Figure 9 depicts that the energy efficiency has definitively a maxima, which is the optimal point a compromise between the number of users within the cell. It can be said that the proposed algorithm may give an optimised solution in terms of the energy efficiency. This is further illustrated in Fig. 9 in a three dimension for the system energy efficiency against the spectral efficiency and the bit error rate. The black dots are the Pareto set front. It should be argued that in assuming random transmitting of the users, some users with 0 dBm transmits power are idle users.

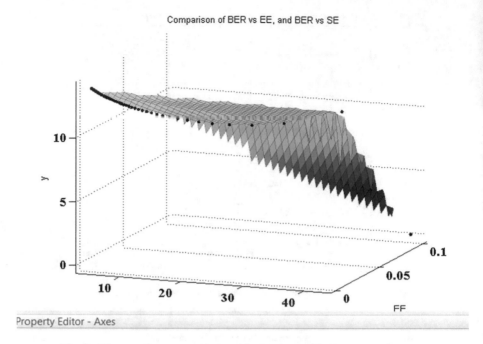

Fig. 9. The system comprised energy efficiency and the spectral efficiency

In Figs. 10, 11, 12, 13 and 14 the wind speed, irradiation solar and the rate of the injected green are considered as function of the number of the small cells.

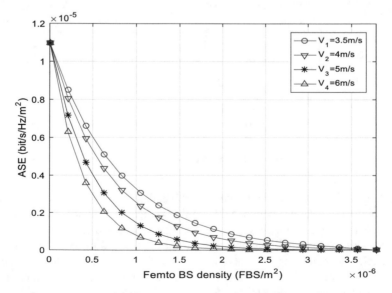

Fig. 10. The system spectral efficiency as function of the wind speed

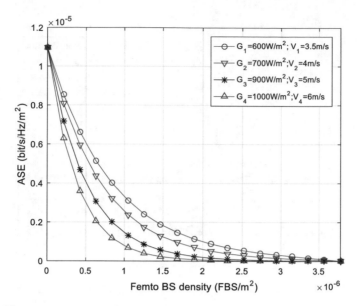

Fig. 11. The system spectral efficiency as function of the wind speed and solar irradiation

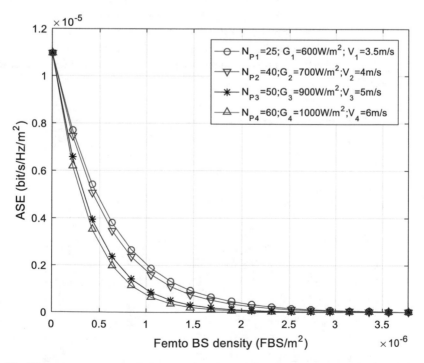

Fig. 12. The system spectral efficiency as function of the wind speed and solar irradiation, for k = 0.5

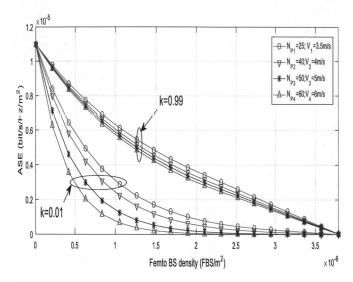

Fig. 13. The system spectral efficiency as function of the wind speed and solar irradiation

In Figs. 13 and 14, it displays that when the value of k is close to 1, the small cells number has no effect since the injected power is sufficient and tends to infinite. However, when the value of k is small, the site may require sufficient solar panels and an acceptable windy condition that will be able to supply that transmitter with a compromise number of small cells. When an ever increase number of small cells are used the spectral efficiency is highly comprised and it is poor, and similar results were obtained in [25].

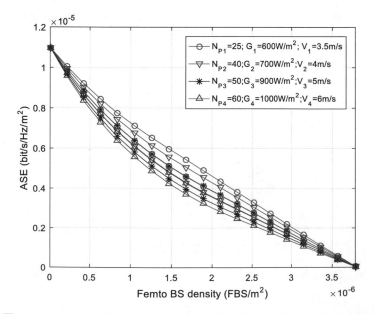

Fig. 14. The system spectral efficiency as function of the wind speed and solar irradiation k = 1.

6 Conclusion

This paper presents the modelling of the integration of renewable energy for a green communication. It also addresses the challenge for this integration with emphasis on measures that should be put in place to cater for their intermittence and maximize their generation efficiency in the system. A multi objective problem on the spectrum and energy efficiency was formulated as function of the injected green energy, and an attempt of resolution with the Pareto methods were proposed.

Moreover, some results will be generated with respect to the number of wind aerogenerator and solar panels that will be needed for the system. It should be indicated that this kind of problem would require the update profile of power demand of the cell area in accordance with the users' distribution, arrival and the exit rate. In this proposed study the energy storage device was not investigated. This is another issue on the size of the storage unit in telecommunication systems for cellular networks. An artificial neural network method also remains an explored approach to predict the cell's status based upon the number of active users.

References

1. Rahnema, M.: UMTS Network Planning, Optimization and Inter-operation with GSM. Wiley (2008)
2. Ajay, R.M.: Fundamentals of Cellular Network Planning and Optimization. Wiley, West Sussex (2004)
3. Yan, C., Oliver, B., Azeddine, G., Antonio, C., Chi-En, W.: Energy saving: scaling network energy efficiency faster than traffic growth. In: IEEE WCNC Workshop on Future green End-to-End Wireless Communication, Shangai (2013)
4. Sunil, V.: Trends in green wireless access. Fujitsu Sci. Tech. **45**(04), 404–408 (2009)
5. Hasan, Z., Boostanimehr, H., Bhargava, V.K.: Green cellular networks: a survey, some research issues and challenges. IEEE Commun. Surv. Tutor. **13**(4), 524–540 (2011)
6. Gunther, A.: How much energy is needed to run a wireless network (2011). https://www.ict-earth.eu/downloads/presentaions/2010-11-article05621969.pdf
7. Han, S., Yang, C., Molisch, A.F.: Spectrum and energy efficient cooperative base station doze (2013)
8. Samuel, C.Y.: CDMA RF Engineering. Artech House Inc., Boston (1998)
9. Parsons, J.D.: The Mobile Radio Propagation Channel, 2nd edn. Wiley, London (1992)
10. Zhuang, W., Zhou, Y.: A survey of cooperative MAC protocols for mobile communication networks (2013)
11. Luis, M., et al.: Challenges and enabling technologies for energy aware in mobile radio network. IEEE Commun. Mag. 66–72 (2010)
12. Touzri, T., Ghorbel, M.B., Hamdaoui, B., Guizani, M., Khalfi, B.: Efficient usage of renewable energy in communication system using dynamic spectrum allocation and collaborative hybrid powering. In: IEEE Global Communications Conference (GLOBE-COM), San Diego (2015). http://web.engr.oregonstate.edu/~hamdaoui/papers/2016/tahatwc-16pdf
13. Bonnefoi, R., Moy, C.M., Palicot, J.: New macrocell downlink energy consumption minimization with cell DTx and power control. HAL, France, hal-01486616 (2017). http://hal.archives-ouvertes.fr

14. The IET Communication: Demand Attentive Networks (2013)
15. Guanyao, D., Ke, X., Dandan, L., Yu, Z., Zhengding, Q.: Transmission power optimization in two-way transmission aware relay-aided cellular networks. In: 4th IET International Conference on Wireless, Mobile & Multimedia Networks (2011)
16. Li, C., Zhang, J., Letaief, K.B.: Energy efficiency analysis in small-cell networks, vol. 1306.6169v1 (2012)
17. Lee, S., Huang, K.: Coverage and economy of cellular networks with many base stations. IEEE Commun. Lett. 16(7), 1038–1040 (2012)
18. Montgomery, C.D., Runger, C.G.: Applied Statistics and Probability for Engineers. Wiley, United States of America (2014)
19. Morosi, S., Piunti, P., Del Re, E.: Energy efficient sleep mode management for cellular networks. In: ETSI Workshop on Energy Efficiency, Genoa (2012)
20. Faruk, N., Ayeni, A.A., Muhammad, M.Y.: Powering cell sites for mobile cellular systems using solar power. Int. J. Eng. Technol. 2(5) (2012)
21. Ali, N.A., Mourad, H.-A.M., Elsayed, H.M., El-Soudani, M., Amer, H.H., Daoud, R.M.: General expressions for downlink signal to interference and noise ratio in homogeneous and heterogeneous LTE-Advanced networks. J. Adv. Res. 6, 923–929 (2016)
22. Feng, Y., Zhang, L.: Scenario analysis of urban energy saving and carbon abatement policies: a case study of Beijing city, China. Procedia Environ. Sci. 13, 632–644 (2012)
23. Boyd, S., Vandenberghe, L.: Convex Optimization. University Press, Cambridge (2009)
24. Karmokar, A., Naeem, M., Anpalagan, A., Jaseemuddin, M.: Energy-efficient power allocation using probabilistic interference model for OFDM-based green cognitive radio networks. Energies 7, 2535–2557 (2014)
25. Mir, T., Dai, L., Yang, Y., Shen, W., Wang, B.: Optimal femtocell density for maximizing throughput in 5G heterogeneous networks under outage constraints. In: Proceedings of IEEE 86th Vehicular Technology Conference (IEEE VTC 2017 Fall), Toronto (2017)

Effect of the Silicon Substrate in the Response of MIS Transistor Sensor for Nano-Watts Light Signal

J. Hernández-Betanzos, A. A. Gonzalez-Fernandez, J. Pedraza, and M. Aceves-Mijares[✉]

Electronics Department, INAOE, Apdo. 51, 72000 Puebla, PUE, Mexico
maceves@inaoep.mx

Abstract. In this paper, the optical response of MOS-like transistors with a Si_3N_4 integrated waveguide as the gate dielectric, and different substrate dopant concentrations is studied. Simulation results show the possibility to integrate a MIS transistor as a detector in an electrophotonic circuit with a compatible bulk CMOS fabrication process and obtaining electrical gain for low power light signals (below 400 nW).

Keywords: Electrophotonics · Integrated photonics · Planar waveguide · Si photonics

1 Introduction

Electrons and photons operating simultaneously in a monolithic chip of silicon is the goal of the new so called electrophotonics [1, 2]. Currently, integrated photonics have evolved from microelectronics, and integrated optical operations are already available in different silicon foundries [3]. Typically, silicon on insulator substrates are used to integrate optical elements, and external light sources must be bonded and/or coupled to the chips [4]. However, such procedures are complicated and not compatible with the standard industrial integration using silicon technologies for electronics, particularly with CMOS. Applications such as lab-on-a-chip or chemical and biological sensing could take maximal advantage of the development of totally integrated silicon electrophotonic chips. One of the most prominent limitations to have a whole silicon electrophotonic chip is the integration of a silicon light source. Consequently, many research groups have been working to develop light sources compatible with silicon fabrication technologies. Two main approaches have been reported: silicon nanocrystals embedded in a dielectric matrix, and reverse biased PN junctions [5–9]. However, the emission power of such devices is low as compared to non-silicon-compatible sources, and faraway of the intensities obtained by lasers.

Despite Pavisi *et al.* have reported optical gain in silicon nanocrystals obtained by Si ion implantation into silicon dioxide as far back as the year 2000 [10], it was only until nowadays that Wang *et al.* reported an optically pumped silicon laser made with silicon nanocrystals [11]. This opens the possibility of having an electrically pumped silicon laser.

K. Arai et al. (Eds.): FTC 2019, AISC 1070, pp. 781–794, 2020.
https://doi.org/10.1007/978-3-030-32523-7_58

Many applications can take better advantage of multiple relatively low power light sources distributed in the chip, instead of one single powerful light source. Electrophotonic systems using relatively low power light emitting devices have been demonstrated in the past [12, 13]. To improve the efficiency of such systems, light detectors with higher sensitivity have been proposed, taking a novel approach which exploits the direct coupling to both the light source and the detector of the waveguides transporting the light [14]. These include bipolar and MOS technologies integrating the waveguides and the detectors to form electrophotonic devices. While heterostructures based on SOI and MOS like transistors have been proposed as light sensors in the past [15–18], the applications and light detection schemes are not suited for distributed integrated light sources, as they are designed for relatively high light intensities and non-fully integrated sources.

In this paper, computer simulations are used to study a device based on a waveguide directly integrated to a MOS-like transistor, which is a completely novel approach to conduct the light to a detector. The oxide in the transistor is replaced by the silicon nitride which is also part of the waveguide transmitting the light to the gate, having effectively an electrophotonic metal-insulator-semiconductor (MIS) structure. The study is limited to identify the effects the substrate concentration, and therefore the depletion region characteristics, on the electrical gain of optical signals with power below 400 nW injected directly from the gate area, in order to aid the fabrication process design of future devices.

The manuscript first describes the structure and modelling of the device in Sect. 2, to afterwards discuss the simulations in Sect. 3, which is divided in two parts: the first corresponding to the electrical behavior, and the second to the optical. Finally, Sect. 4 presents the conclusions of the work.

2 Device Model

The structure of the waveguide integrated with the transistor was simulated with a standard NMOS transistor model to which light was injected from the gate oxide. This was accomplished using the Silvaco simulation platform [19, 20]. Substrate doping concentrations from 1×10^{12} cm^{-3} to 1×10^{15} cm^{-3} were considered. Tables 1 and 2 summarize the values of the constants and characteristics employed for the simulations [20, 21].

Table 1. Model constants used in the simulations.

Constant	Value
N_A: Substrate Concentration	1×10^{12} to 1×10^{15} cm^{-3}
t_{nit}: Nitride thickness	140 nm
L: Channel length	40 μm
W: Transistor Width	1 μm
μ_n: Electron mobility	1000 cm^2/Vs
μ_p: Holes mobility	480 cm^2/Vs

(continued)

Table 1. (*continued*)

Constant	Value
T: Temperature	300 K
k: Boltzmann constant	1.380648×10^{-23} J/K
q: Electron charge	1.602176×10^{19} C
k: State Surface Density	1×10^{10} cm^{-2}
ε_0: Vacuum Permitivity	8.85418×10^{-14} F/cm
ε_{nit}: Silicon Nitride Dielectric Constant	7.5
n_i: Intrinsic Concentration (Room Temperature)	$\approx 1 \times 10^{10}$ cm^{-3}
φ_M: Metal Gate Workfunction (Aluminium)	4.1 eV
χ_{Si}: Silicon Electron Affinity	4.05 eV
ε_{Si}: Silicon Dielectric Constant	11.7
E_g: Bandgap energy	1.12 eV

Table 2. Parameter values for each substrate concentration as obtained by the solution of the theoretical models

Concentration (cm^{-3})	Surface Potential φ_s (V)	Al Work function – Semiconductor φ_{MS} (V)	Depletion Width W_D (μm)	Threshold Voltage V_{TH} (V)
1×10^{12}	0.238	−0.5791	17.544	−0.3688
1×10^{13}	0.357	−0.6386	6.796	−0.2922
1×10^{14}	0.476	−0.6981	2.481	−0.1718
1×10^{15}	0.595	−0.7576	0.877	0.1002

In order of validate the simulations, some results were corroborated comparing them to the following well know theoretical models (see for example [21]).

The threshold voltage is defined as:

$$V_{TH} = \frac{qN_A W_D}{C_{nit}} + \varphi_S(inv) + \varphi_{MS} - \frac{Q_{SS}}{C_{nit}} \qquad (1)$$

Where q is the electron charge, C_{nit} is the capacitance with nitride (as this is the dielectric in the gate in the case of the waveguide-integrated detector), W_D is the depletion width, φ_S the surface potential, φ_{MS} the metal-semiconductor work function, Q_{SS} the interface oxide-semiconductor charge density, and N_A the acceptor substrate concentration.

The surface potential on strong inversion is approximately 2 times the bulk potential φ_B, and can be expressed as follows:

$$\varphi_S(inv) \cong 2\varphi_B = 2\frac{kT}{q}\ln\left(\frac{N_A}{n_i}\right) \qquad (2)$$

where n_i is the silicon intrinsic concentration. The metal–semiconductor Work function is defined as:

$$\varphi_{MS} = \varphi_M - \chi_{Si} - \frac{qE_g}{2} - \varphi_B \tag{3}$$

where φ_M is the metal work function, χ_{Si} the silicon electron affinity, and E_ρ the bandgap energy.

The depletion Width on strong inversion is expressed as:

$$W_D = \sqrt{\frac{2\varepsilon_s \varphi_S(inv)}{qN_A}} \tag{4}$$

where ε_{Si} is the silicon permittivity.

3 Simulation Results and Discussion

3.1 Structural and Electrical Simulation

The simulations were done in two parts. First, the device structure and the technological process were simulated with Athena [19]; and then, the electrical simulation was done with Atlas [20]. The main process steps were:

1. Selection of P type substrates with different acceptor concentrations
2. Deposition of masking oxide: 1 μm-thick
3. Photolithographic definition of drain and source areas
4. Oxide etching to define drain and source windows
5. Implantation of Boron for drain and source. Dose: 8×10^{14} cm^{-2}, Energy: 20 keV.
6. Thermal diffusion: T = 1100 °C, t = 180 min
7. Mask oxide removal
8. Deposition of 140 nm of Si_3N_4 as gate dielectric
9. Etching of silicon nitride
10. Deposition of masking oxide: 1 μm-thick
11. Photolithography to define metal contacts
12. Deposition of aluminum: 1 μm-thick
13. Etching of aluminum
14. Annealing at 350 °C for 120 min in N_2 flow

Figure 1 shows the structure obtained after simulation for a substrate concentration of 1×10^{12} cm^{-3}, and Table 3 compiles the depletion widths for all the substrate concentrations. To validate the simulations, the depletion width was also calculated, as shown in Table 3. As can be seen, the simulated and calculated values have a reasonable agreement.

Table 3. Depletion width for different substrate concentrations as obtained by simulations and the simplified model.

Substrate concentration (cm^{-3})	Simulation W_D (µm)	Calculated W_D (µm)
1×10^{12}	16.663	17.544
1×10^{13}	6.819	6.796
1×10^{14}	2.489	2.481
1×10^{15}	0.876	0.877

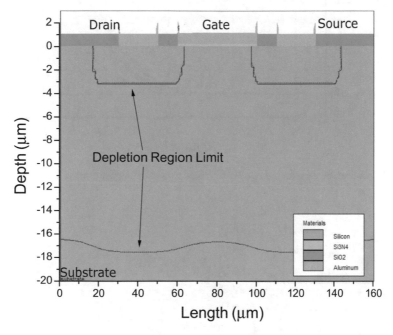

Fig. 1. MOS-like structure obtained after process simulation for a P-substrate concentration of 1×10^{12} cm^{-3}. The depletion width is 16.66 µm, and the gate nitride thickness is 140 nm. The nitride waveguide is the dielectric of the MIS transistor.

Figure 2 shows the drain current versus gate voltage as obtained using Atlas. The drain voltage is 5 V. The threshold voltages were extracted and are shown in the same figure. All transistors are in depletion mode, and they are on at zero gate-source voltage.

Figure 3 shows the characteristic sub-threshold transistor curves for a 1×10^{12} cm^{-3} substrate concentration. The study covers the whole range of channel states: accumulation, weak inversion, and inversion. However, in this work the sub-threshold regime is of higher interest because of its lower drain current at dark condition (no light being injected).

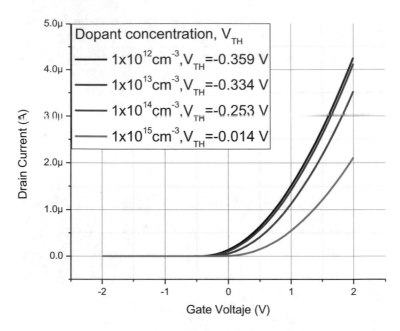

Fig. 2. Drain current vs. gate voltage for different substrate concentrations. The drain voltage is 5 V.

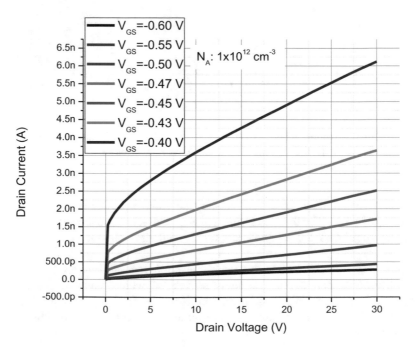

Fig. 3. Drain characteristic curves vs. drain voltage on sub threshold region for different gate voltages.

3.2   Optical Simulation and Discussion

The optical simulations were carried out assuming the light is directly injected from the gate nitride, as it also is the waveguide that would be integrated to the light source. The wavelength used was 600 nm, and the incident optical power was varied from 0 nW (dark condition) to 400 nW. For details on the well-known model used, see for example [22]. The transistor in Fig. 1 was biased in the subthreshold region, and the light was injected directly from the nitride gate, *i.e.*, the waveguide. The gate, source, bulk, and drain were biased as shown in Fig. 4.

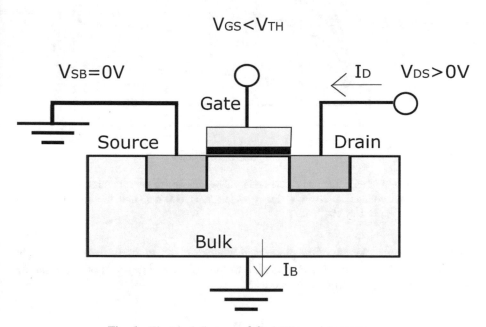

Fig. 4. Electrical diagram of the MIS transistor sensor.

In a semiconductor device, incident light with photon energy equal or larger than the bandgap energy generates electron-hole pairs, which can contribute to a current excess in the presence of an electric field, known as photocurrent.

The ideal incident photocurrent I_{ph} is the number of incident photons per unit of time n_{ph} expressed in electric current units. It depends on light incident power P_{in} and the photon wavelength as follows:

$$I_{ph} = qn_{ph} = q\frac{P_{in}\lambda}{hc} \tag{5}$$

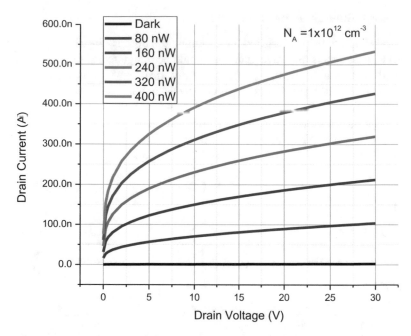

Fig. 5. Drain current as a function of the drain voltage under different illumination conditions. The gate–source voltage is −0.45 V, the wavelength is 600 nm, and the silicon substrate concentration is $N_A = 1 \times 10^{12}$ cm^{-3}.

where, λ is the wavelength of the photons, h Planck's constant, and c the speed of light. The generated photo current I_p is a fraction of this value, and depends on the λ and the material through its quantum efficiency η:

$$\eta(\lambda) = \frac{I_p}{I_{ph}} \tag{6}$$

Figure 5 shows the characteristic curves of I_D as a function of V_D when the transistor is in the subthreshold mode. Each curve corresponds to a different incident optical power. It is clear that the light modulates the drain current. The dark current is 770 pA, and when illuminated with 80 nW at 600 nm wavelength, an increased current up to two orders of magnitude can be achieved.

Figure 6 shows the bulk dark current at subthreshold (I_B) and when the device is illuminated with 400 nW. The I_B under illumination is smaller than the I_{ph} due to the efficiency of generation of carriers per photon. I_D, at dark and under illumination of 400 nW at 600 nm wavelength is also shown in this figure. Under these conditions, according to Eq. (5), the ideal photocurrent is $I_{ph} = 194$ nA. The I_B at this power is equivalent to the current of a silicon photodiode with 80% efficiency.

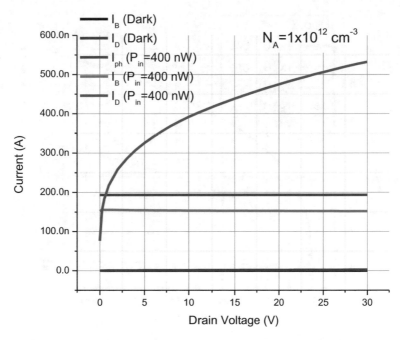

Fig. 6. Transistor's currents (I_D, I_B) in dark and under illumination ($P_{in} = 400$ nW, $\lambda = 600$ nm) vs Drain voltage. It is compared with the photon current. I_D(dark) is in the order of 1 nA and I_B(dark) of 0.1 nA.

On the other hand, the drain current increases as a function of the drain voltage. The current gain is defined as the ratio of the drain current and the bulk current under the same illumination characteristics:

$$G = \frac{I_D}{I_B} \tag{7}$$

Figure 7 shows the current gain of the MIS transistor in the subthreshold regime and under illumination with incident power of 400 nW. At 10 V, the gain is $G = 2.5$.

A remarkable feature of Fig. 7 is the fact that the gain significantly increases with lower dopant concentration in the substrate (higher resistivity). This is most likely caused by the larger depletion region obtained, meaning more photogenerated carriers can contribute to the photocurrent.

Figure 8 shows the photocurrent as a function of the incident optical power, and the substrate concentration. A linear relationship is observed. Then, the responsivity for a 600 nm wavelength under the bias condition used can be extracted from the slope of the curve. This is presented for each substrate concentration in Table 4. The gate-source voltages used for each curve are indicated in the figure.

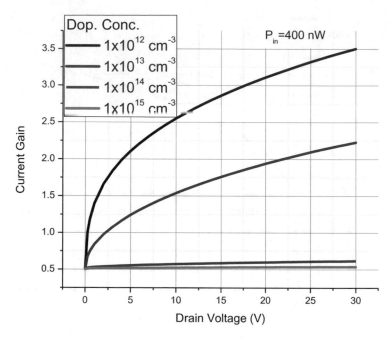

Fig. 7. Current gain of the MIS transistor in the subthreshold regimen and under illumination of 600 nm wavelength and incident power of 400 nW for different dopant concentrations in the substrate.

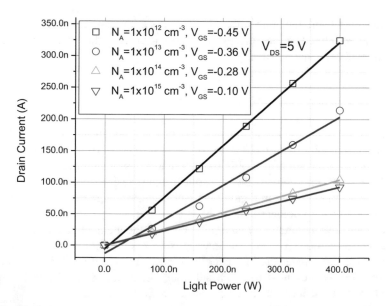

Fig. 8. Drain photocurrent as function of the optical power and the substrate concentration. The substrate concentration varied from 1×10^{12} cm^{-3} to 1×10^{15} cm^{-3}. The points are extracted data. Continuous lines are fitted linear equations as defined in Table 4.

Table 4. Fitted linear relationships between the drain photocurrent and the incident optical power for the different substrates concentration, their slopes, and the coefficient of determination R^2.

Dopant conc. (cm^{-3})	Relationship	m	R^2
1×10^{12}	$I_{ph} = -4.78e-9 + 0.819\ P_{in}$	0.819	0.999
1×10^{13}	$I_{ph} = -1.148e-8 + 0.542\ P_{in}$	0.542	0.985
1×10^{14}	$I_{ph} = 9.25e-10 + 0.262\ P_{in}$	0.262	1
1×10^{15}	$I_{ph} = 3.09e\ 10 + 0.233\ P_{in}$	0.233	1

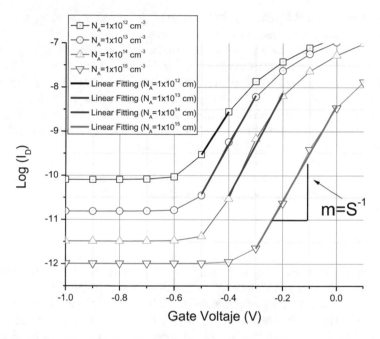

Fig. 9. Sub threshold swing for MIS transistor. The substrate concentration varied from 1×10^{12} cm^{-3} to 1×10^{15} cm^{-3}. Drain voltage is 5 V. The points are extracted data. Continuous lines are fitted linear equations. The slope m as defined in Table 5.

Table 5. Sub threshold swing S obtained for linear fitting. The slope m is the inverse of S.

Dopant conc. (cm^{-3})	Slope m (dec/V)	S (mV/dec)
1×10^{12}	9.68985	103
1×10^{13}	11.18797	89.38
1×10^{14}	11.66766	85.76
1×10^{15}	10.76561	92.93

Figure 9 shows the drain current in log. scale as a gate voltage function. The linear region is the weak inversion region or sub threshold region. This linearity in the plot means that the drain current has an exponential behavior. In order to determine the current decay below the threshold voltage the parameter S is used. S is called sub threshold swing and is expressed as the inverse slope (m) of the drain logarithm curve [23, 24]:

$$S = \left[\frac{d \log_{10} I_D}{dV_{GS}} \right]^{-1} \tag{8}$$

The slope of the different transistors is extracted finding m using linear fitting. Table 5 shows the results and show the S parameter.

The ideal value for S for a bulk MOS transistor is 60 mV/dec [25]. Low values for S are desired. This value increases with the oxide thickness. In the simulated transistors, dielectric thickness is greater than the typical transistor. This obey at the required thickness in the core of the nitride waveguide. However, could be reduced. Values for equivalent transistors using oxide instead of nitride are in the same range. Since nitride must be used in this monolithic photonic integration approach, specific design parameters for this specific application must be taken.

Although the photo response is better in the low concentration transistor, the leakage current is increased as can be seen in Fig. 9. Under this detection scheme, there is a trade-off between optical gain and dark current.

4   Conclusions

A MIS transistor of which the gate dielectric is a waveguide made of silicon nitride was studied through computer simulation. The transistor biased in the subthreshold regime was illuminated directly from the gate considering visible light and power varying from 0 to 400 nW. The results show that the drain current increases as the light intensity increases. In addition, a gain function of the source to drain current is obtained. Such gain increases as the silicon substrate resistivity increases, likely due to the increase of the depletion width as the dopant density reduces. This means that the device can be used to improve the detection of light intensities in the nW range, and in the visible wavelength scope. At the time of writing this document, the experimental process is in preparation. It is important to mention that other integrated devices are currently under study, but this one is totally compatible with MOS and Si-photonics technology. Future simulations and experiments will explore the effects of other parameters such as geometry, drain and source concentrations, and junction depths, in order to reduce the current leakage and improve the optical response.

References

1. Juan-Colás, J., Parkin, A., Dunn, K.E., Scullion, M.G., Krauss, T.F., Johnson, S.D.: The electrophotonic silicon biosensor. Nat. Commun. **7** (2016)
2. Sun, C., et al.: Single-chip microprocessor that communicates directly using light. Nature **528**(7583), 534–538 (2015)
3. Rickman, A.: The commercialization of silicon photonics. Nat. Photonics **8**(8), 579–582 (2014)
4. Bowers, J.E., et al.: Recent advances in silicon photonic integrated circuits. In: Spie Opto, vol. 9774, p. 977402 (2016)
5. Pavesi, L.: Silicon-based light sources for silicon integrated circuits. Adv. Opt. Technol. **2008**(June), 12 (2008)
6. Cabañas-Tay, S.A., et al.: Influence of the gate and dielectric thickness on the electro-optical performance of SRO-based LECs: Resistive switching, IR and deep UV emission. J. Lumin. **192**(March), 919–924 (2017)
7. Muñoz-Rosas, A., Rodríguez-Gómez, A., Alonso-Huitrón, J.: Enhanced electrolumines-cence from silicon quantum dots embedded in silicon nitride thin films coupled with gold nanoparticles in light emitting devices. Nanomaterials **8**(4), 182 (2018)
8. Matsumoto, Y., Dutt, A., Santana-Rodríguez, G., Santoyo-Salazar, J., Aceves-Mijares, M.: Nanocrystalline Si/SiO$_2$ core-shell network with intense white light emission fabricated by hot-wire chemical vapor deposition. Appl. Phys. Lett. **106**(17), 2–7 (2015)
9. Aharoni, H.: Planar light-emitting electro-optical interfaces in standard silicon complemen-tary metal oxide semiconductor integrated circuitry. Opt. Eng. **41**(12), 3230 (2002)
10. Pavesi, L., Negro, L.D., Mazzoleni, C., Franzo, G., Priolo, F.: Optical gain in silicon nanocrystals.pdf. Nature **408**, 440–444 (2000)
11. Wang, D.-C., et al.: An all-silicon laser based on silicon nanocrystals with high optical gains. Sci. Bull. **63**(2), 75–77 (2018)
12. González-Fernández, A.A., Juvert, J., Aceves-Mijares, M., Domínguez, C.: Monolithic integration of a silicon-based photonic transceiver in a CMOS process. IEEE Photonics J. **8** (1), 1–13 (2016)
13. Xu, K., Ning, N., Ogudo, K.A., Polleux, J.-L., Yu, Q., Snyman, L.W.: Light emission in silicon: from device physics to applications, no. November, p. 966702 (2015)
14. Alarcón-Salazar, J., Vázquez, G.V., González-Fernández, A.A., Zaldívar-Huerta, I.E., Pedraza-Chávez, J., Aceves-Mijares, M.: Waveguide-detector system on silicon for sensor application. Adv. Mater. Lett. **9**(2), 116–122 (2018)
15. Abid, K., Khokhar, A.Z., Rahman, F.: High responsivity silicon MOS phototransistors. Sens. Actuators, A **172**(2), 434–439 (2011)
16. Nishiguchi, K., Ono, Y., Fujiwara, A., Yamaguchi, H., Inokawa, H., Takahashi, Y.: Infrared detection with silicon nano-field-effect transistors. Appl. Phys. Lett. **90**(22), 2005–2008 (2007)
17. Averin, S.V., Kuznetzov, P.I., Zhitov, V.A., Alkeev, N.V.: Solar-blind MSM-photodetectors based on Al x Ga 1-x N heterostructures. Opt. Quantum Electron. **39**(3), 181–192 (2007)
18. Zhang, W., Chan, M., Huang, R., Ko, P.K.: High gain gate/body tied NMOSFET photo-detector on SOI substrate for low power applications. Solid State Electron. **44**(3), 535–540 (2000)

19. SILVACO. Inc., "Athena User's Manual," 2013
20. SILVACO. Inc., "Atlas User's Manual." 2015
21. Sze, S.M.: Semiconductor Devices: Physics and Technology, 3rd edn. Wiley, Hoboken (2012)
22. Zimmermann, H.: Integrated Silicon Optoelectronics. Springer, Berlin (2000)
23. Salahuddin, S., Datta, S.: Can the subthreshold swing in a classical FET be lowered below 60 mV/decade?, no. 1, pp. 2–5
24. Jim, J.A., Palma, A.: A simple subthreshold swing model for short channel MOSFETs. Solid-State Electron. **45**, 391–397 (2001)
25. Cheung, K.P.: On the 60 mV/dec @300 K limit for MOSFET subthreshold swing. In: Proceedings of 2010 International Symposium on VLSI Technology, System and Application, vol. 4, pp. 72–73 (2010)

Energy Efficient Balanced Tree-Based Routing Protocol for Wireless Sensor Network (EEBTR)

Rafia Ghoul[1]([⊠]), Jing He[1]([⊠]), Ammar Hawbani[2], and Sana Djaidja[3]

[1] College of Computer Science and Electronic Engineering,
Hunan University, Changsha, Hunan, China
Tigre.eco@live.fr, jhe@hnu.edu.cn
[2] School of Computer Science and Technology,
University of Science and Technology of China, Hefei, Anhui, China
anmande@ustc.edu.cn
[3] Faculty of Medicine, University of Algiers 1, Algiers, Algeria
sanadjal992@gmail.com

Abstract. Standard Tree Routing or General Tree Routing (GTR) is an energy saver protocol designed for low-cost and low-power WSNs to minimize the total energy consumption, its forwarding data process is only via links (Parent-Child). GTR came with improvements such as avoiding flooding in network and providing more energy efficiency, but it does not consider the energy balancing in the network which may lead to network partitioning. The final objective should aim to let all sensors in the same level of tree consume a similar energy in order to maximize the lifetime. However, practically this could be difficult due to the fact that the nodes are located in different communication ranges and only close nodes to BS are eligible to communicate directly with it, and the fact that each sensor consumes different energy rate according to the sensed event frequency and data transmission that lead some sensors die before others. This paper presents an Energy Efficient Balanced Tree-based Routing Protocol (EEBTR) that is designed to achieve a potential balancing of energy usage in each level of the tree such that nodes in the same level have approximately the same number of children. Simulation results show that EEBTR improves GTR performance in terms of both network lifetime and energy-efficiency.

Keywords: Routing Protocol · Hops count · Energy efficiency · Network lifetime

1 Introduction

A network of sensors can be deployed to collect useful information from the field of variety commercial and military for different applications such as surveillance, acoustic and seismic detection, environmental monitoring, it is also growing rapidly in tracking applications, the nodes detect, monitor, and track a target, object, or event [1].

© Springer Nature Switzerland AG 2020
K. Arai et al. (Eds.): FTC 2019, AISC 1070, pp. 795–822, 2020.
https://doi.org/10.1007/978-3-030-32523-7_59

In the literature, a wide range of protocols have been proposed in order to reduce energy consumption of WSN, especially for low-cost and low-power wireless (Micro-sensors) networks used instead of the expensive macro sensors, and to ensure a high quality, the designers have to build a network with a high number of these small devices. Therefore, the protocols for these networks must be designed in such a way that the limited power in the sensor nodes is efficiently used [2]. How to design the routing protocol that reduces the energy dissipation used for communication in each node, that uses minimum number of hops to the sink node and distributes and balances the energy dissipation over all nodes are the most critical issues in WSN [3]. Routing Protocols for Wireless Sensor Networks are divided into four categories; Hierarchical protocols, Data centric protocols, location-based protocol and Network Flow & QoS Aware Protocol.

The Hierarchical (flat) protocols are based on clustering structure that is mainly founded by clusters and the clusters as well are founded by cluster head. The key matter to implement the Hierarchical cluster-based protocols is the selection of the cluster-head, in some algorithms the cluster head is indicated in advance of the deployment, and the energy is without restriction, like the case of Max-min Zpmin algorithm, but in the practical application we can't realize this ideal situation. Thus, the need of algorithm for clustering, in fact many algorithms have been proposed, the most followed algorithm is (DDR) Dynamic, Distributed and Randomized Algorithm, it's especially followed by LEACH. Due to the randomness present in clustering algorithms, number of cluster heads generated varies highly from the optimal count [4]. The Low Energy Adaptive Clustering Hierarchy protocol is a cluster-based routing protocol that came with improvements of the use of nodes, it requires only few nodes unlike the conventional routing protocols that use all nodes, the data transmission via cluster-head utilizes a randomized rotation of local cluster head to evenly distribute the energy load among the nodes sensors in the network, so that make the network has scalability and robustness, and not only that, it also implements data fusion to reduce the amount of information overhead [3], so that the reduce of network traffic and computational complexity. The process is broken up into rounds each consists of two phases; The Setup Phase that organizes and select the cluster heads, and the steady state when the data will be sent to the BS [5]. Later, researchers have proposed many energy efficient routing protocols such as; the Hybrid Energy Efficient Distributed (HEED), Power Efficient Gathering Sensor Information System (PEGASIS), TEEN and BCDCP, and ZRIC. Although the algorithms cited above are effective in scalability and energy saving for WSNs, they are only designed for homogeneous WSN and not suitable for heterogeneous case because of their inability to treat nodes differently and the nodes in heterogeneous WSN have different energy levels. While, the algorithms such as Stable Election Protocol (SEP), Distributed Energy-Efficient Clustering (DEEC), Developed DEEC (DDEEC), Enhanced DEEC (EDEEC) and Threshold DEEC (TDEEC) are designed for heterogeneous WSN. As regards to clustering structure in WSN, the article [6] surveyed energy efficient clustering protocols in heterogeneous wireless sensor networks and compared these protocols on various points like, location awareness, clustering method, heterogeneity level and clustering Attributes.

On the other side, data centric and location-based categories are traditional routing protocols. In the data centric protocol, the BS has to select certain regions to send the queries, and then waits for data from sensors located in the selected regions [7]. The authors in [8] provided the first data-centric called (SPIN) Sensor Protocols for Information via Negotiation that uses the data negotiation scheme among sensor nodes in order to decrease data redundancy and save energy as well. Comparatively than flooding protocol, SPIN gives a factor of 3.5 less of energy dissipation and meta-data negotiation it almost reduces half of data redundancy [7]. But this protocol cannot guarantee the delivery of data. Thus, this protocol is not suitable for such applications that require reliable delivery of data packets within regular intervals.

After that, Directed Diffusion [9] has been developed. The data generated by sensor nodes is named by attribute-value pairs. The base station requests certain type of information by broadcasting interests, and the sensed interest data can be aggregated (through the network hop-by-hop) and then be transmitted back to the base station. Another protocol similar to Directed Diffusion called Energy Aware Routing protocol [10] has been proposed; unlike the directed diffusion protocol it maintains a set of paths instead of one optimal path. Then, many other protocols following similar concept or based in DD (directed diffusion) have been proposed.

The location-based or mobile agent based routing protocols have the goal to know the location of the region to be sent through the sensors located in the selected region in order to diffuse the query to only that region, which will save energy. It can get the information of sensors nodes location via GPS or any estimation algorithms based on received signal strength. After getting the location information, by using power control techniques [11] the consumption of energy could be largely minimized. Some searchers have been focused on the design of energy efficient approaches that achieve full area coverage and maintain the connectivity of the WSN, in [12] the author used some techniques to find the all overlapped covered sub-regions (2 Edges) and (3 Edges) and he ensured the connectivity of the network by finding the maximum covered regions, with this approach he increased the WSN lifetime.

Finally, the Network Flow & QoS Aware Protocol: The provided protocols for this method are developed such that the network traffic is balanced in order to maximize the network lifetime. Many techniques are used at different layers like MAC layer, these techniques avoid collision, overhearing, control packet overrun to minimize the energy consumption.

Researchers have been widely focusing on the design of routing protocols, to forward packets to the sink, along minimum energy consumption, which may cause unbalanced distribution of residual energy among sensor nodes, which may lead to the partition of the network; due to the importance to design an efficient routing protocol that ensure the minimum of the energy consumption and hence the extension of the network's lifetime.

In this work, we assume the following conditions:

- The base station is supposed fixed and located inside the interest area.
- The sensors in the interest area are homogeneous and statics, as shown in Fig. 1.

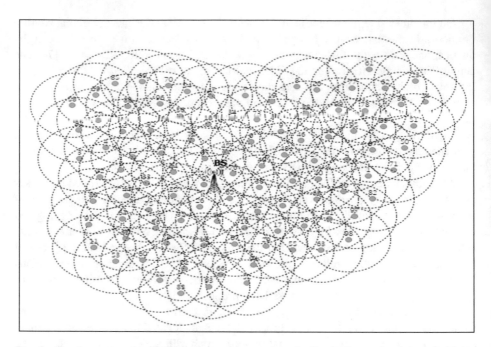

Fig. 1. Random network of 100 homogeneous sensors and a fixed BS near the sink node (0), the dash cycles indicate the communication range for nodes.

Figure 1 shows the appropriate working conditions, in order to achieve longer lifetime for the network, the base station is fixed and located in the center of the interest area, and the sensors are homogeneous and statics. Only nodes close to BS can communicate with it. (sensors have limited communication range). The other nodes should maintain a multi-hop path to communicate with BS. The communication range of each node is represented by dash cycle. The aim of this paper is as following:

- Achieve a potential balancing of energy consumption per level in routing tree such that all nodes at the same level l has approximately the same number of children.
- Provide a performance improvement by the simulation results and comparisons with GTR.

The rest of the paper is organized as follows. Section 2 introduces some related work of energy efficient routing algorithms. In Sect. 3, the Radio Model is presented in details. In Sect. 4, the details of EEBTR algorithm are described. Performance evaluation and comparison are given in Sect. 5, and Sect. 6 concludes this paper.

2   Related Work

Unbalanced energy consumption is an inherent problem in WSNs characterized by the multi-hop routing. Since each node in the network can show different energy usage according to the frequency of event sensing and data transmission, therefore they have different lifetime. This uneven energy dissipation can significantly reduce network lifespan. Energy efficiency and balancing is one of the primary challenges to the successful protocols of WSNs. Significant researches efforts are proposed to develop the energy efficient routing protocols in WSNs, but few of them focused on balancing thus prolonging the lifespan.

At the aim of ensuring the energy efficient and energy balancing, the authors in [13] present a Distance-based Energy Aware Routing (DEAR) based on theoretical deduction and numerical illustration under different energy and traffic models. During the routing process, DEAR treats a distance distribution as the first parameter and the residual energy as the secondary parameter.

An improvement protocol over Tree Routing protocol (TR) is the Enhanced tree routing protocol (ETR) [14], in addition to the parent-child links strategy in TR, it uses links to other one-hop neighbors if it is decided that this will lead to a shorter path. It is shown that such a decision can be made with minimum storage and computing cost by using an address structure, it proved its better energy efficiency comparatively than TR. Some years later, author in [15] designed an Energy-Balanced Routing Protocol (EBRP) by constructing a mixed virtual potential field in terms of depth, energy density, and residual energy. The authors proved by simulation results that the method provides significant improvements in energy balance. Later, [16] came with a tree-based routing protocol called TBRR, which routes packets in the destination node with a highly reliable manner by a process of two stages. TBRR in his process uses a nearly complete set of reliability metrics and keeps searching for the best nodes that satisfy the level of reliability required by data, and then it forwards the packets to them as the next nodes of reliable paths towards the sink. After that, the author in [17] proposed a quadrant-based routing protocol that utilizes the principle of spanning tree. It divided sensor area into four quadrants. Each node selects its parent node by itself based on the distance and residual energy of the parent node. Simulation results shows that the proposed protocol gives better performance than the other protocols with respect to load balancing and network lifetime. Further, author in [18] developed a protocol called EDAL, it aims to generate routes that connect all source nodes with minimal total path cost, under the constraints of packet delay requirements and load balancing needs. The lifetime of the deployed sensor network is also balanced by assigning weights to links based on the remaining power level of individual nodes.

Recently, the authors in [19] developed a new system that processes in three scenarios: power control, non-uniform deployment, and probabilistic switching (ProSwit) routing; especially, ProSwit scenario that has the aim to maximize network lifetime. The paper aims to achieve system-level energy balance in order to prolong network lifetime. Simulation results demonstrate that by exploiting the three schemes, the system-level energy balance with multi-hop routing can be achieved.

Finally, a tree method has also been widely used, in order to maintain the balancing in the protocol, so prolonging the network lifetime. The author in [1] proposed a GLT method which processes in two tiers, the Notification Tree (NT), enhances three mechanisms; the activation, the data cleaning and the energy balancing mechanism, while the second tier; Hierarchical Spanning Tree supports the data reporting mechanism and the lifetime prolonging mechanism.

In this paper, we develop a tree routing protocol called EEBTR to achieve maximum network lifetime, it distributes the energy load evenly among the nodes in the network. Simulation results prove that, the system energy balance-based network lifetime optimization still can be achieved by using this method.

3 Radio Model for EEBTR

This paper, assumes that the radio channel is symmetric such that the energy required to transmit a message of k bits from node x to node y is the same as energy required to transmit the same size message from node y to node x or a given signal to noise ratio (SNR). It assumes a simple model called First Order Radio Model, as demonstrated in Fig. 2. This model is discussed in [13, 20].

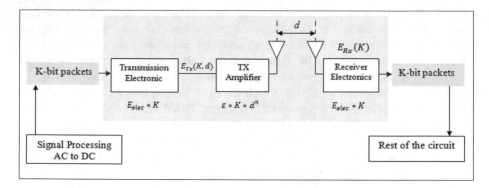

Fig. 2. Radio energy dissipation model

For this model, the energy dissipation for transmit and receive (to run the radio), $E_{elec} = 50\,\text{nJ/bit}$. E_{elec} depends on the factors such as the digital coding, modulation, filtering, and spreading of the signal, it is defined as the energy dissipation to run the radio [20]. The Free space model of transmit er amplifier is $\varepsilon_{fs} = 10\,\text{pJ/bit/m}^2$ and the Multi-path model of transmitter amplifier is $\varepsilon_{mp} = 0.0013\,\text{pJ/bit/m}^4$. Both of Free space and Multi-path models are depending on the distance to the receiver and the acceptable bit-error rate. If the distance is more than a threshold $t_0 = \sqrt{\frac{\varepsilon_{fs}}{\varepsilon_{mp}}}$ (meters), the multi path (mp) model is used; otherwise, the free space (fs) model is used as shown in Eqs. (1) and (2).

A Sensor node will consume $E_{Tx}(k, d)$ of energy to transmit k bits size message over distance d, and consume $E_{Rx}(k)$ of energy to receive transmitted k bits size message.

$$E_{Tx}(d, k) = \begin{cases} (k \cdot E_{elec}) + (k \cdot \varepsilon_{fs} \cdot d^2) & d < t_0 \\ (k \cdot E_{elec}) + (k \cdot \varepsilon_{mp} \cdot d^4) & d \geq t_0 \end{cases} \tag{1}$$

$$E_{Rx}(k) = k \cdot E_{elec} \tag{2}$$

4 EEBTR

EEBTR is a tree-based algorithm that aims to achieve a certain balance between number of hops and energy dissipation in the network, it takes in consideration each two successive levels (l *and* $l+1$) and try to balance between N_l (the number of nodes at level l) and N_{l+1} (the number of nodes at level $l+1$), by trying to give to each node in the level l an average number of children, this average number is calculated by;
$$\mathrm{AV}_l = \left(\frac{N_{l+1}}{N_l}\right).$$

By giving the same number of children to the nodes located at the same level, EEBTR is achieving a potential balance of energy consumption per level in the tree. The proposed algorithm processes in two phases:

Up-Down Balancing Phase (UDB) and **Down-Up Balancing Phase (DUB).** To better explain **EEBTR**, we use the network shown in Fig. 1. Let's assume that the node (0) is the sink (root) node and L is the number of levels in the tree, starting by $l = 1$ which is the location of the sink node, so ($1 \leq l < L$). We analyze the consumption of the energy for each node when transmitting 1000 parquets (1024 bits) from each sensor via a tree topology. The simulation shows that the node (65) is the node that will die first as shown in Fig. 4(a); hence this node has to be saved. In order to prolong the lifetime of this network, and save the node (65), an optimizing strategy should be applied to the tree shown in Fig. 3(a), this strategy will start by **Up-Down Balancing phase (UDB)** then finished by the **Down-Up Balancing phase (DUB)**.

Network Lifetime Definition: It is defined as the time that the first node is declared dead, after an x packets transmission via a designed routing protocol see Table 8. The dead of a node in the network may cause the loss of the functionalities.

Figure 4 shows the simulation of sending 1000 packets of (1024 bits) from each sensor via GTR and EEBTR trees. The assumed sink node is node (0), the communication range is R = 25 m and the initial energy of each sensor is 1 J. The energy dissipation needed to run the radio (E_{elec}) is 50 nJ/bit.

From Fig. 4(a) we can see that the node (65) will die after 220 s by using GTR, while it lives for 92 more rounds via EEBTR as shown in Fig. 4(a) and (c).

Figure 5 shows the residual energy after 220 rounds via GTR tree as shown in Fig. 4(a) and via EEBTR as shown in Fig. 4(b). For example, for node (65), the energy after 220 rounds is: 0% (the node declared dead) in GTR tree as shown in Fig. 4(a), while it is still alive with 40.77% of energy in EEBTR tree as shown in Fig. 4(b), hence the save of (40.77%) of energy for (node 65).

Figure 6 shows the nodes with the residual energy less than 65% for the simulation shown in Fig. 5, for example for node (25) after 220 rounds, it's still alive but with only 9.27% of the initial energy via GTR, while it's still alive with a residual energy of 88.77% via EEBTR, that is to say that EEBTR helped the node (25) to save 79.5% of its energy. The same for the other nodes (EEBTR helped node (27) to save 43.21%, helped node (33) to save 25.4%, helped node (35) to save 28.87%, helped node (38) to save 28.1%, node (65) to save 40.77%, node (76) to save 12.48%, and the node (79) to save 27.09%).

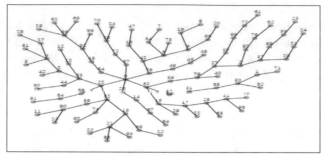

(a)Tree topology for the network represented in **Fig. 1**.

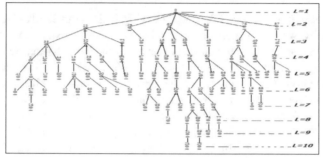

(b) Unbalanced tree (General tree) with 10 Levels. The edges indicate the communication relation between nodes.

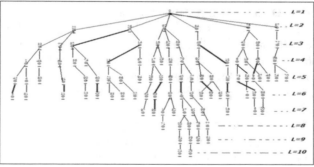

(c): Balancing the tree represented in **Fig. 3(a,b)**, by using **UDB** phase. The blue edges represent the change of position for some sub-trees.

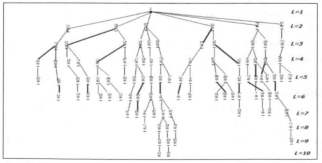

(d): Applying **DUB** to the tree represented in **Fig. 3(c)**. The red edges represent the change of position for some sub-trees.

Fig. 3. Random network with balanced tree and unbalanced tree.

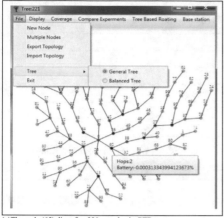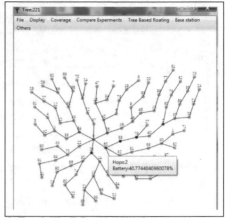

(a)The node (65) dies after 220 rounds via GTR tree.

(b) The node (65) still alive with 40.77% of energy after 220 rounds via EEBTR tree.

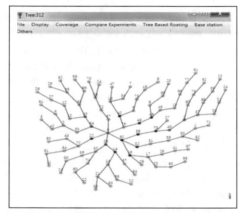

(c)The node (65) dies after 312 rounds via the EEBTR tree

Fig. 4. The situation of node (65) via GTR & EEBTR.

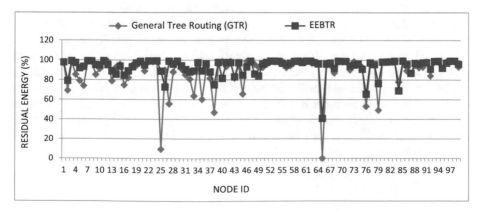

Fig. 5. The residual energy after 220 rounds via GTR and EEBTR

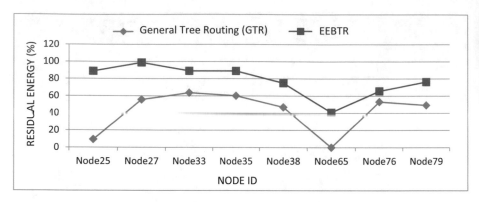

Fig. 6. The nodes with a residual energy less than 56% after 220 rounds via **GTR & EEBTR**

4.1   Tree Routing Building Phase

To build a routing tree it's a very simple process as shown in Algorithm 1. Each node represents a sensor in the tree, these nodes are linked by edges if the sensors can communicate, which means two sensors are linked if they overlapped in the communication rang. Hence, each node can have only one parent, and m children nodes. The node with no child is called a leaf and the sink node in the tree is the root node. To more facilitate the work, we use **exist** (Boolean value) for each node, to show whether the node is already added to the tree, this value is par default false, and becomes true when we add the node to the tree. Each node will select its children; this operation is repeated recursively for each node in the network. This simple algorithm is explained by Algorithm 1 and shown in Fig. 3(a) and (b); the sink node (0) selects and add its 6 children (node (25), node (26), node (76), node (84) and the node (87)) to the tree, then the value of exist for these nodes are true, after that each child node in sequence will add selects its children, the node (65) has three children ((node (16), node (32) and node (38)), and has 32 grand children; the node (76) has 2 children (node (45) and node (49)) and has 13 grandchildren; the node (87) has only one child (node (72)) and 3 grandchildren; the node (25) has 3 children (node (33), node (35) and node (75)) and has 26 grand children; the node (84) has only one child (node (43)) and 6 grand children; and finally the node (26) has only one child (node (14)) and 2 grandchildren. That leads to say; the node (65) is considered as a rooter for 35 nodes; the node (25) is considered as a rooter for 29 nodes; the node (76) is considered as a rooter for 15 nodes, the node (84) is considered as a rooter for 7 nodes; the node (87) is considered as a rooter for 4 nodes and finally the node (26) is considered as a rooter for 3 nodes. Finally, after adding all these nodes to the tree, the tree is already built as shown in Fig. 3(a) and (b). As we can see some nodes have more children than others at the same tree level, that lead to say that the created tree is not balanced. More children mean more network load and then more energy consumption.

ALGORITHM 1: Tree Routing Building Phase

---

1. $R. exist = true;$ // the root node is added to tree

2. $R. parent = null;$ //No parent for the root node

3. $CL = R;$ // current leaves node.

4. $TreeRoutingBuilding(CL)\{$

5. $L = 1;$// the last level of nodes.

6. $foreach\ (node\ in\ L)\{$

7. $if\ (node\ .neighbors\ !=\ null)\{$

8. $for\ each\ (neighboringnode\ .neighbors)\{$

9. $if\ (neighbor\ .exist = false)\{//neighborisnotinthetree.$

10. $node. Childern. Add(neighbor\ );$

11. $neighbor. parent = node;$

12. $CL. add(neighbor);$

13. $neighbor. exist = true;$

14. $\}//end\ if.$

15. $\}//\ end\ for.$

16. $\}//end\ if.$

17. $\}//\ end\ for.$

18. $if\ (CL. Count > 0)$

19. $BuidTree(CL);//$ recursive call to the last level of tree.

20. $\}//\ end.$

4.2   Balanced Tree

Since the created tree in the Algorithm 1 as shown in Fig. 3(b) is not balanced, as we can see, some nodes have more children than others in the same level, more children means more network load and then more energy consumption as each parent acts like a router for its children. Hence, the created tree has to be balanced in order to save the energy. To prolong the lifetime of the network, the proposed algorithm EEBTR with its two phases; **Up-Down Balancing phase (UDB)** and **Down-Up Balancing phase (DUB)** will try to evenly distribute the energy at each level in the tree. At the first phase **Up-Down Balancing**, the balancing starts form the root node located at the level $l = 1$ going down exploring each level and equalizing the number of children for each node in the same level, after finishing **(UDB) phase**, as shown in Fig. 3(c), the **Down-Up Balancing phase** will start, this phase process from the leaves node towards the sink (root) as shown in Fig. 3(d). These two phases are explained below.

4.2.1   Up-Down Balancing (UDB)

The Algorithm **UDB** has the aim to balance between each two successive levels in the tree, in order to give to the nodes located at level l, the same number of children located at level $l+1$, this number is called the average number and it's calculated as: $AV_l = \left(\frac{N_{l+1}}{N_l}\right)$ for each level It starts from the level $l = 2$, as the root node located at $l = 1$ is the only node in this level and has to be linked to all its neighbors (children). The main purpose is to find the node (*NodeOverChild*) which is the node that has a number of children greater than the average number and to find the (*SupporterNode*) which is defined as the node with less children than the average number. #15 is explained as below:

For each level l such that $l < L$ do

1. If any node n_i has m_i children, such that $m_i > AV_l$, then n_i is declared as (*NodeOverChild*).
2. If any node n_i has m_i children, such that $m_i < AV_l$, then n_i is declared as (*SupporterNode*),
3. If any child x of (*NodeOverChild*) is overlapped with (*SupporterNode*) then (*SupporterNode*) will be a new parent for the child x.
4. If any child x of (*NodeOverChild*) is overlapped with any child y of (*SupporterNode*), then the child y will select the node x as its child.

The **Algorithm UDB** is explained by Algorithm 2 and shown in Fig. 3(c).

ALGORITHM 2 : Up-Down Balancing (UDB) Phase

1.	*Up Down Balancing (l){*
2.	*if (l! = 1){*
3.	*int parent = N_l // the number of nodes in l.*
4.	*if (parent > 1){*
5.	*intchildren = N_{l+1};// the number of nodes in l + 1.*
6.	*if (children > 1){*
7.	*intAV = children /parent;*
8.	*for(int n = 0; n < N_l; n + +){*
9.	*NodeOverChild = l[n];// get the nth node in this level.*
10.	*if (NodeOverChild. Children. Count > AV){*
11.	*for(int m = 0; m < N_l; m + +){*
12.	*if (m! = n){*
13.	*SupporterNode = l[m];// get the mth node in this level.*
14.	*if (SupporterNode. Children. Count < AV){*
15.	*parent_child[] = Link();// 0 for parent and 1 for child*
16.	*If (parent_child ! = null){*
17.	*parent = parent_child[0];*
18.	*child = parent_child[1];*
19.	*child. parent = parent;*
20.	*parent. Children. Add(child);*
21.	*nodeOverChild. Children. Remove(child);*
22.	*}// endif.*
23.	*} // endif.*
24.	*}// endif.*
25.	*}// endfor.*
26.	*}// endif.*
27.	*}//endfor*
28.	*}// endif.*
29.	*}//endif.*
30.	*}// endif.*
31.	*}//end.*

From   Fig. 3(b),   we   have $N_1 = 1, N_2 = 6, N_3 = 11, N_4 = 17, N_5 = 24,$ $N_6 = 20, N_7 = 10, N_8 = 5, N_9 = 4$ *and* $N_{10} = 2$, by applying **UDB**, each node in the level $l = 2$, has to have a number $N \cong AV_2 = \left(\frac{11}{6}\right)$ of children, each node at the level $l = 3$, has to have a number $N \cong AV_3 = \left(\frac{17}{11}\right)$ nodes, each node in the level $l = 4$, has to have a number $N \cong AV_4 = \left(\frac{25}{17}\right)$ nodes, each node in the level $l = 5$, has to have a number $N \cong AV_5 = \left(\frac{20}{24}\right)$ nodes, each node in the level $l = 6$, has to have a number $N \cong AV_6 = \left(\frac{10}{20}\right)$ nodes, each node in the level $l = 7$, has to have a number $N \cong AV_7 = \left(\frac{8}{9}\right)$ nodes, each node in the level $l = 8$, has to have a number $N \cong AV_8 = \left(\frac{4}{5}\right)$ nodes, each node in the level $l = 9$, has to have a number $N \cong AV_9 = \left(\frac{2}{4}\right)$ nodes and finally for the nodes in the last level $l = 10$, can't be balanced by using **UDB** because the nodes in the last level are the leaves nodes. Now applying **UDB** for level $= 2$. The node (65) located at $l = 2$, has to be balanced, because it has a number of children greater than the average number, its children are the nodes (node (16), node (32) and node (38)), to balance this node, we have to remove one of its children and link it to another parent located at $l = 2$.

From Fig. 1, the nodes (32) and (16) overlapped with only node (65) in the level $l = 2$, however the node (38) is overlapped with the node (65) and the node (76), but since the node (76) has a number of children equal to the average number $AV_2 = 2$, so the node (76) can't be a new parent for the node (38). For the node (25), it has 3 children and since $3 > AV_2$ then the node (25) needs balance. Its children are (node (33), node (75) and node (35)), for the node (75) it can't be linked to other node, because it is overlapped with only the node (25) in $l = 2$, however for the node (35) it is overlapped also with the node (26), and since the node (26) has only one child and $1 < AV_2$, so this node can be a new parent for the node (35), hence the sub-tree rooted at (35) will be removed from the node (25) and linked to the node (26) as shown in Fig. 3(c). For the third node (33) it is overlapped also with the node (84), so it can be a new parent for the node (33), but since after applying **UDB** for the node (26), the node (26) doesn't need a balance, so the node (33) will not be moved in the **UDB** phase. For the node (87) it has only one child, which means it doesn't need to be balanced at the **UDB phase.** By applying **UDB phase** to the tree shown in Fig. 3(a, b), we will get a balanced tree shown in Fig. 3(c). Applying **UDB phase** for Fig. 3(b) will lead to the following changes;

1. At $l = 3$: The sub-tree rooted at (35) has been removed from the node (25) and linked to the node parent (26).
2. At $l = 4$: The sub-tree rooted at (15) has been removed from the node (33) and linked to the node parent (43).
   The sub-tree rooted at (13) has been removed from the node (35) and linked to the node parent (14).
3. At $l = 5$: The sub-tree rooted at (9) has been removed from the node (6) and linked to the node parent (48).

4. At $l = 6$: The sub-tree rooted at (3) has been removed from the node (4) and linked to the node parent (42).

> The sub-tree rooted at (57) has been removed from the node (4) and linked to the node parent (12).
> The sub-tree rooted at (11) has been removed from the node (90) and linked to the node parent (91).
> The sub-tree rooted at (22) has been removed from the node (21) and linked to the node parent (60).
> The sub-tree rooted at (63) has been removed from the node (28) and linked to the node parent (55).
> The sub-tree rooted at (2) has been removed from the node (27) and linked to the node parent (40).
> The sub-tree rooted at (69) has been removed from the node (86) and linked to the node parent (96).
> The sub-tree rooted at (19) has been removed from the node (9) and linked to the node parent (73).

5. At $l = 7$: The sub-tree rooted at (98) has been removed from the node (41)and linked to the node parent (63).

After balancing the tree shown in Fig. 3(a, b) by using **UDB phase,** we will get a tree shown in Fig. 3(c). From Fig. 5, we can see that some nodes has high energy more than (90%) and others with less than 65%, that lead us to think about modifying the **UDB phase.** Thus, the purpose of **Down-Up Balancing (DUB).**

4.2.2   Down-Up Balancing (DUB)

From Figs. 5 and 6 we can see that some nodes have the value of energy less than 56%, and others still alive with the energy more than 90%, the main objective of this phase is to balance the energy for the nodes that haven't been balanced by **UDB phase**, or still need balance, by processing a new balancing strategy starting from the leaves to the root, by saving the characteristic of **UDB**, each node n_i at level l, has $m_i \leq AV_l$.

Figure 5 shows the list of nodes with a residual energy less than 65%, the nodes are (node (25), node (27), node (33), node (35), node (38), node (65), node (76) and node (79)). The node (76) with residual energy (53.13%), which means that a network load on node (76) is (46.87%), this node has two children; node (45) with residual energy (65.53%) and node (49) with residual energy (92.8%), as we can see, the network load on node (45) is much higher than the network load on node (49), the node (45) has two nodes (node (6) with residual energy (73.91%) and node (18) with residual energy (94.8%), thus the node (6) has much network load, in order to save the node (76) we have to save the node (45), and in order to save node (45) we have to save node (6), to save node (6), node (9) has been removed and linked to node (48) in the **UDB phase,** but since the network load of node (6) is still high, the node (34) will be removed from node (6) and look for another node that overlapped with it, and that has a children count less than its current node parent children count. Since the node (34) overlapped with node (18) and node (18) has only one child and high residual energy (94.8%), the sub-tree (34) will be removed from the node (6) and linked to the node (18) in **DUB phase.**

Following the same strategy, for node (25) with residual energy (9.27%) has three children (node (33) with residual energy (63.6%), node (75) with residual energy (92.8%) and node (35) with residual energy (60.14%). To save the node (25) **UDB phase** has been applied, and the sub-tree rooted at (35) has been removed from the node (25) and linked to the node (26). But, even with this balance, the node (25) is still high loaded, and to save this node, an apply for the second phase is needed. The **DUB phase** has the aim to balance the nodes that haven't been balanced enough, by processing from the leaves toward the sink. To more save the node (25), **DUB phase** removed the sub-tree (33) and linked it to node (84), since node (33) overlapped with node (84) and node (84) has children count less than node (25) children count, also it has a high residual energy (77.87%) comparatively than node (25) with (9.27%). The node (33) with its residual energy(63.6%) also demand a balance, the first balancing process is achieved by using **UDB phase**, while the sub-tree rooted at (15) has been removed from the node (33) and linked to node (43). But since the node (33) has a high network load, removing the node (15) with the residual energy (95.61%) won't save the node (33) for a long time; Due the apply of **DUB phase**, since the node (5) doesn't have another parent node, and the node (44) has another overlapped node parent, the node (75), and node (75) satisfied the requirement of **DUB phase**, node (75) has children count less than node (33) children count, and the residual energy of node (75) is high(92.8), so the node (75) is suitable to take the node (44) as a new child.

Finally, the node (65) is the first node dead, because of its big network load, has been balanced at the **UDB phase**, but the node still need a balanced due the apply for the second phase that will balance it from the leaves, the node (65) has 3 children after that can't be removed, as explained above, let's find the node highly loaded among its children, the node (38) with a residual energy (46.83%), the node (32) with a residual energy (80.74%) and the node (16) with a residual energy (74.77%), from this we have the node (38) that should be saved. This node has only on child node (79), and node (79) has two nodes; the node (27) with residual energy (55.66%) and node (40) with residual energy (98.5), due the node (27) has to be saved, the node (27) as well has two children(node (2) with residual energy (69.06%) and node (10) with residual energy (91.14%), it has been balanced in **UDB phase** by removing the sub-tree rooted at (2) and linked it to the node parent (40). But, the node (27) is not balanced enough and the network load is still high which required another help, due to the application of **DUB phase.**

That will remove the node (10) from the node (27) and link it to node (46) as the node (46) has less children count than the node (27) and has a high residual energy (98.45%). By applying **DUB phase** the tree topology shown in Fig. 3(d) is more efficient. The Algorithm of **DUB phase** is explained in Algorithm 3. It process from the leaves toward the sink, For any level l_i with N_i nodes, if there is a child node $Ch_i \in N_i$ and its parent p_i, if p_i is overlapped with any other node n_j such that $j \neq i$ and p_j is the parent of n_j, if the children count of p_i is less than the children count of p_j. Then p_i can take n_j as a child.

ALGORITHM 3: Down -Up Balancing Phase **(DUB)**

---

1. DownUpBalancechildNode(childNode){

2. ParentOfchild = childNode. parent;

3. if (ParentOfchild ! = null){

4. foreach (nodeIntheSameLevelinTreeLevels[childNode. Level]. Nodes){

5. if (nodeIntheSameLevel. ID ! = childNode. ID){

6. nodeareOverLapped = areOverlapped(ParentOfchild, nodeIntheSameLevel);

7. if(areOverLapped){

8. if(nodeIntheSameLevel. parent. Childern. Count > $ParentOfchild. Childern. Count$){

9. nodeIntheSameLevel. parent. Childern. Remove(nodeIntheSameLevel);

10. nodeIntheSameLevel. parent = ParentOfchild;

11. ParentOfchild. Childern. Add(nodeIntheSameLevel);

12. }// endif

13. }//endif

14. }//endif

15. }// endfor

16. }// endif

17. }// end

---

5 Experimental Results

In our simulation we use OMNET++ 4, and visual studio 2017 c# to evaluate the performance of EEBTR over different time and for different coverage schemes. In each round (1 s), each sensor sends 1000 packet of message 1024 bits to the sink node either direct transmission, if they communicate or multi-hop transmission. Simulation parameters are shown in Table 1 while the network topologies are shown in Table 2 and Fig. 8.

Table 1. Simulation parameters

Parameter	Value
Network size	$500 \times 500 \text{ m}^2$
Radius	30 m for random coverage/20 m for triangular coverage
Data length	1000 packets of (1024) bits
Initial energy	1 J
E_{elec}	50 nJ/bit
ε_{amp}	0.001 pJ/bit/m^2

Table 2. Network topologies

Network topology	Number of nodes
Random coverage scheme	100
Triangular coverage scheme	36

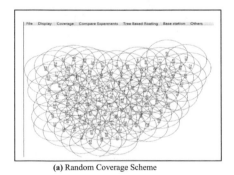

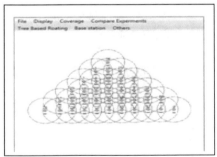

(a) Random Coverage Scheme   (b) Triangular Coverage Scheme

Fig. 7. GTR & EEBTR for two coverage schemes

5.1 Random Coverage Scheme

Figure 8 shows GTR & EEBTR tree routing for the network shown in Fig. 7(a). The node (0) is assumed to be a sink node.

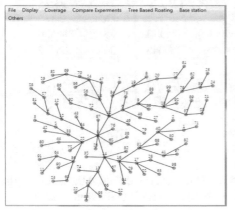
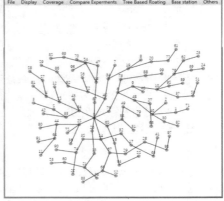

(a) GTR for Random Coverage Scheme   (b) EEBTR for Random Coverage Scheme for EEBTR

Fig. 8. GTR & EEBTR for random coverage

Figure 9 shows the residual energy in each node after 120 rounds for the network shown in Fig. 7(a). In each round, every node transmits (1000) packets of 1024 bits to the sink node, either by directly communication or multi-hop. The simulation shows that EEBTR outperforms GTR in the term of balancing energy. The node (45) is still alive with (18.03%) of energy via GTR topology, while it is still alive with (73.59%) of energy via EEBTR topology.

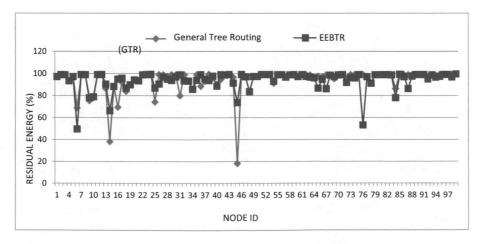

Fig. 9. The residual energy after 120 rounds for network topology shown in Fig. 7(a).

Table 3. Shows simulation results of Fig. 9

Protocol	GTR	EEBTR
Number of nodes	100	100
Density	0.641297656633992	0.641297656633992
Packet rate	1 packet per 0.001 s	1 packet per 0.001 s
Simulation time	120 s	120 s
Total energy consumption	2.90307853415172	2.97515915886536
Average hops/path	3.49959688255845	3.58635451505017
Average redundant transmissions/path	0	0
Average routing distance/path	87.1493414028705	86.6119571872207
Average transmission distance/hop	46.8024447290471	46.1126794547385
Average waiting time/path	0	0
Generated packets	7442	7475

From Table 3, we can see that GTR and EEBTR consumed almost the same energy in the period of t = 120 s, but the residual energy in each node is different, the nodes in EEBTR save more energy than in GTR hence, both energy efficiency and energy balancing is achieved by EEBTR.

Figure 10 shows the GTR & EEBTR routing and the situation of node (45) after 160 rounds. Simulation parameters are shown in Table 1. As we can see the node (45) is almost dead with the residual energy (0.36%) via GTR routing as shown in Fig. 10(a), while it is still alive with residual energy of (70.10%) via EEBTR routing as shown in Fig. 10(b).

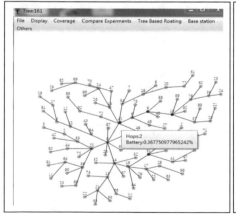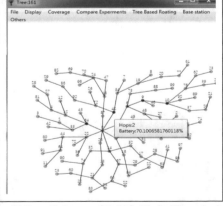

(a) The residual energy of node (45) after 160rounds, via GTR .

(b) The residual energy of the (node 45) after 160 rounds, via EEBTR.

Fig. 10. GTR & EEBTR after 160 rounds and the situation of node (45)

Table 4. The simulation results for network shown in Fig. 10(a, b)

Protocol	GTR	EEBTR
Number of nodes	100	100
Density	0.641297656633992	0.641297656633992
Packet rate	1 packet per 0.001 s	1 packet per 0.001 s
Simulation time	160 s	160 s
Total energy consumption	3.47879997425198	3.44825699516352
Average hops/path	3.49635935924723	3.51093073077839
Average redundant transmissions/path	0	0
Average routing distance/path	87.1255345114673	86.9601673792145
Average transmission distance/hop	46.7715642195868	46.0844506458246
Average waiting time/path	0	0
Generated packets	8927	8919

Figure 11 shows the residual energy in each node after 160 rounds for the network topology shown in Fig. 7(a). From Table 4, the total consumed energy by GTR and by EEBTR is approximately the same at the period t = 160 s, but the residual energy in each node is different, the nodes in EEBTR save more energy than in GTR hence, both energy efficiency and energy balancing is achieved by EEBTR.

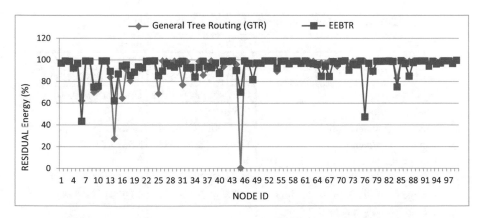

Fig. 11. The residual energy after 160 rounds for network shown in Fig. 10.

Figure 12 shows the round when first node dead appearance via GTR & EEBTR. Sink node is assumed to be node (0), the communication range for each node is 30 m.

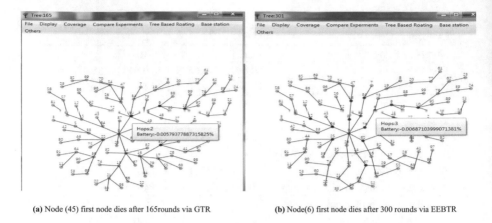

(a) Node (45) first node dies after 165rounds via GTR   **(b)** Node(6) first node dies after 300 rounds via EEBTR

Fig. 12. GTR & EEBTR routing after the round of first node dead

From Table 5, GTR routing generated only 8968 packets after the round of the first node declared dead t = 165 s, while EEBTR generated 15414 because of its longer lifetime t = 300 s, which is to say that EEBTR lives 135 more rounds, and generates 6446 more packets than GTR. Hence, EEBTR proved the improvement of GTR performance.

Table 5. The simulation results for network shown in Fig. 12(a, b)

Protocol	GTR	EEBTR
Number of nodes	100	100
Density	0.641297656633992	0.641297656633992
Packet rate	1 packet per 0.001 s	1 packet per 0.001 s
Simulation time	165 s	300 s
Total energy consumption	3.49400837190394	6.90297437410001
Average hops/path	3.49542818911686	3.56176203451408
Average redundant transmissions/path	0	0
Average routing distance/path	87.1195074356733	86.9524479816732
Average transmission distance/hop	46.7674874190421	45.9094853057864
Average waiting time/path	0	0
Generated packets	8968	15414

5.2 Triangular Coverage Scheme

The triangular coverage scheme can be more efficient when it is used as multiple Triangular patterns in the interest area in order to cover the whole area without overlapping, in order to reduce the sending of redundant data during routing. A part of the interest area may be covered by two or more covered regions [12].

Figure 13 shows GTR & EEBTR tree routing for the network shown in Fig. 7(b). The node (0) is assumed to be a sink node.

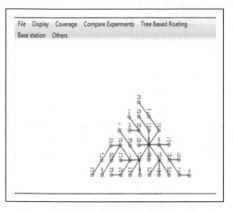
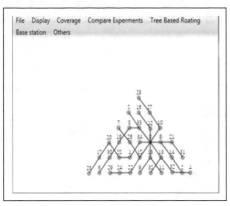

(a)General Tree Routing for Triangular Coverage Scheme   (b) EEBTR for Triangular Coverage Scheme

Fig. 13. GTR & EEBTR tree for triangular coverage scheme

Figure 14 shows the tree routing of 36 nodes for the Triangular coverage scheme shown in Fig. 13(a) and (b) after 120 rounds, with a communication range R = 20 m, and a length packet of 1000 packets for each sensor, each packet will be sent at each t = 0.001 s as explained in Table 1. It shows also that node (15) is the node with higher network load, with a use of (72.53%) energy via GTR as shown in Fig. 14(a) and a use of (28,67%) energy via EEBTR as shown in Fig. 14(b).

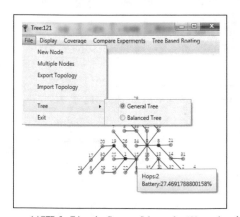
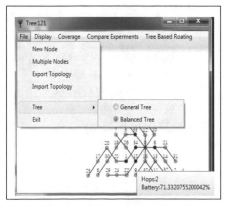

(a)GTR for Triangular Coverage Scheme after 120 rounds and a   (b) EEBTR for Triangular Coverage Scheme after 120 rounds
residual energy of node (15) (27.46%)   and a residual energy of node(15) (71.33%)

Fig. 14. GTR & EEBTR after 120 rounds and the situation of node (15)

Figure 15 shows the residual energy in each node after 120 rounds for the network topology shown in Fig. 7(b). Simulation parameters are shown in Table 1, and the simulation results are shown in Table 6. From Fig. 13 we have, EBTR sends almost the same number of packets with GTR at the period time t = 120 s, and GTR & EEBTR both consume almost the same energy in total but in term of energy balancing, EEBTR shows better performance than GTR.

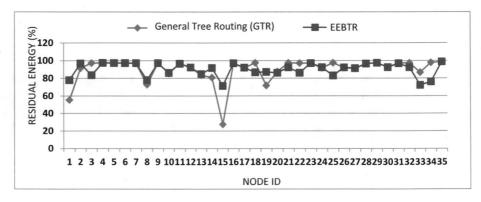

Fig. 15. The residual energy of Fig. 13 after 120 rounds.

Table 6. The simulation results of 36 nodes, after 120 rounds via GTR & EEBTR

Protocol	GTR	EEBTR
Number of nodes	36	36
Density	0.626125576199922	0.614725153184938
Packet rate	1 packet per 0.001 s	1 packet per 0.001 s
Simulation time	120 s	120 s
Total energy consumption	1.76595685375966	1.75974774783994
Average hops/path	2.49636835278859	2.48715991692627
Average redundant transmissions/path	0	0
Average routing distance/path	89.0062410308567	89.4187313514326
Average transmission distance/hop	30.4607048587252	29.7838012527919
Average waiting time/path	0	0
Generated packets	7710	7708

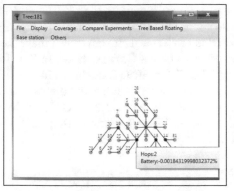

(a) Node(15) dies after 180s via GTR.

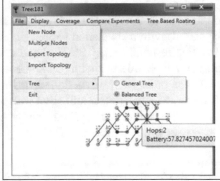

(b) Node(15) is still alive with 57.82% residual energy after 180s via EEBTR.

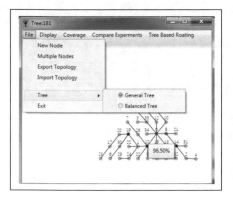

(c) Node (34) is still alive after 180s, with 96.50% residual energy via GTR.

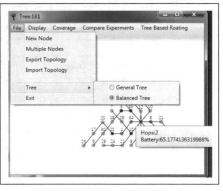

(d) Node(34) still alive after 180s, with 65.17% residual energy via EEBTR.

Fig. 16. GTR & EEBTR after 180 rounds and the situation of node (15) and node (34)

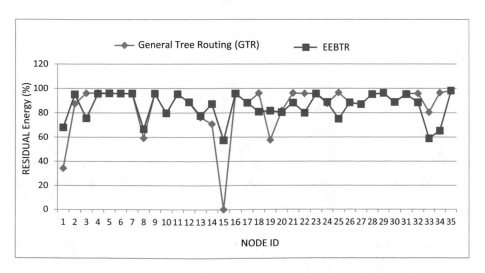

Fig. 17. The residual energy after first node declared dead for GTR & EEBTR

Figure 16 shows the tree routing for the network shown in Fig. 7(b) after 180 rounds, via GTR as shown in Fig. 16(a) and (c), and via EEBTR as shown in Fig. 16 (b) and (d) with a simulation parameters explained in Table 1. It shows also that node (15) with residual energy 0% declared dead after 180 s via GTR as shown in Fig. 16 (a), while it is still alive with residual energy 57.82% via EEBTR as shown in Fig. 16 (b) hence more save of energy for node (15), and longer lifetime for the network. However, for node (34) is still alive with the residual energy 96.50% via GTR, while it is still alive with only residual energy 65.17% via EEBTR, unfortunately EEBTR doesn't show any performance of energy for node (34).

Table 7. The simulation results, after first node declared dead via GTR & EEBTR

Protocol	GTR	EEBTR
Number of nodes	36	36
Density	0.626125576199922	0.687880554069623
Packet rate	1 packet per 0.001 s	1 packet per 0.001 s
Simulation time	180 s	360 s
Total energy consumption	2.59126838271958	4.85514451967947
Average hops/path	2.40966698715147	2.5076885225735
Average redundant transmissions/path	0	0
Average routing distance/path	86.6987346014206	89.5196809776984
Average transmission distance/hop	29.5798431446637	33.5238766000093
Average waiting time/path	0	0
Generated packets	11441	22258

Table 8. Network lifetime using different communication range and coverage schemes

Communication range	Coverage protocol	Nodes	Routing protocol	First node declared dead	Network lifetime (s)
25 m	Random coverage		GTR	65	220
			EEBTR	65	312
30 m	Random coverage	100	GTR	45	165
			EEBTR	6	300
20 m	Triangular	36	GTR	15	180
			EEBTR	15	360

Figure 17 shows the residual energy in each node for the tree routing shown in Fig. 13, after 180 rounds. Node (15) is the first node declared dead via GTR, while it is still alive with residual energy 57.82% via EEBTR, simulation parameters are explained in Table 1. Table 7 shows the simulation results after the first node dead, node (15) dies after 180 s via GTR and dies after 360 s via EEBTR. From Table 7 we can see that EEBTR lives for more 180 rounds and generates more 10817 packets comparatively than GTR, which means EEBTR shows better performance than GTR.

6   Conclusion

This paper proposed Energy Efficient Balanced Tree-Based Routing (EEBTR), a new algorithm for data routing in wireless sensor network that aims to achieve a certain balance between number of hops and energy dissipation in the network by processing in two phases. Up-Down Balancing phase and Down-Up Balancing phase. Hence, the nodes at the same level will have the same number of children, that is to say the nodes will share the same probability to be loaded, then the save of energy in each node comparatively than in GTR.

Acknowledgments. This paper is supported by the National Natural Science Foundation of China under Grants 61775054 and by the Natural Science Foundation of Hunan Province of China (Grant No. 2017JJ2047).

References

1. Hawbani, A., Wang, X., Karmoshi, S., Kuhlani, H., Ghannami, A., Abudukelimu, A., Ghoul, R.: GLT: Grouping based location tracking for object tracking sensor networks. Wirel. Commun. Mob. Comput. **2017**, 19 pages, Article ID 4509697. https://doi.org/10.1155/2017/4509697

2. Manjeshwar, A., Agrawal, D.: Teen: a routing protocol for enhanced efficiency in wireless sensor networks. In: Proceedings of the 15th International Parallel and Distributed Processing Symposium (IPDPS 2001) Workshops, San Francisco, CA, USA, 2001; pp. 2009–2015

3. Robert, A., Sawant, U.: Grid-based coordinated routing in wireless sensor networks. In: Consumer Communications and Networking Conference, CCNC 2007. https://doi.org/10.1109/CCNC.2007.174

4. Batra, P.K., Kant, K.: LEACH-MAC: a new cluster head selection algorithm for wireless sensor networks. Wirel. Netw. **22**(1), 49–60 (2016). https://doi.org/10.1007/s11276-015-0951-y

5. Nikolidakis, S.A., Kandris, D., Vergados, D.D., Douligeris, C.: Energy efficient routing in wireless sensor networks through balanced clustering. Algorithms **6**, 29–42 (2013). https://doi.org/10.3390/a6010029

6. Singh, H., Pawar, L.: A survey on energy efficient clustering protocols in heterogeneous wireless sensor networks. Int. J. Adv. Res. Comput. Commun. Eng. **4**(8) (2015). (2278-1021)

7. Prajapati, G., Parmar, P.: A survey on routing protocols of location aware and data centric routing protocols in wireless sensor network. Int. J. Sci. Res. (IJSR), **2**(1) (2013). (2319-7064)

8. Kulik, J., Heinzelman, W., Balakrishnan, H.: Negotiation-based protocols for disseminating information in wireless sensor networks. Wirel. Netw. **8**, 169–185 (2002). https://doi.org/10.1023/A:1013715909417

9. Intanagonwiwat, C., Govindan, R., Estrin, D.: Directed diffusion: a scalable and robust communication paradigm for sensor networks. In: Proceedings of the 6th Annual ACM/IEEE International Conference on Mobile Computing and Networking (MobiCom 200), Boston, MA, USA, pp. 56–67, November 2000

10. Shah, R.C., Rabaey, J.: Energy aware routing for low energy ad hoc sensor networks. In: IEEE Wireless Communications and Networking Conference (WCNC), 17–21 March 2002, Orlando, FL (2002)
11. Rodoplu, V., Ming, T.H.: Minimum energy mobile wireless networks. IEEE J. Sel. Areas Comm. **17**, 1333–1344 (1999)
12. Ghoul, R., Jing, H., Toure, F.D.M.: Finding the overlapped sub-regions C2, C3 and the maximum covered regions in WSN by using Net Arcs (NA) Method. In: Advances in Computer Science Research, vol. 44, 3rd International Conference on Wireless Communication and Sensor Network (WCSN 2016) (2016). https://doi.org/10.2991/icwcsn-16.2017.140
13. Wang, J., Kim, J.-U., Shu, L., Niu, Y., Lee, S.: A distance-based energy aware routing algorithm for wireless sensor networks. Sensors **10**, 9493–9511 (2010)
14. Qiu, W., Skafidas, E., Hao, P.: Enhanced tree routing for wireless sensor networks. Ad Hoc Netw. **7**(3), 638–650 (2009)
15. Fengyuan, R., Zhang, J., He, T., Chuang, L., Das, S.K.: EBRP: Energy-balanced routing protocol for data gathering in wireless sensor networks. IEEE Trans. Parallel Distrib. Syst. **22**(12), 2108–2125 (2011)
16. Mazinani, S.M., Naderi, A., Jalali, M.: A tree-based reliable routing protocol in wireless sensor networks. In: 2012 International Symposium on Computer, Consumer and Control, 978-0-7695-4655-1/12. https://doi.org/10.1109/IS3C.2012.130
17. Parihar, V., Kansal, P.: Quadrant based routing protocol for improving network lifetime for WSN. In: IEEE INDICON 2015 1570174091
18. Yanjun, Y., Qing, C., Vasilakos, A.V.: EDAL: an energy-efficient, delay-aware, and lifetime-balancing data collection protocol for heterogeneous wireless sensor networks. IEEE/ACM Trans. Netw. **23**(3), 810–823 (2015)
19. Wenyu, Z., Zhenjiang, Z., Han-Chieh, C., Yun, L., Peng, Z.: System-level energy balance for maximizing network lifetime in WSNs. Digital Object Identifier https://doi.org/10.1109/ACCESS.2017.2759093
20. Razieh, S., Jabbehdari, S.: A two-level cluster based routing protocol for wireless sensor networks. Int. J. Adv. Sci. Technol. **45**, 19–30 (2012)

Effect of Traffic Generator and Density on the Performance of Protocols in Mobile Wireless Networks

Amine Kada[✉], Hassan Echoukairi, Khalid Bouragba,
and Mohammed Ouzzif

Laboratory of Computer Networks, Telecommunications and Multimedia,
Higher School of Technology, Hassan II University of Casablanca,
Casablanca, Morocco
a.kada@outlook.com, echoukairi@gmail.com,
bourgba2008@gmail.com, ouzzif@gmail.com

Abstract. In multi-hop mobile wireless networks, it is difficult to distinguish between packets lost through congestion and those lost due to errors or mobility. However, the effectiveness of this type of network depends on the effect of the TCP and CBR traffic source on the overall performance of the routing protocols. In this context, different routing protocols have been evaluated to resolve these issues. In this paper, the performance analysis is carried out on DSR, DSDV and AODV protocols using NS2 simulator for wireless network in order to evaluate and quantify the effects of various factors such as different traffic, density of nodes and the mobility that may influence network performance. The paper also seeks to present analysis of these protocols based on the following performance metrics: packet delivery ratio, throughput and normalized routing load.

Keywords: Mobile wireless networks · TCP · CBR · Routing protocols · Random waypoint · NS-2.34

1 Introduction

Mobile Wireless Networks have become very popular because they are linked to the global Internet network following high demand and user requirements. In this type of network, all the nodes can be dynamically connected arbitrarily and some of them behave like routers in order to auto-discover and maintain the routes.

To do this, a routing protocol allowing the choice of the relay nodes must be defined.

The challenge imposed facing these types of networks is to develop the dynamic routing protocols to find the best routes for communication between nodes. In this context, three types of routing protocols are named: reactive, proactive and hybrid appear. In proactive (table-driven) routing, routes are established before requesting a connection between a source and a destination, and saved in a table with all the appropriate information (time of discovery, number of hops, neighborhood nodes …).

© Springer Nature Switzerland AG 2020
K. Arai et al. (Eds.): FTC 2019, AISC 1070, pp. 823–832, 2020.
https://doi.org/10.1007/978-3-030-32523-7_60

In contrast, in the reactive case (on-demand) routing protocols routes are discovered when they are needed.

However, the hybrid protocols take into account the characteristics of the two previously defined approaches in order to avoid their disadvantages and exploit their advantages [1, 12, 13].

Including this section, this paper has four sections. Section 2 depicts the routing protocols in MANET. Section 3 highlights the simulation configurations, lists the network scenarios and traffic generating. Section 4 describes the performance metrics, discusses and analyses the results which are obtained. Finally, in Sect. 5, we give a conclusion based on our findings.

2 Description of Routing Protocols in Manet

In this work, we focused on the "Destination Sequenced Distance Vector" (DSDV) protocol, as a proactive routing protocol, and on the "Ad Hoc On Demand Distance Vector" (AODV) and on the "Dynamic Source Routing" (DSR) as reactive routing protocols.

2.1 DSDV (Destination Sequenced Distance Vector) Protocol

DSDV is the most familiar proactive protocol [2, 3, 13]. It depends on the distance vector routing utilizing the Bellman-Ford distributed algorithm. In this configuration, each node tries to maintain a routing table that lists all the destinations, the distance required to reach a certain destination and the sequence number that is assigned by destination to avoid the formation of routing loops. The updates of these routing tables are carried out periodically and immediately after a significant change of topology (break of link, new node …).

However, this protocol keeps only the shortest route to each destination. But it is very difficult to maintain routing tables for large networks and does not support multipath because it consumes more bandwidth.

2.2 Ad Hoc on Demand Distance Vector (AODV) Protocol

The AODV protocol is a of reactive routing protocol [4, 13]. The establishment of news routes can be built by the demand from source. When a source node wants to reach a destination node, it broadcasts a specific request packet (RREQ) to its neighbors. The destination node makes again a response message (RREP) which contains the number of hops to achieve it. All nodes keep a routing table containing all data about each destination.

AODV is used for unicast and multicast packets. However, it does not support throughput routing metrics.

2.3   Dynamic Source Routing (DSR) Protocol

DSR is an on demand routing protocol used in multi-hop wireless Ad-hoc networks of mobile nodes [5, 6, 13].

When a node needs a route to destination that is not in its cache, it broadcasts a route request packet. Each node of the network then receives this request; it appends its address in the header of each packet and broadcasts it to the destination by using the newly discovered route.

However, DSR can support fast topology changes because it supports multipath in its design.

In order to quantify and assess the effects of multiple traffic (TCP, CBR) and density in mobile environment on these protocols that may influence network performance, we simulated and compared AODV, DSDV and DSR protocols in terms of packet delivery ration (PDR), throughput and Normalized Routing Protocol (NRL) according to various traffic and node densities under NS2.

3   Simulation Setup and Traffic Generation

3.1   Simulation Setup

The network simulator NS2 (NS2.34) was used in this work. This tool is a suitable simulator for analyzing the routing protocols in a wired as well wireless environment [7–9]. The nodes in our simulation follow a homogeneous configuration. All the nodes are therefore randomly distributed on an area of 10.000 square meters and are given an initial energy of 1000 J. The variables and parameters we are using in our simulation are in order to analyze the performance of routing protocols are reported in Table 1.

Table 1. Simulation parameters setup

Parameter	Value
Channel type	Channel/Wireless Channel
Radio-propagation model	Propagation/ Two Ray Ground
Number of nodes	25, 50, 75, 100
Simulation time	200 s
Routing protocol	DSDV, DSR, AODV
Traffic Type	CBR, TCP
Mobility	Random Waypoint
MAC Type	802.11
Link Layer Type	LL
Interface Type	Queue
Packet Size	512 MB
Max packet in ifq	50 packets
Node speed	5 m/s – 20 m/s
Transmission rate	4 packets/s

We simulated DSDV, DSR, and AODV at different simulation densities from 25 to 100 nodes. Traffic type is CBR and TCP with packet size 512. The model mobility used in this paper is random waypoint. The awk scripts is used to calculate the performance metrics such as PDR, throughput and NRL and the results obtained are illustrated in figures.

3.2   Network Scenario and Traffic Generating

To generate a file of random node positions and their movement, we are using the tool CMU generator under the repertoire ~ /indep-utils/cmu-scen-gen/setdest.

$. /setdest –v <2> -n <nodes> -s <speed type> -m <min speed> -M <max speed> - t <simulation time> -P <pause type> -p <pause time> -x <max X> -y <max Y>> scenario_output_file

For example, we executed the following command:

./setdest -v2 -n 25 -m 5.0 -M 20.0 -t 200 -x 100 -y 100> scen-100x100-200-25-v2

m: 5.0 m/s means the minimum speed

M: 20.0 m/s means the maximum speed

t: 200 s means the simulation time

This syntax means a 100 m × 100 m topology with 25 nodes random distributed and max moving speed of 20.0 m/s, min moving speed of 5.0 m/s.

The output of the generate tcl statements into file whose name is scen-100x100-200-25-v2.

Some fragments of output file are shown in Fig. 1 below:

```
$ns_ at 0.892149302086 "$node_(8) setdest 45.698708576460 36.955170749378 13.376477087687"
$ns_ at 0.893826995195 "$node_(10) setdest 18.543216343268 94.674202006365 18.872398032215"
$ns_ at 2.100903542530 "$node_(7) setdest 63.895787412842 31.630715792810 5.007900782312"
$ns_ at 2.617954076679 "$node_(20) setdest 70.204054634656 82.600753540396 12.393301370974"
$ns_ at 2.762396156433 "$node_(13) setdest 63.385791560692 2.725750642586 16.367517232300"
$ns_ at 2.921308195072 "$node_(21) setdest 31.219818185485 44.693761550892 18.335573649686"
$ns_ at 3.050059567257 "$node_(18) setdest 15.814803771088 22.549579686092 14.917552341440"
$ns_ at 3.193503895087 "$node_(24) setdest 54.771992818586 68.720702522878 5.084373722214"
$ns_ at 3.449278057430 "$node_(19) setdest 71.381472783942 34.687166920614 8.916602159071"
$ns_ at 4.184054560114 "$node_(10) setdest 92.577626965043 1.397107166843 6.446504453400"
$ns_ at 4.451043498129 "$node_(24) setdest 24.820168540486 32.394074215080 11.277534874558"
$ns_ at 4.510173376717 "$node_(8) setdest 47.589832101642 7.111323969242 5.216225839680"
$ns_ at 4.563527088683 "$node_(17) setdest 29.241571594553 25.895467700031 7.512606760166"
$ns_ at 4.832169791532 "$node_(4) setdest 25.501365052602 89.423282793733 18.046573898449"
```

Fig. 1. Fragments of setdest

For the generation of random traffic of TCP and CBR connections between wireless mobile nodes, we are using the traffic-scenario generator script. This script is available under ~ ns/indep-utils/cmu-scen-gen and is called cbrgen.tcl.

The syntax of this script is:

ns cbrgen.tcl [-type cbr|tcp] [-nn nodes] [-seed seed] [-mc connections] [-rate rate] > output.tcl where:

- type: represents the type of traffic CBR or TCP
- nn: represents the number of nodes
- mc: represents the maximum number of connections between source and destination nodes.
- seed: represents a random seed
- rate: represents a rate whose inverse value is used to calculate the interval time between packets sending.

For our simulation, we are used the following command:
$ ns cbrgen.tcl -type cbr -nn 25 -seed 1.0 -mc 10 -rate 4.0 > cbr-traffic-25-10-4
Figure 2 shows a fragment of traffic connection CBR output.

```
# nodes: 25, max conn: 10, send rate: 0.25, seed: 1.0
# 1 connecting to 2 at time 2.5568388786897245
set udp_(0) [new Agent/UDP]
$ns_ attach-agent $node_(1) $udp_(0)
set null_(0) [new Agent/Null]
$ns_ attach-agent $node_(2) $null_(0)
set cbr_(0) [new Application/Traffic/CBR]
$cbr_(0) set packetSize_ 512
$cbr_(0) set interval_ 0.25
$cbr_(0) set random_ 1
$cbr_(0) set maxpkts_ 10000
$cbr_(0) attach-agent $udp_(0)
$ns_ connect $udp_(0) $null_(0)
$ns_ at 2.5568388786897245 "$cbr_(0) start"
# 4 connecting to 5 at time 56.333118917575632
set udp_(1) [new Agent/UDP]
$ns_ attach-agent $node_(4) $udp_(1)
set null_(1) [new Agent/Null]
$ns_ attach-agent $node_(5) $null_(1)
set cbr_(1) [new Application/Traffic/CBR]
$cbr_(1) set packetSize_ 512
$cbr_(1) set interval_ 0.25
$cbr_(1) set random_ 1
$cbr_(1) set maxpkts_ 10000
$cbr_(1) attach-agent $udp_(1)
$ns_ connect $udp_(1) $null_(1)
$ns_ at 56.333118917575632 "$cbr_(1) start"
```

Fig. 2. Fragment of connection CBR

Now, in the wake of adding the creating results to tcl record, we completed one time simulation, however we have to enable numerous tests with various control parameters.

4 Results and Discussion

In order to evaluate and assess the network performance and efficiency of routing protocols using Network Simulator 2, a set of performance metrics must be used. In our work, we focus on the following performance and efficiency metrics described as follow:

Packet Delivery Ratio (PDR): This ratio is obtained by calculating the ratio of the total number of packets successfully received per destination to the total number of packets sent from the source. The greater the value of this metric is, the better performance and the efficiency of the protocol is [11].

Here, we are analyzing the effect in the performance of PDR metric on DSDV, DSR and AODV protocols during various numbers of nodes and different types of traffic (like CBR and TCP), when mobility speed of nodes changes.

In the case of the CBR traffic, as shown in Fig. 3, the PDR value is decreased slightly for all protocols when the number of nodes is increased. It can be seen that with CBR and variable node density, AODV records more PDRs compared to the DSR and DSDV protocols.

In the case of the TCP traffic, as shown in Fig. 4, the PDR value is decreased slightly for DSDV protocols when the number of nodes is increased. It decreases with the increase of the density of nodes to a certain level and then starts increasing again for AODV and DSR. Finally, it decreases with the increase of the number of nodes for all protocols.

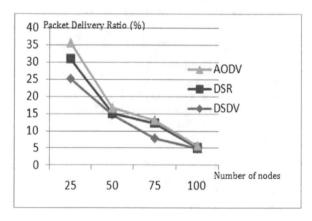

Fig. 3. The PDR compared with the Number of nodes (CBR traffic)

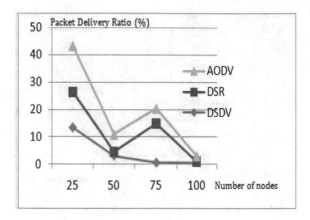

Fig. 4. The PDR compared with the Number of nodes (TCP traffic)

Throughput: This metric measures average rate of successfully data packets delivery from source to destination within a time span. It is represented in bits/bytes per second. The high value of this metric means the better robust network [10].

In CBR traffic, AODV has the best throughput as to the other protocols. DSDV has the lowest performance compared to AODV and DSR. In all protocols, throughput increases as the node density is increased as shown in Fig. 5. In TCP traffic, the throughput for the AODV protocol increases with decreasing the number of nodes. But in DSR and DSDV protocols, it shows high value initially and decreases with the increasing numbers of nodes as shown in Fig. 6.

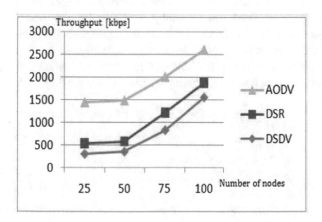

Fig. 5. The throughput compared with the Number of nodes (CBR traffic)

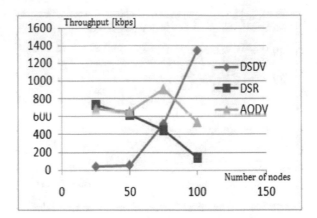

Fig. 6. The Throughput compared with the Number of nodes (TCP traffic)

Normalized Routing Load: This metric represents the ratio of network routing control packets sent from all nodes to all delivered data packets in order to discover and maintain route changes occurred in the simulation [11]. It is given by the following formula:

NRL = Total Routing Packets sent/Total Data Packets Received
This metric evaluate the efficiency of the routing protocol.

In CBR traffic and based on Fig. 5, it is observe that the normalized routing load for DSDV and AODV increases when the number of nodes increases and DSDV produces low results in comparison to AODV. Figure 7 shows also that DSR presents the maximum and minimum values of NRL when varying the number of nodes.

In TCP traffic, NRL increases according to the number of nodes for the AODV protocol. However, DSR and DSDV presents the maximum and minimum values of NRL when varying the number of nodes as shown in Fig. 8.

Therefore, we can conclude that DSDV in CBR traffic and AODV in TCP traffic have better performance in the context of routing overhead because they have the lower NRL.

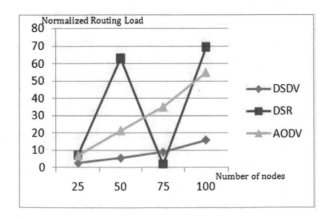

Fig. 7. The Normalized Routing Load compared with the Number of nodes (CBR traffic)

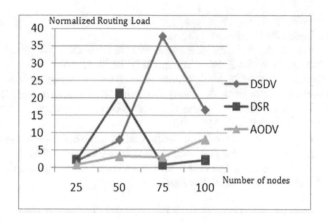

Fig. 8. The Normalized Routing Load compared with the Number of nodes (TCP traffic)

5 Conclusion

In this paper, we evaluated and assessed, in a mobile environment, the performance and execution of three routing protocols; namely, DSDV, DSR and AODV, under both the CBR and TCP traffic classes with a variation in the number of nodes present in the network.

From the results reported in this work, we concluded that the AODV routing protocol performs better and gives better results compared to the DSR and DSDV in PDR for both traffic classes (CBR, TCP) and mobile environment.

AODV becomes the best when throughput metric is considered according to density and CBR traffic. Despite of their dynamic nature while finding the route and configuring it, DSDV in CBR traffic and AODV in TCP traffic present better NRL.

As a result, node density, traffic type and mobility affect the overall performance and behavior of routing protocols in the MANETs.

References

1. Subramanya, B.M., Shwetha, D.: A performance study of proactive, reactive and hybrid routing protocols using qualnet simulator. Int. J. Comput. Appl. **28**(5), 0975–8887 (2011)
2. Perkins, C.E., Bhagwat, P.: Highly dynamic destination-sequenced distance-vector routing (DSDV) for mobile computers. In: SIGCOMM ACM, pp. 234–245 (1994)
3. Jagdale, B.N., Patil, P., Lahane, P., Javale, D.: Analysis and comparison of distance vector, DSDV and AODV protocol of MANET. Int. J. Distrib. Parallel Syst. (IJDPS) **3**(2), 121 (2012)
4. Perkins, C.E., Royer, E.M.: Ad-hoc, "On-demand Distance Vector Routing", draft-ietf-manet-aodv-02.txt (1998)
5. Johnson, D., Hu, Y. C., Maltz, D.: The dynamic source routing protocol (DSR) for mobile ad hoc networks for IPv4 (No. RFC 4728) (2007)
6. Perkins, C.E.: Adhoc Networking. Pearson, US (2000). Chapter-5
7. The Network Simulator-ns2. http://www.isi.edu/nsnam/ns. Accessed 15 Feb 2017

8. Network Simulator 2 for Wireless: My Experience. http://www.winlab.rutgers.edu/~zhibinwu/html/network_simulator_2.html
9. Nayyar, A., Singh, R.: A comprehensive review of simulation tools for wireless sensor networks (WSNs). J. Wirel. Netw. Commun. 5(1), 19–47 (2015)
10. Devi, R., Sumathi, B., Gandhimathi, T. Alaiyarasi, G.: Performance metrics of MANET in multi-hop wireless ad-hoc network routing protocols. Int. J. Comput. Eng. Res. (IJCER)
11, Barakovie, S., Barakovie, J.: Comparative Performance Evaluation of Mobile Ad hoc Routing Protocols. In: MIPRO, May 2010
12. Kaur, H., et al.: A survey of reactive, proactive and hybrid routing protocols in MANET: a review. Int. J. Comput. Sci. Inf. Technol. (IJCSIT) 4(3), 498–500 (2013)
13. Abolhasan, M., Wysocki, T., Dutkiewicz, E.: A review of routing protocols for mobile ad hoc networks. Sci. Direct 2(1), 1–22 (2004)

Development of P2P Educational Service in Russia

Sergey Avdoshin and Elena Pesotskaya[✉]

National Research University Higher School of Economics,
20 Myasnitskaya ulitsa, 101000 Moscow, Russian Federation
{savdoshin, epesotskaya}@hse.ru

Abstract. Education is an important foundation to build on in any area of the modern world. Only quality education allows students to take up the most promising jobs, and for employers to get better trained employees. The purpose of this study is to develop an approach to the development of a new educational peer-to-peer (P2P) platform that can overcome existing barriers to the use of such services and meet the needs of the entire educational ecosystem. To achieve this goal, the research includes a comparative analysis of existing educational services in Russia, identifying their weaknesses and current barriers to the end users. The approach is based on a broad analysis of the needs of all market players and stakeholders. The result: the current paper provides recommendations for the development of an innovative educational P2P service, which will be able to solve the existing problems of the education market.

Keywords: Education · Online platforms · Educational services · P2P education

1 Introduction

The volume of the world education market (EdTech) is about 4.5–5.0 trillion dollars, and in the coming years it will increase to 6–7 trillion. Online education accounts for about 3%, or 165 billion dollars. According to The Global Market Insights hypothesis [1] the EdTech market is likely to grow by 5% between 2016 and 2023, exceeding USD 240 Billion. The USA is the largest and most developed market in EdTech. The rate of its growth has slowed down to approximately 4, 0–4, 4% annually. The second largest region (Southeast Asia, primarily China and India) is developing much faster, at 17%. In 2016, it overtook Western Europe: $11.7 billion compared to $6.8 billion. So far, Eastern Europe, with its $1.2 billion, is lagging behind the West, but is developing faster, at 17% [2].

Over the past decade, society has made a significant breakthrough in technology, which, of course, has affected education, including distance education. Therefore, we can single out the tendency to use large data to improve educational processes, personalize courses, introduce virtual and augmented reality technologies, reduce the volume of courses (micro-teaching and mini-programs), machine learning and artificial intelligence.

© Springer Nature Switzerland AG 2020
K. Arai et al. (Eds.): FTC 2019, AISC 1070, pp. 833–847, 2020.
https://doi.org/10.1007/978-3-030-32523-7_61

The analysis of the EdTech segment, conducted within the framework of this study, allows us to identify the main trends:

- *Blended learning* combines the traditional form with the distance form. Based on a study by Babson Survey Research Group [3], such combined programs are seen by experts to be as effective as full-time ones, and often superior to them. The introduction of blended learning as an innovation leads to a number of advantages: the student acquires a space of freedom and responsibility in which he/she learns to make an informed choice and be responsible for the consequences. The teacher ceases to simply broadcast the material: he/she initiates a dialogue with the students, receives feedback and begins to interact with the audience in a new way.
- *Personality is more important than place*. The importance of a person in online education increases, while the online site becomes less important when choosing an educational service provider. People who want to receive knowledge come to a particular speaker who has managed to gain a reputation in the market, created his own brand which evokes confidence based on numerous ratings.
- *Large IT giants are entering the EdTech market.* Apple, Amazon, Google, Microsoft are competing on the basis of the integration of their products, such as Google Apps for Education and Microsoft Office 365, in the activities of higher education institutions. The EdTech market is growing rapidly, especially in developing countries. At the same time, the quality of educational content and learning as a whole plays an important role.

Russia is the driver of the Eastern European market with an average annual growth, according to various estimates, of 17–25%. At the beginning of 2017, the Russian education market was estimated at 1.8 trillion rubles with a 19.2% share of the private sector and 1.1% in online education. According to the forecasts of the company "Netology Group", by 2021 the market volume will have grown to 2 trillion rubles, the share of the private sector will have decreased to 18.9%, while the share of online education will have grown to 2.6% [4].

The main question posed by the authors of the research is what features and functionality an innovative P2P service should have to bridge the current gaps of online education, attract the maximum number of users to the platform and provide quality services.

To find the solutions the authors carry out a comparative analysis of existing educational services, identify their weaknesses and current barriers for consumers. Further, a study of the needs of participants is conducted and a platform development approach is suggested as well as the ways of the platform monetization. The authors also propose possible prototypes of solutions with a list of recommendations for further development of innovative educational P2P service.

In the framework of this study the authors actually solve two problems. On the one hand, they identify the needs of participants in the educational ecosystem and the existing systems in accordance with the identified requirements - thus identifying gaps and shortcomings of existing solutions. On the other hand, the authors develop their own approach to creating a new innovative P2P service, which includes recommendations on the functionality of such services, the methods of monetization and the architecture of the solution.

Following this introduction, the work is continued in the next Sect. 2 with key barriers to implementing or continuing the use of educational online services along with the demands of the key stakeholders. Section 3 shows an analysis of the interests of consumers and providers of these educational services, the methods of interaction with the platform and means of monetization. In Sect. 4 we discuss recommendations for educational P2P platform services aimed at additional vocational education. Following on from that, Sect. 5 addresses related work and discussion, considering the limitations of the study. Finally, Sect. 6 summarizes our main conclusions and outlines future work.

2   Edtech Market Research and Detection of Existing Barriers

On the market of Russian online education, several segments can be distinguished: preschool education, general and secondary school education, secondary and additional vocational education. Forecasts for market volumes for 2021 are 548 billion rubles, 699 billion rubles and 278 billion rubles, respectively [9]. The growth rates of the segments are high, more than 20% per year. In addition, we can distinguish a segment of language education (31 billion rubles).

The state creates certain barriers for the development of school/preschool online education, so we can talk about a small market share. For example, to get a license for general education, even for an online school, you need to have a gym, a dining room, and a cloakroom, even if they are never used.

The most interesting is additional vocational education, but despite the significant number of players, in-depth market analysis allows us to conclude that there are the gaps in the EdTech industry.

Harvard, which has its own online platform, recently submitted a study, according to which an average of only 6% of the original registered receive a certificate. Russian platforms do not disclose the number of users that reach the end of their education, but most likely the numbers are similar. There are several reasons given by the authors: firstly, often obtaining a certificate costs money; secondly, educational platforms include mostly paid courses (the amount of paid content varies from 60 to 90%, depending on the platform); thirdly, the convenience of training does not always meet the needs of the audience.

The three largest players (Coursera, Universarium, Udemy) have more than 50% of the market. In addition, the audience of the largest Russian educational online platform, Universarium, is 26 times smaller than that of the large western Coursera - 1.5 mln users in comparison with 40 mln (see Fig. 1).

Educational Platform	# of Courses	% of Courses in Rusian	# of Users	Additional vocational education	Career Consultations
COURSERA	2700+	12 %	40 mln.	✓	✗
Udemy	80000+	0,4 %	24 mln.	✗	✗
edX	2000+	0,3 %	17 mln.	✓	✗
Universarium	79	100 %	1,5 mln.	✗	✗
Netology	150+	100 %	260 thous.	✗	✓
Lektorium	70+	100 %	105 thous.	✗	ʌ
Uniweb	56	100 %	11 thous.	✓	✗

Fig. 1. Analysis of the key online educational platforms/players

For a more structural analysis of online education market gaps, the authors identified key barriers to starting/continuing the use of services:

- *Dissatisfaction with the quality of education.* Monitoring of education conducted by HSE [5] showed that 53% of students wanted to get additional professional education.
- *The gap between university education and the needs of employers/companies.* According to Elena Rusanova, head of the HeadHunter Consulting Center [14], companies mostly lack non-standard employees with specific knowledge, skills, qualifications and soft-skills. Many employers believe that young professionals misconstrue their future responsibilities. In their justification, students note the disconnection of education from the demands of the labor market, the lack of necessary subjects, and outdated approaches to educating by some teachers.
- *Lack of practice.* According to the annual survey conducted by Changellenge ≫, the main reason for the dissatisfaction with education is the lack of practice. This factor was indicated by more than 50% of the respondents. Students receive theory, but do not know how to apply it to a working process after employment to a company.

In May 2017 the Department of the Ministry of Labor stated that 72% of graduates of higher education institutions have to pass further study at their workplaces, and employers pay for additional training for a quarter of them.

On the one hand, there is an urgent need for additional education, on the other hand, according to the authors, not all are ready to switch to online education due to a number of constraints.

Firstly, there is the stereotype of the "need for collective communication" imposed by the older generation for the normal development of a child or adolescent which is supposedly much more effective at school than in online courses. Other constraining factors are the lack of trust in Internet education, poor public awareness of existing programs and online opportunities in the field of education, and lack of financial resources. Of course, with the development of the Internet, educational consumers are increasingly willing to trust online services, as evidenced by the growth of EdTech.

Among the barriers to continuing the use of services, it is worth noting the weak involvement and motivation, mainly in services for self-study. The idea of lifelong learning as a basic educational need is lacking, as well as there being a lack of free time.

The existing platforms on the Russian market do not make unique offers in the field of education and do not always satisfy the demands of users. Platforms do not have many key functions, such as communicating and changing the roles of the user. Often, the needs of users of the educational process are not taken into account; the process of interaction with the platform causes difficulties and does not contribute to the continuation of training.

3 The Analysis of Key Stakeholders and Methods of Their Interaction with the Platform

3.1 Interests of Educational Services Consumers and Providers

In order to determine a framework of an educational online platform and design a possible product, it is necessary to understand what the audience needs and what it lacks now, namely, the interests of consumers and service providers.

Consumers can include students and people who graduated from the university. Consumers want to get a good job or get a promotion in their current one. At the same time, they lack practical skills, because they get knowledge but not practice while studying in universities. They may lack experience or connections. According to the research of HeadHunter [6], most consumers want to get a new job, get promotion at their current job, or get additional knowledge. In order to achieve their goals, consumers need to obtain relevant knowledge and get the attention of employers. Finally, consumers may want to change their field of activity.

Service providers include those who have unique knowledge and competencies. They can be employees of companies, teachers of the leading universities, experts from various industries and just people who are deeply versed in specific issues. Providers on behalf of company representatives want to attract quality employees for little money, improve their reputation and their skills. Such providers place priority on increasing the possibility of additional earnings and recruiting. By selecting candidates through training, you can shorten the time of adaptation of a new employee and increase the profit that he/she generates.

Providers also working as university professors tend to improve their status or develop new teaching methods. Among the teachers' motives are the following: to make education accessible, to cooperate with colleagues, to transform traditional education, to allow students to study at a pace convenient for them, to improve their own knowledge, build their reputation, and strengthen their own research. This means that teachers want to strengthen their reputation both within the walls of the university and in the scientific community, and to improve educational programs and methods of teaching. To do this, they need to attract as many people as possible to their courses.

Another group of producers is experts, in other words, people who are proficient in specific issues or who have expertise and are willing to make money. To achieve their goal, they need access to a solvent audience that needs the appropriate knowledge and skills.

It is very important to understand what consumers and producers are guided by while using online platforms, what they want to achieve and what they really need for this.

3.2   Interaction of Platform Users

After identifying the interests of the main stakeholders, it is necessary to understand how participants will interact with the platform and with each other.

The platform should unite producers with consumers and allow them to share value [7]. Any interaction on the platform begins with an exchange of information that helps the interested parties decide whether they want to start the exchange and how they will participate in it. Consumers will not use the service until they see the values

As for the value of the platform for consumers of educational services, it is access to the value created on the platform - courses, training materials, experts, the possibility of collective learning and personal contact with the teacher. All this allows users to attract the attention of employers, and of course accumulate knowledge and develop the necessary competencies.

For educational service providers the value is the access to the educational community, the education market and the personnel market, which is important for them to select the most suitable candidates for the company.

Both consumers and producers get access to tools and services that facilitate interaction.

The educational platform will not have value until users start using it. Most platforms fall apart simply because they have not been able to solve this problem. The business model of a platform should facilitate the exchange of information and offer systems that make the exchange of units of value easy and convenient. The platform needs to solve the problem of initial involvement of participants.

Another important task for the platform is to preserve the interest of users who have logged in or subscribed to the platform. Facebook found out that users consider the platform useful only after subscribing to several other users. Otherwise, they stopped using it. As a result, Facebook changed its marketing tactics from involving the participants to helping in establishing links [8]. There are no bright examples for Russian platforms yet, but such an option may be the entry of users into the industry rating, the expert community or the "personnel reserve" of the company.

3.3   Ways to Monetize Platform Services

An important issue for choosing a platform by users is the ability to access services, the terms of connection. Charging fees from all users is rarely used, because it can scare off users. Rather, the service may charge some participants full cost and provide benefits to those who are sensitive to the price. The category of users who are particularly sensitive to pricing is most likely to leave the platform; it makes sense to offer such users discounts or benefits.

The study identifies the three most popular ways of monetization, which are common among the large service platforms [7]:

First is charging a commission. Often, the interaction between the consumer and the producer involves the exchange of funds when the user selects a paid course or service and the platform acts as an agent.

Second, the platform can have paid access to the professional community for interested parties. Companies in the search for professionals with specific skills and abilities must contribute when placing a job vacancy on the career page.

The third option is payment for advanced access/services. This is when the platform charges for improved services, such as premium courses and master classes with outstanding speakers, or provides a quality assurance for the selection of specialists being a source of significant additional value for companies.

To determine the full cycle of monetization, one needs to clearly understand to whom the platform brings real value, which monetization method to choose, which services are in great demand, and who is the most appropriate audience.

4 Development of Educational P2P Service

In the framework of this study, the authors intend to offer recommendations for educational P2P platform service aimed at additional vocational education. In the opinion of the authors, such a service has a rather large potential, along with low competition in the market. This is confirmed by the study of "Netology-group" [9], according to which the online education market in the area of additional vocational education occupies the largest share and by 2021 will have reached the level of 10.9%, which is about 11 billion rubles (the total market volume is estimated at more than 103 billion rubles).

The target audience is high school students and young professionals aged 18 to 30 years. Specialists over the age of 30 may also experience the need for additional education, but their share among users of online services will be negligible. According to Rosstat, in Russia on January 1, 2017 there were 19,707,000 people between 20–30 years of age. 4,277,000 of them (Masters, Bachelor, specialties) are students of Russian educational institutions [10].

Presuming that half of the students are interested in additional education, and according to the authors' assumption, only 30% will use online educational services. In addition, it is necessary to take into account the conversion rate of educational platforms in the market of educational services - it ranges from 1–4% [11, 12]. This means that only 1–4% will choose the new platform and start using it frequently (become an active user).

Of the remaining 15 million users, suppose that only 7% [13] are interested in additional education, only 30% of them will use the online services, and only 1–4% will choose our P2P platform. Thus, according to the authors' estimates at the initial stage, the platform can attract about 25 thousand users; this number will grow as the platform develops.

In addition, the target audience should include professional teachers 30–45 years old, freelance experts, universities and different companies.

Recommendations will be based on the main problems identified by us and in fact solve them. After analyzing current players in the market and conducting a survey of more than 100 students and graduates, the following reasons for abandoning existing services have been identified:

- Lack of interactive and practical examples in training;

- Lack of originality and involvement of teachers;
- There is not enough motivation and desire to learn;
- The lack of logic in the structure of courses;
- The courses do not meet the expectations of users.

To offer a P2P solution that attracts a large number of users, removes current barriers and restrictions, can keep users and offer high quality of service, three aspects should be worked out in detail: (a) P2P functionality, (b) architecture and technology, (c) marketing and monetization. Let us consider each of them in more detail.

4.1 Functionality

Analyzing all the shortcomings of existing systems' functionality (see Fig. 2), the authors suggest an approach that will allow one to get the maximum effect (quality and quantity of knowledge) from the new platform.

Functionality / Educational Platform	Deep teachers profile	Social networks authorization	Personalized content	Intro test to identify training needs	Quality control based on students constant feedback	Rating of students	Rating of teachers	Forums on topics/industries	Changing roles (student/teacher)	Reward for academic achievement
COURSERA	◐	●	●	◐	●	○	○	◐	○	○
Udemy	●	◐	◐	○	●	○	●	○	●	○
edX	◐	●	◐	○	◐	○	○	◐	○	○
Universarium	○	◐	○	○	◐	○	○	○	○	○
Netology	◐	◐	○	○	○	○	○	○	○	○
Lektorium	◐	◐	○	○	○	◐	○	○	○	◐
Uniweb	◐	◐	○	○	◐	○	○	○	○	○

Full coverage ●   Partial ◐   No coverage ○

Fig. 2. Functionality of the main educational platforms

Within the framework of the approach, all functional requirements are considered from the point of view of process participants and educational content. The functional requirements based on the problems identified within the study should be solved on the new platform.

1. *Working with content.* Under the content/product, we should consider the applied knowledge of a different profile, individual classes and full courses, theoretical studies and materials, webinars, practical works/cases, consultations and courses from companies.

   Among the identified problems, the authors recognized a limited possibility of obtaining individual content and lack of quality control.

   As a solution to this problem, the new platform should provide the possibility of moderating publications, processing user complaints, and organizing technical support services.

   The key solution of the new platform is the ability to select, categorize and filter content. Users should have access to recommended and popular courses. In accordance with the user profile (it is also possible to analyze the profile in social networks), the user can access the section of recommendations - providing the most relevant selections depending on his/her interests. Content categorization can also be implemented by the contribution of other users who tag through professions/branches/functions for more convenient navigation. In the advanced version of the platform, artificial intelligence can be used for categorization. Filtering content will allow a user to search for necessary courses depending on the current preferences and tasks.

Communication on the forum should be available to all users, while only certain interested users can access news of the course, feature articles, expert interviews, various notifications, such as push notifications about the start of classes and key events. The user should be able to view and leave comments, assess the quality and complexity of any content, as well as leave feedback on authors, influence their rating. User profiles can be linked in order to keep up with the progress of group-mates or friends, to be able to purchase courses as a gift, to offer joint training, to share recommendations and knowledge.

2. *Working with the user.* The current problems of working with users can include: (a) Complicated registration of participants; (b) personalized selection of courses; (c) a system of rewards based on ratings of participants.

2.1. To solve the problem of complicated registration, we propose in the new platform to integrate with social networks for a simpler and intuitive log-in. Registration in one click and a further opportunity to share your opinion in social network will attract new users and keep existing ones who are thinking about taking new courses.

2.2. For a more focused and personalized selection of courses, the user can take a test the first time they visit the platform to determine their own training needs. The test can identify current gaps in training, determine the most interesting subject area for the user, and based on the current type of activity, disclose the need for new innovative courses and directions of training. As a result, the user is offered a number of options that he can take into consideration. If the user is firmly confident in choosing a course, the test can help to determine their current level of knowledge and domain ownership in order to choose the optimal level of further study. Based on the results of the course, detailed statistics are available for the user, allowing him to identify his weaknesses, to obtain personal automated recommendations based on data analysis - comparisons with the success of similar profiles, and target values of academic achievement.

2.3. The system of ratings and rewards. For the constant development of the user, increase of involvement and retention on the educational course, better social effect and learning outcomes - it is necessary to implement a rating system of users of the P2P solution. The rating takes into account the number of completed courses, academic performance, timeliness of completion, and feedback from other participants.

For instance, after completing the course and successfully passing the tests, the system will offer the student to become a mentor and assemble a team of "not so good" pupils and help them pass a course. This will help to increase the mentor's rating and add additional features to the account.

For proactive study, users earn points that form a personal rating in the community and provide access to activities that are included in a special incentive program. Incentives can include tournaments with prize-winning places, free access to lectures, tickets to the theater, etc. as a reward for academic achievement. Another option for incentives can be virtual money, which you can earn for active training, passing courses and answers on the forum. A user can spend it on purchasing paid educational content, sub-scribing to new courses and webinars, and promoting his/her questions in the expert community.

3. *Working with provider/lecturer.* In most online platforms, there is insufficient attention to the teacher's personality in the platforms and as a result *a narrow profile elaboration.* Either the teacher's biography is missing, or his competence or scientific achievements are not singled out.

As a solution, the new online platform gives the provider the opportunity to register on the platform and create their own profile, which reflects his biography, competencies and scientific interests and achievements. In order to solve another problem identified, it is not enough to provide quality services. It is proposed to organize a competitive environment on the platform.

To this end, the provider/teacher has a rating, which, on the one hand, creates competition among the providers of educational services, but at the same time gives an incentive to constantly improve the quality of content. The rating gives the teacher the opportunity to earn more money as better content providers. The rating of the teacher is related to the rating of the course and allows users to find the highest quality and popular courses. If the course has a high rating, the teacher is given the right to make it payable.

The problem of insufficient promotion of educational content of providers also has a solution. To improve the level of teaching skills, teachers can use points, depending on the rating and choose courses from partner companies.

Thus, each user has one account, with the ability to switch modes for taking or creating courses, which allows one to combine two roles: the teacher and the student, sharing experience in one industry and acquiring skills in another.

4.2   Architecture

Solution architecture provides the user with constant interaction with the system by carrying out the exchange of data between the client side «front-end» and software and hardware "back-end" (see Fig. 3).

When developing the platform, the authors recommend using a free framework for web applications. For example, Django for Python applications using the Model View Controller (MVC) design pattern. Python has more capabilities than the same PHP (Hypertext Preprocessor) greatly simplifies and speeds up the development process, makes the platform functionality more flexible, and the design is more complete.

The web server needs an interface that will launch the application, send the request from the user, and return the response. To accomplish these tasks, the Web Server Gateway Interface (WSGI) is the standard for interfacing Python programs and the

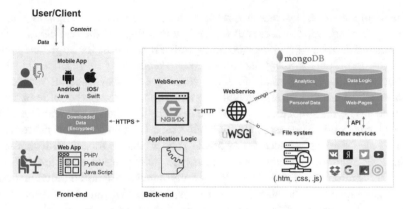

Fig. 3. An example of high level P2P online platform architecture

Web server uWSGI is one of the WSGI implementations. It is stable, flexible in configuration and has enough speed.

As the HTTP server and mail proxy server, nGINX [engine x] can be used, which at the same time has been a long service provider for many heavily used Russian sites.

The most advanced tool for creating interactive e-courses and video lectures at the moment is iSpringSuite. It allows you to create electronic tests, interactive simulators, mount video or create a learning game.

As third-party services you can also use services for payments, video hosting, mail client, mailing list, services for analytics, DDOS Projection, file storage, social Network APIs, Video calls and databases.

For simple usage, the platform functionality can be accessed both in the browser and in specially developed mobile applications. A special application for iOS, Android and PC allows users to view the course lectures in offline mode at a convenient place and time.

To implement the platform, there should be an online payment service (online wallet, banking cards, etc.) to purchase the selected course, as well as API authorization, and the opportunity to share reviews and comments in social networks.

Additionally, cloud LMS can be considered, which will allow access to information from any computer connected to the Internet. The same information can be viewed and edited by different users simultaneously from different devices. However, a permanent connection is required, as well as significant investments in equipment.

Using such technologies as artificial intelligence and Big Data will lead to a huge advantage - adaptive learning systems and control of involvement will significantly raise the quality of education. However, the cost of implementing these technologies is relatively high; the return on investment is possible only after the platform passes the point of self-sufficiency, with stable and dynamic growth.

4.3   Marketing and Monetization

As for the marketing and monetization of the P2P solution, there should be multi-channel promotion and an audience retention funnel.

The main sources of the platform revenue are paid content for users, as well as the employers' commission for access to the tools and ratings of the participants. That is, the platform can act as a training, recruiting and PR platform for large companies.

For instance, paid content includes group webinars with practical blocks and individual trainings. The cost of individual training consists of one-off lectures, practical classes and number of assignments, consultations of industry profile experts and communication with a mentor. Users can be offered "Pro packages" with light content or limited capabilities (for example, without video content). On the platform, there should be also content and lectures available free of charge for all participants.

The P2P solution can earn money from the paid placement of job offers from partners and promotion of partner courses. Development of tailored individual courses (e.g. industry-specific), the ability to add tests, interactive quizzes and other course formats (for the teacher), video content, digitization of the course, the ability to make conspectus, increase course popularity/rating - all these are additional ways to monetize the educational online service.

Collaborating with leading universities (providing a site for the placement of courses) can increase interest among the student audience. Reference partners can base mutually beneficial partnerships with bookstores, online services, coffee houses, and cinema networks on charging a commission for the purchase of recommended goods and services.

5   Related Work and Discussion

We can evaluate the research done in this area from three perspectives, starting with (1) peer-to-peer networking, applications and file-sharing systems, (2) online and distance learning (e-learning), and (3) peer-to-peer education systems as a logical result of the first two.

The very meaning of the term peer-to-peer networking is a process that allows computer hardware and software to function without the need for special central server devices. Peer-to-peer file sharing systems help each user contribute and retrieve information [15].

Peer-to-peer networks became a very popular method of sharing information amongst users between 2000 and 2010. Currently we can see the growing popularity of blockchain technology, which is a decentralized, or peer-to-peer, environment for transactions, where all the transactions are recorded on a public ledger that is visible to everyone [16].

The first studies on the use of peer-to-peer environments for transactions were performed in the early 2000s, with the main focus being on the aspects of the security [17] and transparency [18, 19], for all its users.

An overview of distributed education and its challenges was published by Oblinger, Barone and Hawkins in 2001. The paper was designed to provide college and university presidents with an overview of distance education, e-learning, or what is known as distributed learning. Additionally, a lot of studies were published investigating new environments, roles of learners [20], quality aspects [21] and organizational and technological constraints [22].

Considering our stated main goal, we have attempted to capture peer-to-peer educational systems ontologies that document a conceptualization for this domain. In the process of our research, we discovered that many articles were devoted to the topic of using Blockchain technologies in education. Collaborative and Peer-to-peer learning aspects are very well described by Andrews and Manning in their report "A Guide to Peer-to-Peer Learning" [23]. They answer the question of how to make peer-to-peer support and learning effective in the public sector.

The research of related work shows that the question of monetization of P2P services has not been properly researched, as well as the interests and demands of all interested parties. Additionally, during their research, the authors did not find any architectural and functional proposed solutions, leaving a gap in the research field.

In the research work our aim was to formalize the recommendations for the development of an innovative educational P2P service, which will be able to solve the existing problems of the education market.

We also have several limitations in the current research. Firstly, this study mainly focuses on a single Russian education market, with consideration of global trends. Secondly, we did not consider Blockchain to be peer-to-peer technology while creating the development approach of the P2P educational platform.

This research shows some interesting results on the behavior of the interested parties, the analysis of the potential barriers and their demand on educational services. However, we will be able to discuss the efficiency of the proposed approach only after we conduct the implementation of the proposed architecture and solution and create the distributed education ecosystem.

6 Conclusions and Future Work

The authors solve the problem of creating a development concept and recommendations for P2P platform. The main idea of the platform is not only getting the standard education, but also the exchange of knowledge between people.

To find the solution, the authors carry out a comparative analysis of existing educational services; identify their weak points and current barriers to consumers. The main question posed by the authors of the research is what features and functionality an innovative P2P service should have to solve current educational problems, attract the maximum number of users to its platform, and provide quality services. In order to determine the approach and make the format of a possible product, it is necessary to understand what the audience needs and what it lacks now, and analyze the interests of consumers and producers. As a result of market analysis, the authors have concluded that when developing an educational P2P platform service, one should focus on higher and additional vocational education. Three aspects are considered in detail: (a) the functionality of the P2P solution, (b) architecture and technology, (c) marketing and monetization.

Having analyzed all the shortcomings of the existing systems' functionality, the authors developed an approach that allows one to get the maximum effect from the existing platform. Within the framework of the approach, all functional requirements are considered from the point of view of participants in the process and educational content. As a result, a list of recommendations is developed for further development of the solution. The principle of joint learning, which is proposed in the recommendations, involves the transfer of knowledge not in the form of a lecture, but in the form of direct discussion between the teacher (the knowledge producer) and the student. During the course, everyone gets new knowledge and new values - both the teacher and the students. Such a P2P solution has significant potential and a large target audience that can make the service more large-scale, qualitative and useful due to a constant exchange of knowledge between all participants, and the joint improvement of processes and educational content as demanded in the market.

As an ongoing work in this line of research, we have just started to develop the architecture and functionality of the solution, to be implemented and shared publicly. We can proof the efficiency of the proposed solution only after implementation. This leaves possibilities for future research, where this research can be used as a foundation. The extension of the research on the quality of implemented solutions utilizing the P2P platform could result in educational ecosystems with sets of metrics to measure educational quality, consistency of content and user attrition and satisfaction rate.

References

1. eLearning market trends and forecast 2017–2021. Docebo Press (2017). https://www.docebo.com/resource/elearning-market-trends-and-forecast-2017-2021/. Accessed 27 Apr 2019
2. Shashkina, I.: Obrazovanie na vsu zhizn: kak online prevrashaet obucgenie v element lichnoy infrastruktury. Forbes (2017). https://www.forbes.ru/tehnologii/345797-obrazovanie-na-vsyu-zhizn-kak-onlayn-prevrashchaet-obuchenie-v-element-lichnoy/. Accessed 21 Mar 2019
3. Allen, E., Seaman, J., Poulin, R., Taylor Straut, T.: Online Report Card, Tracking Online Education in the United States. Babson Survey Research Group (2016). https://www.onlinelearningsurvey.com/reports/OnlineReportCard_embargo.pdf. Accessed 18 Mar 2019
4. Online-obrazovanie: rastushaya industriya. Analytical Journal «Universitetskaya Kniga» (2017). http://www.unkniga.ru/vishee/7989-online-obrazovanie-rastuschaya-industriya-1.html/. Accessed 17 Apr 2019
5. Roshina, Ya., Rudakov, V.: Monitoring Ekonomiki Obrazovaniya. NSU HSE, Informazionnii Bulleten 2018-1 [121], p. 46 (2018)
6. Berezina, K.: Chemy Seychas Hotyat Nauchitsa Ludi? HeadHunter (2015). https://www.the-village.ru/village/business/management/213453-uchitsya-i-pereuchivatsya. Accessed 22 Apr 2019
7. Parker, G., Van Alstyne, M., Choudary, S.P.: Platform Revolution: How Networked Markets Are Transforming the Economy and How to Make Them Work for You. Mann, Ivanov, Ferber, p. 17 (2017)
8. Shih, C.: The Facebook Era: Tapping Online Social Networks to Build Better Products, Reach New Audience, and Sell More Stuff. Addison-Wesley Professional, Boston (2011)
9. Bode, M.: Issledovanie rossiyskogo rynka online obrazovaniya I obrazovatelnyh technologiy. Netology Group (2017). https://edmarket.digital/. Accessed 12 Feb 2019
10. Reports FSN No VPO-1 of 2017/18. Ministry of Education and Science of RF (2017). Eis. mon.gov.ru/education/SitePages/ВПО_Формы.aspx. Accessed 30 Apr 2019

11. Rynok online obycheniya yvelichitsa kratno. Business News, Inform Agency (2017). http:// delonovosti.ru/business/4049-rynok-onlayn-obrazovaniya.html. Accessed 20 Apr 2019
12. Krazsova, E.: Millions from Skype: kak za 2 goda sozdat krupneishyu online skoly. RBC (2015). https://www.rbc.ru/own_business/11/03/2015/54fed5c79a7947851f391e22. Accessed 18 Jan 2019
13. Opros vzroslogo naseleniya po voprosam stanovleniya nereryvnogo obrazovaniya v Rossii za 2017 god. NRU HSE. Monitoring of Economic Education (2017). https://memo.hse.ru/ data/2018/03/01/1165054940/ind2017_PO_nepr_7.pdf. Accessed 18 Jan 2019
14. Rusanova, E.: Pochemy horoshih candidatov vsegda malo. HeadHunter (2017). https://hh.ru/ article/530528. Accessed 16 Apr 2019
15. Mahanta, K., Khataniar, G.P.: Peer-to-peer (P2P) file sharing system: a tool for Distance Education. Int. J. Comput. Sci. Inf. Technol. Secur. (IJCSITS) 3(2), 155–158 (2013)
16. Yli-Huumo, J., Ko, D., Choi, S., Park, S., Smolander, K.: Where is current research on Blockchain technology? A systematic review. PLoS One 11, 1–27 (2016)
17. Lachos, V., Androutsellis-Theotokis, S., Spinellis, D.: Security applications of peer-to-peer networks. Technical report 2, New York, NY, USA (2004)
18. Minar, N., Hedlund, M., Shirky, C.: Peer-to-Peer: Harnessing the Power of Disruptive Technologies. O'Reilly, Sebastapol (2001)
19. Ciglaric, M., Vidmar, T.: Position of modern peer-to-peer systems in the distributed systems architecture. In: 11th Mediterranean Conference: Electrotechnical Conference, MELECON 2002 (2002)
20. Rye, S.A.: Dimensions of flexibility - Students, communication technology and distributed education (2008). https://journals.hioa.no/index.php/seminar/article/view/2490. University of Agder. Accessed 11 Jan 2019
21. Antony, S., Gnanam, A.: Quality assurance in distance education: the challenges to be addressed. Higher Education 47(2), 143–160 (2004)
22. Garrison, R.: Theoretical challenges for distance education in the 21st century: a shift from structural to transactional issues. University of Alberta, Canada (2000). http://www.irrodl. org/index.php/irrodl/article/view/2/333. Accessed 23 Feb 2019
23. Andrews, M., Manning, N.: Mapping peer learning initiatives in public sector reforms in development. Effective Institutions Platform (2015). https://www.effectiveinstitutions.org/ media/EIP_Mapping_on_peer_learning_initiatives_in_public_sector_reforms_in_ development.pdf. Accessed 11 Mar 2019

Assessing Collaborative Learning with E-Tools in Engineering and Computer Science Programs

Steven Billis[✉] and Oscar Cubenas

New York Institute of Technology, New York, NY 10023, USA
{sbillis, ocubenas}@nyit.edu

Abstract. Collaborative learning has been gaining popularity as a teaching strategy that creates an active learning environment. Together with the use of e-tools it creates an instructional methodology which places more emphasis on the interaction of the students with each other and with their instructor through group activities. The use of interactive software (e-tools) to supplement the student's in and out of class activities are intended to generate excitement and motivation by providing students with both heads-on and hands-on experiences. This paper will present an example of one such active learning model as described above that was used in a required three semester programming sequence of courses for Computer Science (CS) and Electrical and Computer Engineering (ECE) majors at our university. This paper will report that significant gains in student performance and retention rates were achieved when compared to classes using more traditional teaching methodologies.

Keywords: Collaborative learning · e-Tools · Assessment · t-Test

1 Introduction

The traditional pattern of teaching is one in which the instructor typically assigns students to read textbooks and to work on problem sets outside of class, while listening to lectures and taking tests in class. This is referred to as passive learning [1].

While such lectures can be highly stimulating; and may even provide an efficient way of introducing large numbers of students to a particular field of study, critics of passive learning say that students are not really learning but just memorizing or retaining enough information to pass their next test [1].

Active learning, on the other hand is an instructional methodology which places more emphasis on the interaction of students with their instructor through in class activities that ask them to retrieve, apply, and/or extend the material learned in class [1].

Collaborative learning is a teaching strategy that allows the instructor to create an active learning environment. It is an approach to teaching and learning that has students working together in small groups, in the classroom, to complete a task. The tasks may range from solving a problem to creating a software product or an engineering design.

© Springer Nature Switzerland AG 2020
K. Arai et al. (Eds.): FTC 2019, AISC 1070, pp. 848–854, 2020.
https://doi.org/10.1007/978-3-030-32523-7_62

As such, it has the potential to provide the greatest amount of time for direct student/teacher interaction, and for peer/peer interactions as well. This interactive/ social engagement between instructor and student and among peers provides immediate feedback to both the student and instructor and has been strongly correlated with retention in the discipline [2–4]. In addition, the use of e-tools in and out of class, to supplement instruction, provides hands-on teaching experiences for further student engagement.

In the 2018/19 Academic Year a collaborative learning approach was implemented in a three semester sequence of Java programming courses (CSCI 125 "Programming I, CSCI 185 "Programming II, CSCI 260 "Data Structures") which are required courses for both CS and ECE majors. The courses were assessed with respect to their Learning Outcomes (LOs) using a Faculty Course Assessment Report (FCAR) that all full-time and part-time faculty are required to complete for every course that they teach during the fall and spring semesters. The assessment results were then compared to those sections of the programming sequence that were taught by the same instructor with a more traditional approach (i.e. a lecture oriented approach).

2 The Institute [5]

The Institute is a non-profit independent, private institution of higher education. Led by the President, the Institute is guided by its mission to provide career-oriented professional education, offer access to opportunity to all qualified students, and support applications-oriented research that benefits the larger world. Its students represent nearly all 50 U.S. states and 109 countries. The total number of international students at the various domestic campuses is 1350 (~10% of the total student population), the majority of whom are in the College of Engineering and Computing Sciences (CoECS).

There were 9,930 students enrolled in the class of 2018. The average SAT scores in Math and Critical Reading were 1120 and the average High School GPA was 89. To be admitted to engineering and computer science programs, students must have a minimum combined SAT score of 1000 (critical reading and math only), which includes a minimum of 520 in mathematics. In addition, students should have adequate mathematics skills to permit entry into Calculus I. Regardless of their SAT scores all entering freshman and transfer students who have not received transfer credit for either Mathematics or English courses must take placement exams to determine their skill in English and Mathematics. Their performance on the placement exam determines which course(s) they are eligible to take in their academic major. In general, the students in the computer science and engineering programs are first generation college students and many of them are required to take remedial courses in Mathematics and English. The Center for Teaching and Learning, (CTL), supports faculty members in their work as teacher/scholars by cultivating reflective practice and promoting the scholarship of teaching and learning. As part of the Institute's identity as a partially virtual institution, they serve as a resource for best practices in skillful, appropriate, and effective uses of technology in education. CTL provided assistance with the experimental procedure for the evaluation of the effectiveness of the various instructional methodologies.

3 Collaborative Learning/Zybooks

In the fall 2018 semester the instructor taught two sections of CSCI 125 and one section of CSCI 260. In the past, he had taught each of these courses as a traditional lecture and was not satisfied with the student performance. In the fall 2017 semester he decided to use a collaborative strategy in just one of the two sections of CSCI 125 "Programming 1" and as the classes were essentially the same size (approximately 25 students in each class) compare the student performance for the two sections. The instructor also used a collaborative teaching strategy for the single section of CSCI 260 "Data Structures".

In addition to using a collaborative teaching strategy in one section of CSCI 125, the instructor also used an online interactive book for learning Java from zYBooks: "Programming in Java with zyLabs" and in the CSCI 260 the e-book "Data Structure Essentials". These e-books combine the power of interactive tools, question sets, animations, and embedded homework, so students learn by doing.

In class, the students formed small groups of three or four [6–8]. About two thirds of each class period was dedicated to collaborative learning with students working together on the programming assignments. All program assignments required reading the problem description, designing an algorithm to solve the problem, using object-oriented programming concepts with the appropriate variable types, as well as the Java constructs. The instructor worked with each group to guide them in the application of general software construction principles.

4 Assessment

At the Institute's College of Engineering and Computing Sciences, each program has an assessment process in place to ensure that the LOs of each course have been attained. It is a process that provides data to support continuous course improvement. The course-embedded assessment tool used is the Faculty Course Assessment Report (FCAR). The FCAR concept was created by Dr. John K. Estell of Northern Ohio University [9], and later automated in Excel [10] by our Institute's faculty.

In both fall and spring semesters CoECS faculty members prepare a Faculty Course Assessment Report (FCAR) for each course they teach. The FCAR requires:

- The faculty member to identify course-specific learning outcomes (LO's) for his/her course and to establish appropriate performance tasks (APTs) with appropriate documentation to assess to what extent the LOs are being met. These APTs may be quizzes, exam questions, reports, projects, presentations, etc. Each student's APT is then scored with the method shown below (Table 1), to create an EGMU vector for that specific LO and a corresponding assessment metric.

Table 1. EGMU rubrics

EGMU	Rubric	Score
E – Excellent	Fully demonstrates/accomplishes the attributes and behavior in the rubric	3
G – Good	Mostly demonstrates/accomplishes the attributes and behavior in the rubric	2
M – Minimal	Minimally demonstrates/accomplishes the attributes and behavior in the rubric	1
U – Unsatisfactory	Does not demonstrate/accomplish the attributes and behavior in the rubric	0

The EGMU Vector is obtained as follows:

A typical EGMU vector for a class with 19 students in which the APT was the third problem of the first exam might be (8, 9, 1, 1) which would signify that 8 students demonstrated a complete and accurate understanding, while 9 students applied appropriate strategies, etc. The average score in this case being $43/19 = 2.26$ which is Good.

These course-embedded assessments serve as the primary tools to determine LO achievement and afford a direct link between learning outcomes as one aspect of curriculum change.

The data from FCARs are then evaluated at the spring Faculty Assessment meetings. At these meetings all full-time faculty members and those regular part-time faculty members wishing to participate identify and propose strategies to improve LOs and, hence, our program educational objectives through course work.

The recommendations of the assessment committee meetings are generally of two types: One set of recommendations can be implemented solely through the faculty member making internal changes to the courses (i.e. textbook changes, pedagogical changes). The other set of recommendations would need to be forwarded to the curriculum committees of the CoECS and then to the Academic Senate for adoption (i.e. new course, prerequisite/co-requisite changes, catalog description).

5 Assessment of Learning Outcomes

At the completion CSCI 125 "Programming I, the students will be able to:

LO1. Apply design and development principles in the construction of software systems of varying complexity.

LO2. Read a problem description, **design** an algorithm to solve the problem and document their solution.

LO3. Design, implement and **document** applications using object-oriented programming concepts."

As noted previously, the instructor taught two sections of CSCI 125. Section M01 had 26 students and was taught using a collaborative strategy and the e-tool zyBook. Section M02 had 25 students and was taught as a traditional lecture. Table 2 below shows the mean EGMU score for each of the LOs for the two sections.

Table 2. t-Test, Two-Sample, Unequal Variances: LOs CSCI 125 M01 fall 2018 vs. LOs CSCI 125 M02 fall 2018 (all figures in this and subsequent tables have been rounded to three significant figures after the decimal point).

CSCI 125 M01	CSCI 125 M02
LO1: Mean: 1.98	LO1: Mean: 1.64
Variance: 0.035	Variance: 0.015
P(T <= t) two tail: 0.008	
P < .05, difference in the means is statistically significant	
LO2: Mean: 1.82	LO2: Mean: 1.57
Variance: 0.018	Variance: 0.018
P(T <= t) two tail: 0.016	
P < .05, difference in the means is statistically significant	
LO3: Mean: 1.94	LO3: Mean: 1.71
Variance: 0.022	Variance: 0.031
P(T <= t) two tail: 0.048	
P < .05, difference in the means is statistically significant	

In the fall 2018 semester, the instructor taught one section CSCI 260 M01 "Data Structures" using a collaborative strategy with the e-tool zyBooks. The class had 24 students. Another section of Data Structures CSCI 260 M02 was also offered in the fall semester.

At the completion of CSCI 260, students should be able to:

LO1. Represent lists, stacks, queues, sets, dictionaries, trees, and graphs in the target programming language.

LO2. List the advantages/disadvantages of particular representations of data structures for particular cases.

LO3. Select algorithms appropriate to data structures to analyze behavior of aggregated objects.

Table 3 below shows the mean EGMU score for each of the LOs for the two sections of CSCI 260 M01, M02.

Table 3. t-Test, Two-Sample, Unequal Variances: LOs CSCI 260 M01 fall 2018 vs. LOs CSCI 260 M02 fall 2018 (all figures in this and subsequent tables have been rounded to two significant figures after the decimal point).

CSCI 260 M01	CSCI 260 M02
LO1: Mean: 1.95	LO1: Mean: 1.68
Variance: 0.019	Variance: 0.035
P(T <= t) two tail: 0.029	
P < .05, difference in the means is statistically significant	

(*continued*)

Table 3. (*continued*)

CSCI 260 M01	CSCI 260 M02
LO2: Mean: 1.88	LO2: Mean: 1.62
Variance: 0.023	Variance: 0.019
P(T <= t) two tail: 0.018	
P < .05, difference in the means is statistically significant	
LO3: Mean : 1.82	LO3: Mean: 1.58
Variance: 0.022	Variance: 0.005
P(T <= t) two tail: 0.015	
P < .05, difference in the means is statistically significant	

6 Conclusion

The collaborative teaching strategy together with zyBook's interactive online textbook is one example of an active learning environment that means to improve teaching and learning in the classroom. Active learning puts the responsibility of learning on the students as well as the instructor. Students are no longer just listeners but active participants in and out of the classroom. It is meant to engage students into the process of learning in to provide a more meaningful learning experience. There is a Chinese proverb that concisely describes the collaborative teaching strategy, and active learning in general: "Tell me and I forget, teach me and I remember, involve me and I learn" [11].

In the fall semester, two courses of a three semester programming sequence were offered. Two of the courses were taught using a collaborative teaching strategy, while their counterparts were taught as a traditional lecture. The LOs of all four courses were assessed and the mean EGMU score of each LO in both of the two CSCI 125 were calculated and compared. It was noted that the means of the CSCI 125 and CSCI 260 sections that were taught using a collaborative teaching strategy were higher than their corresponding LOs in the other course sections. The question that needed to be answered was: "Is the difference in the means significant?" and is it due to the difference in how the course was being taught. The t-Test answered both questions in the affirmative.

The department will conduct the experiment in the following spring semester to determine if the results corroborate the findings of the previous fall semester.

References

1. Billis, S., Anid, N.: To flip or not to flip; active and collaborative learning. In: Future Technology Conference (2017)
2. Prince, M.: Does active learning work? a review of the research. J. Eng. Educ. **93**, 223–231 (2004)

3. Hake, R.: Interactive-engagement versus traditional methods: a six-thousand-student survey of mechanics test data for introductory physics courses. Am. J. Phys. **66**, 64 (1998)
4. US Dept. of Educ. Evaluation of Evidence-Based Practices in Online Learning: A Meta-Analysis and Review of Online Learning Studies, Washington DC (2009)
5. www.nyit.edu
6. Kulik, J.A., Kulik, C.-L.C.: Effects of ability grouping on student achievement. Equity Excell. **23**, 22–30 (1987)
7. Kulik, J.A., Kulik, C.-L.C.: Ability grouping and gifted students. In: Colangelo, N., Davis, G.A. (eds.) Handbook of Gifted Education, pp. 178–196. Allyn & Bacon, Boston (1991)
8. Lou, Y., Abrami, P.C., Spence, J.C., Poulsen, C., Chambers, B., d'Apollonia, S.: Within class grouping: a meta-analysis. Rev. Educ. Res. **66**(4), 423–458 (1996)
9. Estell, J.K.: Streamlining the assessment process with the faculty course assessment report. Int. J. Eng. Educ. **25**(5), 941–951 (2009)
10. Beheshti, B., Billis, S.: An iterative approach to program assessment – best practices. In: ICEE
11. www.zybooks.com/

School Leadership Preparation and Technology Implementation: Ensure Successful Change Processes Through Transformative Mind Shifts

Jeff Faust and Ted Price[✉]

Virginia Polytechnic Institute and State University, Blacksburg, VA 24061, USA
pted7@vt.edu

Abstract. Leadership is changing and is now more commonly defined as a shared, collaborative responsibility rather than an individual one. Preparing leaders in the 21st Century, will require a team effort. Those responsible for preparing new leaders will need to consider leader behavior along with new technologies and research based best practices to ensure that preparation programs are in tune with the world as it is now. Mind shifts in the way leaders perceive issues and problems will be essential to success. This issue is of particular interest to principal preparation programs which must focus on developing effective school leaders. One way to begin the process of change will be to redefine popular conceptions of 21st Century leadership. Aspiring leaders and learners will need to be exposed to new ideas, attitudes, and approaches to work —making these shifts will also be required to transition aspiring leaders from teaching to leading. School leaders, and school principals, after teachers, have the most impact on student learning in their schools. However, it is not clear that principals are prepared to make decisions about what works best for staff and students in today's schools, new requirements and new accountability measures are in place, and many principal preparation programs have not changed to reflect what is now needed. In this digital age, being a leader for schools requires adequate preparation, experience, and good decision making: problem solving and technology leadership. An aspiring leader needs to make transformational mind shifts in their approach to leading. Technology changes also raise the bar for school leaders; the stakes are usually higher because of the anticipation around the likely impact of new ideas and newly adopted technologies. The school leader, to be effective, must also shift—their thinking, changing a focus from what happened, the events—to how they respond to what is happening now; from reacting to responding, from contracting to interacting. The way the new leader makes this mind shift and behavior transition; how the leader makes changes becomes a personal choice and responsibility. This article discusses the mind shifts necessary for the new leader to make, first in thought, and later in action changes in his or her personal behavior and awareness, in order to effectively and successfully manage and lead the change process to succeed in today's school environment.

Keywords: Education · Leadership · Technology

K. Arai et al. (Eds.): FTC 2019, AISC 1070, pp. 855–870, 2020.
https://doi.org/10.1007/978-3-030-32523-7_63

1 School Leadership, Technology Leadership and Change Management

There are so many self-help books on leadership. So, why are leaders, who supposedly have the needed skills to lead all aspects of schools, so hard to find? Research suggests three reasons. First, organizations, are often structured in old models that do not reward appropriate leader behavior, appropriate to today's needs, result, and schools haven't developed the leaders we need today. Also, all too often organizations have stifled leadership and leaders by fostering conformity leading to a confused sense of who they are, and what they stand for leaders are more often managers, expected to act like leaders but given managerial tasks, thus have not developed into the kind of leaders needed today. Second, a traditional understanding of leadership is now misunderstood in the new world arena, again the manager vs. the leader, expecting leader behavior but given management tasks new leaders have not emerged as we need them. Thirdly, technology in schools often exists as a separate entity and is not managed seamlessly or integrated into the institutional processes [3]. Consequently, in the new age of technology driven schools leaders have not been trained to manage, lead or integrate new technologies effectively.

As Michael Fullan has suggested, the main body of leadership literature has focused on the characteristics of leaders. He suggests characteristics and behavior can be changed, but first specific leader behaviors now expected need to be taught and understood [4]. Zuiebeck suggests that environment or characteristics alone are capable of creating effective leaders [5]. The traditional assumption has been that leadership is something the leader does to other people. Experts now suggest we should redefine leadership as something we do with other people, for example present a vision that others are willing to follow. Leadership and leaders require followers; it's not about telling people where to go any longer. Today's leader is more about showing the way; presenting a compelling vision of a preferred future. Goffee echoes this thought and suggests leadership should always be viewed as a relationship between the follower and the leader [6]. The challenges around technology proficiency for educational leadership complicates the responsibilities and adds new dimensions which are factors that often narrow the field of candidates as many individuals will not see themselves as being adequately prepared to lead technology changes and challenges in schools.

Principal Preparation Programs could become more effective by focusing on creating successful school leaders by demystifying ideas of 21st Century leadership and instructing students in new ways (attitudes and behaviors), to approach work and workers —making mind shifts—leaders who aspire to lead can make the move from teaching to leading by seeing the new picture and learning the new ways of leading effectively. Preparation programs can demonstrate to educational leaders how to carry out effectively new ways of working by collaborating with colleagues. One way is to change how teaching and learning occur in the modern classroom, away from the sage on the stage and into more of a role as a facilitator and coach; as a change agent. Demonstrating in the classroom new ways of encouraging aspiring leaders to evolve, personally and collectively, in the areas necessary for successful school leadership; collaborating being just one example of the new skill sets leaders must possess. Leaders are in a time of transition, especially those moving from teaching to becoming leaders; school leaders of teachers. This shift will require leaders to embrace aspects of technology [7], either in pieces or as comprehensive and integrated systems. When reviewing these challenges facing school leadership, the following questions come to mind:

What transformative mind shifts are needed in preparing school leadership and aspiring school leaders to make decisions regarding the effective use of technology in their schools? What decisions will fall to a school leader regarding technology usage in their school? How can school leaders measure technology usage and efficacy?

A change to the leader one wants to be, ought to be, and needs to be, can occur when the emergent leader engages in new thinking (and later implements new behaviors) about what makes effective leaders in the new paradigm of leadership. Making mind shifts to embrace a new approach to leadership may quell some of the uncertainty many new aspiring leaders encounter during the transformation from being a person led—a follower, to a person who is now charged with leading—with being the leader. Considering new ideas, new thinking about leadership, new ways of conducting oneself as a leader is a first step in behaving as a different kind of leader. Behavior emanates from thinking. Therefore, many find themselves on a quest for clarity and direction in understanding what "leadership" means and what the new role will look like and be. Shifts in paradigms and mind shifts for new leaders are a first step for the new leader to consider as transformative elements [9]. Seeing leadership in a new way may help an aspiring leader move forward in their career. Clarity in roles and expectations can lead to new behaviors—new leaders need to lead in new ways.

Leading is being redefined as a shared, collaborative responsibility rather than an individual one [1]. As a result, preparation programs, are redesigning, too, both in what the final outcome will be what leaders will look like and be able to do, but also in how training and instruction is being provided; demonstrated to prospective students and aspiring leaders. However, before any redesign occurs new thinking resulting in different mindsets about what leading is and what it should be must change, too. We know the format of programs can be changed—we know the organizational environment can change but only after people begin to think differently about the intentions and the outcomes. By teaching new conceptions of appropriate leadership behavior, new methods to approach and solve problems are learned (Fullan 2014). To instruct future leaders to ensure effectiveness in the 21st Century, new paradigms of leader behavior must be embraced, and shared as the starting point for the redesign of all preparation programs.

Thomas Kuhn's book The Structure of Scientific Revolutions, popularized the word "paradigm." This work was published originally in the latter part of the 20th century, and recently updated [8]. The word "paradigm" traditionally meant a "pattern" or "model", but Joel Barker and others put forward the following extended definition: a system of rules and regulations that first establish boundaries, then offer guidance on how to successfully solve problems within those boundaries. Paradigms, as Barker explains them, offer a problem-solving approach; a model for problem solving [9]. The bottom line of both a shift in paradigm thinking and in leadership thinking is coming up with new and original ways of envisioning problems and acting on them via planning and change management [10]. Leaders must be able to visualize the future and imagine a path to achieving their vision while removing any obstacles. Technology is a catalyst for change that can disrupt the leader's ability to visualize the future, often this change occurs before the industries or individuals that will be affected most are ready to change and in the past few decades digital technologies have changed virtually every aspect of the way humans interact and learn [11].

Change is the only constant; this is evident in the great efforts our schools and communities are making to keep pace with technological forces impacting the behaviors and norms of instruction and school operations. School leadership must embrace these forces and learn to work with them not only to be a successful school leader but also to effectively utilize technology to promote change [12]. One challenge for leaders comes in the form of successfully initiating and managing the change process [13]. Leaders are now expected to move from the passive state of watching events (previously identified as manager behavior but called leadership) by moving to the active state of creating a picture of a hoped-for outcomes and simultaneously creating an environment to afford opportunities for change to occur. Shifting focus from what happens—the events themselves—to the response is a key for transitioning into leadership; leaders who respond and interact are desired, not leaders or managers who react and contract. Ultimately, the way a leader transitions and promotes change is a personal choice and responsibility [2], but responding by interacting is a significantly better method of leading in today's world. Technology systems are invaluable in their ability to collect data and make measurements continuously in ways that would not be possible or sustainable without these systems. School leadership must be efficient in measuring change. When leaders commit to continuous improvement the modus operandi of both leader behavior and school change gets better and does so more efficiently.

General investigations on paradigm shifts and how leadership should be defined and discussed in the 21st Century has been looked at by researchers [14]. Noteworthy is the following: in the field of school leadership, there is a dearth of research around the idea of how personal transformations may be required to transition from classroom teaching to positions as school leaders, at a time when new definitions of leadership are being offered up and program remodels are occurring designed to preparing school leaders to work in schools reformed by recent reform efforts and even legislation. It would be easy to surmise that a new definition of leadership with accompanying revised curriculum models have resulted in major changes in the ways schools are now being led, but such a conclusion is premature [15].

The current definition of leadership and new curriculum modes (digital, individualized, remote, standards-based) were not part of the experiential path our rising leaders experienced or practiced and may in fact be completely foreign to them. Further research studies may need to be initiated to focus on the reasons upcoming leaders need to rethink the work they aspire to do [16]. Sir Ken Robinson has written about the urgent need to transform our industrial model education system. He has suggested that training of teachers and administrators must occur simultaneously to aid in the transformation of schools to a more appropriate model of schooling if we really want to improve student learning. Robinson articulates that the self-transformation of the new leader—a personal behavioral and change process, first in thought, and later in action, is needed to help new leaders make the mind shifts necessary. Society should view these shifts as necessary in order to promote development of effective school leadership.

2   Three Areas of Management for School Leaders

There are three areas of management all leaders must master to successfully create an environment that fosters change—in particular, change that has resulted from making a mind shift that allows for understanding problems and behaviors in new ways. The three areas are:

- **Visioning:** Leading a shared vision accompanied by a focus on staff working collaboratively on personal development, personnel development and organizational development for the whole of the organization.
- **Crucial Conversations:** Open and direct: clearly communicate feedback, including peer review and interaction.
- **Technology innovation:** Tackling assignments, creating products and providing solutions

2.1   Visioning

Leading a shared vision accompanied by a focus on staff working collaboratively on personal development, personnel development and organizational development for the whole of the organization.

In regard to visioning, the following may help clarify: The phrase "vision" means seeing ahead, seeing the future, and must be future focused in a pragmatic and practical sense. A person is described as seeing the future when he or she can understand a situation or concept as a preferred option and does not get bogged down on specific details. So, from a personal perspective what one believes is needed and where their vision leads the organization: it's all about a philosophy, a vision and seeing the "big picture." Leadership is not a technique. It comes from passion and vision–having an idea about what the future can and should be. In having a vision, Irwin suggests that truly engaging others (shared vision) flows from the essence of who we are from our core. Leadership is passionate and personal [17].

2.1.1   Create Goals

Leaders must create their own goals professionally and personally, move from being a follower to becoming a leader.

Self-awareness is the beginning of personal leadership. Many young people are consumed with striving to be perfect, especially to those in the outside world. As a result, many are reluctant to stop and look inward as to what really matters and as to what really motivates us; many of our impulses are noble, while others are not necessarily so. It is not that everyone has been reluctant to face their shadow, but through a lack of self-awareness deny there is a shadow at all [17]. As a leader, one near the top of the ladder or still climbing up, a person can manage dysfunctions and move ahead when they acknowledge some of their own obstacles or even blind spots. To attain a higher level of success leaders are wise to become aware of what others might perceive of, as unconscious annoying habits and this may only occur if one becomes more self-aware of attitudes and behaviors [18]. In a leader role be careful not to misuse power, once used improperly it might be taken away. Aspiring leaders must put self-awareness and self-regulation of behavior at the top of their list of must haves; grow awareness in direct proportion to the power one hopes to exert [19].

2.1.2   Transform Roles

Leaders need to transform from a former teacher with a boss to a leader who others will see as the boss. No longer looking for a person to set the tone or direction to becoming the leader who works with others to set the tone and direction. Being the one now who believes working together with others you can find the answers.

Research suggests leadership failures do not reflect a problem with the leader's skills and abilities. Often failures occur due to a breach of something inside the leader-a character flaw. Because people tend to lead from their core this is not surprising. However, polished a leader may appear to be on the outside, what comes through in decision making is what's on the inside [17]. An analysis of one's abilities is key, but also knowing one's character is often a better predictor of success and or failure. Personal analysis can shed light on how character may emerge in different settings and situations based on each unique individual [20]. A leader who takes charge of his or her growth while leading others accepts the responsibility of leadership [21].

2.1.3   Take Calculated Risks

Go beyond playing it safe; so, take responsible risks. In your growth and development, push yourself. In order to grow and become more than you are now embrace the uncomfortable.

Awareness of one's current situation may promote a willingness to take risks and hopefully grow in the process of changing and in the process to be quick and nimble [22]. Supported leaders feel able to show vulnerability. Being vulnerable can lead to experimenting, taking risks. First comes self-awareness which can lead to an increase in their ability to learn, grow and try new strategies. Buckingham and Coffman's research concludes - great managers share one common trait: They break virtually every rule held sacred by tradition and the previous conventional wisdom (1999). Baldoni's research includes advisement, embrace risk taking behavior, put fear aside and use your experiential experience and intuition to move forward [23]. Create an organization where failure is seen as learning and failure is a step toward success. Tolerate and even encourage risk, trying something different while attending to what works and what has not worked previously, keep the good and throw out the bad can be essential to every breakthrough and is the road to success. Technology usage and adoption presents a risk for many leaders as they are likely to feel unprepared to use it effectively or successfully. In today's schools, leaders must lean into these opportunities as their employees expect them to provide leadership around technology [24].

2.1.4   Become the Person/Leader Who Is Doing for Others,
Not Just for You

To create a sense of we, says Kurtzman, it is necessary to create the specter, the sense of them. Developing a common purpose, can be a path to organized support if it is done without others in mind, including involving others in the process opposition to the cause can quickly arise. Because organizations are primarily focused on goals beyond the individual it is incumbent on the leader to involve others, others point of view and opinions are needed if you want to accomplish shared goals, this can't be done alone. Operationalizing methods of aligning groups of people, so they achieve common goals should be paramount in the leaders thinking and actions [25]. Reading a situation; awareness of the ins and outs of individual concerns is paramount to achieving a balance of the leader's mindset with the thoughts and opinions of others. This balance in expanded awareness comes from having self-knowledge and considering the needs of others and requires a quality such as mindfulness to be present, aware of circumstances, and cognizant that ideas are best shared face-to-face in present time. Servant leadership is really about supporting and serving others, that results in an outward focus toward the organization that considers people and products [26]. The leader who aspires to serve others also inspires others to grow and develop through modeling. Personnel performance improves as a result of personal growth and development encouraged and supported by the leader. The organization benefits all, including the stakeholders when everyone shares. This shared process leads to increased organizational effectiveness and success.

2.1.5 Leading the Group: Go Beyond Being Part of the Group

According to Bill Cohen, leadership is leading a group with integrity and simultaneously lifting individual performance to new levels of achievement while building team spirit; a one for all and all for one attitude. Even a sense of sacrifice for the good of the whole. Heroic leadership, Cohen adds, requires high standards [27]. Michael Maccoby agrees in his book, The Leaders We Need, and says to lead in the global enterprise, the leader needs to know about people, almost to a fault [28]. He means the leader really needs to get down in the weeds with colleagues, both personally and professionally. School leaders have tremendous power to lead. To get the message out these days the leader must use every available digital tool and networks. Means and methods school leaders should be employing to ensure that they are sharing their vision but also listening and hearing what their staff members are saying must be the way of business. Conducting business is communicating effectively with all players; those inside and outside the organization. School leaders, too, need to demonstrate how kindness and caring for others; authentically, can be used to: recognize, motivate unique talents, groom; establish a compassionate environment; accelerate organizational growth; change and adapt; foster risk-taking; and promote the next generation of leaders for the good of their organizations [29]. Organizational development is a by-product of personal and professional development by all members of the organization.

> "Treat people as if they were what they ought to be, and you help them become what they are capable of being."—Johann Wolfgang von Goethe

2.2 Crucial Conversations

2.2.1 Leaders Must Be Open and Direct

According to David Whyte, the core act of leadership is the act of making conversations real. Difficult conversations, now referred to as crucial conversations, are what build relationships. Conversations that create the opportunity for honest feedback. To change a mind a leader must first understand the thinking of another. Conversations about new choices are based on prior knowledge and understanding. The leader who leads is the leader who has at first been a listener. Leaders enable conversations by listening first. In bringing people along for conversations of personal, practical and organizational wellbeing they increase each member's level of engagement, commitment and accountability. Leaders first get at understanding others by asking people to share their own thoughts and feelings, and in process assume personal leadership of their own behaviors. The practice of listening becomes paramount in the process of leading [30].

One doesn't necessarily choose to lead. Leading is an honor given to the leader by those who are willing to follow. Most leaders are humbled by the process and find the decision to accept a leadership role one of the most rewarding decisions a person will make, however with awesome recognition comes increasing responsibility. A leader can use their ability to help others make known, express and fulfill their potential. Leaders influence others by establishing a vision with a clarity of purpose. A clear vision creates opportunities which otherwise might have remained unknown. Leadership is a mindset of service. Leadership is a mindset of life-long learning. Leadership is a mindset of evolution, personally and professionally. The truth of leadership is that leading is a process of creating in that building up others also brings out the best possible version of the leader [31]. Warren Bennis suggests that good leaders are also good followers in his book Reinventing Leadership. Leaders and followers are mutual partners and share similar characteristics: listening, collaborating, and working out issues with colleagues. When the leaders and followers agree on the perspective that there is a shared responsibility for everyone's growth, the whole organization benefits [32].

2.2.2 Watering a Garden Properly Is More Than Adding Water

A culture that is well watered is like a green-house garden where the environment contributes to the effectiveness of watering–water is dispensed in the right amounts and at the right time. So, too, in organizations where people and ideas flourish; everybody in the organization, regardless of position or title, is empowered to act and communicate frankly and openly and is rewarded for sharing ideas about new products, effective systems, more efficient processes, and improved ways to savor and serve customers [22]. Understanding people's strengths, and nurturing them like a gardener with prized plants, pays big dividends to the organization. Developing people, however, in an organization is an organizational commitment that is never fully realized. For the truth is, people can't be developed; they must develop themselves, and so the task for the leader is like that of a head gardener, always trying to figure out what the various microclimates provide to the plants, and figuring out the qualities of the plants that need adjustments in the microclimates. All individuals within an organization (like plants in a green-house garden) must be managed for their strengths and abilities, and if done carefully, may be placed in the most appropriate area for growth and development [33].

2.2.3 From Reactive to Proactive

Leaders must adapt their own perspectives, behaviors and questions from being reactive to proactively driving the conversation forward; the difference between reacting and responding, by adding depth and value after really listening and demonstrating an understanding of the concerns and needs of others; from reacting to responding, from being defensive to being supportive and encouraging. Looking for solutions instead of pointing out problems. Making excuses, venting and sessions focused on complaining are useless.

Herbold writes, In What's Holding You Back? We need courageous leaders who ensure the organization has a clear, understandable, and almost simple to execute game plan for the future. Leaders that can create a culture of curiosity about the future, to encourage people to look for advanced ways for the company or organization to grow and prosper. Leadership must confront problems and challenge people to change and avoid creating a culture focused on pride in the past and the protection of old procedures [34].

2.2.4   Facilitating Discussions and Conversations

Listen more, talk less, and establish processes and procedures that foster learning and growth for everyone. Chris Thurman wrote in The Lies We Believe, and says, some of the causes of unhappiness are the lies people choose to believe in life, what he means is we have to confront misconceptions, we have to dig down to know the truth and hold constant that which we know is true and root out that which we know is not. We can't let people operate apart from reality in that we do things because they have always been done that way. If an individual's interpretation or version of reality isn't in line with what we know to be true, we have to work to change minds.

In the past we have been inclined to choose the status quo as a way of operating over confronting new realities. The resultant old ways have not allowed for organizational growth. Instead of getting the results we now need, one was likely to get reasons, stories, and excuses for why things are not working out, leading to disengagement by followers and ineffective leadership by the leader [35]. A leader needs to foster and facilitate an atmosphere where listening becomes the norm and communication for its honesty and openness is valued. The courageous leader knows what role technology will play can play in communicating effectively with all stakeholders. The new leader also knows what role technology can play and leads in the technology of communication revolution. Leaders can select tools and processes that promote effective communication and information sharing across their organization.

Facilitating and Coaching to Bring Out the Best in Others

Kouzes and Posner raised and interesting idea when they reported that they are asked time again is "What's new in leadership?" Unfortunately, there answer can be disheartening. They often respond to the question by answering, while the context of leadership has changed dramatically, the content of leadership hasn't changed much at all. What they mean is the fundamental behaviors, actions, and practices of leaders have remained essentially the same since their beginning research findings were published, and they have been writing about leadership and leader behavior for over three decades.

At this point we're left with the fundamental truth of leadership development. We haven't developed the kind of leaders we need to lead in the new world. Kouzas and Posner reiterate that leadership requires one to first understand where people are coming from, what they believe or understand. We have not been future focused in our leadership development program. Too much emphasis has been placed on learning content knowledge and not enough on learning how to work with the resources at hand, primarily the people who do the work in the organization. To develop leaders in all contexts and weed out fact from fiction we need to help people see problems and people in new ways [36]. With new paradigms emerging continuously, focusing on the mind shifts and new perspectives, leaders can make to help them move forward while supporting new behaviors and approaches to leading and working with others will help bring out the best across the organization. Inspiring leaders supply the right ingredients, earning individual and team respect. Leaders must blend vision, communication and technology with trust, integrity, partnerships, and affirmations to lead so others will follow [37].

> "I suppose leadership at one time meant muscles; but today it means getting along with people."—Mahatma Gandhi

2.3   Technology Innovation, Tackling Assignments, Creating Products and Providing Solutions

Why are great leaders great? It's not because they are super-human. In fact, they are ordinary in many respects and extraordinary in some others; they are growth-oriented folks with convictions who have chosen to advocate for a bold course of action. A course in the best interest of those they serve despite the odds [38]. Great leaders are both creative and innovative.

2.3.1   Collective Efficacy and Shared Responsibility Means Moving from One Person with the Correct Answers to Many People with Many Answers

Contributions leaders make are to today's bottom line, and also to the development of personnel and institutions that change, prosper, and grow. The impact leaders have on organizational life and culture-and the affect that organizations have on leaders and personnel are problem solving keys [39]. A leader should never take the job of leadership unless they are willing to see beyond personal needs.

Leaders who are motivated to lead by their own needs will ultimately fail, as a result schools and students will suffer [14]. Finding ways to continually innovate, seeking ways to get the job done is the wave and way of the future leader [38]. Leadership must learn from feedback loops and cycles of continuous improvement. This idea is ubiquitous across today's technology and software companies. These methods include Lean, Agile, Scrum, and Kanban. Practices designed to empower the collective voice and shift the focus of change away from glacial progress and towards measurable iterations in cyclical sprints is the way the modern organization succeeds [40].

2.3.2   Getting Things Done

Leaders must avoid asking: Is this what you want? They must instead put forward possible solutions and examples of paths to completion of the tasks and work which lies ahead for their organization.

In stressing the importance of becoming a Performer in leadership development, Authors Chris Brady and Orrin Woodward emphasize the need to create a record of performance. Technology professionals developing a new product tend to embrace the concept of sandboxing for vetting new features and feasibility of new products. Conversely, how are school leaders supporting sandboxing of new ideas, products or opportunities? How is the school leader facilitating conversations around what's possible?

Promoting new ideas and solutions in an organization will not happen by accident and some organizations have even gone so far as to promote intentional practices to ensure that their employees remain focused on the need to innovate and embrace new opportunities. One method of promoting creative solutions and which is usually attributed to Google is to provide employees with 20% time to pursue projects of personal interest that could benefit the organization. School leaders need to investigate how they can facilitate a similar practice to encourage innovation. How can the focus in schools be shifted from high value being placed on knowing and knowledge acquisition to promoting collaboration in pursuit of solutions to already existing, real-word challenges?

If our school leaders keep doing what they have always done, then the product being created will never change. In their book, Launching a Leadership Revolution, Brady and Woodward shared a perspective that the most efficient and guaranteed method of ensuring repeated success as measured by performance is to identify, refine and enculturate any patterns of success already present in the organization. Every leader should strive to become a performer whose success cannot be overlooked by any member of the organization. This will ensure the development of a platform from which they will be effective in helping others to succeed. The route to influence is from past performance [41]. And in Flow: The Psychology of Optimal Experience, Csikszentmihalyi says that by ordering the information that enters our consciousness, we can discover true happiness and greatly improve the quality of our lives while performing at our highest level of efficiency and effectiveness [42]. To arrive at the new, leaders coming from the past, and leading in the future will have an alignment of values, beliefs, expectations, and new performance behaviors/solutions.

2.3.3 Make Your Needs Known

There is no better way to help you grow and learn, except to be proactive in making your needs known. There are no mind readers here. The new leader will help others to move from doing the assignments "as assigned" to suggesting how an assignment will help others to grow and develop the skills needed to more appropriately meet new job expectations, theirs and others.

In his distillation of his own thoughts, Myself and Other More Important Matters, Charles Handy, considers and thinks about the idea of exposing the difficulty in seeing ourselves as others see us [43]. Handy talks about making our blind spots known. He suggests that throughout our lives we all play many parts and, in a sense, become different people. But he asks: Can we become something different from what we are and different from what we see ourselves as being? Can we become a leader? Only by becoming aware of our own perceptions, certainly, but also on the perceptions others share about us, with us. Handy continues these ideas and goes on to state that in psychology scholars ask whether we have a core identity that is sitting there in our inner self, waiting to be revealed, or whether our identity evolves over time. One of the questions that festers in organizations is related to that debate: Are leaders born or made [44]? In connecting with others grounded in empathy and vulnerability consultant David Noer writes about something not realized by all leaders. He says empathy and vulnerability comes as the result of struggles and growth. It is through our struggle that individuals are prepared to help others [45]. Thus the answer to the age old question is: We are born and we are made, we can become a leader at any time.

To see the whole more clearly we need to adopt a peer-review process: solicit feedback and have another pair of eyes on our products prior to submission. Best to develop your skills as a byproduct of "courageous conversations" with one another: listening to feedback and adjusting as appropriate. Leadership in not often tidy or clean. It's messy, and so is the rest of life. To improve work products you have to work successfully with others.

What makes an effective leader is a contradictory collage of motivations and drivers, rewards and costs [1]. In The Lean Startup, Eric Ries advocated the notion of "Fail Fast" not because failure is the end of the path, but rather a necessary part of the learning process [46]. Consider the adage; failure is the first attempt at learning. Leaders must encourage employees to fail fast and support the impending pivots and iterative improvements necessary as part of the organizational response to ineffective or unsuccessful practices. Schools are not agile, but the world around schools is more agile and nimble than ever. This means leadership cannot be taught, at least not in the same manner it has been thus far. Being a good leader is not formulaic or prescriptive. There are not step-by-step directions which end in a successful leader. Instead, it is necessary to understand the complex relationships and situations which will confront all leaders and evaluate how this differs now when compared to leadership traits of the past. By taking a look below the surface, into the heart and pulsing arteries of leadership, we are bound to understand leadership as a process much better [47]. Leadership is a collaborative process that promotes growth, learning and serves as a key indicator of team effectiveness and leader influence [48].

2.3.4 Good Enough Never Is Actually Good Enough

Work in an environment for a leader who advocates for you to his/her boss and uses your work products to demonstrate your high competence and skill.

Behave like a leader, no matter what others do. Be a part of shaping culture. Be inspirational and inspire yourself and others. See everyone as a leader. And still everyone is not the same [33]. Jim Collins has written, good can be the enemy of great [49]. Leaders can become so accepting of thinking of performance as something that develops along evolutionary lines, from poor to good to outstanding, that it takes a while to grasp the notion that competence can actually inhibit achievement. Work must be of the highest quality. Expect excellence. Leaders who embrace the necessity of having do something again and again until they know for sure the product meets the highest standard realize that failing and learning and achieving are all parts on the path to success. When faced with a challenge, leaders must look inwards and outwards.

A leader must be prepared to ask, let go, and wait for an answer. By trusting the process, leaders will eventually shift focus away from looking outside themselves for guidance to intuitively knowing the right way to go as they develop their own potential and achieve success, to lead is to be followed [50].

> "A good leader leads the people from above them. A great leader leads the people from within them."—M. D. Arnold

3 Summary

3.1 Vision

Leaders do not have time to do their job and the job of others, too. Hand holding will not be tolerated while leaders must focus on evaluating and ensuring that organizational objectives are being met. Rather, leaders seek out individuals that will think creatively and approach challenges strategically to assess, create, innovate, implement, evaluate, and repeat the cycle. It's all about crucial conversations, problem identification and decision making in a collaborative environment [38].

3.2 Crucial Conversations

Colleagues and staff, "are looking for vision, a leader to collaborate, be creative and articulate the direction." Staff are looking for a communicator (Martin, R. personal communication, February 5, 2015). Employees requiring too much direction in the processes will inevitably be replaced by someone who can complete the necessary tasks more autonomously.

3.3 Technology Innovations

The best version of you. Any work, all work, must meet one's own highest standards. Putting forward what is unequivocally the best product every time. Leadership connects our actions today to the promise of hope or future life. Great leaders are needed everywhere, but perhaps no single field exemplifies the need as prominently as in education where schools are being asked to do more and be more than ever before with fewer resources and greater accountability [51]. In The Tipping Point Gladwell changed the reader's understanding of the world. In Blink he challenged the readers to rethink thinking. In OUTLIERS he transformed the reader's understanding of success [52]. Now, more than ever before, our schools require innovative leaders with great vision who are ready to embrace the requisite self-transformation to converse successfully with others. In the process of leading self and others the leader will also begin to think differently about the current state and the current vision. The successful leader will know to lead others and will have embraced communication skills and technology innovations as cornerstones of effective leadership behavior.

References

1. Pink, D.H.: Drive: The Surprising Truth About What Motivates Us. Riverhead Books, New York (2009)
2. Brenner, A.: Transitions: How Women Embrace Change and Celebrate Life. CreateSpace, Charleston (2010)
3. Ritzhaupt, A.D., Hohlfeld, T.N., Barron, A.E., Kemker, K.: Trends in technology planning and funding in Florida K-12 public schools. Int. J. Educ. Policy Leadersh. **3**(8), 1–17 (2008)
4. Fullan, M.: The Principal: Three Keys to Maximizing Impact. Jossey-Bass, San Francisco (2014)
5. Zuieback, S.: Leadership Practices for Challenging Times: Principles, Skills and Processes that Work. Synectics LLC, Ukiah (2012)
6. Goffee, R., Jones, G.: Why Should Anyone Be Led by You? What It Takes to Be an Authentic Leader. Harvard Business Press, Boston (2006)
7. Brooks, C.: Locating leadership: the blind spot in Alberta's technology policy discourse. Educ. Policy Anal. Arch. **19**(26) (2011)
8. Kuhn, T.S.: The Structure of Scientific Revolutions, 4th edn. University of Chicago Press, Chicago (2012)
9. Barker, J.A.: Paradigms: The Business of Discovering the Future. Harper Business, New York (1993)
10. Denning, S.: Why the paradigm shift in management is so difficult. Forbes, November 2012. http://www.forbes.com/sites/stevedenning/2012/11/12/why-the-paradigm-shift-in-management-is-so-difficult/
11. Foundation for Excellence in Education: Digital Learning Now (2010). http://digitallearningnow.com/wp-content/uploads/2011/11/Digital-Learning-NowReport-FINAL.pdf
12. Anderson, R.E., Dexter, S.: School technology leadership: an empirical investigation of prevalence and effect. Educ. Adm. Q. **41**(1), 49–82 (2005)
13. White, B.J., Prywes, Y.: The Nature of Leadership: Reptiles, Mammals, and the Challenge of Becoming a Great Leader. AMACOM, New York (2007)
14. Kouzes, J.M., Posner, B.Z.: The Leadership Challenge: How to Make Extraordinary Things Happen in Organizations, 5th edn. Jossey-Bass, San Francisco (2012)
15. Wagner, T., Kegan, R.: Change Leadership: A Practical Guide to Transforming Our Schools. Jossey-Bass, San Francisco (2006)
16. Robinson, K.: Out of Our Minds: Learning to Be Creative, Rev edn. Capstone, Chichester (2011)
17. Irwin, T.: Impact: Great Leadership Changes Everything. BenBella Books, Dallas (2014)
18. Goldsmith, M., Reiter, M.: What Got You Here Won't Get You There: How Successful People Become Even More Successful. Hyperion, New York (2007)
19. Kouzes, J.M., Posner, B.Z.: The Truth About Leadership: The No-Fads, Heart-of-the-Matter Facts You Need to Know. Jossey-Bass, San Francisco (2010)
20. Rath, T.: StrengthsFinder 2.0. Gallup Press, New York (2007)
21. Bell, A.: Great Leadership: What It Is and What It Takes in a Complex World. Davies-Black, Mountain View (2006)
22. Bryant, A.: Quick and Nimble: Lessons from Leading CEOs on How to Create a Culture of Innovation. Times Books, New York (2014)
23. Baldoni, J.: Lead with Purpose: Giving Your Organization a Reason to Believe in Itself. AMACOM, New York (2012)

24. Brown, B., Jacobsen, M.: Principals' technology leadership: how a conceptual framework shaped a mixed methods study. J. Sch. Leadersh. **26**(5), 811+ (2016)
25. Kurtzman, J.: Common Purpose: How Great Leaders Get Organizations to Achieve the Extraordinary. Jossey-Bass, San Francisco (2010)
26. Temes, P.S.: The Power of Purpose: Living Well by Doing Good. Three Rivers Press, New York (2007)
27. Cohen, W.A.: Heroic Leadership: Leading with Integrity and Honor. Jossey-Bass, San Francisco (2010)
28. Maccoby, M.: The Leaders We Need: And What Makes Us Follow. Harvard Business Press, Boston (2007)
29. Baker, W.F., O'Malley, M.: Leading with Kindness: How Good People Consistently Get Superior Results. AMACOM, New York (2008)
30. Whyte, D.: Life at the frontier: the conversational nature of reality [Video File], February 2011. http://youtu.be/5Ss1HuA1hIk
31. Dweck, C.: Mindset: The New Psychology of Success. Ballantine Books, New York (2008)
32. Bennis, W., Townsend, R.: Reinventing Leadership: Strategies to Empower the Organization. HarperCollins, New York (1995)
33. McKinney, M.: Lord Sharman on Helping People Grow [Review of the Book Lessons Learned: Straight Talk from the World's Top Business Leaders, by L. Sharman], 26 September 2007. http://www.leadershipnow.com/leadingblog/2007/09/lord_sharman_on_helping_people.html
34. Herbold, R.J.: What's Holding You Back: 10 Bold Steps that Define Gutsy Leaders. Jossey-Bass, San Francisco (2011)
35. Thurman, C.: The Lies We Believe. Thomas Nelson, Nashville (2003)
36. Kouzes, J.M., Posner, B.Z.: Credibility: How Leaders Gain and Lose It, Why People Demand It, Rev. edn. Jossey-Bass, San Francisco (2011)
37. Blanchard, K., Muchnick, M.: The Leadership Pill: The Missing Ingredient in Motivating People Today. Free Press, New York (2003)
38. Christensen, C.M., Horn, M.B., Johnson, C.W.: Disrupting Class, Expanded Edition: How Disruptive Innovation will Change the Way the World Learns. McGraw-Hill, New York (2010)
39. Kets de Vries, M.F.R.: Reflections on Character and Leadership: On the Couch with Manfred Kets de Vries. Wiley, Hoboken (2010)
40. Schwaber, K., Beedle, M.: Agile Software Development with Scrum, vol. 1. Prentice Hall, Upper Saddle River (2002)
41. Brady, C., Woodward, O.: Launching a Leadership Revolution: Mastering the Five Levels of Influence. Business Plus, New York (2007)
42. Csikszentmihalyi, M.: Flow. HarperCollins, New York (1991)
43. McKinney, M.: Charles Handy: Are Leaders Born or Made? [Review of the Book Myself and Other More Important Matters, by C. B. Handy], 4 February 2008. http://www.leadershipnow.com/leadingblog/2008/02/charles_handy_are_leaders_born.html
44. Handy, C.B.: Myself and Other More Important Matters. AMACOM, New York (2006)
45. Goldsmith, M., Kaye, B., Shelton, K.: Learn Like a Leader: Today's Top Leaders Share Their Learning Journeys. Nicholas Brealey, Boston, MA (2010)
46. Ries, E.: The Lean Startup: How Today's Entrepreneurs Use Continuous Innovation to Create Radically Successful Businesses. Crown Business, New York (2011)
47. Smith, A.F.: The Taboos of Leadership: The 10 Secrets No One Will Tell You About Leaders and What They Really Think. Jossey-Bass, San Francisco (2007)

48. Higgins, M.C., Young, L.V., Weiner, J., Wlodarczyk, S.: Leading teams of leaders: what helps team member learning? Phi Delta Kappan **91**(4), 41–45 (2009)
49. Collins, J.: Good to Great: Why Some Companies Make the Leap … and Others Don't. HarperCollins, New York (2001)
50. Taylor, M.: You are the one you are waiting for: turn to yourself [Web log post], 19 June 2014. http://www.dailyom.com/articles/2014/43916.html
51. Kolditz, T.A.: In Extremis Leadership: Leading As If Your Life Depended On It. Jossey-Bass, San Francisco (2007)
52. Gladwell, M.: Outliers: The Story of Success. Back Bay Books, New York (2011)

Developing Active Personal Learning Environments on Smart Mobile Devices

Brian Whalley[1(✉)], Derek France[2], Julian Park[3], Alice Mauchline[3], and Katharine Welsh[2]

[1] University of Shefficld, Sheffield S10 2TN, UK
b.whalley@sheffield.ac.uk
[2] University of Chester, Chester CH1 4BJ, UK
[3] University of Reading, Reading RG6 6HA, UK

Abstract. 'Tablets' and other 'smart' devices (such as iPads and iPhones) have established themselves as a significant part of mobile technologies used in mobile (m-)learning. Smart devices such as iPads and the Apple Watch not only provide many apps that can be used for a variety of educational purposes; they also allow communication between students and tutors and with the world at large via social media. We argue that 'smart' mobile devices enable personalized learning by adjusting to the educational needs of individuals. We refer to Salmon's quadrat diagram to suggest where using mobile technologies should be of benefit to revising our views of pedagogy, making it much more responsive to students' needs in education as well as the world in general. Smart mobile devices now contain computing power to allow voice and face recognition, augmented reality and machine learning to make them intelligent enough to act as tutors for individual students and adjust and respond accordingly. To take advantage of these facilities on mobile devices, pedagogy must change from an institution-centered to a student-tutor-device focus. This is best done via 'active learning' and incorporating cognitive awareness into an educational operating system that can develop with the owner.

Keywords: Personal learning environments · Smart devices · Quality enhancement · Supercomplexity

1 Educational Scenarios

1.1 Some Quotidian Scenes

Consider the following:

A server in a restaurant takes a break to study for a class
A grandmother sends a letter to her grandson
A farmer in Africa looks for a market for her product
A visitor to a foreign country grapples with translating a conversation
A mother with visual impairment reads a book to her child
A tutor explains a difficult concept to her dyslexic student
A student takes notes from a teacher's class talk.

© Springer Nature Switzerland AG 2020
K. Arai et al. (Eds.): FTC 2019, AISC 1070, pp. 871–889, 2020.
https://doi.org/10.1007/978-3-030-32523-7_64

These are all situations that happen every day somewhere in the world and may well have been occurring for the last hundred years (and more). They are all, in one sense or another, 'educational' and also 'one-to-one' if not 'face-to-face'. Talking and listening, writing and reading; the fundamental ways of information-knowledge transfer. Technology has transformed these in the lifetime of a centenarian; from telegraph to telephone, radio and TV, cinema through to e-mail the internet and the smart devices now familiar around the world. Smartphones have become ubiquitous in todays' information transfer system.

An exception to the list might be the student taking notes from a teacher. The UK's Open University (OU) was set up in 1969 and was an early example of providing flexible distance learning. It originally used short formal lectures or demonstrations delivered via TV and supplemented by local tutors, summer schools and course text-books. For the most part, OU course materials could be studied informally, by the restaurant server for example. Today, an institutional virtual learning environment (VLE) or learning management system (LMS) is the main way in which higher education (HE) is most frequently 'delivered' to students as e-Learning perhaps coupled to traditional lectures. This paper examines recent developments in mobile technologies in HE institutions (HEIs). In particular, we examine how smart devices can promote active and involved learning by students to supplement or replace these traditional forms of educational delivery. We suggest pedagogically sound ways of involving students in their own education, from college and beyond, by developing personal learning environments, formal and informally.

1.2   Approaches

We examine some of the constraints for students' learning in HE, primarily from a United Kingdom perspective, with implications across the international sector. We examine how the concept of 'active learning', especially by way of out-of-classroom or fieldwork activities, can aid individual students. This is especially important for accessibility issues and the delivery of high quality education. The ubiquity of 'smart' devices and the promotion of mobile-learning (m-Learning) shows that smart phones and tablets can contribute to more formal educational scenarios [1]. We suggest that a tutor-based approach, developed via smart devices, involves and integrates students' personalized learning within established structures.

1.3   Institutional Higher Education

The massification and marketisation of HE is evident in many countries [2]. In the UK, universities also need to response to governmental pressures; including monitoring of overseas students, research output (the Research Excellence Framework, REF) and teaching quality (Teaching Excellence Framework, TEF). Economies of scale provide return on investment yet the enlargement of universities and the HE system can result in problems for individual students. The massification of HE has meant that student learn in lecture rooms containing large numbers of students, particularly during their first year.

We *all* live in an uncertain and supercomplex world. Ronald Barnett [3] has argued the central role of the university should be to deal with 'supercomplexity': a world is one in which the very frameworks by which we orient ourselves to the world are themselves contested. Consequently, in their pursuit of quality, educational programs in HE need to adjust to opportunities provided by new technologies to meet the students' requirements of 'active education' and information literacy in today's world. So, how can this be achieved?

A UK report [4], 'Horizon Scanning, what will higher education look like in 2020?' takes an international viewpoint and focuses on the educational importance of MOOCs, where impact 'on pedagogy and university business models will be profound but an evolutionary shift rather than an avalanche of change'. Although undoubtedly significant for mass, open education, MOOCs are still unimportant for most students at institutions where they attend lectures and where lecture theatres are still being built to accommodate increased numbers.

A recent review of next generation of learning environments [5] considered institutional technology enhanced learning (TEL) associated with VLEs, including analytics and user experiences (UX). The section on 'emergent models' includes ideas including; conversational teaching, interfaces and platforms and conversation-led learning, environments and chatbots. The report considers the potential of user experience to facilitate on-line learning environments and describes how 'engagement in online communities is changing dramatically, from a generation of desktop computer users to mobile online-users engaging with new forms of collaboration and interaction'.

Our contribution to this discussion is this article, written as practitioners of fieldwork education and co-designers of active learning opportunities with students. We give first thought to students and their development as individuals and recognize the need for student to work outside the constraints of many aspects of higher education.

2 Delivery of Syllabus and Curriculum

2.1 Classes, Lectures

A traditionalist view of higher education (HE) in the United Kingdom (UK) in the 21st century is still one of massed students listening to lectures given by lecturers ('faculty' in the USA); an 'instructivist' approach. For many students, this 'learning' is aided by study in libraries and constitutes what remains an essentially 'Industrial Education' [6]. The challenge and the opportunity is to harness the power of technology to provide more tailored and individualized personal learning, whereas there is a danger at present that the varied needs of individual students are subsumed via instructivist educational processes due to the larger student numbers. The ultimate goal is the considered use of educational technologies can lead to top down educational processes where the needs of the individual are truly catered for.

In the UK, considerable attention is currently being paid to lecture capture systems as part of institutional responses to quality provision by technology-enhanced education. Recorded lectures do have advantages, especially for students with disabilities or who might be in hospital, but they are still part of an institutional, instructivist, pedagogic template. Bligh's [7] book 'What's the use of lectures?' is mentioned by Phillips [8] in a brief espousal of educational design summarized in Table 1 in which theory-in-use is basically lecture-driven.

Table 1. Learning environments comparing two theories, after Phillips [8].

	Espoused theory	Theory-in-use
Pedagogical philosophy	Constructivist	Instructivist
Approach to learning	Deep	Surface
Approach to teaching	Student-centered	Teacher-centered
Subject design	Outcomes-based	Content-based

Phillips argues that 'it is an important role of the teacher, in a massified tertiary education sector, to assist students to develop... generic, lifelong learning skills'. So the challenge to educators is to consider how this might this be done in a lecture context and to avoid the situation where 'College is a place where a professor's lecture notes go straight to the students' lecture notes, without passing through the brains of either' (a quote attributed to Hamilton Holt and to Mark Twain or Edwin Slosson). Holt promoted a 'Conference Plan' involving one-on-one interaction between tutor and student. This might be likened to the 'Oxford-Cambridge tutorial', based around discussion between two or three students and their tutor. In many HE institutions large student numbers do not allow such low staff-student ratios although technology can support educators in developing key skills and critical thinking.

E-, or online, learning can support lectures by, for example, MOOCs, Spocs (small private on-line courses), visits to Wikipedia, YouTube and iTUNES U, etc. These excursions might be informally suggested by tutors or as part of 'flipped' classes or Just-in-Time teaching. Information literacy needs support and guidance; what are reliable sources? How can students find them and distinguish them from 'conspiracy theory' sites? Tutors should work with students and information professionals to develop meaningful tasks to extend teaching options and opportunities.

Various authors have suggested ways of using the lecture itself to promote smaller chunks of learning ('micro-learning') and to use lecture time for other activities such as problem solving [9, 10] as well as forms of problem-based learning by peer discussion. Such interventions can provide diverse opportunities for student interaction and, as we shall argue, the integrated use of technology and smart devices to support learning, thereby approaching Phillips' 'espoused theory'. MOOCs should indeed be 'designed as challenges', not as online lectures and connectivist eMOOCs provide a more diverse learning scenario [11].

2.2 Some General Attributes of Quality

The report by Davies et al. [12] indicates that the UK's Teaching Excellence Framework (TEF), will, 'focus the attention of university leaders on the opportunities presented by technology-enhanced learning'. However, the report gives no suggestions as to what the technologies will be or how they should be used to aid students' learning.

We now explore Gilly Salmon's ideas [13], via Fig. 1, to examine emergent technologies, missing from Davies et al. [12] that promote a more radical view of change using peripheral, new products, technologies, markets and missions.

Quadrants 1, 2 and 3 provide the basic elements of education as e-learning. Tablets and smart mobile devices are increasingly used by students for informal media consumption or social interaction so literature searching, communicating with tutors, submitting work [14] or interacting with a VLE as recognized by Phipps et al. [5] should encourage their educational use. Indeed, since the introduction of the iPad, a range of educational hardware and software has emerged but there remains a need to place smart devices in a pedagogic setting.

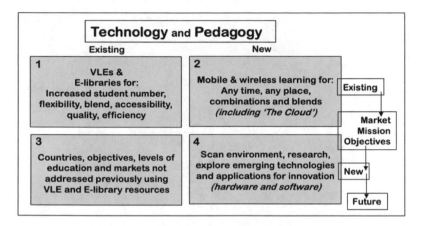

Fig. 1. New and existing practices in technology and pedagogy; modified after Salmon [13]. *Additional terms in italic.*

2.3 Education in a Cognitive and Inter-connected World

There are a wide range of 'learning theories' in education, however, it is *how* educational principles are employed that is of importance. If technology, perhaps delivered with an institutionally-provided VLE, is used *in conjunction* with students' individual technology use then we could reach an optimal learning scenario as envisioned by Beetham and Sharpe [15, 16] and Laurillard [17]. If there is any basis to an overall theory, it should be based on cognitive principles [18], be adaptive to individual needs and use technological affordances wisely. Various books and publications involving metacognitive approaches are available to assist tutors design courses and employ appropriate pedagogic principles [e.g. 19]. We now examine a framework for this provision centered around a student's involvement in education.

Figure 2 uses Beetham's 'student engagement relationships', the pentangle, with types of 'learning'. Importantly, the Personal Learning Environment (PLE), is student-centered and education is appreciative of cognition, especially meta-cognition. The idea of a Personal Learning Environment (PLE) is extended from an educational device linked to a VLE and thus rooted in Salmon's Fig. 1. Dabbagh and Kitsantas [20] discuss how PLEs can 'serve as platforms for both integrating formal and informal learning and fostering self-regulated learning in higher education contexts'. Their framework includes blogs, wikis and social networking leading from personal information management to social interaction and collaboration to information aggregation and management. Our view extends this use of a PLE to facilitate the relationships and methods described by Beetham in Fig. 2 via a students' personalization of smart mobile devices. This is integral to the PLE and incorporates aspects of the instructor/tutors via cognitive and metacognitive assistance.

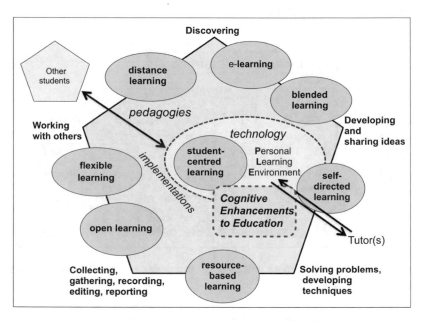

Fig. 2. Beetham's student engagement relationships [21] after France et al. [1].

2.4   Personalization Measures and the Role of Tutors

Individual private tutors, such as Thomasina Coverly's tutor Septimus Hodge in the play *Arcadia* [22], were once commonplace for the nobility and wealthy. The modern-day equivalent can be achieved through co-developing personal learning environments (PLE) with students by tutors, lecturers, research students (graduate students) or post-doctoral researchers. They contribute to this personalization by providing one-to-few educational relationships with their students. Although these will rarely be one-to-one tutorials, a PLE relates various tutorial inputs to the individual student by placing 'student plus PLE' at the heart of individual and diverse learning experiences.

Even 'personalised' education can lead to directed learning and passive responses by students. This may often be in the form of assessments such as end of semester/year examinations rather than active, problem-solving tasks. Further, the metrics used to assess engagement in learning also may need to be re-designed. Rather than using lecture attendance to show that students are 'engaged', the USA the National Survey of

Student Engagement (NSSE; nsse.indiana.edu) focuses on effective educational *practices*. This uses the 'seven principles for good practice in undergraduate education' [23] to promote active learning and student engagement.

These principles can certainly be used by (one-to-many) lecturers and are associated with the questions asked of students about their courses in the NSSE [24] and can be incorporated into the student engagement relationships of Fig. 2 to promote active learning via a PLE. Individual, or perhaps one-to-very few, tuition can certainly promote active learning procedures [25] where tutors can respond to the learning needs of individual students, for instance providing prompt and effective feedback. Examples of the effectiveness of active learning conditions can be found in, for example, Kuh et al. [26] and Healey [27] and promoted generally by McHaney [28].

2.5 Fieldwork and Out-of-Classroom Learning

Fieldwork is an important part of many academic disciplines, typically the earth, environmental and biological sciences as well as archaeology, history and some social sciences. To these may be added the more informal learning experiences when visiting holiday locations or visits to museum and art galleries. Fieldwork allows close tutor to student involvement in a way that lecturing at students does not. Fieldwork allows all of the attributes of Chickering and Gamson's seven principles to be exploited, whether or not technological devices are used. Our project, Enhancing Fieldwork Learning (EFL) has, since 2010, been promoting fieldwork and appropriate technologies designed to enhance learning opportunities for a wide range of students in fieldwork. Our experiences using smart devices, especially iPad tablets, helps promote student involvement and active learning [29–31]. We show that a variety of apps can be used for many learning activities including fieldwork observation, measurement, sketching as well as note-taking and writing.

3 Mobile Technologies and Personalization

Beetham and Sharpe [32 p. 4] point out that digital technologies have, 'profoundly changed how ideas and practices are communicated, and what it means to be a knowledgeable or capable person'. Higher Education institutions still have banks of computers and students are expected to have 'computers', laptops at home or perhaps carried around with them, yet writing essays on laptops and submitting via a VLE is hardly e-learning or active learning pedagogy. In 2005 Salmon [13] indicated that, 'real development beyond projects by innovators has so far been modest'. This view might still be valid nearly 15 years later and still begs the question as to what e-learning actually is and how it differs from 'learning' in an inter-connected world as suggested by our initial scenario list. Currently, it is only necessary to use a tablet or smartphone, with appropriate apps and perhaps a WiFi link, to give convenience to students. Such device ubiquity should be a way of promoting the best pedagogic practice and can be achieved by linking tutors to students. This indicates that mobile or m-learning is just a convenient subset of learning. Students and tutors can be linked via intelligent devices both formally; for course requirements and assessments, and informally; for conversation, keeping in touch and collaboration. This suggests a need to involve students, as co-producers or co-designers [33] in educational processes rather than as 'consumers' or 'investors'. We now make some suggestions within educational contexts to develop active learning with student involvement.

3.1   Tablet Technologies Before and Now

Stemming from the introduction of the iPad in 2010, various books, papers and reports have promoted mobile learning in HE [e.g. 14] with a variety of student-help books. The Enhancing Fieldwork Learning project has shown the importance of mobile devices in fieldwork and out-of-classroom activities [1], bring-your own device (BYOD), and in (mobile) m-learning [30]. We have also shown the significance of the iPad as a 'vade mecum' with respect to students' information gathering, storage and retrieval [34]. The student on the front cover of Macdonald and Creanor's [35] 'Learning with Online and Mobile Technologies' has books, a pencil and notepad but the 'laptop' could now be replaced with a tablet and be located anywhere.

3.2   Cognitive Learning and Technology

Various models of using technology in education have been suggested. Although individual apps and procedures may exist in one or other of these classifications, the development of an educational operating system will itself be an adaptive, or an emergent, process according to the situation of the learner, as for example in Fig. 2. A better way of envisioning e-learning is as an ecosystem of higher education where niches are filled with appropriate pedagogy-technologies supported by apps in the schema of Fig. 2 and delivered by a PLE. We now elaborate on the use of apps to enable the Personal Learning Environment to be tailored for individual students' use.

3.3   Cognition, Recognition and Learning via Apps

Whilst there is a move towards using human cognition in learning [18] this may still tend to be a top-down, how students 'should' be taught, approach. If personalization is to mean anything then it is evident that students should use all the tools available to them, in particular, via smart devices. Our brains are individual, by definition, thus the iPhone (used as a generic term) and ancillary devices will be more important than the VLE, especially if it can be trained or act as a tutor. This follows from Dweck's work on 'mindsets' [36]. The cognitive facilities provided by iPhones extend well beyond scanning text and translating to audio and providing text from audio speech recognition. Even simple apps can be used, once students recognize their importance and utility, to supplement their meta-cognition. Simple examples include making and prioritizing lists in problem solving, using tweets for group communication and feedback, as suggested by Dabbagh and Kitsantas [20]. On a more complex level, graphic organizer tools help concept mapping [37] and visual thinking in general [38]. Image and pattern recognition techniques are increasingly common in apps, not only for recognizing faces but also in photographic collections, tree leaf identification (e.g. Leafsnap) and works of art (e.g. Smartify and Mereasy) and music (e.g. Shazzam or

SoundHound) For those with visual impairments, LookTel provides spoken words for product packages and banknotes. There is little doubt that apps such as these, involving cognitive aspects of AI, will be further developed and when available on tablets will be useful to individual's learning opportunities.

3.4   Tablet Technologies: Looking to the Future

Neil Stephenson's futuristic novel *The Diamond Age* [39] places the reader in a recognizable 'Neo-Victorian' world a few years hence. The nanotech engineer, John Percival Hackworth, was asked to produce an advanced personal primer for a customer. The level of artificial intelligence (AI) supported by this device was used to construct the novel's sub-title, a 'Young Lady's Illustrated Primer'. The Primer is, 'designed to react to its owners' environment and teach them what they need to know to survive and develop' (Wikipedia: en.wikipedia.org/wiki/The Diamond Age). It is thus the ultimate PLE with its own evolving database and responses to the world.

We would not claim that the iPad is such a device, although it has some of the attributes of Stephenson's 'Primer' and could be developed further with existing and developing software technologies. However, iPads, especially in association with other iOS devices, such as iPhone and Apple Watch and linked to MacOS and 'the Cloud', can be developed as true *personal* devices that allow multitasking. They can aid an individual's education at any age. That is, they have attributes of a tutor. Students can already control their own learning, guided by tutors, peers and their own experiences so mobile devices can only assist in that personal development. We envisage even more personalized learning by projecting from the present capabilities of the iPad and iOS to those being developed which include; 'intelligent apps', enhanced hardware, the use of artificial intelligence (AI), virtual reality (VR) and augmented reality (AR) all delivered on high resolution screens.

3.5   Establishing Independent Learning and 'Emergence': Ada and Arthur

Although we make no prescriptive statements about iPads and iPhones, and smartphones in general, HE can look to developing independent learning, especially with the aid of tutors. Students benefits from the tutor's knowledge, experience and wisdom, the tutor benefits from the inquisitiveness of the student (à propos *Arcadia*) perhaps by co-learning. The iPad system contributes nothing unless switched on (an exception is as a tray for a cup of tea and a biscuit). In the sense of the (Young Lady's) primer, an iPad might even replace a tutor. In Tom Stoppard's *Arcadia*'s young tutee was Thomasina, the young lady heroine of *Diamond Age* was Nell, the name of the user of our integrated system PLE is Ada (harking back to Stoppard's character in *Arcadia* who was based on Ada Lovelace).

We generalize the tutor concept by suggesting that Ada asks an on-board addition to Siri that can be called by the individual student, Ada. The operating system for this personal tutor we name 'Arthur', named for Arthur Dent, user of 'The Book' in *The Hitch Hiker's Guide to the Galaxy* [40]; Adaptive Response To HUman Requests. *Arthur* should show emergent behavior by the use of AI (machine learning, etc.), together with its owner, Ada. The term 'emergence' has been used in various ways and includes; new system features and evolution as well as properties of wholeness. Emergent behavior from an iPad (etc.) will not just be the device alone, rather it will be able to respond intelligently to the queries of the young lady and help her make appropriate choices. Decision-making according to experience via creativity, problem solving abilities by Ada with Arthur's assistance.

The geographer William Kirk developed the idea of the 'behavioural environment', how people, individuals or groups', make decisions within the world of 'facts', the phenomenal environment. The behavioural environment is ordered (and sometimes 'disordered' in a supercomplex world) according to the social facts. Arthur, a quasi-intelligent tutor would allow Ada to grow within developing experience.

iPads can act as a *vade mecum* for students using information content (on board and via the Cloud) as a way of personalizing learning needs [34]. We now suggest ways in which developing technologies, already on the horizon, might help in developing personal tutors. For example, Google's 'Pixel Buds', when paired with the Pixel 2 handset can carry out live language translations. A recently launched application, 'Spoke' is a form of workplace tutor. This emergent behavior (of device plus student) perhaps gives the 'Tutor test', when the answer from the personal device tutor, such as Arthur, is indistinguishable from a 'real' tutor.

3.6   Tablets as Inter-communication Devices in Education

From the development of the smartphone-iPhone (from 2007), the iPad (from 2010) there is considerable degree of software commonality via the operating system (iOS) and to MacOS. The Apple Watch (2015) has its own watchOS, based on iOS that enables it to act as a more than a basic communication and media (music, video) replay device. In particular, it can be used to replay conventional TV as well as recorded lectures. Asynchronously, we have podcasts/vodcasts that can be academic as much as social and used for feedback in various ways. More interactively, Skype, FaceTime and Google Duo are videotelephony or Voice-over-Internet protocol (VoIP) products that can be used for effective tutoring (peer-to-peer) via WiFi as well as 4-5G/LTE. Medical (especially general practitioner and paramedic) trials are underway to relieve pressures at surgeries and hospitals. For example, the Welsh trauma surgeon David Nott reports intensive care unit in hospitals under fire in Aleppo (Syria) being monitored 24/7 over a Skype link to a hospital in Washington DC [41] and it has long been known that machine intelligent terminals can offer simple medical advice on an impersonal basis that is acceptable for patients. Finally, various aspects of computer-mediated communication and socio-emotional content have been discussed for many years [42].

Existing developments can aid the user of a smart device in an inter-connected world. Circle of 6 (www.circleof6app.com) was originally designed for college students to prevent, or at least warn, of sexual violence. It could also be useful for students needing to foster safe relationships. This is not relying on AI in the app but rather just using the capabilities of the device. A 'group mind', often referred to as a 'hive mind', offers collective conscience approaches to sharing ideas or information that might be useful to a student class. Showing students a problem image (whether in earth, biological or medical science or history) is a good way of getting engagement by group discussion as much as individual study. This can be used as part of flipped education or Just-in-Time teaching scenarios and to develop observational and interpretative skills. It is one form of citizen science that has been shown to be effective for class-based investigations on a small scale.

Analysis of group information is a way of linking individual responses to grouped data. For example, Reddit provides a news aggregation and content analysis website with discussion. Posts and discussions on Reddit do not themselves comprise a virtual tutor but rather indicate ways in which a tutor could help students' individual analysis of a topic. A student-tutor system, such as Ada-Arthur, might ask the user to be aware of problems of group behavior and perhaps analyze results much in the same way as a phishing scam mail or 'false news' posting might be analyzed. The tutor could offer advice on a simple question-response basis by going through a decision tree asking such questions as, what is the sender's address? is it from the company? are there spelling mistakes? for a suspected phishing e-mail? This provides a simple way of slowing down the quick, but perhaps incorrect, response to an invitation [43] that can be learned by the student from an on-board tutor. In other words, the device + tutor + hive behavior would minimize risks associated with a 'quick-click' response to e-mails. Similarly, Blue's 'practical tips for staying safe on-line' [44] could be made into a personal tutorial and be more user friendly than parent-installed safety devices. Kahneman and Tversky [43] have taught us to 'think twice' before responding to a problem. This is important when dealing with social media, regretting having sent some comment or perhaps rephrasing something, or mailing the wrong person. Some social gaffes may have far-reaching consequences. Machine learning algorithms could be used in various ways to recognize inappropriate responses and provide a learning experience for the student. Apple's predictive text feature, QuickType, in iOS that has a machine learning component that allows the software to build custom dictionaries. Parsing text from these dictionaries could flag a warning to the student similar to grammar and spelling checkers and translation apps operate.

3.7 Virtual and Augmented Reality, Gamification and Fieldwork

Cloud computing (and cloud storage) may be necessary for certain tasks; for example heavy computational use in an educational context is the provision of games. Gamification, is a current topic in learning technology [45] that may be device specific or shared via the internet, in particular related to augmented and virtual reality. The release of ARKit within iOS suggests an Apple VR headset; Virtual Reality (VR) will help in 'gamification' but could also be incorporated into pedagogic practice such as tutorials, fieldwork and laboratory techniques and data visualization. However, it does require additional hardware. Augmented Reality (AR) apps are where educators can take advantage to provide better student experiences. In our own area, fieldwork is likely to be a major recipient of such benefits. Visualization methods from the oil industry and military are likely to have important spin-offs in AR. People with disabilities should be significant benefactors from these technologies. We next examine how all these facilities and affordances can be made operational in an educational system.

4 Cognitive Education

4.1 Cognitive Learning and the Individual

All learning, or more fully, education, is cognitive. The previous discussion tends towards connectivist approaches as exemplified in network creation and an ecological learning system [46] Learners, of whatever stage of development, sit within such networks. What is really needed is better direction regarding individual student's behavior in the system. Good guidance is given by Douglas N Adams definition of 'a learning experience' [47 p. 274]; "You know what a learning experience is? A learning experience is one of those things that says, 'You know that thing you just did'? don't do that". The individual needs to adapt within the educational ecological system. Arthur is viewed as an adaptive system into which individuals pursue their own path as they grow with, even within, the system, as in 'The Young Lady's Illustrated Primer'. To this end the decision-making needs to be informed by previous experience. Some experiences are presented in 'Algorithms to live by' [48] and can be tempered by the findings of Kahneman and Tversky [43] and Adams' definition of 'learning experience'. The ultimate aim may be the development and recognition of competencies rather than excellence alone within a cognitive apprenticeship. There are general cognitive approaches, as presented in Bransford et al. and aspects such as prediction errors, 'nudges' and rewards. However, apps designed for this learning environment must be inclusive [49] and accessibility should be a major feature of developing basic operating systems.

4.2 Empowering Ada, Expanding the Curriculum, Syllabus and Context

Arthur is essentially a personalized operating system (OS) between a device and the user (Ada). It is trained, initially as an app manager and would link to voice, screen or text input. This may well utilize some of the features of the device OS. Some of this will be part of the linkages between OS from Apple and developers should be able to use appropriate APIs. From this viewpoint it would act as a general personal tutor but offering advice and assistance and develop with the student.

Supercomplexity of the educational system, sitting within the modern world, is poorly countered by traditional lectures/tutorials/exam. The strategic view of Salmon (Fig. 1) needs to be developed, not just with one or two apps entered via smart devices, but as an integrated and developing educational ecology. This ecological approach sits within a broad range of educational experiences (Fig. 2) and decision making. Kirk's behavioral environment, adapted by 'intelligent filters', would be a useful way of assisting learning processes for students. False news may be easy to identify and evaluate but misinformation on the web may be insidious in its effects. A 'trust gap' may exist between authorities and pressure groups and explain, for example, why even in some advanced economies vaccination of children is falling. Similarly, climate change denial may have to be countered. Information aggregation and analysis apps could be developed and installed in Arthur to support Ada's analysis and understanding of this complex world.

Assessment is an important part of education; for most students this means 'examinations'. Forms of assessment are varied and the multiple-choice question (MCQ) format, perhaps in the form of quizzes, is popular and simple and can be performed online via mobile devices. Several apps for student collaboration are used successfully in many institutions alongside visualization techniques and polling apps. The need is for instructors to be aware of these apps and to exploit them creatively.

So-called 'e-assessment' is being used to help mark students term papers, essays and reports, whether for summative or formative assessment. However, this is just an update of traditional ('Victorian') systems where a submitted text (often as PDF) is read by an instructor, marked up on the page and seen on a tablet. This may speed up the process but is still conforming to old fashioned practices. Applying several of the principles of Chickering and Gamson [23] would lead to an advance in assessment methods. We envisage a 'Virtorial', sitting within Arthur, to provide new ideas for 'i-Assessment' that encourage student-centered and formative assessment. An enhanced version of MCQs, using metacognition, is given by confidence-based marking (CBM) for MCQs developed by Gardner-Medwin [50]. An ideal would be for an instructor to submit some data (in a general sense) for student comment or analysis. MCQs might be used for this and CBM would be helpful but it would be helpful if an app did the marking of text entries. Message parsers used for text analysis and XML/JSON operations would allow students to examine their own responses to the problems. Tutors would need to supply answers as part of the original question. The tutor's job of marking many individual returns would be done by Virtorial. All sorts of data, numeric, symbolic as well as text could be dealt with in this way. This is a project under development.

4.3   Accessibility and Inclusivity

iPads have a range of on-board affordances that take them well beyond the simple media delivery communication device envisaged by Steve Jobs. The 'Accessibility' tab in settings shows range of assistants that allow the user to control text size and contrast, keyboard switching, and assistive touch to control of Siri. (Again, we should say that most of our work has been with MacOS and iOS systems than that other vendors have similar accessibility features and apps developed for them.) Many features already available may not be known to users or instructors whether or not they have special requirements. For example, VoiceOver (iOS) might be useful to help in fieldwork when hands are full with equipment ('Hey Siri, record these data'). Some control and access to cameras, accelerometers and microphone may already be incorporated in apps. Devices can also be controlled by eye gaze and tracking, using ARKit 2 as mentioned previously. As elsewhere, developers can use features in ways which have yet to be explored by educators.

An app to measure background noise (in dBA) in a room, lab or lecture theatre, may be valuable to indicate working conditions is already in watchOS. The microphone can be used to append audio notes to written material. Similarly, the iOS 'Magnifier' features could be used as a recording microscope for field or laboratory use as well as being helpful for the visually impaired. Students' instructors may be unaware of these possibilities and our project has helped to show lecturers and tutors how they can be used. Arthur could be used, through Siri, to provide reminders or prompts to students or tutors that a feature or app is available. In fieldwork for example, expensive slope measurement devices ($100 for a clinometer) can be replaced by an on-board app for a few cents that each student can obtain.

Care needs to be taken to be inclusive and allow for a wide range of 'impairments' – from which we all suffer to one extent or another, old and young. Base computer operating systems already provide for a range of disabilities, which is where mobile devices score well, but attention needs also to be given to cognitive disadvantages from poor short-term/working memories through cognitive degeneration to SpLDs. James and Linda Nuttall [51] produced 'Dyslexia and the iPad, Overcoming dyslexia with technology'. This gives a personal account of how James became enabled by his device and many ideas and apps are mentioned. Neurodiversity is a general term that covers a range of conditions. These may be unknown to most tutors and educators, especially if they are 'neurotypical', but it is likely that 'educating the educators' with respect to cognitive functions is an important line of research especially where smart devices may well be important in diagnosis and remediation.

Siri, and other personal digital assistants, are complex technologies using neural network technologies. The release of Core ML (machine learning) suggests developers will soon be incorporating natural language and machine vision into apps. It remains to be seen how well these technologies will be integrated into personalized learning tutors. For older people, they may be able to help combat the decline of 'fluid intelligence' (or reasoning) as well as failing memory. Personal Learning Environments coupled to tutorial assistants such as Arthur, are likely to provide important ways for individuals to overcome or combat impairments, physical or cognitive.

5   Discussion and Future Developments

Salmon's schema (Fig. 1) suggests that quadrants 1, 2 and 3 reflect the use of technologies within the core of HE teaching. The term e-learning is still extant in many institutions – as if this was something different from 'good' education despite many years of using the term. The LMS/VLE, and recent technologies, such as video recording of lectures, are still basically top-down structures, often with didactic tendencies. Most undergraduate education is subject based and has to deliver 'content' and certainly lecture-delivery can play a part with educational technologies helping in this delivery. However, education in general tends to be assessment-driven. For many students this means examinations and consequent stresses. Opportunities for lecturers to broaden content may be limited by the necessity to complete a syllabus and teach for the test. Curriculum development and rethinking assessment [52] are part of reviewing and widening educational structures for enhancing quality. Figure 2 provides some ideas for enhancing teaching and considered by over-views such as that by Diana Laurillard [17]. Of themselves, these structures do not consider Barnett's 'supercomplexity'. We suggest that the continually developing capabilities of smart devices, not least linked into the home and the Internet of Things (IoT), needs to be recognized in education, in particular HE. It is here where students are part of Barnett's hazards. In providing quality in HE, institutions have a responsibility to make students fully aware of the digital world and enable them to develop digital capabilities [53]. In developing capabilities, students (and academic staff) need guidance. Such guidance is not just 'pedagogic' or indeed in the use of digital and technological learning (TEL), but in using devices and cognitive capabilities of students [54].

Our experience in fieldwork education, where we encourage active learning and investigation with problem-solving, allows students to work in collaboration and for tutors to work with these small groups. The concept of 'graduateness' [55] and developing students' capabilities [56] are important parts of out-of-class activities that can be explored on campus and in the lecture theatre as well as in remote locations. Our use and promotion of iPads and iPhones is an important part of breaking down barriers, involving students and promoting decision making.

iPads and other smart devices will not, of themselves, make people smart in the same way as Sugata Mitra's 'hole in the wall' made them smart, as opposed to 'knowing some stuff'. But students' capabilities can be extended with assistance from tutors. This assistance needs to include digital awareness and information literacies operating in a complex, indeed supercomplex world. As yet, we do not have Young Lady's primers that can, by themselves, sense the work and its information and evolve for the student. But they can be developed to grow with the student, not only acting as repositories of information, or memory devices but as guide for the best ways to proceed with lifelong learning. AI can help in this just in the same way that the game of Go can be programmed (AlphaGo) to beat 9-dan Go Masters but can also be used inform players. We suggest that iPad virtual tutors, such as *Ada-Arthur*, should be able to assist students in living in the world with more effectiveness and less vulnerability and risk to hazards of learning. Rather than knowledge acquisition for its own sake, the development of intelligent iPads with users can aid our collective power to act wisely, especially in a HE environment. This would accomplish a true 'personal learning environment' (PLE): PLE = tutor + tutee + device + hive. The term 'personalized learning' has been used for many years in the past, usually as an adjunct to e-learning. PLEs have been defined variously as both concept and technology, for example in the review by Fiedler and Väljataga [57]. M-, or mobile, learning [14] has been added to e-learning. It is time to drop the e- and m- now that we should be able to configure portable/wearable devices and make them more adaptable in hardware as well as 'lifewear'. Micro-electro-mechanical systems (MEMS) now allow customizable devices as well as sensors such as near field communication. Haptic screens with audio output, for example, might better integrate knowledge systems with tutees to help further personalizing learning. Keyboard, pencil, oral and eye-controlled data entry and communication may be useful to any user. These attributes can be used and promoted by the development of virtual tutors as suggested in this paper. Active learning and student-centered approaches place more onus on the connected student. Connectedness applies not only with search engines and the internet but with fellow students and tutors, where devices are nodes foe collaboration and co-creation.

Developer features such as ARKit and Core ML in Apple iOS as well as the integration of IoT into Cloud technologies and the use of blockchain concepts for security suggests many advances for the near future. Enhancing Fieldwork Learning continues as project associated with the British Ecological Society and delivers practical short courses on fieldwork teaching and the use of iPads. Our ideas for teaching and student use of mobile devices should be enhanced by these software-hardware developments promoting personal learning environments within educational ecologies.

6   Conclusions

We show that utility of 'smart devices' in the world and across a range of educational practices needs to be better exploited in HE. Tablets and other mobile smart devices, as well as their usability, need to be placed within educationally sound pedagogic environments. We place these principles into a basic ecological structure (Fig. 2) with a personal learning environment at the center. We have also suggested ways that educational principles can be implemented by iPads (for example) to enhance quality in education within inclusive curricula and active learning. Imaginative use of integrated mobile devices (such as watch + tablet) can therefore promote good practice in education' and by empowering the uniqueness of the individual learner's PLE. The Young Lady's Illustrated Primer prototype is already here via the *Ada-Arthur* virtual tutor concept. Its utility can be extended by software and hardware developers working with educators to bring the quotidian to school, college and higher education.

References

1. France, D., Whalley, W.B., Mauchline, A., Powell, V., Welsh, K., Lerczak, A., Park, J., Bednarz, R.: Introduction to tablets and their capabilities. In: Enhancing Fieldwork Learning Using Mobile Technologies. Springer Briefs in Ecology. Springer, London (2015)
2. Calderon, A.: Massification Continues to Transform Higher Education. University World News, 02 September, p. 237 (2012)
3. Barnett, R.: Supercomplexity and the curriculum. Stud. High. Educ. **25**(3), 255–265 (2000)
4. Lawton, W., Mamun, A., Angulo, T., Axel-Berg, A., Burroes, A., Katsomitros, A.: Horizon scanning: what will higher education look like in 2020. In: Horizon Scanning: What will Higher Education Look Like in 2020 Leadership Foundation for Higher Education, International Unit, London? (2013)
5. Phipps, L., Allen, R., Hartland, D., Bryant, P., Hussain, A., Wood, S., Rowett, S., Scott, A.-M., Krohn, A.: Next Generation [Digital] Learning Environments: Present and Future, JISC, Bristol (2018). jisc.ac.uk/rd/projects/next-generation-digital-learning
6. Robinson, K., Aronica, L.: Creative Schools. Allen Lane/Random House, London (2015)
7. Bligh, D.A.: What's the Use of Lectures?. Jossey-Bass, San Francisco (2000)
8. Phillips, R.: Challenging the primacy of lectures: the dissonance between theory and practice in university teaching. J. Univ. Teach. Learn. Pract. **2**(1), 2 (2005). http://ro.uow.edu.au/jutlp/vol2/iss1/2
9. Freeman, S., Eddy, S.L., McDonough, M., Smith, M.K., Okoroafor, N., Jordt, H., Wenderoth, M.P.: Active learning increases student performance in science, engineering, and mathematics. Proc. Natl. Acad. Sci. **111**(23), 8410–8415 (2014)
10. Whalley, W.B.: Lecture discussion tasks: pedagogic and technological approaches to help break down barriers in lectures. In: Kapranos, P. (ed.) Sixth International Symposium for Engineering Education, pp. 93–100. University of Manchester, Sheffield (2016)

11. Jadin, T., Gaisch, M.: eMOOCs for personalised online learning: a diversity perspective. In: Research Track, Proceedings of the European MOOC Stakeholder Summit 2016, pp. 69–80 (2016)
12. Davies, S., Mullan, J., Feldman, P.: Rebooting Learning for the Digital Age: What Next for Technology-Enhanced Higher Education? Higher Education Policy Institute, Report 93, Oxford (2017)
13. Salmon, G.: Flying not flapping: a strategic framework for e-learning and pedagogical innovation in higher education institutions. ALT-J (now Research in Learning Technology) **13**(3), 201–218 (2005)
14. JISC. Mobile technology and m-learning. JISC, Bristol (2015)
15. Beetham, H.: Designing for learning in an uncertain future, In: Beetham, H., Sharpe, R. (eds.) Rethinking Pedagogy for a Digital Age, 2nd edn. pp. 258–281. Routledge, London (2013)
16. Beetham, H., Sharpe, R.: Rethinking Pedagogy for a Digital Age: Designing and Delivering E-learning. Routledge, London (2007)
17. Laurillard, D.: Teaching as a Design Science: Building Pedagogical Patterns for Learning and Technology. Routledge, Abingdon/New York (2012)
18. Bransford, J.D., Brown, A.L., Cocking, R.R.: How People Learn: Brain, Mind, Experience, and School. National Academies Press, Washington DC (2000)
19. Kukulska-Hulme, A., Traxler, J.: Design principles for mobile learning. In: Beetham, H., Sharpe, R. (eds.) Rethinking Pedagogy for a Digital Age: Designing for 21st Century Learning, pp. 244–257. Routledge, London (2013)
20. Dabbagh, N., Kitsantas, A.: Personal learning environments, social media, and self-regulated learning: a natural formula for connecting formal and informal learning. Internet High. Educ. **15**(1), 3–8 (2012)
21. Beetham, H.: Designing for active learning in technology-rich contexts. In: Beetham, H., Sharpe, R. (eds.) Rethinking Pedagogy for a Digital Age, pp. 31–48. Routledge, New York and London (2013)
22. Stoppard, T.: Arcadia. Faber, London (1993)
23. Chickering, A.W., Gamson, Z.F.: Seven principles for good practice in undergraduate education. AAHE Bull. **39**(7), 3–7 (1987)
24. NSSE: using NSSE to assess and improve undergraduate education. Lessons from the field, vol. 1. Indiana University Centre for Postsecondary Research, Bloomington, IN (2009)
25. Livingstone, D., Lynch, K.: Group project work and student-centred active learning: two different experiences. Stud. High. Educ. **25**(3), 325–345 (2000)
26. Kuh, G.D., Kinzie, J., Schuh, J.H., Whitt, E.J.: Student Success in College: Creating Conditions that Matter. Jossey-Bass, San Francisco (2011)
27. Healey, M., Pawson, E., Solem, M.: Active Learning and Student Engagement: International Perspectives and Practices in Geography in Higher Education. Routledge, London (2013)
28. McHaney, R.: The New Digital Shoreline. Stylus, Sterling (2011)
29. Welsh, K.E., Mauchline, A.L., Park, J.R., Whalley, W.B., France, D.: Enhancing fieldwork learning with technology: practitioner's perspectives. J. Geogr. High. Educ. **37**(3), 399–415 (2013)
30. Whalley, W.B., Mauchline, A.L., France, D., Park, J., Welsh, K.: The iPad six years on; progress and problems for enhancing mobile learning with special reference to fieldwork education. In: Crompton, H., Traxler, J. (eds.) Mobile Learning and Higher Education, pp. 8–18. Routledge, London (2018)

31. France, D., Whalley, W.B., Mauchline, A.L., Powell, V., Welsh, K.E., Lerczak, A., Park, J., Bednarz, R.: Enhancing Fieldwork Learning Using Mobile Technologies. Springer Briefs in Ecology. Springer, New York (2015)

32. Beetham, H., Sharpe, R.: Rethinking Pedagogy for a Digital Age: Designing for 21st Century Learning, 2nd edn. Routledge, New York and London (2013)

33. Bovill, C., Cook-Sather, A., Felten, P., Millard, L., Moore-Cherry, N.: Addressing potential challenges in co-creating learning and teaching: overcoming resistance, navigating institutional norms and ensuring inclusivity in student–staff partnerships. High. Educ. **71** (2), 195–208 (2016)

34. Whalley, W.B., France, D., Mauchline, A.L., Welsh, K.E., Park, J.: Everyday student use of iPads: a *vade mecum* for students' active learning. In: Baab, B., Bansavich, J., Souleles, N., Loizides, F., Mavri, A. (eds.) Proceedings of the 2nd International Conference on the use of iPads in Higher Education (ihe2016), pp. 43–61. Cambridge Scholars Publishing, San Francisco (2016)

35. Macdonald, J., Creanor, L.: Learning with Online and Mobile Technologies. A Student Survival Guide. Gower, Farnham (2010)

36. Dweck, C.S.: Motivational processes affecting learning. Am. Psychol. **41**(10), 1040–1048 (1986)

37. Novak, J.D.: Concept maps and vee diagrams: two metacognitive tools to facilitate meaningful learning. Instr. Sci. **19**(1), 29–52 (1990)

38. Bryson, J.M., Ackermann, F., Eden, C., Finn, C.B.: Visible Thinking: Unlocking Causal Mapping for Practical Business Results. Wiley, Chichester (2004)

39. Stephenson, N.: The Diamond Age. Bantam, New York (1995)

40. Adams, D.N.: The Hitchhiker's Guide to the Galaxy. Heinemann, London (1995)

41. Nott, D.: War Doctor: Surgery on the Front Line. Picador, London (2019)

42. Rice, R.E., Love, G.: Electronic emotion: socioemotional content in a computer-mediated communication network. Commun. Res. **14**(1), 85–108 (1987)

43. Kahneman, D.: Thinking, Fast and Slow. Allen Lane, London (2011)

44. Blue, V.: The Smart Girl's Guide to Privacy, Practical Tips for Staying Safe Online. No Starch Press, San Francisco (2015)

45. Wiggins, B.E.: An overview and study on the use of games, simulations, and gamification in higher education. Int. J. Game-Based Learn. **6**(1), 18–29 (2016)

46. Siemens, G.: Knowing Knowledge (2006)

47. Adams, D.N.: The Salmon of Doubt. Macmillan, London (1992)

48. Christian, B., Griffiths, T.: Algorithms to Live By: The Computer Science of Human Decisions. Collins, London (2016)

49. Criado Perez, C.: Invisible Women: Exposing Data Bias in a World Designed for Men. Chatto and Windus, London (2019)

50. Gardner-Medwin, A.R.: Confidence-based marking. In: Bryan, C., Clegg, K. (eds.) Innovative Assessment in Higher Education, pp. 141–149. Routledge, Abingdon (2006)

51. Nuttall, J., Nuttall, L.: Dyslexia and the iPad (2013)

52. Boud, D., Falchikov, N.: Rethinking Assessment in Higher Education: Learning for the Longer Term. Routledge, London (2007)

53. JISC Developing organisational approaches to digital capability (2017). http://www.jisc.ac.uk/full-guide/developing-organisational-approaches-to-digital-capability. Accessed 20 July 2019

54. McGuire, S.Y.: Teach Students How to Learn: Strategies You Can Incorporate Into Any Course to Improve Student Metacognition, Study Skills, and Motivation. Stylus, Sterling (2015)

55. Hill, J., Walkington, H., France, D.: Graduate attributes: implications for higher education practice and policy: introduction. J. Geogr. High. Educ. **40**(2), 155–163 (2016)
56. France, D., Powell, V., Mauchline, A.L., Welsh, K., Park, J., Whalley, W.B., Rewhorn, S.: Ability of students to recognize the relationship between using mobile apps for learning during fieldwork and the development of graduate attributes. J. Geogr. High. Educ. **40**(2), 182–192 (2016)
57. Fiedler, S.H., Väljataga, T.: Personal learning environments: concept or technology? Int. J. Virtual Pers. Learn. Environ. **2**(4), 1–11 (2011)

Effective and Innovative Interactives
for icseBooks

Y. Daniel Liang[(⊠)]

Georgia Southern University, Savannah, GA 31419, USA
y.daniel.liang@gmail.com

Abstract. The term icseBook (pronounced ice book) was coined for Interactive Computer Science Electronic Books. IcseBooks use interactives to engage students. Interactives use animation and visualization to make difficult concepts easy to understand. Interactives enable students to interact with the book to receive hints, tips, and feedback instantaneously. We have created more than 1500 interactives for each of the interactive ebooks on Java, C++, and Python and have received positive reviews. The interactives can be classified into fourteen types. This paper introduces all fourteen types of interactives, discusses their innovative features and pedagogical merits, and compares them with similar interactives used in the icseBooks on the market.

Keywords: Code animation · Interactives · Algorithm visualization ·
Live example · Live exercise · Auto grade · Word match · Parson's problem

1 Introduction

Interactives engage students and improve learning. It has been a dream for many computer science faculty to develop and use interactive textbooks. With the advent of emerging computer technology, a new era of interactive textbooks has arrived. Several computer science interactive ebooks have been created in the recent years [1–3]. The term icseBook (pronounced ice book) was coined for Interactive Computer Science Electronic Books in [11]. As for every book, whether it is in print or in online version, the content is the foundation. The interactives can help better present the contents and help students learn better. The interactives used in all these books are different. Developing interactives is a creative endeavor. We have created fourteen types of interactives. They have been used in the interactive ebooks [4–6] and received positive reviews [7, 8]. In this paper, we will introduce the following interactives used in the Java interactive ebook [4], discuss their innovative features and pedagogical merits, and compare them with similar interactives used in the icseBooks on the market.

1. Code Animation
2. Live Example
3. Interactive Flowchart
4. Algorithm Animation
5. Interactive Figures
6. Interactive Figures with Animation

© Springer Nature Switzerland AG 2020
K. Arai et al. (Eds.): FTC 2019, AISC 1070, pp. 890–903, 2020.
https://doi.org/10.1007/978-3-030-32523-7_65

7. Interactive UML Diagrams
8. Interactive Check Point Questions
9. Interactive Word Matching Exercises
10. Interactive Freestyle Exercises
11. Interactive Order Statement Exercises
12. Interactive Multiple-Choice Questions
13. Live Exercise Sample Run
14. Check Exercise Tool

2 Code Animation

Teaching programming is difficult. Code animation is a great tool for instructors to teach programming effectively. It enables students to see the code execution visually and interactively. As the animation traces the code execution, the tool prompts the questions to check student's understanding of the code.

Students learn programming through examples. The key programming concepts and techniques are best taught through good examples. Many students find that programming textbooks are dry and boring, because the examples in a print book are static, not engaging, and hard to follow. We developed interactive code animation to demonstrate key programming concepts on variables, data types, constants, expressions, selection statements, loops, methods, passing arguments, arrays, multidimensional arrays, and objects. The interactive code animation not only engages students, but also is effective to make the difficult concept easy to teach. We presented in detail on how code animation is used to help students learn variables, selection statements, loops, method, pass by value, arrays, and pass array arguments in [9]. Figure 1 shows an example of the code animation for a program. The example can be viewed from https://liveexample.pearsoncmg.com/codeanimation/TestMax.html.

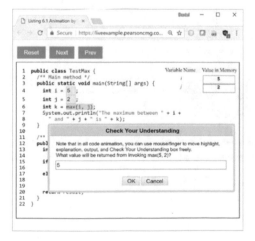

Fig. 1. A code animation example

Note that during the animation, you can move the explanation, code highlight, dialog boxes, and output boxes freely using the mouse.

The best-known code animation tool is developed by Phillip Guo of University of California San Diego [10]. This is a great visualization tool and it is used in many isceBooks, but it does not have the explanation for each line and it does not prompt questions to check student understanding. Good explanations are important for students to understand the code and the "Check Your Understanding" boxes test student understanding of the code.

3   LiveExample

Students learn programming through writing programs. LiveExample enables students to modify, compile, and run the programs live within the book. This encourages students to experiment with examples as they are reading and helps students better understand the code and practice coding. LiveExample leaves some key code for students to complete. This enhances learning and reinforces key concepts. Students can view the missing code by clicking the Answer button. Figure 2 shows an example of LiveExample, which can be accessed from https://liveexample.pearsoncmg.com/LiveRun/faces/LiveExample.xhtml?programName=ComputeAverage&programHeight=380&resultHeight=180&header=off.

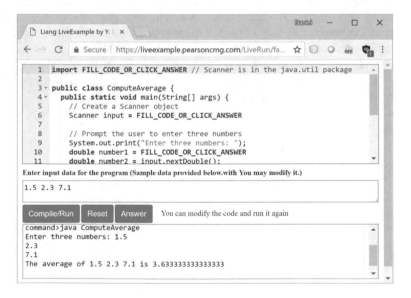

Fig. 2. A LiveExample example

Clicking the Compile/Run button compiles and runs the code. If the code has a compile error, LiveExample can offer hints for fixing the error. Clicking the Answer button displays the answer for FILL_CODE_OR_CLICK_ANSWER. Clicking the Reset button resets the code to its original state. The user can enter the console input from the input box in the middle of user interface.

4   Interactive Flowchart

A flowchart is a diagram that describes an algorithm or process, showing the steps as diamond or rectangle boxes. A diamond box is for testing a condition and a rectangle box is for executing statements. An interactive flowchart is most effective to show students of the flow of the actions. Figure 3 shows an example of interactive flowchart, which can be accessed directly from https://liveexample.pearsoncmg.com/dsanimation/Figure5_1New.html. Through this interactive flowchart, students can clearly see the logic steps for executing a for loop. Students will understand when the initial action takes place, when the actions-after-each-iteration is executed, and when the loop exits.

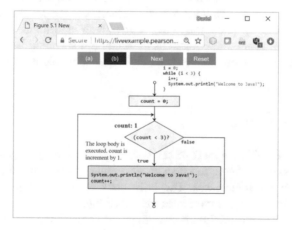

Fig. 3.   Interactive flowchart.

5   Algorithm Animation

Algorithm animation demonstrates how an algorithm works interactively. A picture is worth a thousand words. An interactive picture is worth more than pictures. Figure 4 gives an example of selection sort animation, which can be accessed directly from https://liveexample.pearsoncmg.com/dsanimation/SelectionSorteBook.html.

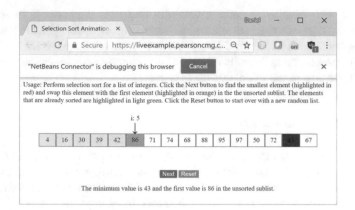

Fig. 4. The selection sort animation.

Through this animation, students will see clearly that the selection sort selects the smallest element and then swaps with the first element in the list. This process is repeated for the remaining list until all the elements are all sorted.

We have developed animations for sorting algorithms (selection sort, insertion sort, bubble sort, merge sort, quick sort, heap sort), graph algorithms (depth-first search, breadth-first search, Hamiltonian path/cycle, minimum spanning trees, shortest paths, travelling salesman problem), convex hull, eight queens problem, closest-pair problem, Sudoku, etc.

6 Interactive Figure

Interactive figures enable students to see various cases of a figure. Students click the buttons on the top of the figure to select cases. Figure 5 shows an interactive figure for finding a minimum spanning tree. The print book uses seven figures and is hard to follow. This interactive figure combines the seven figures into one and colors the new node that is added to the minimum spanning tree. This figure can be accessed from https://liveexample.pearsoncmg.com/dsanimation/Figure29_8Insert.html.

In this interactive figure, the user can click the buttons Initial State to start animation and then click buttons from (a) to (g) to see how each vertex is added to the minimum spanning tree.

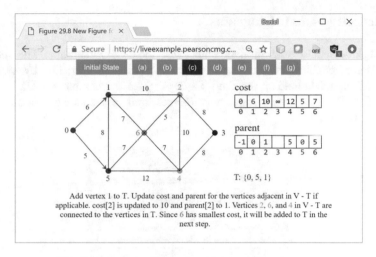

Fig. 5. Interactive figure for minimum spanning tree.

7 Interactive Figure with Animation

The interactive figure with animation provides interactive figures with animation effect to illustrate a concept or a behaviour interactively. Figure 6 shows the Tower of Hanoi animation, which can be accessed directly from https://liveexample.pearsoncmg.com/dsanimation/TowerOfHanoieBook.html. The user can see how discs are moved from tower A to B with the help of C from this animation.

In this figure, the user can select the number of discs and click the Start button to see the movement of discs in animation.

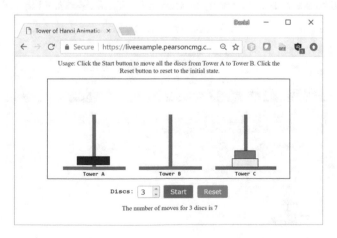

Fig. 6. Animation for tower of Hanoi.

8   Interactive UML Diagrams

Many object-oriented programming books use the UML diagrams. The interactive UML diagram enables the user to view the description of each line by moving the mouse over on a data field or a method. Figure 7 gives an example of a UML diagram, which can be accessed from https://liveexample.pearsoncmg.com/dsanimation/ Figure15_10v2.html.

Fig. 7. An interactive UML diagram example.

In the figure, the user can also click javafx.scene.input.MouseEvent to view the complete documentation for the MouseEvent class.

9   Interactive Check Point Questions

The check points provide review questions to help students review what they have learned in a section. It is presented interactively to enable the user to expand/shrink a question and to see/hide the answer, as shown in Fig. 8 (also see https://liveexample.

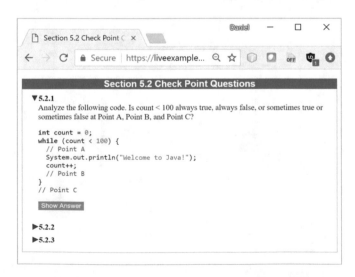

Fig. 8. An interactive check point questions example.

pearsoncmg.com/checkpoint/Section5_2.html). As shown in this figure, the user can click the Show Answer button to view the answer for the question and click 5.2.2 and 5.2.3 to expand the questions to see the descriptions.

10   Word Match Exercises

Word Match Exercises let students match the key terms with their descriptions. Figure 9 shows an example, which can be accessed from https://liveexample.pearsoncmg.com/wordmatch/Section5_2new.html. The user drags a term to match a description. If a term is dragged to a wrong place, it will be automatically bounced back to the key term list. After all terms are matched, a dialog box is displayed and the user may reset to practice the exercise again.

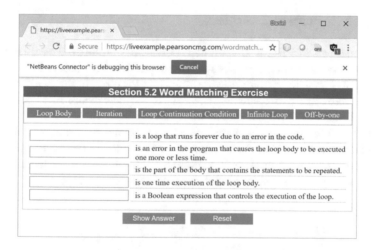

Fig. 9. A word match exercise example.

11   Freestyle Exercises

Freestyle exercises let the student type an answer to a question and get a response. Figure 10 shows an example of a freestyle exercise, which can be accessed directly from https://liveexample.pearsoncmg.com/freestyle/Section5_2.html. If the user enters a wrong answer, the tool displays a hint.

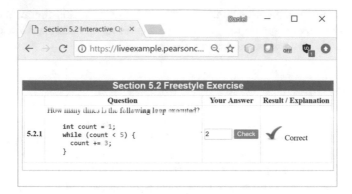

Fig. 10. A freestyle exercise example.

12   Order Statement Exercises

An order statement exercise is also known as Parson's problem. It gives a code with statements in a random order and asks students to put the statements in the correct order by dragging the statements up or down to the right location. In the isceBooks such as [1], this exercise has two columns. The left is the statements in a random order. Students will need to move the statements from the left column to the right column. Our interactive uses only one column, which is suitable for a mobile device with a narrow width. Figure 11 shows an example of the order statement exercise (https://liveexample.pearsoncmg.com/wordmatch/Order5_2.html.) After the correct order is achieved, a dialog box is displayed and user can reset the exercise.

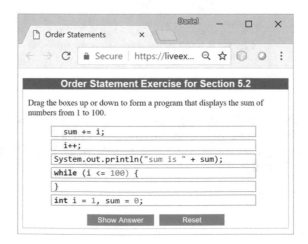

Fig. 11. An order statement exercise example.

13   Interactive Multiple-Choice Questions

Multiple-choice questions are placed at the end of a section. Here is an example for multiple-choice questions in Fig. 12 (https://liveexample-ppe.pearsoncmg.com/selftest/OneSectionJava?sectionNo=5.2). The user answers one question and then clicks the Next button to see the next question. After all the questions are answered, the user can view all the questions and answers in one page. For each question, the interactive gives a hint if the student makes a wrong choice. As shown in Fig. 12, you can also click the Get Statistics button to get the statistics on the question that shows the correct number of attempts and the total number of attempts by all users.

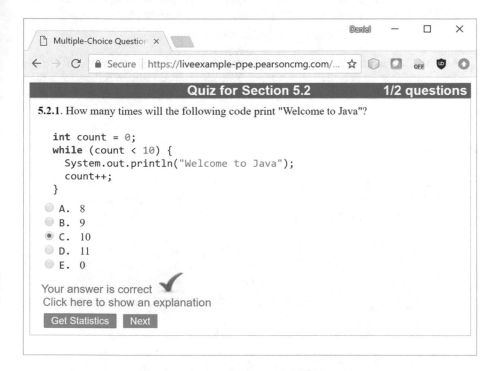

Fig. 12. An interactive multiple-choice quiz.

14   Live Exercise Sample Run

Students learn programming by practice. The interactive ebook provides a wide variety of problems at various levels of difficulty to motivate students. To appeal to students in all majors, the problems cover many application areas, including math, science, business, financial, gaming, animation, and multimedia. Live exercise sample runs show students the input and output from a sample run. Students can modify the input and see the expected output. Figure 13 shows an example (https://liveexample.pearsoncmg.com/LiveRun/faces/LiveExampleResultOnly.xhtml?programName=Exercise03_01&programHeight=0&resultHeight=100&compilerVisible=false).

Fig. 13. A live exercise example.

As shown in the interactive, the sample run for the exercises is provided. But the user can enter new input and click the Show Sample Output Using the Preceding Input button to run the exercise using the specified input. The user can also click the Reset button to display the default input and output.

15   Check Exercise Tool

Students can check the correctness of their programs using the CheckExerciseTool, as shown in Fig. 14. The tool can be accessed from https://liveexample.pearsoncmg.com/CheckExercise/faces/CheckExercise.xhtml?chapter=3&programName=Exercise03_01.

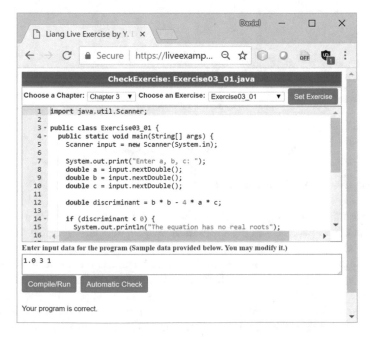

Fig. 14. Check exercise tool.

Students can select a chapter and the exercise. The exercises are programming exercises in the book. The user can enter the complete code and compile/run to see the output by clicking the Compile/Run. Students can also check the code by clicking the Automatic Check button. If the code is wrong, the tool highlights where the error starts in the output.

16 Instructor and Student Feedback

We have used the interactive ebook for our CS1 and CS2 courses for more than three years. We like the interactive ebook for a very simple reason: It helps students learn better. The static contents of the interactive ebook are just like a print book. However, the interactive ebook has more than 1500 interactives. The interactives are extremely helpful for students to learn difficult concepts. Our students use the code animation to see how an example work step by step visually and interactively with explanations and dialog boxes that prompt the user to answer questions. Our students use LiveExample to fill in the missing code and run the program to see the result without leaving the book. Our students get to know how an exercise behaves by using the LiveExercise interactive before writing the code for the exercise. Our students use the Check Point questions to review the concepts learned from each section. Our students use the WordMatch interactives to study and review for the key terms. Our students use the Freestyle exercise interactives and the multiple-choice question interactives to test their knowledge. Our students use the OrderStatement interactives to practice Parson's puzzles. Our students use the CheckExercise tool to check if their code for the end-of-the-book programming exercises is correct.

Two studies were conducted by the Central Michigan University and the University of Louisiana [7, 8] to evaluate the effectiveness of the interactive Java book. The studies cover many features of book including the interactives. The studies show that the interactives help improve student learning.

Here is an excerpt from the Central Michigan University study [8] on the interactives conducted in 2015:

- *62% of students agreed or strongly agreed that the Animations in the interactive ebook that stepped line by line through code, showing what was happening in the program, helped them better understand how to replicate the coding on their own.*
- *64% of students agreed or strongly agreed that the use of Interactives in the interactive ebook (LiveExamples) helped them practice and modify running live code for retention of key programming concepts.*

Here is an excerpt from the University of Louisiana study [7] on the interactives conducted in 2016:

- *87% of students who used the Animations strongly agreed or agreed that they helped them better understand how to replicate coding on their own.*
- *85% of students who used the Interactives strongly agreed or agreed that they helped them run live code for the retention of key programming concepts.*

The Java interactive ebook was first launched in 2015. The Central Michigan study was conducted in 2015. We substantially improved the interactives based on instructor and student feedback and created a new version in 2016 after receiving feedback from instructors and students. As shown in the University of Louisiana study conducted in 2016, the result is much better than the University of Central Michigan study.

17   Future Work

Many of the interactives were first created using Java applets. Due to security restrictions, Java applets are prohibited from running in a Web browser or a mobile device. To enable the interactives to run from a Web browser and from any device, we have recreated all the interactives using HTML5, CSS, and JavaScript on the client side. Additionally, we also use JavaServer Faces to develop LiveExample, LiveExercise, and CheckExercise tool and use MySQL to store data on the server side. The code animation is customized for every program. It is tedious and time-consuming to create a code animation. Our future work is to create a code animation generator that can automatically generate the HTML script for animating the code.

18   Conclusions

This paper presented fourteen types of interactives used in the interactive ebooks for Java, C++, and Python. Each book has more than 1500 interactives. The interactives are tailored to the contents and they are extremely helpful for students to learn programming. The studies conducted by University of Louisiana and Central Michigan University indicate that majority of students agree or strongly agree that interactives help them learn programming.

References

1. Runestone. http://runestoneinteractive.org/. Accessed 16 May 2019
2. Zyante. https://zybooks.zyante.com. Accessed 16 May 2019
3. OpenDSA. https://opendsa-server.cs.vt.edu/home/books. Accessed 16 May 2019
4. Daniel Liang, Y.: REVEL™ for introduction to java programming and data structures. Pearson Education (2016). ISBN-13: 978-0134167008
5. Daniel Liang, Y.: REVEL™ for introduction to C++ programming and data structures. Pearson Education (2018). ISBN-13: 978-0134669854
6. Daniel Liang, Y.: REVEL™ for introduction to python programming and data structures. Pearson Education (2018). ISBN-13: 978-0135187753
7. REVEL™ educator study observes homework and exam grades at University of Louisiana, Spring (2016). http://www.pearsoned.com/results/revel-educator-study-observes-homework-exam-grades-university-louisiana/. Accessed 16 May 2019
8. REVEL educator study assesses quiz, exam, and final course grades at Central Michigan University, Fall (2015). http://www.pearsoned.com/results/revel-educator-study-assesses-quiz-exam-final-course-grades-central-michigan-university/. Accessed 16 May 2019

9. Daniel Liang, Y.: Teaching and learning programming using code animation. In: The 14th International Conference on Frontiers in Education: Computer Science and Computer Engineering (FECS 2018), 30 July–2 August 2018, Las Vegas, USA, pp 24–30 (2018)

10. Guo, P.: Online python tutor: embeddable web-based program visualization for cs education. In: Proceedings of 44th ACM Technical Symposium on Computer Science Education (2013)

11. Korhonen, A., Naps, T., Boisvert, C., Karavirta, V., Mannila, L., Miller, B., Morrison, B., Rodger, S.H., Ross, R., Shaffer, A.: Requirements and design strategies for open source interactive computer science eBooks. In: ITICSE 2013 Working Group Reports, 29 June–3 July 2013, Canterbury, UK (2013)

Learning Analytics as a Sociotechnical System

Marcel Simonette[1]([⊠]), Mario Magalhães[2], and Edison Spina[1]

[1] Universidade de São Paulo, Cidade Universitária, São Paulo, Brazil
{marceljs, spina}@usp.br
[2] Centro de Estudos Sociedade e Tecnologia, Cidade Universitária, São Paulo, Brazil
m.e.magalhaes@ieee.org

Abstract. In the current education, there are several analytics strategies to support educators in the monitoring and analysis of course events. The systems used to explore the raw data about facts and processes in courses contexts include humans and Information and Communication Technology. These sociotechnical systems are environments that enable collaborative work among humans and between humans and computer agents. These environments have features that dare technological determinism in order to develop innovative situation awareness and decision support approaches. This work is about System Thinking and Meta-design to support sociotechnical environments in which the management process of the Learning Analysis considers humans factors, adding value to learning processes.

Keywords: Learning Analytics · Systems Thinking · Meta-design · Sociotechnical systems

1 Introduction

Each breakthrough in Information and Communication Technology (ICT) is a possible pinnacle. The present pinnacle is data analysis, which makes it possible to go beyond the raw data and make decisions and recommendations based on these data. In the educational context, it is tempting to think that it is possible to use data about courses to describe and to understand facts and processes that happen in course environments, in which learners and educators share the use of ICT to interact with each other and to create course contexts.

Learning Analytics supports educators and learners, allowing the analysis of behaviors and decisions, monitoring processes, and reaction to course events. The management of this sociotechnical process must-have features that are not limited to the ones of technological determinism, it must consider the people involved. People interaction mediated by ICT brings challenges to any system. Technologies currently available do not allow recreating the human experience as it occurs in the physical space, in which human factors, such as purposes, values, and emotions, have significant influence on creativity. This scenario impacts two essential elements of Learning Analytics systems:

1. The development of innovative situation awareness.

© Springer Nature Switzerland AG 2020
K. Arai et al. (Eds.): FTC 2019, AISC 1070, pp. 904–912, 2020.
https://doi.org/10.1007/978-3-030-32523-7_66

2. The approaches to decision support.

This work is about Systems Thinking and Meta-design to developed methods and tools for helping the management processes of Learning Analytics. It is a proposal to deal with Learning Analytics systems considering them as sociotechnical systems. Systems in which the management process of Learning Analytics considers human aspects, promoting a collaborative environment in which both humans and ICT engage in the process of general situation awareness and recommendation through Leaning Analytics.

This paper is structured into six sections. In the next section, we briefly characterize Learning Analytics. The third section is about the management process and its inherent complexity. In section four, we present Systems Thinking and Meta-design concepts. In section five, we compound Systems Thinking, Meta-design and, ICT. Finally, the last section presents our conclusions and future work; followed by the references.

2 Learning Analytics

In the current education, changes in learning environments are related to ICT and with the result of small and successive changes both in human behavior and in social organization. Learning environments give rise to a massive volume of data, which enables education institutions to take administrative decisions, such as marketing decisions to promote the institution, cultivate donors, and efficient use of budgets and staff time. Moreover, as pointed in [1], data analysis of the educational environment has the potential to guide students, educators, institutional administrators, and funders in making learning-related decisions.

Business leaders know data and analytics as ways to gain insights about their business. The term "business intelligence" describes the relationship between data and insights, as a way to obtain information from the interactions between workers or consumers with information, with technology, and with organizations, taking decisions and recommendations [2].

When applied to education, data and analytics about the learners and educators' context are called Learning Analytics. However, Learning Analytics is a term that has a broad meaning, sometimes referring to predictive models, and at other times, referring to routine tasks, such as classroom allocation. These two meanings are related to the spectrum of learning, which involves: education, workplace learning, formal learning, informal learning, network learning, and the social context in which learning is inserted. The Society for Learning Analytics emphasizes the learner in its definition: "Learning analytics is the measurement, collection, analysis, and reporting of data about learners and their contexts, for the purposes of understanding and optimizing learning and the environments in which it occurs" [3, 4].

We understand Learning Analytics as a way to provide stakeholders with better information and support decisions to deal with the factors of the learning process that contribute to learner success. However, to reach the goal of providing decisions makers with the needed information and assisting learning, it is necessary to manage the Learning Analytics process, to prevent delays in data capture, data definition, data analysis, and decisions. Furthermore, the management process needs to deal with the fact that was pointed out by Siemens [5]: learning and interaction between learners and teachers are increasingly distributed and ubiquitous, using different tools and environments.

3 Management Process

Management consists of planning, organizing, staffing, leading, and controlling an organization or initiative to accomplish a goal. Management process is the process of planning, controlling, organizing, and leadership of the execution of any organizational activity [6]. It is a complex process because of the increasing rate of changes happening in the world, which are directly and indirectly affecting all organizations. Moreover, the management process is complex because it is necessary to consider people, i.e., human factors, such as values, ethics, loyalty, honor, dignity, discipline, personality, goodwill, kindness, and mood; human factors that bring complexity to endeavors.

3.1 Complexity

It is natural to want a kind of cookbook approach to deal with all the internal and external changes that affect organizations. Nevertheless, due to the diversity of events happening in society, there is no easy recipe to follow in management processes. Advice and strategies that worked for one situation may not work with the same situation in different contexts.

Managing the diversity of dimensions that impact a management process is more than only paying attention to local and external changes. People differences must be acknowledged, recognizing the importance and impact of these differences; notwithstanding that, it is hard to recognize the relation among people diversity and management processes. The latter has an inherent complexity related to the need for a design process to meet the interests, values, and intentions of the people involved in the process.

3.2 Volatility

The diversity of dimensions that impact management processes confers high volatility to these processes. Therefore, process models must be developed to deal with the volatility and with the increasing rate of changing conditions associated with it.

Modeling is a critical activity in any system design, and it is essential to involve all stakeholders in this activity and, at the same time, respect the stakeholders' diversity. If these people are not involved in process modeling, their needs, ideas, and intentions will not be present in the process; the process will fail due to the weak requirement definitions and no capacity to deal with volatility.

3.3 Collaborative Modeling

Due to the volatility and complexity of management processes, the development of a common understanding of the processes is critical. Usually, the designer's or manager's view about processes is, in many ways, different from a collaborative understanding of the same processes.

Collaborative modeling allows a shared understanding and joint vision about the management process design. Collaborative effort triggers communication and problem understanding [7]. The outcome of collaborative effort is a process model that represents the multiple stakeholders' views. It is relevant to highlight that the process model may change from time to time due to changes in requirements.

4 Systems Thinking and Meta-design

The Cartesian way of handling problems is to break them down into as many separate simple parts as possible, which is the most successful technique used by engineering. Hitchins [8] states that System Thinking came to the attention of engineers that were facing difficulties in applying the Cartesian approach to systems including humans. People are different from each other, and "one-size-fits-all" solution that does not meet all the interests, values and intentions of the people involved in the system design process [9, 10].

4.1 Systems Thinking

System Thinking allows understanding the system as a whole, instead of focusing on small events that occur in it. This approach allows identifying the relationships between systems elements and how this relationship influences both the system behavior patterns and events that derive from these relationships, including events that can affect other events, even if the second event occurs long after the first [8, 11, 12].

4.2 Meta-design

Fischer and Giaccardi [13] state that the creation of technical and social conditions to allow broad participation of stakeholders in design activities is as important as creating the artifact itself.

The environments created by Meta-design enable the stakeholders' participation as co-designers at all the stages of artifact creation. Meta-design characterizes the objectives, techniques, and processes necessary to create environments that allow stakeholders to act as designers. These environments are sociotechnical environments that empower stakeholders to engage actively in the continuous development of systems [14], and the most important and far-reaching impact of these environments is the possibility of transforming cultures by empowering people to be contributors in personally meaningful activities [15].

5   Application Domain

Learning Analytics is a system that operates with different purposes, according to the different users of the system. Users may be learners looking for information about them, teachers making pedagogical decisions about groups, leaders making decisions about improvement strategies for the organization, policymakers studying the impact of policies, and researchers.

The analysis and use of datasets in education, and in other sectors, are supposed to facilitate judgments, predict outcomes, and support interventions. There are sets of data that represent common concepts among all educational institutions. Also, there are several sets of data that are subjective and dependent on the culture of each institution. These data are difficult to understand, and the data interpretation is frequently subjective due to the lack of understanding of individuals about the data relationship and about how to interpret and purposefully use educational data to support learning. The essential element in the Learning Analytics endeavor is the ability to make sense of the data and, at the same time, to make effective use of the data, which are abilities that have a direct relationship with the management process of the Learning Analytics life cycle, from conception to grave.

5.1   Comprehensive Framework

A comprehensive framework is necessary to allow creating environments in which educational institutions stakeholders can develop relationships and share information about their proposals and activities, adding value to Learning Analytics systems. The authors of this work are studying and developing a framework based on System Thinking and Meta-design to present a response that considers the complexity, volatility, and collaborative understanding of the management process of Learning Analytics (see Fig. 1).

Meta-design is an enabler of environments that create conditions for stakeholders' participation in collaborative design activities. These collaborative activities are as important as the creation of the Learning Analytics system itself [13]. The result of this collaborative process is the socially enabled management process of the Learning Analytics life cycle, representing the multiple views of the stakeholders. Meta-design allows natural and emergent behavior by the stakeholders, which are not bothered with technical issues, but with specific tasks and challenges of the Learning Analytics domain.

Systems Thinking is an approach for seeing and working with systems as a whole, rather than with the systems individual parts. It promotes the integration of people, purpose, and process [8, 12]. The approach includes several strategies for knowledge construction and modeling, also allowing both hierarchical and dynamic process modeling, thinking in loops and layers, allowing the identification and construction of a range of Learning Analytics systems elements, considering both the properties and relationships [16, 17]. System Thinking facilitates sensemaking, which is the process by which people give meaning to their collective experiences [18].

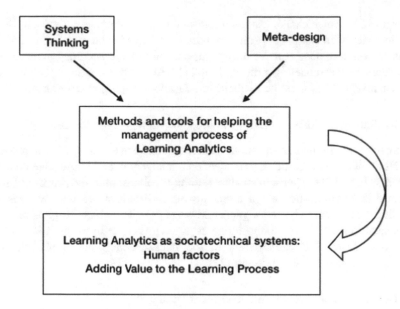

Fig. 1. Key elements involved in developing a management process of Learning Analytics systems.

5.2 Dealing with the Complexity of Management Process

Management process is not a single act performed a single time. The process of planning, controlling, organizing, and leading is interrelated and dependent on each other. To deal with the complexity of this process, we, the authors, consider the work in Systems Thinking and complexity developed by Blockley [16, 17], in which identifying the purpose of the system to be developed provides the lens through which it is possible to identify system boundaries and the architecture of the social, organizational, and technical components of the system.

The purpose of Learning Analytics is to support institutional administrators, educators, and learners through the analysis of behaviors and decisions. The purpose of formative support to the development of learners' competencies means that the educational institution desires that its learners develop a set of capabilities, which intrinsically include educators' and learners' human factors.

The management process needs to deal with another important characteristic of complex systems as Learning Analytics, the feedback loops, which influence the system because the analysis of behaviors and decisions demand the continuous monitoring of the processes and reaction to course events.

5.3 Dealing with the Volatility of Management Process

Process models have structural entities and attributes. Typical entities are activities, logic flow, events, conditions, and decision nodes; attributes are costs and resources allocation. Volatility is related to the need for process adaptation to an increasing rate of changes [19], which challenges the traditional process of modeling, which follows a linear approach with sharply separated phases [20]. Incremental approaches and spiral models can offer an answer to volatility, supporting flexible and easy participation of stakeholders in the modeling process [19–21]. However, there is a premise to this approach: stakeholders must be resilient to adapt to changing environments.

5.4   Dealing with Collaborative Modeling of Management Process

Collaborative environments promote stakeholder's engagement in interest-driven activities allowing democratic design decisions, innovation, and knowledge creation. It is a process in which the participants have direct involvement in the process of solving problems. The democratic design shifts power and controls to real systems users, supporting them to act as both designers and users [22]. The Meta-design approach is a kind of catalyzer, allowing the development of environments in which it is possible to have active stakeholder participation in the management process, articulating the challenges of collaborative modeling [23].

5.5   Learning Analytics as Sociotechnical Systems

The stakeholders' diversity implies that people taking part in Learning Analytics systems have different levels of skills and knowledge. Both Systems Thinking and Meta-design allow knowledge sharing face-to-face and through ICT tools, such as mailing lists, chat rooms, content management, and other technological tools to exchange knowledge. This knowledge supports stakeholders in using the design facilities, enabling the finding of relevant processes of Learning Analytics and fostering practice sharing within a community.

Both the technical and social elements of Learning Analytics systems are relevant. However, even if the best technological tools are adopted and established, there is the need for these tools to be easy to use by institutional administrators, educators, and learners. Otherwise, few stakeholders will be able to make use of the analyses to make data-enabled decisions; therefore, Learning Analytics will be relegated to technical and scientific circles, having relatively little impact on the majority of learners and educators.

6   Conclusion

Learning Analytics as sociotechnical systems highlight the fact that these systems must not be technology-oriented. It is necessary to consider stakeholders throughout all the systems life cycle, dealing with humans factors, which brings complexity to endeavors, dealing with these factors following the advice posted by Meadows and Wright [24]: "We can't control systems or figure them out. But we can dance with them!

Several knowledge domains take part in Learning Analytics systems, such as business intelligence, HCI, assessment, and evaluation. These domains have in common the dependency of data quantity and quality. Data are captured as learners engage in the learning process, which highlights the relevance of the management process of

Learning Analytics systems, which must consider technical, social and human elements present in the Learning Analytics systems life cycle, from their conception to grave. The authors of this paper are studying and developing this framework; a framework based on Systems Thinking and Meta-design to offer responses that consider Learning Analytics complexity, supporting the management process of these systems.

6.1 Future Work

Learning Analytics may be used to predict and to evaluate learners' and institutions performance and trajectories. Also, it may be used by institutional administrators to compel educators to apply learning methodologies that had a certain threshold of success rates, limiting educator's ability to innovate. Another issue of the use of Learning Analytics is learners having their difficulty exposed to others than their educators.

Several ethical and privacy issues can arise and need to be considered by the management process of Learning Analytics, an issue to be treated in the next steps of this research.

References

1. Campbell, J.P., DeBlois, P.B., Oblinger, D.G.: Academic analytics, a new tool for a new era. EDUCAUSE Rev. **42**(4), 40–57 (2007). http://www.educause.edu/ero/article/academic-analytics-new-tool-new-era

2. Rud, O.: Business Intelligence Success Factors: Tools for Aligning Your Business in the Global Economy. Wiley, Hoboken (2009). ISBN 9780470392409

3. Siemens, G.: Learning analytics: envisioning a research discipline and a domain of practice. In: Proceedings of the 2nd International Conference on Learning Analytics and Knowledge (LAK 2012), Vancouver, Canada, 2012, pp. 4–8. (2012). https://doi.org/10.1145/2330601.2330605

4. SOLAR, Society for Learning Analytics Research. http://www.solaresearch.org/about/. Accessed Apr 20 2019

5. Siemens, G.: Connectivism: creating a learning ecology in distributed environment. In: Hug, T. (ed.) Didactics of Microlearning: Concepts, Discourses, and Examples, pp. 53–68. Waxmann Verlag, New York (2007)

6. Samson, D., Daft, R.L.: Management. Cengage Learning, South Victoria (2012). ISBN 9780170192705

7. Renger, M., Kolfschoten, G., Vreede, G.: Challenges in collaborative modeling: a literature review. In: Advance in Enterprise Engineering I. Lecture Notes in Business Information Processing, vol. 10, pp. 61–77, Springer, Berlin (2008). https://doi.org/10.1007/978-3-540-68644-6_5

8. Hitchins, D.K.: Systems Engineering: A 21st Century Systems Methodology. Wiley, Chichester (2008)

9. Meadows, D.: Thinking in Systems: A Primer. Earthscan, London (2009). Edited by D. Wright

10. Appelo, J.: How to Change the World: Change Management 3.0. eBook, version 1.01, Rotterdam, The Netherlands (2012)

11. Sweeney, L.B.: When A Butterfly Sneezes: A Guide for Helping Kids Explore Interconnections in our World Through Favorites Stories, 1st edn. Pegasus Communications, Waltham (2001)
12. Checkland, P.: Soft Systems Methodology: A 30-Year Retrospective. Wiley, New York (1999)
13. Fischer, G., Giaccardi, E.: Meta-design: a framework for the future of end-user development. In: Lieberman, H., Paterno, F., Wulf, V. (eds.) End User Development, pp. 427–457. Springer, Dordrecht (2006)
14. Fischer, G., Giaccardi, E., Ye, Y., Sutcliffe, A., Mehandjiev, N.: Meta-design: a manifesto for end-user development. Commun. ACM **47**(9), 33–37 (2004). https://doi.org/10.1145/1015864.1015884
15. Fischer, G., Fogli, D., Piccinno, A.: Revisiting and broadening the meta-design framework for end-user development. In: Paternò, F., Wulf, V. (eds.) New Perspectives in End-User Development. Springer, Cham (2017). https://doi.org/10.1007/978-3-319-60291-2_4
16. Blockley, D.: The importance of being process. Civil Eng. Environ. Syst. **27**(3), 189–199 (2010). https://doi.org/10.1080/10286608.2010.482658
17. Blockley, D.: Process modelling from reflective practice for engineering quality. Civil Eng. Environ. Syst. **16**(4), 287–313 (1999). https://doi.org/10.1080/02630259908970268
18. Weick, K.E., Sutcliffe, K.M., Obstfeld, D.: Organizing and the process of sensemaking. Organ. Sci. **16**(4), 409–421 (2005). https://doi.org/10.1287/orsc.1050.0133
19. Dadam, P., Reichert, M.: The ADEPT project: a decade of research and development for robust and flexible process support – challenges and achievements. Comput. Sci. Res. Dev. **23**(2), 81–97 (2009). https://doi.org/10.1007/s00450-009-0068-6
20. Weske, M.: Business Process Management – Concepts, Languages, Architectures. Springer, Heidelberg (2007)
21. Davies, I., Green, P., Rosemann, M.: How do practitioners use conceptual modeling in practice? J. Data Knowl. Eng. **58**, 358–380 (2006). https://doi.org/10.1016/j.datak.2005.07.007
22. von Hippel, E.: Democratizing Innovation. MIT Press, Cambridge (2005)
23. Erol, S., Mödritscher, F., Neumann, G.: A Meta-design approach for collaborative process modeling. In: Proceedings of the 2nd International Workshop on Open Design Spaces (ODS 2010) International Reports on Socio-Informatics, vol. 7, no. 2 (2010), pp. 46–62. https://www.researchgate.net/publication/249991052_A_Meta-Design_Approach_for_Collaborative_Process_Modeling
24. Meadows, D., Wright, D.: Thinking in Systems: A Primer. Chelsea Green Pub, Hartford (2008)

Cluster and Sentiment Analyses of YouTube Textual Feedback of Programming Language Learners to Enhance Learning in Programming

Rex P. Bringula[1,2(✉)], John Noel Victorino[2], Marlene M. De Leon[2], and Ma. Regina Estuar[2]

[1] University of the East, Manila, Philippines
rexbringula@gmail.com
[2] Ateneo de Manila University, Manila, Philippines

Abstract. This study intends to determine the clusters and sentiments of feedback of YouTube users in learning to program in Python and C++. Toward this goal, a total of 2,583 feedback on introductory video tutorials about Python and C++ were collected. It is found that the words "*thanks*" and "*thank*" were the most frequently occurring word in both YouTube videos – indicating appreciation and helpfulness of the video tutorials. The results of k-means cluster analyses further disclosed that groups of feedback are similar across the two languages, i.e., confirmation, helpfulness, gratitude, and recommendation. YouTube users expressed positive sentiments towards the tutorial videos. Implications to teaching programming and YouTube video content development are presented. Limitations of the study are also offered.

Keywords: Learning · Programming · Sentiment analysis · Videos · YouTube

1 Introduction

Programming has been consistently found to be a desirable skill [7, 14, 19]. According to [7], it is expected that in 2024 there will be an increase of 12% in the demand for Information Technology professionals programming and other technical skills. Unfortunately, learning to program is a difficult educational task [12, 15] because it requires exceptional perfection [13, 16]. This poses a big challenge for novice learners like students. When students could not cope with difficulty in programming, they tend to shift to other degree programs [10, 18].

Educators and higher education institutions (HEIs) are aware of these situations. In order to address these learning challenges, HEIs provide print and digital educational resources. Educators are backed up by literature that attempt to provide better teaching strategies on programming (e.g., [9]).

Despite these efforts, there is an empirical evidence showing that students tend not to use electronic resources (e.g., e-books, digital library collections) [3] when learning to program. Instead, students use blogs, forums, YouTube, and educational websites to learn programming [2]. While these studies provided evidence on the use of these

© Springer Nature Switzerland AG 2020
K. Arai et al. (Eds.): FTC 2019, AISC 1070, pp. 913–924, 2020.
https://doi.org/10.1007/978-3-030-32523-7_67

online resources, it is still unclear how students respond to the content to these resources. Understanding the textual feedback (subsequently referred as feedback) is important since it ensures that learning takes place [22]. In particular, it was shown that YouTube is one of the online resources that are utilized by the students in order to learn programming [2]. While there are prior studies on the use of YouTube (e.g., [6]), very few studies are conducted about its use in terms of learning programming (e.g., [2, 4]).

This study intends to contribute to the existing threads of discussions about the use of YouTube in learning programming. Understanding the feedback of the learners may serve as basis for the educators, YouTube content developers, and higher education institutions to improve educational resources or programs that could help them achieve learners' educational goals. This would further inform educators and HEIs to continuously develop educational materials suited to the educational needs of the students. Toward this goal, the study gathered the feedback of the YouTube users in programming tutorial videos in Python and C++ - two of the most popular programming languages in 2018 [5]. The threads of feedback were then analyzed through sentiment and cluster analyses. Specifically, the study aims to answer the following questions.

1. How can the feedback of YouTube users in the two programming languages be described in terms of number of feedback, most frequently used words, classifications of feedback, and sentiments of the feedback?
2. How can the clusters of and sentiments towards the feedback be utilized in developing educational video?

The paper was subdivided into five main sections in order to answer the above questions. In Literature Review section, studies conducted related to this study were presented. The Methodology section presented how data were collected and analyzed. The Results were presented and followed immediately by the Discussion section. The summary of the study was presented in Conclusions, Implications, Limitations, and Recommendations.

2   Literature Review

YouTube is a useful resource for knowledge [1] and learning [8]. It aids students to solve problem solving tasks through the development of their cognitive and social abilities [21]. The suitability of YouTube as an educational resource can be attributed to its compatibility between students' preferred learning styles and video content, wide variety of relevant, up-to-date, and sufficient videos, sense of clarity of content, and easy-to-follow topic discussions across multiple domains [8]. Moghavvemi et al. [11] confirmed that entertainment, seeking information, and academic learning are the factors that motivate the students to use YouTube.

One of the capabilities of YouTube is allowing users to provide feedback (i.e., comments and ratings) about the video. Zher Ng and Maznah Raja Hussain [22] conducted an experiment where their students uploaded videos in YouTube. The target audience of the video was secondary school students or teachers. The effectiveness of the videos was rated based on the ratings, watch count, and comments given by their peers. Their findings showed that students were engaged in learner development which allowed them to monitor, manage, and evaluate their own learning. The authors acknowledged the possible biased of the study since the evaluation was done among their peers.

The educational value of YouTube has been consistently found in various disciplines. There is a growing body of research on the use of YouTube in the field of medicine. Akgun et al. [1] determined the usefulness and accuracy of YouTube videos relating to electrocardiogram (ECG). Excluding non-English language and non-educational videos in their analysis, the researchers developed a checklist that evaluated the usefulness of the 119 videos they collected. To determine the accuracy of the videos, they utilized a medical textbook. They found that 47% ($n = 56$) of the videos were very useful. On one hand, however, there were misleading videos ($n = 16$, 13.4%). They concluded that utmost care should be exercised when selecting YouTube videos relating to ECG. In the same discipline, [20] conducted a study on the use of YouTube regarding seizures and epilepsy. According to the authors, 98% of the 100 videos they collected were easy to understand. It was also reported that the majority (85%) of the videos were sympathetic towards people with these medical conditions. Similar to the findings of [1], it was revealed that there were accurate (51%) as well as misleading videos (9%).

The educational use YouTube is particularly helpful for procedural learning [17]. This is confirmed in the study of [6]. In this study, a consensus on the advantages of the use YouTube as teaching and learning tool in the performing arts was determined. Using fuzzy Delphi technique, it was shown that YouTube was a useful instructional tool in the areas of music, creative writing, theatre, television film, dance, animation, and fine arts. Additionally, as a platform that can capture the current trends in the performing arts, YouTube was able to sustain students' interest and learning achievement [6].

Carlisle [4] utilized YouTube to engage students' class participation. Students watched 21 YouTube videos for introductory programming in Java in the laboratory class. While teachers reduced their lecture time, students increased laboratory time. The experiment showed that students read more educational materials after watching the videos and would prefer not to have more lecture sessions. Even after the experiment, students watched videos relating to the topic before going to their class.

3   Methodology

3.1   Data Collection Procedure, Sample Size, and Sampling Design

Users' feedback in the YouTube tutorial videos (or simply videos) were collected, processed, and analyzed using the R statistical software. The keyword "*Python Programming*" was used to search for tutorial videos for the Python programming. The R software returned 642 YouTube videos search results using that keyword. Afterwards, the first-100 videos were selected using the keyword "*beginner*". The threads of feedback were then collected using the unique YouTube IDs of the first-100 videos and stored in a spreadsheet. The same processes were repeated for the C++ programming except that the keyword used was "*C++*" to search for videos in the said language. There were 1,356 and 1,227 feedbacks in Python and C++ videos, respectively.

3.2   Data Cleaning and Data Preprocessing

The collected threads of feedback, which comprised the dataset of the study, were subjected to data cleaning and data preprocessing. In data cleaning, all special characters relating to web links were removed (e.g., "\", "http"). Afterwards, the dataset was subjected to data preprocessing. Stopwords comprising of English standard terms (e.g., "the", "a", "an", etc.), keyword terms (e.g., "programming", "python", "language", "program", "tutorial"), and noise words (e.g., alphanumeric, whitespace, punctuations, and unrecognizable characters) were removed in the dataset. All words in the dataset were transformed into lowercase. The same were tokenized and made to comprise the corpus. The corpus in Python contained 3,109 words while C++ had 2,894 words.

3.3   Data Analysis

The corpus was analyzed using clustering technique and sentiment analysis. Cluster analysis using *k*-means algorithm and sentiment analysis were employed using the R software. In cluster analysis, words that appeared in less than 1% in the dataset were removed in the corpus. This step resulted to 20 and 19 words in Python and in C++, respectively. To determine the optimal number of clusters, the elbow method was used. The sentiments were based on the *loughran* sentiments classifications (i.e., positive, negative, and uncertainty). For the sake of image clarity, words with less than 5% frequency were eliminated in the corpus.

4   Results

Excluding the "*cout*" keyword in C++, the most frequent words appearing in the feedback in both programming languages are "*thanks*" and "*thank*". It is worth noting that both words have almost the same frequencies. Furthermore, the words "*please*", "*help*", and "*like*" appear in both languages. As shown in Tables 1 and 2, the word "*please*" denotes two reactions. First, it signifies a polite request of seeking help from the developers' community (e.g., "*Brother could you please tell me how can I enlarge the text while writing the code?*"). It further signifies a kind recommendation: "*please make a video on python django*").

From the extracted feedback, the word "*help*" may have two meanings. The feedback "*Thx for the help*" indicates that the YouTube tutorials are useful in learning programming. The second implication is that YouTube becomes a platform to seek further information (e.g., "*Hi I need your help*"). This confirms the findings of [11]. YouTube users provide feedback that they enjoy the videos, as indicated by these feedbacks: "*Hi corey, I like the way you teach*" and "*I absolutely like your tutorials*". The word "*like*" and "*make*" are used to further suggest and enumerate other topics they want to learn more from the videos. The feedback "*plss make language video.....like C++, HTML*" demonstrates this finding.

Table 1. Ten most frequently used words in Python YouTube tutorials

Rank	Python	Frequency	Sample feedback
1	thanks	129	Hey dude everything looks great thanks so much. Just subtitles pls and, then it will be greatest
2	thank	120	Thank you for this easy to understanding tutorial
3	learn	104	My friends just started to learn from your channel as well and they are excited and motivated
4	please	99	Brother could you please tell me how can I enlarge the text while writing the code??
5	great	91	Awesome, awesome presentation. This is a good jump in learning a great programming language. :)
6	much	87	I just want to say, thank you so much for your time and effort in making this tutorial
7	good	83	Awesome, awesome presentation. This is a good jump in learning a great programming language. :)
8	help	79	Hi I need your help // thank you for your kind help sir
9	like	77	You plss make language video…..like c++, HTML, python // Hi corey, I like the way you teach
10	really	75	Really great help and learned a lot from your videos

Table 2. Ten most frequently used words in C++ YouTube tutorials

Rank	C++	Frequency	Sample feedback
1	cout	111	I fixed a problem with the code it needed to be std::cout << Name "\n";
2	thanks	110	Awesome, would you continue this tutorial to stl? Thanks
3	thank	109	great job thank you so much
4	int	108	What does Int Mean? (I know it is a stupid question)
5	like	98	I can't seem to find an organized playlist….the playlists are named things like "arrays" // I absolutely like your tutorials
6	help	83	Thx for the help
7	use	81	But can we use any other header files besides the iostream like string, ctype and etc.
8	make	74	Please make a video on python django
9	please	73	Please make a video on python django
10	value	70	One more question, When I do not initialize variables values

Feedback that contains the word "*much*" is mostly found in the Python videos but not in the C++ videos. This word is usually combined with the phrase "*thank you*" to indicate the intensity of gratitude. The words "*good*", "*great*", and "*really*" are signs that the users of Python video tutorials are appreciated. Meanwhile, the keyword "*cout*" and the variable type "*int*" appear in the C++ video tutorials feedback. Users seek help to demonstrate the use of functions and keywords, and the possible values associated with the code. As such, the words "*use*" and "*values*" appeared mostly in the feedback on C++ videos.

Cluster analyses using the R language displayed two figures (one each for Python and C++) that categorized the words in the corpus. The clustered words were manually converted into tables for better interpretations. Table 3 shows that the feedback in the Python and C++ videos can be categorized into 4 groups. The first group of words is called "*Confirmation*" since users' feedback validates the usefulness of the videos. The second group of words was labeled as "*Helpfulness*". Cluster analysis had two distinct groupings for the words "thanks" and "*thank*". The word "thanks" was labeled as "*Gratitude with Recommendations*" while "*thank*" was labeled "*Gratitude*". The distinction between the two groups is that the former provides encouragement for the YouTube content developers to make more videos of different topics while the latter only indicates appreciation towards the videos.

Table 3. Result of cluster analysis on the feedback on YouTube videos about Python and C++

Types of feedback	Python feedback words	C++ feedback words
Cluster 1 – Confirmation	"awesome", "best", "get", "learning", "like", "make", "much", "one", "really", "use", "videos", "want", "way"	"code", "error", "get", "int", "know", "learn", "much", "need", "run", "videos", "work"
Cluster 2 – Helpfulness	"good", "great", "help", "please", "learn"	"great", "like", "use", "please", "help", "make"
Cluster 3 – Gratitude with Recommendations	"thanks"	"thanks"
Cluster 4 – Gratitude	"thank"	"thank"

The groupings were based on the actual feedback shown in Tables 1 and 2. Visual inspections on the results of the elbow method analysis (Figs. 1 and 2) confirm that there are 4 clusters on each group of feedback. Therefore, the number of clusters is retained.

Figures 3 and 4 shows the results of the sentiment analysis. Generally, both tutorial videos received positive sentiments (Table 4). Python ($n = 332$, 60%) received more positive feedback than C++ ($n = 174$, 45%). One-way χ^2 test shows that there is an unequal number of positive, negative, and uncertain sentiments in both languages (Python, $\chi^2(2) = 192.72$; C++, $\chi^2(2) = 46.95$). The results are unlikely to have arisen from sampling error ($p < .05$).

The word "*great*" is the dominating positive sentiment in both languages. Both languages had the same prominent negative sentiment relating to the word "*error*". This can be explained by the fact that programming is usually prone to different errors. Upon inspection of the feedback, YouTube users are seeking help to fix an error

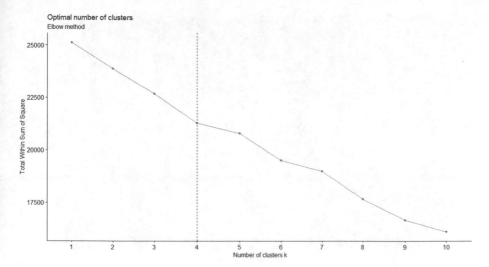

Fig. 1. Optimal clusters for Python words using the elbow method

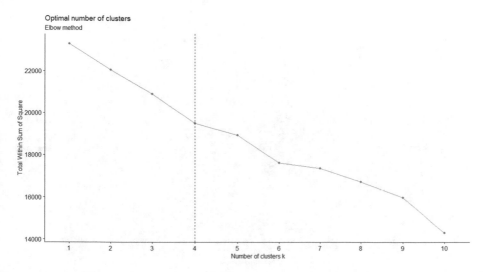

Fig. 2. Optimal clusters for C++ words using the elbow method

(or technically, a bug) in the programming they are doing. For example, one feedback said "*Pls help me with this error: Failed to launch the Python Process, please validate the path.*" This is similar to the case of C++ video tutorials. One user said "*I have compiling error I don't know how to fix it. It says there was problem in compiling.*"

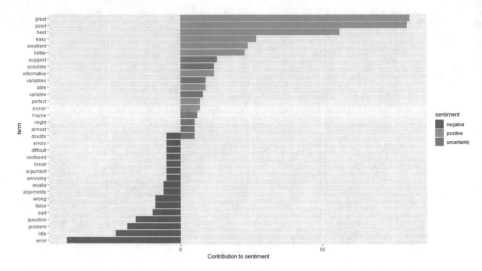

Fig. 3. Result of sentiment analysis on the feedback on YouTube videos about Python

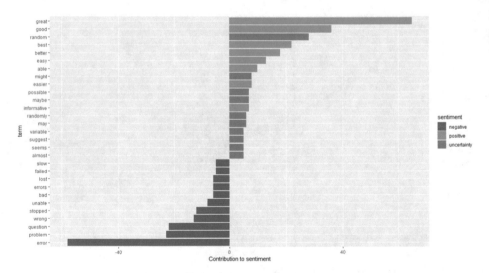

Fig. 4. Result of sentiment analysis on the feedback on YouTube videos about C++

Table 4. Polarity of Words

Polarity	Python	χ^2 (df = 3, α = 0.05)	C++	χ^2 (df = 3, α = 0.05)
Positive	332 (60%)		174 (45%)	
Negative	157 (28%)	χ^2 = 192.72	148 (38%)	χ^2 = 46.95
Uncertain	69 (12%)		68 (17%)	
TOTAL	558		390	

5 Discussion

This study analyzed the feedback of YouTube users in Python and C++ programming tutorials. It aims to understand their feedback through cluster and sentiments analyses. The results of these analyses may serve as basis for the educators, YouTube content developers, and the higher education institutions to improve educational resources or programs that could help them achieve learners' educational goals.

The results of the study suggest that YouTube users appreciated the video tutorials. They find the content of the video helpful. This conforms to the findings of [2] that YouTube programming tutorials are helpful to learn programming. The helpfulness of the video can be explained by the fact that learners can repeat the discussions at their convenient time and place. Unlike in classroom or laboratory settings where lectures are only delivered once, learners can easily replay the video until they understand the topic. Moreover, YouTube videos are more liberating since they can easily switch from one video to another if they do not understand the content delivery of the tutorial. Another explanation of the helpfulness of YouTube videos is the reproducibility of the codes. Learners can easily type the codes and can instantly verify the correctness of the code.

The users' feedback disclosed that they do not only learn from the videos but also validate the educational content of the videos (e.g., "*awesome, awesome presentation. This is a good jump in learning a great programming language. :)*"). The YouTube platform allows users to provide comments and informs the content developers whether they are able to meet the learning needs of the users. In other words, the YouTube platform exhibits a teach-evaluation loop which is similar in an educational setting. (The teach-evaluation loop is described as learners provide feedback about the teacher and use that feedback to improve teaching.)

Alongside with conformance with and usefulness of the content, gratitude and recommendations to make other videos are discovered. Cluster analysis made a clear distinction between the words "*thanks*" and "*thank*". While they both signify appreciation, the latter does not provide suggestions. The comment "*Awesome, would you continue this tutorial to stl? Thanks*" demonstrate this finding. This feedback signals that the users are grateful with the video and found it useful. Such gratification drives a user to encourage content developers to make another video.

It is interesting to note that all types of feedback are positive. This is because learners can switch to another video and would not give constructive criticisms on how to improve a video. As a result, learners will only provide positive feedback to the videos they deemed suitable to their programming needs.

It can be gleaned from Table 2 that C++ keyword "*cout*" and the variable type "*int*" are frequently used words. This is because the questions raised in the comment sections usually contain codes with the standard input/output keyword and the data type integer. Meanwhile, the word "*use*" is one of the visible words in the feedback. In the context of this study, the word "*use*" denotes two usages for the C++ dataset. First, it denotes that C++ video users are seeking more particular information. One user sought help by asking the question "*So a question I have that I haven't been able to find is how do I make a game with the most efficient method and which libraries to use that use the least amount of overhead.*" Learners further seek help on the usage of special characters, keyboard shortcuts, functions, and arguments. The second usage denotes recommendation to use alternative codes, such as in "*but can we use any other header files besides the iostream like string, ctype and etc.*".

Using loughran sentiments classification, it is revealed that the videos mostly received positive feedback. The results of one-way chi-square test revealed that this finding is unlikely to have arisen from sampling error. Thus, videos in both programming languages received favorable feedback from the users. The words "*error*" and "*errors*" are the prevalent negative sentiments in both videos. This is understandable since programming languages are closely tied up with the discussion of programming errors. Furthermore, it must be emphasized that the words classified as negative sentiments are not necessarily negative in the context of this study. One example is the word "*argument*" since this has a different meaning in programming parlance.

The negative sentiments provide another perspective of YouTube content development. They may serve as basis of the development of new videos. Content developers may compile the errors encountered by the users in their previous videos and may develop videos addressing these errors. Another possible way is that content developers may already include the possible errors in their videos.

6   Conclusions, Implications, Limitations, and Recommendations

This study contributes to the existing body of literature by uncovering other perceptions towards the use of YouTube which were reflected through learners' feedback. It reveals that learners are able to confirm the suitability of the video content to their learning needs. It also confirms that the videos are helpful and that users are able to express their gratitude towards these videos. As satisfied learners, they further encourage content developers to make more tutorial videos since this study confirms that the YouTube video tutorials in programming are valuable resources for beginners. It shows that YouTube is an educational platform to seek help and information about programming. Overall, users find that YouTube tutorial videos in Python and C++ programming elicit positive sentiments.

This study shows that YouTube videos in programming receive favorable responses from its users. Higher education institutions (HEIs) can capitalize on this finding to initiate student learning support program. HEIs can institutionalize a multidisciplinary collaboration among its departments in developing the content of YouTube video tutorials. Videos can be constructed based on the syllabus. In particular, teachers may focus their content on the uses of keywords, data types, variables, functions, and arguments in C++.

To realize this recommendation, it is advised that teachers be given incentives for the content development. Teachers may be given an equivalent of a 3-unit course in lieu of their video development. This can be integrated in the existing policy of an institution in terms of content development of learning management system materials. It is advised that teachers may include YouTube videos as one of the learning materials in their syllabus. These strategies may help the students cope with programming. The chances of students transferring to other programs may be lowered.

While this study provided empirical evidence on the usefulness of the YouTube programming tutorials, the extent of the effectiveness of these videos in terms of learning is still unknown. Future researchers may compare and contrast the effectiveness of YouTube videos with other learning materials in teaching programming. Furthermore, the study is delimited by the type of feedback provided by the users. Constructive criticisms are not present in the study because YouTube users may simply ignore a video and look for another one. Thus, only positive feedback will reflect in the study.

As shown in the study, only positive types of feedback are reflected. Video developers may invite users to provide their constructive criticisms. A mixed-method study can be initiated to capture the constructive criticisms about the videos. Thus, it is recommended that YouTube may provide the length of stay of users and number of likes in the video to further analyze the usage of the users. These data could be correlated to the quality of feedback of the users.

Moreover, it is recommended that the types of feedback found in this study be empirically tested. Researchers may investigate the relationship of the types of feedback and users' learning experiences in YouTube. Future studies may compare the types of errors encountered in a laboratory setting and in using YouTube tutorials. Lastly, this study can be replicated using other programming languages.

Acknowledgment. The authors are indebted to the University of the East and Ateneo de Manila University for funding this study.

References

1. Akgun, T., Karabay, C.Y., Kocabay, G., Kalayci, A., Oduncu, V., Guler, A., Pala, S., Kirma, C.: Learning electrocardiogram on YouTube: how useful is it? J. Electrocardiol. **47**(1), 113–117 (2014)
2. Bringula, R.: Influence of usage of e-books, online educational materials, and other programming books and students' profiles on adoption of printed programming textbooks. Program **51**(4), 441–457 (2017)
3. Bringula, R., Aborot, A., Lim, P.J.G., Canlas, K.C., Amador, S.M.: HCI group: why computing students are not using e-resources? evidence from the university of the east. In: Liu, X. (ed.) WCCCE 2014 Proceedings of the Western Canadian Conference on Computing Education. ACM, New York (2014). Article 2
4. Carlisle, M.C.: Using You Tube to enhance student class preparation in an introductory Java course. In: Proceedings of the 41st ACM Technical Symposium on Computer Science Education, pp. 470–474. ACM, New York (2010)
5. Cass, S.: The 2018 top programming languages. https://spectrum.ieee.org/at-work/innovation/the-2018-top-programming-languages. Accessed 11 Oct 2018
6. DeWitt, D., Alias, N., Siraj, S., Yaakub, M.Y., Ayob, J., Ishak, R.: The potential of YouTube for teaching and learning in the performing arts. Procedia Soc. Behav. Sci. **103**, 1118–1126 (2013)
7. Forbes technology council (2018). https://www.forbes.com/sites/forbestechcouncil/2017/12/21/13-top-tech-skills-in-high-demand-for-2018/#4d4ea1bb1e5c

8. Lee, D.Y., Lehto, M.R.: User acceptance of YouTube for procedural learning: an extension of the technology acceptance model. Comput. Educ. **61**, 193–208 (2013)
9. Mannila, L.: What about a simple language? analyzing the difficulties in learning to program. Comput. Sci. Educ. **3**, 211–227 (2006)
10. Merida, R.A., Torres, R., Bringula, R.: Push and pull of institutional image indicators and computing degree programs viewed through the lens of shifters and transferees at the University of the East. Manage. Educ. Int. J. **16**(3), 13–27 (2016)
11. Moghavvemi, S., Sulaiman, A., Jaafar, N.I., Kasem, N.: Social media as a complementary learning tool for teaching and learning: The case of youtube. Int. J. Manage. Educ. **16**, 37–42 (2018)
12. Pendergast, M.O.: Teaching introductory programming to IS students: Java problems and pitfalls. J. Inf. Technol. Educ. **5**, 491–595 (2006). http://jite.org/documents/Vol5/v5p491-515Pendergast128.pdf. Accessed 8 Jan 2016
13. Perkins, D.N., Martin, F.: Fragile knowledge and neglected strategies in novice programmers. In: Soloway, E., Iyengar, S. (eds.) Empirical Studies of Programmers, First Workshop, pp. 213–229. Ablex, Norwood (1986)
14. Poe, E.: Programmers have most in-demand IT skills of 2015 (2015). Fiercecio.Com, http://rizal.lib.admu.edu.ph:2048/login?url=http://search.ebscohost.com/login.aspx?direct=true&db=edsbig&AN=edsbig.A414111425&scope=site. Accessed 1 Oct 2018
15. Robins, A., Rountree, J., Rountree, N.: Learning and teaching programming: a review and discussion. Comput. Sci. Educ. **13**(2), 137–172 (2003)
16. Rogerson, C., Scott, E.: The fear factor: how it affects students learning to program in a tertiary environment. J. Inf. Technol. Educ. **9**, 147–171 (2010). http://jite.org/documents/Vol9/JITEv9p147-171Rogerson803.pdf. Accessed 08 Apr 2010
17. Rössler, B., Lahner, D., Schebesta, K., Chiari, A., Plöchl, W.: Medical information on the internet: quality assessment of lumbar puncture and neuroaxial block techniques on YouTube. Clin. Neurol. Neurosurg. **114**(6), 655–658 (2012)
18. Teague, D., Roe, P.: Learning to program: going pair-shaped. Innov. Teach. Learn. Inf. Comput. Sci. **6**(4), 4–22 (2015). https://doi.org/10.11120/ital.2007.06040004
19. Torres, C.: Demand for programmers hits full boil as U.S. job market simmers. https://www.bloomberg.com/news/articles/2018-03-08/demand-for-programmers-hits-full-boil-as-u-s-job-market-simmers. Accessed 4 Sep 2018
20. Wong, V.S.S., Stevenson, M., Selwa, L.: The presentation of seizures and epilepsy in YouTube videos. Epilepsy Behav. **27**(1), 247–250 (2013)
21. Zahn, C., Pea, R., Hesse, F.W., Rosen, J.: Comparing simple and advanced video tools as supports for complex collaborative design processes. J. Learn. Sci. **19**(3), 403–440 (2010)
22. Zher Ng, H., Maznah Raja Hussain, R.: Empowering learners as the owners of feedback while YouTube-ing. Interact. Technol. Smart Educ. **6**(4), 274–285 (2009)

Method for Estimation of Multiple Reflection, Scattering and Absorption in Mountainous Areas of Remote Sensing Satellite Data

Kohei Arai[✉]

Saga University, Saga City 8408502, Japan
arai@is.saga-u.ac.jp

Abstract. Method for estimation of multiple reflection, scattering and absorption in mountainous areas of remote sensing satellite data is proposed. An experiment of TOA: Top of the Atmosphere radiance evaluation for miniature forest with slope grades, tree species, tree canopy shapes in modeled V-shape mountainous areas is conducted. Through experiments, it is found that the influences due to multiple reflection between two slope surfaces are quite significant; the maximum radiance rates are about 4% to 5%. It is also found that the multiple reflection can reveal important differences in clarifying radiative transfer process in atmosphere. The results are also simulated by means of Monte-Carlo based radiative transfer model with forestry surface model.

Keywords: Multiple reflection · Experiment with miniature forest model · Slope surfaces · Interaction between two surfaces · Monte-Carlo model · Top of the atmosphere radiance

1 Introduction

Solar reflection channels (ultra-violet to shortwave infrared, in general) of optical sensors onboard on remote sensing satellites have problem on the estimation of surface radiance with observed radiance of the sensor data of Digital Number: DN. This problem is caused by absorption and scattering in the atmosphere and multiple reflection, absorption and scattering between land surface as well as the atmosphere. The latter is called adjacency effect and the former is called atmospheric effect. In particular, V-shaped mountainous areas (valley areas) have a relatively great multiple reflection, absorption and scattering between land surface as well as the atmosphere. Therefore, it is highly required to correct such influence to get the actual radiance from the surface in concern accurately.

Monte-Carlo simulation[1] for radiative transfer modeling of forested areas and the atmosphere is reported precisely [1]. In particular, layered atmospheric model with consideration of surface model is not easy to formulate. Reflectance model of

---

[1] https://www.investopedia.com/terms/m/montecarlosimulation.asp.

K. Arai et al. (Eds.): FTC 2019, AISC 1070, pp. 925–935, 2020.
https://doi.org/10.1007/978-3-030-32523-7_68

vegetation in forested areas taking into account multiple reflection between trees is also well reported [2]. Forest parameter estimation by means of Monte-Carlo simulations with experimental considerations as well as estimation of multiple reflection among trees depending on forest parameters are discussed [3]. The use of a suitable and rigor method to accurately estimate forest biomass is significant.

On the other hand, Monte-Carlo simulation is well studied and established already. Adjacency effect taking into account layered clouds based on Monte Carlo method is proposed and well reported [4]. Also, adjacency effect of layered clouds taking into account phase function of cloud particles and multi-layered plane parallel atmosphere based on Monte-Carlo Method is investigated [5]. Then, adjacency effect of layered clouds estimated with Monte-Carlo simulation is published [6].

Meanwhile, estimation of tree type, shape, tree distance using radiative transfer model based on Monte-Carlo model is discussed [6]. After that, non-linear mixture model of mixed pixels in remote sensing satellite images based on Monte-Carlo simulation is proposed [7]. Furthermore, forest parameter estimation by means of Monte-Carlo simulations with experimental considerations is proposed in conjunction with estimation of multiple reflection among trees depending on forest parameters [8]. Forest parameter estimation, by means of Monte-Carlo simulations with experimental consideration of estimation of multiple reflection among trees depending on forest parameters is discussed [9].

Meanwhile, micro traffic simulation with unpredictable disturbance based on Monte-Carlo simulation is proposed for evaluation of effectiveness of the proposed agent cars of Sidoarjo Indonesia hot mudflow disaster [3]. Similarly, micro traffic simulation with unpredictable disturbance based on Monte-Carlo simulation and effectiveness of the proposed agent cars of Sidoarjo Indonesia hot mudflow disaster is well reported as one of applications of Monte-Carlo simulation [3].

Monte-Carlo ray tracing simulation for bi-directional reflectance distribution function and grow index of tealeaves estimation is proposed [10]. Then Monte-Carlo ray tracing simulation for bi-directional reflectance distribution function and grow index of tealeaves estimations is also discussed [11]. The quality of a Monte Carlo ray tracer (based on path tracing algorithms) is much more realistic than a distributed (stochastic) engine[2].

In conjunction with the above application of MCRT: Monte-Carlo Ray Tracing simulation, MCRT simulation for bi-directional reflectance distribution functions and grow index of tealeaves estimations is proposed [12]. Also, Monte-Carlo simulation of polarized atmospheric irradiance for determination of refractive index of aerosols is discussed [13]. Moreover, MCRT simulation of polarization characteristics of sea water which contains spherical and non-spherical shapes of suspended solid and phytoplankton is proposed [14].

---

[2] https://computergraphics.stackexchange.com/questions/6/why-does-monte-carlo-ray-tracing-perform-better-than-distributed-ray-tracing.

MCRT based non-linear mixture model of earth observation satellite imagery pixel data is investigated [15] together with MCRT based sensitivity analysis of the atmospheric and oceanic parameters on the top of the atmosphere radiance. Furthermore, MCRT based nonlinear mixture model of mixed pixels in Earth observation satellite imagery data is proposed and discussed [16].

In this paper, a method for estimation of multiple reflection, scattering and absorption in forested areas in remote sensing satellite data is proposed. An experimental study of multiple reflection and scattering as well as absorption in V-Shaped mountainous areas is conducted. The experimental results are validated based on MCRT to solve the aforementioned problem of solar reflection channels radiometers onboard remote sensing satellites.

The following section describes experimental configurations followed by preliminary results from the experiments. Then the experimental results is validated with MCRT after that followed by conclusion with some discussions.

2   Experiments

2.1   Experimental Configuration

Simulated forested areas are intended to create for experimental study on multiple reflection and adjacency effects. Commercial lawn and two coniferous trees were used for the experiment. Tree No. 1 is line gold (scientific name: *Thuja occidentalis* 'Rheingold', tree shape: circular, lateral diameter: 13 cm, longitudinal diameter: 12 cm), No. 2 is gold crest (scientific name: *Cupressus macrocarpa* 'Goldcrest' Shape: conical, bottom diameter: 11 cm, height: 26 cm). Through these, a simulated forest modeling the V-shaped mountain topography was constructed. There are 5 cases (No. 1 to No. 5) depending on composition of the simulated forest (existence of trees · species of trees) and are shown in Table 1 and Fig. 1. The A and B plants are respectively 60 cm by 60 cm. When the trees are in the grassland, eight trees were used on one side and two-dimensionally arranged. Also, to reduce reflection from the bare ground around the experiment area, the surroundings were covered with black cloth.

Outlook of the experiment configuration is shown in Fig. 2. The experiment site was selected at the parking lot of Saga University (33° 14′N, 130° 17 ′E) and was carried out on May 3, 2006 (fine weather). The data used here was acquired from 11: 29 to 15: 24. Measurement of radiance at the upper end of the simulated forest was carried out using a spectroradiometer (MS 720[3] manufactured by EKO Seiki Co., Ltd.) installed using two tripods. The installation height of the radiometer was 128 cm, and the installation direction was set just under. By using this forest simulation model, multiple reflection effect can be confirmed and evaluated.

---

[3] http://eko.co.jp/meteorology/met_products/0015.html.

(a)Case No.1   (b)Case No.2

(c)Case No.3   (d)Case No.4

(e)Top view of case No.3

Fig. 1. The experiment's cases

Table 1. Specification of experiment's cases

No. of case	A surface		B surface	
	Slope angle	Surface situation	Slope angle	Surface situation
No. 1	26.7°fixed	Grass ground	0°~36°changeable	Grass ground
No. 2	23.3°fixed	Tree No.1	0°~24.5° changeable	Tree No. 1
No. 3	23.3°fixed	Tree No.1	0°~26.8° changeable	Grass ground
No. 4	0°~24.2°changeable	Tree No.1	12.5°fixed	Tree No. 2
No. 5	4.4°~23.9° changeable	Tree No.2	10.0°fixed	Tree No. 2

2.2   Experimental Set-up

The radiance was measured while one surface of the V-shaped slope was fixed at an angle and the inclination angle of the other surface was changed stepwise. The change of sunlight at this measurement time (10 to 12 min) was monitored by a spectrora-diometer (Spectra Vista Corp., GER 2600)[4]. Also, shadow areas are calculated and compensated from the measured up-welling radiance. The slope angle is changed along with the line perpendicular to the solar direction.

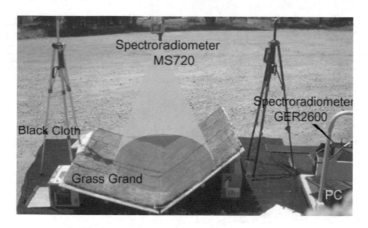

Fig. 2. Outlook of the experiment

2.3   Measuring Instruments

Two spectral radiometers, MS-720 and GER-2600 are used for experiments. The wavelength coverage MS-720 ranges from 350 to 1050 nm while that of GER-2600 ranges from 350 nm to 1050 nm with 1.5 nm interval and from 1050 nm to 2500 nm with 11.5 nm interval.

Wavelength coverages of the MS-720 and GER-2600 are overlapped so that it is possible to check a confidentiality of the measuring instruments. Also, down-welling solar irradiance can be measured and confirmed with these two measuring instruments.

2.4   Experimental Results

Four wavelengths of 750 nm, 800 nm, 900 nm, and 1000 nm in the near infrared region where the reflectance is large are selected, and the spectral radiance associated with the change in the tilt angle is obtained and shown in Fig. 3(a) to (e) for the Case No. 1 to 5, respectively. These wavelengths are common to the MS-720 and GER-2600. Generally, up-welling radiance is increased with increasing of the slope angle and then it is decreased with increasing of the slope angle.

---

[4] https://spectravistacorp.fm.alibaba.com/product/11012876-10723430/Spectroradiometer_GER_2600.html.

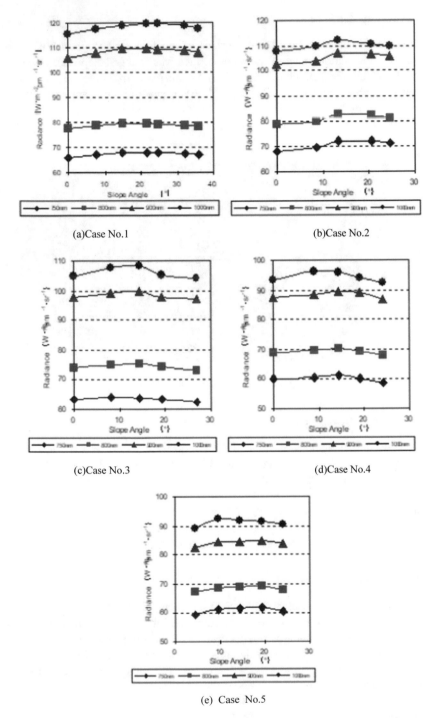

(a)Case No.1

(b)Case No.2

(c)Case No.3

(d)Case No.4

(e) Case No.5

Fig. 3. TOP radiance change depending on slope angles in five experiment cases

Table 2. Maximum radiance change ratio (%)

No.	750 nm	800 nm	900 nm	1000 nm
Case 1	2.61	2.27	3.05	3.22
Case 2	5.40	4.88	4.37	3.96
Case 3	2.84	2.88	1.94	3.92
Case 4	3.81	3.28	2.84	3.69
Case 5	3.66	2.86	2.86	3.75

Not so large up-welling radiance change is observed for Case No. 1 because both of planes are almost flat grasses. Within slope angle ranges from 0 to 25°, multiple reflection between plane A and B is getting large. Therefore, up-welling radiance is also getting large. Meanwhile, it is getting down because multiple absorption is greater than that of reflection for the slope angle ranges from 25 to 36°. Although these trends almost similar for all the cases, change ratios are different.

Maximum radiance change ratio of each case is shown in Table 2. Maximum radiance change ratio of the case No. 2 is greatest followed by No. 4, No. 5, No. 3, and No. 1. Influence due to multiple reflection and absorption for the case No. 2 is greatest. Therefore, it is concluded that the spherical shape of trees has much great multiple reflection effect rather than that of coniferous cone shaped trees.

2.5 Validation of Experimental Results with Monte Carlo Ray Tracing: MCRT

In order to validate the experimental results, MCRT simulation is conducted. MCRT allows simulation of polarization characteristics of ground surface with designated parameters of the atmospheric conditions and ground surface and the conditions of ground cover targets. Illustrative view of MCRT is shown in Fig. 4.

Photons from the sun are input from the top of the atmosphere (the top of the simulation cell) with the solar constant[5] of 1361 W/m^2. Travel length of the photon is calculated with optical depth[6] of the atmospheric molecule[7] and that of aerosol. Aerosol is a system of particles of ultra-microscopic size dispersed in a gas. There are two components in the atmosphere; molecule and aerosol particles. Two inclined

---

[5] https://en.wikipedia.org/wiki/Solar_constant The solar constant includes all types of solar radiation, not just the visible light. It is measured by satellite as being 1.361 kilowatts per square meter (kW/m^2) at solar minimum and approximately 0.1% greater (roughly 1.362 kW/m^2) at solar maximum.

[6] https://en.wikipedia.org/wiki/Optical_depth n physics, optical depth or optical thickness, is the natural logarithm of the ratio of incident to transmitted radiant power through a material, and spectral optical depth or spectral optical thickness is the natural logarithm of the ratio of incident to transmitted spectral radiant power through a material. Optical depth is dimensionless, and in particular is not a length, though it is a monotonically increasing function of optical path length, and approaches zero as the path length approaches zero. The use of the term ``optical density' for optical depth is discouraged.

[7] https://www.eurekalert.org/multimedia/pub/9482.php.

planes are situated and spherical shape and cone shape of trees and grasses are also situated on the planes as shown in Table 1. The number of photons is corresponding to the radiance (up-welling) and the irradiance (down-welling).

When the photon meets molecule or aerosol (the meeting probability with molecule and aerosol depends on their optical depth), then the photon scattered in accordance with scattering properties of molecule and aerosol. The scattering property is called as phase function 1. In the visible to near infrared wavelength region, the scattering by molecule is followed by Rayleigh scattering law[8] while that by aerosol is followed by Mie scattering law[9].

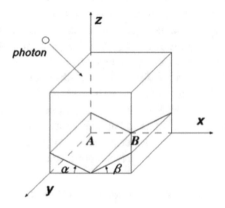

Fig. 4. Definition of MCRT

For simplifying the calculations of the atmospheric influences, it is assumed that the atmosphere containing only molecules and aerosols. Thus the travel length of the photon at once, L is expressed with Eq. (1).

$$L = L_0 \, \text{RND}(i) \tag{1}$$

$$L_0 = Z_{max}/\tau \tag{2}$$

where Z_{max}, τ, RND(i) are maximum length, altitude of the atmosphere, optical depth, and i-th random number, respectively.

Mersenne Twister method[10] is used for generation of random number. Pseudo-random numbers are mainly generated and used on computers. Therefore, the speed generated on the computer, the simplicity of programming, and the size of the storage area used are the points of evaluation of the random number generator, as well as the

---

[8] https://en.wikipedia.org/wiki/Rayleigh_scattering.

[9] https://en.wikipedia.org/wiki/Mie_scattering.

[10] http://random.ism.ac.jp/info01/random_number_generation/node6.html.

quality and period length of the random numbers generated. Mersenne Twister has a high reputation for these being well-balanced. In this equation, τ is optical depth of molecule or aerosol.

The photon meets molecule when the random number is greater than τ. Meanwhile, if the random number is less than τ, then the photon meats aerosol. The photon is scattered at the molecule or aerosol to the direction which is determined with the aforementioned phase function and with the rest of the travel length of the photon.

Table 3 shows MCRT simulation condition.

Table 3. MCRT simulation conditions

Reflectance A	0.08
Reflectance B	1.14
Optical Depth (Aerosol)	0.35
Optical Depth (Molecule)	0.14
Optical Depth (Ozone)	0.009
Optical Depth (Water)	0.001
Wavelength	750 nm

Table 4 shows the simulation and the experimental results of the Top of the Atmosphere: TOA radiance (at sensor radiance).

Table 4. TOA radiance derived from the MCRT simulation and the experiment

Case	TOA (Multiple reflection/absorption)	TOA (Direct)	TOA (Experiment)
No. 1	56.8	56.2	66.0
No. 2	57.4	68.1	68.0
No. 5	55.7	52.2	59.0

The Case No. 1, 2 and 5 are simulated based on MCRT. Namely, there are three types of cases, grass plane and grass plane (No. 1), No. 1 tree plane and No. 1 tree plane (No. 2) as well as No. 2 tree plane and No. 2 tree plane (No. 5). One of the merits of MCRT is that direct component, multiple reflection and absorption components can be discriminated. Therefore, Table 4 shows three of TOA radiance derived from the experiment, MCRT simulation derived TOA radiance due to direct component and MCRT simulation derived TOA radiance taking into account of multiple reflection and absorption component. The difference of TOA radiance derived from the experiment, MCRT simulation with and without consideration of multiple reflection and absorption on the ground ranges from 0.1 to 9%. Also, it is found that the MRCT simulation allows component based estimation of TOA radiance, namely, direct component and multiple reflection and absorption can be discriminated.

3 Conclusion

The radiance at the top of the simulated forest modeling V-shaped mountain topography is measured. The influence of the inclination angle of the ground surface on the atmospheric radiance is quite obvious, and the maximum rate of change is about 4 to 5%. Therefore, it would be better to take into account these effects when visible to near infrared radiometer data of forested areas is analyzed. It is thought that it is mainly due to multiple reflection between both sides. Estimation of vegetation parameters by radiative transfer model based on Monte Carlo method of simulated sloping forest based on the experiment result is continued this research.

The difference of TOA radiance derived from the experiment, MCRT simulation with and without consideration of multiple reflection and absorption on the ground ranges from 0.1 to 9%. Also, it is found that the MRCT simulation allows component based estimation of TOA radiance, namely, direct component and multiple reflection and absorption can be discriminated.

Future Works
Although the proposed method is validated with MCRT simulation, further study is required for validation of the proposed method with the actual visible to near infrared radiometer onboard remote sensing satellites.

Acknowledgements. Author would like to thank Dr. Ding YaLiu for her valuable discussions and preparation of experimental data.

References

1. Ding, Y., Arai, K.: Monte-Carlo simulation for radiative transfer modeling of forested areas and the atmosphere. Res. Note Sci. Eng. Fac. Saga Univ. **35**(1), 35–40 (2006)
2. Arai, K., Ding, Y.: Reflectance model of vegetation in forested areas taking into account multiple reflection between trees. J. Jpn. Soc. Photogramm. Remote Sens. **45**(6), 56–66 (2006)
3. Ding, Y., Arai, K.: Forest parameter estimation by means of Monte Carlo simulations with experimental considerations - estimation of multiple reflections among trees depending on forest parameters. Adv. Space Res. **43**(3), 438–447 (2009)
4. Arai, K., Kawaguchi, T.: Adjacency effect taking into account layered clouds based on Monte Carlo method. J. Remote Sens. Soc. Jpn. **21**(2), 179–185 (2001)
5. Arai, K., Kawaguchi, T.: Adjacency effect of layered clouds taking into account phase function of cloud particles and multi-layered plane parallel atmosphere based on Monte Carlo method. J. Jpn. Soc. Photogramm. Remote Sens. **40**(6), 38–47 (2001)
6. Arai, K.: Adjacency effect of layered clouds estimated with Monte-Carlo simulation. Adv. Space Res. **29**(19), 1807–1812 (2002)
7. Arai, K., Ding, Y.: Reflection model of forest vegetation considering the influence of multiple reflections among trees. Photogramm. Remote Sens. **45**(6), 56–66 (2006)
8. Ding, Y., Arai, K.: Estimation of tree type, shape, tree distance using radiative transfer model based on Monte Carlo model. J. Remote Sens. Soc. Jpn. **27**(2), 141–152 (2007)
9. Arai, K.: Non-linear mixture model of mixed pixels in remote sensing satellite images based on Monte Carlo simulation. Adv. Space Res. **41**(11), 1715–1723 (2008)

10. Arai, K., Harsono, T., Basuki, A.: Micro traffic simulation with unpredictable disturbance based on Monte Carlo simulation: effectiveness of the proposed agent cars of Sidoarjo hot mudflow disaster. J. Emitter **1**(1), 1–10 (2010)
11. Arai, K.: Monte Carlo ray tracing simulation for bi-directional reflectance distribution function and grow index of tealeaves estimation. Int. J. Res. Rev. Comput. Sci. **2**(6), 1313–1318 (2011)
12. Arai, K.: Monte Carlo simulation of polarized atmospheric irradiance for determination of refractive index of aerosols. Int. J. Res. Rev. Comput. Sci. **3**(4), 1744–1748 (2012)
13. Arai, K., Terayama, Y.: Monte Carlo ray tracing simulation of polarization characteristics of sea water which contains spherical and non-spherical shapes of suspended solid and phytoplankton. Int. J. Adv. Comput. Sci. Appl. **3**(6), 85–89 (2012)
14. Arai, K.: Monte Carlo based non-linear mixture model of earth observation satellite imagery pixel data. Int. J. Adv. Comput. Sci. Appl. **3**(8), 18–22 (2012)
15. Arai, K.: Monte Carlo ray tracing based sensitivity analysis of the atmospheric and oceanic parameters on the top of the atmosphere radiance. Int. J. Adv. Comput. Sci. Appl. **3**(12), 7–13 (2012)
16. Arai, K.: Monte Carlo ray tracing based nonlinear mixture model of mixed pixels in Earth observation satellite imagery data. Int. J. Adv. Comput. Sci. Appl. **4**(1), 148–152 (2013)

A Method to Input Secret Information Using an Eye Tracking Device

Hazuki Owada, Daiki Kamitai, Chinayo Shonen Inoue,
and Manabu Okamoto(✉)

Kanagawa Institute of Technology, Atsugi, Kanagawa, Japan
manabu@nw.kanagawa-it.ac.jp

Abstract. Passwords are an authentication method used for identity verification in secure information services. However, people with physical disabilities may struggle to input a password because they cannot use a normal keyboard to input a password. Other forms of password input exist, but these come with security drawbacks. People with disabilities can use voice recognition to enter their password, but this is vulnerable to bystanders eavesdropping on them. Additionally, people who cannot move their arms can enter their password by pressing the touchscreen with their toes. When you use a touchscreen, a person nearby can look at the display and recognize what you input. Instead, we propose an input method using an eye-tracking system to keep the passwords secret. Eye-tracking enables a person to control a device using their eye movements. Using this method, people who cannot move their limbs can input a password securely by gazing and blinking. This method of password protection is robust against shoulder surfing, which can occur when the person turns their back on a bystander.

Keywords: Security · Password · Eye-tracking · Authentication · People with physical disabilities

1 Introduction

Many web services require authentication for access. A correctly inputted user ID and password are commonly required to use the service or to access content. However, for people with physical disabilities who cannot use a regular keyboard, entering an ID and a password can be difficult.

The onscreen keyboard [1] enables a user to input keystrokes without a physical keyboard by pressing the image of a keyboard key displayed on the screen. People with physical disabilities who cannot move either of their arms can move a pointer by gripping and manipulating a "mouth stick" [2] bar with their mouth. Other similar support devices such as the use of voice recognition, chin input devices, detection of eyelid movements, or moving a trackball by foot have been proposed for people with physical disabilities.

Unfortunately, these password input methods are vulnerable to observation by bystanders, who might be able to see what a user inputs by observing the pointer moving on the screen. This would also be true for input using voice recognition

© Springer Nature Switzerland AG 2020
K. Arai et al. (Eds.): FTC 2019, AISC 1070, pp. 936–943, 2020.
https://doi.org/10.1007/978-3-030-32523-7_69

wherein the bystander can listen to what a user says. When people with physical disabilities enter secret or private information (such as a password) using such assistive technologies, they must ask bystanders to leave the room to create a secure environment wherein no one is watching or listening. However, asking unknown strangers to leave a communal area such as a hospital is not always feasible. Another issue with inputting passwords is that one has to enter it themselves. For other PC operations, individuals with disabilities may be able to ask someone else to type it for them, but for inputting secret information, this can render them vulnerable to security risks.

In this paper, we propose a new method of password input for use by individuals with disabilities. Our proposed method and eye-tracking system can be used by people who cannot fully use their arms or feet or those who for some other reason cannot use a regular keyboard. An eye-tracking system measures eye activity such as looking, ignoring, or blinking, which can be used to interact with a PC. Using our method, a user can input secret information such as a password without the risks associated with other assistive technologies. It is a simple input method that people with physical disabilities can learn and perform easily. Using this method of input, the user is safe from observation by a bystander who might look at the display from behind the user as the information on the display does not reveal any information about a password. Additionally, this method is effective against keyloggers.

2   Method

In this section, we present our proposed eye tracking method, which we also proposed at the Future Technologies Conference 2018 (FTC2018) [3], the International Conference on Soft Computing, Artificial Intelligence and Applications (SAI2018) [4], and SAI 2016 [5]. In the method described herein, the user can securely input a password using a two-choice input device. We use an eye-tracking system as the selection device, which allows the input of a four-digit personal identification number (PIN) as a password. The basic structure of our proposed method is identical to those described in our previous papers [3] and [4], which we quote below.

First, the PC or server generates a 4 × 4 random matrix table and displays it to the user. The matrix comprises the characters A–F, 0, and 1–9 as shown in Fig. 1. The sequence of the characters varies each time. A four-digit hexadecimal number can be used as a PIN.

The matrix table is automatically divided into two sections delineated by a red line: left or right. The user can recognize the two sections, as shown in the image on the left in Fig. 2. The annotations "Section 1" or "Section 2" are not displayed, but the user can imagine them easily. The user selects the section containing the first character in the password by gazing at the section followed by a blinking action as follows. As the user stares at a section as sensed by the eye-tracking system, the user selects Section 1 by blinking the right eye while closing the left eye. To select Section 2, the user blinks the left eye while closing the right eye.

After selection, these two sections move left and right to create a space between them. After selecting 1 or 2, the matrix takes on the appearance of the image on the right in Fig. 2.

Fig. 1. Base matrix.

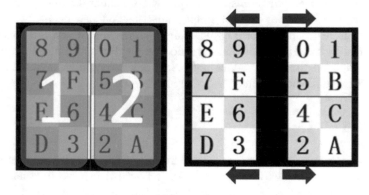

Fig. 2. Action 1.

Each section is further subdivided into two smaller sections by a red line, as shown in Fig. 3. The user selects Section 1 (up) or 2 (bottom) containing the next character in the password to be inputted, after which each subsection moves up and down to create a space between them. In the same manner, the user continues to select the section in which a letter of the password is located, as shown in Figs. 4 and 5.

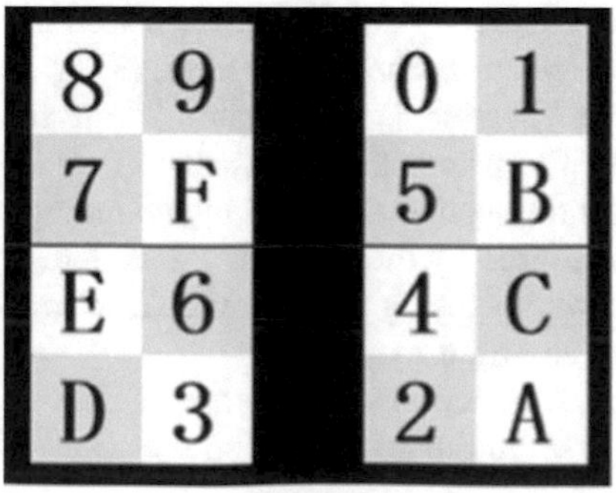

Fig. 3. Action 2.

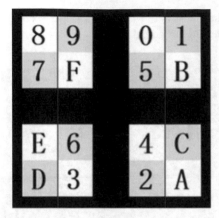

Fig. 4. Action 3.

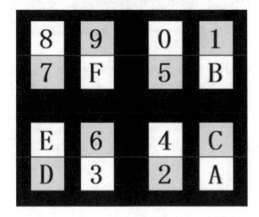

Fig. 5. Action 4.

In this example, when the user wants to input "0," which is the first letter of the password, they select a section such as 2-1-1-1. To input the next letter of the password, the server generates another 4 × 4 random matrix table. The sequences of the characters in the new matrix are different from those of the previous one. When the user finishes entering all characters included in the password and does not need to input any additional characters, they only need to press the Enter key to complete authentication. Because pressing the enter key does not disclose or require any confidential information, the user can ask someone else to press the Enter key instead. Alternatively, instead of the pressing an Enter key, another recognizable eye movement such as closing both eyes for a fixed time can be applied.

Figure 6 shows an example of the input operation. The circle displayed on the screen indicates the location where the user is looking, and the circle in the middle indicates that one of the eyes is closed. This visual clue is useful for training users as they can easily see what the system is tracking. For security purposes, these visual clues can be easily removed for actual input.

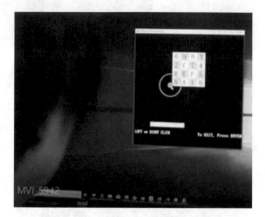

Fig. 6. Example of the input operation.

In this example, we describe a method whereby an eye-tracking system is used as an input device. However, the selection of matrix sections only requires the use of a two-selection input such as up/down, left/right, long-push/short-push, single/double, on/off, or input/nothing (time-out), making this selection method versatile for other input methods.

3   Results

3.1   User Testing

We conducted user tests using our prototype software. The user test conditions are as follows. These tests were also conducted in [3, 4, 5] under similar conditions.

- The subjects comprised eight students at our university without disabilities
- The subject created their four-digit PIN under no time limit.
- The experiment measured the time required to input the password and the authentication success rate.
- Subjects inputted the password for a total of three times, once every other week.
- We measured the time between the moment the character matrix was displayed and the moment the "Enter" key was pressed to indicate the end of the password input.
- Before the actual test, we allowed subjects to practice inputting ten times.
- If subjects made a mistake, we did not measure the time used and allowed them to try again.
- We used the Tobii Eye Tracker [6] as our input device and the Synchrogazer [7] as the driver software.
- We setup these devices at the bottom of the display, as shown in Fig. 7.

Fig. 7. Tobii eye tracker.

Table 1. Input times for a four-digit PIN (unit: s)

Subject	1st	2nd	3rd	Average
A	45.213	43.998	35.868	41.693
B	77.835	65.032	60.194	67.687
C	44.418	49.356	43.121	45.631
D	46.997	50.321	45.213	47.510
E	46.752	69.002	43.998	53.250
F	57.200	63.857	59.197	60.084
G	87.843	77.093	43.121	69.352
H	54.318	48.122	39.622	47.354
Average	57.572	58.347	46.291	54.070

Table 2. Success rate

Rounds	1st	2nd	3rd	Average
Success rate	62.50%	50.00%	75.00%	62.50%

As indicated in Tables 1 and 2, the proposed technique was clearly more time-consuming than inputting with a conventional keyboard. However, we may expect that many users will gradually improve their input skills. Although we could only conduct the test for a short period of time, the input time may be shortened by testing over a longer period.

The issue with our proposed method is its low success rate. We expect that users will gradually improve their skill with the device with practice. However, we found that some people are not good at winking and many are unaware of their lack of winking ability. A person who is good at winking can easily enter their password successfully, but the behavior of those not skilled at winking was often misrecognized. As a result, we think that we need to consider other eye behaviors for selection.

This method enables the use of an eye-tracking device for password input to assist people with physical disabilities who cannot move their limbs and are unable to use conventional keyboards. Even people with severe disabilities can often move their eyes. Therefore, a wide range of people with disabilities can use this method.

This method is a robust way to defend against shoulder surfing, which is the act of stealing private information by watching someone inputting their password, because it prevents the attacker from obtaining the password by simply watching the display, which contains no hints as to what is being selected. By turning toward the screen and away from the observer, bystanders cannot even use the eye motion of the user to figure out the password.

To test the physical security of our method, we conducted an attack test as follows:

- A pair of testers comprising an attacker and a subject.
- The subject entered his or her secret four-digit PIN using our method.
- The attacker could see both the display and the subject's face and could take notes.
- After the subject entered the PIN, the attacker was asked to guess what the subject input.
- The subject indicated whether the guess was correct.
- A total of six attackers/subject pairs took the test.

No attackers were able to correctly guess the subject's PIN because it was difficult for an attacker to see both the screen and the subject's face simultaneously. We expect that is especially true if the user turns their back to the attacker, as the position of the eyes is completely hidden. Of course, if both the screen and the face are captured and analyzed in a video, this method is no longer completely secure. To verify this, we need to perform video capture and examine the results in a future study.

In addition to protection against over the shoulder surfing, this method provides strong protection against keylogging. A keylogger saves only the operations that are used to select a number, not the number or action itself. This makes figuring out a password especially difficult when the matrix in Fig. 1 is changed each time to disassociate numerical values from locations on the screen.

As an additional advantage of this approach, we can use a typical password that includes English letters and numbers. In this study, we use a 4×4 matrix, as shown in Fig. 1; however, we can expand it to be larger, such as an 8×8 matrix comprising both upper-case and lower-case letters and numerals. However, with an 8×8 matrix, our method requires six actions to input one character. Thus, to input a password comprising eight characters, we need to use 48 actions, which is much slower than a standard keyboard and the 4×4 matrix discussed earlier.

4   Conclusion

We have proposed a method to input secret information such as a password with an eye-tracking system to help secure the password against keylogging and shoulder surfing. In our test, we use gaze detection and eye-blinking to input a password. This offers protection against shoulder surfing, as users can easily turn their backs against a bystander and hide their face to prevent the attacker from obtaining the password by capturing the display, which contains no hints.

In a future study, we aim to use various other eye movements to improve this method, such as gazing somewhere other than the matrices on the screen to benefit people who are not good at winking. In addition, we need to conduct long-term testing to confirm the usefulness of this method.

Acknowledgment. This work was supported by JSPS KAKENHI Grant Number JP17K00194.

References

1. Microsoft screen keyboard. https://www.microsoft.com/ja-jp/enable/products/windows7/onscreenkeyboard.aspx. Accessed 13 Feb 2019
2. Soft Fit mouse stick. https://toksoamor-com.ssl-xserver.jp/32_427.html. Accessed 13 Feb 2019
3. Manabu, O.: Input password method with simple device. IN: FTC 2018, Vancouver (2018)
4. Momose, R., Miyu, S., Manabu, O.: Password input method using simple device. In: SAI 2018, "Intelligent Computing," pp. 1244–1252. Springer, London (2018)
5. Ito, A., Kumazawa, Y., Okamoto, M.: Input password method for handicapped people. In: SAI Computing Conference (SAI) (2016)
6. Tobii Eye Tracker. https://www.tobiipro.com. Accessed 13 Feb 2019
7. Synchrogazer software. https://www.baku-dreameater.net/synchrogazer. Accessed 13 Feb 2019

Web-Based Learning for Enhancing CSL Learners' Language Proficiency in Singapore

Liu May[✉]

Singapore Centre for Chinese Language, Nanyang Technological University,
Singapore, Singapore
may.liu@sccl.sg

Abstract. In language learning, the cognitive development of second language learners is often higher than their language level. How to bridge the gap between the two is a topic worthy of further discussion. This paper will demo how to use collaborative technologies such as collaborative post-it, mind mapping and co-writing. It takes Singaporean primary five students in learning Chinese as an example. When applying the aforementioned methods, this paper shows how to design a teaching sequence among the three collaborative tools. While implementation, this paper also shows how to interact among teacher-students and students-students. The scaffolding instruction strategies will be provided to meet the CSL learners' need. This paper helps language teachers to design web-based learning effectively using collaborative technologies and tools to enhance CSL learners' both cognitive level and language proficiency.

Keywords: Chinese as a Second Language (CSL) · Collaborative technologies · Primary students

1 Introduction

Collaborative learning has acquired a new dimension with the widespread use of information and communication technologies [1]. The main advantages of using technology for language learning are a greater exposure to authentic language, access to a wide range of sources of information and to different varieties of language, opportunities for interaction and communication and more intensive learner participation [2]. As ACTFL [3] states: "The use of technology is not a goal in and of itself; rather technology is one tool that supports language learners as they use the target language in culturally appropriate ways to accomplish authentic tasks. Further, all language learning opportunities whether facilitated through technology or in a classroom setting should be standards-based, instructor-designed, learner-centered, and aimed at developing proficiency in the target language through interactive, meaningful, and cognitively engaging learning experiences" [3]. In this sense, students learn alongside someone they could talk to and collaborate with – the web-based collaborative technology can help students do their best learning with their peers.

© Springer Nature Switzerland AG 2020
K. Arai et al. (Eds.): FTC 2019, AISC 1070, pp. 944–949, 2020.
https://doi.org/10.1007/978-3-030-32523-7_70

2   Purpose, and Background/Significance

Purpose: In this paper, the author examines how to use collaborative technologies and tools such as collaborative post-it (sticky notes, e.g., Padlet), mind mapping (e.g., Coggle, FreeMind and Sketchboard) and co-writing (e.g., Google Docs, Draft) as well as how the steps of the process (the teaching sequence) are related in order to develop CSL learners' both cognitive ability and Chinese language.

Background/Significance: In language learning, the cognitive development of second language learners is often higher than their language level. How to bridge the gap between the two is a topic worthy of further discussion. Some researchers [4–6] explain that collaborative activities make students more responsible in collaborative writing tasks, especially because they have joint responsibility. Storch [4] points out that "This may promote a sense of co-ownership and hence encourage students to contribute to the decision making on all aspects of writing: content, structure, and language." Although it is shown in literature that there is much evidence to support the occurrence of learning in connection with collaboration, both in face-to-face and online teaching situations, but much less research on how such collaboration develops over the duration of a course. The learning process entails concepts such as collective learning, information acquisition, technological skills and processes of meaning negotiation. This is the reason why collaborative activities should take advantage of technology, as it facilitates collaborative activities. This paper aims at showing how to integrate collaborative technologies and tools into CSL learning at primary school in Singapore.

3   Research Questions

1. How to design a web-based learning via the collaborative technologies and tools for CSL learners?
2. Do students enhance their writing ability through web-based collaborative learning?

4   Research Methodology

4.1   Experimental Design

This study will use a mixed method design involving both quantitative and qualitative approaches. For quantitative approach, it uses the Pretest-Posttest Equivalent Group Design as shown in Table 1. Data in terms of writing will be collected. It is assumed that the Experimental and Comparison Groups will be equivalent in writing ability to begin with. It is expected that $O1 < O3$ and that $O2 = O4$. Effect sizes will be estimated and evaluated by using Cohen's criteria. As randomization is unlikely in the school context, in case the groups are found to be non-equivalent in the beginning, gains-analysis will be used.

Table 1. Two groups, Nonrandom Selection, Pre-, Post-test

Group	Pre-test	Treatment	Post-test
Experimental group = E	O1	X	O3
Comparison Group = C	O2		O4

4.2   Samples

Cluster sampling will be used to select the students as random sampling is impractical in the school context. The sample size (N = 120 each group) is adequate for an experimental study, with an expected effect size of 0.5, statistical power of 0.8, and probability of 0.05. The samples will be taken from the Government Schools as these are the main bulk of schools, and the students will be distributed to as many schools as practicably possible to ensure representativeness. An effort will also be made to keep a balance between boys and girls as it is well-known that there is a gender difference in language development as well as interest in favor of girls.

4.3   Research Design

This intervention addresses to assist online collaborative writing tasks in Primary five students. It aims to improve both students' meaningful learning of Chinese language and student's skills of individual and collaborative writing. For so doing, the intervention combines in diverse manners the use of different kinds of supports and scaffolds (post-it, and mind map) all along the process of collaborative writing.

Step 1. Brainstorming via the Collaborative Post-it
Firstly, the teacher creates a collaborative post-it for each small group with a title and introduction. The students in a small group brainstorm their creative ideas and post them onto the collaborative post-it. In the space of collaborative post-it, student can post the multimedia files, such as text, images (photos), URLs of video and website. Students can also change the font size, font color and background color to facilitate classification and help learners' awareness and memory (see Fig. 3).

When engaging in interactive activities, teachers can follow the principle of "first student-student interaction (see Fig. 1) and then teacher-student interaction" (see Fig. 2).

Fig. 1. Student-student interaction via collaborative post-it.

Fig. 2. Teacher-student interaction via collaborative post-it.

The advantage of collaborative post-it is to mobilize the knowledge of the second-language learners through brainstorming; for learners who have no prior knowledge, they can also supplement their knowledge by learning from each other. Therefore the collaborative post-it plays an important role in the learning of L2 learners, helping them to collectively brainstorm and after that, it can help students to transit to individual divergent thinking.

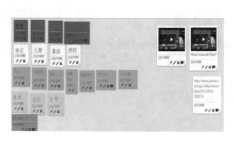

Fig. 3. Changing the background color.

Fig. 4. The strategy scaffolding via collaborative post-it.

Figure 4 shows the strategy scaffolding via collaborative post-it. Taking the using mobile phone as an example, the teacher uses the video clip as a stimulus and attaches it along with the guiding questions. The student can use the "comment" function on the post-it to answer the questions.

Step 2. Sorting out their Creative Ideas via the Mind-map
After brainstorming, students in the same small group post their creative ideas onto the mind-map to sort out the ideas with hierarchical distinction and interrelatedness (see Figs. 5 and 6).

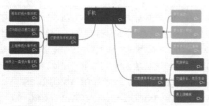

Fig. 5. Sorting out the creative ideas via the mind-map.

Fig. 6. The strategy scaffolding via collaborative mind-map.

When students visualize their own thinking and share their experiences via mind-map, it becomes easier for them to mutual learning between peers.

The most difference of the functions between the collaborative post-it and the mind map is that the former can be used to generate creative ideas, while the later is used to sort out the ideas. With the support of collaborative technologies and tools, everyone's thinking becomes visual and can be shared with each other.

Step 3. Completing Articles via Co-writing

Finally, students complete their article via co-writing. Collaborative writing allows a group of students to add content, edit or comment on documents in real time. Using this feature to design a writing experience for students, it can help students to build knowledge together, and develop writing skills and social interactions.

The functions of the web-based co-writing tool can display the writing histories, the co-writers and the contributors, as well as save the changes, and export PDF files, etc.

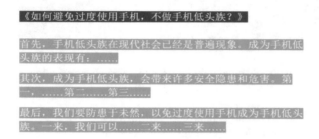

Fig. 7. The strategy scaffolding via co-writing.

Figure 7 shows the strategy scaffolding via co-writing. Taking the using mobile phone as an example, the teacher writes the title and the opening sentence of the three paragraphs, as well as the specific requirements. For instance, students are requires to write the situations of using mobile phone at the first paragraph; and then to write the safety concerns and harms at the second paragraph; finally to write the suggestions to using mobile phone at the last paragraph. There are three strategies for collaborative writing: (1) free selecting writing tasks: the team members are free to choose the part to write in accordance with the above-mentioned writing framework; (2) teacher's assignment according to the levels of Chinese proficiency of students as well as the difficulty of the paragraph; (3) jigsaw method: The teacher sets up a panel of experts (homogeneous grouping) in the collaborative writing space. For example, those with lower Chinese proficiency only need to write the first paragraph together, and so on. After completing the task in each expert group, the teacher resets the study group (heterogeneous grouping) in the collaborative writing space, and the team members complete the whole article together. After the practice of co-writing, the teacher can ask each student to write an article by him/herself. Differentiated instruction will be used for assigning the writing task. For instance, for the students with lower Chinese ability, they can complete their own article by imitating the co-writing article. For students with medium Chinese ability, they can rewrite the article. For students with higher Chinese ability, they can create a related article in a new way. Therefore, the affordance of collaborative technology tools allow each student to develop his/her Chinese language proficiency according one's ability and by one's pace under the strategy scaffolding.

5   Results and Conclusions

The results of this paper reveal how to design the collaborative activities and technology for language learning as a way to enhance the writing ability of primary five students in Singapore. Learners will be engaged in the meaningful collaborative activities and tasks that require the use of technological tools and in which students are able to communicate with their peers in an interactive and specific context.

This paper helps language teachers to design web-based learning effectively using collaborative technologies and tools to enhance CSL learners' both cognitive level and language proficiency. This paper can be used in pre-, and in service teachers' training program. It also can serve as a guide for designing and implementing web-based learning for language teachers.

References

1. McCafferty, S.G., Jacobs, G.M., DaSilva Iddings, A.C. (eds.): Cooperative Learning and Second Language Teaching. Cambridge University Press, New York (2006)
2. Carrió-Pastor, M.L.: Enhancing learner-teacher collaboration through the use of on-line activities. In: González-Pueyo, C.I., Foz-Gil, M., Siso, J., Luzón Marco, M.J. (eds.) Teaching Academic and Professional English Online. Peter Lang, Berlin (2009)
3. ACTFL   Homepage.   https://www.commonsense.org/education/top-picks/best-student-collaboration-tools. Accessed 31 May 2019
4. Stoch, N.: Collaborative writing: product, process, and students' reflections. J. Second Lang. Writ. **14**, 153–173 (2005)
5. Woodrich, M., Fan, Y.: Google docs as a tool for collaborative writing in the middle school classroom. J. Inf. Technol. Educ. Res. **16**, 391–410 (2017)
6. Humphris, R.: Developing students as writers through collaboration. Changing English: Studies in Culture and Education **17–2**, 201–214 (2010)

Deep Siamese Networks with Bayesian Non-parametrics for Video Object Tracking

Anthony D. Rhodes[1,2(✉)] and Manan Goel[2]

[1] Portland State University, Portland, OR, USA
arhodes@pdx.edu
[2] Intel Corporation, Santa Clara, CA, USA

Abstract. We present a novel algorithm utilizing a deep Siamese neural network as a general object similarity function in combination with a Bayesian optimization (BO) framework to encode spatio-temporal information for efficient object tracking in video. In particular, we treat the video tracking problem as a dynamic (i.e. temporally-evolving) optimization problem. Using Gaussian Process priors, we model a dynamic objective function representing the location of a tracked object in each frame. By exploiting temporal correlations, the proposed method queries the search space in a statistically principled and efficient way, offering several benefits over current state of the art video tracking methods.

Keywords: Video tracking · Gaussian processes · Bayesian optimization · Siamese network · Computer vision · Deep learning

1 Introduction

The problem of tracking an arbitrary object in video, where an object is identified by a single bounding-box in the first frame, requires both a robust similarity function and an efficient method for querying plausible locations of the object in subsequent frames. Early video tracking approaches have included feature-based approaches and template matching algorithms [1] that attempt to track specific features of an object or even the object as a whole.

Feature-based approaches use local features, including points and edges, keypoints [2], SIFT features [3], HOG features [4] and deformable parts [5]. Conversely, template-based methods take the object as a whole – offering the potential advantage that they treat complex templates or patterns that cannot be modeled by local features alone. Through the course of a video, an object can potentially undergo a variety of different visual transformations, including rotation, occlusion, changes in scale, illumination changes, etc., that pose significant challenges for tracking. In order to obtain a robust template matching for video tracking, researchers have developed a host of methods, including mean-shift [6] and cross-correlation filtering which entails convolving a template over a search region; significant advances to cross-correlation filtering for video tracking include MOSSE [7] adaptive correlation filter and the MUSTer algorithm [8] which draws influence from cognitive psychology in the design

© Springer Nature Switzerland AG 2020
K. Arai et al. (Eds.): FTC 2019, AISC 1070, pp. 950–958, 2020.
https://doi.org/10.1007/978-3-030-32523-7_71

of a flexible object representation using long and short-term memory stored by means of an integrated correlation filter.

More recently, deep learning models have been applied to video tracking to leverage the benefits of learning complex functions from large data sets. While deep models offer the potential of improved robustness for tracking, they have nevertheless presented two significant challenges to tracking research to date. First, many deep tracking models are too slow for practical use due to the fact that they require online training, and, second, many deep trackers, when trained offline, are based on classification approaches, so that they are limited to class-specific searches and frequently require the aggregation of many image patches (and thus many passes through the network) in order to locate the object [9]. In light of these difficulties, several contemporary state of the art deep learning-based tracking models have been developed as generic object trackers in an effort to obviate the need for online training and to also improve the generalizability of the tracker. Author in [10] applies a regression-based approach to train a generic tracker, GOTURN, offline to learn a generic relationship between appearance and motion; several deep techniques additionally incorporate motion and occlusion models, including particle filtering methods [11] and optical flow [12].

Author in [13] demonstrated the power of deep Siamese networks (see Sect. 2.1) based on [14], achieving a new state of the art for generic object matching for video tracking. Remarkably, the SINT algorithm delivered state of the art performance despite the fact that it was not equipped with any model updating, no occlusion detection, and no explicit geometric or feature matching components. Authors in [15, 18] extended this work to achieve state of the art Siamese-based tracking while operating at frame rates beyond real-time by exploiting a fully-convolutional network structure.

Even with these recent successes in video object tracking, there nevertheless exists a void in state-of-the-art video tracking workflows that fully integrate deep learning models with classical statistics and machine learning approaches. Most state of the art video trackers lack – for instance – a capacity to generate systematic "belief" states (e.g. through explicit error and uncertainty measures), or ways to seamlessly incorporate contextual and scene structure, or to adaptively encode temporal information (e.g. by imposing intelligent search stopping conditions and bounds) and the ability to otherwise directly and inferentially control region proposal generation or sampling methods in a precise and principled way. To this end, we believe that the fusion of deep models with classical approaches can provide a necessary incubation for "intelligent" computer vision systems capable of high-level vision tasks in the future (e.g. scene and behavior understanding).

In the current work we present the first integrated dynamic Bayesian optimization framework in conjunction with deep learning for object tracking in video.

2   Siamese Networks

We adopt the Siamese network-based approach for one-shot image recognition from [15] to learn a generic, deep similarity function for object tracking. The network learns a function $f(z, x)$ that compares an exemplar crop z to a candidate crop x and returns a high score if the two images depict the same object and a low score otherwise. For computer vision tasks, a natural candidate for the similarity function f is a deep conv-net [16, 17]. Following [14, 15], a Siamese network applies an identical transformation φ to both input image crops and then combines their representations using another function g that is trained to learn a general similarity function on the deep conv-net features, so that $f(z, x) = g(\varphi(z), \varphi(x))$.

The network is trained on positive and negative pairs, using logistic loss:

$$l(y, v) = log(1 + exp(-yv)) \tag{1}$$

where v is the real-valued score of an exemplar-candidate pair and $y \in \{-1, +1\}$ is its ground-truth label. The parameters of the conv-net θ are obtained by applying Stochastic Gradient Descent (SGD) to:

$$argmin_\theta E_{(z,x,y)}[l(y, f(z, x; \theta))] \tag{2}$$

where the expectation is computed over the data distribution.

Pairs of image crops were obtained using annotated videos from the 2015 edition of ImageNet for Large Scale Visual Recognition Challenge (ILSVRC); images were extracted from two different frames, at most a distance of T frames apart; positive image exemplars were defined as a function of their center offset distance from the ground-truth and the network stride length. Image sizes were normalized for consistency during training.

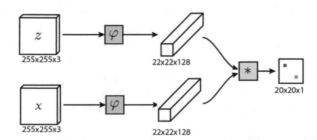

Fig. 1. The Siamese network φ takes the exemplar image z and search image x as inputs. We then convolve (denoted by *) the output tensors to generate a similarity score. Similarity scores for a batch of sample search images are later rendered in a 20 × 20 × 1 search grid using a Gaussian Process.

We use a five-layer conv-net architecture, with pooling layers after the first and second layers, and stride lengths of 2 and 1 throughout. The final network output is a $22 \times 22 \times 128$ tensor, as shown in Fig. 1.

3   Dynamic Bayesian Optimization

Author in [19] define object tracking in video as a dynamic optimization problem (DOP):

$$DOP = \{\max f(\boldsymbol{x}, t) \, s.t. \, x \in F(t) \subseteq S, t \in T\} \tag{3}$$

where $S \in \mathbb{R}^D$, with S in the search space; $f : S \times T \to \mathbb{R}$ is the temporally-evolving objective function which yields a maximum when the input x matches the ground-truth of the target object; $F(t)$ is the set of all feasible solutions $x \in F(t) \subseteq S$ at time t.

Bayesian optimization is a sequential framework for optimizing an unknown, noisy and/or expensive objective function $f(\boldsymbol{x}, t)$. BO works in two key stages: first, we generate a surrogate model to learn a latent objective function from collected samples; next, we determine plausible points to sample from the objective function in the search space. In the present work we use Gaussian Process Regression (GPR) to render the surrogate model. The second phase involves a secondary optimization of a surrogate-dependent *acquisition function* $a(\boldsymbol{x}, t)$, which strikes a balance between exploring new regions in the search space and exploiting information obtained from previous samples of the objective function. Common choices of acquisition functions include *expected-improvement* (EI) and *probability of improvement* (PI) functions [20]. We devise a novel acquisition function, which we call *memory-score expected-improvement* (MS-EI) that demonstrated superior performance to EI and PI on our experimental data. We define MS-EI as:

$$MS - EI(\boldsymbol{x}) = (\mu(\boldsymbol{x}) - f(\boldsymbol{x}^*) - \xi)\Phi(Z) + \sigma(\boldsymbol{x})\rho(Z) \tag{4}$$

where $Z = \frac{\mu(\boldsymbol{x}) - f(\boldsymbol{x}^*) - \xi}{\sigma(\boldsymbol{x})}$, $\boldsymbol{x}^* = argmax f(\boldsymbol{x})$, Φ and ρ denote the pdf and cdf of the standard normal distribution respectively. We define $\xi = \left(\alpha \cdot meanf(\boldsymbol{x})_D \cdot n^q\right)^{-1}$; where α and q are tunable parameters that depend on the scale of the objective function (we use $\alpha = 1$, $q = 1.1$); D denotes the sample data set, and n is the sample iteration number, with $|D| = n$; $meanf(\boldsymbol{x})_D$ is the sample mean of the previously observed values. Here ξ serves to balance the exploration-exploitation trade-off to the specificity of a particular search. In this way, MS-EI employs a cooling schedule so that exploration is encouraged early in the search; however, the degree of exploration is conversely dynamically attenuated for exploitation as the search generates sample points with larger output values.

3.1   Gaussian Processes

A Gaussian Process (GP) defines a prior distribution over functions with a joint Normality assumption. We denote $\hat{f}$, the realization of the Gaussian process: $\hat{f} \sim GP(\mu, K)$. Here the GP is fully specified by the mean $\mu : X \to \mathbb{R}$ and covariance $K : X \times X \to \mathbb{R}$, $K((x,t),(x',t')) = E[(\hat{f}(x,t) - \mu(x,t))(\hat{f}(x',t') - \mu(x',t'))]$, where $K(\cdot,\cdot) \leq 1$ and $X = S \times T$. See [20] for further details.

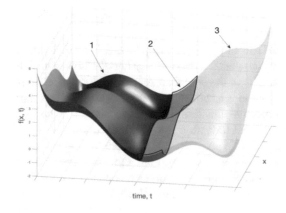

Fig. 2. Illustration of $\hat{f}(x,t)$ for DOP: Region (1) shows previous sample instances for time instances prior to time t; region (2) depicts the bounded region of the search at time t; region (3) represents future time slices. Image credit: [22].

3.2   Dynamic Gaussian Processes

Following [21], we model a DOP $f(x,t)$ as a spatio-temporal GP where the objective function at time t represents a slice of f constrained at t (see Fig. 2). This dynamic GP model will therefore encapsulate statistical correlations in space and time; furthermore, the GP can enable tracking the location of an object, expressed as the temporally-evolving maximum of the objective function $f(x,t)$.

Let $\hat{f}(x,t) \sim GP(0, K(\{x,t\},\{x',t'\}))$, where $(x,t) \in \mathbb{R}^3$ (x is the bounding-box spatial location), and K is the covariance function of the zero-mean spatio-temporal GP. For simplicity, we assume that K is both stationary and separable of the form [21]:

$$K(\hat{f}(x,t),\hat{f}(x',t')) = K_S(x,x') \cdot K_T(t,t') \tag{5}$$

where K_S and K_T are the spatio and temporal covariance functions, respectively. We use Matérn kernel functions [20] in experiments and train the spatial and temporal covariance functions independently, following our separable assumption.

3.3 Siamese-Dynamic Bayesian Tracking Algorithm

We now present the details of our *Siamese-Dynamic Bayesian Tracking Algorithm* (SDBTA). The algorithm makes use of the previously-described deep Siamese convnet. In the first step, we train the dynamic GP model. Then, for each current frame t in the video containing T total frames (consider $t = 0$ the initial frame containing the ground-truth bounding-box for the target object), we render the GPR approximation over a resized search grid of size $d \times d$ (we use $d = 20$ for computational efficiency), and then subsequently apply upscaling (e.g. cubic interpolation) over the original search space dimensions. In order to allow our algorithm to handle changes in the scale of the target object, each evaluation of an image crop is rendered by the Siamese network as a triplet score, where we compute the similarity score for the current crop compared to the exemplar at three scales: $\{1.00 - p, 1.00, 1.00 + p\}$, where we heuristically set $p = 0.05$. The remaining algorithm steps are straightforward and detailed below.

4 Experimental Results

We tested our algorithm using a subset of the VOT14 [27] and VOT16 [28] data sets, the "CFNET" video tracking data set [18], against three baseline video tracking models: template matching using normalized cross correlation (TM) [26] the MOSSE tracker algorithm [7], and ADNET (2017, CVPR), a state of the art, deep reinforcement learning-based video tracking algorithm [25]. For our algorithm, we fixed the number of samples per frame at 80 (cf. region proposal systems commonly rely on thousands of image queries [9]). We report the search summary statistics for IOU (intersection over union) for each model in Table 1. Beyond these strong quantitative tracking results, we additionally observed that the comparison models suffered from either significant long-term tracking deterioration or episodic instability (see Fig. 3). The SDBTA algorithm in general did not exhibit this behavior based on our experimental trials (Algorithm 1).

Table 1. Experimental results for SDBTA on the CFNET video tracking dataset.

	TM	MOSSE	ADNET	SDBTA (ours)
Mean IOU	0.26	0.10	0.47	**0.56**
Std IOU	0.22	0.25	0.23	**0.17**

Algorithm 1 Siamese-Dynamic Bayesian Tracking Algorithm

Train Dynamic GP model
for i **do** = 1,2,...T frames **do**
    for j **do** = 1,2,...{Max iterations per frame} **do**
        Calculate $\{x_i, t_i\}$ = arg max$_{x,t}$ $MS\text{-}EI(x, t)$
        Query Siamese network $y_i \leftarrow f(x_i, t_i)$
        Augment new point to the data
        Render GPR with set $\{y\}$ over $d \times d$ grid
        Upsample grid data to dim. of search space S
        Update current location of optimum over S
    end for
end for

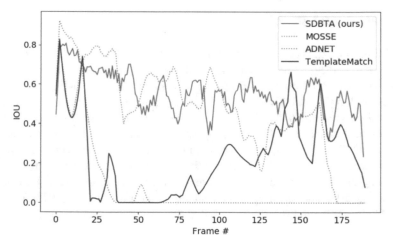

Fig. 3. The graph shows the general stability of the SDBTA tracker for a representative test video, 'tc_boat_ce1' (T = 200 frames); IOU is represented by the vertical axis and the frame number corresponds with the horizontal axis. By comparison, the MOSSE tracker essentially fails to track after frame 30; TM fails to track for nearly half of the duration of the video (frames 25–100); and ADNET fails to track after frame 170.

5   Future Work

While the present algorithm has already demonstrated its effectiveness in video tracking, we nevertheless believe it can be further improved in the near future. We intend to expand the current approach to accommodate the following enhancements: (1) GP-enabled multiscaling (so the GP is generated in five dimensions, including space, size and time); (2) adaptive Bayesian optimization (ABO) which adaptively alters the bounds and sample constraints at each frame for optimizing the acquisition function based on the learned time-related length-scale parameter [22]; (3) we anticipate furthermore that incorporating a fully-convolutional [15] architecture into the Siamese conv-net with our current pipeline will yield faster than real-time video tracking with the added benefits of BNP. Following [23] the current research can be augmented to include visual context models for structured image and video types to be used with video scene and behavior recognition; various numerical optimization techniques can further improve the efficiency and speed of our GP-based video tracking, including [24]. We believe that the current research has significant potential for widespread real-world use, including applications to surveillance, high-level scene understanding in computer vision systems, and a myriad of commercial and consumer-based applications.

References

1. Brunelli, R.: Template Matching Techniques in Computer Vision: Theory and Practice. Wiley Publishing, Hoboken (2009)
2. Lowe, D.G.: Object recognition from local scale-invariant features. In: Proceedings of the International Conference on Computer Vision-Volume 2 - Volume 2 (ICCV 1999), vol. 2, pp. 1150–1157. IEEE Computer Society, Washington, DC (1999)
3. Nebehay, G., Pflugfelder, R.P.: Consensus-based matching and tracking of keypoints for object tracking. In: (IEEE) Winter Conference on Applications of Computer Vision, Steamboat Springs, CO, USA, pp. 862–869 (2014)
4. Dalal, A., Triggs, B.: Histograms of oriented gradients for human detection. In: Proceedings of the 2005 IEEE Computer Society Conference on Computer Vision and Pattern Recognition (CVPR 2005) - Volume 1 - Volume 01 (CVPR 2005), vol. 1, pp. 886–893. IEEE Computer Society, Washington, DC (2005)
5. Nebehay, G., Pflugfelder, R.P.: Clustering of static-adaptive correspondences for deformable object tracking. In: (IEEE) Conference on Computer Vision and Pattern Recognition (CVPR), Boston, MA, USA, pp. 2784–2791 (2015)
6. Comaniciu, D., Meer, P.: Mean shift: a robust approach toward feature space analysis. IEEE Trans. Pattern Anal. Mach. Intell. 24(5), 603–619 (2002). https://doi.org/10.1109/34. 1000236
7. Draper, B.A., Bolme, D.S., Beveridge, J.R., Lui, Y.M.: Visual object tracking using adaptive correlation filters. In: 2010 IEEE Computer Society Conference on Computer Vision and Pattern Recognition (CVPR), San Francisco, CA, USA, pp. 2544–2550 (2010). https://doi. org/10.1109/cvpr.2010.5539960
8. Hong, Z., Chen, Z., Wang, C., Mei, X., Prokhorov, D., Tao, D.: MUlti-Store Tracker (MUSTer): a cognitive psychology inspired approach to object tracking. In: IEEE Conference on Computer Vision and Pattern Recognition (CVPR), Boston, USA (2015)
9. Girshick, R.: Fast R-CNN. In: Proceedings of the 2015 IEEE International Conference on Computer Vision (ICCV), ICCV 2015, pp. 1440–1448. IEEE Computer Society, Washington, DC (2015). https://doi.org/10.1109/ICCV.2015.169
10. Held, D., Thrun, S., Savarese, S.: Learning to track at 100 FPS with deep regression networks. In: European Conference on Computer Vision (ECCV). Springer (2017)
11. Fan, Z., Ji, H., Zhang, Y.: Iterative particle filter for visual tracking. Image Commun. 36(C), 140–153 (2015)

12. Weng, S.-K., Kuo, C.-M., Tu, S.-K.: Video object tracking using adaptive Kalman filter. J. Vis. Comun. Image Represent. **17**(6), 1190–1208 (2006)
13. Tao, R., Gavves, E., Smeulders, A.: Siamese instance search for tracking. In: Computer Vision and Pattern Recognition (CVPR) (2016)
14. Koch, G., Zemel, R., Salakhutdinov, R.: Siamese neural networks for one-shot image recognition. Paper Presented at the Meeting of the 32nd International Conference on Machine Learning (2015)
15. Bertinetto, L., et al.: Fully-convolutional Siamese networks for object tracking. In: Computer Vision ECCV 2016 Workshops, pp. 850–865 (2016). Crossref. Web
16. Krizhevsky, A., Sutskever, I., Hinton, G.E.: ImageNet classification with deep convolutional neural networks. In: Pereira, F., Burges, C.J.C., Bottou, L., Weinberger, K.Q. (eds.) Proceedings of the 25th International Conference on Neural Information Processing Systems - Volume 1 (NIPS 2012), vol. 1, pp. 1097–1105. Curran Associates Inc., Red Hook (2012)
17. LeCun, Y., Bengio, Y.: Convolutional networks for images, speech, and time series. In: Arbib, M.A. (ed.) The Handbook of Brain Theory and Neural Networks, pp. 255–258. MIT Press, Cambridge (1998)
18. Bertinetto, L., Valmadre, J., Henriques, J., Vedaldi, A., Torr, P.: Fully-convolutional siamese networks for object tracking. In: ECCV (2016)
19. Branke, J.: Evolutionary Optimization in Dynamic Environments. Kluwer Academic Publishers, Norwell (2001)
20. Rasmussen, C.E., Williams, C.K.I.: Gaussian Processes for Machine Learning (Adaptive Computation and Machine Learning). The MIT Press, Cambridge (2005)
21. Jebara, T., Kondor, R., Howard, A.: Probability product kernels. J. Mach. Learn. Res. **5**, 819–844 (2004)
22. Nyikosa, F.M., Osbore, M.A., Roberts, S.J.: Bayesian optimization for dynamic problems. arXiv:1803.03432 (2018)
23. Rhodes, A., Witte, J., Jedynak, B., Mitchell, M.: Gaussian processes with context-supported priors for active object localization. In: International Joint Conference on Neural Networks (IJCNN) (2018)
24. Flaxman, S., Wilson, A., Neill, D., Nickisch, H., Smola, A.: Fast Kronecker inference in Gaussian processes with non-Gaussian likelihoods. In: Proceedings of the 32nd International Conference on Machine Learning (2015)
25. Yun, S., Choi, J., Yoo, Y., Yun, K., Choi, J.Y.: Action-decision networks for visual tracking with deep reinforcement learning. In: The IEEE Conference on Computer Vision and Pattern Recognition (CVPR 2017) (2017)
26. Briechle, K., Hanebeck, U.: Template matching using fast normalized cross correlation. In: Proceedings of SPIE - The International Society for Optical Engineering, vol. 4387 (2001). https://doi.org/10.1117/12.421129
27. Kristan, M., et al.: The visual object tracking VOT2014 challenge results. In: ECCV 2014 Workshops (2015)
28. Kristan, M., et al.: The visual object tracking VOT2016 challenge results. In: ECCV 2017 Workshops (2017)

A Distributed Ledger Based Cyber-Physical Architecture to Enforce Social Contracts: Paper Cup Recycling

Tarun Goel[1], Yingqi Gu[1], Francesco Pilla[2], and Robert Shorten[1(✉)]

[1] Department of Electrical and Electronic Engineering, University College Dublin,
Dublin, Ireland
{tarun.goel,yingqi.gu,robert.shorten}@ucd.ie
[2] Department of Planning and Environmental Policy, University College Dublin,
Dublin, Ireland
francesco.pilla@ucd.ie

Abstract. In this paper, we describe a distributed ledger re-cycling system to encourage responsible disposal of paper cups. A complete working prototype is described. Real measurements are presented to illustrate the potential suitability of the IOTA based distributed ledger for this application.

Keywords: Sharing economy · Distributed ledgers · Social compliance · Waste reduction · Paper cups

1   Introduction

Driven by an agenda to make our cities more efficient, and more enjoyable places for citizens to live and work, the general *Smart City* research theme is driving advances across a broad sweep of scientific areas ranging from combinatorial optimisation, battery chemistry, internet of things, to smart wearable materials and many others. Broadly speaking, the Smart City agenda is concerned with making better use of resources in our cities, and there are already many notable examples of cities where the quality of life for citizens has improved immeasurably as a result of technological interventions. Unfortunately, while it without question is true that cities have become *smarter* as a result of these interventions, they are not necessarily *fairer* places to live. Poorly thought-out interventions, while leading to some societal benefits, often lead to unintended consequences, that can limit or exclude communities from accessing resources. One only has to think of AirBnB, and the effect that it is having in property prices in many cities to realise that what is ostensibly a good thing (making use of under used capacity), may in fact be a bad thing. A less obvious, concerns deposits on bottles/plastics to encourage recycling and more responsible environmental behaviour. This seemingly good initiative has led to entire sub-economies emerging in many countries. Here, bottle searchers, sort through rubbish bins to recover bounty from discarded bottles. While such an initiative may

© Springer Nature Switzerland AG 2020
K. Arai et al. (Eds.): FTC 2019, AISC 1070, pp. 959–967, 2020.
https://doi.org/10.1007/978-3-030-32523-7_72

do environmental good, and may provide economic benefit to some individuals, it is extremely exploitative whereby the good of society may in fact depend on exploiting the (perhaps existential) needs of the few. Here, the people searching for plastic bottles perform the essential function of *sorting rubbish into recyclable and non-recyclable parts*. Yet, these people are often poorly paid for the task they perform. For example, typically, *Pfandsammler* (as they are known in Germany) would have to collect a large number of bottles per hour, just to meet minimum wage. Clearly, while deposits on plastic products are a useful tool, they are far from ideal for two main reasons. First, the deposit is not set at an expensive enough level to encourage consumers to respect social contracts (for example, disposing of cups responsibly). Second, the deposit is effectively added at point of manufacture and is associated with the product rather than the consumer. This effectively creates a currency whereby value can be redeemed by anyone. We believed that distributed ledgers may be useful in addressing both of these points.

At a more general societal level, recycling rubbish is one of the most pressing issues facing municipalities. For example, according to estimates, only 5% of the value of plastic packaging material retains in the economy, the rest is lost after a very short first-use. The associated annual bill for this in Europe is estimated to be between €70 and €105 billion[1]. Moreover, as the reader is no doubt aware, plastic takes hundreds of years to break down, with the million or so tonnes of plastic litter that end up in the oceans every year being one of the most visible manifestations of this problem. Even when consumers make the effort, much of the packaging of consumed goods is difficult to recycle (such as plastic lined paper coffee cups). To many, compostable cups are seen as the solution to this problem. Yet, at closer inspection, the introduction of biodegradable cups is only a small part of the solution; namely, not only must we use such cups when drinking our coffee, we must also put our biodegradable cup in the correct bin (compost bin) or else risk contaminating other waste. It is in this context that the present work is placed. The work presented in this paper proposes an innovative use of Distributed Ledger Technology (DLT), the agnostic term for Blockchain and related technologies, to tackle the issue of wrongly disposed paper-cups, without creating sub-economies in the process. Since the Blockchain DLT was first introduced in Nakamoto's white paper in 2008 [6], the technology has been used primarily as an immutable record keeping tool or as a means to enable financial transactions based on peer-to-peer trust [1,2,8,12,13]. Unlike much of the literature on the topic, this paper demonstrates a new, innovative, use of DLT's, as a tool for designing cyber-physical systems [3,4]. In particular, we demonstrate the use of DLT to nudge people towards more sustainable behaviours, and in specific, to dispose of single-use and biodegradable coffee cups in the dedicated collector. We hope that such technologies will be a helpful element in the transition of Europe towards a Low-Carbon and Circular economy, making a tangible contribution to reaching the 2030 Sustainable Development Goals [5] and the Paris Climate Agreement objectives [9].

---

[1] European commission memo/18/6: An European strategy for plastics (2018).

2   Technology Bricks

Our basic idea is to use digital tokens that are associated with individual coffee cups to enforce a deposit-and-return based system. In our envisaged system, when purchasing a coffee, a high-value digital token is transferred from the purchasers digital wallet to the *cup* and only returned when the exact cup is disposed of responsibly. This solution builds on two components: first the notion of a digital ledger, second, their use in enforcing social contracts. Before proceeding, we now briefly comment on both these concepts. A more comprehensive review can be found in [4] upon which the following section is based.

2.1   A Primer on Distributed Ledgers

Distributed Ledger Technology (DLT) is a term that describes Blockchain and a suite of related technologies. At a very high level, a DLT is nothing more than a ledger, held in multiple places, and a mechanism for agreeing on the contents of the ledger; the so-called consensus mechanism. While a detailed review of DLT's is beyond the scope of this present paper (and because there are many existing reviews and tutorials on the topic), we note here that roughly speaking, two types of DLT architectures, roughly classified according to the consensus mechanism employed. Architectures such as blockchain operate a competitive consensus mechanism enabled via mining, whereas architectures such as the IOTA Tangle [11] based on graph structures often operate a cooperative consensus mechanism. In this work, we shall make use of the IOTA DLT. Our interest in IOTA stems from the fact that the architecture is designed to facilitate high-frequency micro-trading. In particular, the architecture places a low computational and energy burden on devices using IOTA; it is highly scalable; there are no transaction fees; and transactions are pseudonymous [7]. As we shall see, these latter two points will be important in our discussion.

2.2   Social Contracts in Smart Cities and Distributed Ledgers

Despite widespread interest in DLT, to the best of the authors knowledge, its use as a tool to orchestrate behaviours in cyber-physical systems and encourage more sustainable behaviours has only recently been discussed [3,4], and its implementation as part of a real-life application has yet remained unexplored. Our interest, as anticipated in an earlier section, stems from the possible applications of DLT's in a smart city environment, using the digital tokens as a way to enforce the desired level of compliance in the resource sharing interactions between humans and machines. Before proceeding, it is worth noting that the idea of using price signals in related areas is not new; see the discussion in [4] and the references therein. The subtle, but substantial difference between this pricing work and our "deposit" based system is that a user is forced to deposit a certain amount of tokens that will be returned once certain criteria are met. Thus, in the latter application, it is a risk that is being priced, rather than a form of demand management.

3   Architecture

The overall conceptual model of our proposed system is illustrated in Fig. 1. Our motivation in this use case is to encourage the responsible disposal of coffee cups. In particular, we wish to ensure that coffee cups are disposed of in the bin using a refundable deposit based scheme that is uniquely linked to the purchaser of the coffee cup. To this end, we assume that the customers participating in the scheme have purchased IOTA based tokens that are pegged at fixed value against a FIAT currency (pegging avoids exposure to price volatility). For the purpose of this paper, we call these tokens, CFtokens. For example, one CFtokens token might have a fixed value of $10. Furthermore, we also assume that each customer has a digital wallet on his/her phone that can store CFtokens, and perhaps other digital currencies. A more elaborate scheme may allow deposits to be priced in a number of tokens, thereby allowing a form of dynamic pricing to be implemented. To understand how this system might work consider a scenario in which a customer coming to a coffee shop wishes to purchase a cup of coffee.

- The customer purchases the coffee using traditional cash or using some other form of digital currency; and the customer transfers one CFtoken to the merchant.
- After serving coffee to the customer, the merchant of the coffee shop registers the ID of the cup to a cloud-based database which centrally records and manages all coffee cups served for this application. The cup, the public address of the customer's digital wallet, and the CFtoken are associated in this database.
- With this in place, the database can accept further querying sent by the smart bin when a customer attempts to dispose the cup to a bin. If the querying is successful and the user is disposing the coffee cup to the correct category of the bin, then the cloud will send notifications to both the smart bin and coffee shop for further actions. Upon receiving a notification, the smart bin will open for the user and the coffee shop will activate the functions to return the CFtoken to the customer based on his/her wallet address that was previously recorded.

This basic system is depicted in Fig. 1.

Comment 1: Note that each cup is uniquely associated to one digital wallet. Cups are worthless when associated with any other digital wallet and consequently do not incentivise secondary (*Pfandsammler*) sorting behaviour. Thus, the value of each CFtoken must be high enough to sufficiently encourage social compliance.

Comment 2: For this application, the IOTA Tangle is of great utility due to fee-less transactions. Transactive methods in which a portion of the coin is kept as a fee by some intermediary would not be functional to enable a deposit based system.

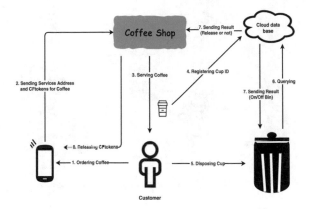

Fig. 1. Conceptual model of the smart bin scheme.

Comment 3: Pseudo-anonymity is very important in our use case. Digital currencies usually leave a digital trace of transactions. Who purchased what, when and where, is usually available to a central intermediary. In contrast, DLT enabled transactions, are much more like a traditional coin, with far superior privacy preserving properties.

Comment 4: By pricing deposits as a number of tokens, one may enforce dynamic pricing to regulate sustainable behaviour. See [4] for examples of pricing models and their stability.

4   Implementation

The physical components of the prototype system implementation is illustrated in Fig. 2. The proposed smart bin system mainly consists of five components, namely a Raspberry Pi, a motor driver circuit board, a waste bin with a built-in bar code scanner, a cloud database, and a mobile phone with the IOTA App for transactions. We shall elaborate the functionality of each component next.

1. Raspberry Pi: We used a Raspberry Pi 3, Model B+. It is the third-generation single-board computer with two rows of General Purpose Input Output (GPIO) pins, which can be used as connections between the Raspberry Pi and the real world. In our use case, the main program that reads from the cloud database to check whether a cup being recycled was previously purchased or not, writes to the cloud database when a correctly identified cup was scanned via the bar code scanner connected to the Raspberry Pi, as well as controls the bin motor via the GPIO pins is written on the Raspberry Pi using Python 3.5 scripts.

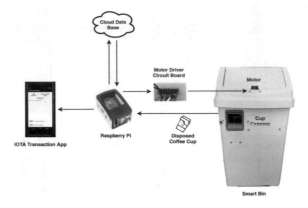

Fig. 2. System implementation of the smart bin scheme.

2. Motor driver-circuit board: The motor driver-circuit board connects the GPIO pins on the Raspberry Pi to the motor on the bin lid. The motor driver we used is the SN754410, which is a quadruple high-current half-H driver designed to provide bidirectional drive currents up to 1 A at voltages from 4.5 V to 36 V. Thus, the motor equipped on the smart bin can be controlled by algorithms written in the Raspberry Pi.
3. Smart bin with built-in bar code scanner: The cup ID reader fitted to the smart bin is a 2D omnidirectional automatic bar code scanner which can easily capture the cup ID tag. The bar code scanner is connected with the Raspberry Pi in order to forward the bar code sequence on each coffee cup to Raspberry Pi for further processing. The bin also includes an attached DC motor on the fore-side of the bins' lid that allows for the locking and unlocking of the lid depending on the state of the motor, such as whether it is in forward or reverse motion.
4. Cloud database: All the coffee cup IDs are stored on the cloud database when each cup of coffee is sold to customers. The Raspberry Pi can be connected to the cloud database using WiFi or 3G. After receiving the ID sequence from the bar code scanner, the Raspberry Pi will forward it to the cloud for further ID matching. If the cup ID is matched with one of the records in the database (if this cup is recyclable) an unlock command will be sent to the smart bin; In the meanwhile, an IOTA transaction for CFtoken return will be initiated through the mobile phone.
5. Mobile phone: In our architecture, we developed a specific App on a Samsung Galaxy S4 smartphone running the Android Oreo operating system (version 8.0) to demonstrate the ease-to-use IOTA client user interface to customers. Customers can make use of the App for IOTA transaction with the coffee shop directly and easily whether this involves transferring or retrieving a token.

A concern with digital ledgers such as blockchain are processing time for both complete transactions, and for transactions to be selected by miners. Real measurements from the IOTA tangle attachment times showed in Fig. 3. These suggest a real practical applicability of the proposed system in real-life scenarios with citizens purchasing hot beverages served in compostable cups. The vast majority of the 2000 transactions in our pilot happened within 10 s (around 74%) and almost the totality (around 94%) within 20 s. While tangle transaction verification times can be longer, these measured attachment times suggest that transactions can be initiated while the beverage is prepared, with completion shortly after the cup is scanned and returned to the smart bin. The flexibility of the cyber-physical framework and its replicability makes it a suitable approach to encourage compliance in a wide range of applications in the arena of smart cities and sustainable communities.

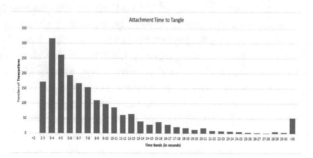

Fig. 3. Tangle attachment time for 2000 transactions.

5   Comments on Pricing Models and Pricing Analytics

Even though it is beyond the scope of the current paper we briefly mention pricing models that can be used to determine the price of CFtokens. As in [4], the price of a token can be used to regulate the level of social compliance. As with traditional pricing models, a number of issues arise in any such strategy. For example, with a series of interconnected activities, string-stability issues may arise whereby disturbances in one process can propagate through the interconnected activities in a destructive manner. A further issue arises when a single malevolent actor can effectively attack the entire community by purposely misbehaving in order to increase prices. While both of these issues arise, and are often neglected, in traditional pricing models, the use of CFtokens potentially offers new tools to address these issues. In particular, the use of tokens attached to individual digital wallets, offers the possibility of pricing strategies tailored to individuals. Specifically, by exploiting an analogy with packets in a communication network, and an item to be recycled, it may be possible to reuse some pricing strategies developed in this context to ensure fair pricing. We anticipate ideas, originally developed in the context of router design [10], will be of great utility in exploring this direction.

6   Discussion and Conclusion

The objective of this project is to use Distributed Ledger Technologies (DLT) to encourage people towards a more sustainable behaviour and in specific to dispose of paper cups in the dedicated collector. The project contributes to the objectives of the following Sustainable Development Goals (SDG): (i) SDG11 "Sustainable Cities and Communities" as it will contribute to make cities more sustainable and ensure sustainable consumption in them; (ii) SDG13 "Climate Action" as by recycling disposable cups, it will reduce their disposal in landfill and thus protect the terrestrial ecosystem and reduce production of greenhouse gases. The proposed solution is based on a digital form of a coin or token that is used to manage, for example, shopping trolleys. As mentioned above, it may appear that this problem could be successfully tackled with compostable cups. However, it has to be noted though that compostable cups would only partially solve the problem as citizens would not only have us to use these cups, but they would also have to dispose of them in the correct bin or else risk contaminating other waste. The IOTA-based DLT framework was chosen for the implementation detailed in this paper because it possesses certain properties which are useful in a large-scale cyber-physical setting: (i) the architecture is be scalable and be able to deal with large amounts of operations; (ii) transactions are free of transaction costs; and (iii) tokens can be created that are fixed against a stable currency to avoid volatility and hoarding. Furthermore, a key factor in this application is that full (pseudo) anonymity is preserved such that there is no traceback and a user. If users comply to the rules, they will not be charged any unnecessary transaction fees. Future work will consider adapting our architecture to batter management problems and other smart city compliance problems.

References

1. Banerjee, M., Lee, J., Raymond Choo, K.-K.: A blockchain future for internet of things security: a position paper. Digit. Commun. Netw. **4**(3), 149–160 (2018)
2. Conoscenti, M., Vetro, A., De Martin, J.C.: Blockchain for the internet of things: a systematic literature review. In: 2016 IEEE/ACS 13th International Conference of Computer Systems and Applications (AICCSA), pp. 1–6. IEEE (2016)
3. Cruse, J.P., Cantwell, D., Xu, M., Hardy, C., Schwarz, B., Collins, K.S., Nguyen, A., Sui, Z., Lee, E.: Methods for processing substrates in process systems having shared resources, 30 July 2013. US Patent 8,496,756
4. Ferraro, P., King, C., Shorten, R.: IOTA-based Directed Acyclic Graphs without Orphans. arXiv e-prints, page arXiv:1901.07302, December 2018
5. Griggs, D., Stafford-Smith, M., Gaffney, O., Rockström, J., Öhman, M.C., Shyamsundar, P., Steffen, W., Glaser, G., Kanie, N., Noble, I.: Policy: Sustainable development goals for people and planet. Nature **495**(7441), 305 (2013)
6. Nakamoto, S.: Bitcoin: A peer-to-peer electronic cash system (2008)
7. Popov, S., Saa, O., Finardi, P.: Equilibria in the tangle. arXiv preprint arXiv:1712.05385 (2017)
8. Puthal, D., Malik, N., Mohanty, S.P., Kougianos, E., Das, G.: Everything you wanted to know about the blockchain: its promise, components, processes, and problems. IEEE Consum. Electron. Mag. **7**(4), 6–14 (2018)

9. Rogelj, J., Elzen, D.M., Höhne, N., Fransen, T., Fekete, H., Winkler, H., Schaeffer, R., Sha, F., Riahi, K., Meinshausen, M.: Paris agreement climate proposals need a boost to keep warming well below 2 °C. Nature **534**(7609), 631 (2016)
10. Stanojevic, R.: Router-based algorithms for improving internet quality of service. PhD thesis, Hamilton Institute, Maynooth University (2007)
11. Wang, W., Hoang, D.T., Xiong, Z., Niyato, D., Wang, P., Hu, P., Wen, Y.: A survey on consensus mechanisms and mining management in blockchain networks. arXiv preprint arXiv:1805.02707
12. Yli-Huumo, J., Ko, D., Choi, S., Park, S., Smolander, K.: Where is current research on blockchain technology? A systematic review. PloS one **11**(10), e0163477 (2016)
13. Zheng, Z., Xie, S., Dai, H., Chen, X., Wang, H.: An overview of blockchain technology: architecture, consensus, and future trends. In: 2017 IEEE International Congress on Big Data (BigData Congress), pp. 557–564. IEEE (2017)

3D Design and Manufacturing Analysis of Liquid Propellant Rocket Engine (LPRE) Nozzle

Samuel O. Alamu[✉], Marc J Louise Caballes, Yulai Yang,
Orlyse Mballa, and Guangming Chen

Department of Industrial and Systems Engineering,
Morgan State University, Baltimore, MD 21251, USA
olala22@morgan.edu

Abstract. The innovation and development of 3D printing technology has benefitted aerospace industries in the design and fabrication of parts and model structure of aerospace products such as building rocket parts from engines to propellant tanks. The structure can be created with little or no generation of industrial waste, thereby reducing overall cost and shortening the production time. Morgan State University (MSU) has been awarded a grant from BASE 11 for building a rocketry program. The MSU students have an opportunity to participate in the design and fabrication of rockets. This project seeks to adopt the 3D printing method to design a liquid propellant rocket engine (LPRE) nozzle. A 3D model of the rocket engine nozzle was developed using a CAD software (AutoCAD and Autodesk Inventor) and a prototype was printed with LPA filaments using a 3D printer. The design geometry specifications and flow conditions were simulated to determine the performance of the LPRE nozzle. The overall design specifications and investigated Ni-based superalloys materials contained in this research findings will be utilized in the overall ongoing design of liquid propellant rocket program awarded to the MSU Rocketry Team.

Keywords: Additive manufacturing · Liquid Propellant Rocket Engine (LPRE) · 3D printer · Nozzle

1 Purpose

The purpose of this research is to design and print a 2D and 3D Model of the LPRE Nozzle that will be used as a guide and visualization for the MSU Rocketry project. Furthermore, it is also to show how efficient and convenient additive manufacturing process is, not only in academia but as well as in the manufacturing and technology industry. Lastly, after gathering the optimal parameters of the scaled design, a brief simulation was conducted to find out how the angles, spacings, and sizing of the nozzle affects the overall performance of the liquid propellant rocket.

© Springer Nature Switzerland AG 2020
K. Arai et al. (Eds.): FTC 2019, AISC 1070, pp. 968–980, 2020.
https://doi.org/10.1007/978-3-030-32523-7_73

2   Background/Significance

Robert Goddard successfully patented the very first working fueled rocket in 1914. Thus, making him the Godfather of modern rocketry. One of the mainframes of a rocket is the rocket nozzle. It is a cylindrical shape-like system with a variable cross-sectional area where its main purpose is not only to control different variables, such as - flow rate, speed, direction, mass, shape, etc. but as well as to pressurized the steam that emerges. Furthermore, aside from converting the chemical and thermal energies into kinetic energy in the combustion chamber, the LPRE nozzle will also convert the low velocity with a high temperature and pressure into a high velocity with a lower temperature and pressure. With that said, the general range of exhaust velocity can be easily achieved, which is 2 to 4.5 km/s [1]. The convergent and divergent type of nozzle, known as DE-LAVAL nozzle is being modeled and manufactured in this research.

The total thrust produced by the liquid rocket engine depends on its mass flowrate, the exit velocity of the exhaust, and the pressure at the nozzle exit. All these parameters depend on the nozzle geometry. Figure 1 represents the theoretical concept of a liquid fuel rocket consisting of three major parts - the combustion chamber, the nozzle, and the injector. The throat of a nozzle is the smallest cross-sectional area through which the hot exhaust leaving the combustion chamber converges. Thus, the flow becomes choked resulting in a sonic flow with Mach number 1.0 (i.e., Me = 1). The mass flowrate of LPR engine is determined by the moving fluid density, velocity and the cross-sectional area (i.e., the throat). The mass flowrate is maximum at sonic condition where the hot flow is choked [2].

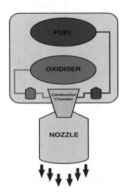

Fig. 1. Theoretical design of a liquid propellant rocket.

2.1   Material Selection

For all rocket propulsion systems that use gas expansion as the mechanism for ejecting matter at high velocities, the fundamental working principle and thermodynamic relations are similar. The choice of materials for inner chamber walls in combustion chamber and nozzle throat regions, the most critical locations, is crucial for the success of a launch mission, which is influenced by hot gas composition, maximum allowable wall temperatures, heat transfer rates, and duty cycles. A large variety of materials have been utilized. Table 1 lists some typical materials selected for several applications [3].

Table 1. Typical materials used in several common liquid rocket propellant thrust

Application	Propellant	Components	Cooling method	Typical materials
Bipropellant TC, high pressure (booster or upper stage)	Oxygen – hydrogen	C, N, E	F	Copper alloy
		I	F	Transpiration Cooled porous stainless-steel face. Structure is stainless steel
		Alternate E	R	Carbon fiber in a carbon matrix or niobium
		Alternate E	T	Steel shell with ablative inner liner
Bipropellant TC, high pressure (booster or upper stage)	Oxygen – hydrocarbon or storable propellant	C, N, E, I	F	Stainless steel with tubes or milled slots
		Alternate E	R	Carbon fiber in a carbon matrix or niobium
		Alternate E	T	Steel shell with ablative inner liner
Experimental TC (very limited duration – only a few seconds)	All types	C, N, E	U	Low-carbon steel
Small bipropellant TC	All types	C, N, E	R	Carbon fiber in a carbon matrix, rhenium, niobium
		Alternate E	T	Steel shell with ablative inner liner
		I	F	Stainless steels, titanium
Small monopropellant TC	Hydrazine	C, N, E	R	Inconel, alloy steels
		I	F	Stainless steel
Cold gas TC	Compressed air, nitrogen	C, N, E, I	U	Aluminum steel, or plastic

Additive manufacturing (AM) or commonly called as 3D printing, is a method and process of manufacturing and fabricating parts layer by layer. Nowadays, AM is widely used in different major industries and fields such as – automotive, aerospace, defense, research, and health industry. As the technology industries are booming fast, AM gives the manufacturers several advantages but not limited to (1) freedom to design and innovate without penalties; (2) increased supply chain proficiency with "3D faxing"; (3) support of green manufacturing initiatives; (4) bottom line improvements through factory physics; and (5) faster production.

Steel is the most widely used Engineering material in the field of additive manufacturing as at today [9]. The available stainless steel material includes: Austenitic
stainless steels (AISI 316L/EN: 1.4404/X2CrNiMo17-12-2 [4], AISI 304L/EN:
1.4306/X2CrNi19-11) [5], maraging steel (18Ni-300/1.2709/X3NiCoMoTi18-9-5) [6],
precipitation hardenable stainless steels (17-4 PH/EN: 1.4542/X5CrNi-CuNb16-
4/AISI: 630 [7] and 15e5 PH/EN: 1.4545/X5CrNiCu15-5) [8]. The above materials
can satisfy the requirements of general-purpose applications. In addition, the Aluminum (Al) alloys available for additive manufacturing is very limited when compared
to other materials. There is low commercial advantage to manufacture Al parts by
additive manufacturing in that fabrication/machining of Aluminum is easier, and the
costs of its parts are comparatively low. On the other hand, Al has a low melting point,
so it is not an ideal material for rocket nozzle. Additive manufacturing parts fabricated
from the a-b alloy Ti-6Al-4V have been investigated and widely used for commercial
fabrication [9]. Nickel-based superalloys for high-temperature applications, such as
Inconel 625 [10] and Inconel 718 used in LBM [11] and EBM [12] can achieve good
performance.

According to Table 2, Maraging steel 300, Inconel 625, have high strength and
high melting points. They are found suitable for the designing LPRE nozzle. Combustion temperatures of rocket propellants are generally higher than the melting points
of common metal alloys and refractory materials up to 3600 K. The strength of most
materials declines rapidly at high temperatures. For rocket engine applications, the
maximum allowable wall temperature is well below the material melting point. Enough
heat must be removed to keep material at a designed temperature. The Rocket engine
cooling mechanical is very important [14].

Table 2. Mechanical properties of several stainless-steel grades by different AM processes

Alloy	Process	YS (Mpa)	UTS (Mpa)	Melting point	Reported by
316L	LBM	590 ± 17	705 ± 15	1390–1440 °C	[21]
Maraging steel 300	SLM	1214 ± 99	1290 ± 114	1412.8 °C	[6]
Maraging steel 300	SLM + aging (5 h, 480 °C)	1998 ± 32	2217 ± 73		[6]
Inconel 625	SLM	800 ± 20 (horizontal samples)	1030 ± 50 (horizontal samples)	1290–1350 °C	[10]
Inconel 625	SLM	720 ± 30 (vertical samples)	1070 ± 60 (vertical samples)		[10]
Inconel 718	SEBM, +homogenizing (1060 °C, 2H)+ annealing (720 °C and 620 °C 8 H each)	1030 MPa	1275 MPa	1210–1344 °C	[13]

3   Methodology

3.1   CAD Designing of the LPRE Nozzle

A 2D Design with all the correct parameters was created alongside the 3D Printed product using Autodesk Inventor. In the 21st century era, most of the engineers designed products with 3D CAD tools and analyze it from all engineering aspects with CAE tools to make it better functioning. Production with 3D printing has many advantages. It can potentially reduce the cost, making the product light, and is more efficient to create a finished since no welds or joints will be needed. Based on the materials or filaments that will be used to print, there is a higher chance that the finished product can withstand extreme temperature and make the pressure fluctuations better. The reduced weight means that the rocket can carry much more weight during the mission. No welds or joins will reduce the overall failure rate of the rocket.

Besides the above unique advantage, the 3D printing technique reduces the production of waste. 3D printing can produce parts significantly quicker than conventional manufacturing methods. With this kind of advancement, it will only be a matter of time when it will not be a question of limitations anymore, but the possibilities of the things it can print. Sooner or later, printing the whole system of aircraft designs will no longer be impossible. In addition, with this kind of approach, it will make the designers and manufacturers' work optimal and efficiently, since designers will be able to create highly complex designs and manufacturers will be able to produce a complete product in a single unit rather than multiple parts and needs to be assembled. Even though additive manufacturing is just starting to flourish, it will still have a long way to go and more rooms to be improved to become the main face of the manufacturing industry. With that said, 3D manufacturing produces products/objects in higher quality in terms of precision and reliability than the traditional method. Figure 2 represents the 2D Design of the LPRE Nozzle and the 3D CAD design was shown in Fig. 3. Each part is divided into different sections to visualize the hidden lines and gap inside the hollow nozzle.

In the industry, rocket nozzles and combustion chambers are very expensive to manufacture not only because of its complexity of materials but as well as the environment of which it will be fabricated because it needs to expose in extreme heat and pressures from the combustion process. Now, if most of the crucial part will be developed in 3D printing, it will solve most of the problems because the 3D printer machines in the industry that is intended to do this kind of work use a free-form directed energy wire deposition. Thus, it has the ability to reduce build time from several months to several weeks [15].

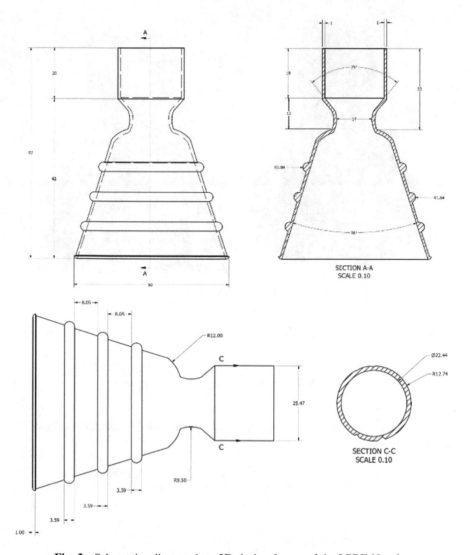

Fig. 2. Schematics diagram in a 2D design format of the LPRE Nozzle.

3.2   LPRE Nozzle Design Equations

The overall design was based on Isentropic condition. Isentropic flows occur when the change in flow variables is small and gradual, such as the ideal flow through the nozzle. Mach number is the ratio of the gas velocity to the local speed of sound. The Mach number at the nozzle exit is given by a perfect gas expansion expression as:

$$Exit\,Mach\,Number:\; Me^2 = \frac{2}{\gamma - 1}\left[\left(\frac{Pc}{Patm}\right)^{\left(\frac{\gamma-1}{\gamma}\right)} - 1\right] \tag{1}$$

Fig. 3. 3D design of the LPRE nozzle.

The exit area of the nozzle which correspond to the exit Mach number is expressed as:

$$Exit\ area\ of\ the\ nozzle:\ A_e = \frac{At}{Me}\left[\frac{1 + \frac{\gamma-1}{2}Me2}{\frac{\gamma+1}{2}}\right]^{\frac{\gamma+1}{2(\gamma-1)}} \tag{2}$$

The exit pressure **Pe** and temperature **Te** of the nozzle is calculated by the given Eqs. 3 and 4:

$$Exit\ Pressure:\ P_e = Pt\left(1 + \frac{\gamma-1}{2}Me2\right)^{\left(\frac{-\gamma}{\gamma-1}\right)} \tag{3}$$

$$Exit\ Temperature:\ T_e = Tt\left(1 + \frac{\gamma-1}{2}Me2\right)^{(-1)} \tag{4}$$

The exit velocity can be calculated from the resulting temperature exit and gas constant as:

$$Exit\ Velocity:\ V_e = M_e\sqrt{\gamma RTe} \tag{5}$$

$$Mass\ flowrate:\ \dot{m} = \frac{A*Pt}{\sqrt{Tt}}*\sqrt{\frac{\gamma}{R}}*Me\left(1 + \frac{\gamma-1}{2}Me2\right)^{-\left(\frac{\gamma+1}{2(\gamma-1)}\right)} \tag{6}$$

Having calculated the exit velocity and the mass flow rate, then we can determine the thrust of the nozzle with the expression in Eq. 7:

$$Thrust: \; F_0 = \dot{m}\,V_e + (P_e - P_0)A_e \tag{7}$$

Where R – Universal Gas constant (1545.32 ft-lb/lb(deg)R, 8.314 J/mol.K), γ – Ratio of specific heat (assumed value of 1.2), Pc – Chamber Pressure and Patm – atmospheric pressure (14.7 psi), Tc – Chamber Temperature.

3.3   3D Printing Materials

A filament is also called a fiber which is a long chain of polymers lines up in the same direction. The straighter they are, the stronger the fiber will be. In the production process, the filaments used were packed together in a straight manner. There are different types of filaments (standard, flexible, composite, specialty, support and many others) used while performing 3D printing. Each one of them possesses different properties depending on the nature and specifications of a project. For the scope of this paper, only the standard filaments will be used. Table 3 displays the sub filaments used in standard filament and their properties.

Table 3. Types of filament and some of their properties

Types of filament	Properties
PRO	Hard but brittle
	Low to no shrinkage
PLA	Stiff
	Odorless
	Flexible
NYLON	Shatter resistant
	Rigid
	Sensitive to moisture
	Strong
ABS	Ductile
	Heat tolerance

3.4   3D Printing Machine

In this research, a Flashforge Finder 3D printer (Fig. 4) was used for printing the designed LPRE nozzle from the CAD software. One of the most common approached in the additive manufacturing industry is called the Fused Filament Fabrication (FFF). It is the trademarked term for 3D modeling and printing. The working principle of the FFF is melting the plastic filament, either PLA or ABS, in layers at a very high temperature. Thus, because of the heat, the layers of the filament in where it is printed will be "fuse" together. Once the cool down process is finished, the solidification of the filament occurs instantaneously after it is extruded of the printer's nozzle. The machine has a filament diameter of 1.75 mm and nozzle diameter of 0.4 mm with a build-volume of 140L × 140H × 140W mm. It is easy, flexible and affordable for students learning 3D printing. Flashforge Finder uses only PLA filaments which are non-toxic, biodegradable plastic that is recyclable and safe for the environment.

Fig. 4. Flashforge Finder 3D printer.

4   Results and Discussion

4.1   Nozzle Simulation

The design parameters for simulating the behavioral pattern and performance of the LPRE nozzle were shown in Table 4. The design specifications given by the CAD was scaled down to 10:1 ratio for the input geometry to the simulation software. The flow inputs (Pc and Tc) were also specified.

Table 4.   Design parameters for LPRE Nozzle simulation

Design parameters	Values
Chamber Pressure (Pc)	160 kPa
Chamber Temperature (Tc)	400 K
Design Throat diameter	35.56 mm
Exit Diameter	64.69 mm

Due to the shape of the rocket nozzle which is converging-diverging, the hot exhaust can efficiently leave the combustion chamber while converging it down to the throat. Furthermore, the throat size was carefully analyzed that it can able to withstand the temperature and can optimally choke the flow. In addition to that, it would be able to set the mass flow rate. According to previous works, the flow of the throat is sonic, which in this case, Me = 1. Moreover, three important variables of the nozzle which are - throat's downstream, geometric application (diverges), and its flow will have a continuous isentropic expansion due to the state of the supersonic value, which is Me > 1. Thus, expansion of supersonic flow is heavily dependent on the ratio of the throat's exit is. With that said, this event usually causes a continual decrease in both temperature and pressure from the throat to the exit. With that, the amount of expansion also determines the exit pressure and temperature.

The exit temperature determined the exit speed of sound, which further determined the exit velocity. The amount of thrust produced by the nozzle was determined by the exit velocity and pressure and the mass flowrate through the nozzle as shown in Table 5. The exit velocity of 3.38 km/s conforms with the range specified by [1].

Table 5. Simulation output of LPRE nozzle design

Exit Mach number (Me)	Area ratio (Ae/At)	Exit pressure ratio (Pe/Pc)	Exit temperature ratio (Te/Tc)	Exit velocity (Ve) (m/s)	Mass flowrate (Kg/s)	Thrust (N)
2.5	3.31	0.0094	0.459	3384.98	0.174	593.6

4.2 Additive Manufacturing

Additive manufacturing (AM) is regarded as an incremental layer by layer manufacturing process which contrasts with the conventional subtractive methods [16]. The manufacturing process involves melting of powder or wire by a focused heat source and then followed by subsequent cooling in order to form a part [17]. According to [18], AM is characterized for its five any's: use of almost *any* material to fabricate *any* part, in *any* quantity and *any* location, for *any* industrial field. AM has advantages when compared with subtractive manufacturing in that it can reduce the amount of supply chain management; reduces process time, and it is cost effective by minimizing material waste. AM can make lightweight, improved and complex geometries products which could lead to fuel savings in the aerospace industry [19].

The structure of thrust chamber is usually very complex, which makes the design and manufacture process challenge and time consuming. The final product will have many joints, which will increase the overall failure rate since if one of the joint failures the system will fail. This prompted the need for additive manufacturing which helps to minimize material waste and time in complex designs.

4.3 3D Printing and Manufacturing

The CAD document was extracted into the Flashforge Finder 3D printer. The design specifications shown in Fig. 2 were scaled down to 20:1 ratio for easy printing. The total duration for the printing is 2 h 47 min at extruder temperature of 220 °C.

Figure 5 represents the finished product of the 3D Model of the LPRE Nozzle. Based on research, even though nozzles look very simple, but actually they are quite complex components despite their appearance. Author in [20] fabricated a nozzle using a 3D printer that was able to improve the efficiency of the propulsion, meaning it does not require much gas. 3D Printing of the nozzle saved weight and costs while taking into considerations all the crucial factors and constraints.

Fig. 5. Finished product of the 3D model of the LPRE nozzle.

4.4   Limitation

3D Printing is becoming more popular day by day due to its ability to be a custom-made production. Thus, this process is less wasteful when compared to other methods due to the fact that carving or cutting the away the material is not needed to build an object. Even though 3D Printing and Manufacturing make lives easier and faster it still has its own several disadvantages such as (1) energy consumption; (2) limited materials; (3) it takes time to manufacture; (4) production of dangerous weaponry; and lastly (4) copyright infringements. Overall, one cannot deny how 3D printing easily provides countless benefits in the field of industry. 3D printing system has been widely used in areas where product customization is important ranging from printing hearing aids and dental implants to printing a miniature of the happy couple for their wedding cake. However, as of now, it is still not enough to replace traditional manufacturing. In the future, there is a huge need to consider the disadvantages mentioned in the product development phase in order to complement the process of design and manufacturing.

5 Conclusion

Additive manufacturing has proven to be very advantageous when compared with conventional manufacturing processes. Complex LPRE parts such as nozzle and combustion chamber can be successfully manufactured using additive manufacturing process to save cost and production time. In this research, AM methodology was adopted to produce a 3D model of LPRE nozzle using Autodesk CAD software and a Flashforge Finder 3D printer. The design geometry was imported into simulation software. The simulation output values indicate the overall performance of the nozzle based on the input flow condition. The LPRE nozzle prototype was manufactured using PLA filaments. On research findings, the best material for manufacturing purposes of LPRE nozzle is Ni-based superalloys (Inconel 625 or 718) since they can withstand high-temperature applications. These superalloys will be used when manufacturing the LPRE nozzle and combustion chamber by the Morgan State University Rocketry Team using the design model developed in this study. Even though there are some limitations, the future of 3D printing and manufacturing is bright due to its capabilities and varieties of uses. With the advancement of the technology industry, especially in the aerospace field, it would be more beneficial for developing quantities of utilized cases and verifiable designs and projects that demonstrates additive manufacturing can be the standard in assembling innovation. What should be a theoretical concept before can be now the standard in the manufacturing process.

6 Scope of Future Work

In pursuit of the developing and launching of liquid propellant rocket by the MSU Rocketry team, the research work continues having the study carried out in this research paper as benchmark. As shown in Table 2, Inconel 718 will be chosen for our designed LPRE nozzle because of its high temperature strength, tensile strength, and melting point. But different raw material (different kind of metal powder), 3D metal printer, manufacture process and final heat treatment will result in different final performance of nozzle. Design of experiment will be used to choose the best factors to achieve the ideal product. A comparison between the simulated and experimented nozzle thrust will be conducted to validate the results obtained and for optimum performance of the LPRE nozzle. The use of 3D printer will be further explored for manufacturing complex parts of the LP rocket engine to save cost and production time. All these put together will help the MSU Rocketry team to develop the most cost-effective LP rocket engine.

References

1. Mishra, N.K., Prasad, S.S., Padania, M.A.: Modeling & simulation of rocket nozzle. Int. J. Adv. Eng. Glob. Technol. **2**, 988–995 (2014)
2. Snyder, C.A.: NASA Chemical Equilibrium with Applications (CEA) (2016). https://www.grc.nasa.gov/www/CEAWeb/

3. Sutton, G.P., Biblarz, O.: Rocket Propulsion Elements. Wiley, Hoboken (2016)
4. Badrossamay, M., Childs, T.H.C.: Further studies in selective laser melting of stainless and tool steel powders. Int. J. Mach. Tools Manuf **47**(5), 779–784 (2007)
5. Guan, K., Wang, Z., Gao, M., Li, X., Zeng, X.: Effects of processing parameters on tensile properties of selective laser melted 304 stainless steel. Mater. Des. **50**, 581–586 (2013)
6. Kempen, K., Yasa, E., Thijs, L., Kruth, J.P., Humbeeck, J.V.: Microstructure and mechanical properties of Selective Laser Melted 18Ni-300 steel. Phys. Procedia **12**, 255–263 (2011)
7. Murr, L.E., Martinez, E., Hernandez, J., Collins, S., Amato, K.N., Gaytan, S.M., Shindo, P. W.: Microstructures and properties of 17-4 PH stainless steel fabricated by selective laser melting. J. Mater. Res. Technol. **1**(3), 167–177 (2012)
8. Islam, M., Purtonen, T., Piili, H., Salminen, A., Nyrhilä, O.: Temperature profile and imaging analysis of laser additive manufacturing of stainless steel. Phys. Procedia **41**, 835–842 (2013)
9. Tan, X., Kok, Y., Tan, Y.J., Descoins, M., Mangelinck, D., Tor, S.B., Chua, C.K.: Graded microstructure and mechanical properties of additive manufactured Ti–6Al–4V via electron beam melting. Acta Mater. **97**, 1–16 (2015)
10. Yadroitsev, I., Thivillon, L., Bertrand, P., Smurov, I.: Strategy of manufacturing components with designed internal structure by selective laser melting of metallic powder. Appl. Surf. Sci. **254**(4), 980–983 (2007)
11. Nie, P., Ojo, O.A., Li, Z.: Numerical modeling of microstructure evolution during laser additive manufacturing of a nickel-based superalloy. Acta Mater. **77**, 85–95 (2014)
12. Dehoff, R.R., Kirka, M.M., Sames, W.J., Bilheux, H., Tremsin, A.S., Lowe, L.E., Babu, S. S.: Site specific control of crystallographic grain orientation through electron beam additive manufacturing. Mater. Sci. Technol. **31**(8), 931–938 (2015)
13. Körner, C., Helmer, H., Bauereiß, A., Singer, R.F.: Tailoring the grain structure of IN718 during selective electron beam melting. In: MATEC Web of Conferences, vol. 14, p. 08001. EDP Sciences (2014)
14. Pizzarelli, M.: Regenerative cooling of liquid rocket engine thrust chambers (2017). https://doi.org/10.13140/RG.2.2.30668.92804
15. Urbano, A., Selle, L., Staffelbach, G., Cuenot, B., Schmitt, T., Ducruix, S., Candel, S.: Exploration of combustion instability triggering using large eddy simulation of a multiple injector liquid rocket engine. Combust. Flame **169**, 129–140 (2016)
16. Emmelmann, C., Kranz, J., Herzog, D., Wycisk, E.: Laser additive manufacturing of metals. In: Schmidt, V., Belegratis, M. (eds.) Laser Technology in Biomimetics, pp. 143–162. Springer, Heidelberg (2013)
17. Brandl, E., Palm, F., Michailov, V., Viehweger, B., Leyens, C.: Mechanical properties of additive manufactured titanium (Ti–6Al–4V) blocks deposited by a solid-state laser and wire. Mater. Des. **32**(10), 4665–4675 (2011)
18. Lu, B., Li, D., Tian, X.: Development trends in additive manufacturing and 3D printing. Engineering **1**(1), 85–89 (2015)
19. Joshi, S.C., Sheikh, A.A.: 3D printing in aerospace and its long-term sustainability. Virtual Phys. Prototyping **10**(4), 175–185 (2015)
20. Moorthy, C.V., Srinivas, V., Prasad, V.V.S.H., Vanaja, T.: Computational analysis of a CD nozzle with 'SED' for a rocket air ejector in space applications. Int. J. Mech. Prod. Eng. Res. Dev. (IJMPERD) **7**(1), 53–60 (2017)
21. Carlton, H.D., Haboub, A., Gallegos, G.F., Parkinson, D.Y., MacDowell, A.A.: Damage evolution and failure mechanisms in additively manufactured stainless steel. Mater. Sci. Eng. A **651**, 406–414 (2016)

Author Index

A

Abegaz, Tamirat, 368
Aceves-Mijares, M., 781
Adeitan, Ayodeji Dennis, 652
Agbenyeku, Emmanuel Emem-Obong, 652
Aigbavboa, Clinton, 652
Al-Abri, Dawood, 1
Al-Ahmadi, Saleh, 122
Alahmadi, Saad, 338
Alamu, Samuel O., 968
Algarni, Fahad, 13
Alharbi, Sarah, 338
Aloufi, Khalid, 13
Alsmadi, Izzat, 542
Aman, Waqas, 1
Andrew, Changa, 717
Antoun, Sherine, 275
Arai, Kohei, 925
Arora, Geetika, 511
Avdoshin, Sergey, 833

B

Baldi, Vania, 176
Barkouti, Wahid, 122
Bedja, Koffi-Sa, 758
BenMessaoud, Fawzi, 596
Billis, Steven, 848
Blum, David, 100
Boakye, Andrews, 40
Boranbayev, Askar, 324
Boranbayev, Seilkhan, 324
Bouragba, Khalid, 823
Bringula, Rex P., 913

Brown, DeWayne, 710
Bulega, Tonny, 717
Burggräf, Peter, 692

C

Caballes, Marc J Louise, 968
Chabchoub, Abdelkader, 122
Chakraborty, Ishani, 582
Chandler, Jordan, 100
Chaovavanich, Korakot, 589
Chatwiriyachai, Sornchai, 589
Chaurasia, Nisha, 670
Chen, Guangming, 968
Chen, Haixian, 293
Chen, Qifeng, 484
Cheng, Xu, 293
Cherif, Adnen, 122
Chou, Annie Y. H., 605
Chowdhury, Brinta, 710
Cotton, Chase, 306
Cubenas, Oscar, 848

D

Daniel Liang, Y., 890
Dave, Rushit, 710
De Leon, Marlene M., 913
Demidov, Roman, 454
Diallo, Moussa, 745
Dias, Norman, 435
Diop, Idy, 745
Dioum, Ibra, 745
Djaidja, Sana, 795
Dotche, Koffi A., 758

© Springer Nature Switzerland AG 2020
K. Arai et al. (Eds.): FTC 2019, AISC 1070, pp. 981–983, 2020.
https://doi.org/10.1007/978-3-030-32523-7

Doumbia, Mamadou, 293
Dromard, Juliette, 397

E
Easttom, Chuck, 542
Echoukairi, Hassan, 823
Eddin, Anas Salah, 379
El Naga, Halima, 379
El-Hadedy, Mohamed, 379
Elmounchid, F., 87

F
Farssi, Sidi Mohamed, 745
Faust, Jeff, 855
Fayad, Mohamed E., 641
Flood, Ian, 519
France, Derek, 871

G
García-Sierra, Rodolfo, 49
Ghoul, Rafia, 795
Gil, Santiago, 49
Gnanasekaran, Lavanya, 379
Goel, Manan, 950
Goel, Tarun, 959
Gonzalez-Fernandez, A. A., 781
Gregory, Meredith, 100
Gu, Yingqi, 959
Gueye, Ibrahima, 745
Gupta, Phalguni, 511

H
Haddara, Moutaz, 552
Hall, Dwight William, 596
Hamouda, Ali, 122
Hamu, David, 641
Handlon, Holly Nichole, 596
Hatami, Mohsen, 519
Hawbani, Ammar, 795
He, Jing, 795
Heljakka, Katriina, 625
Hellam, S., 87
Hernández-Betanzos, J., 781
Hosseini, Changiz, 552
Huang, Ching-Chun (Jim), 353
Husted, Taryn Elizabeth, 596
Hwang, C. Jinshong, 511

I
I. Sarwat, Arif, 493
Ihamäki, Pirita, 625
Inoue, Chinayo Shonen, 936
Islam, Md Saiful, 338

J
Jankowski, Charles, 204

K
Kada, Amine, 823
Kalra, Agastya, 188
Kamitai, Daiki, 936
Karsenti, Thierry, 528
Kausar, Firdous, 1
Keita, K. Wane, 745
Kodjo, Koffi M., 758
Kshirsagar, Niranjan Valmik, 596
Kuppa, Gaurav, 641

L
Lawson, Thomas, 260
Li, Lichun, 731
Li, Yutian, 40
Lim, Hock Chuan, 284
Lorber, Carolin, 692

M
Magalhães, Mario, 904
Martinez, Gina, 661
Mauchline, Alice, 871
May, Liu, 944
Mballa, Orlyse, 968
Mbarki, S., 87
Mekonnen, Yemeserach, 493
Miao, Xuhong, 40
Mizanoor Rahman, S. M., 224, 244
Montellano, Jerwin M., 132
Mostowsky, Zachary, 100
Moud, Hashem Izadi, 519
Mruthyunjaya, Vishwas, 204

N
Nanda, Aparajita, 670
Narang, Mahima, 670
Ndiaye, Papis, 745
Ngamarunchot, Bank, 589
Ngamkajornwiwat, Potiwat, 589
Nigam, Charu, 670
Ninyawee, Nutchanon, 589
Niu, Li, 40
Nobre, Helena, 176
Nurbekov, Askar, 324
Nurusheva, Assel, 324

O
Ocasio, Kendall, 100
Okamoto, Manabu, 936
Okopa, Michael, 717
Omari, Safwan, 661

Ostritsova, Valeriia, 682
Oulahrir, Y., 87
Ouzzif, Mohammed, 823
Owada, Hazuki, 936
Owezarski, Philippe, 397
Öztürk, Kağan, 417

P
Park, Julian, 871
Parvez, Imtiaz, 493
Pataranutaporn, Pat, 589
Payne, Bryson, 368
Pechenkin, Alexander, 454
Pedraza, J., 781
Pereira, João, 682
Pereira, João Paulo, 682
Perlovsky, Leonid, 162
Pesotskaya, Elena, 833
Peterson, Ben, 188
Pilla, Francesco, 959
Pournaghshband, Hassan, 619
Price, Ted, 855
Priyadarshini, Ishaani, 306
Promersberger, John, 162
Pyka, Andreas, 692

R
Raji, Rafiu King, 40
Rawashdeh, Ahmad, 563
Reck, Ryan, 275
Reeja, S. R., 435
Regina Estuar, Ma., 913
Rhodes, Anthony D., 950
Richey, Melonie, 100
Rolo, Ema, 176
Rubin, Izhak, 66

S
Sadiq, A., 87
Salami, Adekunlé A., 758
Salimi, Abi, 368
Sanfilippo, James, 368
Schmeelk, Suzanna, 467
Seitkulov, Yerzhan, 324

Sekyere, François, 758
Semwal, Sudhanshu Kumar, 162
Shojaei, Alireza, 519
Shorten, Robert, 959
Simonette, Marcel, 904
Sowells-Boone, Evelyn, 710
Spina, Edison, 904
Sunyoto, Yulia, 66
Surareungchai, Werasak, 589

T
Tiwari, Kamlesh, 511
Tseng, Frank S. C., 605

U
Ullah, Azmat, 13

V
Victorino, John Noel, 913

W
Wagner, Johannes, 692
Wan, Ailan, 40
Wang, Haining, 306
Wei, Feng, 731
Wei, Shugang, 390
Wei, Wilson, 484
Weißer, Tim, 692
Welsh, Katharine, 871
Whalley, Brian, 871

Y
Yang, Yanyan, 260
Yang, Yulai, 968
Yılmaz, Mustafa Berkay, 417
Yuan, Zih-shiuan (Spin), 353

Z
Zapata-Madrigal, Germán D., 49
Zhang, Shiyu, 484
Zhang, Xun, 519
Zhang, Youshan, 112